Operations Research

Operations Research

An Introduction

Fourth Edition

Hamdy A. Taha

Department of Industrial Engineering
University of Arkansas, Fayetteville

MACMILLAN PUBLISHING COMPANY
New York

COLLIER MACMILLAN PUBLISHERS
London

Macmillan Publishing Company
866 Third Avenue, New York, New York 10022

Collier Macmillan Canada, Inc.

Library of Congress Cataloging-in-Publication Data

Taha, Hamdy A.
  Operations research.

  Includes index.
  1. Operations research.   2.   Programming (Mathematics)
I. Title.
T57.6.T3   1987   001.4'24   86–5366
ISBN 0–02–418940–5 (Hardcover Edition)
ISBN 0–02–946750–0 (International Edition)

Printing: 1 2 3 4 5 6 7 8   Year: 7 8 9 0 1 2 3 4 5 6

ISBN 0-02-418940-5

To Karen

Los ríos no llevan agua,
el sol las fuentes secó . . .

¡Yo sé donde hay una fuente
que no ha de secar el sol!

La fuente que no se agota
es mi propio corazón . . .

V. Ruiz Aguilera (1862)

Preface

As in the first three editions, the goal of the fourth edition continues to be the improvement of the readability of the book. Although the third edition has contributed significantly toward achieving this goal, the changes introduced in the fourth edition should further enhance the utility of the text. As always, the new changes are designed to best serve the interest of the student.

The most visible addition to the book is the introduction of short cases at the end of most chapters. These cases deviate from familiar "storytelling" case presentations in that the facts of the situation are given in a more straightforward manner. However, they differ from regular textbook problems in that the amount of information conveyed by the case may not be tailored exactly to what the decision maker would need to solve the problem. In many cases, the presentations of the cases are made deliberately ambiguous. The objective, of course, is to get the student to think about the solution to the problem in a nonregimented manner. Since some of the cases are taken from published literature, the student is encouraged to consult the original sources (which are cited in the text) whenever possible.

Two chapters have been rewritten completely: the introductory chapter (Chapter 1) and the simulation chapter (Chapter 17). The presentation of the introductory chapter deviates from the narrative style followed in the previous editions. Chapter 1 now presents the elements of the decision-making process in a more concrete manner. It also provides an overview of the use and applicability of operations research. Simple, but instructive, examples are used to introduce the main ideas of the chapter.

The simulation chapter has been rewritten to stress the practical aspects of this important tool. In particular, the chapter provides a presentation of the different methods of simulation. It also gives a summary of available simulation languages.

Other changes have also been introduced in many parts of the book. For example, I have abandoned the use of networks as a vehicle for introducing the basic concepts of dynamic programming. The feedback I received from the students regarding the effectiveness of this method was not totally favorable. Along the same lines, I have changed the presentation in the network chapter (Chapter 6) to include the labeling procedure for the maximal flow problem.

The organization of the material in the fourth edition continues to provide the same flexibility experienced in the preceding editions. The book may be used to support serious courses in specialized areas, particularly in the area of linear programming. It may also be used in survey courses both at the undergraduate and graduate levels.

ACKNOWLEDGMENTS

Many colleagues have helped generously by preparing detailed criticisms of the third edition. I am deeply indebted to all of them. In particular, my sincere gratitude goes to Professors Guy Curry (Texas A&M University), John Littschwager (University of Iowa), Allen C. Schuermann (Oklahoma State University), David Finkel (Bucknell University), Bernard Rasof (Illinois Institute of Technology), and Ron Barnes (University of Houston).

I wish to express my appreciation to Professor C. R. Emerson, Head of the Industrial Engineering Department at the University of Arkansas for his support and encouragement during the preparation of the fourth edition. I also wish to acknowledge that a portion of this revision was completed during my sabbatical leave at the Kuwait Institute for Scientific Research (KISR).

I feel fortunate that Mrs. Elaine Wetterau, who ably supervised the production of the third edition, was again assigned to continue the same quality work with the fourth edition. I am deeply thankful to Mrs. Wetterau for her superior editorial work and assistance.

H. A. T.

Contents

PART II: PROBABILISTIC MODELS

Chapter 10   Review of Probability Theory   385

Chapter 11   Decision Theory and Games   427

Chapter 12   Project Scheduling by PERT–CPM   469

Chapter 13 Inventory Models 503

Chapter 14 Markovian Decision Process 570

Chapter 15 Queueing Theory (with Miniapplications) 595

APPENDIXES

Operations Research

Chapter 1

Decision Making in Operations Research

1.1 THE ART AND SCIENCE OF OPERATIONS RESEARCH

Operations research (OR) seeks the determination of the best (optimum) course of action of a decision problem under the restriction of limited resources. The term **operations research** quite often is associated almost exclusively with the use of **mathematical techniques** to model and analyze decision problems. Although mathematics and mathematical models represent a cornerstone of OR, there is more to problem solving than the construction and solution of mathematical models. Specifically, decision problems usually include important intangible factors that cannot be translated directly in terms of the mathematical model. Foremost among these factors is the presence of the human element in almost every decision environment. Indeed, decision situations have been reported where the effect of human behavior has so influenced the decision problem that the solution obtained from the mathematical model is deemed impractical. A good illustration of these cases is a version of the widely circulated *elevator problem*. In response to tenants' complaints about the slow elevator service in a large office building, a solution based on analysis by waiting line theory was found unsatisfactory. After studying the system further, it was discovered that the tenants' complaints were more a case of boredom, since in reality the actual waiting time was quite small. A solution was proposed whereby full-length mirrors were installed at the entrances of the elevators. The complaints disappeared because the elevator users were kept occupied watching themselves and others while waiting for the elevator service.

The elevator illustration underscores the importance of viewing the mathematical aspect of operations research in the wider context of a decision-making process whose elements cannot be represented totally by a mathematical model. Indeed, this point was recognized by the British scientists who pioneered the first OR activities during World War II. Although their work was concerned primarily with the optimum allocation of the limited resources of war materiel, the team included scientists from such fields as sociology, psychology, and behavioral science in recognition of the importance of their contribution in considering the intangible factors of the decision process.

As a problem-solving technique, OR must be viewed as both a science and an art. The science aspect lies in providing mathematical techniques and algorithms for solving appropriate decision problems. Operations research is an art because success in all the phases that precede and succeed the solution of a mathematical model largely depends on the creativity and personal ability of the decision-making analysts. Thus gathering of the data for model construction, validation of the model, and implementation of the obtained solution will depend on the ability of the OR team to establish good lines of communication with the sources of information as well as with the individuals responsible for implementing recommended solutions.

It must be emphasized that a successful OR team is expected to exhibit adequate ability in the science and art aspects of OR. Emphasis on one aspect and not the other is apt to impede the effective utilization of OR in practice.

1.2 A SIMPLE DECISION MODEL

A decision model should be thought of as merely a vehicle for "summarizing" a decision problem in a manner that allows systematic *identification and evaluation* of all decision alternatives of the problem. A decision is then reached by selecting the alternative that is judged to be the "best" among all available options.

A simple, but instructive, example for demonstrating the function of a model is the situation in which a production department manager must decide on the question of acquiring an automatic, as opposed to a semiautomatic, machine. The two machines produce a specific part in batches. The setup cost per batch and the variable unit production cost are

| | Cost in Dollars ||
	Semiautomatic	Automatic
Set up cost per batch	20.0	50.0
Unit variable cost	.6	.4

To formalize the situation as a decision mode, we must

1. Identify the decision alternatives.
2. Design a criterion for evaluating the "worth" of each alternative.
3. Use the developed criterion as a basis for selecting the best of the available alternatives.

The statement of the problem tells us that there are two alternatives:

1. Buy an automatic machine.
2. Buy a semiautomatic machine.

The evaluation of these two alternatives can be appropriately based on the operating cost of the machine consisting of a fixed setup cost and a variable production cost. Our objective is to select the alternative with the smaller cost.

To formalize the cost criterion, let x represent the number of units to be produced in a single batch (that is, before a new setup is effected). The cost function then becomes

$$\text{production cost per batch} = \text{setup cost} + (\text{variable unit cost})x$$
$$= \begin{cases} 50 + .4x, \text{ for automatic machine} \\ 20 + .6x, \text{ for semiautomatic machine} \end{cases}$$

We can state now the complete decision model:

Select one of the alternatives:
  1. Buy an automatic machine.
  2. Buy a semiautomatic machine.
The chosen alternative must yield the smaller production cost per batch.

The next step after developing the model is to obtain the solution, that is, make a decision. We can achieve this by employing a **break-even chart.**

Let the x-axis represent the batch size and define the y-axis to represent the production cost. The associated cost functions are then plotted as straight lines as shown in Figure 1–1. The two alternatives cost exactly the same at $x = 150$ units. For batch sizes less than 150 units, the semiautomatic machine is cheaper. The opposite is true for batches greater than 150 units. Thus a general solution based on the model is

1. Buy semiautomatic machine if batch size is less than 150 units.
2. Buy automatic machine if batch size is greater than 150 units.
3. Buy either machine if the batch size is equal to 150 units.

The solution implicitly assumes that both machines produce parts at the same rate so that batch sizes corresponding to a given production period are necessarily equal. Suppose that in reality the hourly production rates for the automatic and semiautomatic machines are 25 and 15 units, respectively. Suppose further that the factory operates on a daily 8-hour single-shift basis. The produced parts are used in an assembly elsewhere in the factory at the daily rate of 100 units. However, a possible future expansion may raise the daily demand to 150 units.

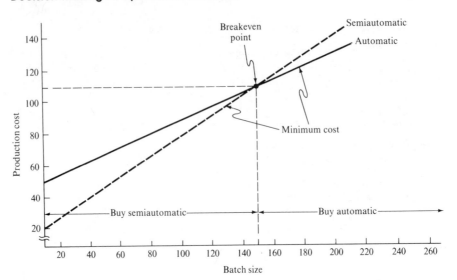

Figure 1-1

The new information adds *restrictions* that were not accounted for in the model. Since the factory operates on a single-shift 8-hour basis, the maximum batch size for the automatic and semiautomatic machines are limited to 200(= 25 × 8) and 120(= 15 × 8), respectively. Given this information, the model for the new situation is changed as shown in Figure 1–2.

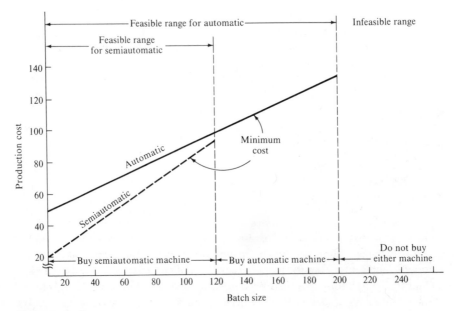

Figure 1-2

Notice the effect of the 8-hour single-shift restriction. If the batch size does not exceed 120 units, the decision problem has two alternatives, of which the semiautomatic machine is the better choice. On the other hand, for batch sizes between 120 and 200, the semiautomatic machine is an *infeasible* alternative, leaving the automatic machine as the only *feasible* choice. Finally, for batch sizes greater than 200 units, both alternatives are infeasible.

Exercise 1.2–1

Determine the optimum decision in Figure 1–2 assuming that the shift length extends to 10 hours.

[*Ans.* Same as in Figure 1–1, provided that batch sizes stay below 250.]

We can now see that restrictions in a decision problem act to limit our choices by eliminating the infeasible alternatives. A more restricted decision problem usually yields a worse solution (in term of the evaluation criterion). For example, a comparison of Figures 1–1 and 1–2 shows that in the absence of the 8-hour shift restriction, a batch size in the range 120 to 150 would lead to selecting the semiautomatic machine as compared with the choice of the automatic machine when the 8-hour restriction is instituted. We can also see from Figures 1–1 and 1–2 that the production cost in the restricted case is worse than in the unrestricted case. As a general rule, a restricted solution can never be better (in terms of the evaluation criterion) than the unrestricted one.

The simple example serves to introduce all the essential ingredients of *any* decision model. In essence, a decision model must include three elements:

1. Decision **alternatives** from which a selection is made.
2. **Restrictions** for excluding infeasible alternatives.
3. **Criteria** for evaluating, and hence ranking, feasible alternatives.

Note that there is a difference between constructing the model and obtaining its solution. Normally, the first step in the decision-making process is to build the model. Following this step, the decisionmaker must find a method for solving the model. In some cases, there may be more than one way to solve the model. In other cases, the resulting model may be so complex that it may be difficult to obtain an exact solution. In such a situation, one must be satisfied with obtaining an *approximate* solution to the problem. We will readdress this point later in the chapter.

The general procedure of constructing a decision model and then seeking its solution represents the core of the decision-making process in the field of operations research. OR, however, employs a somewhat different terminology. Thus we speak of **decision variables** in place of decision alternatives. We also seek to determine the "value" of the decision variables by *optimizing* (cost *minimizing* or profit *maximizing*) an **objective function,** a procedure that is exactly equivalent to ranking the decision alternatives. The optimization process is normally confined to the feasible values of the decision variables satisfying all the restrictions of the model.

1.3 ART OF MODELING

In Section 1.2, the decision-making process in OR is shown to consist of constructing a decision model and then solving it to determine the optimum decision. The model is defined as an objective function and restrictions expressed in terms of the decision variables (alternatives) of the problem.

Model construction and development represents the crucial step in the implementation of OR, and, indeed, any systematic decision-making procedure. Thus a solution to a model, albeit accurate and exact, will not be useful unless the model itself provides an adequate representation of the true decision situation. For example, the exact solution provided in Figure 1–1 for the machine selection problem can be useless in practice since it does not take into account the reality of the situation that sets an 8-hour one-shift limit on the factory operation.

Although a real situation may involve a substantial number of variables and constraints, usually only a small fraction of these variables and constraints truly dominates the behavior of the real system. Thus the simplification of the real system for the purpose of constructing a model should concentrate primarily on identifying the dominant variables and constraints as well as other data pertinent to decision making.

Figure 1–3 depicts the levels of abstraction of a real-life situation that lead to the construction of a model. The **assumed real system** is abstracted from the real situation by concentrating on identifying the dominant factors (variables, constraints, and parameters) that control the behavior of the real system. The model, being an abstraction of the *assumed* real system, then identifies the pertinent relationships of the system in the form of an objective and a set of constraints.

The following example is introduced to gain an appreciation of the significance of the different levels of abstraction.

Example 1.3-1. A manufactured product typically undergoes a number of operations from the time it is conceived by the designer until it reaches the consumer. After the design is approved, a production order is issued to the production department, which in turn requests the necessary materials from the materials department. The materials department either satisfies the request from its stocks or contacts the purchasing department to initiate a purchase order. After the final product is completed, the sales department, in conjunction

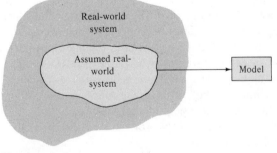

Figure 1–3

with the marketing department, assumes the responsibility for distributing it to the consumer.

Suppose that it is desired to determine the "best" level of production in the plant manufacturing the product. Looking at the overall system, we can see that a large number of factors can influence the production level. The following are some examples:

1. *Production department.* Available machine hours, specific sequencing of operations on machines, in-process inventory, number of defective items produced, and inspection rate.

2. *Materials department.* Available stock of material, rate of delivery of purchased material, and storage limitations.

3. *Marketing department.* Sales forecast, intensity of advertising campaign, capacity of distribution facilities, and effect of competition.

If each of these factors is to be considered explicitly in a model that determines the level of production, we would be faced with a staggering task indeed. For example, we can consider explicitly such variables as the assignment of machine hours, the assignment of labor hours, and the inspection rate. As for the constraints, we can include capacities of the machines, limit on labor hours, limit on in-process inventory, limit on demand, and storage limitation. Already you can see the complexity of the relationships that express the level of production in terms of such detailed variables as those exemplified here.

The definition of the "assumed real" system for the foregoing situation entails looking at the system as an entity rather than initially concerning ourselves with the finer details of the problem. In essence, we can look at the entire system in a general sense from the standpoint of the producer and the consumer. With some reflection, we can see that the producer's side can be expressed in terms of the **production rate** whereas the consumer's side can be represented by a **consumption rate.**

Naturally, the production rate is a function of such factors as the availability of machine and labor hours, sequencing of operations, and availability of raw material. Similarly, the consumption rate is based on the limitation of the distribution system and the sales forecast. In essence, the simplification from the "real" to the "assumed real" system is effected by "lumping" several factors in the assumed system.

It is easier now to think in terms of the assumed real system. The desired model would now seek the determination of the stock level in terms of the production and consumption rates. A proper objective could be to select the stock level that balances the cost of carrying excess inventory against the cost of running out of stock when the product is needed.

We must keep in mind, however, that the degree of complexity of the model is always an inverse function of the degree of simplification of the assumed real system as abstracted from the real system. For example, we can assume that the production and consumption rates are constant or change as functions of time. The latter case should lead to a more complex model, naturally.   ◄

In general, there are no fixed rules for effecting the levels of abstraction cited in Figure 1–3. The reduction of the factors controlling the system to a relatively small number of dominant factors and the abstraction of a model from the assumed real system is more an art than a science. The validity of the model in representing the real system depends primarily on the creativity, insight, and imagination of the OR team. Such personal qualities cannot be regulated by the establishment of fixed rules for constructing models.

Although it is not possible to present fixed rules about *how* a model is constructed, it may be helpful to present ideas about possible types of models, their general structures, and their characteristics. This is the subject of the next section.

1.4   TYPES OF OR MODELS

The discussions in Sections 1.2 and 1.3 stress the fact that the model construction phrase comes first, followed by solving the model to secure a desired solution. The solution methods are usually devised to take advantage of the special structures of the resulting models. As such, the wide variety of models associated with existing real systems gives rise to a corresponding number of solution techniques. Hence are the familiar names of linear, integer, dynamic, and nonlinear programming that represent algorithms for solving special classes of OR models.

In most OR applications, it is assumed that the objective and constraints of the model can be expressed quantitatively or mathematically as functions of the decision variables. In such a case, we say that we are dealing with a **mathematical model.**

Unfortunately, in spite of the impressive advances in mathematical modeling, a considerable number of real situations still lies well beyond the capabilities of presently available mathematical techniques. For one thing, the real system may be too involved to allow an "adequate" mathematical representation. Alternatively, even when a mathematical model can be formulated, it may prove to be too complex to be solved by available solution methods.

A different approach to modeling (complex) systems is to use **simulation.** Simulation models differ from mathematical models in that the relationships between input and output are not explicitly stated. Instead, a simulation model breaks down the modeled system into basic or elemental modules that are then linked to one another by well-defined logical relationships (in the form of IF/ THEN). Thus starting from the input module, the computations will move from one module to another until an output result is realized.

The following two examples are designed to make you appreciate the applicability of mathematical models vis-à-vis simulation modeling.

Example 1.4-1 (A Mathematical Model). Farmers Coop has two central warehouses that supply corn seeds to three regional stores for distribution to farmers. The monthly supply available at the two warehouses is estimated to be 1000 and 2000 sacks of corn seeds. The demand at the three regional stores is estimated at 1500, 750, and 750 sacks, respectively. The cost in cents per

sack for transporting the seed from the warehouses to the stores can be summarized as given in Table 1–1.

Table 1-1

		Store		
		1	2	3
Warehouse	1	50	100	60
	2	30	20	35

The goal of the Coop is satisfy the monthly demand at the regional stores at
the least possible transportation cost.

To facilitate the modeling process, we can look at the problem graphically
as shown in Figure 1–4. The figure shows that warehouses 1 and 2 can supply
1000 and 2000 sacks of seed, whereas stores 1, 2, and 3 have the respective
demands of 1500, 750, and 750 sacks. The lines joining the supply and demand
points represent possible transportation routes. The unit transportation cost
(per sack) is shown on each route.

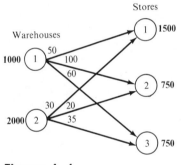

Figure 1-4

What are the decision variables (alternatives) of the problem? We are interested in determining the number of sacks that should be transported *on each
route*. For example, the number of sacks from warehouse 1 to store 2 can be
anywhere from 0 to 750 sacks. If we apply the same reasoning to all the routes,
we will discover that the overall number of possibilities that must be considered
to obtain the solution is indeed "extremely" large (compare this with the simple
model in Section 1.2 where we dealt with two alternatives only). This is why
we need a model that can express the decision problem in a manner that allows
a systematic determination of the best values of the decision variables.

The model, as we stated earlier, includes an objective and a set of constraints.
The objective in this case is straightforward, namely, the minimization of the
transportation costs. What about the constraints? With some reflection, we notice
that since the total supply of $1000 + 1500 = 2500$ sacks equals the total demand
of $1500 + 750 + 750 = 2500$, then each warehouse will ship out its exact
supply to some or all of the three stores. Simultaneously, each store will receive

its exact demand from one or both warehouses. Our model thus seeks to determine the amounts to be transported on each route to

---

minimize total transportation cost on all routes subject to

amount shipped from a warehouse = its supply
amount received by a store = its demand

We can go ahead and translate the model mathematically in terms of the decision variables. However, to dampen the (possible) initial "discomfort" of dealing with mathematical notation, we will present the model in an elementary form. Let us define the decision variables as x_{11}, x_{12}, x_{13}, x_{21}, x_{22}, and x_{23} to represent the amounts to be shipped from warehouse 1 to store 1, warehouse 1 to store 2, and so on. Then the model can be summarized as shown in Table 1–2.

In Table 1–2, each cell (or square) represents a route with a decision variable. Thus cell (1, 1) represents x_{11}, the amount to be shipped from warehouse 1 to store 1. The top right corner of each cell records the unit transportation cost. The amounts of supply are shown to the far right of each row, and the amounts of demand are given at the bottom of each column. Now, the mathematical functions of the model can be expressed more readily. Namely, the objective is to minimize the total transportation cost expressed as the sum of the decision variables multipled by the unit cost. The constraints reduce to saying that the sum of the variables in each row must equal the associated supply (exactly), and the sum of the variables in each column must equal the associated demand. All this information may now be translated into an "authentic" mathematical model as follows:

Minimize
   transportation cost = $50x_{11} + 100x_{12} + 60x_{13} + 30x_{21} + 20x_{22} + 35x_{23}$

subject to

$$x_{11} + x_{12} + x_{13} = 1000 \quad \text{(warehouse 1 supply)}$$
$$x_{21} + x_{22} + x_{23} = 2000 \quad \text{(warehouse 2 supply)}$$
$$x_{11} + x_{21} = 1500 \quad \text{(store 1 demand)}$$
$$x_{12} + x_{22} = 750 \quad \text{(store 2 demand)}$$
$$x_{13} + x_{23} = 750 \quad \text{(store 3 demand)}$$

All decision variables are nonnegative.

The mathematical formulation of the problem is now complete. The next step is to find the solution. As we stated earlier, the real drawback of mathematical modeling is that the resulting model may not be solvable analytically. Luckily, a general solution method is available for this model as we will show in Chapter 5.

Table 1–2

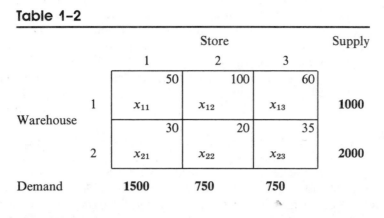

			Store		Supply
		1	2	3	
Warehouse	1	50 x_{11}	100 x_{12}	60 x_{13}	**1000**
	2	30 x_{21}	20 x_{22}	35 x_{23}	**2000**
Demand		**1500**	**750**	**750**	

Exercise 1.4–1

Consider the formulation of Table 1–2. Attempt to find a *feasible* schedule and express the solution in terms of the decision variables. Next, devise a "logical" procedure for finding a *better* solution.

Example 1.4–2 (A Simulation Model). A bank must decide on the number of tellers to be installed to serve its customers. If the number of tellers is not sufficient, a customer may wait too long for service. On the other hand, installing too many tellers could represent needless additional costs. The objective then is to balance the "cost" of waiting for service against the cost of (installing and) operating the tellers.

Our decision variable in this problem is the number of tellers to be installed. The optimum value of the decision variable is determined by minimizing the sum of the cost of operating the tellers and the cost of waiting for service. Figure 1–5 summarizes the situation graphically. The waiting cost is a function

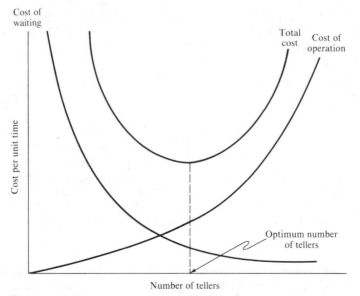

Figure 1–5

of the number of customers queueing for service, whereas the cost of operation will depend on the number of installed tellers. For simplicity, we have ignored the fact that the number of tellers is an integer variable.

Although this model has no explicit constraints, it has the basic structure of a mathematical model. In fact, if we stipulate that the bank cannot house more than four tellers because of space limitation, then the model would reduce to determining the number of tellers, not to exceed four, that minimizes the total cost function.

Why is this model then different from any other mathematical model? Insofar as the model itself is concerned, there really is no difference. However, when it comes to the *implementation* of the model, we will find ourselves faced with the problem of how to determine the cost of waiting. To determine such a cost, we need to determine the (average) number of customers waiting as a function of the number of tellers. And except for special cases that can be handled by queueing theory (Chapter 15), there are no direct methods for securing this information. This is where simulation modeling proves useful.

The bank simulation model can be viewed as shown in Figure 1–6. Customers arrive at the bank from a SOURCE. If all the tellers are busy, the arriving customer will wait in a QUEUE. Otherwise, the customer will proceed to TELLER for service. When a customer completes service, TELLER becomes available and waiting customers, if any, will leave QUEUE to seek service. Otherwise, TELLER remains idle.

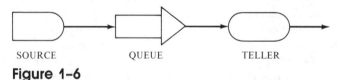

SOURCE   QUEUE   TELLER

Figure 1–6

What simulation does is to "send" individual customers from SOURCE into the system, and then keep track of the changes that take place in the length of QUEUE as a function of time. After "observing" the system a sufficient number of times, we can then compute the (average) queue length. The process is repeated for a different number of tellers to determine the desired data. These data can then be used within the cost model to determine the optimum number of tellers.

Although we have not discussed how *simulated* customers "arrive" from SOURCE or how long they "stay" in TELLER for service, our objective here is to show that simulation basically provides the desired information literally through *observation*. If we were to obtain the queue length mathematically as in queueing theory (Chapter 15), we would simply use the relationship between how customers arrive from SOURCE and how they are served in TELLER to obtain a *closed-form* formula for the queue length. This is not the case with simulation.   ◀

Simulation models, when compared with mathematical models, do offer greater flexibility in representing complex systems. The main reason for this flexibility is that simulation views the system from a basic elemental level. Mathematical modeling, on the other hand, tends to consider the system from a less detailed level of representation.

The flexibility of simulation is not without drawbacks. The development of a simulation model is usually quite costly both in time and resources. Additionally, the execution of a simulation model, even on the fastest computer, could incur considerable cost. On the other hand, successful mathematical models are usually manageable computationally.

We must keep in mind, however, that simulation is not a substitute to OR mathematical modeling. Rather, simulation is used to estimate data that otherwise are not readily available analytically. Once these data are determined, they can be used within the framework of a mathematical model to secure an optimum solution. The bank model of Example 1.4–2 serves to illustrate this point.

1.5 EFFECT OF DATA AVAILABILITY ON MODELING

Models of any kind, regardless of sophistication and accuracy, may prove of little practical value if they are not supported by reliable data. Take, for example, the transportation model presented in Example 1.4–1. Although the model is well defined, the quality of the solution is obviously dependent on how well we can estimate the unit transportation costs. If the estimates are distorted, the obtained solution, though optimum in a mathematical sense, will actually be of inferior quality from the standpoint of the real system.

In some situations, data may not be known with certainty. Rather, they are estimated by probability distributions. For example, in the transportation model of Example 1.4–1, it is assumed that the monthly demand at the stores is known with certainty, that is, exactly. If, however, it is discovered that the monthly demand fluctuates, it may then necessary to estimate the demand in the form of a probability distribution. More important, it would be necessary to change the structure of the model to accommodate the probabilistic nature of the demand. This gives rise to the so-called **probabilistic** or **stochastic models** as opposed to **deterministic models.**

Sometimes a model is constructed under the assumption that certain data can be secured, but later search may prove that such information is difficult to obtain. In this case, it may be necessary to reconstruct the model to reflect the lack of data. Thus the degree of data availability may also affect the accuracy of the model. As an illustration, consider an inventory model in which the stock level of a certain item is determined such that the total cost of holding excess inventory and not satisfying all demand is minimized. This requires estimating a holding cost per excess unit held in stock and a shortage cost per unsatisfied unit of demand. The holding cost, which depends on storage expenses and cost of capital, may be relatively simple to estimate. But if the shortage cost accounts for the loss in customer's goodwill, it may be difficult to quantify such an intangible factor. Under such conditions, the model may have to be changed so that the shortage cost is not spelled out explicitly. For example, one may have to specify an acceptable upper limit on the shortage quantity at any time. In essence the specified upper limit implies a certain estimate of shortage cost. But it appears much simpler to determine such a limit than to estimate a shortage cost.

The gathering of data may actually be the most difficult part of completing a model. Unfortunately, no rules can be suggested for this procedure. While accumulating experience in modeling in an organization, the OR analyst should also develop means for gathering and documenting data in a manner useful for both present and future projects.

1.6 COMPUTATIONS IN OR

In OR there are two distinct types of computations: those involving simulation and those dealing with mathematical models. In simulation models, computations are typically voluminous and mostly time consuming. Yet, in simulation one is always assured that the desired results will definitely be secured. It is simply a matter of providing sufficient computer time!

Computations in OR mathematical models, on the other hand, are typically **iterative** in nature. By this we mean that the optimum solution of a mathematical model usually is not available in a closed form. Instead, the final answer is reached in steps or **iterations,** with each new iteration bringing the solution closer to the optimum. In this respect, we say that the solution *converges* iteratively to the optimum. As an illustration, in the transportation model of Example 1.4–1, we start with a transportation schedule that is feasible (satisfies the supply and demand conditions). Then by testing the resulting solution, we can decide whether or not the solution is optimum. If not, we find a better solution. Logically, if every new iteration should improve (reduce) the transportation cost, eventually a terminal iteration will be reached where the costs can no longer be reduced. This is the optimum!

Unfortunately, not all OR mathematical models possess solution algorithms (methods) that always converge to the optimum. There are two reasons for this difficulty:

1. The solution algorithm may be proven to converge to the optimum, but only in a theoretical sense. Theoretical convergence says that there is a finite upper ceiling on the number of iterations, but it does not say how high this ceiling may be. Thus one can consume hours of computer time without reaching the final iteration. Worse still, if the iterations are stopped prematurely before reaching the optimum, one is usually unable to measure the quality of the obtained solution relative to the true optimum. (Notice the difference between this situation and that of simulation. In simulation, one has control over the computational time simply by reducing the observation period of the model. In mathematical models, the number of iterations is a function of the efficiency of the solution algorithm and the specific structure of the model, both of which may not be controllable by the user.)

2. The complexity of the mathematical model may make it impossible to devise a solution algorithm. In this case, the model may remain computationally unsolvable.

The apparent difficulties in mathematical model computations have forced practitioners to seek alternative computational methods. These methods are also iterative in nature, but they do not guarantee optimality. Instead, they simply seek a *good* solution to the problem. Such methods are usually known as **heuristics** because their logic is based on rules of thumb that are conducive to obtaining a good solution. The advantage of heuristics is that they normally involve less computations when compared with exact algorithms. Also, because they are based on rules of thumb, they normally are easier to explain to users who are not mathematically oriented.

In OR, heuristics are generally employed for two different purposes:

1. They can be used within the context of an exact optimization algorithm to speed up the process of reaching the optimum. The need for "beefing up" the optimization algorithm becomes more evident with large-scale models.

2. They are simply used to find a "good" solution to the problem. The resulting solution is not guaranteed to be optimum, and, in fact, its quality relative to the true optimum may be difficult to measure.

We will illustrate these two cases by examples. For the first case, consider the transportation model of Example 1.4–1. The model is repeated in Table 1–3 for convenience.

Table 1–3

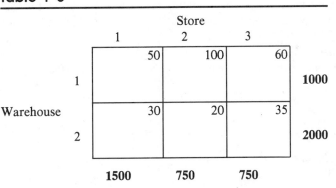

Although the transportation model possesses an efficient exact algorithm, we can employ a heuristic to make it even more efficient. This is achieved by selecting the initial transportation schedule intelligently. The idea of the heuristic is to assign as much as possible to the cheapest route in the table without violating the supply and demand restrictions. An exhausted supply row or a satisfied demand column is then crossed out, and the remaining supply and demand quantities are adjusted. The process of selecting the next cheapest route is then repeated until all the demands are satisfied. Applying this to Table 1–3, we assign 750 sacks to route (2, 2), that is, $x_{22} = 750$, which crosses out column 2 and leaves 1250 at warehouse 2. Next, we assign 1250 sacks to route

(2, 1), followed by 250 sacks to route (1, 1) and 750 sacks to route (1, 3). The initial solution is thus given as

From/To	Quantity (sacks)	Unit Cost (¢)	Cost ($)
1/1	250	50	125.00
1/3	750	60	450.00
2/1	1250	30	375.00
2/2	750	20	150.00

Exercise 1.6-1

Devise a heuristic that will improve on the preceding solution by using a form of perturbation in the schedule.

To illustrate the second type of heuristics, consider the problem of a traveling salesperson who must travel to five cities, with *each city visited exactly once* before returning back to his or her hometown. Figure 1–7 summarizes the distances (in miles) between all cities. The objective of the salesperson is to minimize the total travel distance.

This problem can be formulated as an exact mathematical model. However, obtaining the exact optimum solution to this problem has proven to be formidable. However, a "good" solution can be obtained by using a heuristic that calls for traveling from the present city to the closest unvisited city. Thus, starting from city 1, the salesperson will travel to 4 (distance = 3 miles), then from 4 to 5, followed by 5 to 3, and then 3 to 2, from which the trip is completed by returning back to 1. The total traveled distance of the tour is 18 miles, which is not optimal since the route 1–2–3–4–5–1 is shorter by 3 miles.

Exercise 1.6-2

Devise rules of thumb that may improve on the solution obtained by the foregoing heuristic.

1.7 PHASES OF OR STUDY

An OR study cannot be conducted and controlled by the OR analyst alone. Although he or she may be the expert on modeling and model solution techniques, the analyst cannot possibly be an expert in all the areas where OR problems arise. Consequently, an OR team should include members of the organization directly responsible for the functions in which the problem exists as well as for the execution and implementation of the recommended solution. In other words, an OR analyst commits a grave mistake by not seeking the cooperation of those who will implement recommended solutions.

The major phases through which the OR team would proceed to effect an OR study include

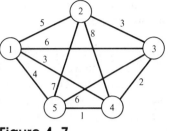

Figure 1-7

1. Definition of the problem.
2. Construction of the model.
3. Solution of the model.
4. Validation of the model.
5. Implementation of the final results.

Although this sequence is by no means standard, it seems generally acceptable. Except for the "model solution" phase, which is based generally on well-developed techniques, the remaining phases do not seem to follow fixed rules. This stems from the fact that the procedures for these phases depend on the type of problem under investigation and the operating environment in which it exists. In this respect, an operations research team would be guided in the study principally by the professional experiences of its members rather than by fixed rules.

In spite of the apparent difficulties in setting fixed rules for the execution of these phases, it seems desirable to establish general guidelines. The remainder of this section is thus devoted to providing an orientation of the main points involved in an operations research study.

The first phase of the study deals with the **problem definition.** From the viewpoint of operations research, this indicates three major aspects: (1) a description of the goal or the objective of the study, (2) an identification of the decision alternatives of the system, and (3) a recognition of the limitations, restrictions, and requirements of the system.

The second phase of the study deals with the **model construction.** Depending on the definition of the problem, the operations research team should decide on the most suitable model for representing the system. Such a model should specify quantitative expressions for the objective and the constraints of the problem in terms of its decision variables. If the resulting model fits into one of the common mathematical models (e.g., linear programming), a convenient solution may be obtained by using mathematical techniques. If the mathematical relationships of the model are too complex to allow analytic solutions, a simulation model may be more appropriate. Some cases may require the use of a combination of mathematical, simulation, and heuristic models. This, of course, is largely dependent on the nature and the complexity of the system under investigation.

The third phase of the study deals with the **model solution.** In mathematical models, this is achieved by using well-defined optimization techniques and the model is said to yield an optimum solution. If simulation or heuristic models are used, the concept of optimality is not as well defined, and the solution in these cases is used to obtain approximate evaluations of the measures of the system.

In addition to the (optimum) solution of the model, one must also secure, whenever possible, additional information concerning the behavior of the solution due to changes in the system's parameters. This is usually referred to as **sensitivity analysis.** In particular, such an analysis is needed when the parameters of the system cannot be estimated accurately. In this case, it is important to study the behavior of the optimal solution in the neighborhood of these estimates.

The fourth phase calls for checking the **model validity.** A model is valid if, despite its inexactness in representing the system, it can give a reasonable prediction of the system's performance. A common method for testing the validity of a model is to compare its performance with some past data available for the actual system. The model will be valid if under similar conditions of inputs, it can reproduce the past performance of the system. The problem here is that there is no assurance that future performance will continue to duplicate past behavior. Also, since the model is based on careful examination of past data, the comparison should always reveal favorable results. In some instances this problem may be overcome by using data from trial runs of the system.

It must be noted that such a validation method is not appropriate for nonexisting systems, since data will not be available for comparison. In some cases, if the original system is investigated by a mathematical model, it may be feasible to construct a simulation model from which data are obtained to carry out the comparison.

The final phase of the study deals with the **implementation** of the tested results of the model. The burden of executing these results lies primarily with the operations researchers. It involves the translation of these results into detailed operating instructions issued in an understandable form to the individuals who will administer and operate the recommended system. The interaction between the operations research team and the operating personnel will reach its peak in this phase. Communication between the two groups can be improved by seeking the participation of the operating personnel in developing the implementation plan. In fact, this participation should be sought throughout all phases of the study. In this way no practical consideration that might lead to system failure will be overlooked. Meanwhile, possible modifications or adjustments in the system may be checked for feasibility by the operating personnel. In other words, it is imperative that the implementation phase be executed through the cooperation of both the operations research team and those who will be responsible for managing and operating the system.

1.8   ABOUT THIS BOOK

This chapter opened with the strong statement that operations research is both an *art* and a *science.* As you advance through the remaining chapters, you will get the impression that the book puts more emphasis on the scientific aspect of OR. In essence, the topics of the book are classified according to the well-known mathematical models of OR (e.g., linear programming, integer programming, inventory, and queueing theory). There are two reasons for this. First, as may be expected, there are no definite rules that can be prescribed for the art aspect of OR. The diversity of the situations where OR can be applied makes attempts in this direction almost futile. Second, we believe that it is essential that the OR practitioner acquire an adequate understanding of the capabilities and limitations of the mathematical techniques of OR. The viewpoint that an OR user need not learn about the mathematics of OR because the computer can "take care" of solving the problem is dangerous. We must keep in mind that the computer solves the model as presented by the user. If the user is unaware of the limitations of the model being used, the quality of the solution will reflect this deficiency.

We must point out that the book does not completely neglect the art aspect of OR. The numerous examples and minicases presented throughout the book should provide insight into the art of OR modeling. The short cases introduced in most chapters are taken from published real situations and are designed to stimulate thinking about finding ways to solve practical problems. We have also stressed topics that are most useful in the analysis of practical problems. One example is the topic of sensitivity analysis, which plays important roles in the study of OR problems.

We believe that a first course in OR should give students a good foundation in the mathematics of OR, coupled with meaningful application examples and minicases. This plan will provide OR users with the kind of *confidence* that normally would be missing if they direct their principal training toward the philosophical and artistic aspects of decision making. Once a fundamental knowledge of the mathematical foundation of OR is acquired, students can increase their capacity as decision makers by studying reported case studies and actually working on real-life problems.

SELECTED REFERENCES

ACKOFF, R. L., *The Art of Problem Solving*, Wiley, New York, 1978.
OSBORN, A. F., *Applied Imagination*, 3rd rev. ed., Scribner, New York, 1963.
SALE, KIRKPATRICK, *Human Scale*, Coward, New York, 1980.

REVIEW QUESTIONS

True (T) or False (F)?

1. ____ The main components of a decision model include the decision alternatives, the constraints, and the objective criterion.

2. ____ The most important step toward solving a decision problem is the construction of an adequate model.

3. ____ A typical decision problem must have at least two decision alternatives.

4. ____ The number of alternatives in a decision problem is typically finite.

5. ____ A decision problem that has infinity of alternatives is usually unsolvable.

6. ____ A feasible alternative in a decision model must satisfy all the restrictions of the problem.

7. ____ An optimum solution is the best solutions in terms of the objective criterion among all the feasible *and* infeasible alternatives of the model.

8. ____ The more restrictions we add to the decision model, the more likely that we worsen the optimum value of the objective criterion.

9. ____ Most mathematical OR models are solved by using iterative procedures.

10. ____ Simulation models collect information by actually observing the simulated system.

11. ____ A simulation model, unlike a mathematical model, does not provide a solution to the decision problem.

12. ____ The accuracy of the results obtained from a simulation model depends on how long we observe the simulated system.

13. ____ Heuristic procedures do not guarantee the attainment of an optimum solution.

14. ____ Heuristics can be used to find a good starting solution and hence lead to a faster attainment of the optimum solution of a mathematical model.

[*Ans.* **1**—T, **2**—T, **3**—T, **4**—F, **5**—F, **6**—T, **7**—F, **8**—T, **9**—T, **10**—T, **11**—T, **12**—T, **13**—T, **14**—T.]

PROBLEMS

☐ **1–1** Consider the machine selection problem in Section 1.2. Suppose that both machines can operate for two shifts a day. Each new shift must reset the machine. Determine the optimum decision.

☐ **1–2** Repeat Problem 1–1 assuming that the unit variable cost of the semiautomatic machine is $.45 instead of $.60.

☐ **1–3** Suppose in Problem 1–1 that a second automatic machine can be secured from another manufacturer. The new machine has a setup cost of $65 and its unit variable cost is $.30. The production rate of the new machine is 30 units per hour. All machines will operate on the basis of one shift per day. State the decision alternatives, their feasible domain, and the criterion employed to select the best alternative; then determine the optimum decision.

☐ **1–4*** During the summer months, a professor has a consulting commitment in another city. She flies there every Monday and returns on Wednesday of the same week. The commitment will last for five weeks. A roundtrip ticket that is bought on Monday for return on Wednesday of the same week will cost 20% more than a ticket that includes a weekend. How should the professor buy the tickets during the five-week consulting commitment?

☐ **1–5** A school board is holding a public meeting to discuss the need for constructing two additional schools. The two schools will serve four communities. The following table summarizes the mileages between the communities as well as the number of students in each community:

Community	1	2	3	4	Number of Students
1	.0	2.3	2.4	3.1	500
2	2.3	.0	3.5	3.8	600
3	2.4	3.5	.0	5.1	700
4	3.1	3.8	5.1	.0	550

In which communities should the schools be located?

---

* This problem and the next can be solved by using sophisticated and exact mathematical models. Naturally, you are not expected to do so at this stage. Instead, the problems are chosen so that the *optimum* can be determined by using simple logical arguments. You will find it helpful, however, to look at the problems from the standpoint of optimization and not as puzzles.

PART I

LINEAR, INTEGER, AND DYNAMIC PROGRAMMING

CHAPTERS 2 THROUGH 7 describe linear programming. Integer programming is presented in Chapter 8. Dynamic programming is developed in Chapter 9.

The material in Chapters 2 through 7 is sufficient to support a serious course in linear programming. The material may be augmented with the closely related topic of integer programming (Chapter 8) and game theory (Section 11.4).

We have designed Chapter 2 so that the important concept of *sensitivity analysis* in linear programming can be appreciated even when linear programming is presented at the elementary level of graphical solution. Chapter 3 also shows how sensitivity analysis can be carried out algebraically from the simplex tableau. The basis for sensitivity analysis is explained in Chapter 4 through the use of duality theory. We believe that the first course in operation research should include Chapters 2 through 4. However, the materials in Chapters 2 and 3 are designed to equip the student with a good background regarding the potential use and interpretation of linear programming results.

Chapter 5 on the transportation model can be studied independently of Chapters 2 through 4, although a knowledge of the simplex method should further enhance understanding of the transportation model. Chapter 6 on networks is a natural sequel to Chapter 5 but may be studied independently.

Chapter 7 on advanced linear programming cannot be studied without a good understanding of Chapters 3 and 4. The prerequisites for Chapter 8 on integer programming are the simplex method (Chapter 3) and the dual simplex method (Section 4.4). Chapter 9 on dynamic programming may be studied independently.

Chapter 2

Linear Programming: Formulations and Graphical Solution

You will notice that this book starts with a detailed coverage of linear programming (LP). Our reason is more than a simple adherence to the tradition followed in most OR books. We are actually motivated by the fact that LP provides important foundations for the development of solution methods of other OR techniques, such as integer, stochastic, and nonlinear programming. In addition, after three decades of experimentation and scrutiny, LP has been applied with impressive success to problems ranging from the familiar cases in industry, military, agriculture, economics, transportation, and health systems to the extreme cases in behavioral and social sciences. For example, an unusual application of LP is the *marriage* problem, whose optimum solution shows that monogamy is the best type of marriage.

This chapter presents various LP applications. The graphical solution of a two-variable problem provides a concrete foundation in the details of the optimization process. It also addresses the important question of how sensitivity analysis is applied to the optimum solution. The chapter closes with the framework for an economic interpretation of the general LP model.

2.1   A SIMPLE LP MODEL AND ITS GRAPHICAL SOLUTION

This section provides the data needed to construct and solve a mathematical model—a linear program in this situation. As you study the material, pay special attention to the assumptions of the model and their possible effects on implementing the solution. Concurrently, try to draw general conclusions about how the procedures used with this example can be applied to other situations.

Example 2.1-1 (The Reddy Mikks Company). The Reddy Mikks Company owns a small paint factory that produces both interior and exterior house paints for wholesale distribution. Two basic raw materials, A and B, are used to manufacture the paints. The maximum availability of A is 6 tons a day; that of B is 8 tons a day. The daily requirements of the raw materials *per ton* of interior and exterior paints are summarized in the following table.

	Tons of Raw Material per Ton of Paint		Maximum Availability (tons)
	Exterior	Interior	
Raw material A	1	2	6
Raw material B	2	1	8

A market survey has established that the daily demand for interior paint cannot exceed that of exterior paint by more than 1 ton. The survey also shows that the maximum demand for interior paint is limited to 2 tons daily.

The wholesale price per ton is $3000 for exterior paint and $2000 for interior paint.

How much interior and exterior paints should the company produce daily to maximize gross income?

Construction of the Mathematical Model

The construction of a mathematical model can be initiated by answering the following three questions:

1. What does the model seek to determine? In other words, what are the **variables** (unknowns) of the problem?

2. What **constraints** must be imposed on the variables to satisfy the limitations of the modeled system?

3. What is the **objective** (goal) that needs to be achieved to determine the optimum (best) solution from among all the *feasible* values of the variables?

An effective way to answer these questions is to give a verbal summary of the problem. In terms of the Reddy Mikks example, the situation is described as follows.

The company seeks to determine the *amounts* (in tons) of interior and exterior paints to be produced to *maximize* (increase as much as is feasible) the total gross income (in thousands of dollars) while satisfying the *constraints* of demand and raw materials usage.

The crux of the mathematical model is first to identify the variables and then to express the objective and constraints as mathematical functions of the variables. Thus, for the Reddy Mikks problem, we have the following.

Variables. Since we desire to determine the amounts of interior and exterior paints to be produced, the variables of the model can be defined as

x_E = tons produced daily of exterior paint
x_I = tons produced daily of interior paint

Objective Function. Since each ton of exterior paint sells for $3000, the gross income from selling x_E tons is $3x_E$ thousand dollars. Similarly, the gross income from x_I tons of interior paint is $2x_I$ thousand dollars. Under the assumption that the sales of interior and exterior paints are independent, the total gross income becomes the sum of the two revenues.

If we let z represent the total gross revenue (in thousands of dollars), the objective function may be written mathematically as $z = 3x_E + 2x_I$. The goal is to determine the (feasible) values of x_E and x_I that will maximize this criterion.

Constraints. The Reddy Mikks problem imposes restrictions on the usage of raw materials and on demand. The usage restriction may be expressed verbally as

$$\begin{pmatrix} \text{usage of raw material} \\ \text{by both paints} \end{pmatrix} \leq \begin{pmatrix} \text{maximum raw material} \\ \text{availability} \end{pmatrix}$$

This leads to the following restrictions (see the data for the problem):

$$x_E + 2x_I \leq 6 \qquad \text{(raw material A)}$$
$$2x_E + x_I \leq 8 \qquad \text{(raw material B)}$$

The demand restrictions are expressed verbally as

$$\begin{pmatrix} \text{excess amount of interior} \\ \text{over exterior paint} \end{pmatrix} \leq 1 \text{ ton per day}$$

$$(\text{demand for interior paint}) \leq 2 \text{ tons per day}$$

Mathematically, these are expressed, respectively, as

$$x_I - x_E \leq 1 \qquad \text{(excess of interior over exterior paint)}$$
$$x_I \leq 2 \qquad \text{(maximum demand for interior paint)}$$

An implicit (or "understood-to-be") constraint is that the amount produced of each paint cannot be negative (less than zero). To avoid obtaining such a solution, we impose the **nonnegativity restrictions,** which are normally written as

$$x_I \geq 0 \quad \text{(interior paint)}$$
$$x_E \geq 0 \quad \text{(exterior paint)}$$

The values of the variables x_E and x_I are said to constitute a **feasible solution** if they satisfy *all* the constraints of the model.

The complete mathematical model for the Reddy Mikks problem may now be summarized as follows:

Determine the tons of interior and exterior paints, x_I and x_E, to be produced to

maximize $z = 3x_E + 2x_I$ (objective function)

subject to

$$x_E + 2x_I \leq 6$$
$$2x_E + x_I \leq 8$$
$$-x_E + x_I \leq 1 \quad \text{(constraints)}$$
$$x_I \leq 2$$
$$x_E \geq 0, \quad x_I \geq 0$$

◄

What makes this model a linear program? Technically, it is a linear program because all its functions (constraints and objective) are *linear*. Linearity implies that both the **proportionality** and **additivity** properties are satisfied.

1. *Proportionality* requires that the contribution of each variable (i.e., x_E and x_I) in the objective function or its usage of the resources be *directly proportional* to the level (value) of the variable. For example, if the Reddy Mikks Company grants quantity discounts by selling a ton of exterior paint for $2500 when the sales exceed 2 tons, it will no longer be true that each ton produced will bring a revenue of $3000. Rather, it will bring $3000 per ton for $x_E \leq 2$ tons and $2500 per ton for $x_E > 2$ tons. This situation does not satisfy the condition of *direct* proportionality with x_E.

2. *Additivity* requires that the objective function be the *direct sum* of the individual contributions of the different variables. Similarly, the left side of each constraint must be the sum of the individual usages of each variable from the corresponding resource. For example, in the case of two *competing* products, where an increase in the sales level of one product adversely affects that of the other, the two products do not satisfy the additivity property.

Exercise 2.1–1

The following questions apply to the model just described.
(a) Rewrite each of the constraints below under the stipulated conditions.
   (1) The daily demand for interior paint exceeds that of exterior paint by *at least* 1 ton.
      [*Ans.* $x_I - x_E \geq 1$.]
   (2) The daily usage of raw material A is *at most* 6 tons and *at least* 3 tons.
      [*Ans.* $x_E + 2x_I \leq 6$ and $x_E + 2x_I \geq 3$.]

(3) The demand for interior paint cannot be less than the demand for exterior paint.
[*Ans.* $x_I - x_E \geq 0$.]

(b) Check whether the following solutions are feasible.
(1) $x_E = 1$, $x_I = 4$; (2) $x_E = 2$, $x_I = 2$; (3) $x_E = 3\frac{1}{3}$, $x_I = 1\frac{1}{3}$; (4) $x_E = 2$, $x_I = 1$; (5) $x_E = 2$, $x_I = -1$.
[*Ans.* All solutions are feasible except (1) and (5).]

(c) Consider the feasible solution $x_E = 2$, $x_I = 2$. Determine
(1) Slack (unused) amount of raw material A.
[*Ans.* Zero.]
(2) Slack amount of raw material B.
[*Ans.* 2 tons.]

(d) Determine the best solution among all the feasible solutions in part (b).
[*Ans.* (2) $z = 10$; (3) $z = 12\frac{2}{3}$; (4) $z = 8$. Solution (3) is the best.]

(e) Can you guess the number of *feasible* solutions the Reddy Mikks problem may have?
[*Ans.* Infinity of solutions, which makes it futile to use the enumeration procedure in part (d) and points to the need for a more "selective" technique, as we show in the following section.]

2.1.1 GRAPHICAL SOLUTION OF LP MODELS

In this section we consider the solution of Reddy Mikks LP model. The model can be solved graphically because it has only two variables. For models with three or more variables, the graphical method is either impractical or impossible. Nevertheless, we shall be able to draw general conclusions from the graphical method that will serve as the basis for the development of the general solution method in Chapter 3.

The first step in the graphical method is to plot the feasible solutions, or the (feasible) **solution space,** which satisfies all the constraints *simultaneously.* Figure 2–1 depicts the required solution space. The nonnegativity restrictions $x_E \geq 0$ and $x_I \geq 0$ confine all the feasible values to the first quadrant (which is defined by the space above or on the x_E-axis and to the right or on the x_I-axis). The space enclosed by the remaining constraints is determined by first replacing ($\leq$) by ($=$) for each constraint, thus yielding a straight-line equation. Each straight line is then plotted on the (x_E, x_I) plane, and the region in which each constraint holds when the inequality is activated is indicated by the direction of the arrow on the associated straight line. The resulting solution space is shown in Figure 2–1 by the area *ABCDEF*. (You must satisfy yourself that the arrows on each constraint actually represent the associated inequality.)

Each point within or on the boundary of the solution space *ABCDEF* satisfies all the constraints and hence represents a *feasible* point. Although there is an *infinity* of feasible points in the solution space, the **optimum solution** can be determined by observing the direction in which the objective function $3x_E + 2x_I$ increases. Figure 2–2 illustrates this result. The parallel lines representing the objective function are plotted by assigning (arbitrary) increasing values to $z = 3x_E + 2x_I$ to determine both the slope and the direction in which total revenue (objective function) increases. In Figure 2–2 we used $z = 6$ and $z = 9$. (Verify!)

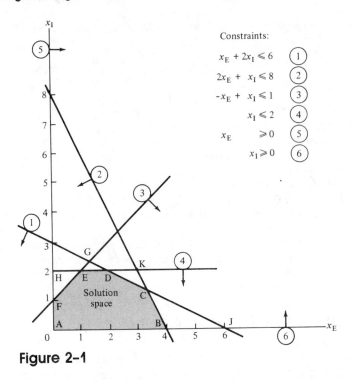

Figure 2-1

To find the optimum (maximum) solution, we move the revenue line "uphill" to the point where any further increase in revenue would render an infeasible solution. Figure 2–2 shows that the optimum solution occurs at point C. Since C is the intersection of lines ① and ② (see Figure 2–1), the values of x_E and x_I are determined by solving the following two equations simultaneously:

$$x_E + 2x_I = 6$$
$$2x_E + x_I = 8$$

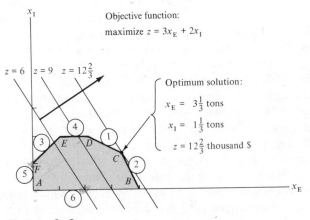

Figure 2-2

The two equations yield $x_E = 3\frac{1}{3}$ and $x_I = 1\frac{1}{3}$. The solution thus says that the daily production should be $3\frac{1}{3}$ tons of exterior paint and $1\frac{1}{3}$ tons of interior paint. The associated revenue is

$$z = 3(3\tfrac{1}{3}) + 2(1\tfrac{1}{3}) = \$12\tfrac{2}{3} \text{ thousand}$$

Exercise 2.1-2

(a) Identify the solution space and the optimum solution for the Reddy Mikks problem if each of the following changes is effected separately. Assume that each change replaces, rather than augments, an existing condition in the model and that the remaining information of the model remains unchanged. (Refer to Figure 2–1 for identifying and checking the answers.)

   (1) The maximum demand for interior paint is 3 tons daily.
   [*Ans.* The solution space is the area *ABCGF;* the optimum solution remains at *C.*]
   (2) Demand for interior paint is *at least* 2 tons daily.
   [*Ans.* Area *EDG;* optimum $x_E = x_I = 2$, $z = 10$ at *D.*]
   (3) Demand for interior paint is *exactly* 1 ton higher than for exterior paint.
   [*Ans.* Line segment *EF;* optimum $x_E = 1$, $x_I = 2$, $z = 7$ at *E.*]
   (4) Daily availability of raw material B is *at least* 8 tons.
   [*Ans.* Area *BCJ;* optimum $x_E = 6$, $x_I = 0$, $z = 18$ at *J.*]
   (5) Availability of raw material B is *at least* 8 tons per day and the demand for interior paint exceeds that of exterior paint by *at least* 1 ton.
   [*Ans.* The problem has no feasible solution space.]

(b) Identify the optimum solution in Figure 2–2 if the objective function is changed as shown:

   (1) $z = 3x_E + x_I$
   [*Ans.* $x_E = 4$, $x_I = 0$ at *B.*]
   (2) $z = 3x_E + 1.5x_I$
   [*Ans.* Any point on the line segment joining the points *B* and *C*.]
   (3) $z = x_E + 3x_I$
   [*Ans.* $x_E = 2$, $x_I = 2$ at *D.*]
   (4) In case (2) the problem has more than one optimum solution. What is the value of the objective function at all these optima?
   [*Ans.* The value of the objective function remains *the same* and equal to 12. We refer to such solutions as **alternative optima** because they yield the *same* optimum objective value.]

Exercise 2.1–2(b) reveals the interesting observation that the optimum solution can always be identified with one of the feasible **corner** (or **extreme**) **points** of the solution space: *A, B, C, D, E,* and *F* in Figure 2–2. The choice of the specific corner point depends in the first place on the slope (coefficients) of the objective function. Notice that even in case (2), where the optimum solution need not occur at a corner point, all the alternative optima are known once the corner points *B* and *C* are determined.

We shall show in Chapter 3 that the observation just discussed is the key idea to solving linear programs in general. Indeed, we can see that we no longer have to concern ourselves with the fact that the solution space has an infinity of solutions because we can now concentrate on a *finite* number of *corner* points.

2.1.2 SENSITIVITY OR POSTOPTIMAL ANALYSIS—AN ELEMENTARY PRESENTATION

Sensitivity analysis is a procedure that is normally carried out *after* the optimum solution is obtained. It determines how sensitive the optimum solution is to making certain changes in the original model. For example, in the Reddy Mikks model we may be interested in studying changes in the optimum solution as a result of increasing or decreasing demand and/or availability of raw materials. We may also want to know how the optimum solution is changed if market prices are changed.

This type of analysis is always considered an *integral* part of solving linear programming models and, in fact, any operations research model. It gives the model a *dynamic* characteristic that allows the analyst to check changes in the optimum solution that could result from possible future changes in the data of the model. The dynamic characteristics of models actually typify most business activities. The absence of procedures capable of studying the way the optimum solution responds to changes may render a (static) solution obsolete before there is a chance to implement it.

In this section we utilize graphical methods to carry out sensitivity analysis. The procedure, by necessity, is thus not refined. However, the discussion will lead to general conclusions that are used in the development of powerful and efficient methods of postoptimal analysis in Chapters 3 and 4.

Sensitivity Problem 1. How much increase or decrease in resources?

After the determination of the optimum, it is logical to study possible changes in the optimum solution that could result from adjusting the level of resources used in the constraints. In particular, we are interested in two types of analysis:

1. By how much can a resource be *increased* to *improve* the optimum value of the objective function, z?

2. By how much can a resource be *decreased* without causing a change in the current optimum?

Since the level of a resource is usually given by the right-hand side of the constraints, it is common to refer to this type of analysis as **right-hand sensitivity.**

Before answering these questions, we first classify the constraints of a linear program as either **binding** or **nonbinding.** A binding constraint must pass through the optimum solution point. If it does not, it is nonbinding. In Figure 2–1 of the Reddy Mikks model, only constraints ① and ② are binding. These are the constraints associated with materials A and B.

Logically, if a constraint is *binding,* we may regard it as a **scarce resource,** since it has been used completely. On the other hand, a *nonbinding* constraint represents an **abundant resource.** Thus, for right-hand sensitivity, we are interested in knowing the amounts by which *scarce* resources can be increased to improve the solution. Similarly, we would like to determine the amounts by which *abundant* resources can be decreased without affecting the current optimum. The latter information will be especially useful if the surplus amounts of abundant resource can be diverted for use with other activities.

You may ask: Aren't we also interested in studying the effect of *increasing*

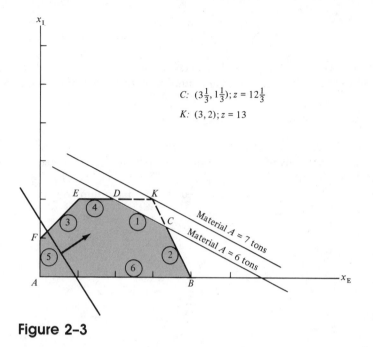

Figure 2–3

abundant resources and *decreasing* scarce resources? Well, the first change is unnecessary, since we would be trying to make an already abundant resource "more" abundant, with the expected result of not affecting the solution. On the other hand, there may be merit to studying the effect of *decreasing* a *scarce* resource in anticipation of a future shortage in supply. However, keep in mind that *a decrease in a scarce resource can never improve the value of the objective function.* (Study of this problem is postponed until Chapter 4.)

We now turn to specifics. In the Reddy Mikks model, materials A and B (constraints ① and ②) represent scarce resources. Let us consider material A first. Figure 2–3 shows that as the level of material A increases, line ① (or *CD*) moves upward parallel to itself, gradually incorporating corresponding portions of the triangle *CDK*. (Sides *CK* and *DK* of triangle *CDK* are extensions of constraints ② and ④.) When point *K* is reached, constraints ② and ④ become binding, with the optimum solution occurring at *K* and the solution space defined by *ABKEF*. At *K*, constraint ① (material A) becomes redundant, since any further increase in the level of material A will have no effect on the *solution space* or the *optimum solution*.† Thus we say that the level of material A should not be increased beyond the level that will make its constraint ① *just redundant,* that is, pass through the *new optimum* point *K*. This level is determined as follows. First, determine point *K*, which is the intersection of lines ② and ④; that is, solve the equations

---

† Notice the distinct difference between *abundance* and *redundancy*. The *deletion* of a redundant constraint has no effect whatsoever on the *solution space* or the *optimum,* whereas the constraint associated with an abundant resource does affect the *solution space* but not necessarily the optimum.

$$2x_E + x_I = 8 \qquad \text{(line ②)}$$
$$x_I = 2 \qquad \text{(line ④)}$$

This yields $x_E = 3$ and $x_I = 2$. Next, substitute K in the left-hand side of constraint ① to determine the maximum allowable level of material A. This yields

$$x_E + 2x_I = 3 + 2(2) = 7 \text{ tons}$$

Figure 2–4 represents the case where the scarce resource ② (material B) is increased. Point J gives the new optimum, which is the intersection of line ⑥ and line ①; $x_I = 0$ and $x_E + 2x_I = 6$. This yields $x_E = 6$ and $x_I = 0$, and material B can be increased to a maximum of $2x_E + x_I = 2(6) + 1(0) = 12$ tons.

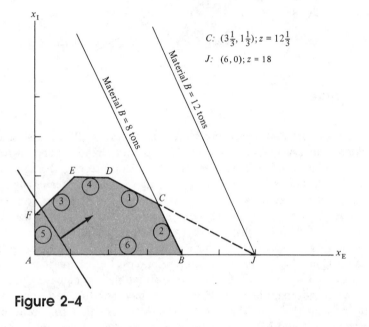

Figure 2–4

We consider next a *decrease* in the right-hand side of the nonbinding constraints. Constraint ④, $x_I \leq 2$, represents the maximum limit on the demand for interior paint. Figure 2–2 shows that line ④ (*ED*) can be lowered until it passes through the optimum C without affecting the solution. Since point C has $x_E = 3\frac{1}{3}$ and $x_I = 1\frac{1}{3}$, the maximum limit on demand for interior paint can be reduced to $x_I = 1\frac{1}{3}$ tons without changing the current optimum solution at C.

Now consider constraint ③, $-x_E + x_I \leq 1$, which represents the relationship between the demands for interior and exterior paints. Again, the right-hand sides of the constraints can be decreased until line ③ (*EF*) passes through point C. Thus the right-hand side of constraint ③ becomes equal to $-x_E + x_I = (-3\frac{1}{3}) + (1\frac{1}{3}) = -2$, or constraint ③ would read as $-x_E + x_I \leq -2$, which is equivalent to $x_E - x_I \geq 2$. This means that the current optimum solution at C will not be affected even if the exterior paint exceeds the interior paint by 2 or more tons.

These results can be summarized in the following tabular form.

Resource	Type	Maximum Change in Level of Resource (tons)	Maximum Change in Revenue z (thousands of dollars)
1	Scarce	$7 - 6 = +1$	$13 - 12\frac{2}{3} = +\frac{1}{3}$
2	Scarce	$12 - 8 = +4$	$18 - 12\frac{2}{3} = +5\frac{1}{3}$
3	Abundant	$-2 - 1 = -3$	$12\frac{2}{3} - 12\frac{2}{3} = 0$
4	Abundant	$1\frac{1}{3} - 2 = -\frac{2}{3}$	$12\frac{2}{3} - 12\frac{2}{3} = 0$

Exercise 2.1–3

(a) Consider Figure 2–1. Indicate the binding, nonbinding, and redundant constraints under each of the following conditions. Refer to the summary table.
  (1) Resource 2 is increased to its maximum limit (= 12 tons).
    [*Ans.* ② redundant; ① and ⑥ binding; all others nonbinding.]
  (2) Resource 3 is decreased to its minimum limit.
    [*Ans.* ①, ④, and ⑤ redundant; ② and ③ binding; ⑥ nonbinding.]
  (3) Resource 4 is decreased to its minimum limit.
    [*Ans.* ① redundant; ② and ④ binding; all others nonbinding.]
(b) In Figure 2–3 suppose that constraint ① (material A) is given by $2x_E + 3x_I \leq 10$ instead of $x_E + 2x_I \leq 6$. All the remaining information remains unchanged. Determine the resulting optimum and the maximum allowable increase in resource A together with the increase in revenue z.
  [*Ans.* Current optimum: $x_E = 3\frac{1}{2}$, $x_I = 1$, $z = 12\frac{1}{2}$; A is increased from 10 to 12 tons and z from $12\frac{1}{2}$ to 13.]

Sensitivity Problem 2. Which resource to increase?

In Sensitivity Problem 1 we studied the effect of increases in scarce resources (binding constraints). Under limited budget consideration, which is normally the case in most economic situations, we would like to know which resources should receive higher priorities in the allocation of funds. Linear programming analysis provides this information in a neat manner by considering the *worth* (in terms of the optimum value of the objective function) of *one* additional unit of the scarce resource. This information is obtained directly from the summary table of right-hand sensitivity given at the end of Sensitivity Problem 1.

Let y_i be the worth per unit of resource i; then y_i is determined from the formula

$$y_i = \frac{\text{maximum change in optimum } z}{\text{maximum allowable increase in resource } i}$$

For example, for constraint ① (material A), the summary table given earlier shows that

$$y_1 = \frac{13 - 12\frac{2}{3}}{7 - 6} = \frac{1}{3} \text{ thousand dollars per ton of A}$$

We can thus summarize the worth per unit for all the resources as follows:

Resource	Type	Value of y_i (thousand dollars/ton)
1	Scarce	$y_1 = 1/3$
2	Scarce	$y_2 = 4/3$
3	Abundant	$y_3 = 0$
4	Abundant	$y_4 = 0$

This information shows that resource 2 (material B) should receive first priority in allocation of funds, followed by resource 1 (material B). Abundant resources 3 and 4 should not be increased—as one should expect.

Exercise 2.1–4

In the Reddy Mikks model, suppose that constraint ① (material A) is changed from $x_E + 2x_I \leq 6$ to $2x_E + 3x_I \leq 10$, with all the remaining information unchanged. Compute y_i for all the resources.
[*Ans.* $y_1 = 1/4$; $y_2 = 5/4$; $y_3 = y_4 = 0$.]

Sensitivity Problem 3. How much change in the objective function coefficients?

A change in the objective function coefficients can affect the *slope* of the straight line representing it. We have seen in Section 2.1.1 that the choice of a specific corner point as the optimum solution depends primarily on the slope of the objective function. This means that changes in the objective function coefficients can change the set of binding constraints and hence the status (abundant or scarce) of a given resource. Sensitivity analysis of the objective coefficients may thus involve the following typical situations:

1. By how much can a coefficient be changed (increased or decreased) without causing any change in the optimal corner point?
2. By how much must a coefficient be changed to reverse the status of a given resource from abundant to scarce, and vice versa?

We shall demonstrate the answer to these questions by using the Reddy Mikks model.

To see how changes in the coefficients of the objective function can affect the optimal solution of the Reddy Mikks model, let c_E and c_I be the respective revenues per ton of exterior and interior paints. Thus the objective function may be written

$$z = c_E x_E + c_I x_I$$

Figure 2–5 shows that as c_E increases *or* c_I decreases, the objective function z will rotate in a *clockwise* direction (pivoting at C). On the other hand, a decrease in c_E *or* an increase in c_I causes z to rotate in a *counterclockwise* direction. Thus the point C will remain optimum as long as the slope of z varies between the slopes of constraints ① and ②. When the slope of z coincides with that

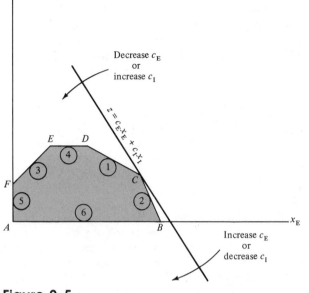

Figure 2–5

of ①, we have two *alternative* corner optima at C and D. Similarly, when it coincides with that of ②, we obtain *alternative* corner optima at B and C. (Alternative optima mean that the value of z remains the same even though the value of the variable may change. See part (b-4) of Exercise 2.1–2.) Any small change outside the range just defined for c_E will result in a new solution at B *or* at D.

We illustrate the procedure by determining the allowable range for c_E that will maintain the optimum at C. The value of c_I will be fixed at 2, its current value. Figure 2–5 shows that c_E can be increased until z coincides with ② or decreased until z coincides with ①. Thus the minimum and maximum values of c_E can be determined by equating the slope of z to the slopes of ① and ②, respectively. Since the slope of z is $c_E/2$ while those of ① and ② are 1/2 and 2/1, it follows that the minimum value of c_E is determined from

$$\frac{c_E}{2} = \frac{1}{2} \qquad \text{or} \qquad \text{minimum } c_E = 1$$

Similarly, maximum c_E is determined from

$$\frac{c_E}{2} = \frac{2}{1} \qquad \text{or} \qquad \text{maximum } c_E = 4$$

The range of c_E for which point C remains strictly the *unique* optimum is given by

$$1 < c_E < 4$$

When $c_E = 1$, the optimum *corner* solution occurs either at C or D. If c_E is decreased just below 1, the optimum is shifted to D. A similar interpretation can be developed if c_E is equal to 4 or increased just above 4.

We can now see that if c_E is just below 1, resource 2 becomes abundant and resource 4 becomes scarce. In terms of the Reddy Mikks problem, this means that if the revenue per ton of exterior paint is dropped below $1000, it is more profitable to produce the maximum limit of interior paint (i.e., $x_I = 2$ tons). In this case the use of raw material B (constraint ②) will drop, which will make resource 2 abundant. Similar conclusions can be drawn if c_E is increased just above 4.

Exercise 2.1–5

(a) In Figure 2–5, find the range of variation for c_I that will keep the optimum solution uniquely at C. The value of c_E remains fixed at 3.
   [*Ans.* $3/2 < c_I < 6$.]
(b) In part (a) identify the optimum corner points when c_I is increased just above its maximum limit.
   [*Ans.* Point D.]
(c) Suppose that the objective function is originally given as $z = 3x_E + x_I$ (instead of $z = 3x_E + 2x_I$). The associated optimum solution will then occur at B, where $x_E = 4$ and $x_I = 0$. This means that no interior paint will be produced. By how much should the profit per ton of exterior paint be adjusted to start producing positive quantities of interior paint?
   [*Ans.* Decrease c_E from $3000 to at least $2000.]

The foregoing sensitivity problems present the basic ideas underlying **postoptimal analysis.** Chapter 4 gives a more complete presentation of the topic as it applies to the general LP problem.

2.2   LP FORMULATIONS

In this section we present two LP examples. You will notice that the decision variables in the two examples are easy to define. Later, in Section 2.3, we present other formulations in which the identification and/or use of the decision variables are not as straightforward.

Example 2.2–1 (Bank Loan Policy). A financial institution, the Thriftem Bank, is in the process of formulating its loan policy for the next quarter. A total of $12 million is allocated for that purpose. Being a full-service facility, the bank is obligated to grant loans to different clientele. The following table provides the types of loans, the interest rate charged by the bank, and the possibility of bad debt as estimated from past experience:

Type of Loan	Interest Rate	Probability of Bad Debt
Personal	.140	.10
Car	.130	.07
Home	.120	.03
Farm	.125	.05
Commercial	.100	.02

Bad debts are assumed unrecoverable and hence produce no interest revenue.

Competition with other financial institutions in the area requires that the bank allocate at least 40% of the total funds to farm and commercial loans. To assist the housing industry in the region, home loans must equal at least 50% of the personal, car, and home loans. The bank also has a stated policy specifying that the overall ratio for bad debts on all loans may not exceed .04.

Mathematical Model

The variables of the model can be defined as follows:

$x_1 =$ personal loans (in millions of dollars)
$x_2 =$ car loans
$x_3 =$ home loans
$x_4 =$ farm loans
$x_5 =$ commercial loans

The objective of the Thriftem Bank is to maximize its net return comprised of the difference between the revenue from interest and lost funds due to bad debts. Since bad debts are not recoverable, both as principal and interest, the objective function may be written as

$$\text{maximize } z = .14(.9x_1) + .13(.93x_2) + .12(.97x_3) + .125(.95x_4)$$
$$+ .1(.98x_5) - .1x_1 - .07x_2 - .03x_3 - .05x_4 - .02x_5$$

This function simplifies to

$$\text{maximize } z = + .026x_1 + .0509x_2 + .0864x_3 + .06875x_4 + .078x_5$$

The problem has five constraints:

1. *Total funds*

$$x_1 + x_2 + x_3 + x_4 + x_5 \leq 12$$

2. *Farm and commercial loans*

$$x_4 + x_5 \geq .4 \times 12$$

or

$$x_4 + x_5 \geq 4.8$$

3. *Home loans*

$$x_3 \geq .5\,(x_1 + x_2 + x_3)$$

4. *Limit on bad debts*

$$\frac{.1x_1 + .07x_2 + .03x_3 + .05x_4 + .02x_5}{x_1 + x_2 + x_3 + x_4 + x_5} \leq .04$$

or

$$.06x_1 + .03x_2 - .01x_3 + .01x_4 - .02x_5 \leq 0$$

5. *Nonnegativity*

$$x_1 \geq 0, \, x_2 \geq 0, \, x_3 \geq 0, \, x_4 \geq 0, \, x_5 \geq 0$$ ◄

Example 2.2–2 (Land Use and Development). The Birdeyes Real Estate Co. owns 800 acres of prime, but undeveloped, land on a scenic lake in the heart of the Ozark Mountains. In the past, little or no regulation was applied to new developments around the lake. The lake shores are now lined with clustered vacation homes. Because of the lack of sewage service, septic tanks, mostly improperly installed, are in extensive use. Over the years, seepage from the septic tanks has resulted in a severe water pollution problem.

To curb further degradation in the quality of water, county officials introduced and approved some stringent ordinances applicable to all future developments.

1. Only single-, double-, and triple-family homes can be constructed, with the single-family homes accounting for at least 50% of the total.
2. To limit the number of septic tanks, minimum lot sizes of 2, 3, and 4 acres are required for single-, double-, and triple-family homes.
3. Recreation areas of 1 acre each must be established at the rate of one area per 200 families.
4. To preserve the ecology of the lake, underground water may not be pumped for house or garden use.

The president of Birdeyes Real Estate is studying the possibility of developing the company's 800 acres on the lake. The new development will include single-, double-, and triple-family homes. He estimates that 15% of the acreage will be consumed in the opening of streets and easements for utilities. He also estimates his returns from the different housing units:

Housing Units	Single	Double	Triple
Net return per unit ($)	10,000	15,000	20,000

The cost of connecting water service to the area is proportionate to the number of units constructed. However, the county stipulates that a minimum of $100,000 must be collected for the project to be economically feasible. Additionally, the expansion of the water system beyond its present capacity is limited to 200,000 gallons per day during peak periods. The following data summarize the cost of connecting water service as well as the water consumption assuming an average size family:

Housing Unit	Single	Double	Triple	Recreation
Water service cost per unit ($)	1000	1200	1400	800
Water consumption per unit (gal/day)	400	600	840	450

Mathematical Model

The company must decide upon the number of units to be constructed of each housing type together with the number of recreation areas satisfying county ordinances. Define

x_1 = number of units of single-family homes
x_2 = number of units of double-family homes
x_3 = number of units of triple-family homes
x_4 = number of recreation areas

An apparent objective of the company is to maximize total return. The objective function is thus given as

$$\text{maximize } z = 1000x_1 + 15{,}000x_2 + 2000x_3$$

The constraints of the problem include

1. Limit on land use.
2. Limit on the requirements for single-family homes relative to other styles.
3. Limit on the requirements for recreation areas.
4. Capital requirement for connecting water service.
5. Limit on peak-period daily water consumption.

These constraints are expressed mathematically as follows:

1. *Land use*

$$2x_1 + 3x_2 + 4x_3 + 1x_4 \leq 680$$

2. *Single-family homes*

$$\frac{x_1}{x_1 + x_2 + x_3} \geq .5$$

or

$$.5x_1 - .5x_2 - .5x_3 \geq 0$$

3. *Recreation areas*

$$x_4 \geq \frac{x_1 + 2x_2 + 3x_3}{200}$$

or

$$200x_4 - x_1 - 2x_2 - 3x_3 \geq 0$$

4. *Capital*

$$1000x_1 + 1200x_2 + 1400x_3 + 800x_4 \geq 100{,}000$$

5. *Water consumption*

$$400x_1 + 600x_2 + 840x_3 + 450x_4 \leq 200{,}000$$

6. *Nonnegativity*

$$x_1 \geq 0, \; x_2 \geq 0, \; x_3 \geq 0, \; x_4 \geq 0$$

◄

2.3   ADDITIONAL LP FORMULATIONS

In Section 2.2 we presented two LP formulations in which the definitions of the decision variables as well as the construction of the objective and constraint functions are almost straightforward. In this section we present three additional formulations that are characterized by a degree of subtlety in the way the variables are defined and used in the model. The objective, of course, is to expose you to new ideas in model development.

Example 2.3-1 (Bus Scheduling Problem). Progress City is studying the feasibility of introducing a mass transit bus system that will alleviate the smog problem by reducing in-city driving. The initial study seeks the determination of the minimum number of buses that can handle the transportation needs. After gathering necessary information, the city engineer noticed that the minimum number of buses needed to meet demand fluctuates with the time of the day. Studying the data further, she discovered that the required number of buses can be assumed constant over successive intervals of 4 hours each. Figure 2–6 summarizes the engineer's findings. It was decided that to carry out the required daily maintenance, each bus could operate only 8 successive hours a day.

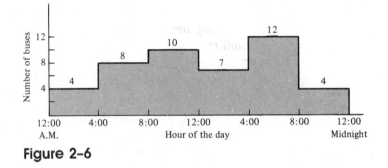

Figure 2–6

Mathematical Representation

It is required to determine the *number of buses to operate during different shifts* (variables) that will *meet the minimum demand* (constraints) while *minimizing the total number of daily buses in operation* (objective).

You may already have noticed that the definition of the variables is ambiguous. We know that each bus will run 8-hour shifts, but we do not know when a shift should start. If we follow a normal three-shift schedule (8:01 A.M.–4:00 P.M., 4:01 P.M.–12:00 midnight, and 12:01 A.M.–8:00 A.M.) and assume that x_1, x_2, and x_3 are the number of buses starting in the first, second, and third shifts, we can see from Figure 2–6 that $x_1 \geq 10$, $x_2 \geq 12$, and $x_3 \geq 8$, with the corresponding minimum number equal to $x_1 + x_2 + x_3 = 10 + 12 + 8 = 30$ buses daily.

This solution is acceptable only if the shifts *must* coincide with the normal three-shift schedule. It may be advantageous, however, to allow the optimization

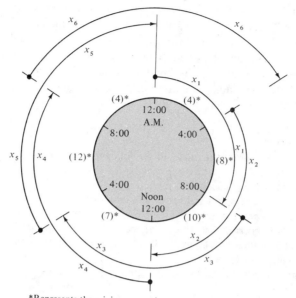

*Represents the minimum requirement in the 4-hour period.

Figure 2–7

process to choose the "best" starting time for a shift. A reasonable way to accomplish this is to allow a shift to start every 4 hours. Figure 2–7 illustrates this concept where (overlapping) shifts may start at 12:01 A.M., 4:01 A.M., 8:01 A.M., 12:01 P.M., 4:01 P.M., and 8:01 P.M., with each shift continuing for 8 consecutive hours. We are now ready to define the variables:

x_1 = number of buses starting at 12:01 A.M.
x_2 = number of buses starting at 4:01 A.M.
x_3 = number of buses starting at 8:01 A.M.
x_4 = number of buses starting at 12:01 P.M.
x_5 = number of buses starting at 4:01 P.M.
x_6 = number of buses starting at 8:01 P.M.

The mathematical model (see Figure 2–8) is thus written as

$$\text{minimize } z = x_1 + x_2 + x_3 + x_4 + x_5 + x_6$$

subject to

$$
\begin{aligned}
x_1 \qquad\qquad\qquad\quad + x_6 &\geq 4 \quad \text{(12:01 A.M.–4:00 A.M.)}\\
x_1 + x_2 \qquad\qquad\qquad &\geq 8 \quad \text{(4:01 A.M.–8:00 A.M.)}\\
x_2 + x_3 \qquad\qquad\quad &\geq 10 \quad \text{(8:01 A.M.–12:00 noon)}\\
x_3 + x_4 \qquad\quad &\geq 7 \quad \text{(12:01 P.M.–4:00 P.M.)}\\
x_4 + x_5 \quad &\geq 12 \quad \text{(4:01 P.M.–8:00 P.M.)}\\
x_5 + x_6 &\geq 4 \quad \text{(8:01 P.M.–12:00 A.M.)}\\
x_j \geq 0, \quad j = 1, 2, \ldots, 6
\end{aligned}
$$

You will be interested to know that the optimum solution to this model requires the use of 26 buses only, with 10 buses starting at 4:01 A.M. (x_2), 12 at 12:01 P.M. (x_4), 4 at 8:01 P.M. (x_6), and no shifts scheduled to start at 12:01 A.M., 8:01 A.M., and 4:01 P.M. (i.e., $x_1 = x_3 = x_5 = 0$). Thus, by allowing the solution to choose the starting time for the shifts (as compared with sticking to the normal shift schedule), we are able to reduce the daily number of buses from 30 to 26.   ◀

Exercise 2.3–1

(a) Study the application of the model to the following situations:
   (1) Number of nurses in a hospital.
   (2) Number of police officers in a city.
   (3) Number of waiters and waitresses in a 24-hour cafeteria.
   (4) Number of operators in a telephone exchange center.
(b) If the operating cost for buses that *start* between 8:01 A.M. and 8:00 P.M. is about 80% that of buses that *start* between 8:01 P.M. and 8:00 A.M., how can this information be incorporated in the model?
   [*Ans.* Change the objective function to minimize $z = .8(x_3 + x_4 + x_5) + (x_1 + x_2 + x_6)$.]

Example 2.3–2 (Trim-Loss or Stock-Slitting Problem).

The Pacific Paper Company produces paper rolls with a standard width of 20 feet each. Special customer orders with different widths are produced by slitting the standard rolls. Typical orders (which may vary from day to day) are summarized in the following table:

Order	Desired Width (ft)	Desired Number of Rolls
1	5	150
2	7	200
3	9	300

In practice, an order is filled by setting the slitting knives to the desired widths. Usually, there are a number of ways in which a standard roll can be slit to fill a given order. Figure 2–8 shows three possible knife settings for the 20-foot roll. Although there are other feasible settings, we limit the discussion for the moment to considering settings *A, B,* and *C* in Figure 2–8. We can combine the given settings in a number of ways to fill orders for widths 5, 7, and 9 feet. The following are two examples of feasible combinations:

1. Slit 300 (standard) rolls using setting *A* and 75 rolls using *B*.
2. Slit 200 rolls using setting *A* and 100 rolls using setting *C*.

Which combination is better? We can answer this question by considering the "waste" that each combination will produce. In Figure 2–8 the shaded portion represents surplus rolls not wide enough to fill the required orders. These surplus rolls are referred to as *trim loss.* We can thus evaluate the "good-

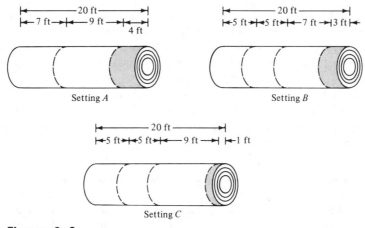

Setting *A*   Setting *B*

Setting *C*

Figure 2–8

ness" of each combination by computing its trim loss. However, since the surplus rolls may have different widths, we should base the evaluation on the *area* of trim loss rather than the number of surplus rolls. Thus, assuming that the standard roll has a length L feet, we can compute the trim-loss area as follows (see Figure 2–8):

$$\text{combination 1:}\quad 300\,(4 \times L) + \ 75\,(3 \times L) = 1425L \text{ ft}^2$$
$$\text{combination 2:}\quad 200\,(4 \times L) + 100\,(1 \times L) = \ 900L \text{ ft}^2$$

These areas account only for the shaded portions in Figure 2–8. Note, however, that any surplus production of the 5-, 7-, and 9-foot rolls must be considered in the computation of the trim-loss area. Thus, in combination 1, setting A will produce a surplus of $300 - 200 = 100$ extra 7-foot rolls while setting B will produce 75 extra 7-foot rolls. Thus the additional "waste" area is $175\,(7 \times L) = 1225L$ ft². Combination 2 does not produce surplus rolls of the 7- and 9-foot rolls. Setting C, however, does produce $200 - 150 = 50$ extra 5-foot rolls, with an added waste area of $50\,(5 \times L) = 250L$ ft². As a result we have

$$\binom{\text{total trim loss area}}{\text{for combination 1}} = 1425L + 1225L = 1650L \text{ ft}^2$$

$$\binom{\text{total trim loss area}}{\text{for combination 2}} = \ 900L + \ 250L = 1150L \text{ ft}^2$$

Combination 2 is thus better because it yields a smaller trim-loss area.

To obtain the optimum solution to the problem, it would be necessary first to determine all possible knife settings and then generate *all* the feasible combinations. Although the determination of all the settings may not be too difficult, generating all feasible combinations may be a formidable task. The need for a systematic approach is thus evident. This is what the LP model will accomplish.

Mathematical Representation

We seek to determine the *knife setting combinations* (variables) that will *fill the required orders* (constraints) with the *least trim-loss area* (objective).

The definition of the variables as given must be translated in a way that the mill operator can use. By studying the way we constructed the two combinations, we note that the variables should be defined as *the number of standard rolls to be slit according to a given knife setting*. This definition obviously requires identifying all possible knife settings as summarized in the following table. Settings 1, 2, and 3 are given in Figure 2–8. You should convince yourself of the validity of the remaining settings and that no "promising" settings have been forgotten. Keep in mind that a promising setting cannot yield a trim-loss roll of width 5 feet or higher.

Required Width (ft)	Knife Settings						Minimum Number of Rolls
	1	2	3	4	5	6	
5	0	2	2	4	1	0	150
7	1	1	0	0	2	0	200
9	1	0	1	0	0	2	300
Trim loss per foot of length	4	3	1	0	1	2	

To express the model mathematically, we define the variables as

x_j = number of standard rolls to be slit according to setting j
$\quad j = 1, 2, \ldots, 6$

The constraints of the model deal directly with satisfying the minimum number of rolls ordered. Thus if all the settings exhibited in the table are used, we get

number of 5-ft rolls produced $= 2x_2 + 2x_3 + 4x_4 + x_5$
number of 7-ft rolls produced $= x_1 + x_2 + 2x_5$
number of 9-ft rolls produced $= x_1 + x_3 + 2x_6$

These expressions represent the actual number of rolls produced with widths of 5, 7, and 9 feet, and thus must equal at least 150, 200, and 300 rolls, respectively. These are all the constraints of the model.

To construct the objective function, let y_1, y_2, and y_3 be the number of surplus rolls of the 5-, 7-, and 9-foot rolls, respectively. Thus,

$$y_1 = 2x_2 + 2x_3 + 4x_4 + x_5 - 150$$
$$y_2 = x_1 + x_2 + 2x_5 - 200$$
$$y_3 = x_1 + x_3 + 2x_6 - 300$$

A general expression for measuring the total trim-loss area is thus given by

$$L(4x_1 + 3x_2 + x_3 + x_5 + 2x_6 + 5y_1 + 7y_2 + 9y_3) \text{ ft}^2$$

Since L, the length of the standard roll, is a common factor, we can divide the entire expression by L without affecting the optimization of the objective function.

Thus the general model can be written as

$$\text{minimize } z = 4x_1 + 3x_2 + x_3 + x_5 + 2x_6 + 5y_1 + 7y_2 + 9y_3$$

subject to

$$
\begin{aligned}
2x_2 + 2x_3 + 4x_4 + x_5 \quad\quad\quad - y_1 &= 150 \quad \text{(5-ft rolls)} \\
x_1 + x_2 \quad\quad\quad + 2x_5 \quad\quad - y_2 &= 200 \quad \text{(7-ft rolls)} \\
x_1 \quad\quad + x_3 \quad\quad\quad + 2x_6 - y_3 &= 300 \quad \text{(9-ft rolls)} \\
x_j \geq 0, \quad j = 1, 2, \ldots, 6 & \\
y_i \geq 0, \quad i = 1, 2, 3 &
\end{aligned}
$$

◄

Exercise 2.3-2

(a) Using the table of knife settings given in Example 2.3–2, express each of the following feasible solutions in terms of the variables x_j and compute the trim-loss area in each case.
   (1) 200 rolls using setting 1 and 100 rolls using setting 3.
      [*Ans.* $x_1 = 200$, $x_3 = 100$, trim-loss area $= 1150L$ ft².]
   (2) 50 rolls using setting 2, 75 rolls using setting 5, and 150 rolls using setting 6.
      [*Ans.* $x_2 = 50$, $x_5 = 75$, $x_6 = 150$, trim-loss area $= 650L$ ft².]
(b) Suppose that the only available standard roll is 15 feet wide. Generate all possible knife settings for producing 5-, 7-, and 9-foot rolls and compute the associated trim loss per foot length.
   [*Ans.* Settings: (3, 0, 0), (0, 2, 0), (1, 1, 0), and (1, 0, 1). Trim loss per foot for the four settings: (0, 1, 3, 1).]

Example 2.3–3 (Goal Programming). In the previous examples, the constraints represent permanent relationships; that is, the left and right sides of each constraint are related by one of three relationships: (1) $\leq$, (2) $\geq$, or (3) $=$. Practical situations exist, however, where it may be advantageous to violate a constraint, possibly at the expense of incurring a penalty. For example, a company entertaining a number of business ventures usually operates under limited capital restrictions, but may elect to exceed that limit by borrowing additional money. The penalty incurred in this case is the cost of borrowed money (interest). Naturally, a loan can be justified on an economic basis only if the new business ventures are profitable. This type of modeling is sometimes referred to as **goal programming,** since the model automatically adjusts the level of certain resources to satisfy the goal of the decision maker.

We illustrate the goal programming model by a simple example. Two products are manufactured by passing sequentially through two different machines. The time available for the two products on each machine is limited to 8 hours daily but may be exceeded by up to 4 hours on an overtime basis. Each overtime hour will cost an additional $5. The production rates for the two products together with their profits per unit are summarized in the table that follows. It is required to determine the production level for each product that will maximize the net profit.

	Production Rate (units/hr)	
Machine	Product 1	Product 2
1	5	6
2	4	8
Profit per unit	$6	$4

Mathematical Representation

It is required to determine the *number of units of each product* (variables) that *maximizes net profit* (objective) provided that the *maximum allowable machine hours are exceeded only on an overtime basis* (constraints).

Let

$$x_j = \text{number of units of product } j, \quad j = 1, 2$$

In the absence of the overtime option, the constraints of the model are written as

$$x_1/5 + x_2/6 \leq 8 \qquad \text{(machine 1)}$$
$$x_1/4 + x_2/8 \leq 8 \qquad \text{(machine 2)}$$

To include the overtime option, we can rewrite the constraints as

$$x_1/5 + x_2/6 - y_1 \quad\;\; = 8$$
$$x_1/4 + x_2/8 \quad\;\; - y_2 = 8$$

where the variables y_1 and y_2 are unrestricted in sign for the following reason. If y_i is *negative,* the 8-hour limit on the capacity of the machine is not exceeded and no overtime is used. If it is *positive,* the used machine hours will exceed the daily limit and y_i will thus represent the overtime hours.

We have accounted for the overtime option by letting y_i assume unrestricted values. Next, we need to limit the daily use of overtime to 4 hours and also to include the cost of overtime in the objective function. Since y_i is positive only when overtime is used, the constraints

$$y_i \leq 4, \qquad i = 1, 2$$

will provide the desired restriction on the use of overtime. Note that the constraint becomes redundant when $y_i < 0$ (no overtime).

We now consider the objective function. Our goal is to maximize the net profit that equals the total profit from the two products *less* the additional cost of overtime. The expression for the total profit is given directly by $6x_1 + 4x_2$. To include the overtime cost, we note that it is incurred only when $y_i > 0$. Thus a suitable way for expressing the overtime cost is

$$\text{overtime cost} = \text{cost per hour} \times \text{overtime hours}$$
$$= 5 \, (\max \{0, y_i\})$$

Note that $\max \{0, y_i\} = 0$ when $y_i < 0$, which yields zero overtime cost, as desired.

The complete model can thus be written as

$$\text{maximize } z = 6x_1 + 4x_2 - 5(\max\{0, y_1\} + \max\{0, y_2\})$$

subject to

$$
\begin{aligned}
x_1/5 + x_2/6 - y_1 \quad &= 8 \\
x_1/4 + x_2/8 \quad\quad - y_2 &= 8 \\
y_1 \quad\quad &\leq 4 \\
y_2 &\leq 4
\end{aligned}
$$

$x_1, x_2 \geq 0$

y_1, y_2 unrestricted in sign

To convert the model to a linear program, we use the substitution

$$w_i = \max\{0, y_i\}$$

which is equivalent to

$$w_i \geq y_i \quad \text{and} \quad w_i \geq 0$$

because the *negative* coefficient of w_i in the objective function will force it to assume the smallest possible *nonnegative* value: zero or y_i. Thus the LP model can be written as

$$\text{maximize } z = 6x_1 + 4x_2 - 5(w_1 + w_2)$$

subject to

$$
\begin{aligned}
x_1/5 + x_2/6 - y_1 \quad\quad\quad &= 8 \\
x_1/4 + x_2/8 \quad\quad - y_2 \quad\quad &= 8 \\
y_1 \quad - w_1 \quad\quad &\leq 0 \\
y_2 \quad\quad - w_2 &\leq 0 \\
y_1 \quad\quad\quad &\leq 4 \\
y_2 \quad\quad\quad &\leq 4
\end{aligned}
$$

$x_1, x_2, w_1, w_2 \geq 0$

y_1, y_2 unrestricted in sign   ◀

Exercise 2.3–3

(a) Suppose that a feasible solution to the model of Example 2.3–3 yields $y_1 = 2$ and $y_2 = -1$, what does this mean in terms of the use of regular time and overtime?
[*Ans.* Machine 1 will use 2 hours of overtime and machine 2 will have 1 hour of unused regular time.]

(b) In part (a), compute the additional cost due to overtime using the objective function as defined in terms of w_i.
[*Ans.* $y_1 = 2$ will make $w_1 \geq 2$. By the optimization process $w_1 = 2$ and the associated term in the objective function is $-5 \times 2 = -\$10$. For $y_2 = -1$, $w_2 \geq -1$, but since $w_2 \geq 0$ by definition, the optimization process will force w_2 to equal zero. Hence its term in the objective function will be $-5 \times 0 = 0$.]

In closing this section, we remark that in some of the models we have presented, the physical properties of the problem actually require the variables to assume *integer* values. For example, in the *trim-loss problem* (Example 2.3–2) the variables represent the *number* of rolls that must be slit according to a given knife setting. We must point out that LP solutions are not guaranteed to be integers except in special cases. As a result, the only possible remedy at this point is to *round* the continuous optimum solution. This procedure may be acceptable, particularly if the optimum values of the variables are relatively large. In any event, Chapter 10 will present exact methods for solving integer linear programs. The major drawback is that available integer programming methods are generally inefficient computationally.

2.4 THE LP MODEL AND RESOURCE ALLOCATION

Linear programming often represents **allocation problems** in which limited resources are allocated to a number of economic activities. To provide this interpretation, we write the LP model as follows:

$$\text{maximize } z = c_1 x_1 + c_2 x_2 + \cdots + c_n x_n$$

subject to

$$a_{11} x_1 + a_{12} x_2 + \cdots + a_{1n} x_n \leq b_1$$
$$a_{21} x_1 + a_{22} x_2 + \cdots + a_{2n} x_n \leq b_2$$
$$\vdots$$
$$a_{m1} x_1 + a_{m2} x_2 + \cdots + a_{mn} x_n \leq b_m$$
$$x_1, x_2, \ldots, x_n \geq 0$$

From the economic standpoint, linear programming seeks the *best* allocation of *limited* resources to specific economic activities. In the general LP model, there are n activities whose levels (unknown values) are represented by x_1, $x_2, \ldots, x_n$. There are also m resources whose maximum availabilities are given by $b_1, b_2, \ldots, b_m$. Each unit of activity j consumes an amount a_{ij} of resource i. This means that the quantity $\sum_{j=1}^{n} a_{ij} x_j$ represents the total usage by all n activities of resource i and hence cannot exceed b_i.

The objective function $\sum_{j=1}^{n} c_j x_j$ represents a measure of the contribution of the different activities. In the maximization case, c_j represents the profit per unit of activity j, whereas in the case of minimization, c_j represents the cost per unit. We note that the "worth" of an activity cannot be judged in terms of the objective coefficient c_j only: The activity's consumption of the limited resources is also an important factor. Because all the activities of the model are competing for limited resources, the *relative* contribution of an activity (with respect to other activities) depends on both its objective coefficient c_j and its consumption of the resources a_{ij}. Thus an activity with very high unit profit may remain at the zero level because of its excessive use of limited resources.

2.5 SUMMARY

Linear programming is a *resources allocation model* that seeks the best allocation of limited resources to a number of competing activities. LP has been applied with considerable success to a multitude of practical problems.

The suitability of the graphical LP solution is limited to two-variable problems. However, the graphical method reveals the important result that for solving LP problems it is only necessary to consider the *corner* (or *extreme*) points of the solution space. This result is the key point in the development of the **simplex method,** which is an algebraic procedure designed to solve the general LP problem.

Sensitivity analysis should be regarded as an integral part of solving any optimization problem. It gives the LP solutions dynamic characteristics that are absolutely necessary for making sound decisions in a constantly changing decision-making environment.

SELECTED REFERENCES

BAZARAA, M., and J. JARVIS, *Linear Programming and Network Flows,* Wiley, New York, 1977.

DANTZIG, G. B., *Linear Programming and Extensions,* Princeton University Press, Princeton, N.J., 1963.

MURTY, KATTA, *Linear Programming,* Wiley, New York, 1983.

REVIEW QUESTIONS

True (T) or False (F)?

1. ____ The proportionality property in LP is not satisfied when the per unit contribution of a variable in the objective function is dependent on the value of the variable.

2. ____ In an LP model, replacing ≤ or ≥ by = in the constraints can improve the value of the objective function.

3. ____ Replacing ≤ by = in a constraint can result in a more restrictive solution space.

4. ____ A constraint of the type ≥ can be made more restrictive by decreasing its right-side constant.

5. ____ A two-dimensional space with two equality constraints can include at most one feasible point provided that the resulting two lines do not coincide (i.e., the two equations are independent).

6. ____ A two-dimensional solution space with two equality constraints can include an infinity of feasible points only if the two lines coincide (i.e., the two equations are dependent).

7. ____ The optimum LP solution, when finite, can always be determined from a knowledge of *all* the extreme points of the solution space.

8. ____ In a two-dimensional LP solution, the objective function can assume the *same* value at two distinct extreme points.

9. ____ In an LP model, the *feasible* solution space can be affected when *redundant* constraints are deleted.

10. ____ In an LP model, the *feasible* solution space can be changed when *nonbinding* constraints are deleted.

11. ____ In an LP model, the *optimal solution* can be changed when *nonbinding* constraints are deleted.

12. ____ *Redundant* constraints represent *abundant* resources.

13. ____ Changes in the availability of a *scarce* resource will definitely affect the optimal values of both the objective function and the variables.

14. ____ Changes in the coefficients of the objective function will definitely result in changing the optimal values of the variables.

15. ____ Changes in the coefficients of the objective function can change the status of a resource from abundant to scarce, and vice versa.

16. ____ In real-life situations, the variables of an LP model can be unrestricted in sign.

17. ____ In an LP model, the variable representing the activity with the largest profit *per unit* in the objective function will always appear at positive level in the optimal solution.

18. ____ A nonpromising activity can be made profitable by reducing its consumption of the limited resources.

[*Ans.* **1**—T, **2**—F, **3**—T, **4**—F, **5**—T, **6**—T, **7**—T, **8**—T, **9**—F, **10**—T, **11**—F, **12**—F, **13**—T, **14**—F, **15**—T, **16**—T, **17**—F, **18**—T.]

PROBLEMS

Section	Assigned Problems
2.1.1	2–1 to 2–19
2.1.2	2–20 to 2–25
2.2, 2.3	2–26 to 2–36

☐ **2–1** A small furniture factory manufactures tables and chairs. It takes 2 hours to assemble a table and 30 minutes to assemble a chair. Assembly is carried out by four workers on the basis of a single 8-hour shift per day. Customers usually buy at least four chairs with each table, meaning that the factory must produce at least four times as many chairs as tables. The sale price is $150 per table and $50 per chair. Determine the daily production mix of chairs and tables that would maximize the total daily revenue to the factory and comment on the significance of the obtained solution.

☐ **2–2** A farmer owns 200 pigs that consume 90 lb of special feed daily. The feed is prepared as a mixture of corn and soybean meal with the following compositions:

Feedstuff	Pounds per Pound of Feedstuff			Cost (dollars/lb)
	Calcium	Protein	Fiber	
Corn	.001	.09	.02	.20
Soybean meal	.002	.60	.06	.60

The dietary requirements of the pigs are

1. At most 1% calcium.
2. At least 30% protein.
3. At most 5% fiber.

Determine the daily minimum-cost feed mix.

☐ **2–3** A small bank is allocating a maximum of $20,000 for personal and car loans during the next month. The bank charges an annual interest rate of 14% for personal loans and 12% for car loans. Both types of loans are repaid over three-year periods. The amount of car loans should be at least twice as much as that of personal loans. Past experience has shown that bad debts amount to 1% of all personal loans. How should the funds be allocated?

☐ **2–4** The Popeye Canning Company is contracted to receive 60,000 lb of ripe tomatoes at 7¢/lb from which it produces both canned tomato juice and tomato paste. The canned products are packaged in cases of 24 cans each. A single can of juice requires 1 lb of fresh tomatoes whereas that of paste requires 1/3 lb only. The company's share of the market is limited to 2000 cases of juice and 6000 cases of paste. The wholesale prices per case of juice and paste stand at $18 and $9, respectively. Devise a production schedule for Popeye.

☐ **2–5** A radio assembly plant produces two models, HiFi-1 and HiFi-2, on the same assembly line. The assembly line consists of three stations. The assembly times in the workstations are

Workstation	Minutes per Unit of	
	HiFi-1	HiFi-2
1	6	4
2	5	5
3	4	6

Each workstation has a maximum availability of 480 minutes per day. However, the workstations require daily maintenance, which amounts to 10%, 14%, and 12% of the 480 minutes daily availability for stations 1, 2, and 3, respectively. The company wishes to determine the daily units to be assembled of HiFi-1 and HiFi-2 to minimize the sum of unused (idle) times at all three workstations.

☐ **2-6** An electronic company manufacturers two radio models, each on a separate rate production line. The daily capacity of the first line is 60 radios and that of the second is 75 radios. Each unit of the first model uses 10 pieces of a certain electronic component, whereas each unit of the second model requires 8 pieces of the same component. The maximum daily availability of the special component is 800 pieces. The profit per unit of models 1 and 2 is $30 and $20, respectively. Determine the optimum daily production of each model.

☐ **2-7** Two products are manufactured by passing sequentially through three machines. Time per machine allocated to the two products is limited to 10 hours per day. The production time and profit per unit of each product are

	Minutes per Unit			
Product	Machine 1	Machine 2	Machine 3	Profit
1	10	6	8	$2
2	5	20	15	$3

Find the optimal mix of the two products.

☐ **2-8** A company can advertise its product by using local radio and TV stations. Its budget limits the advertisement expenditures to $1000 a month. Each minute of radio advertisement costs $5, and each minute of TV advertisement costs $100. The company would like to use the radio at least twice as much as the TV. Past experience shows that each minute of TV advertisement will usually generate 25 times as many sales as each minute of radio advertisement. Determine the optimum allocation of the monthly budget to radio and TV advertisements.

☐ **2-9** A company produces two products, A and B. The sales volume for product A is at least 60% of the total sales of the two products. Both products use the same raw material, of which the daily availability is limited to 100 lb. Products A and B use this raw material at the rates of 2 lb/unit and 4lb/unit, respectively. The sales price for the two products are $20 and $40 per unit. Determine the optimal allocation of the raw material to the two products.

☐ **2-10** A company produces two types of cowboy hats. Each hat of the first type requires twice as much labor time as does each hat of the second type. If all hats are of the second type only, the company can produce a total of 500 hats a day. The market limits daily sales of the first and second types to 150 and 200 hats. Assume that the profit per hat is $8 for type 1 and $5 for type 2. Determine the number of hats of each type to produce to maximize profit.

☐ **2–11** Determine the solution space graphically for the following inequalities.

$$x_1 + x_2 \le 4$$
$$4x_1 + 3x_2 \le 12$$
$$-x_1 + x_2 \ge 1$$
$$x_1 + x_2 \le 6$$

$$x_1, x_2 \ge 0$$

Which constraints are redundant? Reduce the system to the smallest number of constraints that will define the same solution space.

☐ **2–12** Write the constraints associated with the solution space shown in Figure 2–9 and identify all redundant constraints.

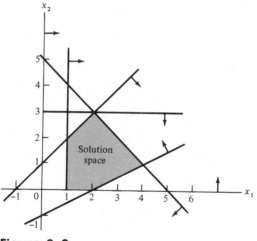

Figure 2–9

☐ **2–13** Consider the following problem:

$$\text{maximize } z = 6x_1 - 2x_2$$

subject to

$$x_1 - x_2 \le 1$$
$$3x_1 - x_2 \le 6$$
$$x_1, x_2 \ge 0$$

Show graphically that at the optimal solution the variables x_1 and x_2 can be increased indefinitely while the value of the objective function remains constant.

☐ **2–14** Consider the following linear programming problem:

$$\text{maximize } z = 4x_1 + 4x_2$$

subject to

$$2x_1 + 7x_2 \leq 21$$
$$7x_1 + 2x_2 \leq 49$$
$$x_1, x_2 \geq 0$$

Find the optimal solution (x_1^*, x_2^*) graphically. What are the ranges of variation of the coefficients of the objective function that will keep (x_1^*, x_2^*) optimal?

□ **2–15** Solve the following problem graphically:

$$\text{maximize } z = 5x_1 + 6x_2$$

subject to

$$x_1 - 2x_2 \geq 2$$
$$-2x_1 + 3x_2 \geq 2$$
$$x_1, x_2 \text{ unrestricted in sign}$$

□ **2–16** Consider the following problem:

$$\text{maximize } z = 3x_1 + 2x_2$$

subject to

$$2x_1 + x_2 \leq 2$$
$$3x_1 + 4x_2 \geq 12$$
$$x_1, x_2 \geq 0$$

Show graphically that the problem has no *feasible* extreme points. What can one conclude concerning the solution to the problem?

□ **2–17** Solve the following problem graphically:

$$\text{maximize } z = 5x_1 + 2x_2$$

subject to

$$x_1 + x_2 \leq 10$$
$$x_1 = 5$$
$$x_1, x_2 \geq 0$$

□ **2–18** In Problem 2–17, identify numerically the extreme points of the solution space. If the constraint $x_1 = 5$ is changed to $x_1 \leq 5$, determine all the *feasible* extreme points and find the optimum by evaluating the objective function numerically at each point. Show that the answer agrees with the graphical solution. Repeat the procedure with $x_1 = 5$ replaced by $x_1 \geq 5$.

□ **2–19** Consider the solution space in Figure 2–9 (Problem 2–12). Determine the optimum solution assuming that the objective function is as given:
  (a) Minimize $z = 2x_1 + 6x_2$.
  (b) Maximize $z = -3x_1 + 4x_2$.
  (c) Minimize $z = 3x_1 + 4x_2$.
  (d) Minimize $z = x_1 - 2x_2$.
  (e) Minimize $z = x_1$.
  (f) Maximize $z = x_1$.

□ **2–20** Consider the following linear program:

$$\text{maximize } z = 5x_1 + 3x_2$$

subject to

$$x_1 + x_2 \leq 4 \qquad \text{(resource 1)}$$
$$5x_1 + 2x_2 \leq 10 \qquad \text{(resource 2)}$$

$$x_1, x_2 \geq 0$$

Determine the following.
  (a) The *increase* in resource 1 that will make its constraint just redundant and the associated change in the value of z.
  (b) The *decrease* in resource 2 that will make constraint 1 just redundant and the associated change in the value of z.

□ **2–21** In Problem 2–6, the optimum solution shows that the constraints associated with the availability of the special component and the capacity of production line 1 are both scarce whereas those of line 2 are abundant.
  (a) Determine the maximum *increase* in the capacity of line 1 beyond which the objective value will not improve.
  (b) Determine the maximum *decrease* in the capacity of line 2 that will leave the current optimum unchanged.
  (c) Determine the maximum increase in the daily availability of the special component beyond which the objective value will cease to improve.
  (d) Basing your decision on the worth per resource unit, determine the scarce resource that must be given higher priority for level increase.

□ **2–22** Consider Problem 2–7, where three machines are used to manufacture two products.
  (a) Identify the machine(s) with abundant capacity at the optimum solution, then determine the amount of unused capacity (in machine hours).
  (b) For each machine with full utilization, determine the worth per unit increase in its capacity.
  (c) Which of the three machines should be given the highest priority for capacity increase?

□ **2–23** Consider Problem 2–8. Determine the maximum increase in the monthly budget beyond which any further increase will not affect the optimal solution. Analyze the result from the standpoint of its implementation in real-life situations.

□ **2–24** Consider Problem 2–6. The optimum solution specifies the daily production as 60 radios of model 1 and 25 of model 2, with an optimum profit of $2300.
  (a) Determine the range of change in the per unit profit of model 1 that will keep the optimum solution unchanged.
  (b) Repeat part (a) for model 2.
  (c) Determine the per unit profit of model 1 that can result in an optimum

solution in which both constraints representing the line capacities become *nonbinding*. The per unit profit of model 2 is assumed fixed at $20.

(d) Suppose that we vary the per unit profits of models 1 and 2 *simultaneously*. Determine a range of the ratio of the two profit coefficients that will keep the current optimum solution unchanged.

□ **2–25** Consider Problem 2–8. Determine the increase in sales per minute of radio advertisement that will make it more attractive to assign the entire monthly budget to radio advertisement only.

□ **2–26** Four products are processed successively on two machines. The manufacturing times in hours per unit of each product are tabulated for the two machines:

	Time per Unit (hr)			
Machine	Product 1	Product 2	Product 3	Product 4
1	2	3	4	2
2	3	2	1	2

The total cost of producing 1 unit of each product is based directly on the machine time. Assume that the cost per hour for machines 1 and 2 is $10 and $15, respectively. The total hours budgeted for all the products on machines 1 and 2 are 500 and 380. If the sales price per unit for products 1, 2, 3, and 4 are $65, $70, $55, and $45, formulate the problem as a linear programming model to maximize total net profit.

□ **2–27** A manufacturer produces three models (I, II, and III) of a certain product. He uses two types of raw material (A and B), of which 4000 and 6000 units are available, respectively. The raw material requirements per unit of the three models are

Raw Material	Requirements per Unit of Given Model		
	I	II	III
A	2	3	5
B	4	2	7

The labor time for each unit of model I is twice that of model II and three times that of model III. The entire labor force of the factory can produce the equivalent of 1500 units of model I. A market survey indicates that the minimum demand for the three models is 200, 200, and 150 units, respectively. However, the ratios of the number of units produced must be equal to 3:2:5. Assume that the profit per unit of models I, II, and III is $30, $20, and $50, respectively. Formulate the problem as a linear programming model to determine the number of units of each product that will maximize profit.

☐ **2–28** A business executive has the option of investing money in two plans. Plan A guarantees that each dollar invested will earn 70 cents a year hence, and plan B guarantees that each dollar invested will earn $2.00 two years hence. In plan B, only investments for periods that are multiples of 2 years are allowed. How should the executive invest $100,000 to maximize the earnings at the end of 3 years? Formulate the problem as a linear programming model.

☐ **2–29** Suppose that the *minimum* number of buses required at the ith hour of the day is b_i, $i = 1, 2, \ldots, 24$. Each bus runs 6 consecutive hours. If the number of buses in period i exceeds the minimum required b_i, an excess cost c_i per bus hour is incurred. Formulate the problem as a linear programming model so as to minimize the total excess cost incurred.

☐ **2–30** Consider the problem of assigning three types of aircraft to four routes. The table gives the pertinent data:

Aircraft Type	Capacity (passengers)	Number of Aircraft	Number of Daily Trips on Route			
			1	2	3	4
1	50	5	3	2	2	1
2	30	8	4	3	3	2
3	20	10	5	5	4	2
Daily number of customers			100	200	90	120

The associated costs are

Aircraft Type	Operating Cost Per Trip on Given Route ($)			
	1	2	3	4
1	1000	1100	1200	1500
2	800	900	1000	1000
3	600	800	800	900
Penalty cost per lost customer	40	50	45	70

Formulate the problem as a linear programming model.

☐ **2–31** In the trim-loss problem of Example 2.3–2, show that the expression for the objective function can be put in the more simplified form

$$\text{minimize } z = 20(x_1 + x_2 + \cdots + x_6)$$

☐ **2–32** Two alloys, A and B, are made from four different metals, I, II, III, and IV, according to the following specifications.

Alloy	Specifications
A	At most 80% of I At most 30% of II At least 50% of IV
B	Between 40 and 60% of II At least 30% of III At most 70% of IV

The four metals are extracted from three different ores:

Ore	Maximum Quantity (tons)	Constituents (%)					Price ($/ton)
		I	II	III	IV	Others	
1	1000	20	10	30	30	10	30
2	2000	10	20	30	30	10	40
3	3000	5	5	70	20	0	50

Assuming that the selling prices of alloys A and B are $200 and $300 per ton, formulate the problem as a linear programming model.

[*Hint:* Let X_{ijk} be the tons of the ith metal obtained from the jth ore and allocated to the kth alloy.]

☐ **2–34** A gambler plays a game that requires dividing bet money among four different choices. The game has three outcomes. The following table gives the corresponding gain (or loss) per dollar deposited in each of the four choices for the three outcomes.

Outcome	Gain (or Loss) per Dollar Deposited in Given Choice			
	1	2	3	4
1	−3	4	−7	15
2	5	−3	9	4
3	3	−9	10	−8

Assume that the gambler has a total of $500, which may be played only once. The exact outcome of the game is not known a priori, and in face of this uncertainty the gambler decided to make the allocation that would maximize the *minimum* return. Formulate the problem as a linear programming model. [*Hint:* The gambler's return may be negative, zero, or positive.]

☐ **2–35** A manufacturing company produces a final product that is assembled from three different parts. The parts are manufactured within the company by two different departments. Because of the specific setup of the machines, each department produces the three parts at different rates. The following table provides the production rates together with the maximum number of hours the two departments can allocate weekly to manufacturing the three parts.

Department	Maximum Weekly Capacity (hr)	Production Rate (units/hr)		
		Part 1	Part 2	Part 3
1	100	8	5	10
2	80	6	12	4

It would be ideal if the two departments could adjust their production facilities to produce equal quantities of the three parts, as this would result in perfect matches in terms of the final assembly. This objective may be difficult to accomplish because of the variations in production rates. A more realistic goal would be to maximize the number of final assembly units, which in essence is equivalent to minimizing the mismatches resulting from shortages in one or more parts. Formulate the problem as an LP model.

☐ **2–36** In Example 2.3–3, we can use the following substitution for the unrestricted variable y_i.

$$-y_i = y_i' - y_i''$$

where y_i' and y_i'' are *nonnegative* variables. This substitution has the property that when $y_i' > 0$, $y_i'' = 0$ and when $y_i'' > 0$, $y_i' = 0$. We can thus think of y_i' as the amount of unused resource. In this case, y_i'' will represent the amount by which the available resource is exceeded. Show how this substitution can be implemented in Example 2.3–3.

SHORT CASES

Case 2–1†

The Hi-C Company manufactures and cans three orange extracts: juice concentrate, regular juice, and jam. The products, which are intended for commercial use, are manufactured in 5-gallon cans. Jam uses Grade I oranges, and the remaining two products use Grade II. The following table lists the usages of orange as well as next year's demand.

Product	Orange Grade	Pounds of Oranges per 5-Gallon Can	Maximum Demand (cans)
Jam	I	5	10,000
Concentrate	II	30	12,000
Juice	II	15	40,000

† This case is motivated by "Red Brand Canners," *Stanford Business Cases 1965,* Graduate School of Business, Stanford University.

A market survey shows that the demand for regular juice is at least twice as high as that for the concentrate.

In the past, Hi-C bought Grade I and Grade II oranges separately at the respective prices of 25 cents and 20 cents per pound. This year, an unexpected frost forced the growers to harvest and sell the crop early without being sorted to Grade I and Grade II. It is estimated that 30% of the 3,000,000-lb crop falls into Grade I and that only 60% into Grade II. For this reason, the crop is being offered at the uniform discount price of 19 cents per pound. Hi-C estimates that it will cost the company about 2.15 cents per pound to sort the oranges into Grade I and Grade II. The below-standard oranges (10% of the crop) will be discarded.

For the purpose of cost allocation, the Accounting Department uses the following argument to estimate the cost per pound of Grade I and Grade II oranges. Since 10% of the purchased crop will fall below Grade II standards, the effective average cost per pound can be computed as $(19 + 2.15)/.9 = 23.5$ cents. Now, since the ratio of Grade I to Grade II in the purchased lot is 1 to 2, the corresponding average cost per pound based on the old prices is $(20 \times 2 + 25 \times 1)/3 = 21.67$ cents. Thus, the increase in the average price $(= 23.5$ cents $- 21.67$ cents $= 1.83$ cents) should be reallocated to the two grades by using the $1:2$ ratio, yielding a Grade I cost per pound of $25 + 1.83(1/3) = 25.61$ cents and a Grade II cost of $20 + 1.83(2/3) = 21.22$ cents. By using this information, the Accounting Department complies the following profitability sheet for the three products.

	Product (5-gallon can)		
	Jam	Concentrate	Juice
Sales price	$15.50	$30.25	$20.75
Variable costs	9.85	21.05	13.28
Allocated fixed overhead	1.05	2.15	1.96
Total cost	$10.90	$23.20	$15.24
Net profit	4.60	7.05	5.51

Based on the information given, it is desired to establish a production plan for the Hi-C Company.

Case 2–2†

A steel company operates a foundry and two mills. The foundry casts three types of steel rolls that are machined in its machine shop before being shipped to the mills. Machined rolls are used by the mills for manufacturing various products.

At the beginning of each quarter, the mills prepare their monthly needs

---

† This case is based on S. Jain, K. Stott, and E. Vasold, "Orderbook Balancing Using a Combination of Linear Programming and Heuristic Techniques," *Interfaces*, Vol. 9, no. 1, November 1978, pp. 55–67.

of rolls and submit them to the foundry. The foundry manager then draws a production plan that is essentially constrained by the machining capacity of the shop. Shortages are covered by direct purchase at a premium price from outside sources. A comparison between the cost per roll when acquired from the foundry and its outside purchase price is given in the table that follows. However, management points out that such shortage is not frequent and can be estimated to occur about 5% of the time.

Roll Type	Weight (lb)	Internal Cost (dollars per roll)	External Purchase Price (dollars per roll)
1	800	90	108
2	1200	130	145
3	1650	180	194

Processing times on the four different machines in the machine shop are

Machine Type	Processing Time per Roll			Number of Machines	Available Hours per Machine per Month
	Roll 1	Roll 2	Roll 3		
1	1	5	7	10	320
2	0	4	6	8	310
3	6	3	0	9	300
4	3	6	9	5	310

The demand for rolls by the two mills over the next three months is

	Demand in Number of Rolls					
	Mill 1			Mill 2		
Month	Roll 1	Roll 2	Roll 3	Roll 1	Roll 2	Roll 3
1	50	20	40	20	10	0
2	0	30	50	30	20	20
3	10	0	30	0	40	20

Devise a production schedule for the machine shop.

Chapter 3

Linear Programming: Algebraic Solution

A graphical solution of LP models with more than two variables is at best cumbersome. The need for an algebraic solution method is thus evident. This chapter introduces the general method called the **simplex algorithm,** which is designed to solve any linear program.

The information that can be secured from the simplex method goes beyond determining the optimum values of the variables. Indeed, it provides important economic interpretations of the problem and shows how sensitivity analyses, similar to those presented in Section 2.1.2, can be carried out algebraically.

The simplex method solves linear programs in **iterations** where the same computational steps are repeated a number of times before the optimum is reached. The nature of computations thus requires the use of the digital computer as an essential tool for solving linear programs.

The simplex method is a perfect example of the iterative process that characterizes computations in most optimization models. This chapter should thus provide a basic understanding of the use of iterations in solving OR models.

3.1 THE STANDARD FORM OF THE LP MODEL

We have seen in Chapter 2 that an LP model may include constraints of the types $\leq$, $=$, and $\geq$. Moreover, the variables may be nonnegative or unrestricted in sign. To develop a general solution method, the LP problem must be put in a common format, which we call the **standard form.** The properties of the standard LP form are

1. All the constraints are equations with *nonnegative right-hand side.*
2. All the variables are nonnegative.
3. The objective function may be maximization or minimization.

We now show how any LP model can be put in the standard format.

Constraints

1. A constraint of the type $\leq$ ($\geq$) can be converted to an equation by adding a **slack** variable to (subtracting a **surplus** variable from) the left side of the constraint.

For example, in the constraint

$$x_1 + 2x_2 \leq 6$$

we add a slack $s_1 \geq 0$ to the left side to obtain the equation

$$x_1 + 2x_2 + s_1 = 6, \qquad s_1 \geq 0$$

If the constraint represents the limit on the usage of a resource, s_1 will represent the *slack* or *unused amount* of the resource.

Next, consider the constraint

$$3x_1 + 2x_1 - 3x_3 \geq 5$$

Since the left side is not smaller than the right side, we subtract a *surplus* variable $s_2 \geq 0$ from the left side to obtain the equation

$$3x_1 + 2x_1 - 3x_3 - s_2 = 5, \qquad s_2 \geq 0$$

2. The right side of an equation can always be made nonnegative by multiplying both sides by -1.

For example, $2x_1 + 3x_2 - 7x_3 = -5$ is mathematically equivalent to $-2x_1 - 3x_2 + 7x_3 = +5$.

3. The direction of an inequality is reversed when both sides are multiplied by -1.

For example, whereas $2 < 4$, $-2 > -4$. Thus, the inequality $2x_1 - x_2 \leq -5$ can be replaced by $-2x_1 + x_2 \geq 5$.

Exercise 3.1–1

Convert the following inequalities to equations with nonnegative right-hand sides by using two procedures: (1) Multiply both sides by -1 and then augment the slack or surplus variable. (2) Convert the inequalities to equations first and then multiply both sides by -1. Does it make a difference which procedure is followed?

(a) $x_1 - 2x_2 \geq -2$.

(b) $-2x_1 + 7x_2 \leq -1$.

[*Ans.* (a) $-x_1 + 2x_2 + s_1 = 2$, $s_1 \geq 0$. (b) $2x_1 - 7x_2 - s_1 = 1$, $s_1 \geq 0$. The two procedures are exactly equivalent.]

Variables

An *unrestricted* variable y_i can be expressed in terms of two *nonnegative* variables by using the substitution

$$y_i = y_i' - y_i'' \qquad y_i', y_i'' \geq 0$$

The substitution must be effected throughout *all* the constraints and in the objective function.

The LP problem is normally solved in terms of y_i' and y_i'', from which y_i is determined by reverse substitution. An interesting property of y_i' and y_i'' is that in the optimal (simplex) LP solution only *one* of the two variables can assume a positive value, but never both. Thus, when $y_i' > 0$, $y_i'' = 0$, and vice versa. In the case where (unrestricted) y_i represents both slack and surplus, we can think of y_i' as a *slack* variable and of y_i'' as a *surplus* variable since only one of the two can assume a positive value at a time. This observation is used extensively in *goal programming* (see Example 2.3–3) and, indeed, is the basis for the conversion idea introduced in Problem 2–36.

Exercise 3.1–2

The substitution $y = y' - y''$ is used in an LP model to replace unrestricted y by the two nonnegative variables y' and y''. If y assumes the respective values -6, 10, and 0, determine the associated optimal values of y' and y'' in each case.

[*Ans.* (1) $y' = 0$, $y'' = 6$; (2) $y' = 10$, $y'' = 0$; (3) $y' = y'' = 0$.]

Objective Function

Although the standard LP model can be of either the maximization or the minimization type, it is sometimes useful to convert one form to the other.

The maximization of a function is equivalent to the minimization of the *negative* of the same function, and vice versa. For example,

$$\text{maximize } z = 5x_1 + 2x_2 + 3x_3$$

is mathematically equivalent to

$$\text{minimize } (-z) = -5x_1 - 2x_2 - 3x_3$$

Equivalence means that for the same set of constraints the *optimum* values of x_1, x_2, and x_3 are the same in both cases. The only difference is that the values of the objective functions, although equal numerically, will appear with opposite signs.

Example 3.1-1. Write the following LP model in the standard form

$$\text{minimize } z = 2x_1 + 3x_2$$

subject to

$$x_1 + x_2 = 10$$
$$-2x_1 + 3x_2 \leq -5$$
$$7x_1 - 4x_2 \leq 6$$
$$x_1 \text{ unrestricted}$$
$$x_2 \geq 0$$

The following changes must be effected.

1. Multiply the second constraint by -1 and subtract a surplus variable $s_2 \geq 0$ from the left side.
2. Add a slack variable $s_3 \geq 0$ to the left side of the third constraint.
3. Substitute $x_1 = x_1' - x_1''$, where x_1', $x_1'' \geq 0$, in the objective function and all the constraints.

Thus we get the standard form as

$$\text{minimize } z = 2x_1' - 2x_1'' + 3x_2$$

subject to

$$x_1' - x_1'' + x_2 = 10$$
$$2x_1' - 2x_1'' - 3x_2 - s_2 = 5$$
$$7x_1' - 7x_1'' - 4x_2 + s_3 = 6$$
$$x_1', x_1'', x_2, x_3, s_2, s_3, \geq 0 \qquad \blacktriangleleft$$

3.2 THE SIMPLEX METHOD

In the graphical solution in Section 2.1, we observed that the optimum solution is always associated with a *corner* (or *extreme*) point of the solution space. The simplex method is based fundamentally on this idea.

Lacking the visual advantage associated with the graphical representation of the solution space, the simplex method employs an iterative process that starts at a *feasible* corner point, normally the origin, and systematically moves from one *feasible* extreme point to another until the optimum point is eventually reached.

The general idea of the method will be illustrated in terms of the Reddy Mikks model, whose solution space is depicted in Figure 3–1. The simplex algorithm starts at the origin (point *A* in Figure 3–1), which is usually referred to as the **starting solution.** It then moves to an **adjacent corner point,** which could be either *B* or *F.* The specific choice of either point depends on the coefficients of the objective function. Since x_E has a larger coefficient than x_I and since we are *maximizing,* the solution will move in the direction in which x_E increases until extreme point *B* is reached. At *B,* the process is repeated to see if there is another extreme point that can improve the value of the objective function. Again, by using the information in the objective function we can decide whether or not such a point exists. Eventually, the solution will stop at *C,* the *optimum* extreme point.

There are two rules that govern the selection of the next extreme point in the simplex method:

1. The *next* corner point must be *adjacent* to the current one. For example, in Figure 3–1, the solution cannot move from *A* to *C* directly. Instead, it must follow the (border) **edges** of the solution space: $A \rightarrow B$ and then $B \rightarrow C$.

2. The solution can never go back to a previously considered extreme point. For example, in Figure 3–1 the solution cannot "regress" from *B* to *A*.

To summarize the ideas of the simplex method, we notice that the method always starts at a *feasible* corner point and always moves to an *adjacent* feasible corner point, checking each point for optimality before moving to a new one. In the Reddy Mikks model where the method starts at *A,* passes through *B,* and locates the optimum at *C,* we say that it took three iterations (*A, B,* and *C*) to reach the optimum.

The solution space of a linear program is represented in the simplex method by the standard form of the model. Thus, in Reddy Mikks model, the solution space is represented by

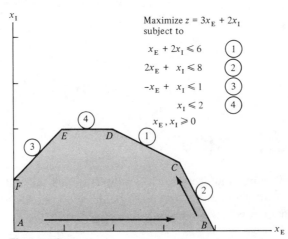

Maximize $z = 3x_E + 2x_I$
subject to

$x_E + 2x_I \leqslant 6$ ①

$2x_E + x_I \leqslant 8$ ②

$-x_E + x_I \leqslant 1$ ③

$x_I \leqslant 2$ ④

$x_E, x_I \geqslant 0$

Figure 3–1

$$\text{maximize } z = 3x_E + 2x_I + 0s_1 + 0s_2 + 0s_3 + 0s_4$$

subject to

$$
\begin{aligned}
x_E + 2x_I + s_1 \quad\quad\quad\quad &= 6 \\
2x_E + x_I \quad\quad + s_2 \quad\quad\quad &= 8 \\
-x_E + x_I \quad\quad\quad\quad + s_3 \quad &= 1 \\
x_I \quad\quad\quad\quad\quad\quad + s_4 &= 2
\end{aligned}
$$

$$x_E, \ x_I, \ s_1, \ s_2, \ s_3, \ s_4 \geq 0$$

The model consists of four equations and six unknowns that completely define all the points of the solution space. In general, the standard model will include m equations and n unknowns ($m < n$).

To determine the corner (or extreme) points directly form the standard form, we observe that a corner point is uniquely identified geometrically as an intersection of the boundary planes of the solution space. Since the standard form has more unknowns ($= n$) than equations ($= m$), the corner points are identified by setting ($n - m$) variables equal to zero and then solving for the remaining m unknowns. A mandatory requirement for the selection of ($n - m$) variables to be set equal to zero is that the remaining m variables have a *unique nonnegative* solution (otherwise, they would not represent a corner point).

To illustrate the idea, in Figure 3–1 representing the Reddy Mikks solution space, corner point A has $x_E = x_I = 0$, which yields $s_1 = 6$, $s_2 = 8$, $s_3 = 1$, and $s_4 = 2$ as the desired point (convince yourself of what it means to have the indicated slack values at point A). Similarly, at corner point B, $s_2 = 0$ and $x_I = 0$, which yield $x_E = 4$, $s_1 = 2$, $s_3 = 5$, and $s_4 = 2$.

Algebraically, the *unique* solutions resulting from setting ($n - m$) variables equal to zero are called **basic solutions.** If a basic solution satisfies the nonnegativity restrictions, it is called a **feasible basic solution.** The variables set equal to zero are called **nonbasic variables;** the remaining ones are called **basic variables.**

The simplex methods deals only with basic (feasible) solutions in the sense that it moves from one basic solution to another exactly as illustrated in Figure 3–1. Each basic solution is associated with an **iteration.** As a result, the maximum number of iterations in the simplex method cannot exceed the number of basic solutions of the standard form. We can thus conclude that the maximum number of iterations cannot exceed $C_m^n = n! / [m!(n - m)!]$.

Before presenting the details of the simplex algorithm, we need to show how we can move from one basic solution to another (i.e., from one corner point to another). We use the Reddy Mikks example as a vehicle of explanation. In Figure 3–1, we see that corner points A and B are adjacent. The following table lists the nonbasic and basic variables associated with the two points.

Extreme Point	Nonbasic (Zero) Variable	Basic Variable
A	x_E , x_I	s_1, s_2 , s_3, s_4
B	s_2 , x_I	s_1, x_E , s_3, s_4

The table shows that (adjacent) point B can be generated from point A by swapping exactly two variables. Specifically, nonbasic x_E takes the place of basic s_2 in A, thus yielding the adjacent point B in which x_E is basic and s_2 is nonbasic.

The basic–nonbasic swap gives rise to two suggestive names: The **entering variable** is a current nonbasic variable that will "enter" the set of basic variables at the next (adjacent extreme point) iteration. The **leaving variable** is a current basic variable that will "leave" the basic solution in the next iteration. Thus, in moving from A to B, the entering and leaving variables are x_E and s_2, respectively.

Exercise 3.2–1

In Figure 3–1, determine the solution space in each of the following cases:
(a) $s_1 = 0$.
  [*Ans.* Entire line ① given by $x_E = 2x_I = 6$.]
(b) $s_1, s_2, s_3, s_4, x_E, x_I \geq 0$.
  [*Ans.* Area *ABCDEF*.]
(c) $s_2 \geq 0, x_E \geq 0, x_I = 0$.
  [*Ans.* Line segment *AB*.]
(d) $s_4 < 0$.
  [*Ans.* Entire halfspace above the line passing through D and E.]

3.2.1 COMPUTATIONAL DETAILS OF THE SIMPLEX ALGORITHM

The steps of the simplex algorithm follow.

Step 0: Using the standard form, determine a **starting basic feasible solution** by setting $n - m$ appropriate (nonbasic) variables at zero level.

Step 1: Select an **entering variable** from among the current (zero) nonbasic variables which, when increased above zero, can improve the value of the objective function. If none exists, stop; the current basic solution is optimal. Otherwise, go to step 2.

Step 2: Select a **leaving variable** from among the current basic variables that must be set to zero (become nonbasic) when the entering variable becomes basic.

Step 3: Determine the new basic solution by making the entering variable basic and the leaving variable nonbasic. Go to step 1.

The details of the simplex algorithm will be explained by using the Reddy Mikks example. This will require expressing the objective function and all the constraints of the standard form as

$$
\begin{aligned}
z - 3x_E - 2x_I \qquad\qquad\qquad &= 0 \quad \text{(objective equation)} \\
x_E + 2x_I + s_1 \qquad\qquad &= 6 \\
2x_E + x_I \quad + s_2 \qquad\qquad &= 8 \\
-x_E + x_I \qquad + s_3 \qquad &= 1 \\
x_I \qquad\qquad + s_4 &= 2
\end{aligned}
\quad
\begin{pmatrix} \text{constraint} \\ \text{equations} \end{pmatrix}
$$

As mentioned, the starting solution is determined from the *constraint* equations by setting two ($= 6 - 4$) variables equal to zero, provided that the resulting solution is *unique* and *feasible*. It is evident that by putting $x_E = x_I = 0$, we immediately obtain $s_1 = 6$, $s_2 = 8$, $s_3 = 1$, and $s_4 = 2$ (point A in Figure 3–1). We can thus use this point as a starting feasible solution. The corresponding value of z is zero, since both x_E and x_I are zero. As a result, by changing the objective equation so that its right side is equal to zero, we can see that the right sides of the objective and constraint equations will automatically yield the complete starting solution. This is always the case when the starting solution consists of all *slack* variables.

We can summarize the foregoing information in a convenient tableau form as follows:

Basic	z	x_E	x_I	s_1	s_2	s_3	s_4	Solution	
z	1	−3	−2	0	0	0	0	0	z-equation
s_1	0	1	2	1	0	0	0	6	s_1-equation
s_2	0	2	1	0	1	0	0	8	s_2-equation
s_3	0	−1	1	0	0	1	0	1	s_3-equation
s_4	0	0	1	0	0	0	1	2	s_4-equation

The information in the tableau reads as follows. The "basic" column identifies the current basic variables s_1, s_2, s_3, and s_4, whose values are given in the "solution" column. This implicitly assumes that the nonbasic variables x_E and x_I (those not present in the "basic" column) are at the zero level. The value of the objective function is $z = 3 \times 0 + 2 \times 0 + 0 \times 6 + 0 \times 8 + 0 \times 1 + 0 \times 2 = 0$, as shown in the solution column.

How do we know if the current solution is the best (optimum)? By inspecting the z-equation, we notice that the current zero variables, x_E and x_I, both have *negative* coefficients, which is equivalent to having *positive* coefficients in the original objective function. Since we are *maximizing*, the value of z can thus be improved by increasing either x_E or x_I above zero level. However, we always select the variable with the *most* negative objective coefficient because computational experience has shown that such a selection is more likely to lead to the optimum solution rapidly.

This observation is the basis for what we call the **optimality condition** of the simplex method. It states that, in the case of maximization, if *all* the *non*basic variables have *nonnegative* coefficients in the z-equation of the current tableau, the current solution is optimal. Otherwise, the nonbasic variable with the *most negative* coefficient is selected as the entering variable.

Applying the optimality condition to the starting tableau, we select x_E as the **entering variable**. At this point, the leaving variable must be one of the current basic variables s_1, s_2, s_3, and s_4. This is achieved by using the **feasibility condition** that selects the **leaving variable** as a current basic variable whose value will reach zero level when the entering variable x_E reaches its maximum value at the adjacent extreme point. Of course, we need to do this without the use of the graphical solution. However, the graphical solution can assist us to develop the *feasibility condition* algebraically.

Consider the solution space of the Reddy Mikks model given in Figure 3–2. The maximum feasible value of x_E equals the *smallest positive* intercept of the constraints with the x_E-axis. Algebraically, each of these intercepts is equal to the **ratio** of the right-hand side of the constraint equation to the associated *positive* coefficient of the entering variable x_E. If the coefficient of x_E is negative or zero, the associated constraint does not intersect x_E in the positive direction. This is evident in Figure 3–2, where constraint ③ with negative x_E-coefficient ($= -1$) intersects the x_E-axis in the negative direction and constraint ④ with zero x_E-coefficient is parallel to the x_E-axis. On the other hand, the intercepts of constraints ① and ② are given by $x_E = 6/1$ and $x_E = 8/2 = 4$. Thus x_E reaches its maximum value of 4 at B, at which point s_2 will be the leaving variable.

The ratios (intercepts) just defined and the leaving variable can be determined directly from the simplex tableau. First, identify the column under the entering variable x_E and cross out all its negative and zero elements in the constraint equations. Then, excluding the objective equation, take the ratios of the right-side elements of the equations to the uncrossed-out elements under the entering variable. The leaving variable is the current basic variable associated with the minimum ratio.

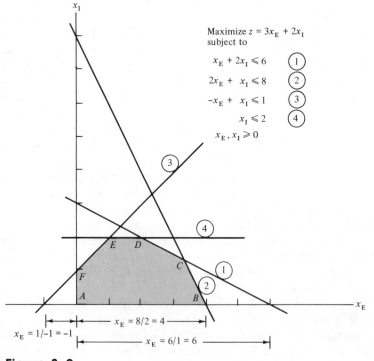

Figure 3–2

The starting tableau of the Reddy Mikks model is repeated after applying the *feasibility condition* (i.e., computing the ratios and identifying the leaving variable). For the purpose of computing the next iteration, we identify the column under the entering variable as the **entering column.** The row associated with the leaving variable will be called the **pivot equation** and the element at the intersection of the entering column and the pivot equation will be called the **pivot element.**

Entering
column
↓

Basic	z	x_E	x_I	s_1	s_2	s_3	s_4	Solution	Ratio
z	1	−3	−2	0	0	0	0	0	
s_1	0	1	2	1	0	0	0	6	$6/1 = 6$
s_2	0	2	1	0	1	0	0	8	$8/2 = ④$
s_3	0	⧸	1	0	0	1	0	1	—
s_4	0	⧸	1	0	0	0	1	2	—

Pivot
equation ← s_2

└─Pivot
  element

After determining the entering and the leaving variables (by applying the *optimality* and *feasibility conditions*), the next iteration (new basic solution) is determined by applying the **Gauss–Jordan** method. The method effects a **change in basis** by using two types of computations:

1. Type 1 (pivot equation):

   new pivot equation = old pivot equation ÷ pivot element

2. Type 2 (all other equations, including z):

$$\text{new equation} = \text{old equation} - \begin{pmatrix} \text{its } \textit{entering} \\ \textit{column} \\ \text{coefficient} \end{pmatrix} \times \begin{pmatrix} \text{new pivot} \\ \text{equation} \end{pmatrix}$$

Type 1 computations make the pivot element equal to 1 in the new pivot equation, whereas type 2 computations create zero coefficients everywhere else in the *entering column.* This is essentially equivalent to solving for the new basic solution by *substituting out* the entering variable in all but the pivot equation.

Applying type 1 to the starting tableau, we divide the s_2-equation by the pivot element 2. Since x_E takes the place of s_2 in the basic column, type 1 will lead to the following changes in the starting tableau:

Basic	z	x_E	x_I	s_1	s_2	s_3	s_4	Solution
z								
s_1								
x_E	0	1	1/2	0	1/2	0	0	$8/2 = 4$
s_3								
s_4								

Notice that the "solution" column yields the new value of x_E ($= 4$), which equals the minimum ratio of the feasibility condition.

To complete the tableau, we carry out the following type 2 computations.

1. z-equation:

$$
\begin{array}{rl}
\text{old } z\text{-equation:} & (1 \quad -3 \quad -2 \quad 0 \quad 0 \quad 0 \quad 0 \quad 0) \\
-(-3) \times \text{new pivot equation:} & (0 \quad 3 \quad 3/2 \quad 0 \quad 3/2 \quad 0 \quad 0 \quad 12) \\
= \text{new } z\text{-equation:} & (1 \quad 0 \quad -1/2 \quad 0 \quad 3/2 \quad 0 \quad 0 \quad 12)
\end{array}
$$

2. s_1-equation:

$$
\begin{array}{rl}
\text{old } s_1\text{-equation:} & (0 \quad 1 \quad 2 \quad 1 \quad 0 \quad 0 \quad 0 \quad 6) \\
-(1) \times \text{new pivot equation:} & (0 \quad -1 \quad -1/2 \quad 0 \quad -1/2 \quad 0 \quad 0 \quad -4) \\
= \text{new } s_1\text{-equation:} & (0 \quad 0 \quad 3/2 \quad 1 \quad -1/2 \quad 0 \quad 0 \quad 2)
\end{array}
$$

3. s_3-equation:

$$
\begin{array}{rl}
\text{old } s_3\text{-equation:} & (0 \quad -1 \quad 1 \quad 0 \quad 0 \quad 1 \quad 0 \quad 1) \\
-(-1) \times \text{new pivot equation:} & (0 \quad 1 \quad 1/2 \quad 0 \quad 1/2 \quad 0 \quad 0 \quad 4) \\
= \text{new } s_3\text{-equation:} & (0 \quad 0 \quad 3/2 \quad 0 \quad 1/2 \quad 1 \quad 0 \quad 5)
\end{array}
$$

4. s_4-equation. The new s_4-equation is the same as the old s_4-equation because its *entering column* coefficient is zero.

The complete new tableau thus looks as follows:

Basic	z	x_E	x_I	s_1	s_2	s_3	s_4	Solution	Ratio
z	1	0	−1/2	0	3/2	0	0	12	
s_1	0	0	3/2	1	−1/2	0	0	2	$\dfrac{2}{3/2} = \left(\dfrac{4}{3}\right)$
x_E	0	1	1/2	0	1/2	0	0	4	$\dfrac{4}{1/2} = 8$
s_3	0	0	3/2	0	1/2	1	0	5	$\dfrac{5}{3/2} = \dfrac{10}{3}$
s_4	0	0	1	0	0	0	1	2	$2/1 = 2$

The new solution gives $x_E = 4$ and $x_I = 0$ (point B in Figure 3–2). The value of z has increased from 0 to 12. The increase follows because each unit increase in x_E contributes 3 to the value of z; thus, total increase in z is $3 \times 4 = 12$.

Notice that the new tableau has the same properties as the preceding one; namely, once the nonbasic variables x_I and s_2 are set equal to zero, the values of the basic variables are immediately given in the solution column. This is precisely what the Gauss–Jordan method accomplishes.

Examining the last tableau, the optimality condition selects x_I as the entering variable because its z-coefficient is $-1/2$. The feasibility condition then shows that s_1 is the leaving variable. The ratios shown in the last tableau indicate that x_I enters the basic solution at the value 4/3 (= minimum ratio), thus improving the value of the objective function by $(4/3) \times (1/2) = 2/3$.

The following Gauss–Jordan operations will produce the new tableau:

(i) New pivot (s_1) equation = old s_1-equation $\div$ (3/2).
(ii) New z-equation = old z-equation $-$ $(-1/2) \times$ new pivot equation.
(iii) New x_E-equation = old x_E-equation $-$ $(1/2) \times$ new pivot equation.
(iv) New s_3-equation = old s_3-equation $-$ $(3/2) \times$ new pivot equation.
(v) New s_4-equation = old s_4-equation $-$ $(1) \times$ new pivot equation.

These computations lead to the following tableau.

Basic	z	x_E	x_I	s_1	s_2	s_3	s_4	Solution
z	1	0	0	1/3	4/3	0	0	$12\frac{2}{3}$
x_I	0	0	1	2/3	$-1/3$	0	0	4/3
x_E	0	1	0	$-1/3$	2/3	0	0	10/3
s_3	0	0	0	-1	1	1	0	3
s_4	0	0	0	$-2/3$	1/3	0	1	2/3

The solution yields $x_E = 3\frac{1}{3}$ and $x_I = 1\frac{1}{3}$ (point C in Figure 3–2). The value of z has increased from 12 in the preceding tableau to $12\frac{2}{3}$. The increase $(12\frac{2}{3} - 12) = 2/3$ is the result of x_I increasing from 0 to 4/3, with each unit increase accounting for 1/2 in the objective function. The total increase in z thus equals $(4/3) \times (1/2) = 2/3$.

The last tableau is optimal because *none* of the *non*basic variables has a negative coefficient in the z-equation. This completes the simplex method computations.

The simplex algorithm is applied to a maximization problem. In considering a minimization problem, we need only change the optimality condition so that the entering variable is selected as the variable having the most *positive* coefficient in the z-equation. The feasibility condition is the same for both problems. We summarize the two conditions here for convenience.

Optimality Condition. The entering variable in maximization (minimization) is the *non*basic variable with the most negative (positive) coefficient in the z-equation. A tie is broken arbitrarily. When *all* the nonbasic coefficients in the z-equation are nonnegative (nonpositive), the optimum is reached.

Feasibility Condition. For *both* the maximization and minimization problems, the leaving variable is the basic variable having the smallest ratio (with positive denominator). A tie is broken arbitrarily.

Exercise 3.2–2

(a) In the application of the simplex method to a *maximization* model, suppose that the entering variable is selected to be *any* of the nonbasic variables with a negative coefficient in the z-equation. Will the simplex method eventually reach the optimum solution?

[*Ans.* Yes, because a negative z-coefficient of a nonbasic variable indicates that the value of z can increase when this variable is increased above zero. The only advantage of selecting the nonbasic variable with the most negative coefficient is that it has the potential to reach the optimum in the smallest number of iterations.]

(b) Consider the graphical solution of the Reddy Mikks model given in Figure 3–2. Starting at the origin (point A), suppose that x_1 is selected as the entering variable. Determine the ratios of the feasibility condition directly from the graphical space; then compare them with those obtained from the starting simplex tableau. Which current basic variable should leave the basic solution?

[*Ans.* The ratios (intercepts) are 3, 8, 1, and 2. Also, s_3 leaves the solution.]

(c) In part (b) determine the increase in the value of z when x_1 enters the solution without carrying out the Gauss–Jordan computations.

[*Ans.* Increase in $z = 2 \times 1 = 2$.]

(d) Examine Figure 3–2 and determine how many iterations will be needed to reach the optimum at C if x_1 is selected as the entering variable in the starting tableau.

[*Ans.* Five iterations, corresponding to extreme points A, F, E, D, and C.]

(e) In the optimum tableau of the Reddy Mikks model, determine the leaving variable and the corresponding change (increase or decrease) in the value of z if the entering variable is given by

(1) s_1. [*Ans.* x_1 leaves and z *decreases* by $2 \times 1/3 = 2/3$.]

(2) s_2. [*Ans.* s_4 leaves and z *decreases* by $2 \times 4/3 = 8/3$.]

(f) What conclusion can be drawn from the computations in part (e)?

[*Ans.* Any attempt to make a nonbasic variable with a positive coefficient in the z-equation enter the basic solution can never increase the value of z. This is why we say that the *maximum* value of z has been attained when all the nonbasic coefficients of the z-equation have become nonnegative.]

3.2.2 ARTIFICIAL STARTING SOLUTION

In our presentation of the simplex method we have used the *slack* variables as the *starting* basic solution. However, if the original constraint is an equation or of the type ($\geq$), we no longer have a *ready* starting basic *feasible* solution. We illustrate this point by the following example:

$$\text{minimize } z = 4x_1 + x_2$$

subject to

$$3x_1 + x_2 = 3$$
$$4x_1 + 3x_2 \geq 6$$
$$x_1 + 2x_2 \leq 4$$
$$x_1, x_2 \geq 0$$

The standard form is obtained by augmenting a surplus x_3 and adding a slack x_4 to the left sides of constraints 2 and 3. Thus, we have

$$\text{minimize } z = 4x_1 + x_2$$

subject to

$$3x_1 + x_2 \qquad\qquad = 3$$
$$4x_1 + 3x_2 - x_3 \qquad = 6$$
$$x_1 + 2x_2 \qquad + x_4 = 4$$
$$x_1, x_2, x_3, x_4 \geq 0$$

We have three equations and four unknowns, which means that one variable must be nonbasic at zero in any basic solution. Unlike the case where we have slack variables in every equation, we cannot be sure that by setting one variable equal to zero, the resulting basic variables will be nonnegative. Of course, we can use trial and error by setting one variable at a time equal to zero. In addition to being time consuming, trial and error is not suitable for automatic computations. We must thus resort to a more direct method for finding a starting basic feasible solution.

The idea of using **artificial variables** is quite simple. It calls for *adding* a *nonnegative* variable to the left side of each equation that has no obvious *starting* basic variables. The added variable will play the same role as that of a slack variable, in providing a starting basic variable. However, since such artificial variables have no physical meaning from the standpoint of the original problem (hence the name "artificial"), the procedure will be valid only if we force these variables to be zero when the optimum is reached. In other words, we use them only to start the solution and must subsequently force them to be zero in the final solution; otherwise, the resulting solution will be infeasible.

We accomplish this result by using information feedback, which, through the optimization process, will ultimately force the artificial variables to be zero in the final solution *provided that a feasible solution exists*. A logical way to do this is to *penalize* the artificial variables in the objective function. Two (closely related) methods based on the idea of using penalties are devised for this purpose: (1) the M-technique or method of penalty and (2) the two-phase technique. We shall illustrate each technique using the example described earlier.

A.   The M-Technique (Method of Penalty)

Let us consider the standard form of the example just given:

$$\text{minimize } z = 4x_1 + x_2$$

subject to

$$3x_1 + x_2 \qquad\qquad = 3$$
$$4x_1 + 3x_2 - x_3 \qquad = 6$$
$$x_1 + 2x_2 \qquad + x_4 = 4$$
$$x_1, x_2, x_3, x_4 \geq 0$$

The first and second equations do not have variables that play the role of a slack. Hence we augment the two artificial variables R_1 and R_2 in these two equations as follows:

$$3x_1 + x_2 \quad + R_1 \quad = 3$$
$$4x_1 + 3x_2 - x_3 \quad + R_2 = 6$$

We can *penalize* R_1 and R_2 in the objective function by assigning them very large positive coefficients in the objective function. Let $M > 0$ be a very large constant; then the LP with its artificial variables becomes

$$\text{minimize } z = 4x_1 + x_2 + MR_1 + MR_2$$

subject to

$$3x_1 + x_2 \quad + R_1 \qquad = 3$$
$$4x_1 + 3x_2 - x_3 \quad + R_2 \quad = 6$$
$$x_1 + 2x_2 \qquad + x_4 = 4$$
$$x_1, x_2, x_3, R_1, R_2, x_4 \geq 0$$

Notice the reason behind the use of the artificial variables. We have three equations and six unknowns. Hence the starting basic solution must include $6 - 3 = 3$ zero variables. If we put x_1, x_2, and x_3 at zero level, we immediately obtain the solution $R_1 = 3$, $R_2 = 6$, and $x_4 = 4$, which is the required starting *feasible* solution.

Now, observe how the "new" model automatically forces R_1 and R_2 to be zero. Since we are minimizing, by assigning M to R_1 and R_2 in the objective function, the optimization process that is seeking the *minimum* value of z will eventually assign zero values to R_1 and R_2 in the *optimum* solution. Notice that the intermediate iterations preceding the optimum iteration are of no importance to us. Consequently, it is immaterial whether or not they include artificial variables at positive level.

How does the M-technique change if we are maximizing instead of minimizing? Using the same logic of penalizing the artificial variable, we must assign them the coefficient $-M$ in the objective function ($M > 0$), thus making it unattractive to maintain the artificial variable at a positive level in the optimum solution.

Exercise 3.2–3

In the foregoing example, suppose that the problem is of the maximization type. Write the objective function after augmenting the artificial variables.
[*Ans.* Maximize $z = 4x_1 + x_2 - MR_1 - MR_2$.]

Having constructed a starting feasible solution, we must "condition" the problem so that when we put it in tabular form, the right-side column will render the starting solution directly. This is done by using the constraint equations to substitute out R_1 and R_2 in the objective function. Thus

$$R_1 = 3 - 3x_1 - x_2$$
$$R_2 = 6 - 4x_1 - 3x_2 + x_3$$

The objective function thus becomes

$$z = 4x_1 + x_2 + M(3 - 3x_1 - x_2) + M(6 - 4x_1 - 3x_2 + x_3)$$
$$= (4 - 7M)x_1 + (1 - 4M)x_2 + Mx_3 + 9M$$

and the z-equation now appears in the tableau as

$$z - (4 - 7M)x_1 - (1 - 4M)x_2 - Mx_3 = 9M$$

Now you can see that at the starting solution, given $x_1 = x_2 = x_3 = 0$, the value of z is $9M$, as it should be when $R_1 = 3$ and $R_2 = 6$.

The sequence of tableaus leading to the optimum solution is shown in Table 3–1. Observe that this is a *minimization* problem so that the entering variable must have the most *positive* coefficient in the z-equation. The optimum is reached when *all* the nonbasic variables have *non*positive z-coefficients. (Remember that M is a very large positive constant.)

The optimum solution is $x_1 = 2/5$, $x_2 = 9/5$, and $z = 17/5$. Since it contains no artificial variables *at positive level,* the solution is feasible with respect to the original problem before the artificials are added. (If the problem has no feasible solution, at least one artificial variable will be positive in the optimum solution. This case is treated in the next section.)

Exercise 3.2–4

(a) Write the z-equation for the preceding example as it appears in the starting tableau when each of the following changes occurs independently.
  (1) The third constraint is originally of the type $\geq$.
      [*Ans.* $z + (-4 + 8M)x_1 + (-1 + 6M)x_2 - Mx_3 - Mx_4 = 13M$. Use artificial variables in all three equations.]
  (2) The second constraint is originally of the type $\leq$.
      [*Ans.* $z + (-4 + 3M)x_1 + (-1 + M)x_2 = 3M$. Use an artificial variable in the first equation only.]
  (3) The objective function is to maximize $z = 4x_1 + x_2$.
      [*Ans.* $z + (-4 - 7M)x_1 + (-1 - 4M)x_2 + Mx_3 = -9M$. Use artificial variables in the first and second equations.]
(b) In each of the following cases, indicate whether it is *absolutely necessary* to use artificial variables to secure a starting solution. Assume that all variables are nonnegative.
  (1) Maximize $z = x_1 + x_2$
      subject to

$$2x_1 + 3x_2 = 5$$
$$7x_1 + 2x_2 \leq 6$$

  [*Ans.* Yes, use R_1 in the first equation and a slack in the second.]
  (2) Minimize $z = x_1 + x_2 + x_3 + x_4$
      subject to

$$2x_1 + x_2 + x_3 \quad\quad = 7$$
$$4x_1 + 3x_2 \quad\quad + x_4 = 8$$

  [*Ans.* No, use x_3 and x_4; but first substitute them out in the z-function using $x_3 = 7 - 2x_1 - x_2$ and $x_4 = 8 - 4x_1 - 3x_2$.]

Table 3-1[a]

Iteration	Basic	x_1	x_2	x_3	R_1	R_2	x_4	Solution
0 (starting)	z	$-4+7M$	$-1+4M$	$-M$	0	0	0	$9M$
x_1 enters	R_1	3	1	0	1	0	0	3
R_1 leaves	R_2	4	3	-1	0	1	0	6
	x_4	1	2	0	0	0	1	4
1	z	0	$\dfrac{1+5M}{3}$	$-M$	$\dfrac{4-7M}{3}$	0	0	$4+2M$
x_2 enters	x_1	1	1/3	0	1/3	0	0	1
R_2 leaves	R_2	0	5/3	-1	$-4/3$	1	0	2
	x_4	0	5/3	0	$-1/3$	0	1	3
2	z	0	0	1/5	$8/5-M$	$-1/5-M$	0	18/5
x_3 enters	x_1	1	0	1/5	3/5	$-1/5$	0	3/5
x_4 leaves	x_2	0	1	$-3/5$	$-4/5$	3/5	0	6/5
	x_4	0	0	1	1	-1	1	1
3 (optimum)	z	0	0	0	$7/5-M$	$-M$	$-1/5$	17/5
	x_1	1	0	0	2/5	0	$-1/5$	2/5
	x_2	0	1	0	$-1/5$	0	3/5	9/5
	x_3	0	0	1	1	-1	1	1

[a] We have eliminated the z column for convenience, since it never changes. We shall follow this convention throughout the book.

B. The Two-Phase Technique

A drawback of the M-technique is the possible computational error that could result from assigning a very large value to the constant M. To illustrate this point, suppose that $M = 100,000$ in the M-technique example of Section 3.2.3. Then, in the starting tableau (see Table 3–1), the coefficients of x_1 and x_2 in the z-equation become $(-4 + 700,000)$ and $(-1 + 400,000)$. The effect of the original coefficients (4 and 1) is now too small relative to the large numbers created by the multiples of M. Because of round-off error, which is inherent in any digital computer, the computations may become insensitive to the relative values of the *original* objective coefficients of x_1 and x_2. The dangerous outcome is that x_1 and x_2 may be treated as having zero coefficients in the objective function.

The two-phase method is designed to alleviate this difficulty. Although the artificial variables are added in the same manner employed in the M-technique, the use of the constant M is eliminated by solving the problem in two phases (hence the name "two-phase" method). These two phases are outlined as follows:

Phase I. Augment the artificial variables as necessary to secure a starting solution. Form a new objective function that seeks the *minimization* of the *sum* of the artificial variables subject to the constraints of the original problem modified by the artificial variables. If the *minimum* value of the new objective function is zero (meaning that all artificials are zero), the problem has a feasible solution space. Go to phase II. Otherwise, if the minimum is positive, the problem has no feasible solution. Stop.

Phase II. Use the optimum basic solution of phase I as a starting solution for the original problem.

We illustrate the procedure using the M-technique example in Section 3.2.2.

Phase I. Since we need artificials R_1 and R_2 in the first and second equations, the phase I problem reads as

$$\text{minimize } r = R_1 + R_2$$

subject to

$$
\begin{aligned}
3x_1 + x_2 \quad\;\; + R_1 \qquad\qquad\;\; &= 3 \\
4x_1 + 3x_2 - x_3 \quad\;\; + R_2 \quad\; &= 6 \\
x_1 + 2x_2 \qquad\qquad\quad + x_4 &= 4 \\
x_1, x_2, x_3, R_1, R_2, x_4 &\geq 0
\end{aligned}
$$

Because R_1 and R_2 are in the starting solution, they must be substituted out in the objective function (compare with the M-technique) as follows:

$$
\begin{aligned}
r &= R_1 + R_2 \\
&= (3 - 3x_1 - x_2) + (6 - 4x_1 - 3x_2 + x_3) \\
&= -7x_1 - 4x_2 + x_3 + 9
\end{aligned}
$$

The *starting* tableau thus becomes

Basic	x_1	x_2	x_3	R_1	R_2	x_4	Solution
r	7	4	-1	0	0	0	9
R_1	3	1	0	1	0	0	3
R_2	4	3	-1	0	1	0	6
x_4	1	2	0	0	0	1	4

The *optimum* tableau is obtained in *two* iterations and is given by (verify)

Basic	x_1	x_2	x_3	R_1	R_2	x_4	Solution
r	0	0	0	-1	-1	0	0
x_1	1	0	1/5	3/5	$-1/5$	0	3/5
x_2	0	1	$-3/5$	$-4/5$	3/5	0	6/5
x_4	0	0	1	1	-1	1	1

Since the minimum $r = 0$, the problem has a feasible solution and we move to phase II.

Phase II. The artificial variables have now served their purpose and must be dispensed with in all subsequent computations. This means that the equations of the optimum tableau in phase I can be written as

$$x_1 \quad + \frac{1}{5}x_3 \qquad = 3/5$$

$$x_2 - \frac{3}{5}x_3 \qquad = 6/5$$

$$x_3 + x_4 \quad = 1$$

These equations are *exactly equivalent* to those in the standard form of the original problem (before artificials are added). Thus the original problem can be written as

$$\text{minimize } z = 4x_1 + x_2$$

subject to

$$x_1 \quad + \frac{1}{5}x_3 \qquad = 3/5$$

$$x_2 - \frac{3}{5}x_3 \qquad = 6/5$$

$$x_3 + x_4 = 1$$

$$x_1, x_2, x_3, x_4 \geq 0$$

As you can see, the principal contribution of the phase I computations is to provide a ready starting solution to the original problem. Since the problem has three equations and four variables, by putting $4 - 3 = 1$ variable, namely,

x_3, equal to zero, we immediately obtain the starting basic feasible solution $x_1 = 3/5$, $x_2 = 6/5$, and $x_4 = 1$.

To solve the problem, we need to substitute out the basic variables x_1 and x_2 in the objective function as we did in the M-technique. This is accomplished by using the constraint equations as follows:

$$z = 4x_1 + x_2$$
$$= 4\left(3/5 - \frac{1}{5}x_3\right) + \left(6/5 + \frac{3}{5}x_3\right)$$
$$= -\frac{1}{5}x_3 + 18/5$$

Thus the starting tableau for phase II becomes

Basic	x_1	x_2	x_3	x_4	Solution
z	0	0	1/5	0	18/5
x_1	1	0	1/5	0	3/5
x_2	0	1	−3/5	0	6/5
x_4	0	0	1	1	1

The tableau is not optimal, since x_3 must enter the solution. If we carry out the simplex computations, we shall obtain the optimum solution in one iteration (verify).

It is interesting to note that the number of iterations in the M-technique and the two-phase method are necessarily the same. In fact, there is a one-to-one correspondence between the tableaus of both methods. The advantage of the two-phase method rests in the elimination of the constant M.

It must be noted that in phase II, the artificial variables are removed only when they are nonbasic at the end of phase I (as in the foregoing example). It is possible, however, that an artificial variable remain *basic* at *zero* level at the end of phase I. In this case, provisions must be made to ensure that the artificial variable never becomes positive during phase II computations. The details of the procedure can be found in Dantzig [(1963), Sec. 5–2].

Exercise 3.2–5

(a) In phase I, what do the artificial variables signify? Specifically, why is the sum of the artificial variables minimized?

[*Ans.* Artificials may be regarded as a "measure of infeasibility" in the constraints. When the minimum sum of the (nonnegative) artificial variables is zero, each of these variables must be zero. Thus the resulting basic solution must be feasible for the original problem.]

(b) If the original linear program is of the maximization type, do we maximize the sum of the artificial variables in phase I?

[*Ans.* No, we should always minimize in phase I, since this is equivalent to penalizing the artificial variables.]

3.3 SPECIAL CASES IN SIMPLEX METHOD APPLICATION

In this section we consider special cases that can arise in the application of the simplex method, which include

1. Degeneracy.
2. Alternative optima.
3. Unbounded solutions.
4. Nonexisting (or infeasible) solutions.

Our interest in studying these special cases is twofold: (1) to present a *theoretical* explanation for the reason these situations arise and (2) to provide a *practical* interpretation of what these special results could mean in a real-life problem.

3.3.1 DEGENERACY

In Section 3.2 we indicated that in the application of the feasibility condition, a tie for the minimum ratio may be broken arbitrarily for the purpose of determining the leaving variable. When this happens, however, one or more of the *basic* variables will necessarily equal zero in the next iteration. In this case, we say that the new solution is **degenerate**. (In all the LP examples we have solved so far, the basic variables always assumed strictly positive values.)

There is nothing alarming about dealing with a degenerate solution, with the exception of a small theoretical inconvenience, which we shall discuss shortly. From the practical standpoint, the condition reveals that the model has at least one *redundant* constraint. To be able to provide more insight into the practical and theoretical impacts of degeneracy, we consider two numeric examples. The graphical illustrations should enhance the understanding of ideas underlying these special situations.

Example 3.3-1 (Degenerate Optimal Solution)

$$\text{Maximize } z = 3x_1 + 9x_2$$

subject to

$$x_1 + 4x_2 \leq 8$$
$$x_1 + 2x_2 \leq 4$$
$$x_1, x_2 \geq 0$$

Using x_3 and x_4 as slack variables, we list the simplex iterations for the example in Table 3–2. In the starting iteration, a tie for the leaving variable exists between x_3 and x_4. This is the reason the basic variable x_4 has a zero value in iteration 1, thus resulting in a degenerate basic solution. The optimum is reached after an additional iteration is carried out.

What is the practical implication of degeneracy? Look at Figure 3–3, which provides the graphical solution to the model. Three lines pass through the optimum ($x_1 = 0$, $x_2 = 2$). Since this is a two-dimensional problem, the point is

Table 3–2

Iteration	Basic	x_1	x_2	x_3	x_4	Solution
0 (starting)	z	-3	-9	0	0	0
x_2 enters	x_3	1	4	1	0	8
x_3 leaves	x_4	1	2	0	1	4
1	z	$-3/4$	0	$9/4$	0	18
x_1 enters	x_2	$1/4$	1	$1/4$	0	2
x_4 leaves	x_4	$1/2$	0	$-1/2$	1	0
2 (optimum)	z	0	0	$3/2$	$3/2$	18
	x_2	0	1	$1/2$	$-1/2$	2
	x_1	1	0	-1	2	0

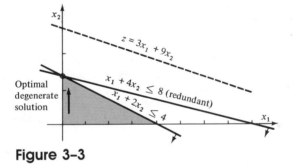

Figure 3–3

said to be *overdetermined*, since we only need two lines to identify it. For this reason, we conclude that one of the constraints is redundant. In practice, the mere knowledge that some resources are superfluous can prove valuable during the implementation of the solution. Such information may also lead to discovering irregularities in the construction of the model. Unfortunately, there are no reliable techniques for identifying redundant constraints directly from the tableau. In the absence of graphical representation, we may have to rely on other means to identify redundancy in the model.

From the theoretical standpoint, degeneracy has two implications. The first deals with the phenomenon of **cycling** or **circling.** If you look at iterations 1 and 2 in Table 3–2, you will find that the objective value has not improved ($z = 18$). It is thus conceivable, in general, that the simplex procedure would repeat the *same sequence* of iterations, never improving the objective value and never terminating the computations. Although there are methods for handling this situation so that cycling does not happen, these methods could lead to a drastic slowdown in computations. For this reason, most LP codes do not include provisions for cycling, relying on the fact that the percentage of LP problems having this complication is usually too small to warrant a routine implementation of the cycling procedures.

The second theoretical point arises in the examination of iterations 1 and 2. Both iterations, although differing in classifying the variables as basic and nonbasic, yield identical values of all the variables and objective value, namely,

$$x_1 = 0, \quad x_2 = 2, \quad x_3 = 0, \quad x_4 = 0, \quad z = 18$$

An argument thus arises as to the possibility of stopping the computations at iteration 1 (when degeneracy first appears), even though it is not optimum. This argument is not valid because, in general, a solution may be *temporarily* degenerate (see Problem 3–24).　　　　　　　　　　　　　　　◀

Exercise 3.3–1

(a) If, in a two-dimensional problem such as Example 3.3–1, we have three zero basic variables at a simplex iteration, how many redundant constraints exist at this corner point?
   [*Ans.* 3.]
(b) In an n-dimensional problem, how many planes (constraints) must pass through a corner point to produce a degenerate situation?
   [*Ans.* At least $n + 1$ planes.]
(c) Judging from the foregoing discussion, is the number of basic solutions larger than the number of corner points under degenerate conditions?
   [*Ans.* Yes, definitely.]
(d) Assuming that cycling will not occur, what is the ultimate effect of degeneracy on computations as compared to the case where redundant constraints are removed, that is, degeneracy is removed?
   [*Ans.* The number of iterations until the optimum is reached may be larger under degeneracy, since a corner point may be represented by more than one basic solution.]

3.3.2 ALTERNATIVE OPTIMA

When the objective function is parallel to a *binding* constraint (i.e., a constraint that is satisfied in the equality sense by the optimal solution), the objective function will assume the *same optimal value* at more than one solution point. For this reason they are called **alternative optima.** The next example shows that normally there is an *infinity* of such solutions. The example also demonstrates the practical significance of encountering alternative optima.

Example 3.3–2 (Infinity of Solutions)

$$\text{Maximize } z = 2x_1 + 4x_2$$

subject to

$$x_1 + 2x_2 \le 5$$
$$x_1 + x_2 \le 4$$
$$x_1, x_2 \ge 0$$

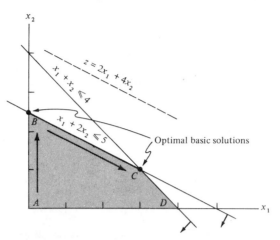

Figure 3-4

Figure 3-4 demonstrates how alternative optima can arise in LP model when the objective function is parallel to a binding constraint. Any point on the *line segment BC* represents an alternative optimum with the same objective value $z = 10$.

Algebraically, we know that the simplex method is capable of encountering corner-point solutions only. Table 3-3 shows that the optimum ($x_1 = 0$, $x_2 = 5/2$, $z = 10$) is encountered in iteration 1 at point *B*. How do we know from this tableau that alternative optima exist? Look at the coefficients of the *non*basic variables in the z-equation of iteration 1. The coefficient of (nonbasic) x_1 is zero, indicating that x_1 can enter the basic solution without changing the value of z, but causing a change in the values of the variables. Iteration 2 does just

Table 3-3

Iteration	Basic	x_1	x_2	x_3	x_4	Solution
0 (starting)	z	-2	-4	0	0	0
x_2 enters	x_3	1	2	1	0	5
x_3 leaves	x_4	1	1	0	1	4
1 (optimum)	z	0	0	2	0	10
x_1 enters	x_2	1/2	1	1/2	0	5/2
x_4 leaves	x_4	1/2	0	$-1/2$	1	3/2
2 (alternate optimum)	z	0	0	2	0	10
	x_2	0	1	1	-1	1
	x_1	1	0	-1	2	3

that—letting x_1 enter the basic solution, which will force x_4 to leave. This results in the new solution point at C ($x_1 = 3$, $x_2 = 1$, $z = 10$).

As expected, the simplex method determines only the two corner points B and C. Mathematically, we can determine all the points $(\hat{x}_1, \hat{x}_2)$ on the line segment BC as a nonnegative weighted average of the points B and C. That is, given $0 \leq \alpha \leq 1$ and

$$B: \quad x_1 = 0, \quad x_2 = 5/2$$
$$C: \quad x_1 = 3, \quad x_2 = 1$$

then all the points on the line segment BC are given by

$$\hat{x}_1 = \alpha(0) + (1 - \alpha)(3) = 3 - 3\alpha$$
$$\hat{x}_2 = \alpha(5/2) + (1 - \alpha)(1) = 1 + 3\alpha/2$$

Observe that when $\alpha = 0$, $(\hat{x}_1, \hat{x}_2) = (3, 1)$, which is point C. When $\alpha = 1$, $(\hat{x}_1, \hat{x}_2) = (0, 5/2)$, which is point B. For values of α between 0 and 1, $(\hat{x}_1, \hat{x}_2)$ lies between B and C.   ◀

In practice, knowledge of alternative optima is useful because it gives management the opportunity to choose the solution that best suits their situation without experiencing any deterioration in the objective value. In the example, for instance, the solution at B shows that only activity 2 is at a positive level, whereas at C both activities are positive. If the example represents a product-mix situation, it may be advantageous from the standpoint of sales competition to produce two products rather than one. In this case the solution at C would be recommended.

Exercise 3.3–2
(a) What is the value of α that will locate $(\hat{x}_1, \hat{x}_2)$ halfway on the line segment BC? one-third of the way from B? Compute $(\hat{x}_1, \hat{x}_2)$ in each case.
   [*Ans.* $\alpha = 1/2$ gives $(3/2, 7/4)$; $\alpha = 2/3$ gives $(1, 2)$.]
(b) Show that for $(\hat{x}_1, \hat{x}_2) = (3 - 3\alpha, 1 + 3\alpha/2)$, the value of z ($= 2\hat{x}_1 + 4\hat{x}_2$) equals 10 for all α such that $0 \leq \alpha \leq 1$.

3.3.3   UNBOUNDED SOLUTION

In some LP models, the values of the variables may be increased indefinitely without violating any of the constraints, meaning that the solution space is **unbounded** in at least one direction. As a result, the objective value may increase (maximization case) or decrease (minimization case) indefinitely. In this case we say that both the solution space and the "optimum" objective value are unbounded.

Unboundedness in a model can point to one thing only. The model is poorly constructed. Having a model that produces an "infinite" profit is obviously nonsensical. The most likely irregularities in such models are (1) one or more nonredundant constraints are not accounted for, and (2) the parameters (constants) of some constraints are not estimated correctly.

The following examples show how unboundedness, both in the solution space and the objective value, can be recognized in the simplex tableau.

Example 3.3–3 (Unbounded Objective Value)

$$\text{Maximize } z = 2x_1 + x_2$$

subject to

$$x_1 - x_2 \leq 10$$
$$2x_1 \quad\;\; \leq 40$$
$$x_1, x_2 \geq 0$$

Starting Iteration

Basic	x_1	x_2	x_3	x_4	Solution
z	-2	-1	0	0	0
x_3	1	-1	1	0	10
x_4	2	0	0	1	40

In the starting tableau, both x_1 and x_2 are candidates for entering the solution. Since x_1 has the most negative coefficient, it is normally selected as the entering variable. Notice, however, that *all* the *constraint* coefficients under x_2 are *negative* or *zero,* meaning that x_2 can be increased indefinitely without violating any of the constraints. Since each unit increase in x_2 will increase z by 1, an infinite increase in x_2 will also result in an infinite increase in z. Thus we conclude without further computations that the problem has no bounded solution. This result can be seen in Figure 3–5. The solution space is unbounded in the direction of x_2 and the value of z can be increased indefinitely.　　　◄

The general rule for recognizing unboundedness is as follows. If at any iteration the constraint coefficients of a *nonbasic* variable are nonpositive, then the *solution space* is unbounded in that direction. If, in addition, the objective coefficient of that variable is negative in the case of maximization or positive in the case of minimization, then the *objective value* also is unbounded.

Exercise 3.3–3

(a) In Example 3.3–3, carry out the simplex iterations, always selecting the nonbasic variable with the *most* negative coefficient to enter the solution (i.e., start with x_1 as the entering variable). Would this procedure also lead to the conclusion that the objective value is unbounded?
[*Ans.* Yes, the second iteration will show that x_3 is the entering variable but with all zero or negative constraint coefficients.]
(b) Explain why negative or zero constraint coefficients of a nonbasic variable indicate that the variable can be increased indefinitely without violating feasibility.
[*Ans.* Negative or zero constraint coefficients indicate that *no* constraints intersect the *positive* direction of the axis representing the entering variable; see Figure 3–2.]

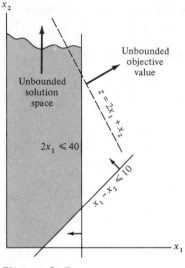

Figure 3–5

(c) Can the condition of unboundedness *always* be detected from the *starting* simplex iteration?
[*Ans.* No, it may become evident for the first time at a later iteration.]

3.3.4 INFEASIBLE SOLUTION

If the constraints cannot be satisfied simultaneously, the model is said to have no feasible solution. This situation can never occur if *all* the constraints are of the type $\leq$ (assuming nonnegative right-side constants), since the slack variable always provides a *feasible* solution. However, when we employ the other types of constraints, we resort to the use of artificial variables which, by their very design, do not provide a feasible solution to the *original* model. Although provisions are made (through the use of penalty) to force the artificial variables to zero at the optimum, this can occur only if the model has a feasible space. If it does not, at least one artificial variable will be *positive* in the optimum iteration. This is our indication that the problem has no feasible solution.

From the practical standpoint, an infeasible space points to the possibility that the model is not formulated correctly, since the constraints are in conflict. It is also possible that the constraints are not meant to be satisfied simultaneously. In this case, a completely different model structure that does not admit all constraints simultaneously may be needed. Examples of such models using the so-called **either–or constraints** are presented in Chapter 8 as applications of integer programming.

The following example illustrates the case of infeasible solution space.

Example 3.3–4 (Infeasible Solution Space)

$$\text{Maximize } z = 3x_1 + 2x_2$$

subject to

$$2x_1 + x_2 \le 2$$
$$3x_1 + 4x_2 \ge 12$$
$$x_1, x_2 \ge 0$$

The simplex iterations in Table 3–4 show that the artificial variable R is *positive* ($= 4$) in the optimal solution. This is an indication that the solution space is infeasible. Figure 3–6 demonstrates the infeasible solution space. The simplex method, by allowing the artificial variable to be positive, in essence has reversed the direction of the inequality from $3x_1 + 4x_2 \ge 12$ to $3x_1 + 4x_2 \le 12$. (Can you explain how?) The result is what we may call the **pseudo-optimal solution,** as shown in Figure 3–6.   ◄

Table 3–4

Iteration	Basic	x_1	x_2	x_4	x_3	R	Solution
0 (starting)	z	$-3 - 3M$	$-2 - 4M$	M	0	0	$-12M$
x_2 enters	x_3	2	1	0	1	0	2
x_3 leaves	R	3	4	-1	0	1	12
1 (pseudo-optimum)	z	$1 + 5M$	0	M	$2 + 4M$	0	$4 - 4M$
	x_2	2	1	0	1	0	2
	R	-5	0	-1	-4	1	4

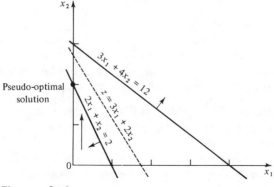

Figure 3–6

Exercise 3.3–4

Apply phase I of the two-phase method (Section 3.2.2) to Example 3.3–4. How does phase I conclude that the problem has no feasible solution space?

[*Ans.* The optimum objective value in phase I is positive ($= 4$).]

3.4   INTERPRETING THE SIMPLEX TABLEAU—SENSITIVITY ANALYSIS

The preceding two sections presented the details of the simplex method. You will probably need to carry out the simplex computations by hand at least once. This done, you can expect the computer to relieve you forever of the burden of having to solve a linear program by hand computations. The convenience of accessing a large variety of LP codes that allow you to use batch processing or to interact with the computer through terminals makes it unnecessary to solve a linear program by hand computations.† Indeed, as you become involved in solving real-life LP models, you will discover that most of your time is spent in formulating the model, gathering the data, and preparing the input information to suit the LP code. Once you have done this, the computer will take over and provide you with the optimal solution to the problem.

We can now see that our attention must be directed to reading, interpreting, and analyzing the computer results. It is erroneous, however, to assume that one can interpret the computer printout without having adequate knowledge of *how* and *why* the simplex method works. This background was presented to you in Sections 3.2 and 3.3; and while we are minimizing the need for solving an LP model by hand computations, we wish to underscore the importance of gaining a fundamental insight into why the solution method is valid. Without this understanding, you are likely to reach erroneous decisions that could lead eventually to a "no-confidence" vote.

It will be disappointing to you to think that all we can get out of the optimum simplex tableau is a list of variables and their optimum values. The fact is that the simplex tableau is "loaded" with important information, the least of which are the optimum values of the variables. The following list summarizes the information that can be obtained from the simplex tableau either directly or with simple additional computations

1. The optimum solution.
2. The status of the resources.
3. The unit worth of each resource.
4. The sensitivity of the optimum solution to changes in availability of resources, coefficients of the objective function, and usage of resources by activities.

The first three items are readily available in the optimum simplex tableau. The fourth item requires additional computations that are based on the information in the optimum solution.

To demonstrate these items, we use the Reddy Mikks model, which we repeat here for convenience. (You will find it helpful to review the description of the model in Example 2.1–1.)

$$\text{Maximize } z = 3x_E + 2x_I \quad \text{(profit)}$$

---

† For your convenience, the LP computer program in Appendix C is designed to reproduce the same computational format followed in this book.

subject to

$$
\begin{aligned}
x_E + 2x_I + s_1 &= 6 \quad \text{(raw material A)} \\
2x_E + x_I + s_2 &= 8 \quad \text{(raw material B)} \\
-x_E + x_I + s_3 &= 1 \quad \text{(demand)} \\
x_I + s_4 &= 2 \quad \text{(demand)} \\
x_E, x_I, s_1, s_2, s_3, s_4 &\geq 0
\end{aligned}
$$

The optimum tableau is given as

Basic	x_E	x_I	s_1	s_2	s_3	s_4	Solution
z	0	0	1/3	4/3	0	0	$12\frac{2}{3}$
x_I	0	1	2/3	$-1/3$	0	0	$1\frac{1}{3}$
x_E	1	0	$-1/3$	2/3	0	0	$3\frac{1}{3}$
s_3	0	0	-1	1	1	0	3
s_4	0	0	$-2/3$	1/3	0	1	2/3

3.4.1 OPTIMUM SOLUTION

From the standpoint of *implementing* the LP solution, the mathematical classification of the variables as basic and nonbasic is of no importance and should be totally ignored in reading the optimum solution. The variables *not* listed in the "basic" column necessarily have zero values. The rest have their values in the solution column. In terms of the optimum solution of the Reddy Mikks model, we are interested primarily in the product mix of exterior and interior paint, that is, the decision variables x_E and x_I. From the optimum tableau we have the following summary.

Decision Variable	Optimum Value	Decision
x_E	$3\frac{1}{3}$	Produce $3\frac{1}{3}$ tons of exterior paint
x_I	$1\frac{1}{3}$	Produce $1\frac{1}{3}$ tons of interior paint daily
z	$12\frac{2}{3}$	Resulting profit is $12\frac{2}{3}$ thousand dollars

Notice that $z = 3x_E + 2x_I = 3x(3\frac{1}{3}) + 2(1\frac{1}{3}) = 12\frac{2}{3}$, as given in the optimum tableau.

3.4.2 STATUS OF RESOURCES

In Section 2.1.2 we classified constraints as *scarce* or *abundant,* depending respectively on whether or not the optimum solution "consumes" the entire available amount of the associated resource. Our objective is to secure this information from the optimum tableau. First we must clarify one point. Speaking of a *resource* implies that there is a *maximum* limit on its availability, which means that the constraint must originally be of the type ≤. Thus constraints of the type ≥ cannot physically represent a resource restriction; rather, they imply that the solution must meet certain requirements, such as satisfying minimum demand or minimum specification.

In the Reddy Mikks model we have four constraints of the type ≤. The first two (representing raw materials usage) are "authentic" resources restrictions. The third and fourth constraints deal with demand limitations imposed by the market conditions. We can think of these constraints as limited "resources," since increasing demand limits is equivalent to expanding the company's share in the market. Monetarily, this has the same effect as increasing availability of physical resources (such as raw materials) through allocation of additional funds.

Following this discussion, the status of the resources (abundant or scarce) in any LP model can be secured directly from the optimum tableau by observing the values of the *slack* variables. In the Reddy Mikks model we have the following summary.

Resource	Slack	Status of Resource
Raw material A	$s_1 = 0$	Scarce
Raw material B	$s_2 = 0$	Scarce
Limit on excess of interior over exterior paint	$s_3 = 3$	Abundant
Limit on demand for interior paint	$s_4 = 2/3$	Abundant

A positive slack means that the resource is not used completely, thus is abundant, whereas a zero slack indicates that the entire amount of the resource is consumed by the activities of the model. In the summary we see that demand limitations (resources 3 and 4) are "abundant"; hence any increase in their maximum limit will simply make them "more" abundant without affecting the optimum solution.

The resources that can be increased for the purpose of improving the solution (increasing profit) are raw materials A and B, since the optimum tableau shows they are scarce. A logical question would naturally arise: Which of the scarce resources should be given priority in the allocation of additional funds to improve profit most advantageously? We answer this question when we consider the *per unit* worth of the different resources.

Exercise 3.4-1

In the Reddy Mikks model, give a physical interpretation of the optimal slack values $s_3 = 3$ and $s_4 = 2/3$.

[*Ans.* Since $x_E - x_I = s_3 - 1$, s_3 is the excess amount of exterior over interior paint inflated by 1 ton. Thus $s_3 = 3$ means that the amount of exterior paint exceeds that of interior paint by 2 tons; s_4 is the amount by which interior paint production can be increased before exceeding the maximum market requirement.]

3.4.3 UNIT WORTH OF A RESOURCE

A *unit worth* of a resource is the rate of improvement in the optimum value of z as a result of *increasing* the available amount of that resource. This point was analyzed graphically in Section 2.1.2 (Sensitivity Problem 2) as an application to the Reddy Mikks model. (You should review Section 2.1.2 before proceeding with this presentation.) The graphical analysis shows that the unit worths of resources 1, 2, 3, and 4 are

$\quad y_1 = 1/3$ thousand dollars/additional ton of material A

$\quad y_2 = 4/3$ thousand dollars/additional ton of material B

$\quad y_3 = 0$

$\quad y_4 = 0$

This information is readily available in the optimum simplex tableau. Look at the coefficients in the z-equation under the *starting* basic variables s_1, s_2, s_3, and s_4, which we reproduce here for convenience.

Basic	x_E	x_I	s_1	s_2	s_3	s_4	Solution
z	0	0	1/3	4/3	0	0	$12\frac{2}{3}$

These coefficients (1/3, 4/3, 0, 0) exactly equal y_1, y_2, y_3, and y_4. The theory of linear programming tells us that it is always possible to secure the unit worth of a resource from coefficients of the *starting* basic variables in the *optimum* z-equation. There should be no confusion as to which coefficient applies to which resource, since s_i is *uniquely* associated with resource i.

Although Section 2.1.2 has provided the rationale behind the definition of unit worth of a resource, we can deduce the same result directly from the optimum simplex tableau. Consider the optimal z-equation of the Reddy Mikks model:

$$z = 12\tfrac{2}{3} - (1/3s_1 + 4/3s_2 + 0s_3 + 0s_4)$$

If we increase s_1 from its current zero level to a positive level, the value of z will decrease at the *rate* 1/3 thousand dollars/ton. But an increase in s_1 is actually equivalent to *reducing* resource 1 (raw material A), as could be seen from the first constraint:

$$x_E + 2x_I + s_1 = 6$$

We thus conclude that a reduction in the first resource will reduce the objective function at the rate 1/3 thousand dollars/ton. Since we are dealing with *linear* functions, we can reciprocate the argument by concluding that an *increase* in the first resource (equivalently $s_1 < 0$, a *surplus* variable) will *increase* z at the rate 1/3 thousand dollars/ton. A similar argument applies to resource 2.

Turning to resources 3 and 4, we find that their unit worth is zero ($y_3 = y_4 = 0$). This should be expected, since the two resources are already abundant. This will always be the case when the slack variables are positive.

In spite of the fact that we have associated dollar values with the unit worth variables y_i, we should not think of them in the same terms as, for example, the real price we may pay to *buy* the resource. Instead, they are economic measures that quantify the unit worth of a resource from the viewpoint of the optimal objective value. The value of these economic measures will vary as we change the constraints even though we may be utilizing the same *physical* resources. For this reason, economists prefer to use the apt terms **shadow price, imputed price,** or, more technically, **dual price** (see Chapter 4) to describe the unit worth of a resource.

Exercise 3.4-2

In the Reddy Mikks model, if the objective function is changed from $z = 3x_E + 2x_I$ to $z = 2x_E + 5x_I$ while maintaining the same constraints, the simplex computations will yield the following optimum z-equation:

$$z + 2s_1 + s_4 = 14$$

The associated optimal values of the variables are $x_E = x_I = 2$, $s_2 = 2$, $s_3 = 1$. All others are zero.

(a) Classify the status of the four resources of the model.
  [*Ans.* 1 and 4: scarce; 2 and 3: abundant.]
(b) Find the shadow prices for the four resources.
  [*Ans.* $y_1 = 2$, $y_2 = y_3 = 0$, $y_4 = 1$.]
(c) Can the optimum value of z improve by increasing the availability of raw material B?
  [*Ans.* No, because $y_2 = 0$, which implies that the resource is abundant.]
(d) Since $y_4 = 1$, an increase in the fourth resource will improve the optimum value of z. Give a physical interpretation of what it means to increase the fourth resource.
  [*Ans.* The fourth constraint represents the maximum limit on demand. The fourth resource can be increased by expanding the company's share in the market.]
(e) Which of the four resources should be given priority in the allocation of new funds?
  [*Ans.* Raw material A, because it has the largest shadow price, $y_1 = 2$.]

Observe that the definition of shadow prices (unit worth of a resource) gives us the *rate* of improvement in optimum z. It does not specify the *amount* by which a resource can be increased while maintaining the same rate of improvement. Logically, in most situations we would expect an upper limit beyond which any increase in the resource would make its constraint *redundant,* with the result that a new basic solution, and hence new shadow prices, must be sought. The following presentation addresses the point of determining the maximum change in the availability of a resource before its constraint becomes redundant.

3.4.4 MAXIMUM CHANGE IN RESOURCE AVAILABILITY

We normally use the *shadow prices* to decide which resources should be expanded. Our goal is to determine the range of variation in the availability of a resource that will yield the unit worth (shadow price) encountered in the optimum tableau. To achieve this, we need to perform additional computations. We shall first demonstrate how the procedure works and then show how the same information can be secured from the optimum tableau.

Suppose that we consider changing the first resource in the Reddy Mikks model by the amount Δ_1, meaning that available raw material A is $6 + \Delta_1$ tons. If Δ_1 is positive, the resource increases; if it is negative, the resource decreases. Although we normally shall be interested in the case where the resource increases ($\Delta_1 > 0$), we shall present both cases for the sake of generalization.

How is the simplex tableau changed by effecting the change Δ_1? The simplest way to answer the question is to augment Δ_1 to the right side of the first constraint in the starting tableau and then apply the same arithmetic operations that were used to develop the successive iterations. If we keep in mind that the right-side constants are never used as *pivot elements*, it is evident that the change Δ_1 will affect only the right side of each iteration. You should verify that the successive iterations of the model will change as shown in Table 3–5.

Actually, the changes in the right sides resulting from Δ_1 can be obtained *directly* from the successive tableaus. First, notice that in each iteration the elements of the new right side consist of two components: (1) a constant and (2) a linear term in Δ_1. The constant components exactly equal the right side of the iteration *before* Δ_1 is added. The coefficients of the linear term in Δ_1 are essentially those under s_1 in the same iteration. For example, in the optimum iteration the constants ($12\frac{2}{3}$, 4/3, 10/3, 3, 2/3) represent the right side of the optimum tableau before Δ_1 is effected, and (1/3, 2/3, −1/3, −1, −2/3) are the coefficients under s_1 in the same tableau. Why s_1? Because it is uniquely

Table 3–5

		Right-Side Elements in Iteration	
Equation	0 (starting)	1	2 (optimum)
z	0	12	$12\frac{2}{3} + \frac{1}{3}\Delta_1$
1	$6 + \Delta_1$	$2 + \Delta_1$	$\frac{4}{3} + \frac{2}{3}\Delta_1$
2	8	4	$\frac{10}{3} - \frac{1}{3}\Delta_1$
3	1	5	$3 - 1\Delta_1$
4	2	2	$\frac{2}{3} - \frac{2}{3}\Delta_1$

associated with the first constraint. In other words, for right-side changes in the second, third, and fourth constraints, we should use the coefficients under s_2, s_3, and s_4, respectively.

What does this information mean? Since we have concluded that the change Δ_1 will affect only the right side of the tableau, it means that such a change can affect only the *feasibility* of the solution. Thus Δ_1 should not be changed in a manner that will make any of the (basic) *variables* negative. This means that Δ_1 must be restricted to the range that will maintain the *nonnegativity* of the right side of the constraint equations in the optimum tableau. That is,

$$x_I = \frac{4}{3} + \frac{2}{3}\Delta_1 \geq 0 \tag{1}$$

$$x_E = \frac{10}{3} - \frac{1}{3}\Delta_1 \geq 0 \tag{2}$$

$$s_3 = 3 - \Delta_1 \geq 0 \tag{3}$$

$$s_4 = \frac{2}{3} - \frac{2}{3}\Delta_1 \geq 0 \tag{4}$$

To determine the admissible range for Δ_1, we consider two cases.

Case 1: $\Delta_1 > 0$. Relation (1) is always satisfied for $\Delta_1 > 0$. Relations (2), (3), and (4), on the other hand, produce the following respective limits: $\Delta_1 \leq 10$, $\Delta_1 \leq 3$, and $\Delta_1 \leq 1$. Thus, all four relations are satisfied for $\Delta_1 \leq 1$.

Case 2: $\Delta_1 < 0$. Relations (2), (3), and (4) are always satisfied for $\Delta_1 < 0$, whereas relation (1) yields the limit $\Delta_1 \geq -2$.

By combining cases 1 and 2, we see that

$$-2 \leq \Delta_1 \leq 1$$

will always result in a feasible solution. Any change outside this range (i.e, *decreasing* raw material A by more than 2 tons or *increasing* it by more than 1 ton) will lead to infeasibility and a new set of basic variables (see Chapter 4).

Exercise 3.4–3

Consider the Reddy Mikks model.

(a) Given $\Delta_1 = 1/2$ ton, find the new optimum solution.
  [*Ans.* $z = 12\frac{5}{6}$, $x_E = 3\frac{1}{6}$, $x_I = 1\frac{2}{3}$, $s_1 = s_2 = 0$, $s_3 = 2\frac{1}{2}$, $s_4 = 1/3$.]

(b) Determine the optimum right side of the constraint equations resulting from *independently* changing resources 2, 3, and 4 by Δ_2, Δ_3, and Δ_4.

  [*Ans.* Δ_2: $4/3 - \Delta_2/3$, $10/3 + 2\Delta_2/3$, $3 + \Delta_2$, $2/3 + \Delta_2/3$
  Δ_3: $4/3$, $10/3$, $3 + \Delta_3$, $2/3$
  Δ_4: $4/3$, $10/3$, 3, $2/3 + \Delta_4$.]

(c) Determine the feasible ranges for Δ_2, Δ_3, Δ_4 in part (b).
  [*Ans.* $-2 \leq \Delta_2 \leq 4$, $-3 \leq \Delta_3 < \infty$, $-2/3 \leq \Delta_4 < \infty$.]

(d) Determine the ranges in optimal z resulting from the changes in part (c).
  [*Ans.* Δ_2: $10 \leq z \leq 18$. Δ_3 and Δ_4: $z = 12\frac{2}{3}$ regardless of the values of Δ_3 and Δ_4. The answer agrees with earlier analysis on unit worth of resources.]

(e) Are the ranges in part (d) correct if the changes Δ_2, Δ_3, and Δ_4 are effected *simultaneously*?

[*Ans.* No, because *simultaneous* changes will make the right-side elements functions of Δ_2, Δ_3, and Δ_4. The analysis above is correct only when we consider one resource at a time.]

3.4.5 MAXIMUM CHANGE IN MARGINAL PROFIT/COST

Just as we did in studying the permissible ranges for changes in resources, we are also interested in studying the permissible ranges for changes in marginal profits (or costs). We have shown graphically in Section 2.1.2 (Sensitivity Problem 3) that the objective function coefficients can change within limits without affecting the optimum values of the variables (the optimum value of z will change, though). In this presentation, we show how this information can be obtained from the optimum tableau.

In the present situation, as in the case of resource changes, the objective equation is never used as a pivot equation. Thus any changes in the coefficients of the objective function will affect only the objective equation in the optimum tableau. This means that such changes can have the effect of making the solution nonoptimal. Our goal is to determine the range of variation for the object coefficients (one at a time) for which the current optimum remains unchanged.

To illustrate the computations, suppose that the marginal profit of x_E in the Reddy Mikks model is changed from 3 to $3 + \delta_1$, where δ_1 represents either positive or negative change. Thus the objective function reads as

$$z = (3 + \delta_1)x_E + 2x_I$$

If we use this information in the starting tableau and carry out the *same* arithmetic operations used to produce the optimum tableau, the optimum z-equation will appear as

Basic	x_E	x_I	s_1	s_2	s_3	s_4	Solution
z	0	0	$\frac{1}{3} - \frac{1}{3}\delta_1$	$\frac{4}{3} + \frac{2}{3}\delta_1$	0	0	$12\frac{2}{3} + \frac{10}{3}\delta_1$

The objective coefficients of *basic* x_E, x_1, s_3, and s_4 will always remain equal to zero.

The resulting equation is the same as the optimum z-equation *before* the change δ_1 is effected, modified by terms of δ_1. The coefficients of δ_1 are essentially those in the x_E-equation of the optimum tableau, which are given by

Basic	x_E	x_I	s_1	s_2	s_3	s_4	Solution
x_E	1	0	$-1/3$	$2/3$	0	0	$10/3$

We choose the x_E-equation because x_E is the variable whose objective coefficient is being changed by δ_1.

The change δ_1 will not affect the optimality of the problem as long as all the z-equation coefficients of the *nonbasic* variables remain nonnegative (maximization). That is,

$$1/3 - \delta_1/3 \geq 0 \tag{1}$$

$$4/3 + 2\delta_1/3 \geq 0 \tag{2}$$

Relation (1) shows that $\delta_1 \leq 1$, and relation (2) yields $\delta_1 \geq -2$. Both relations limit c_1 by $-2 \leq \delta_1 \leq 1$. This means that the coefficient of x_E can be as small as $3 + (-2) = 1$ or as large as $3 + 1 = 4$ without causing any change in the optimal values of the variables (compare with Sensitivity Problem 3, Section 2.1.2). The optimal value of z, however, will change according to the expression $12\frac{2}{3} + \frac{10}{3}\delta_1$, where $-2 \leq \delta_1 \leq 1$.

Exercise 3.4-4
Suppose that the coefficient of x_I in the Reddy Mikks model is changed to δ_2. Define the range of δ_2 that will keep the current optimum unchanged.
[Ans. $-1/2 \leq \delta_2 \leq 4$, $z = 12\frac{2}{3} + \frac{4}{3}\delta_2$.]

The foregoing discussion assumes that the variable whose coefficient is being changed has an equation in the constraints. This is true only if the variable is basic (such as x_E and x_I). If it is nonbasic, it will not appear in the basic column.

The treatment of a nonbasic variable is straightforward. Any change in the objective coefficient of a nonbasic variable will affect only that coefficient in the optimal tableau. To illustrate this point, consider changing the coefficient of s_1 (the first slack variable) from 0 to $0 + \delta_3$. If you carry out the arithmetic operations leading to the optimum tableau, the resulting z-equation becomes

Basic	x_E	x_I	s_1	s_2	s_3	s_4	Solution
z	0	0	$1/3 - \delta_3$	$4/3$	0	0	$12\frac{2}{3}$

It shows that the only change occurs in the coefficient of s_1, where it is decreased by δ_3. As a general rule, then, all we have to do in the case of a nonbasic variable is to decrease the z-coefficient of the nonbasic variable by the amount by which the original coefficient of the variable is increased.

3.5 SUMMARY

The simplex method theory shows that a corner point is essentially identified by a basic solution of the standard form of linear programming. The optimality and feasibility conditions of the simplex method ensure that, starting from a feasible corner point (basic solution), the simplex method will move to an adjacent corner point which has the potential to improve the value of the objective function. The maximum number of iterations (basic solutions or corner points) until the optimum is reached is limited by $n!/[(n-m)!m!]$ in an n-variable m-equation standard LP model.

The optimum tableau offers more than just the optimum values of the variables. It gives the status and worth (shadow prices) of the different resources. Additionally, the sensitivity study shows that the resources can be changed within certain limits while maintaining the same activity mix in the solution. Also, marginal profits/costs can be changes within certain ranges without changing the optimum value of the variables.

SELECTED REFERENCES

BAZARAA, M., and J. JARVIS, *Linear Programming and Network Flows,* Wiley, New York, 1977.

DANTZIG, G., *Linear Programming and Extensions,* Princeton University Press, Princeton, N.J., 1963.

HADLEY, G., *Linear Programming,* Addison-Wesley, Reading, Mass., 1962.

REVIEW QUESTIONS

True (T) or False (F)?

1. ____ Every equality constraint can be replaced equivalently by two inequalities.

2. ____ The maximization of a function f subject to a set of constraints is exactly equivalent to the minimization of $g = -f$ subject to the same set of constraints, except that min $g = -\max f$.

3. ____ In an LP with m constraints, a simplex iteration may include more than m positive basic variables.

4. ____ A simplex iteration (basic solution) may not necessarily coincide with a feasible extreme point of the solution space.

5. ____ In the simplex method, all variables must be nonnegative.

6. ____ Given the three extreme points A, B, and C of an LP, if A is adjacent to B and B is adjacent to C, then A can be determined from C by interchanging exactly two basic and two nonbasic variables.

7. ____ Every feasible point in a *bounded* LP solution space can be determined from its feasible extreme points.

8. ____ In the simplex method, the optimality conditions for the maximization and minimization problems are different.

9. ____ In the simplex method, the feasibility condition for the maximization and minimization problems are different.

10. ____ If the leaving variable does not correspond to the minimum ratio, at least one basic variable will definitely become negative in the next iteration.

11. ____ The selection of the entering variable from among the current nonbasic variables as the one with the most negative objective coefficient guarantees the most increase in the objective value in the next iteration.

12. ____ In the simplex method, the volume of computations increases primarily with the number of constraints.

13. ____ The optimality condition always guarantees that the next solution will have a *better* objective value than in the immediately preceding iteration.

14. ____ In a simplex iteration, the pivot element can be zero or negative.

15. ____ An artificial variable column can be dropped all together from the simplex tableau once the variable becomes nonbasic.

16. ____ The two-phase method and the *M*-technique require the same number of iterations for solving a linear program.

17. ____ Degeneracy can be avoided if redundant constraints can be deleted.

18. ____ The simplex method may not move to an adjacent extreme point if the current iteration is degenerate.

19. ____ If a current iteration is degenerate, the next iteration will necessarily be degenerate.

20. ____ If the solution space is unbounded, the objective value always will be unbounded.

21. ____ An LP may have a feasible solution even though an artificial appears at a positive level in the optimal iteration.

22. ____ In an LP, it is possible to change the coefficients of the objective function without changing the optimal values of the variables.

23. ____ Changes in the availability of resources can only affect the feasibility of the optimal LP solution.

24. ____ If a unit worth of a resource is positive, the resource must necessarily be scarce.

25. ____ If the slack variable associated with a resource is positive, the unit worth of the resource may not equal zero.

[*Ans.* 1—T, 2—T, 3—F, 4—F, 5—T, 6—T, 7—T, 8—T, 9—F, 10—T, 11—F, 12—T, 13—F, 14—F, 15—T, 16—T, 17—T, 18—T, 19—F, 20—F, 21—F, 22—T, 23—T, 24—T, 25—F.]

PROBLEMS

Section	Assigned Problems
3.1	3–1, 3–2; 3–35 to 3–38
3.2.1	3–3 to 3–15
3.2.2	3–16 to 3–23
3.3	3–24 to 3–29
3.4	3–30 to 3–34

□ **3–1** Convert the following LP to the standard form:

$$\text{maximize } z = 2x_1 + 3x_2 + 5x_3$$

subject to

$$x_1 + x_2 - x_3 \geq -5$$
$$-6x_1 + 7x_2 - 9x_3 \leq 4$$
$$x_1 + x_2 + 4x_3 = 10$$
$$x_1, x_2 \geq 0$$
$$x_3 \text{ unrestricted}$$

□ **3–2** Repeat Problem 3–1 for each of the following independent changes.
  (a) The first constraint is $x_1 + x_2 - x_3 \leq -5$.
  (b) The second constraint is $-6x_1 + 7x_2 - 9x_3 \geq 4$.
  (c) The third constraint is $x_1 + x_2 + 4x_3 \geq 10$.
  (d) $x_3 \geq 0$.
  (e) x_1 unrestricted.
  (f) The objective function is minimize $z = 2x_1 + 3x_2 + 5x_3$.
  (g) Changes (a), (b), and (c) are effected simultaneously.
  (h) Changes (c), (d), (e), and (f) are effected simultaneously.

□ **3–3** Consider the three-dimensional LP solution space shown in Figure 3–7 with its feasible extreme points identified by $A, B, C, \ldots, J$. The coordinates of each point are shown in the figure.

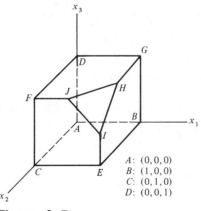

A: $(0,0,0)$
B: $(1,0,0)$
C: $(0,1,0)$
D: $(0,0,1)$

Figure 3–7

·(a) Indicate whether or not the following pairs of extreme points are adjacent:
    (1) A, B; (2) B, D; (3) E, H; (4) A, I.
 (b) From the standpoint of the simplex method, suppose that the solution starts at A and that the optimum occurs at H. Indicate whether or not the simplex iterations from A to H can be identified by the following sequences of extreme points, and state the reason.
    (1) $A \rightarrow B \rightarrow G \rightarrow H$.
    (2) $A \rightarrow C \rightarrow F \rightarrow J \rightarrow H$.

(3) $A \rightarrow C \rightarrow I \rightarrow H$.
(4) $A \rightarrow I \rightarrow H$.
(5) $A \rightarrow D \rightarrow G \rightarrow H$.
(6) $A \rightarrow D \rightarrow A \rightarrow B \rightarrow G \rightarrow H$.
(7) $A \rightarrow C \rightarrow F \rightarrow D \rightarrow A \rightarrow B \rightarrow G \rightarrow H$.

☐ **3–4** In Figure 3–7, all the constraints associated with the solution space are of the type $\leq$. Let s_1, s_2, s_3, and s_4 represent the slack variables associated with the constraints represented by the planes *CEIJF*, *BEIHG*, *DFJHG*, and *HIJ*, respectively. Identify the basic and nonbasic variables associated with each feasible extreme point.
[*Note:* The problem implicitly assumes that x_1, x_2, $x_3 \geq 0$.]

☐ **3–5** In Problem 3–4, identify the entering and leaving variables when the solution moves between the following pairs of extreme points: (a) $A \rightarrow B$, (b) $E \rightarrow I$, (c) $F \rightarrow J$, (d) $D \rightarrow G$.

☐ **3–6** Consider the following problem:

$$\text{maximize } z = 2x_1 - 4x_2 + 5x_3 - 6x_4$$

subject to

$$x_1 + 4x_2 - 2x_3 + 8x_4 \leq 2$$
$$-x_1 + 2x_2 + 3x_3 + 4x_4 \leq 1$$
$$x_1, x_2, x_3, x_4 \geq 0$$

Determine:
(a) The maximum number of possible basic solutions.
(b) The feasible extreme points.
(c) The optimal basic feasible solution.

☐ **3–7** In Figure 3–7, suppose that the simplex method starts at A. Determine the entering variable in the *first* iteration together with its value and the improvement in the objective value given that the objective function is defined as follows.
(a) Maximize $z = x_1 - 2x_2 + 3x_3$.
(b) Maximize $z = 5x_1 + 2x_2 + 4x_3$.
(c) Maximize $z = -2x_1 + 7x_2 + 2x_3$.
(d) Maximize $z = x_1 + x_2 + x_3$.

☐ **3–8** Consider the two-dimensional solution space in Figure 3–8. Suppose that the objective function is given by

$$\text{maximize } z = 3x_1 + 6x_2$$

(a) Determine the optimum extreme point graphically.
(b) Suppose that the simplex method starts at A, identify the successive extreme points that will lead to the optimum obtained in part (a).
(c) Determine the entering variable and the ratios of the feasibility condition

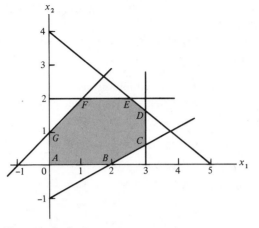

Figure 3-8

assuming that the simplex solution is at A and the objective function is given by

$$\text{maximize } z = 4x_1 + x_2$$

(d) Repeat part (c) when the objective function is replaced by
$$\text{maximize } z = x_1 + 2x_2$$

(e) In parts (c) and (d), determine the resulting improvements in the value of z.

☐ **3–9** Consider the following system of equations:

$$\begin{aligned}
x_1 + 2x_2 - 3x_3 + 5x_4 + x_5 &= 4 \\
5x_1 - 2x_2 + 6x_4 + x_6 &= 8 \\
2x_1 + 3x_2 - 2x_3 + 3x_4 + x_7 &= 3 \\
-x_1 + x_3 + 2x_4 + x_8 &= 0 \\
x_1, x_2, \ldots, x_8 &\geq 0
\end{aligned}$$

Let $(x_5, \ldots, x_8)$ be a given initial basic solution. If x_1 becomes basic, which of the current basic variables must become nonbasic at zero level in order for all the variables to remain nonnegative, and what would be the value of x_1 in the new basic solution? Repeat this procedure for x_2, x_3, and x_4.

☐ **3–10** The following tableau represents a specific simplex iteration.

Basic	x_1	x_2	x_3	x_4	x_5	x_6	x_7	x_8	Solution
z	0	−5	0	4	−1	−10	0	0	620
x_8	0	3	0	−2	−3	−1	5	1	12
x_3	0	2	1	3	1	0	3	0	6
x_1	1	−1	0	0	6	−4	0	0	0

(a) Determine the leaving variable if the entering variable is (1) x_2, (2) x_4, (3) x_5, (4) x_6, (5) x_7.
(b) For each of the cases in part (a), determine the resulting increase *or* decrease in z.

☐ **3–11** Solve the following sets of simultaneous linear equations by using the row operations (Gauss–Jordan) method introduced with the simplex method (Section 3.2.2).

(a)
$$-3x_1 + 2x_2 + 5x_3 = 5$$
$$4x_1 + 3x_2 + 2x_3 = 8$$
$$x_1 - x_2 + 3x_3 = 10$$

(b)
$$x_2 + x_3 = 5$$
$$2x_1 + x_2 - x_3 = 12$$
$$x_1 + 3x_2 + 4x_3 = 10$$

☐ **3–12** Consider the following set of constraints:

$$x_1 + 7x_2 + 3x_3 + 7x_4 \leq 46$$
$$3x_1 - x_2 + x_3 + 2x_4 \leq 8$$
$$2x_1 + 3x_2 - x_3 + x_4 \leq 10$$

Solve the problem by the simplex method assuming that the objective function is given as follows:
(a) Maximize $z = 2x_1 + x_2 - 3x_3 + 5x_4$.
(b) Maximize $z = -2x_1 + 6x_2 + 3x_3 - 2x_4$.
(c) Maximize $z = 3x_1 - x_2 + 3x_3 + 4x_4$.
(d) Minimize $z = 5x_1 - 4x_2 + 6x_3 + 8x_4$.
(e) Minimize $z = 3x_1 + 6x_2 - 2x_3 + 4x_4$.

☐ **3–13** Solve the following problem by inspection and justify the method of solution in terms of the simplex method.

$$\text{maximize } z = 5x_1 - 6x_2 + 3x_3 - 5x_4 + 12x_5$$

subject to

$$x_1 + 3x_2 + 5x_3 + 6x_4 + 3x_5 \leq 90$$
$$x_1, x_2, x_3, x_4, x_5 \geq 0$$

☐ **3–14** Consider the following LP:

$$\text{minimize } z = x_1 - 3x_2 - 2x_3$$

subject to

$$3x_1 - x_2 + 2x_3 \leq 7$$
$$-2x_1 + 4x_2 \leq 12$$
$$-4x_1 + 3x_2 + 8x_3 \leq 10$$
$$x_1, x_2, x_3 \geq 0$$

(a) Solve the problem by the simplex method, where the entering variable is the nonbasic variable with the *most* positive objective coefficient.

(b) Resolve the problem by the simplex method, always selecting the entering variable as the nonbasic variable with the *least* positive coefficient.

(c) Compare the number of iterations in parts (a) and (b).

(d) Suppose that the sense of optimization is changed to maximization by multiplying the minimization objective function by -1, in which case we must use the maximization optimality condition. How would this change affect the simplex computations?

☐ **3–15** Solve the following problem by using x_4, x_5, and x_6 for the starting basic (feasible) solution:

$$\text{maximize } z = 3x_1 + x_2 + 2x_3$$

subject to

$$12x_1 + 3x_2 + 6x_3 + 3x_4 \qquad\qquad = 9$$
$$8x_1 + x_2 - 4x_3 \qquad + 2x_5 \quad = 10$$
$$3x_1 \qquad\qquad\qquad\qquad - x_6 = 0$$
$$x_1, x_2, \ldots, x_6 \geq 0$$

☐ **3–16** Consider the following set of constraints:

$$-2x_1 + 3x_2 = 3 \tag{1}$$
$$4x_1 + 5x_2 \geq 10 \tag{2}$$
$$x_1 + 2x_2 \leq 5 \tag{3}$$
$$6x_1 + 7x_2 \leq 3 \tag{4}$$
$$4x_1 + 8x_2 \geq 5 \tag{5}$$

Assuming that x_1, $x_2 \geq 0$, determine the starting objective equation in each of the following cases after the artificial variables are substituted out in the M-technique.

(a) Maximize $z = 5x_1 + 6x_2$ subject to constraints (1), (3), and (4).

(b) Maximize $z = 2x_1 - 7x_2$ subject to constraints (1), (2), (4), and (5).

(c) Minimize $z = 3x_1 + 6x_2$ subject to constraints (3), (4), and (5).

(d) Minimize $z = 4x_1 + 6x_2$ subject to constraints (1), (2), and (5).

(e) Minimize $z = 3x_1 + 2x_2$ subject to constraints (1) and (5).

☐ **3–17** Consider the following set of constraints:

$$x_1 + x_2 + x_3 = 7$$
$$2x_1 - 5x_2 + x_3 \geq 10$$
$$x_1, x_2, x_3 \geq 0$$

Solve by using the M-technique, assuming that the objective function is given as follows:

(a) Maximize $z = 2x_1 + 3x_2 - 5x_3$.

· (b) Minimize $z = 2x_1 + 3x_2 - 5x_3$.

(c) Maximize $z = x_1 + 2x_2 + x_3$.

(d) Minimize $z = 4x_1 - 8x_2 + 3x_3$.

☐ **3–18** Consider the problem

$$\text{maximize } z = x_1 + 5x_2 + 3x_3$$

subject to

$$x_1 + 2x_2 + x_3 = 3$$
$$2x_1 - x_2 = 4$$
$$x_1, x_2, x_3 \geq 0$$

Let R be an artificial variable in the second constraint equation. Solve the problem by using x_3 and R for a starting basic solution.

☐ **3–19** Consider the problem

$$\text{maximize } z = 2x_1 + 4x_2 + 4x_3 - 3x_4$$

subject to

$$x_1 + x_2 + x_3 = 4$$
$$x_1 + 4x_2 + x_4 = 8$$
$$x_1, x_2, x_3, x_4 \geq 0$$

Find the optimum solution by using (x_3, x_4) as the starting basic solution.

☐ **3–20** Solve the following problem by using x_3 and x_4 as a starting basic feasible solution:

$$\text{minimize } z = 3x_1 + 2x_2 + 3x_3$$

subject to

$$x_1 + 4x_2 + x_3 \geq 7$$
$$2x_1 + x_2 + x_4 \geq 10$$
$$x_1, x_2, x_3, x_4 \geq 0$$

☐ **3–21** In Problem 3–16, write the objective function for phase I in each case.

☐ **3–22** Solve Problem 3–17 by the two-phase method and compare the resulting number of iterations with those in the M-technique.

☐ **3–23** Consider the following linear program:

$$\text{maximize } z = c_1 x_1 + c_2 x_2$$

subject to

$$a_{11}x_1 + a_{12}x_2 \leq b_1$$
$$a_{21}x_1 + a_{22}x_2 \leq b_2$$
$$x_1, x_2 \geq 0$$

where $1 \le c_1 \le 3$, $4 \le c_2 \le 6$, $-1 \le a_{11} \le 3$, $2 \le a_{12} \le 5$, $8 \le b_1 \le 12$, $2 \le a_{21} \le 5$, $4 \le a_{22} \le 6$, and $10 \le b_2 \le 14$. Find the upper and lower bounds on the optimum value of z.

[*Hint:* A more restrictive solution space yields a smaller value of z.]

☐ **3–24** Show that the following LP problem is *temporarily* degenerate:

$$\text{maximize } z = 3x_1 + 2x_2$$

subject to

$$4x_1 + 3x_2 \le 12$$
$$4x_1 + x_2 \le 8$$
$$4x_1 - x_2 \le 8$$
$$x_1, x_2 \ge 0$$

☐ **3–25** Consider the problem

$$\text{maximize } z = x_1 + 2x_2 + 3x_3$$

subject to

$$x_1 + 2x_2 + 3x_3 \le 10$$
$$x_1 + x_2 \qquad \le 5$$
$$x_1 \qquad\qquad \le 1$$
$$x_1, x_2, x_3 \ge 0$$

Find at least three alternative optimal basic solutions and then write a general expression for all the *non* basic optimal solutions comprised by the basic optima obtained.

☐ **3–26** Consider the following linear programming problem:

$$\text{maximize } z = 2x_1 - x_2 + 3x_3$$

subject to

$$x_1 - x_2 + 5x_3 \le 10$$
$$2x_1 - x_2 + 3x_3 \le 40$$
$$x_1, x_2, x_3 \ge 0$$

Show that the problem has alternative solutions that are all *non*basic. What could one conclude concerning the solution space and the objective function? Show that the values of the optimal basic variables can be increased indefinitely while the value of z remains constant.

☐ **3–27** Consider the problem

$$\text{maximize } z = 3x_1 + x_2$$

subject to

$$x_1 + 2x_2 \qquad \le 5$$
$$x_1 + x_2 - x_3 \le 2$$
$$7x_1 + 3x_2 - 5x_3 \le 20$$
$$x_1,\, x_2,\, x_3 \ge 0$$

Show that the optimal solution is degenerate and that there exist alternative solutions that are all nonbasic.

☐ **3–28** In the problem

$$\text{maximize } z = 20x_1 + 10x_2 + x_3$$

subject to

$$3x_1 - 3x_2 + 5x_3 \le 50$$
$$x_1 \qquad + x_3 \le 10$$
$$x_1 - x_2 + 4x_3 \le 20$$
$$x_1,\, x_2,\, x_3 \ge 0$$

in which direction is the solution space unbounded? Without further computations, what could one conclude concerning the optimal solution to the problem?

☐ **3–29** Consider the problem

$$\text{maximize } z = 3x_1 + 2x_2 + 3x_3$$

subject to

$$2x_1 + x_2 + x_3 \le 2$$
$$3x_1 + 4x_2 + 2x_3 \ge 8$$
$$x_1,\, x_2,\, x_3 \ge 0$$

By using the *M*-technique, show that the optimal solution can include an artificial basic variable at the *zero* level. Hence conclude that a feasible optimal solution exists.

☐ **3–30** Consider the following LP allocation model:

$$\text{maximize } z = 3x_1 + 2x_2 \qquad \text{(profit)}$$

subject to

$$4x_1 + 3x_2 \le 12 \qquad \text{(resource 1)}$$
$$4x_1 + x_2 \le 8 \qquad \text{(resource 2)}$$
$$4x_1 - x_2 \le 8 \qquad \text{(resource 3)}$$
$$x_1,\, x_2 \ge 0$$

The optimum tableau of the model is given by

Basic	x_1	x_2	x_3	x_4	x_5	Solution
z	0	0	5/8	1/8	0	17/2
x_2	0	1	1/2	−1/2	0	2
x_1	1	0	−1/8	3/8	0	3/2
x_5	0	0	1	−2	1	4

(a) Determine the status of each resource.

(b) Determine the unit worth of each resource.

(c) Based on the unit worth of each resource, which resource should be given priority for an increase in level?

(d) Determine the maximum range of change in the availability of the first resource that will keep the current solution feasible.

(e) Repeat part (d) for resource 2.

(f) In parts (d) and (e), determine the associated change in the optimal value of z.

(g) Determine the maximum change in the profit coefficient of x_1 that will keep the solution optimal.

(h) Repeat part (g) for x_2.

☐ 3–31 Consider the following LP allocation model:

$$\text{maximize } z = 2x_1 + 4x_2 \quad \text{(profit)}$$

subject to

$$x_1 + 2x_2 \le 5 \quad \text{(resource 1)}$$
$$x_1 + x_2 \le 4 \quad \text{(resource 2)}$$
$$x_1, x_2 \ge 0$$

The optimal tableau is given by:

Basic	x_1	x_2	x_3	x_4	Solution
z	0	0	2	0	10
x_2	1/2	1	1/2	0	5/2
x_4	1/2	0	−1/2	1	3/2

(a) Classify the two resources as scarce or abundant.

(b) Determine the maximum range of change in the availability of each resource that will keep the solution optimal.

(c) Compute the range of optimal z associated with the results in part (b).

(d) Compute the maximum change in the unit profit of x_1 that will keep the solution optimal.

(e) Repeat part (d) for x_2.

☐ 3–32 Consider the following LP allocation model:

$$\text{maximize } z = 3x_1 + 2x_2 + 5x_3 \qquad \text{(profit)}$$

subject to

$$
\begin{aligned}
x_1 + 2x_2 + x_3 &\le 430 & \text{(resource 1)} \\
3x_1 + + 2x_3 &\le 460 & \text{(resource 2)} \\
x_1 + 4x_2 &\le 420 & \text{(resource 3)} \\
x_1, x_2, x_3 &\ge 0
\end{aligned}
$$

The optimal tableau of the model is given by

Basic	x_1	x_2	x_3	x_4	x_5	x_6	Solution
z	4	0	0	1	2	0	1350
x_2	$-1/4$	1	0	1/2	$-1/4$	0	100
x_3	3/2	0	1	0	1/2	0	230
x_6	2	0	0	-2	1	1	20

(a) In each of the following cases, indicate whether the given solution remains feasible. If feasible, compute the associated values of x_1, x_2, x_3, and z.
   (1) Resource 1 availability is increased to 500 units.
   (2) Resource 1 availability is decreased to 400 units.
   (3) Resource 2 availability is decreased to 450 units.
   (4) Resource 3 availability is increased to 440 units.
   (5) Resource 3 availability is decreased to 380 units.
(b) In each of the following cases, indicate whether the given solution remains optimal.
   (1) The profit coefficient of x_1 is decreased to 2.
   (2) The profit coefficient of x_1 is increased to 9.
   (3) The profit coefficient of x_2 is increased to 5.
   (4) The profit coefficient of x_3 is reduced to 1.

☐ **3–33** In Problem 3–30, determine the optimal values of x_1 and x_2 when resource 1 is increased by 2 units and, simultaneously, resource 2 is decreased by 1 unit.

☐ **3–34** In Problem 3–31, suppose that resource 1 and resource 2 are changed simultaneously by the quantities Δ_1 and Δ_2. Determine the relationship between Δ_1 and Δ_2 that will always keep the solution optimal.

☐ **3–35** Show that the *m equalities*

$$\sum_{j=1}^{n} a_{ij}x_j = b_i, \qquad i = 1, 2, \ldots, m$$

are equivalent to the $m + 1$ *inequalities*

$$\sum_{j=1}^{n} a_{ij}x_j \le b_i, \quad i = 1, 2, \ldots, m$$

$$\sum_{j=1}^{n} \left(\sum_{i=1}^{m} a_{ij}\right) x_j \ge \sum_{i=1}^{m} b_i$$

☐ **3-36** In linear programming problems in which there are several unrestricted variables, a transformation of the type $x_j = x_j' - x_j''$ will double the corresponding number of nonnegative variables. Show that it is possible, in general, to replace k unrestricted variables with exactly $k + 1$ nonnegative variables and develop the details of the substitution method.
[*Hint:* Let $x_j = x_j' - w$, where $x_j', w \ge 0$.]

☐ **3-37** Show how the following inequality in absolute form can be replaced by two regular inequalities.

$$|\sum_{j=1}^{n} a_{ij}x_j| \le b_i, \qquad b_i > 0$$

☐ **3-38** Show how the following objective function can be linearized:

$$\text{minimize } z = \max \left\{ \left|\sum_{j=1}^{n} c_{1j}x_j\right|, \ldots, \left|\sum_{j=1}^{n} c_{mj}x_j\right| \right\}$$

SHORT CASES

Case 3-1

A small canning company produces five types of canned goods that are extracted from three types of fresh fruit. The manufacturing process utilizes two production departments. These departments are originally designed with surplus capacities to accommodate possible future expansion. In fact, the company operates currently on a one-shift basis and can easily expand to two or three shifts per day to meet future increases in demand. The real restriction for the time being appears to be the limited availability of fresh fruit. Because of the limited refrigeration capacity on the company's premises, fresh fruit must be brought in daily.

A young operations researcher has just joined the company. After analyzing the production situation for a few weeks, she decides to study the possibility of optimizing the company's product mix. With LP still fresh in her mind, she formulates a master LP model for the plant and immediately comes to realize that the LP model has five decision variables (for the products) and three constraints (for the raw materials). With three constraints and five variables, LP theory tells her that the optimum product mix cannot include more than three products. "Aha," she says, "obviously the company is not operating optimally!"

The operations researcher holds a meeting with the plant manager to discuss the situation. She explains the LP model and the manager seems to follow the modeling concept quite well. He also agrees that, in spite of his limited training in OR, the representation probably is close to reality.

The operations researcher then goes on to explain that according to LP theory, since the model has only three constraints, the optimal mix can never exceed three products. As such, it may be worthwhile to consider discontinuing two of the current products.

The manager listens attentively and then tells the operations researcher that the company is committed to producing all five products because of competition and in no way can the company discontinue any of the products. The operations researcher snaps back by saying that the only way out of the situation is to add at least two more constraints, in which case the optimal LP solution will most likely include all five products.

At this point, the plant manager becomes really confused. To him, the idea of having to add more restrictions to be able to produce more products does not suggest optimality. "That is what the LP theory says" is the answer he gets from the young operations researcher.

What is your opinion of this "paradox"?

Case 3–2

The Reddy Mikks Company is preparing a future expansion plan. A study of the market indicates that the company can increase its sales by about 25%. The following proposals are being studied for the development of an action plan. (Refer to the LP formulation and solution in Section 3.4.)

The first proposal calls for increasing the daily availability of raw material A by about 1 ton and raw material B by about 1.5 tons. The reasoning behind this proposal is that each ton of interior and exterior paints uses an *average* of 1.5 tons each of raw materials A and B.† Thus, to increase the output by 25%, it is necessary to increase the availability of raw materials A and B by $1.5 \times 25\% = 37.5\%$, which yields an increase of $6 \times .375 = 2.25$ tons for A and $8 \times .375 = 3$ tons for B. The purchase price of A and B currently stands at $1200 and $800, respectively.

The second proposal is made by a member of management who has some training in linear programming. She estimates that a 25% expansion would roughly equal a $3165 increase in profit. She also figures that the worth per unit of raw material A is 1/3 and that of B is 4/3 (in thousands of dollars). As a result, the desired increase in production can be achieved by increasing raw material A by $[3.165/(1/3)] = 9.5$ tons or raw material B by $[3.165/(4/3)] = 2.4$ tons. A combined increase in A and B can be effected by basing the estimates on the average worth per unit, which is 5/6. In this case, both A and B can each be increased by $[3.165/(5/6)] = 3.8$ tons.

Do you agree with these proposals? If not, how would you go about solving the problem?

---

† Estimate of average usage $= [(6 + 8)/(4/3 + 10/3)]/2 = 1.5$.

Chapter 4

Linear Programming: Duality and Sensitivity Analysis

This chapter introduces the new topic of duality in linear programming. Aside from its immense theoretical interest, duality is the key concept in the development of the important practical topic of sensitivity analysis. We have dealt with this topic at an elementary level in Sections 2.1.2 and 3.4. You will now fully appreciate the practical significance of sensitivity analysis as we develop it through the use of duality. In addition, the theory will enhance your understanding of some new and efficient computational techniques, which we present throughout the remainder of the book.

4.1 DEFINITION OF THE DUAL PROBLEM

The **dual** is an auxiliary LP problem defined directly and systematically from the original or **primal** LP model. In most LP treatments, the dual is defined for various forms of the primal depending on the types of the constraints, the signs of the variables, and the sense of optimization. Our experience indicates that beginners often become confused by the details of these definitions. Additionally, the use of duality theory does not really necessitate a knowledge of the various definitions in an explicit manner.

In this book we introduce a *single* definition of the dual that automatically accounts for *all* forms of the primal. It is based on the fact that any LP problem must be put in the *standard form* (Section 3.1) before the model is solved by the simplex method. Since all the primal–dual computations are obtained directly from the simplex tableau, it is logical to define the dual in a way that is consistent with the standard form of the primal. You will see later that by defining the dual in this manner, we automatically account for the signs of the dual variables, which are often a source of confusion. Keep in mind, however, that the single definition we give here is *general* in the sense that it automatically accounts for *all* forms of the primal.

The general *standard* form of the primal is defined as

$$\text{maximize or minimize } z = \sum_{j=1}^{n} c_j x_j$$

subject to

$$\sum_{j=1}^{n} a_{ij} x_j = b_i, \quad i = 1, 2, \ldots, m$$

$$x_j \geq 0, \quad j = 1, 2, \ldots, n$$

Note that the n variables, x_j, include the surplus and slacks. For the purpose of constructing the dual, we arrange the coefficients of the primal schematically as shown in Table 4–1.

The diagram shows that the dual is obtained symmetrically from the primal according to the following rules:

1. For every primal constraint there is a dual variable.
2. For every primal variable there is a dual constraint.
3. The constraint coefficients of a primal variable form the left-side coefficients of the corresponding dual constraint; and the objective coefficient of the same variable becomes the right side of the dual constraint. (See, e.g., the tinted column under x_j.)

These rules indicate that the dual problem will have m variables ($y_1, y_2, \ldots, y_m$) and n constraints (corresponding to $x_1, x_2, \ldots, x_n$).

We now turn our attention to determining the remaining elements of the dual problem: the sense of optimization, the type of constraints, and the sign of the dual variables. This information is summarized in Table 4–2 for the maximization and minimization types of the standard form. Recall again that

Table 4-1

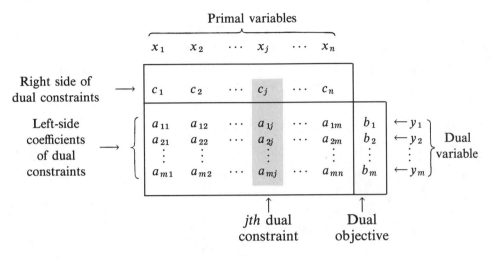

Table 4-2

Standard Primal Objective[a]	Dual		
	Objective	Constraints	Variables
Maximization	Minimization	$\geq$	Unrestricted
Minimization	Maximization	$\leq$	Unrestricted

[a] All primal constraints are equations with nonnegative right-hand sides and all variables are nonnegative.

the standard primal form requires all constraints to be equations (with nonnegative right side) and all the variables to be nonnegative. Thus the only difference in the two primals occurs in the sense of optimization.

The following examples are designed to illustrate the use of these rules and, more important, to show that our definition incorporates all forms of the primal.

Example 4.1-1

Primal

$$\text{Maximize } z = 5x_1 + 12x_2 + 4x_3$$

subject to

$$x_1 + 2x_2 + x_3 \leq 10$$
$$2x_1 - x_2 + 3x_3 = 8$$
$$x_1, x_2, x_3 \geq 0$$

Standard Primal

$$\text{Maximize } z = 5x_1 + 12x_2 + 4x_3 + 0x_4$$

subject to

$$x_1 + 2x_2 + x_3 + x_4 = 10$$
$$2x_1 - x_2 + 3x_3 + 0x_4 = 8$$
$$x_1, x_2, x_3, x_4 \geq 0$$

Notice that x_4 is a slack in the first constraint; hence, it has zero coefficients in the objective function and the second constraint.

Dual

$$\text{Minimize } w = 10y_1 + 8y_2$$

subject to

$$x_1: \quad y_1 + 2y_2 \geq 5$$
$$x_2: \quad 2y_1 - y_2 \geq 12$$
$$x_3: \quad y_1 + 3y_2 \geq 4$$
$$x_4: \quad y_1 + 0y_2 \geq 0 \qquad \text{(implies that } y_1 \geq 0)$$
$$y_1, y_2 \text{ unrestricted}$$

Observe that "y_1 unrestricted" is dominated by $y_1 \geq 0$, the dual constraint associated with x_4. Thus, eliminating the redundancy, the dual problem should read as

$$\text{minimize } w = 10y_1 + 8y_2$$

subject to

$$y_1 + 2y_2 \geq 5$$
$$2y_1 - y_2 \geq 12$$
$$y_1 + 3y_2 \geq 4$$
$$y_1 \geq 0$$
$$y_2 \text{ unrestricted}$$ ◀

Exercise 4.1-1

Indicate the *changes* in the dual shown if its primal is minimization instead of maximization.

[*Ans.* Changes are: Maximize w, first three constraints are of the type $\leq$, and $y_1 \leq 0$.]

Example 4.1-2

Primal

$$\text{Minimize } z = 5x_1 - 2x_2$$

subject to

$$-x_1 + x_2 \geq -3$$
$$2x_1 + 3x_2 \leq 5$$
$$x_1, x_2 \geq 0$$

Standard Primal

$$\text{Minimize } z = 5x_1 - 2x_2$$

subject to

$$x_1 - x_2 + x_3 \qquad = 3$$
$$2x_1 + 3x_2 \qquad + x_4 = 5$$
$$x_1, x_2, x_3, x_4 \geq 0$$

Dual

$$\text{Maximize } w = 3y_1 + 5y_2$$

subject to

$$y_1 + 2y_2 \leq \ \ 5$$
$$-y_1 + 3y_2 \leq -2$$
$$y_1 \qquad \leq \ \ 0$$
$$y_2 \leq \ \ 0$$

y_1, y_2 unrestricted   (redundant)   ◀

Exercise 4.1-2

Indicate the changes in the dual shown if the second primal constraint is of the type "$\geq$" instead of "$\leq$."
[*Ans.* $y_2 \geq 0$ instead of $y_2 \leq 0$.]

Example 4.1-3

Primal

$$\text{Maximize } z = 5x_1 + 6x_2$$

subject to

$$x_1 + 2x_2 = 5$$
$$-x_1 + 5x_2 \geq 3$$
$$4x_1 + 7x_2 \leq 8$$
$$x_1 \text{ unrestricted}$$
$$x_2 \geq 0$$

Standard Primal

Let $x_1 = x_1' - x_1''$, where $x_1', x_1'' \geq 0$. Then the standard primal becomes

$$\text{maximize } z = 5x_1' - 5x_1'' + 6x_2$$

subject to

$$x_1' - \ x_1'' + 2x_2 \qquad \qquad = 5$$
$$-x_1' + \ x_1'' + 5x_2 - x_3 \qquad = 3$$
$$4x_1' - 4x_1'' + 7x_2 \qquad + x_4 = 8$$
$$x_1', x_1'', x_2, x_3, x_4 \geq 0$$

Dual

$$\text{Minimize } w = 5y_1 + 3y_2 + 8y_3$$

subject to

$$y_1 - y_2 + 4y_3 \geq 5 \brace -y_1 + y_2 - 4y_3 \geq -5 \quad \text{(imply that } y_1 - y_2 + 4y_3 = 5)$$

$$2y_1 + 5y_2 + 7y_3 \geq 6$$

$$- y_2 \geq 0 \quad \text{(implies that } y_2 \leq 0)$$

$$y_3 \geq 0$$

y_1 unrestricted

y_2, y_3 unrestricted   (redundant)

Observe that the first and second dual constraints can (*but need not*) be replaced by the equation $y_1 - y_2 + 4y_3 = 5$ (see Section 3.1). This will always be the case when the primal variable is unrestricted, meaning that an unrestricted primal variable will always lead to a dual *equation* (rather than inequality). The result is true whether the primal represents maximization or minimization. ◀

Exercise 4.1-3

Indicate the changes in the dual just shown if the objective is minimization and the first constraint is of the type "≥."
[*Ans.* Maximize w, first three constraints are of the type "≤," and $y_1 \geq 0$, $y_2 \geq 0$, $y_3 \leq 0$.]

If you investigate the foregoing examples carefully, you will be able to devise all the general rules that are traditionally presented in conjunction with the definition of the dual. This, however, will lead to a variety of conditions, particularly in connection with the signs of the dual variables (as you already saw in the examples). However, if you follow the two simple rules we gave in Table 4-2, you will never concern yourself with such problems. Also, remember that the preparation of the standard form does not really represent additional work, since it is always used in the starting simplex iteration.

Exercise 4.1-4

Show that the dual constraint associated with an *artificial* variable (R_i) in the standard form of the primal is always redundant. Hence it is never necessary to consider the dual constraint associated with an artificial variable.
[*Ans.* The dual constraint of artificial variable R_i is $y_i \geq -M$ in case of primal *maximization* and $y_i \leq M$ in case of primal *minimization*. Both are redundant, since M can assume as large a value as desired.]

4.2 PRIMAL–DUAL RELATIONSHIPS

In this section we show the close relationships between the primal and dual problems. Indeed, the optimal solution of one problem is immediately available (without any computations) from the optimal simplex tableau of the other problems. Moreover, in preparation for developing sensitivity analysis techniques, we present certain dual computations that allow the detection of changes in the optimal primal solution as a result of making changes in the original model.

4.2.1 OPTIMAL DUAL SOLUTION IN THE SIMPLEX TABLEAU

We use Example 4.1–1 to demonstrate how the optimal dual solution can be obtained directly from the optimal simplex tableau of the primal. We repeat the two problems here for convenience.

<table>
<tr><th>Primal</th><th>Dual</th></tr>
<tr>
<td>

Maximize $z = 5x_1 + 12x_2 + 4x_3$
subject to

$$x_1 + 2x_2 + x_3 \le 10$$
$$2x_1 - x_2 + 3x_3 = 8$$
$$x_1, x_2, x_3 \ge 0$$

</td>
<td>

Minimize $w = 10y_1 + 8y_2$
subject to

$$y_1 + 2y_2 \ge 5$$
$$2y_1 - y_2 \ge 12$$
$$y_1 + 3y_2 \ge 4$$
$$y_1 \ge 0, y_2 \text{ unrestricted}$$

</td>
</tr>
</table>

We now solve the two problems *independently* by the simplex method. A comparison of the optimal primal and dual tableaus would show how the optimal solution of one problem can be obtained directly from the other. Table 4–3 presents the simplex iterations leading to the optimal primal solution. The iterations are presented in full for later use in the section. The starting and optimal iterations of the dual are given in Table 4–4. The use of artificial variables is necessary to secure a starting solution. Also, the unrestricted variable y_2 in the dual is substituted out as $y_2 = y_2' - y_2''$, where y_2' and y_2'' are nonnegative.

The optimal dual solution *obtained directly by the simplex method* is given in Table 4–4 as

$$w = 54\tfrac{4}{5}, \quad y_1 = 29/5, \quad y_2 = y_1' - y_1'' = 0 - 2/5 = -2/5$$

This information can be obtained directly from the optimal primal tableau (Table 4–3) by using the following equation:

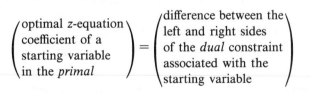

$$\begin{pmatrix} \text{optimal } z\text{-equation} \\ \text{coefficient of a} \\ \text{starting variable} \\ \text{in the } primal \end{pmatrix} = \begin{pmatrix} \text{difference between the} \\ \text{left and right sides} \\ \text{of the } dual \text{ constraint} \\ \text{associated with the} \\ \text{starting variable} \end{pmatrix}$$

In Table 4–3, the starting solution variables are x_4 and R, and their respective coefficients in the *optimal z*-equation are $29/5$ and $-2/5 + M$. The dual constraints associated with x_4 and R are $y_1 \ge 0$ and $y_2 \ge -M$. This information is organized in the following convenient form.

Table 4-3

Iteration	Basic	x_1	x_2	x_3	x_4	R	Solution
0 (starting)	z	$-5-2M$	$-12+M$	$-4-3M$	0	0	$-8M$
x_3 enters	x_4	1	2	1	1	0	10
R leaves	R	2	-1	3	0	1	8
1	z	$-7/3$	$-40/3$	0	0	$\frac{4}{3}+M$	32/3
x_2 enters	x_4	1/3	7/3	0	1	$-1/3$	22/3
x_4 leaves	x_3	2/3	$-1/3$	1	0	1/3	8/3
2	z	$-3/7$	0	0	40/7	$-\frac{4}{7}+M$	368/7
x_1 enters	x_2	1/7	1	0	3/7	$-1/7$	22/7
x_3 leaves	x_3	5/7	0	1	1/7	2/7	26/7
3 (optimal)	z	0	0	3/5	29/5	$-\frac{2}{5}+M$	$54\frac{4}{5}$
	x_2	0	1	$-1/5$	2/5	$-1/5$	12/5
	x_1	1	0	7/5	1/5	2/5	26/5

Starting Primal Variables	x_4	R
Optimal z-equation coefficient	29/5	$-2/5+M$
Left minus right sides of the dual constraint associated with the starting primal variable	$y_1 - 0$	$y_2-(-M)$

Application of the foregoing equation thus yields

$$29/5 = y_1 - 0 \quad \text{and} \quad -2/5+M = y_2-(-M)$$

We thus obtain $y_1 = 29/5$ and $y_2 = -2/5$, which are the same values as those obtained directly from the optimal dual tableau.

We now show that the optimal *dual* tableau (Table 4-4) also yields the optimal primal solution directly and by using the equation stated earlier. Observe,

Table 4-4

Iteration	Basic	y_1	y_2'	y_2''	y_3	y_4	y_5	R_1	R_2	R_3	Solution
0 (starting)	w	$-10+4M$	$-8+4M$	$8-4M$	$-M$	$-M$	$-M$	0	0	0	$21M$
	R_1	1	2	-2	-1	0	0	1	0	0	5
	R_2	2	-1	1	0	-1	0	0	1	0	12
	R_3	1	3	-3	0	0	-1	0	0	1	4
4 (optimal)	w	0	0	0	$-26/5$	$-12/5$	0	$26/5-M$	$12/5-M$	$-M$	$54\frac{4}{5}$
	y_5	0	0	0	$-7/5$	$1/5$	1	$7/5$	$-1/5$	-1	$3/5$
	y_2''	0	-1	1	$2/5$	$-1/5$	0	$-2/5$	$1/5$	0	$2/5$
	y_1	1	0	0	$-1/5$	$-2/5$	0	$1/5$	$2/5$	0	$29/5$

first, that x_1, x_2, and x_3 are respectively associated with the first, second, and third *dual* constraints and, hence, with the artificials R_1, R_2, and R_3.†

Starting Dual Variables	R_1	R_2	R_3
Optimal *w*-equation coefficients	$26/5 - M$	$12/5 - M$	$-M$
Left minus right sides of the primal constraint associated with the starting dual variable	$x_1 - M$	$x_2 - M$	$x_3 - M$

Thus we get

$$x_1 = (26/5 - M) + M = 26/5$$
$$x_2 = (12/5 - M) + M = 12/5$$
$$x_3 = (-M) + M = 0$$

which is the same solution obtained directly in the optimal primal tableau (Table 4–3).

Why should we be interested in obtaining the optimal solution of the primal by solving the dual? The answer is that it may be more advantageous computationally to solve the dual rather than the primal. Recall that the computational effort in linear programming depends more on the number of constraints than on the number of variables. Thus, if the dual happens to have a smaller number of constraints than the primal, generally it will be more efficient to solve the dual, from which the optimal primal solution can then be obtained.

Exercise 4.2–1
Find the optimal dual solution from the optimal primal tableau of each of the following examples in Chapter 3.
(a) Example 3.3–1, Table 3–2.
  [*Ans.* $y_1 = y_2 = 3/2$.]
(b) Example 3.3–2, Table 3–3.
  [*Ans.* $y_1 = 5/8$, $y_2 = 1/8$, $y_3 = 0$.]
(c) The example in Table 3–1, Section 3.2.3–A.
  [*Ans.* $y_1 = 7/5$, $y_2 = 0$, $y_3 = -1/5$.]

The primal and dual solutions in Tables 4–3 and 4–4 reveal two interesting results:

1. At the optimum iteration we have

$$\max z = \min w = 54\tfrac{4}{5}$$

This is always true and, indeed, should be consistent with the *optimal* values of the variables in both problems, namely,

$$z = 5x_1 + 12x_2 + 4x_3 = 5 \times 26/5 + 12 \times 12/5 + 4 \times 0 = 54\tfrac{4}{5}$$
$$w = 10y_1 + 8y_2 = 10 \times 29/5 + 8 \times (-2/5) = 54\tfrac{4}{5}$$

---

† This statement is based on the (almost) obvious observation that *the dual of the dual is the primal.* You should verify this by considering the dual as the *given* problem. Then use x_1, x_2, x_3 as its "dual" variables. By applying the rules in Table 4–2, you will find that the resulting "dual" is the original primal!

2. In the *maximization* problem (Table 4–3) the objective value starts at $z = -8M$ and *increases* successively until it reaches the optimum value at $z = 54\frac{4}{5}$. On the other hand, in the *minimization* problem (Table 4–4), the objective value starts at $w = 21M$ and *decreases* successively until the optimum is reached at $w = 54\frac{4}{5}$. This information is summarized in Figure 4–1. It implies that the objective value for any *feasible* solution in the *minimization* problem always acts as an *upper bound* on the objective value for any *feasible* solution in the *maximization* problem. This condition necessarily requires that the maximization and minimization processes reach an **equilibrium point** beyond which neither problem can improve its objective value. This equilibrium point is reached when both objective values are equal, and it represents the optimum solutions.

These results can be generalized for any pair of primal and dual problems as follows:

1. For any pair of *feasible* primal and dual solutions,

$$\begin{pmatrix} \text{objective value in} \\ \textit{maximization} \text{ problem} \end{pmatrix} \leq \begin{pmatrix} \text{objective value in} \\ \textit{minimization} \text{ problem} \end{pmatrix}$$

2. At the optimum solution for both problems,

$$\begin{pmatrix} \text{objective value in} \\ \textit{maximization} \text{ problem} \end{pmatrix} = \begin{pmatrix} \text{objective value in} \\ \textit{minimization} \text{ problem} \end{pmatrix}$$

Observe closely that these results say nothing about which problem is primal and which is dual. All that matters is the sense of optimization (maximization or minimization) in each problem. Tables 4–3 and 4–4 illustrate the case where the primal is maximization and the dual is minimization. The following example illustrates the opposite case.

Example 4.2–1

Primal	Dual
Minimize $z = 5x_1 + 2x_2$ subject to $\quad\quad x_1 - x_2 \geq 3$ $\quad\quad 2x_1 + 3x_2 \geq 5$ $\quad\quad x_1, x_2 \geq 0$ **Feasible solution:** $\quad\quad x_1 = 3, x_2 = 0$ **Objective value:** $\quad z = 5 \times (3) + 2 \times (0) = 15$	Maximize $w = 3y_1 + 5y_2$ subject to $\quad\quad y_1 + 2y_2 \leq 5$ $\quad\quad -y_1 + 3y_2 \leq 2$ $\quad\quad y_1, y_2 \geq 0$ **Feasible solution:** $\quad\quad y_1 = 3, y_2 = 1$ **Objective value:** $\quad w = 3 \times (3) + 5 \times (1) = 14$

This example shows that for the given feasible solutions, the objective value in the maximization problem (dual) is less than the objective value in the minimization problem (primal).

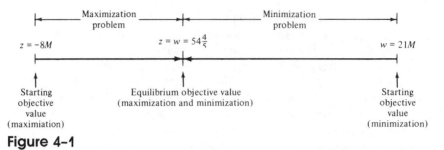

Figure 4-1

Keep in mind that the *feasible* solutions shown here are obtained by using the process of trial and error and that any pair of *feasible* solutions will yield the same conclusion. (To test your comprehension of this point, try to find other feasible solutions for this example.)   ◀

The preceding discussion reveals the following practical result: We can estimate the optimum objective value of an LP problem by judiciously selecting (good) feasible solutions for the primal and dual. These will yield the range in which the optimal objective lies. Indeed, if the feasible solutions happen to yield equal objective values, both solutions are optimal. This result is illustrated in Example 4.2–1, where $z = 15$ and $w = 14$, meaning that the optimum objective values (min z and max w) lie in the range

$$14 \leq (\text{min } z = \text{max } w) \leq 15$$

Since the range is narrow (14 to 15), we can actually think of the feasible solutions in Example 4.2–1 as *near-optimal* solutions. The practical usefulness of this result is now evident, since it can be used to test the "goodness" of feasible solutions.

Exercise 4.2–2
In Example 4.2–1, determine whether or not the following solutions are optimal.
(a) $(x_1 = 3, x_2 = 1; y_1 = 4, y_2 = 1)$
   [*Ans.* Even though $z = w = 17$, the solutions are not optimal because they are not feasible.]
(b) $(x_1 = 4, x_2 = 1; y_1 = 1, y_2 = 0)$
   [*Ans.* No, because $z = 22 \neq w = 3$, even though the solutions are feasible.]
(c) $(x_1 = 3, x_2 = 0; y_1 = 5, y_2 = 0)$
   [*Ans.* Yes, because the solutions are feasible *and* $z = w = 15$.]

4.2.2   IMPORTANT PRIMAL–DUAL COMPUTATIONS

In carrying out sensitivity analysis, we are interested in only one result: Will the changes in the problem's coefficients affect the feasibility or the optimality of the current optimum solution? We have seen in Sections 3.4.4 and 3.4.5 that all we have to do is to recompute the right-side or the objective equation coefficients in the current optimum tableau, and then check optimality and feasibility. We observed also that recomputing the new coefficients can be achieved by using directly the information in the optimum tableau.

In this section we expound on this idea in a more sophisticated manner by exploiting the primal–dual relationships. The computations we present will lead to interesting concepts that will be the basis of an important economic interpretation of duality and the development of the new solution technique called the **dual simplex method.**

It is convenient to present the *primal–dual computations* by using matrix manipulations. We really need only the elementary definitions of (row vector by matrix) and (matrix by column vector) multiplications.† (If you are already familiar with these definitions, you may skip directly to the primal–dual computations.)

Definition. An $(m \times n)$ matrix is a rectangular array of numbers with m rows and n columns. A row vector of size n is a $(1 \times n)$ matrix and a column vector of size m is an $(m \times 1)$ matrix. Thus a matrix of size $(m \times n)$ consists of m row vectors of size n each and n column vectors of size m each. For example, the matrix

$$A = \begin{pmatrix} 3 & 2 \\ 4 & -1 \\ 1 & 0 \end{pmatrix}$$

is of size (3×2). It has the two column vectors

$$\begin{pmatrix} 3 \\ 4 \\ 1 \end{pmatrix} \quad \text{and} \quad \begin{pmatrix} 2 \\ -1 \\ 0 \end{pmatrix}$$

each of size 3 and the three row vectors $(3, 2)$, $(4, -1)$, and $(1, 0)$, each of size 2.

Row Vector × Matrix Multiplication

Given the row vector $\mathbf{V}$ and the matrix $\mathbf{A}$, the product $\mathbf{V} \times \mathbf{A}$ is defined if, and only if, the size of $\mathbf{V}$ is equal to the number of rows of $\mathbf{A}$. Thus, if

$$\mathbf{V} = (v_1, v_2, \ldots, v_m) \quad \text{and} \quad \mathbf{A} = \begin{pmatrix} a_{11} & a_{12} & \cdots & a_{1n} \\ a_{21} & a_{22} & \cdots & a_{2n} \\ \vdots & \vdots & & \vdots \\ a_{m1} & a_{m2} & \cdots & a_{mn} \end{pmatrix}$$

then

$$\mathbf{V} \times \mathbf{A} = (v_1, v_2, \ldots, v_m) \begin{pmatrix} a_{11} & a_{12} & \cdots & a_{1n} \\ a_{21} & a_{22} & \cdots & a_{2n} \\ \vdots & \vdots & & \vdots \\ a_{m1} & a_{m2} & \cdots & a_{mn} \end{pmatrix}$$

$$= \left(\sum_{i=1}^{m} v_i a_{i1}, \sum_{i=1}^{m} v_i a_{i2}, \ldots, \sum_{i=1}^{m} v_i a_{in} \right)$$

---

† Appendix A gives a more complete summary of vectors and matrices. However, this section does not require more than the two elementary multiplications indicated.

The result is a row vector of size n. To illustrate this numerically, consider

$$(11, 22, 33)\begin{pmatrix} 3 & -1 \\ 4 & 8 \\ 6 & 9 \end{pmatrix} = \begin{matrix} (3 \times 11 + 4 \times 22 + 6 \times 33, -1 \times 11 + 8 \times 22 \\ + 9 \times 33) \end{matrix}$$
$$= (319, 462)$$

Matrix × Column Vector Multiplication

The product $\mathbf{A} \times \mathbf{P}$ of the matrix $\mathbf{A}$ and a column vector $\mathbf{P}$ is defined if and only if the number of columns of $\mathbf{A}$ is equal to the size of $\mathbf{P}$. Thus, if

$$\mathbf{P} = \begin{pmatrix} p_1 \\ p_2 \\ \vdots \\ p_n \end{pmatrix}$$

then, for the matrix $\mathbf{A}$ just defined,

$$\mathbf{A} \times \mathbf{P} = \begin{pmatrix} a_{11} & a_{12} & \cdots & a_{1n} \\ a_{21} & a_{22} & \cdots & a_{2n} \\ \vdots & \vdots & & \vdots \\ a_{m1} & a_{m2} & \cdots & a_{mn} \end{pmatrix} \begin{pmatrix} p_1 \\ p_2 \\ \vdots \\ p_n \end{pmatrix} = \begin{pmatrix} \sum_{j=1}^{n} a_{1j}p_j \\ \sum_{j=1}^{n} a_{2j}p_j \\ \vdots \\ \sum_{j=1}^{n} a_{mj}p_j \end{pmatrix}$$

which is a column vector of size m. Numerically, this can be illustrated by

$$\begin{pmatrix} 1 & 5 & 7 \\ 2 & 0 & 8 \\ 3 & 6 & 9 \\ -1 & 4 & -2 \end{pmatrix} \begin{pmatrix} 11 \\ 22 \\ 33 \end{pmatrix} = \begin{pmatrix} 11 \times 1 + 22 \times 5 + 33 \times 7 \\ 11 \times 2 + 22 \times 0 + 33 \times 8 \\ 11 \times 3 + 22 \times 6 + 33 \times 9 \\ 11 \times -1 + 22 \times 4 + 33 \times -2 \end{pmatrix} = \begin{pmatrix} 352 \\ 286 \\ 462 \\ 11 \end{pmatrix}$$

Primal–Dual Computations

The special arrangement introduced in Chapter 3 of the starting simplex tableau is important in the development of the primal–dual computations. Namely, the *starting m* basic variables are always associated with the last m columns of the left side of the starting tableau; also, their constraint coefficients are arranged in the form of an identity matrix.

Under this arrangement, important information can be extracted from the simplex tableau. Figure 4–2 gives a schematic representation of the simplex tableau at *any* iteration. The top row represents the objective equation, and the columns are associated with the constraints. The matrix under the starting variables is called the **inverse** and its position will always be as shown in Figure 4–2.

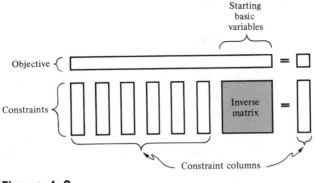

Figure 4–2

Given the inverse, *all* the remaining elements of the tableau can be computed from this inverse and the coefficients in the *original* model. Why are we interested in this type of computation? Recall that sensitivity analysis deals with changing the coefficients in the *original* model. Our goal is to check whether such changes will affect the optimality or the feasibility of the *current* optimal solution. By using the associated inverse together with the "new" (changed) coefficients of the original model to recompute the new elements of the current tableau, we can tell whether the optimality or feasibility of the current solution is affected.

We can divide the computations of the simplex elements into two types (see Figure 4–2): (1) constraint column computations and (2) objective row computations. Constraint column computations are made without the use of the primal–dual relationships, whereas the objective row computations are based on the dual properties.

We explain these computations using the following pair of primal and dual problems whose simplex iterations are given in Tables 4–3 and 4–4 (Section 4.2.1). The primal is presented in the standard form to facilitate the presentation. The reason for including the artificial R in the primal and its corresponding (redundant) dual constraint will be explained shortly.

Primal	Dual
Maximize $z = 5x_1 + 12x_2 + 4x_3 - MR$ subject to $x_1 + 2x_2 + x_3 + x_4 \quad = 10 \quad (y_1)$ $2x_1 - x_2 + 3x_3 \quad + R = 8 \quad (y_2)$ $x_1, x_2, x_3, x_4, R \geq 0$	Minimize $w = 10y_1 + 8y_2$ subject to $y_1 + 2y_2 \geq 5 \quad (x_1)$ $2y_1 - y_2 \geq 12 \quad (x_2)$ $y_1 + 3y_2 \geq 4 \quad (x_3)$ $y_1 \quad \geq 0 \quad (x_4)$ $y_2 \geq -M \quad (R)$ y_2 unrestricted

Constraint Column Computations

At any simplex iteration (primal or dual), the elements of a left or a right column of the constraints of the tableau are computed from

$$\begin{pmatrix} \text{column in} \\ \text{iteration } i \end{pmatrix} = \begin{pmatrix} \text{inverse in} \\ \text{iteration } i \end{pmatrix} \times \begin{pmatrix} \text{original} \\ \text{model column} \end{pmatrix}$$

To illustrate the use of this formula, consider the primal just shown. The starting basic variables are x_3 and R. In Table 4–3, the inverse matrix *associated with each iteration* is shown by tint overlay. Consider iteration 1 and its x_1-constraints column

$$\begin{pmatrix} x_1\text{-column in} \\ \text{iteration 1} \end{pmatrix} = \begin{pmatrix} \text{inverse in} \\ \text{iteration 1} \end{pmatrix} \times \begin{pmatrix} \text{original} \\ x_1\text{-column} \end{pmatrix}$$

$$= \begin{pmatrix} 1 & -1/3 \\ 0 & 1/3 \end{pmatrix} \begin{pmatrix} 1 \\ 2 \end{pmatrix} = \begin{pmatrix} 1/3 \\ 2/3 \end{pmatrix}$$

which is the same as given in iteration 1, Table 4–3.

Now, consider iteration 2 and its right column

$$\begin{pmatrix} \text{right column} \\ \text{in iteration 2} \end{pmatrix} = \begin{pmatrix} \text{inverse in} \\ \text{iteration 2} \end{pmatrix} \times \begin{pmatrix} \text{original} \\ \text{right column} \end{pmatrix}$$

$$= \begin{pmatrix} 3/7 & -1/7 \\ 1/7 & 2/7 \end{pmatrix} \begin{pmatrix} 10 \\ 8 \end{pmatrix} = \begin{pmatrix} 22/7 \\ 26/7 \end{pmatrix}$$

Exercise 4.2–3

Verify that the preceding formula will yield the indicated result in each of the following cases:

(a) Right column in iteration 1, Table 4–3.
(b) x_2-column in iteration 2, Table 4–3.
(c) Right column in iteration 3, Table 4–3.
(d) y_4-column in the optimal iteration, Table 4–4.
(e) Right column in the optimal iteration, Table 4–4.

Objective Row Computations

At *any* simplex iteration of the primal, the elements in the objective *equation* of the variable x_j is computed from

$$\begin{pmatrix} \text{element of } x_j \text{ in} \\ \text{the objective} \\ \text{equation} \end{pmatrix} = \begin{pmatrix} \text{left side of} \\ \text{corresponding} \\ \text{dual constraint} \end{pmatrix} - \begin{pmatrix} \text{right side of} \\ \text{corresponding} \\ \text{dual constraint} \end{pmatrix}$$

Applying this formula to the pair of primal and dual problems shown earlier, we get the following equations (see the constraints of the dual):

$$z\text{-coefficient of } x_1 = y_1 + 2y_2 - 5$$
$$z\text{-coefficient of } x_2 = 2y_1 - y_2 - 12$$
$$z\text{-coefficient of } x_3 = y_1 + 3y_2 - 4$$
$$z\text{-coefficient of } x_4 = y_1 - 0$$
$$z\text{-coefficient of } R = y_2 - (-M) = y_2 + M$$

To compute these coefficients numerically, we need numeric values for the dual variables y_1 and y_2. Since the objective coefficients vary with the iterations, we expect the values of y_1 and y_2 to change from one iteration to the next. The following formula computes the dual variables at any iteration.

$$\begin{pmatrix} \text{values of } dual \\ \text{variables at} \\ \text{iteration } i \end{pmatrix} = \begin{pmatrix} \text{original objective} \\ \text{coefficients of } primal \\ \text{basic variables in} \\ \text{iteration } i \end{pmatrix} \times \begin{pmatrix} \text{inverse in} \\ \text{iteration } i \end{pmatrix}$$

The original objective coefficients of primal basic variables are arranged in a *row vector* whose elements are taken in the same order in which the basic variables appear in the "basic" column of the simplex iteration. For example, in Table 4–3, the successive row vectors associated with the formula are (*note the order in each case*):

$$\begin{aligned} &\textit{iteration 0:} \quad (\text{coefficients of } x_4, R) &= (0, -M) \\ &\textit{iteration 1:} \quad (\text{coefficients of } x_4, x_3) &= (0, \quad 4) \\ &\textit{iteration 2:} \quad (\text{coefficients of } x_2, x_3) &= (12, \quad 4) \\ &\textit{iteration 3:} \quad (\text{coefficients of } x_2, x_1) &= (12, \quad 5) \end{aligned}$$

Exercise 4.2–4

Determine the row vector representing the objective coefficients of the basic variables associated with the optimal iteration in Table 4–4.
[*Ans.* (Coefficients of y_5, y_2'', y_1) $= (0, -8, 10)$.]

Next, we illustrate the use of the computation by determining the coefficients in the z-equation of iteration 3 (optimal) of Table 4–3:

$$\text{dual values} = (\text{objective coefficients of } x_2, x_1) \times (\text{inverse})$$

$$= (12, 5) \begin{pmatrix} 2/5 & -1/5 \\ 1/5 & 2/5 \end{pmatrix} = (29/5, -2/5) = (y_1, y_2)$$

Thus, in iteration 4,

$$\begin{aligned} z\text{-coefficient of } x_1 &= y_1 + 2y_2 - 5 = 29/5 + 2(-2/5) - 5 = 0 \\ z\text{-coefficient of } x_2 &= 2y_1 - y_2 - 12 = 2(29/5) - (-2/5) - 12 = 0 \\ z\text{-coefficient of } x_3 &= y_1 + 3y_2 - 4 = 29/5 + 3(-2/5) - 4 = 3/5 \\ z\text{-coefficient of } x_4 &= y_1 - 0 = 29/5 - 0 = 29/5 \\ z\text{-coefficient of } R &= y_2 - (-M) = -2/5 - (-M) = -\tfrac{2}{5} + M \end{aligned}$$

You may already have noticed that the *dual values* associated with the *optimal* primal iteration yield the optimal dual solution directly (see Section 4.2.1). At any iteration, the associated dual values are sometimes referred to as the **simplex multipliers.** Thus the simplex multipliers of the optimal iteration are the optimal dual values.

Exercise 4.2–5

Compute the simplex multipliers in each of the following cases; then compute the associated z-coefficients in each case and verify that they coincide with those appearing in Tables 4–3 and 4–4.

(a) Iteration 1, Table 4–3.
   [*Ans.* $(y_1, y_2) = (0, 4/3)$.]
(b) Iteration 2, Table 4–3.
   [*Ans.* $(y_1, y_2) = (40/7, -4/7)$.]
(c) Optimal iteration, Table 4–4.
   [*Ans.* $(x_1, x_2, x_3) = (26/5, 12/5, 0)$.]

Summary of Primal–Dual Computations

The primal–dual computations at any interation involve three steps:

1. Compute the elements of each column in the constraints by postmultiplying the current inverse by the original column vector.
2. Compute the dual values or simplex multipliers by premultiplying the original primal objective coefficients of the current basic vector by the current inverse.
3. Compute the (left-side) elements of the objective equation as the difference between the left and right sides of the corresponding dual constraints.

Our experience has shown that the best way to remember the primal–dual computations is to do the following. Select any of the linear programs solved in this chapter or in Chapter 3 and extract the inverse matrix from one of the simplex iterations. Now close the book and attempt to generate the entire simplex tableau using the inverse matrix and the data in the original linear program. You should end up with the same tableau given in the book, of course. This exercise should greatly facilitate your comprehension of the primal–dual computations.

The next two sections introduce an economic interpretation of duality and a new solution method called the dual simplex technique. Both topics are inspired by the presentation in this section. The economic interpretation and the dual simplex method are used directly in the development of sensitivity analysis in Section 4.5.

4.3 ECONOMIC INTERPRETATION OF DUALITY

In the preceding section, we have the following two results that directly link the primal and dual problems:

1. At the optimum,

$$\begin{pmatrix} \text{objective value} \\ \text{in the primal} \end{pmatrix} = \begin{pmatrix} \text{objective value} \\ \text{in the dual} \end{pmatrix}$$

2. At any iteration of the primal,

$$\begin{pmatrix} \text{objective equation} \\ \text{coefficient of the} \\ \text{variable } x_j \end{pmatrix} = \begin{pmatrix} \text{left side of} \\ \text{corresponding} \\ \text{dual constraint} \end{pmatrix} - \begin{pmatrix} \text{right side of} \\ \text{corresponding} \\ \text{dual constraint} \end{pmatrix}$$

These two results lead to important economic interpretations of duality and the dual variables. To present these interpretations formally, we consider the primal as an allocation model in which the objective function is maximized (see Section 2.3). The general primal and dual problems are

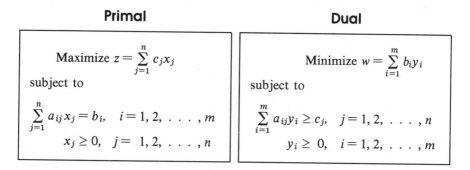

Primal	**Dual**

Maximize $z = \sum_{j=1}^{n} c_j x_j$ Minimize $w = \sum_{i=1}^{m} b_i y_i$

subject to   subject to

$\sum_{j=1}^{n} a_{ij} x_j = b_i, \quad i = 1, 2, \ldots, m$ $\sum_{i=1}^{m} a_{ij} y_i \geq c_j, \quad j = 1, 2, \ldots, n$

$x_j \geq 0, \quad j = 1, 2, \ldots, n$ $y_i \geq 0, \quad i = 1, 2, \ldots, m$

We can think of the primal model in this manner. The coefficient c_j represents the *profit* per unit output of activity j. The available amount of resource i, b_i, is allocated at the rate of a_{ij} units of resource i per unit output of activity j.

Let us now consider the economic interpretations of the two results we summarized at the beginning of this section.

4.3.1 ECONOMIC INTERPRETATION OF DUAL VARIABLES

The equality of the primal and dual objective values can be written in terms of the general definitions given above as

$$z = w = \sum_{i=1}^{m} b_i y_i$$

We analyze this equation dimensionally. Since z represents dollars (profit) and b_i represents the units (amount) of resource i, then dimensionally y_i must represent dollars per unit of resource i because

$$\text{dollars (profit)} = \sum_{i} (\text{units of resource } i)(\text{dollars/unit of resource } i)$$

The dual variables, y_i, then represent the *worth per unit* of resource i. (They are also referred to as **shadow prices.**) This interpretation was obtained in Sections 2.1.2 and 3.4.3 without the use of duality.

The dual values can be used to rank the resources according to their contributions to the value of the objective function. For instance, in the primal and dual problems used in Section 4.2.1, Table 4–3, shows that $y_1 = 29/5$ and $y_2 = -2/5$. Hence, to increase the optimum value of z, the first resource is the only one that can be increased. (In fact, the second must be *decreased* to increase z.) Suppose that Δ_1 is the amount of increase in the first resource. We can determine the maximum value of Δ_1 (as we did in Section 3.4.4) as follows:

$$\begin{pmatrix} x_2 \\ x_1 \end{pmatrix} = \begin{pmatrix} 2/5 & -1/5 \\ 1/5 & 2/5 \end{pmatrix} \begin{pmatrix} 10 + \Delta_1 \\ 8 \end{pmatrix} = \begin{pmatrix} 12/5 + 2\Delta_1/5 \\ 26/5 + \Delta_1/5 \end{pmatrix} \geq \begin{pmatrix} 0 \\ 0 \end{pmatrix}$$

The example shows that Δ_1 can be increased indefinitely, with each unit increase contributing $29/5$ to the objective function. In general, there will be a limit on the amount of increase, but this depends on the technology of the model.

Exercise 4.3-1

If the right side of the second constraint is *decreased* by Δ_2 (recall that $y_2 = -2/5$), determine the maximum value of Δ_2 and the associated total *increase* in optimum z.
[*Ans.* $\Delta_2 \leq 13$ and increase in $z = 26/5$.]

The dimensional analysis given previously leads to an interesting observation. We have noted in Section 4.2.1 that for nonoptimal *feasible* solutions, we have

$$z < w$$

or expressing the relationship verbally in terms of the economic interpretation described earlier, we write

$$(\text{profit}) < (\text{worth of resources})$$

This inequality says that as long as the profit is less than the worth of resources, the solution is not optimal. Optimality is reached when profit equals worth of resources. Alternatively, we can think of the LP model as an input–output system. The resources represent the input and the profit stands for the output. The system remains *unstable* as long as the input exceeds output. Stability is reached when the profit equals the worth of resources.

4.3.2 ECONOMIC INTERPRETATION OF DUAL CONSTRAINTS

In any simplex iteration, the objective coefficient of primal variable x_j is the difference between the left and right sides of the jth dual constraint (Section 4.2.2). By using the general definitions of the primal and dual problems given earlier, this result can be expressed mathematically as

$$\text{objective coefficient of } x_j = \sum_{i=1}^{m} a_{ij}y_i - c_j$$

We analyze this information dimensionally. Note that c_j is the profit per unit of the output activity x_j. This means that c_j has the dimension dollars per unit. For consistency, $\sum_{i=1}^{m} a_{ij}y_i$ also must have the dimension dollars per unit.

Now consider the equation. Since c_j represents profit, $\sum_{i=1}^{m} a_{ij}y_i$, which appears with opposite sign, must stand for "cost." Also, since a_{ij} represents the amount of resource i, used by 1 unit of activity j, y_i must represent the **imputed cost** per unit of resource i and we can think of $\sum_{i=1}^{m} a_{ij}y_i$ as the imputed cost of all resources used to produce 1 unit of activity j. Dimensionally, this information leads to expressing the equation as

$$\$(\text{profit or cost})/\text{unit} = \$(\text{cost})/\text{unit} - \$(\text{profit})/\text{unit}$$

Observe that the left side may represent cost or profit, depending on whether the right side is positive or negative.

The *optimality condition* (maximization) of the simplex method (Section 3.2.2) says that the level of a currently unused activity j (i.e., nonbasic $x_j = 0$) should be increased above zero only if its objective coefficient is negative. If the coefficient is zero or positive, an increase in the level of the activity cannot improve the objective value. Let us interpret this condition economically using dimensional analysis. An unused activity j should be included in the solution if $\Sigma_{i=1}^m a_{ij}y_i - c_j < 0$; that is,

$$\begin{pmatrix} \text{imputed cost of} \\ \text{used resources} \\ \text{per unit of } j \end{pmatrix} - \begin{pmatrix} \text{profit per} \\ \text{unit of } j \end{pmatrix} < 0$$

or

$$\begin{pmatrix} \text{imputed cost of} \\ \text{used resources} \\ \text{per unit of } j \end{pmatrix} < \begin{pmatrix} \text{profit per} \\ \text{unit of } j \end{pmatrix}$$

Thus, as long as the profit exceeds the imputed cost, the activity level should be increased above zero.

You will notice that when we admit an activity into the solution (make its variable basic), we increase its level to the point where its objective equation coefficient becomes zero. This is equivalent to exploiting the profitability of the activity to the fullest extent, since any increase beyond this point will result in increasing the imputed cost above the potential profit. Obviously, this is not beneficial from the viewpoint of optimization.

Finally, you can see why in maximization models an unused activity with positive objective equation coefficient should remain at the zero level. The fact that its unit imputed cost of resources exceeds its unit profit makes it unattractive.

To familiarize you with the notation often used in LP literature, we introduce the following definition:

$$z_j = \sum_{i=1}^m a_{ij}y_i$$

$$= \text{imputed cost of used resources} \atop \text{per unit of } j$$

The notation $(z_j - c_j)$ is the objective coefficient of x_j in the simplex tableau and is often referred to as the **reduced cost** of activity j. Indeed, some books utilize

$$z_j - c_j = \sum_{i=1}^m a_{ij}y_i - c_j$$

to compute the objective equation coefficients directly rather than use the Gauss–Jordan row operations introduced in Chapter 3. The use of $(z_j - c_j)$ in the simplex computations is part of a version of the simplex method, called the **revised simplex method,** which is developed primarily to alleviate problems associated with the use of the digital computer as an essential tool for solving linear programs. We introduce this method in Chapter 7.

4.3.3 APPLICATIONS BASED ON ECONOMIC INTERPRETATIONS OF DUALITY

The LP model may be represented as an input–output system with the following principal elements:

> input = limited resources
>
> output = activity's contribution to the objective function
>
> model = conversion of limited resources to activities.

We are interested primarily in finding ways to improve the performance (output) of the LP model. As we can see from the definition of the input–output system, the model's performance can be improved by considering one or more of the following points (assume a maximization model):

1. Increase the activity's contribution to the objective function.
2. Increase the availability of limited resources.
3. Decrease the activity's consumption of the limited resources.

The first point deals with factors that are normally outside the management's sphere of control. For example, profit margins are normally dictated by market conditions, where competition may not make it possible to increase prices as a means of improving the profit margin. Thus the first point may not prove effective in increasing the model's output.

The increase of limited resources is a more controllable factor, since it normally entails committing new funds for the acquisition of additional quantities of the limited resources. The values of the dual variables (shadow prices) can be used to set priorities for allocation of available funds to the different resources. We have seen in Section 4.3.1 that the dual variables represent the *worths per unit* of the respective resources. Thus the higher the dual value, the higher should be the priority of its resource for receiving additional funds (see also Section 3.4.3).

The last point deals with factors that are normally under management control. It deals with reducing the consumption of resources by an activity. This can be accomplished through improvement of the process utilizing the resources to produce the activities.

As in the case of setting priorities for allocation of resources funds, the dual values can be used to point out which part of the activity should be paid the most attention to produce the "most" improvement. Specifically, to make activity j more profitable, its total imputed cost, $\sum_{i=1}^{m} a_{ij} y_i$, must be reduced. This is normally done by paying attention to reducing the consumption rate a_{kj} associated with the largest dual variable y_k. The technical aspect of how a_{kj} is reduced depends on the system under study.

Example 4.3-1. Consider the product-mix problem in which each of three products is processed on three different operations. The limits on the available time for the three operations are 430, 460, and 420 minutes daily and the profits per unit of the three products are \$3, \$2, and \$5. The times in minutes per unit on the three operations are given as follows:

	Product 1	Product 2	Product 3
Operation 1	1	2	1
Operation 2	3	0	2
Operation 3	1	4	0

The LP model is written as

$$\text{maximize } z = 3x_1 + 2x_2 + 5x_3 \qquad \text{(daily profit)}$$

subject to

$$
\left.
\begin{array}{ll}
\text{operation 1:} & 1x_1 + 2x_2 + 1x_3 \le 430 \\
\text{operation 2:} & 3x_1 + 0x_2 + 2x_3 \le 460 \\
\text{operation 3:} & 1x_1 + 4x_2 + 0x_3 \le 420
\end{array}
\right\}
\quad
\left(
\begin{array}{l}
\text{limits on} \\
\text{daily usages} \\
\text{of operations}
\end{array}
\right)
$$
$$x_j \ge 0, \quad j = 1, 2, 3$$

The optimal tableau of the model is given by

Basic	x_1	x_2	x_3	x_4	x_5	x_6	Solution
z	4	0	0	1	2	0	1350
x_2	−1/4	1	0	1/2	−1/4	0	100
x_3	3/2	0	1	0	1/2	0	230
x_6	2	0	0	−2	1	1	20

We notice that the optimal mix does not include product 1 ($x_1 = 0$). This means that product 1 is not profitable, which occurs because the imputed cost of product 1 is larger than its unit profit; that is, $z_1 > c_1$. Since $z_1 = 1y_1 + 3y_2 + 1y_3$ and $c_1 = \$3$, we can make x_1 profitable by decreasing the value of z_1. This result can be accomplished by reducing the usages by product 1 of the three operations times. These usages are given by the coefficients of y_1, y_2, and y_3 in the expression for z_1.

The optimal tableau gives $y_1 = 1$, $y_2 = 2$, and $y_3 = 0$. This means that a reduction in the usage of operation 3 will not be effective, since its imputed cost per unit, y_3, is zero. Thus, considering operations 1 and 2, we notice that $y_2 (= 2)$ is larger than $y_1 (= 1)$. As a result, it is more attractive to give higher priority to reducing the usage of operation 2.

Suppose that we are interested in determining the amount of reduction in the usage of operation 2 that will make product 1 just profitable. To do so, let r_2 represent reduction in minutes per unit of product 1 on operation 2. In this case,

$$z_1 = y_1 + (3 - r_2)y_2 + y_3 = 1(1) + (3 - r_2)(2) + 1(0)$$
$$= 7 - 2r_2$$

Product 1 becomes just profitable when c_1 just exceeds z_1; that is, $z_1 < c_1$ or $7 - 2r_2 < 3$, which yields $r_2 > 2$. This means that the usage of operation 2 must be reduced by more than 2 minutes to make product 1 profitable. ◀

Exercise 4.3-2

In Example 4.3–1, suppose that the per unit usage of operation 2 cannot be reduced by more than 1.75 minutes. Determine the additional reduction in the usage of operation 1 that will make x_1 just profitable.
[*Ans.* .5 minute.]

4.4 DUAL SIMPLEX METHOD

In Section 4.2 we showed that at any primal iteration $z_j - c_j$, the objective-equation coefficient of x_j, equals the difference between the left and right sides of the associated dual constraint. When, in the case of maximization, the primal iteration is not optimal, $z_j - c_j < 0$ for at least one variable. Only at the optimum do we have $z_j - c_j \geq 0$ for *all j*.

Looking at this condition from the standpoint of duality, we have

$$z_j - c_j = \sum_{i=1}^{m} a_{ij}y_i - c_j$$

Thus, when $z_j - c_j < 0$, $\sum_{i=1}^{m} a_{ij}y_i < c_j$, which means that the dual is infeasible when the primal is nonoptimal. On the other hand, when $z_j - c_j \geq 0$, $\sum_{i=1}^{m} a_{ij}y_i \geq c_j$, which means that the dual becomes feasible when the primal reaches optimality.

These results suggest a new solution method for linear programs, which starts *infeasible* but (better than) *optimal*. (Compare with the regular simplex method, which starts *feasible* but *nonoptimal*.) The new method, called the **dual simplex,** will then maintain optimality and the successive iterations will work to clear the infeasibility. When feasibility is reached, the process terminates since the solution is both feasible and optimal. This type of problem is encountered in certain LP models; but more important, it is used directly in carrying out sensitivity analysis, as we shall show in the following section.

Example 4.4-1

$$\text{Minimize } z = 2x_1 + x_2$$

subject to

$$3x_1 + x_2 \geq 3$$
$$4x_1 + 3x_2 \geq 6$$
$$x_1 + 2x_2 \leq 3$$
$$x_1, x_2 \geq 0$$

The initial step requires converting all the constraints to "$\leq$" inequalities and then augmenting the slack variables. Thus we get

$$\text{minimize } z = 2x_1 + x_2$$

subject to

$$
\begin{aligned}
-3x_1 - x_2 + x_3 &&&= -3 \\
-4x_1 - 3x_2 &&+ x_4 &= -6 \\
x_1 + 2x_2 &&+ x_5 &= 3
\end{aligned}
$$

$$x_1, x_2, x_3, x_4, x_5 \geq 0$$

If you try to express this problem as a starting simplex tableau, you will notice that the slack variables (x_3, x_4, x_5) do not provide a *feasible* starting solution. Since this is a *minimization* problem and all the objective-equation coefficients are ≤ 0, the starting basic solution $x_3 = -3$, $x_4 = -6$, $x_5 = 3$ is optimal but infeasible. This problem is typical of the type that can be handled by the dual simplex method (see also Problem 4–28).

The starting optimal but infeasible tableau is thus given as

Basic	x_1	x_2	x_3	x_4	x_5	Solution
z	-2	-1	0	0	0	0
x_3	-3	-1	1	0	0	-3
x_4	-4	-3	0	1	0	-6
x_5	1	2	0	0	1	3

As in the regular simplex method, the method of solution is based on the optimality and feasibility conditions. The optimality condition guarantees that the solution remains always optimal, whereas the feasibility condition forces the basic solutions toward the feasible space.

Feasibility Condition

The leaving variable is the basic variable having the most negative value. (Break ties arbitrarily.) If all the basic variables are nonnegative, the process ends and the feasible (optimal) solution is reached.

Optimality Condition

The entering variable is selected from among the nonbasic variables as follows. Take the ratios of the left-hand-side coefficients of the z-equation to the corresponding coefficients in the equation associated with the leaving variable. Ignore the ratios associated with positive or zero denominators. The entering variable is the one with the smallest ratio if the problem is minimization, or the smallest *absolute value* of the ratios if the problem is maximization. (Break ties arbitrarily.) If all the denominators are zero or positive, the problem has no feasible solution.

After selecting the entering and the leaving variables, row operations are applied as usual to obtain the next iteration of the solution.

The leaving variable in the preceding tableau is x_4 ($= -6$), since it has the most negative value. For the entering variable, the ratios are given by

Variable	x_1	x_2	x_3	x_4	x_5
z-equation	-2	-1	0	0	0
x_4-equation	-4	-3	0	1	0
Ratios	1/2	1/3	—	—	—

The entering variable is x_2, since it corresponds to the smallest ratio 1/3. By applying row operations as usual, we find the new tableau:

Basic	x_1	x_2	x_3	x_4	x_5	Solution
z	$-2/3$	0	0	$-1/3$	0	2
x_3	$-5/3$	0	1	$-1/3$	0	-1
x_2	4/3	1	0	$-1/3$	0	2
x_5	$-5/3$	0	0	2/3	1	-1

The new solution is still optimal but infeasible ($x_3 = -1$, $x_5 = -1$). If x_3 is arbitrarily selected to leave the solution, x_1 becomes the entering variable. This gives

Basic	x_1	x_2	x_3	x_4	x_5	Solution
z	0	0	$-2/5$	$-1/5$	0	12/5
x_1	1	0	$-3/5$	1/5	0	3/5
x_2	0	1	4/5	$-3/5$	0	6/5
x_5	0	0	-1	1	1	0

which is now optimal and feasible.

The application of the dual simplex method is especially useful in sensitivity analysis. This occurs, for example, when a new constraint is added to the problem after the optimal solution is obtained. If this constraint is not satisfied by the optimal solution, the problem remains optimal but it becomes infeasible. The dual simplex method is then used to clear the infeasibility in the problem. ◄

4.5 SENSITIVITY OR POSTOPTIMAL ANALYSIS

In Chapters 2 and 3 we emphasized the fact that there is more to linear programming and the simplex method than the determination of the optimum values of the variables. The LP solution must provide *dynamic* information. A *static* optimum solution will become obsolete as soon as the conditions under which the model is constructed change. Sensitivity analysis is concerned with studying possible changes in the available optimal solution as a result of making changes in the *original* model.

To gain an appreciation of the potential advantages of sensitivity analysis, we consider its application to the Reddy Mikks model, which we summarize here for convenience. The model deals with the production of exterior and interior paints subject to the limitations of demand and two raw materials. Given that x_1 and x_2 represent the tons of exterior and interior paints, the Reddy Mikks model seeking the maximization of total profit is given as (see Example 2.1–1)

$$\text{maximize } z = 3x_1 + 2x_2 \quad \text{(profit)}$$

subject to

$$x_1 + 2x_2 \leq 6 \quad \text{(raw material A)}$$
$$2x_1 + x_2 \leq 8 \quad \text{(raw material B)}$$
$$-x_1 + x_2 \leq 1 \quad \text{(limit on demand)}$$
$$x_2 \leq 2 \quad \text{(limit on demand)}$$
$$x_1, x_2 \geq 0$$

After obtaining the optimum solution, the Reddy Mikks top management may be interested in exploring the following example situations in response to inquiries or information provided by the company's various departments.

1. In next week's production schedule, the *production department* wants to divert 2 tons daily of raw material B to another process, but will compensate for this reduction in B by increasing the allocation of raw material A by 3 tons daily.

2. The *marketing department* believes that over the next 6 months changes in the market conditions will increase their market share of interior paint from a maximum of 2 tons to $3\frac{1}{2}$ tons daily.

3. The *R&D department* discovered a new process that will reduce usage of raw materials A and B per ton of exterior paint from 1 and 2 tons, respectively, to .8 and 1.7 tons.

4. The *finance department* anticipates that competition for a limited market will reduce the marginal profits of exterior and interior paint to $2500 and $1500 per ton.

5. The *OR analyst*, after analyzing the sales records over the past year, discovered that the market never absorbed more than 3 tons of exterior paint daily. He thus believes that the Reddy Mikks market share of exterior paint cannot exceed 3 tons daily.

6. The *marketing department* is toying with the idea of introducing a cheaper brand of exterior paint to satisfy a certain segment of the market.

The foregoing list is typical of the situations that can be investigated by sensitivity analysis. They may be considered individually or combined, depending on the situation under study. The ultimate result of the analysis should answer the question: Does the *current* optimum change? If so, what is the new optimum?

Study the preceding list carefully. Then ask yourself the question: How can each situation change the current *optimum* solution? With some reflection, your answers will fall in one of the following two categories:

1. The *current* solution can become *infeasible.*
2. The *current* solution can become *nonoptimal.*†

The two categories are based on the results of the primal–dual computations presented in Section 4.2. If you review them again at this point, you will discover the following:

1. *Infeasibility* of the current solution can arise only if we change the availability of resources (*right side* of constraints) and/or add new constraints (e.g., situations 1, 2, and 5 in the preceding list).
2. *Nonoptimality* of the current solution can occur only if we change the objective function and/or *certain* elements of the *left sides* of the constraints (e.g., situations 3 and 4). It can also occur if a new activity is added to the model (e.g., situation 6).

Based on the foregoing discussion, the general procedure for carrying out sensitivity analysis can be summarized as follows:

Step 1: Solve the original LP model and obtain its optimal simplex tableau. Go to step 2.

Step 2: For the proposed change(s) in the original model, recompute the new elements of the current optimal tableau by using the primal–dual computations in Section 4.2. Go to step 3.

Step 3: If the *new* tableau is nonoptimal, go to step 4. If it is infeasible, go to step 5. Otherwise, record the solution in the new tableau as the new optimum. Stop.

Step 4: Apply the *regular* simplex method to the *new* tableau to obtain a new optimal solution (or indicate that the solution is unbounded). Stop.

Step 5: Apply the *dual* simplex method to the new tableau to recover feasibility (or indicate that no feasible solution exists). Stop.

We now turn to specifics and show how sensitivity analysis is applied to the Reddy Mikks model. To that end, we summarize here the primal and dual problems of the model, together with the (current) optimal primal tableau. This is all the information needed to start the analysis (see step 1).

---

† A third category also exists in which the current solution can become both *nonoptimal* and *infeasible.* This can occur if we effect changes in the left side of the constraints that in turn change the *inverse* matrix. As the inverse is used directly in computing the objective equation *and* the right side of the simplex tableau, both optimally and feasibility can be affected. We shall discuss this point later, primarily to help you recognize such situations. However, we emphasize that this type of change does not lend itself neatly to sensitivity analysis.

<table>
<tr><td>

Reddy Mikks Primal

Maximize $z = 3x_1 + 2x_2$
subject to

$$x_1 + 2x_2 \le 6$$
$$2x_1 + x_2 \le 8$$
$$-x_1 + x_2 \le 1$$
$$x_2 \le 2$$
$$x_1, x_2 \ge 0$$

</td><td>

Reddy Mikks Dual

Minimize $w = 6y_1 + 8y_2 + y_3 + 2y_4$
subject to

$$y_1 + 2y_2 - y_3 \ge 3$$
$$2y_1 + y_2 + y_3 + y_4 \ge 2$$
$$y_1, y_2, y_3, y_4 \ge 0$$

</td></tr>
</table>

Optimal Primal Tableau

Basic	x_1	x_2	x_3	x_4	x_5	x_6	Solution
z	0	0	1/3	4/3	0	0	38/3
x_2	0	1	2/3	−1/3	0	0	4/3
x_1	1	0	−1/3	2/3	0	0	10/3
x_5	0	0	−1	1	1	0	3
x_6	0	0	−2/3	1/3	0	1	2/3

(As you proceed through the remainder of this section, it will be convenient to keep in front of you a copy of the information given. Henceforth, we refer to the optimal primal solution as the *current* solution.)

The remainder of this section is organized in two parts. The first part deals with changes in the original model that could lead to infeasibility. The second part treats the cases where the changes could lead to nonoptimality.

4.5.1 CHANGES AFFECTING FEASIBILITY

There are two types of changes that could affect the feasibility of the current solution: (1) changes in resources availability (or right side of the constraints) and (2) addition of new constraints. We consider each case separately.

Changes in the Right Side of Constraints

Suppose that in the Reddy Mikks model, the daily availability of raw material A is changed to 7 tons instead of the current 6 tons. How is the current solution affected?

We know from the primal–dual computations that changes in the right side of the constraints can affect only the right side of the optimal tableau; that is, it can affect feasibility only. Thus all we have to do is recompute the *new* right side of the optimal tableau using the computations given in Section 4.2. (The inverse matrix is shown by a tinted overlay in the optimal tableau.)

The new (basic) solution of the problem is (see Section 4.2.2)

$$\begin{pmatrix} x_2 \\ x_1 \\ x_5 \\ x_6 \end{pmatrix} = \begin{pmatrix} 2/3 & -1/3 & 0 & 0 \\ -1/3 & 2/3 & 0 & 0 \\ -1 & 1 & 1 & 0 \\ -2/3 & 1/3 & 0 & 1 \end{pmatrix} \begin{pmatrix} 7 \\ 8 \\ 1 \\ 2 \end{pmatrix} = \begin{pmatrix} 2 \\ 3 \\ 2 \\ 0 \end{pmatrix} = \begin{pmatrix} \text{new right} \\ \text{side of} \\ \text{the tableau} \end{pmatrix}$$

Since the right-side elements remain nonnegative, the current basic variables remain unchanged. Only their new values become $x_1 = 3$, $x_2 = 2$, $x_5 = 2$, $x_3 = x_4 = x_6 = 0$. The new value of z is $3(3) + 2(2) = 13$.

Exercise 4.5-1

Suppose that, in addition to the change in availability of material A, the maximum limit on the daily demand of interior paint (x_2) is increased from 2 tons to 3 tons. Check if the current (basic) variables remain feasible and find the new values of the variables.

[*Ans*. The basic variables remain unchanged. $x_1 = 3$, $x_2 = 2$, $x_5 = 2$, $x_6 = 1$, $x_3 = x_4 = 0$.]

Let us now consider an example of what happens when the current basic variable becomes infeasible. Suppose that the daily availabilities of materials A and B are 7 tons and 4 tons instead of 6 tons and 8 tons. The right side of the tableau is computed as follows:

$$\begin{pmatrix} x_2 \\ x_1 \\ x_5 \\ x_6 \end{pmatrix} = \begin{pmatrix} 2/3 & -1/3 & 0 & 0 \\ -1/3 & 2/3 & 0 & 0 \\ -1 & 1 & 1 & 0 \\ -2/3 & 1/3 & 0 & 1 \end{pmatrix} \begin{pmatrix} 7 \\ 4 \\ 1 \\ 2 \end{pmatrix} = \begin{pmatrix} 10/3 \\ 1/3 \\ -2 \\ -4/3 \end{pmatrix} = \begin{pmatrix} \text{new right} \\ \text{side of} \\ \text{the tableau} \end{pmatrix}$$

The changes will make x_5 and x_6 negative, meaning that the current solution is infeasible. We must thus use the dual simplex method to recover feasibility.

The following optimal primal tableau shows the *new* right side with tinted overlay. All the remaining elements are unchanged.

Basic	x_1	x_2	x_3	x_4	x_5	x_6	Solution
z	0	0	1/3	4/3	0	0	23/3
x_2	0	1	2/3	-1/3	0	0	10/3
x_1	1	0	-1/3	2/3	0	0	1/3
x_5	0	0	-1	1	1	0	-2
x_6	0	0	-2/3	1/3	0	1	-4/3

Notice that the new value of z is $3(1/3) + 2(10/3) = 23/3$.

The tableau is optimal (all coefficients in the z-equation are ≥ 0) but infeasible (at least one basic variable is negative). The application of the *dual* simplex method shows that the leaving and entering variables are x_5 and x_3. This leads to the following tableau:

Basic	x_1	x_2	x_3	x_4	x_5	x_6	Solution
z	0	0	0	5/3	1/3	0	7
x_2	0	1	0	1/3	2/3	0	2
x_1	1	0	0	1/3	−1/3	0	1
x_3	0	0	1	−1	−1	0	2
x_6	0	0	0	−1/3	−2/3	1	0

This tableau is both optimal and feasible. The new solution is $x_1 = 1$, $x_2 = 2$, and $z = 7$. Although the feasible solution was recovered in one iteration, in general the dual simplex method may require more than one iteration to reach feasibility.

Exercise 4.5–2

Consider the Reddy Mikks model as *originally* given. Suppose that the daily availability of raw material A is 3 tons (instead of 6 tons). Show that the current solution becomes infeasible and determine the entering and leaving variables in the first dual simplex iteration.
[*Ans.* New right-side elements = $(−2/3, 13/3, 6, 8/3) = (x_2, x_1, x_5, x_6)$. Dual simplex: x_2 leaves and x_4 enters.]

Addition of a New Constraint

The addition of a new constraint can result in one of two conditions:

1. The constraint is satisfied by the current solution, in which case the constraint is either *nonbinding* or *redundant* and its addition will thus not change the solution.

2. The constraint is not satisfied by the current solution. It will thus become *binding* and the new solution is obtained by using the dual simplex method.

To illustrate these cases, suppose that the daily demand on exterior paint does not exceed 4 tons. A new constraint of the form

$$x_1 \leq 4$$

must be added to the model. Since the current solution ($x_1 = 10/3$, $x_2 = 4/3$) obviously satisfies the new constraint, it is labeled nonbinding, and the current solution remains unchanged.

Now suppose that the maximum demand on exterior paint is 3 tons instead of 4 tons. Then the new constraint becomes $x_1 \leq 3$, which is not satisfied by the current solution $x_1 = 10/3$ and $x_2 = 4/3$.

Here is what we do to recover feasibility. First, put the new constraint in the standard form by augmenting a slack or a surplus variable if necessary. Then substitute out any of the current basic variables in the constraint in terms of the (current) nonbasic variables. The final step is to augment the "modified" constraint to the current optimum tableau and apply the dual simplex to recover feasibility.

Using x_7 as a slack, we find that the standard form of $x_1 \leq 3$ is

$$x_1 + x_7 = 3, \qquad x_7 \geq 0$$

Now, in the current solution x_1 is a basic variable and we must substitute it out in terms of the nonbasic variables. In the x_1-equation of the current optimal tableau, we have

$$x_1 - (1/3)x_3 + (2/3)x_4 = 10/3$$

Thus the new constraint expressed in terms of the current nonbasic variables becomes

$$(10/3) + (1/3)x_3 - (2/3)x_4 + x_7 = 3$$

or

$$(1/3)x_3 - (2/3)x_4 + x_7 = -1/3$$

(The negative right side indicates infeasibility since, given $x_3 = x_4 = 0$, $x_7 = -1/3$, which violates the requirement $x_7 \geq 0$.)

The modified constraint is now augmented to the current optimal tableau as given next. The augmentation to the tableau is shown by tinted overlay. The remaining elements are taken directly from the current optimum tableau:

Basic	x_1	x_2	x_3	x_4	x_5	x_6	x_7	Solution
z	0	0	1/3	4/3	0	0	0	38/3
x_2	0	1	2/3	−1/3	0	0	0	4/3
x_1	1	0	−1/3	2/3	0	0	0	10/3
x_5	0	0	−1	1	1	0	0	3
x_6	0	0	−2/3	1/3	0	1	0	2/3
x_7	0	0	1/3	−2/3	0	0	1	−1/3

By the dual simplex, x_7 leaves the solution and x_4 enters. This yields the following optimal feasible tableau:

Basic	x_1	x_2	x_3	x_4	x_5	x_6	x_7	Solution
z	0	0	1	0	0	0	2	12
x_2	0	1	1/2	0	0	0	−1/2	3/2
x_1	1	0	0	0	0	0	1	3
x_5	0	0	−1/2	0	1	0	3/2	5/2
x_6	0	0	−1/2	0	0	1	1/2	1/2
x_4	0	0	−1/2	1	0	0	−3/2	1/2

The new solution has a worse optimum value of z than before the constraint is augmented. This is always expected since the addition of a new *binding* constraint can never improve the value of z.

Exercise 4.5-3

Suppose that each of the following constraints is added to the Reddy Mikks model. Determine whether the added constraint is binding. If so, express the binding constraint in terms of the nonbasic variables of the current solution and identify the entering and leaving variables of the dual simplex method.

(a) $x_2 \leq 2$

[*Ans.* The constraint is nonbinding.]

(b) $x_1 + x_2 \leq 4$

[*Ans.* $-(1/3)x_3 - (1/3)x_4 + x_7 = -2/3$. x_7 leaves and x_3 enters.]

(c) $2x_1 + 3x_2 \geq 11$

[*Ans.* $(4/3)x_3 + (1/3)x_4 + x_7 = -1/3$. No feasible solution exists.]

The idea of adding constraints, one at a time, to a current optimum tableau is sometimes used to reduce the computational burden in solving linear programs. As we noted in Chapter 3, the amount of computations in the simplex method depends primarily on the number of constraints. We can effectively reduce the number of constraints in a model as follows. First, we identify the **secondary constraints,** which are the constraints we "feel" are the least restrictive in the optimum solution. The model is then solved subject to the remaining (primary) constraints. The secondary constraints are then augmented to the resulting optimum tableau, one at a time, until a solution is encountered that satisfies all secondary constraints not augmented in the tableau. A computational advantage is realized when a large number of secondary constraints are not used in the computations.

4.5.2 CHANGES AFFECTING OPTIMALITY

The current solution will cease to be optimal only if the coefficients of the objective equation violate the optimality condition. We have seen in Section 4.3 that the objective coefficients directly equal the difference between the left and right sides of the dual constraint. Since the left side of the dual constraint is determined from the primal columns associated with the activities, we can conclude that optimality can be affected if changes occur in the objective function or in the resource usage by each activity. We first consider each case separately. A combination of the two situations is then presented as an application of the augmentation of new activities to the model. Keep in mind that all that is needed in any of these cases is to check optimality by recomputing the new objective equation coefficients.

Changes in the Objective Function

Recall in the primal–dual computations that the dual values are computed using the coefficients of the *basic* variables in the objective function. These dual values are in turn substituted in the dual constraint to compute the z-equation coefficients. Two important points thus arise:

1. If the changes in the objective function involve the coefficients of a *current basic* variable, determine the *new* dual values and then use them to recompute the new z-equation coefficients.

2. If the changes involve *nonbasic* variables only, use the current dual values (directly from the current tableau) and recompute the z-equation coefficients of the involved nonbasic variables only. No other changes will occur in the tableau.

Suppose that in the Reddy Mikks model the objective function is changed from $z = 3x_1 + 2x_2$ to $z = 5x_1 + 4x_2$. The changes involve both x_1 and x_2, which happen to be basic in the current solution. Thus we must determine the new dual values (see Section 4.2.2). Notice that the order of the basic variables in the current tableau is (x_2, x_1, x_5, x_6).

$$(y_1, y_2, y_3, y_4) = (4, 5, 0, 0) \begin{pmatrix} 2/3 & -1/3 & 0 & 0 \\ -1/3 & 2/3 & 0 & 0 \\ -1 & 1 & 1 & 0 \\ -2/3 & 1/3 & 0 & 1 \end{pmatrix}$$

$$= (1, 2, 0, 0)$$

The next step is to recompute the z-equation coefficients by taking the difference between the left and right sides of the dual constraints. This is done as follows. (Observe that the right sides of the dual now must be equal to the *new* objective function coefficients.)

$$x_1\text{-coefficient} = y_1 + 2y_2 - y_3 - 5$$
$$= 1(1) + 2(2) - (0) - 5 = 0$$
$$x_2\text{-coefficient} = 2y_1 + y_2 + y_3 + y_4 - 4$$
$$= 2(1) + 2 + 0 + 0 - 4 = 0$$
$$x_3\text{-coefficient} = y_1 - 0 = 1 - 0 = 1$$
$$x_4\text{-coefficient} = y_2 - 0 = 2 - 0 = 2$$
$$x_5\text{-coefficient} = y_3 - 0 = 0 - 0 = 0$$
$$x_6\text{-coefficient} = y_4 - 0 = 0 - 0 = 0$$

Since all the z-equation coefficients are ≥ 0 (maximization), the changes indicated in the objective function will *not* change the optimum variables or their values. About the only change is the value of z, which is now given by $5 \times (10/3) + 4 \times (4/3) = 22$.

Exercise 4.5–4

Recompute the z-equation coefficients in each of the following cases. What general observation do you have regarding the objective coefficients of the basic variables? Specifically, could they ever differ from zero?

(a) $z = 6x_1 + 4x_2$

  [*Ans.* z-equation coefficients = (0, 0, 2/3, 8/3, 0, 0).]

(b) $z = 5x_1 + 5x_2$

[*Ans.* z-equation coefficients = (0, 0, 5/3, 5/3, 0, 0). The general observation is that the coefficients of the basic variables *always* remain zero, indicating that the corresponding *dual* constraint must be satisfied in equation form. This observation means that it is necessary to recompute the coefficients of the *nonbasic* variables only.]

Let us consider another example in which changes in the objective function will result in nonoptimality. Suppose that

$$z = 4x_1 + x_2$$

then

$$(y_1, y_2, y_3, y_4) = (1, 4, 0, 0) \begin{pmatrix} 2/3 & -1/3 & 0 & 0 \\ -1/3 & 2/3 & 0 & 0 \\ -1 & 1 & 1 & 0 \\ -2/3 & 1/3 & 0 & 1 \end{pmatrix}$$

$$= (-2/3, 7/3, 0, 0)$$

You can verify that the new z-equation becomes

Basic	x_1	x_2	x_3	x_4	x_5	x_6	Solution
z	0	0	-2/3	7/3	0	0	44/3

Since x_3 has a negative coefficient, x_3 must enter the solution and optimality is recovered by applying the regular simplex method. Table 4–5 shows that the new optimum is reached in one iteration. The first iteration is the same as the current optimum iteration, with the exception of the z-equation.

Table 4–5

Iteration	Basic	x_1	x_2	x_3	x_4	x_5	x_6	Solution
1 (starting)	z	0	0	-2/3	7/3	0	0	44/3
x_3 enters x_2 leaves	x_2	0	1	2/3	-1/3	0	0	4/3
	x_1	1	0	-1/3	2/3	0	0	10/3
	x_5	0	0	-1	1	1	0	3
	x_6	0	0	-2/3	1/3	0	1	2/3
2 (optimal)	z	0	1	0	2	0	0	16
	x_3	0	3/2	1	-1/2	0	0	2
	x_1	1	1/2	0	1/2	0	0	4
	x_5	0	3/2	0	1/2	1	0	5
	x_6	0	1	0	0	0	1	2

Exercise 4.5–5

Consider the optimum in Table 4–5, which yields the optimum solution given $z = 4x_1 + x_2$ subject to the original constraints of the Reddy Mikks model.

(a) Recompute the z-equation given $z = 4x_1 + (3/2)x_2$.
    [*Ans.* Since the change occurs only in x_2, which is nonbasic, the dual values remain $(0, 2, 0, 0)$ and only the coefficient of x_2 will change in the z-equation. Its value is $1/2$, which means that x_2 will remain zero and the solution will be unchanged.]
(b) Will the given solution remain optimal when $z = 3x_1 + 2x_2$?
    [*Ans.* Recompute dual values, since the coefficient of basic x_1 changes. This yields $(y_1, y_2, y_3, y_4) = (0, 3/2, 0, 0)$. The z-equation coefficients of nonbasic x_2 and x_4 are $-1/2$ and $3/2$. All others equal zero. Thus x_2 enters and x_3 leaves.]

Changes in Activity's Usage of Resources

A change in an activity's usage of resources can affect only the optimality of the solution, since it affects the left side of its dual constraint. However, we must restrict this statement to activities that are currently *nonbasic*. A change in the constraint coefficients of basic activities will affect the inverse and could lead to complications in the computations. We shall thus restrict our presentation to changes in nonbasic activities. The easiest way to handle changes in basic activities is to solve the problem *anew*. Although methods exist for handling changes in a *single* constraint coefficient of a basic activity, the "quality" of the information they yield is not on a par with what one gets from other sensitivity analysis procedures.

Let us consider the Reddy Mikks model with $z = 4x_1 + x_2$. Its optimal solution is shown in Table 4–5. Activity x_2 is nonbasic, and we can consider modifying its constraint coefficients. Suppose that the usages by activity 2 of raw materials A and B are 4 and 3 tons instead of 2 and 1 tons. The associated dual constraint is

$$4y_1 + 3y_2 + y_3 + y_4 \geq 1$$

(Note that the right side equals the coefficient of x_2 in $z = 4x_1 + x_2$.) Since the objective function remains unchanged, the dual values remain the same as given in Table 4–5. We thus have in the z-equation.

$$\text{new } x_2\text{-coefficient} = 4(0) + 3(2) + 1(0) + 1(0) - 1 = 5$$

Since it is ≥ 0, the proposed change does not affect the optimum solution in Table 4–5.

Exercise 4.5–6

Indicate whether the following changes in the constraint coefficients of x_2 will affect the optimal solution in Table 4–5.

(a) Usage of raw materials A and B is 4 and 2 tons.
    [*Ans.* No, because the new x_2-coefficient in the z-equation is 3.]
(b) Usage of raw materials A and B is 2 and 1/4 tons.
    [*Ans.* Yes, because the new x_2-coefficient is $-1/2$.]

Instead of considering separately how the new optimum solution is obtained when the change results in nonoptimality (such as in case b), we present next the case of adding a completely new activity. Its treatment will automatically include this situation.

Addition of a New Activity

We can think of adding a new activity as a nonbasic activity that started originally in the model with *all zero* coefficients in the objective and constraints. The coefficients of the new activity will then represent the changes from zero to the new values.

In the original Reddy Mikks model (with $z = 3x_1 + 2x_2$), suppose that we are interested in producing a cheaper brand of exterior paint, which uses 3/4 ton of each of raw materials A and B per ton of the new paint. The relationship between interior and exterior paints as expressed in constraint ③ will remain binding except that now *both* types of exterior paint must be considered in the new constraint. The profit per ton of the new paint is $1\frac{1}{2}$ (thousand dollars).

Let x_7 equal the tons of new paint produced. The original model is modified as follows:

$$\text{maximize } z = 3x_1 + 2x_2 + (3/2)x_7$$

subject to

$$x_1 + 2x_2 + (3/4)x_7 \leq 6$$
$$2x_1 + x_2 + (3/4)x_7 \leq 8$$
$$-x_1 + x_2 - x_7 \leq 1$$
$$x_2 \leq 2$$
$$x_1, x_2, x_7 \geq 0$$

The addition of a new activity is equivalent to combining the analysis of making changes in the objective and the resource usages. We can think of x_7 as if it were part of the original model with *all zero* coefficients, which are now changed as shown in the foregoing model. This case is equivalent to saying that x_7 is nonbasic.

The first thing to do is to check the corresponding dual constraint:

$$(3/4)y_1 + (3/4)y_2 - y_3 \geq 3/2$$

Since x_7 is regarded as a nonbasic variable in the original tableau, the dual values remain unchanged. Thus, the coefficient of x_7 in the current optimal tableau is

$$(3/4)(1/3) + (3/4)(4/3) - (1)(0) - 3/2 = -1/4$$

This means that the current solution will improve if x_7 becomes positive.

The *current optimal tableau* is modified by creating an x_7-column in the left side with its z-equation coefficient equal to $-1/4$. The associated constraint coefficients are computed as follows (see Section 4.2.2):

$$\begin{pmatrix} 2/3 & -1/3 & 0 & 0 \\ -1/3 & 2/3 & 0 & 0 \\ -1 & 1 & 1 & 0 \\ -2/3 & 1/3 & 0 & 1 \end{pmatrix} \begin{pmatrix} 3/4 \\ 3/4 \\ -1 \\ 0 \end{pmatrix} = \begin{pmatrix} 1/4 \\ 1/4 \\ -1 \\ -1/4 \end{pmatrix}$$

Table 4–6

Iteration	Basic	x_1	x_2	x_7	x_3	x_4	x_5	x_6	Solution
1 (starting)	z	0	0	$-1/4$	1/3	4/3	0	0	38/3
x_7 enters	x_2	0	1	1/4	2/3	$-1/3$	0	0	4/3
x_2 leaves	x_1	1	0	1/4	$-1/3$	2/3	0	0	10/3
	x_5	0	0	-1	-1	1	1	0	3
	x_6	0	0	$-1/4$	$-2/3$	1/3	0	1	2/3
2 (optimal)	z	0	1	0	1	1	0	0	14
	x_7	0	4	1	8/3	$-4/3$	0	0	16/3
	x_1	1	-1	0	-1	1	0	0	2
	x_5	0	4	0	5/3	$-1/3$	1	0	25/3
	x_6	0	1	0	0	0	0	1	2

Table 4–6 gives the iterations for the new solution.

Exercise 4.5–7
Suppose that the objective and constraint coefficients of x_7 were given respectively as 1, 1/2, 1, -1, and 0. Will it be profitable to produce the new product?
[*Ans.* No, because the z-equation coefficient of x_7 is 1/2.]

You may have already noticed that a new activity cannot be admitted into the solution unless it improves the objective value (e.g., in Table 4–6 the optimum value of z increased from $12\frac{2}{3}$ to 14 as a result of incorporating x_7 in the optimum solution). This result is in contrast to the addition of a new constraint (Section 4.5.1), where a new constraint can never improve the optimum objective value. Indeed, if the additional constraint is binding, it must worsen the optimum value of z.

4.6 SUMMARY

This chapter has introduced the concept of duality in linear programming. The presentation indicates the following uses of duality:

1. Since the optimal primal solution can be obtained directly from the optimal dual tableau (and vice versa), it will be advantageous computationally to solve the dual when it has fewer constraints than the primal.
2. Duality provides an economic interpretation that sheds light on the *unit worth* or *shadow price* of the different resources. It also explains the condition

of optimality by introducing the new economic definition of imputed costs for each activity.

3. Duality plays an important role in the development of the sensitivity analysis techniques.

4. A by-product of duality is the development of the new computational technique called the *dual simplex* method. The method is of immense importance in some sensitivity analysis procedures.

The chapter closes with a presentation of the important techniques of sensitivity analysis, which gives linear programming the dynamic characteristic of modifying the optimum solution to reflect changes in the model. The main results to be drawn from the sensitivity analysis presentation are

1. Changes in the different coefficients of the model can affect either the optimality or feasibility of the current solution and will lead to one of three situations regarding the optimum solution.

(a) The variables and their optimum values remain essentially unchanged.

(b) The variables remain the same, but their optimum values change.

(c) The variables and their values change completely.

In cases (a) and (b), sensitivity analysis methods are very efficient. Even in the third case, the new optimal solution is normally recovered in a few additional iterations.

2. The addition of a new constraint can *never* improve the value of the objective function.

3. The addition of a new variable can *never* worsen the value of the objective function.

SELECTED REFERENCES

DANTZIG, G. B., *Linear Programming and Extensions*, Princeton University Press, Princeton, N.J., 1963.

LUENBERGER, D. G., *Introduction to Linear and Nonlinear Programming*, Addison-Wesley, Reading, Mass., 1973.

MURTY, KATTA, *Linear Programming*, Wiley, New York, 1983.

REVIEW QUESTIONS

True (T) or False (F)?

1. _____ If the *standard* LP primal is minimization, the dual is maximization with ≤ constraints and unrestricted variables.

2. _____ The primal problem must always be of the maximization type.

3. _____ If no slack or surplus variable is needed to convert a primal constraint to an equation (standard form), the corresponding dual variable will necessarily be unrestricted.

4. _____ If a *slack* variable is used to convert a primal constraint to an equation (standard form), the corresponding dual variable will be restricted to nonnegative values if the primal is maximization and nonpositive values if the primal is minimization.

5. _____ If a *surplus* variable is used to convert a primal constraint to an equation (standard form), the corresponding dual variable will be unrestricted regardless of whether the primal is maximization or minimization.

6. _____ An unrestricted primal variable will have the effect of yielding an *equality* dual constraint.

7. _____ The dual of the dual is the primal.

8. _____ The dual constraints associated with artificial variables in the primal necessarily lead to redundant constraints in the dual.

9. _____ The optimal primal (dual) solution is readily available from the optimal dual (primal) tableau.

10. _____ If the number of primal variables is much smaller than the number of constraints, it is more efficient to obtain the solution of the primal by solving its dual.

11. _____ For a feasible pair of primal and dual solutions, the objective value in the primal can never exceed that of the dual, regardless of which problem is maximization and which is minimization.

12. _____ Equality of the objective values in the primal and dual is the *only* condition required to prove the optimality of the chosen values of the primal and dual variables.

13. _____ The entire simplex tableau of a linear program can be computed from knowledge of the associated inverse matrix (and the original model).

14. _____ Changes in the right side of the constraints of a linear program can affect only the right side of its optimal tableau, that is, feasibility.

15. _____ Changes in *any* coefficients of the linear program other than the right side of its constraints can affect only the optimality of the solution.

16. _____ If a variable is nonbasic in the optimal tableau, changes in its original objective coefficient can affect only its objective equation coefficient in the optimal tableau, and nothing else.

17. _____ Changes in the objective coefficient of a basic variable in the optimal solution can change *all* the coefficients of nonbasic variables in the optimal tableau.

18. _____ When the primal problem is nonoptimal, the dual problem is automatically infeasible.

19. _____ If the primal has unbounded optimal, the dual is always infeasible.

20. _____ If the primal is infeasible, the dual *always* has unbounded optimum (see Problem 4–19).

21. _____ At the optimal solution of a maximization problem, the optimal profit must equal the worth of used resources.

22. _____ If at the optimal solution the imputed cost of an activity exceeds its marginal profit, the level of the activity will be zero.

23. _____ In the dual simplex method, the LP starting solution must be optimal and infeasible.

24. _____ In the dual simplex method, if none of the constraint coefficients associated with the leaving variable equation is negative, the LP problem has no feasible solution.

25. _____ The addition of a new activity can improve the objective value.

26. _____ The addition of a new constraint can improve the objective value.

27. _____ If we change both the right side of the constraints and the coefficients of the objective function, we can destroy both the optimality and feasibility of the solution.

[*Ans.* 1—T, 2—F, 3—T, 4—T, 5—F, 6 to 10—T, 11—F, 12—F, 13—T, 14—T, 15—F, 16—T, 17—T, 18—T, 19—T, 20—F, 21 to 25—T, 26—F, 27—T.]

PROBLEMS

Section	Assigned Problems
4.1	4–1 to 4–3
4.2.1	4–4 to 4–11
4.2.2	4–12 to 4–21
4.3	4–22 to 4–25
4.4	4–26 to 4–28
4.5.1	4–29 to 4–35
4.5.2	4–36 to 4–49

☐ **4–1** Write the duals of the following problems.

(a) Maximize $z = -5x_1 + 2x_2$
subject to

$$-x_1 + x_2 \leq -3$$
$$2x_1 + 3x_2 \leq 5$$
$$x_1, x_2 \geq 0$$

(b) Minimize $z = 6x_1 + 3x_2$
subject to

$$6x_1 - 3x_2 + x_3 \geq 2$$
$$3x_1 + 4x_2 + x_3 \geq 5$$
$$x_1, x_2, x_3 \geq 0$$

(c) Maximize $z = 5x_1 + 6x_2$
subject to

$$x_1 + 2x_2 = 5$$
$$-x_1 + 5x_2 \geq 3$$
$$x_1 \text{ unrestricted}$$
$$x_2 \geq 0$$

(d) Minimize $z = 3x_1 + 4x_2 + 6x_3$
   subject to

$$x_1 + x_2 \geq 10$$
$$x_1, x_3 \geq 0$$
$$x_2 \leq 0$$

(e) Maximize $z = x_1 + x_2$
   subject to

$$2x_1 + x_2 = 5$$
$$3x_1 - x_2 = 6$$
$$x_1, x_2 \text{ unrestricted}$$

☐ **4–2** Consider the following problem:

$$\text{minimize } z = x_1 + 2x_2 - 3x_3$$

subject to

$$-x_1 + x_2 + x_3 = 5$$
$$12x_1 - 9x_2 + 9x_3 \geq 8$$
$$x_1, x_2, x_3 \geq 0$$

The simplex solution of this problem requires the use of artificial variables in both constraints. Show that the dual problems obtained from the standard primal before and after the artificials are added are exactly the same. As a result, the artificial variables can lead only to redundant constraints.

☐ **4–3** Repeat Problem 4–2 given that the objective function is replaced by

$$\text{maximize } z = 5x_1 - 6x_2 + 7x_3$$

☐ **4–4** Consider the following linear program:

$$\text{maximize } z = 5x_1 + 2x_2 + 3x_3$$

subject to

$$x_1 + 5x_2 + 2x_3 = 30$$
$$x_1 - 5x_2 - 6x_3 \leq 40$$
$$x_1, x_2, x_3 \geq 0$$

The optimal solution is given by:

Basic	x_1	x_2	x_3	R	x_4	Solution
z	0	23	7	$5+M$	0	150
x_1	1	5	2	1	0	30
x_4	0	−10	−8	−1	1	10

Write the dual problem and find its optimal solution from the optimal primal tableau.

☐ **4-5** Consider the following linear program:

$$\text{maximize } z = x_1 + 5x_2 + 3x_3$$

subject to

$$x_1 + 2x_2 + x_3 = 3$$
$$2x_1 - x_2 = 4$$
$$x_1, x_2, x_3 \geq 0$$

Using a starting solution consisting of x_3 and an artificial variable R in the second constraint, we obtain the following optimum tableau:

Basic	x_1	x_2	x_3	R	Solution
z	0	2	0	$-1+M$	5
x_3	0	5/2	1	−1/2	1
x_1	1	−1/2	0	1/2	2

Write the dual problem and find its optimal solution from the optimal primal tableau.

☐ **4-6** Consider the following linear program:

$$\text{maximize } z = 2x_1 + 4x_2 + 4x_3 - 3x_4$$

subject to

$$x_1 + x_2 + x_3 = 4$$
$$x_1 + 4x_2 + x_4 = 8$$
$$x_1, x_2, x_3, x_4 \geq 0$$

By using x_3 and x_4 as the starting variables, the optimum tableau is given by

Basic	x_1	x_2	x_3	x_4	Solution
z	2	0	0	3	16
x_3	3/4	0	1	−1/4	2
x_2	1/4	1	0	1/4	2

Write the dual problem and find its solution from the optimal primal tableau.

☐ **4–7** Find the *optimal objective value* of the following problem by inspecting only its dual. (Do not solve the dual by the simplex method.)

$$\text{minimize } z = 10x_1 + 4x_2 + 5x_3$$

subject to

$$5x_1 - 7x_2 + 3x_3 \geq 50$$
$$x_1, x_2, x_3 \geq 0$$

☐ **4–8** Find a solution to the following set of inequalities by using the dual problem.

$$2x_1 + 3x_2 \leq 12$$
$$-3x_1 + 2x_2 \leq -4$$
$$3x_1 - 5x_2 \leq 2$$
$$x_1 \text{ unrestricted}$$
$$x_2 \geq 0$$

[*Hint:* Augment the trivial objective function maximize $z = 0x_1 + 0x_2$ to the inequalities, then solve the dual.]

☐ **4–9** Solve the following problem by considering its dual:

$$\text{minimize } z = 5x_1 + 6x_2 + 3x_3$$

subject to

$$5x_1 + 5x_2 + 3x_3 \geq 50$$
$$x_1 + x_2 - x_3 \geq 20$$
$$7x_1 + 6x_2 - 9x_3 \geq 30$$
$$5x_1 + 5x_2 + 5x_3 \geq 35$$
$$2x_1 + 4x_2 - 15x_3 \geq 10$$
$$12x_1 + 10x_2 \geq 90$$
$$x_2 - 10x_3 \geq 20$$
$$x_1, x_2, x_3 \geq 0$$

Compare the number of constraints in the two problems.

☐ **4–10** Consider the primal problem:

$$\text{maximize } z = 3x_1 + 2x_2 + 5x_3$$

subject to

$$x_1 + 2x_2 + x_3 \leq 500$$
$$3x_1 + 2x_3 \leq 460$$
$$x_1 + 4x_2 \leq 420$$
$$x_1, x_2, x_3 \geq 0$$

Write the dual problem for the primal. Without carrying out the simplex method computations on either the primal or the dual problem, estimate a range for the optimum value of the objective function.

☐ **4–11** For the following pairs of primal–dual problems, determine whether the listed solutions are optimal:

Primal

Minimize $z = 2x_1 + 3x_2$
subject to

$$2x_1 + 3x_2 \leq 30$$
$$x_1 + 2x_2 \geq 10$$
$$x_1 - x_2 \geq 0$$
$$x_1, x_2 \geq 0$$

Dual

Maximize $w = 30y_1 + 10y_2$
subject to

$$2y_1 + y_2 + y_3 \leq 2$$
$$3y_1 + 2y_2 - y_3 \leq 3$$
$$y_1 \leq 0, \quad y_2 \geq 0, \quad y_3 \geq 0$$

(a) $(x_1 = 10, x_2 = 10/3; y_1 = 0, y_2 = 1, y_3 = 1)$
(b) $(x_1 = 20, x_2 = 10; y_1 = 1, y_2 = 4, y_3 = 0)$
(c) $(x_1 = 10/3, x_2 = 10/3; y_1 = 0, y_2 = 5/3, y_3 = 1/3)$

☐ **4–12** Consider the following linear program expressed in the standard form:

$$\text{maximize } z = 3x_1 + 2x_2 + 5x_3$$

subject to

$$x_1 + 2x_2 + x_3 + x_4 = 30$$
$$3x_1 + 2x_3 + x_5 = 60$$
$$x_1 + 4x_2 + x_6 = 20$$
$$x_1, x_2, \ldots, x_6 \geq 0$$

The following matrices represent the inverses and their corresponding basic variables associated with different simplex iterations of the problem. Compute the associated *constraint equations* of each iteration and determine the corresponding basic variables and their values.

(a) $(x_4, x_3, x_6);$ $\begin{pmatrix} 1 & -1/2 & 0 \\ 0 & 1/2 & 0 \\ 0 & 0 & 1 \end{pmatrix}$

(b) $(x_2, x_3, x_1);$ $\begin{pmatrix} 1/4 & -1/8 & 1/8 \\ 3/2 & -1/4 & -3/4 \\ -1 & 1/2 & 1/2 \end{pmatrix}$

(c) $(x_2, x_3, x_6);$ $\begin{pmatrix} 1/2 & -1/4 & 0 \\ 0 & 1/2 & 0 \\ -2 & 1 & 1 \end{pmatrix}$

☐ **4–13** Consider the following linear program expressed in the standard form:

$$\text{maximize } z = 4x_1 + 14x_2$$

subject to

$$
\begin{aligned}
2x_1 + 7x_2 + x_3 &= 21 \\
7x_1 + 2x_2 \quad\quad + x_4 &= 21 \\
x_1, x_2, x_3, x_4 &\geq 0
\end{aligned}
$$

Each of the following cases provides an inverse matrix and its corresponding basic variables for the LP given. Determine whether or not each basic solution is feasible.

(a) (x_2, x_4); $\begin{pmatrix} 1/7 & 0 \\ -2/7 & 1 \end{pmatrix}$

(b) (x_2, x_3); $\begin{pmatrix} 0 & 1/2 \\ 1 & -7/2 \end{pmatrix}$

(c) (x_2, x_1); $\begin{pmatrix} 7/45 & -2/45 \\ -2/45 & 7/45 \end{pmatrix}$

(d) (x_1, x_4); $\begin{pmatrix} 1/2 & 0 \\ -7/2 & 1 \end{pmatrix}$

☐ **4–14** In Problem 4–12, determine the simplex multipliers (dual values) associated with the given inverses and their basic variables.

☐ **4–15** Consider the following linear program expressed in the standard form:

$$\text{minimize } z = 2x_1 + x_2$$

subject to

$$
\begin{aligned}
3x_1 + x_2 - x_3 &= 3 \\
4x_1 + 3x_2 \quad\quad - x_4 &= 6 \\
x_1 + 2x_2 \quad\quad\quad + x_5 &= 3 \\
x_1, x_2, x_3, x_4, x_5 &\geq 0
\end{aligned}
$$

(a) Compute the entire simplex tableau associated with the following inverse matrix using the primal–dual relationships.

$$\text{basic variables} = (x_1, x_2, x_5); \quad \text{inverse} = \begin{pmatrix} 3/5 & -1/5 & 0 \\ -4/5 & 3/5 & 0 \\ 1 & -1 & 1 \end{pmatrix}$$

(b) Determine whether the iteration in part (a) is optimal and feasible.

☐ **4–16** Consider the following linear program expressed in the standard form:

$$\text{maximize } z = 5x_1 + 12x_2 + 4x_3$$

subject to

$$x_1 + 2x_2 + x_3 + x_4 = 10$$
$$2x_1 - x_2 + 3x_3 = 2$$
$$x_1, x_2, x_3, x_4 \geq 0$$

Each of the following inverses is associated with a *basic feasible* solution and only one of them represents the optimal solution. Show how you can identify the optimal solution by using the primal–dual relationships.

(a) (x_4, x_3); $\begin{pmatrix} 1 & -1/3 \\ 0 & 1/3 \end{pmatrix}$

(b) (x_2, x_1); $\begin{pmatrix} 2/5 & -1/5 \\ 1/5 & 2/5 \end{pmatrix}$

(c) (x_2, x_3); $\begin{pmatrix} 3/7 & -1/7 \\ 1/7 & 2/7 \end{pmatrix}$

☐ **4–17** The final optimal tableau of a maximization linear programming problem with three constraints of type ($\leq$) and two unknowns (x_1, x_2) is

Basic	x_1	x_2	s_1	s_2	s_3	Solution
z	0	0	0	3	2	?
s_1	0	0	1	1	−1	2
x_2	0	1	0	1	0	6
x_1	1	0	0	−1	1	2

s_1, s_2, and s_3 are slack variables. Find the value of the objective function z in *two* different ways by using the primal–dual relationships.

☐ **4–18** Given that a linear programming problem has an unbounded solution, why is it that its dual must necessarily be infeasible?

☐ **4–19** Consider the problem:

$$\text{maximize } z = 8x_1 + 6x_2$$

subject to

$$x_1 - x_2 \leq 3/5$$
$$x_1 - x_2 \geq 2$$
$$x_1, x_2 \geq 0$$

Show that both the primal and the dual problems have no feasible space. Hence it is not always true that when one problem is infeasible, its dual is unbounded. (Notice the important difference between the argument in this problem and the one in Problem 4–18.)

☐ **4–20** Consider the following problem:

$$\text{minimize } w = y_1 - 5y_2 + 6y_3$$

subject to

$$2y_1 \quad\quad + 4y_3 \geq 50$$
$$y_1 + 2y_2 \quad\quad \geq 30$$
$$y_3 \geq 10$$
$$y_1, y_2, y_3 \text{ unrestricted}$$

Show that the solution to this problem is unbounded by showing that the dual is infeasible and the primal is feasible. Suppose that the primal problem is not checked for feasibility. Would it be possible to make this conclusion? Why?

☐ **4–21** Consider the following primal problem:

$$\text{maximize } z = -2x_1 + 3x_2 + 5x_3$$

subject to

$$x_1 - x_2 + x_3 \leq 15$$
$$x_1, x_2, x_3 \geq 0$$

Show by inspection that the dual is infeasible. What can be said about the solution to the primal?

☐ **4–22** Consider the product-mix problem of Example 4.3–1, where each of three products is processed on three different operations. The LP of the problem is given by

$$\text{maximize } z = 3x_1 + 2x_2 + 5x_3 \quad\quad \text{(profit)}$$

subject to

$$x_1 + 2x_2 + x_3 \leq 430 \quad\quad \text{(operation 1)}$$
$$3x_1 \quad\quad + 2x_3 \leq 460 \quad\quad \text{(operation 2)}$$
$$x_1 + 4x_2 \quad\quad \leq 420 \quad\quad \text{(operation 3)}$$
$$x_1, x_2, x_3 \geq 0$$

Let x_4, x_5, and x_6 represent the slacks in the three constraints. The optimal basic variables and the associated inverse are given by

$$(x_2, x_3, x_6); \quad \begin{pmatrix} 1/2 & -1/4 & 0 \\ 0 & 1/2 & 0 \\ -2 & 1 & 1 \end{pmatrix}$$

(a) Determine the worth of increasing the maximum allowable time of each operation by 1 minute.
(b) Rank the three operations in order of priority for increase in time allocation.
(c) Determine the maximum allowable increase in the time for each operation that will maintain the worth per minute computed in part (a).
(d) In part (c), determine the associated change in the optimal value of the profit.

☐ **4–23** The optimum simplex tableau for a maximization problem with all ($\leq$) constraints is

Basic	x_1	x_2	s_1	s_2	s_3	Solution
z	0	0	1/4	1/4	0	5
x_2	0	1	1/2	−1/2	0	2
x_1	1	0	−1/8	3/8	0	3/2
s_3	0	0	1	−2	1	4

where x_1 and x_2 are the decision variables and s_1, s_2, and s_3 are the slack variables. Suppose that it is decided to increase the right-hand side of one of the constraints. Which one do you recommend for expansion and why? What is the maximum amount of increase in this case? Find the corresponding new value of the objective function.

☐ **4–24** Consider the problem

$$\text{maximize } z = 5x_1 + 2x_2 + 3x_3$$

subject to

$$x_1 + 5x_2 + 2x_3 \leq b_1$$
$$x_1 - 5x_2 - 6x_3 \leq b_2$$
$$x_1, x_2, x_3 \geq 0$$

where b_1 and b_2 are constants. For specific values of b_1 and b_2, the optimal solution is

Basic	x_1	x_2	x_3	s_1	s_2	Solution
z	0	a	7	d	e	150
x_1	1	b	2	1	0	30
s_1	0	c	−8	−1	1	10

where a, b, c, d, and e are constants. Determine
(a) The values of b_1 and b_2 that yield the given optimal solution.
(b) The optimal dual solution.
(c) The values of a, b, and c in the optimal tableau.
(d) If it is required to increase optimum z, should b_1 or b_2 be increased and by how much?

☐ **4–25** In the product-mix model in Problem 4–22, the first product is not in the optimal mix even though x_1 has a higher marginal profit than x_2. Market conditions do not permit increasing the marginal profit of x_1. Thus it is decided to improve the profitability of product 1 by reducing its usages of the times of the three operations. Indicate whether the following changes will improve the profitability of x_1.

(a) Change the usages per unit of product 1 from (1, 3, 1) to (1, 2, 3).

(b) Change the usages to (1/2, 7/2, 2).

(c) Change the usages to (1, 1, 5).

☐ **4–26** Solve the following problems by the dual simplex method.

(a) Minimize $z = 2x_1 + 3x_2$

subject to

$$2x_1 + 3x_2 \leq 30$$
$$x_1 + 2x_2 \geq 10$$
$$x_1, x_2 \geq 0$$

(b) Minimize $z = 5x_1 + 6x_2$

subject to

$$x_1 + x_2 \geq 2$$
$$4x_1 + x_2 \geq 4$$
$$x_1, x_2 \geq 0$$

(c) The bus scheduling problem of Example 2.3–1.

☐ **4–27** Consider the following problem:

$$\text{minimize } w = 6y_1 + 7y_2 + 3y_3 + 5y_4$$

subject to

$$5y_1 + 6y_2 - 3y_3 + 4y_4 \geq 12$$
$$y_2 - 5y_3 - 6y_4 \geq 10$$
$$2y_1 + 5y_2 + y_3 + y_4 \geq 8$$
$$y_1, y_2, y_3, y_4 \geq 0$$

(a) Indicate *three* different methods for solving this problem, and give the complete starting tableau in each case.

(b) Compute the maximum number of possible iterations in each of the three cases.

(c) Which of the foregoing methods would you use, and why?

☐ **4–28 (Dual Simplex Method).** Consider the following problem:

$$\text{maximize } z = 2x_1 - x_2 + x_3$$

subject to

$$2x_1 + 3x_2 - 5x_3 \geq 4$$
$$-x_1 + 9x_2 - x_3 \geq 3$$
$$4x_1 + 6x_2 + 3x_3 \leq 8$$
$$x_1, x_2, x_3 \geq 0$$

The starting basic solution for this problem consisting of all slacks is infeasible; that is, $s_1 = -4$, $s_2 = -3$, and $s_3 = 8$. However, the dual simplex method cannot be applied directly since neither x_1 nor x_3 satisfies the optimality condition for a maximization problem. Show that by augmenting the *artificial constraints* $x_1 + x_3 \leq M$ (where $M > 0$ is sufficiently large that it is not restrictive with respect to the original constraints) and then using the augmented constraint as a pivot row, the selection of x_1 as the entering variable will render an objective row that is all optimal. Hence it is possible to carry out the regular dual simplex calculations. (Notice that generally, in a maximization problem, the entering variable is the one having the largest objective coefficient among the variables that do not satisfy the optimality condition.)

☐ **4–29** In the product-mix model of Example 4.3–1, determine the optimal solution when the time limits (in minutes) of the daily usages of the operations are changed as shown by the following vectors.

$$\text{(a)} \begin{pmatrix} 420 \\ 460 \\ 440 \end{pmatrix} \quad \text{(b)} \begin{pmatrix} 500 \\ 400 \\ 600 \end{pmatrix} \quad \text{(c)} \begin{pmatrix} 300 \\ 800 \\ 200 \end{pmatrix} \quad \text{(d)} \begin{pmatrix} 300 \\ 400 \\ 150 \end{pmatrix}$$

☐ **4–30** Consider Problem 4–4. Suppose that the right side of the constraints becomes $(30 + \theta, 40 - \theta)$, where θ is a nonnegative parameter. Determine the values of θ for which the basic solution in Problem 4–4 remains feasible.

☐ **4–31** In Problem 4–30, find the new optimal solution when $\theta = 10$.

☐ **4–32** In Problem 4–29, suppose that it is necessary to add a fourth operation to all the products of the product-mix problem. The maximum production rate based on 480 minutes a day is *either* 120 units of product 1, 480 units of product 2, *or* 240 units of product 3. Determine the optimal solution assuming that the daily capacity of the fourth operation is limited by (a) 570 minutes; (b) 548 minutes.

☐ **4–33** Consider Problem 4–5. Check whether each of the following constraints will affect the current optimum solution. If so, find the new solution:
    (a) $x_1 + x_2 \leq 2$.
    (b) $2x_1 + 4x_2 \geq 10$.
    (c) $2x_1 + x_2 = 6$.
    (d) $x_1 + x_2 + x_3 \leq 2$.

☐ **4–34** Consider Problem 4–6. Check whether each of the following constraints will affect the current solution. If so, find the new solution:
    (a) $x_1 + x_2 + x_3 \leq 5$.
    (b) $2x_1 + x_2 - x_4 \geq 4$.
    (c) $x_1 + 2x_2 + x_3 + x_4 \leq 4$.
    (d) $x_1 + x_4 = 1$.

☐ **4–35** Consider the feed-mix problem of Example 2.2–2. The poultry industry charts show that the weekly feed per broiler varies according to the following schedule.

Week	1	2	3	4	5	6	7	8
Pounds per broiler	.26	.48	.75	1.0	1.3	1.6	1.9	2.1

Develop the optimal weekly feed mix for each of the eight weeks by using sensitivity analysis.

☐ **4–36** Consider the product-mix model of Example 4.3–1. Determine the optimal solution when the profit function is changed as follows:
  (a) $z = 4x_1 + 2x_2 + x_3$.
  (b) $z = 3x_1 + 4x_2 + 2x_3$.
  (c) $z = 3x_2 + x_3$.
  (d) $z = 2x_1 + 2x_2 + 8x_3$.
  (e) $z = 5x_1 + 2x_2 + 5x_3$.

☐ **4–37** In the product-mix model of Example 4.3–1, determine the necessary increase in the profit coefficient of product 1 that will make it profitable.

☐ **4–38** In Problem 4–5, find the new optimum assuming that the objective function is changed to

$$\text{maximize } z = 12x_1 + 5x_2 + 2x_3$$

☐ **4–39** In Problem 4–38, suppose that the objective is changed to

$$\text{minimize } z = 2x_2 - 5x_3$$

Show how the new optimum is obtained by using sensitivity analysis.

☐ **4–40** In Problem 4–4, suppose that the objective function is

$$z = (5 - \theta)x_1 + (2 + \theta)x_2 + (3 + \theta)x_3$$

where θ is a nonnegative parameter. Find the values of θ that maintain the optimality of the solution obtained at $\theta = 0$.

☐ **4–41** In Problem 4–23, let c_1 and c_2 be the coefficients of x_1 and x_2 in the objective function. Find the range of the ratio c_1/c_2 that will always keep the solution in Problem 4–23 optimal.

☐ **4–42** In the product-mix model of Example 4.3–1, suppose that a fourth product is scheduled on the original three operations. The new product has the following data:

Operation	1	2	3
Minutes per unit	3	2	4

Determine the optimal solution when the profit per unit of the new product is given by (a) $5, (b) $10.

☐ **4–43** Consider Problem 4–5. Check whether a new variable x_4 will improve the optimum value of z assuming that its objective and constraint coefficients are as given. If so, find the new optimum solution:

(a) (5; 2, 2).
(b) (4; 1, 1).
(c) (3; 4, 6).
(d) (5; 3, 3).

☐ **4–44** Consider Problem 4–6. Check whether a new variable x_5 will improve the optimum value of z assuming that its objective and constraint coefficients are as given below. If so, find the new optimum solution:

(a) (5; 1, 2).
(b) (6; 2, 3).
(c) (10; 2, 5).
(d) (15; 3, 3).

☐ **4–45** Consider Problem 4–4. Suppose that the technological coefficients of x_2 are $(5 - \theta, -5 + \theta)$ instead of $(5, -5)$, where θ is a nonnegative parameter. Find the values of θ for which the solution to Problem 4–4 remains optimal.

☐ **4–46** Find the values of θ that satisfy the conditions of Problems 4–30 and 4–40 *simultaneously*.

☐ **4–47** Find the values of θ that satisfy the conditions of Problems 4–30, 4–40, and 4–45 *simultaneously*.

☐ **4–48** Consider the problem

$$\text{maximize } z = 2x_2 - 5x_3$$

subject to

$$
\begin{aligned}
x_1 \quad\ + x_3 &\geq 2 \\
2x_1 + x_2 + 6x_3 &\leq 6 \\
x_1 - x_2 + 3x_3 &= 0 \\
x_1, x_2, x_3 &\geq 0
\end{aligned}
$$

(a) Write the dual.
(b) Solve the primal; then find the solution to the dual.
(c) Suppose that the right-hand side of the primal is changed from (2, 6, 0) to (2, 10, 5). Find the new optimal solution.
(d) Suppose that the coefficients of x_2 and x_3 in the objective function are changed from (2, −5) to (1, 1). Find the new solution.

☐ **4–49** Show how the optimal values of the slack variables in the dual problem can be obtained *directly* from the optimal primal tableau. Apply the procedure to Example 4.1–1, and check the result by substituting the optimal dual variables in the dual constraints. (The optimal solution of Example 4.1–1 is given in Table 4–3.)

SHORT CASES

Case 4–1†

A manufacturing company produces three products P1, P2, and P3. The production process utilizes two raw materials, R1 and R2, which are processed on two facilities, F1 and F2. The maximum daily availability of R1, R2, F1, and F2 and the utilizations per unit of P1, P2, and P3 are listed in the following table:

Resource	Units	Utilization per Unit of			Maximum Daily Availability
		P1	P2	P3	
F1	minutes	1	2	1	430
F2	minutes	3	0	2	460
R1	pounds	1	4	0	420
R2	pounds	1	1	1	300

The minimum daily demand for product P2 is 70 units, whereas that of P3 cannot exceed 240 units. It is estimated that the per unit profit contributions of P1, P2, and P3 are $3, $2, and $5, respectively.

A management meeting is held to discuss ways of improving the financial image of the company. The following are the most prominent proposals.

1. The per unit profit of P3 can be increased by 20%. However, the market demand will drop to a maximum of 210 units instead of the present limit of 240 units.

2. Raw material R2 appears to be a critical factor in limiting current production. Additional units can be secured from a different supplier at a unit price that is $3 higher than that of the present supplier.

3. The capacities of F1 and F2 can be increased by up to 40 minutes daily, each at an additional cost of $35 per day.

4. The chief buyer of product P2 is requesting that its daily supply be increased from the present 70 units to 100 units.

5. The per unit processing time of P1 on F2 can be cut from 3 minutes to 2 minutes at an additional cost of $4 daily.

Discuss the feasibility of these proposals. Keep in mind that (some of) the proposals are not mutually exclusive.

Case 4–2

The Beaver Furniture Company manufactures and assembles chairs, tables, and bookshelves. The manufacturing plant produces semifinished products that are assembled in the company's assembling facility.

---

† This case is based on D. C. S. Sheran, "Postoptimal Analysis in Linear Programming—The Right Example," *IIE Transactions,* Vol. 16, no. 1, March 1984, pp. 99–102.

The manufacturing plant, being highly automated, has predictable output volume that can be estimated as

Product	Production Capacity in Semifinished Units per Month
Chairs	3000
Table	1000
Bookshelf	580

The assembling facility, being labor intensive, employs 150 assembly workers. The facility operates on the basis of two 8-hour shifts per day, five days a week. The average assembly time for each assembly unit is

Products	Labor Minutes per Assembly Unit
Chair	20
Table	40
Bookshelf	15

The size of the labor force in the assembly plant fluctuates because of annual leaves taken by the employees. The estimated changes in the labor force over the next four months are given next. However, depending on work needs, the company can postpone the annual leaves of some or all of the employees.

	May	June	July	July
Number of employees	130	125	105	140

The sales forecast for the three products for the same period of time was estimated by the marketing department as

Product	Sales Forecast in Number of Units				Inventory at the end of April
	May	June	July	August	
Chair	2800	2300	3350	2950	30
Table	500	800	1400	700	100
Bookshelf	320	300	600	520	50

The production cost and selling price for the products are

Product	Production Cost ($)	Selling Price ($)
Chair	25	45
Table	65	100
Bookshelf	10	20

If a unit is not sold in the same month in which it is manufactured, it will be held over for possible sale in a later month. The storage cost is estimated at about 2% of the production cost.

Should the Beaver Company approve the proposed annual leaves?

Chapter 5

Linear Programming: Transportation Model

This chapter presents the transportation model and its variants. In the obvious sense, the model deals with the determination of a minimum-cost plan for transporting a single commodity from a number of sources (e.g., factories) to a number of destinations (e.g., warehouses). The model can be extended in a direct manner to cover practical situations in the areas of inventory control, employment scheduling, personnel assignment, cash flow, scheduling dam reservoir levels, and many others. The model also can be modified to account for multiple commodities.

The transportation model is basically a linear program that can be solved by the regular simplex method. However, its special structure allows the development of a solution procedure, called the transportation technique, that is computationally more efficient.

The transportation technique can be, and often is, presented in an elementary manner that appears completely detached from the simplex method. However, we must emphasize that the "new" technique essentially follows the *exact* steps of the simplex method.

The transportation model can be extended to cover a number of important applications, including the assignment model and the transshipment model. However, the transportation problem and its extensions are also special cases of network models. Chapter 6 introduces additional network models that are of immense practical importance.

5.1 DEFINITION AND APPLICATION OF THE TRANSPORTATION MODEL

In this section we present the standard definition of the transportation model. We then describe variants of the model that extend its scope of application to a wider class of real-life problems.

In the direct sense, the transportation model seeks the determination of a transportation plan of a *single* commodity from a number of sources to a number of destinations. The data of the model include

1. Level of supply at each source and amount of demand at each destination.
2. The *unit* transportation cost of the commodity from each source to each destination.

Since there is only one commodity, a destination can receive its demand from one or more sources. The objective of the model is to determine the amount to be shipped from each source to each destination such that the total transportation cost is minimized.

The basic assumption of the model is that the transportation cost on a given route is directly proportional to the number of units transported. The definition of "unit of transportation" will vary depending on the "commodity" transported. For example, we may be speaking of a unit of transportation as each of the steel beams needed to build a bridge. Or we may use a truckload of the commodity as a unit of transportation. In either case, the units of supply and demand must be consistent with our definition of a "transported unit."

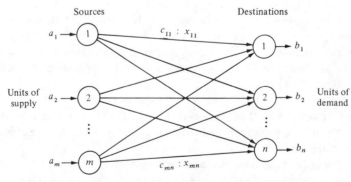

Figure 5-1

Figure 5–1 depicts the transportation model as a network with m sources and n destinations. A *source* or a *destination* is represented by a **node.** The **arc** joining a source and a destination represents the route through which the commodity is transported. The amount of supply at source i is a_i and the demand at destination j is b_j. The *unit* transportation cost between source i and destination j is c_{ij}.

Let x_{ij} represent the amount transported from source i to destination j; then the LP model representing the transportation problem is given generally as

$$\text{minimize } z = \sum_{i=1}^{m} \sum_{j=1}^{n} c_{ij} x_{ij}$$

subject to

$$\sum_{j=1}^{n} x_{ij} \le a_i, \qquad i = 1, 2, \ldots, m$$

$$\sum_{i=1}^{m} x_{ij} \ge b_j, \qquad j = 1, 2, \ldots, n$$

$$x_{ij} \ge 0, \qquad \text{for all } i \text{ and } j$$

The first set of constraints stipulates that the sum of the shipments *from* a source cannot exceed its supply; similarly, the second set requires that the sum of the shipments *to* a destination must satisfy its demand.

The model just described implies that the total supply $\Sigma_{i=1}^{m} a_i$ must at least equal total demand $\Sigma_{j=1}^{n} b_j$. When the total supply *equals* the total demand ($\Sigma_{i=1}^{m} a_i = \Sigma_{j=1}^{n} b_j$), the resulting formulation is called a **balanced transportation model.** It differs from the model only in the fact that all constraints are equations; that is,

$$\sum_{j=1}^{n} x_{ij} = a_i, \qquad i = 1, 2, \ldots, m$$

$$\sum_{i=1}^{m} x_{ij} = b_j, \qquad j = 1, 2, \ldots, n$$

In real life it is not necessarily true that supply equal demand or, for that matter, exceed it. However, a transportation model can always be balanced. The balancing, in addition to its usefulness in modeling certain practical situations, is important for the development of a solution method that fully exploits the special structure of the transportation model. The following two examples present the idea of balancing as well as its practical implications.

Example 5.1–1 (Standard Transportation Model). The MG Auto Company has plants in Los Angeles, Detroit, and New Orleans. Its major distribution centers are located in Denver and Miami. The capacities of the three plants during the next quarter are 1000, 1500, and 1200 cars. The quarterly demands at the two distribution centers are 2300 and 1400 cars. The train transportation cost per car per mile is approximately 8 cents. The mileage chart between the plants and distribution centers is as follows:

	Denver	Miami
Los Angeles	1000	2690
Detroit	1250	1350
New Orleans	1275	850

The mileage chart can be translated to cost per car at the rate of 8 cents per mile. This yields the following costs (rounded to the closest dollar), which represent c_{ij} in the general model:

		Denver (1)	Miami (2)
Los Angeles	(1)	80	215
Detroit	(2)	100	108
New Orleans	(3)	102	68

Using numeric codes to represent the plants and distribution centers, we let x_{ij} represent the number of cars transported from source i to destination j. Since the total supply ($= 1000 + 1500 + 1200 = 3700$) happens to equal the total demand ($= 2300 + 1400 = 3700$), the resulting transportation model is balanced. Hence the following LP model presenting the problem has all *equality* constraints:

$$\text{minimize } z = 80x_{11} + 215x_{12} + 100x_{21} + 108x_{22} + 102x_{31} + 68x_{32}$$

subject to

$$
\begin{aligned}
x_{11} + x_{12} &\phantom{+x_{21}} &\phantom{+x_{31}} &= 1000 \\
&+ x_{21} + x_{22} & &= 1500 \\
& &+ x_{31} + x_{32} &= 1200 \\
x_{11} &+ x_{21} &+ x_{31} &= 2300 \\
+ x_{12} &+ x_{22} &+ x_{32} &= 1400 \\
x_{ij} &\geq 0, \quad \text{for all } i \text{ and } j
\end{aligned}
$$

A more compact method for representing the transportation model is to use what we call the **transportation tableau.** It is a matrix form with its rows representing the sources and its columns the destination. The cost elements c_{ij} are summarized in the northeast corner of the matrix cell (i, j). The MG model can thus be summarized as shown in Table 5–1.

We shall see in the next section that the transportation tableau is the basis for the development of the special simplex-based method for solving the transportation problem. ◀

Exercise 5.1–1

Suppose that it is desired not to ship any cars from the Detroit plant to the Denver distribution center. How can this condition be incorporated in the MG model?

Table 5-1

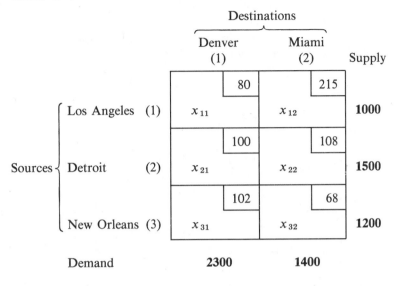

		Denver (1)	Miami (2)	Supply
		Destinations		
Los Angeles	(1)	80 x_{11}	215 x_{12}	**1000**
Detroit	(2)	100 x_{21}	108 x_{22}	**1500**
New Orleans	(3)	102 x_{31}	68 x_{32}	**1200**
Demand		**2300**	**1400**	

Sources: { Los Angeles, Detroit, New Orleans }

[*Ans.* Assign a very high unit transportation cost, *M,* to the Detroit–Denver route; compare the penalty method, Section 3.2.3.]

The MG model happened to have equal supply and demand. It also deals with a single commodity (single-car model), which is an apparent restriction of the standard transportation model. Example 5.1–2 demonstrates how a transportation model can always be balanced and Example 5.1–3 illustrates how multicommodities can be handled. Keep in mind that the main reason for wanting to balance the transportation problem (i.e., converting all the constraints to equations) is that it allows the development of an efficient computational procedure based on the tableau representation illustrated in Table 5–1.

Example 5.1-2 (Balanced Transportation Model). In Example 5.1–1, suppose that the Detroit plant capacity is 1300 cars (instead of 1500). The situation is said to be **unbalanced** because the total supply (= 3500) does not equal the total demand (= 3700).

Stated differently, this unbalanced situation means that it will not be possible to fill *all* the demand at the distribution centers. Our objective is to reformulate the transportation model in a manner that will distribute the shortage quantity (= 3700 − 3500 = 200 cars) optimally among the distribution centers.

Since demand exceeds supply, a fictitious or **dummy source** (plant) can be added with its capacity equal to 200 cars. The dummy plant is allowed, under normal conditions, to ship its "production" to all distribution centers. Physically, the amount shipped to a destination from a dummy plant will represent the shortage quantity at that destination.

The only information missing for completion of the model is the unit "transportation" costs from the dummy plant to the destinations. Since the plant does not exist, no physical shipping will occur and the corresponding unit transportation cost is zero. We may look at the situation differently, however, by saying that a penalty cost is incurred for every unsatisfied demand unit at the distribution centers. In this case the unit transportation costs will equal the unit penalty costs at the various destinations.

Table 5–2 summarizes the balanced model under the new capacity restriction of the Detroit plant. The dummy plant (shown by a tint overlay) has a capacity of 200 cars.

Table 5-2

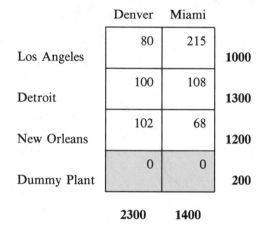

	Denver	Miami	
Los Angeles	80	215	1000
Detroit	100	108	1300
New Orleans	102	68	1200
Dummy Plant	0	0	200
	2300	1400	

In a similar manner, if the supply exceeds the demand, we can add a fictitious or **dummy destination** that will absorb the difference. For example, suppose in Example 5.1–1 that the demand at Denver drops to 1900 cars. Table 5–3 summarizes the model with the dummy distribution center. Any cars shipped from a plant to a dummy distribution center represent a *surplus* quantity at that plant. The associated unit transportation cost is zero. However, we can charge a *storage* cost for holding the car at the plant, in which case the unit transportation cost will equal the unit storage cost. ◄

Table 5-3

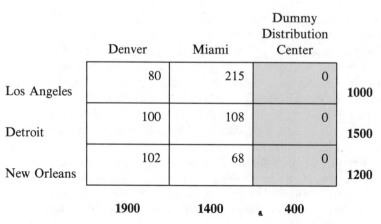

	Denver	Miami	Dummy Distribution Center	
Los Angeles	80	215	0	1000
Detroit	100	108	0	1500
New Orleans	102	68	0	1200
	1900	1400	400	

Exercise 5.1–2

(a) Will it ever be necessary to add both a dummy source and a dummy destination to produce a balanced transportation model?
[*Ans.* No.]

(b) In Table 5–2 suppose that the penalty costs for each unreceived car at Denver and Miami are $200 and $260. Change the model to include this information.
[*Ans.* The unit transportation cost in the dummy row should be 200 and 260 instead of 0.]

(c) Interpret the solution in Table 5–2 if the number of cars "shipped" from the dummy plant to Denver and Miami are 150 and 50, respectively.
[*Ans.* The orders at Denver and Miami will be 150 and 50 cars short.]

(d) In Table 5–3, suppose that the Detroit plant must ship out *all* its production of 1500 cars. How can we implement this restriction?
[*Ans.* Assign a very high cost M to the Detroit–dummy distribution center route.]

(e) In each of the following cases, indicate whether a dummy source or destination should be added to balance the model.
   (1) $a_1 = 10, a_2 = 5, a_3 = 4, a_4 = 6$
   $b_1 = 10, b_2 = 5, b_3 = 7, b_4 = 9$
   [*Ans.* Add a dummy source with capacity of 6 units.]
   (2) $a_1 = 30, a_2 = 44$
   $b_1 = 25, b_2 = 30, b_3 = 10$
   [*Ans.* Add a dummy destination with demand 9 units.]

Examples 5.1–1 and 5.1–2 illustrate the case where only one commodity is involved in the transportation system. The next example shows how the transportation model can be modified to include more than one commodity.

Example 5.1–3 (Multicommodity Transportation Model). The MG Company produces four different models, which we refer to for simplicity as M1, M2, M3, and M4. The Detroit plant produces models M1, M2, and M4. Models M1 and M2 only are produced in New Orleans. The Los Angeles plant manufactures models M3 and M4. The capacities of the various plants and the demands of the distribution centers are given here according to model type.

	M1	M2	M3	M4	Totals
Plant					
Los Angeles	—	—	700	300	1000
Detroit	500	600	—	400	1500
New Orleans	800	400	—	—	1200
Distribution center					
Denver	700	500	500	600	2300
Miami	600	500	200	100	1400

For simplicity, we assume that the transportation rate remains 8 cents per car per mile for all models.

To account for the multiple-car models, we view the transportation problem in the following manner. Instead of considering each plant as one source, we now subdivide it into a number of sources equal to the number of models it produces. Similarly, each distribution center may be viewed as consisting of four receiving stations representing the four models. The end result for this situation is that we have seven sources and eight destinations. The model is depicted in Figure 5-2.

Table 5-4 provides a complete representation of the transportation tableau. Note that certain routes are not admissible since, as the model now stands, car models cannot be substituted for one another. For example, we cannot ship from an M1-source to an M4-destination. In Figure 5-2, a closed route is indicated by a missing arc. These routes are represented in Table 5-4 by the assignment of a very high unit cost M.

If you study Table 5-4 carefully, you will notice that it is really unnecessary to represent the problem by a *single* transportation model. Because of the complete independence of the different models, we should be able to represent the problem for each car model by a separate, yet much smaller transportation tableau. In essence, Table 5-4 can be decomposed into the following for transportation models:

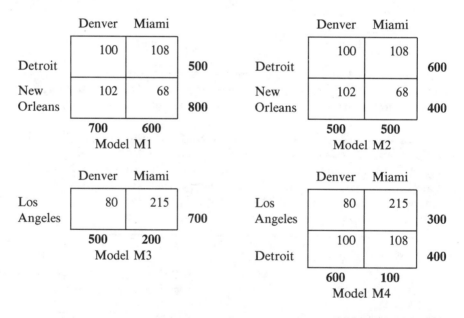

The combined optimal solution of these four transportation models is exactly the same as the optimum solution of Table 5-4. From the computational standpoint, solving the smaller subproblems will be much more efficient than solving the combined model of Table 5-4.

Of what use then is the combined model of Table 5-4? Remember that the main reason for being able to decompose Table 5-4 is that the various car models are completely independent of one another. If interaction exists between the car models (e.g., one car model can substitute for another), however, it will not be possible in general to decompose the combined model in the straightforward manner we used. Problem 5-5 is designed to illustrate this point. ◀

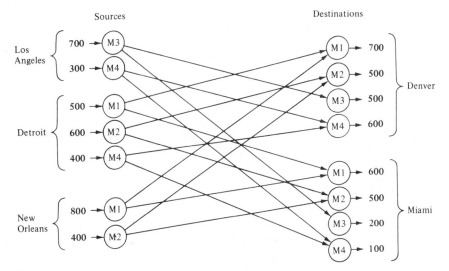

Figure 5–2

Table 5–4

		Denver				Miami				
		M1	M2	M3	M4	M1	M2	M3	M4	
Los Angeles	M3	M	M	80	M	M	M	215	M	700
	M4	M	M	M	80	M	M	M	215	300
Detroit	M1	100	M	M	M	108	M	M	M	500
	M2	M	100	M	M	M	108	M	M	600
	M4	M	M	M	100	M	M	M	108	400
New Orleans	M1	102	M	M	M	68	M	M	M	800
	M2	M	102	M	M	M	68	M	M	400
		700	500	500	600	600	500	200	100	

Exercise 5.1–3

If an extra car model M5 is produced in all three plants and is needed only at the Miami distribution center, how many sources and destinations will the new transportation model have?

[*Ans.* 10 sources and 9 destinations.]

The application of the transportation model is not limited to the problem of "transporting" commodities between geographical sources and destinations. The following example illustrates an unrelated problem in the area of **production inventory control.** The next section presents the assignment model, which deals with the assignment of jobs to machines or personnel. (See Problems 5–6 and 5–7 for other applications.)

Example 5.1–4 (Production–Inventory Model).

A company is developing a master plant for the production of an item over a 4-month period. The demands for the four months are 100, 200, 180, and 300 units, respectively. A current month's demand may be satisfied by

1. Excess production in an earlier month held in stock for later consumption.
2. Production in the current month.
3. Excess production in a later month backordered for preceding months.

The variable production cost per unit in any month is $4.00. A unit produced for later consumption will incur a holding (or storage) cost at the rate of $.50 per unit per month. On the other hand, backordered items incur a penalty cost of $2.00 per unit per month.

The production capacity for producing the item varies monthly depending on the other items being manufactured. The estimates for the next four months are 50, 180, 280, and 270 units, respectively.

The objective is to devise the minimum-cost production–inventory plan.

This problem can be formulated as a "transportation" model. The equivalence between the elements of the production and the transportation systems is established as follows:

Transportation System	Production System
1. Source i	1. Production period i
2. Destination j	2. Demand period j
3. Supply at source i	3. Production capacity of period i
4. Demand at destination j	4. Demand per period j
5. Transportation cost from source i to destination j	5. Production and inventory cost from period i to j

Table 5–5 summarizes the problem as a transportation model. The unit "transportation" cost from period i to period j is

$$c_{ij} = \begin{cases} \text{production cost in } i, & i = j \\ \text{production cost in } i + \text{holding cost from } i \text{ to } j, & i < j \\ \text{production cost in } i + \text{penalty cost from } i \text{ to } j, & i > j \end{cases}$$

Table 5–5

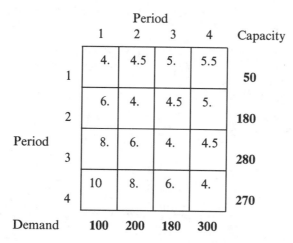

		1	2	3	4	Capacity
	1	4.	4.5	5.	5.5	**50**
	2	6.	4.	4.5	5.	**180**
Period	**3**	8.	6.	4.	4.5	**280**
	4	10	8.	6.	4.	**270**
Demand		**100**	**200**	**180**	**300**	

The definition of c_{ij} indicates that production in period i for the same period $(i = j)$ will result in production cost only. If period i produces for future periods $(i < j)$, an additional holding cost is incurred. Similarly, production in i to fill backorders $(i > j)$ incurs an additional penalty cost. For example,

$$c_{11} = \$4$$
$$c_{24} = 4 + (.5 + .5) = \$5$$
$$c_{41} = 4 + (2 + 2 + 2) = \$10 \qquad \blacktriangleleft$$

Exercise 5.1–4

In Table 5–5, suppose that the holding costs per unit change with the periods and are given by $.4, and $.3, and $.7 for periods 1, 2, and 3. The penalty costs remain unchanged. Recompute c_{ij} in Table 5–5.

[*Ans.* c_{ij} by row are (4, 4.4, 4.7, 5.4), (6, 4, 4.3, 5.0), (8, 6, 4, 4.7), and (10, 8, 6, 4).]

5.2 SOLUTION OF THE TRANSPORTATION PROBLEM

In this section we introduce the details for solving the transportation model. The method uses the steps of the simplex method directly and differs only in the details of implementing the optimality and feasibility conditions.

5.2.1 THE TRANSPORTATION TECHNIQUE

The basic steps of the transportation technique are

Step 1: Determine a starting feasible solution.

Step 2: Determine an entering variable from among the nonbasic variables. If all such variables satisfy the optimality condition (of the simplex method), stop; otherwise, go to step 3.

Step 3: Determine a leaving variable (using the feasibility condition) from among the variables of the current basic solution; then find the new basic solution. Return to step 2.

These steps will be considered in detail. The vehicle of explanation is the problem in Table 5–6. The unit transportation cost c_{ij} is in dollars. The supply and demand are given in number of units.

Table 5–6

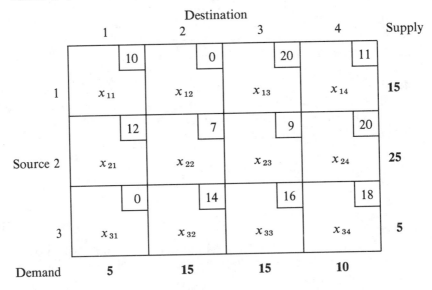

Determination of the Starting Solution

The general definition of the transportation model in Section 5.1 requires that $\sum_{i=1}^{m} a_i = \sum_{j=1}^{n} b_j$. This requirement results in one dependent equation, which means that the transportation model has only $m + n - 1$ independent equations. Thus, as in the simplex method, a starting basic feasible solution must include $m + n - 1$ basic variables.

Normally, if the transportation model is formulated as a simplex tableau, it would be necessary to utilize artificial variables to secure a starting basic solution. However, when the transportation tableau is used, a starting basic feasible solution can be obtained easily and directly. We present a procedure called the **northwest-corner rule** for this purpose. Two other procedures, called **least-cost** method and **Vogel's approximation**, are presented in Section 5.2.2. These procedures usually provide better starting solutions in the sense that the associated values of the objective function are smaller.

The *northwest-corner* method starts by allocating the maximum amount allowable by the supply and demand to the variable x_{11} (the one in the northwest corner of the tableau). The satisfied column (row) is then crossed out, indicating that the remaining variables in the crossed-out column (row) equal zero. If a column and a row are satisfied simultaneously, *only one* (either one) may be crossed out. (This condition guarantees locating *zero* basic variables, if any, automatically; see Table 5–8 for an illustration.) After adjusting the amounts of supply and demand for all uncrossed-out rows and columns, the maximum feasible amount is allocated to the first uncrossed-out element in the new column (row). The process is completed when *exactly* one row *or* one column is left uncrossed out.

The procedure just described is now applied to the example in Table 5–6.

1. $x_{11} = 5$, which crosses out column 1. Thus no further allocation can be made in column 1. The amount left in row 1 is 10 units.
2. $x_{12} = 10$, which crosses out row 1 and leaves 5 units in column 2.
3. $x_{22} = 5$, which crosses out column 2 and leaves 20 units in row 2.
4. $x_{23} = 15$, which crosses out column 3 and leaves 5 units in row 2.
5. $x_{24} = 5$, which crosses out row 2 and leaves 5 units in column 4.
6. $x_{34} = 5$, which crosses out row 3 *or* column 4. Since only one row or one column remains uncrossed out, the process ends.

The resulting starting basic solution is given in Table 5–7. The *basic* variables are $x_{11} = 5$, $x_{12} = 10$, $x_{22} = 5$, $x_{23} = 15$, $x_{24} = 5$, and $x_{34} = 5$. The remaining variables are *nonbasic* at zero level. The associated transportation cost is $5 \times 10 + 10 \times 0 + 5 \times 7 + 15 \times 9 + 5 \times 20 + 5 \times 18 = \$410$.

Table 5–7

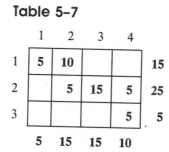

When both a column and a row are satisfied simultaneously, the next variable to be added to the basic solution will necessarily be at the zero level. Table 5–8 illustrates this point. Column 2 and row 2 are satisfied simultaneously. If

Table 5–8

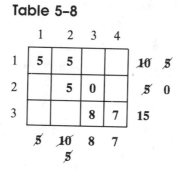

column 2 is crossed out, x_{23} becomes basic at zero level in the next step, since
the remaining supply for row 2 is now zero. (This case is shown in Table 5–
8.) If, instead, row 2 is crossed out, x_{32} would be the zero basic variable.

The starting solutions in Tables 5–7 and 5–8 include the proper number of
basic variables, namely, $m + n - 1 = 6$. The northwest-corner rule always
yields the proper number of basic variables.

Determination of Entering Variable (Method of Multipliers)

The entering variable is determined by using the optimality condition of
the simplex method. The computations of the objective equation coefficients
is based on the primal–dual relationships presented in Section 4.3. We first
present the mechanics of the method and then provide a rigorous explanation
of the procedure based on duality theory. Another method, called the **stepping-
stone** procedure, is also available for determining the entering variable. Although
the computations in the two methods are exactly equivalent, the stepping-stone
method gives the impression that the procedure is completely unrelated to the
simplex method.

In the method of multipliers we associate the multipliers u_i and v_j with
row i and column j of the transportation tableau. For each *basic* variable x_{ij}
in the current solution, the multipliers u_i and v_j must satisfy the following
equation:

$$u_i + v_j = c_{ij}, \quad \text{for each } basic \text{ variable } x_{ij}$$

These equations yield $m + n - 1$ equations (because there are only $m +
n - 1$ basic variables) in $m + n$ unknowns. The values of the multipliers can be
determined from these equations by assuming an *arbitrary* value for *any
one* of the multipliers (usually u_1 is set equal to zero) and then solving the $m +
n - 1$ equations in the remaining $m + n - 1$ unknown multipliers. Once this
is done, the evaluation of each nonbasic variable x_{pq} is given by

$$\bar{c}_{pq} = u_p + v_q - c_{pq}, \quad \text{for each } nonbasic \text{ variable } x_{pq}$$

(These values will be the same regardless of the arbitrary choice of the value
of u_1.) The entering variable is then selected as the nonbasic variable with the
most positive $\bar{c}_{pq}$ (compare with the minimization optimality condition of the
simplex method).

If we apply this procedure to the nonbasic variables in Table 5–7 (current
solution), the equations associated with the basic variables are given as

$$x_{11}: \quad u_1 + v_1 = c_{11} = 10$$
$$x_{12}: \quad u_1 + v_2 = c_{12} = 0$$
$$x_{22}: \quad u_2 + v_2 = c_{22} = 7$$
$$x_{23}: \quad u_2 + v_3 = c_{23} = 9$$
$$x_{24}: \quad u_2 + v_4 = c_{24} = 20$$
$$x_{34}: \quad u_3 + v_4 = c_{34} = 18$$

By letting $u_1 = 0$, the values of the multipliers are successively determined as $v_1 = 10$, $v_2 = 0$, $u_2 = 7$, $v_3 = 2$, $v_4 = 13$, and $u_3 = 5$. The evaluations of the nonbasic variables are thus given as follows:

$$x_{13}: \quad \bar{c}_{13} = u_1 + v_3 - c_{13} = 0 + 2 - 20 = -18$$
$$x_{14}: \quad \bar{c}_{14} = u_1 + v_4 - c_{14} = 0 + 13 - 11 = 2$$
$$x_{21}: \quad \bar{c}_{21} = u_2 + v_1 - c_{21} = 7 + 10 - 12 = 5$$
$$x_{31}: \quad \bar{c}_{31} = u_3 + v_1 - c_{31} = 5 + 10 - 0 = \boxed{15}$$
$$x_{32}: \quad \bar{c}_{32} = u_3 + v_2 - c_{32} = 5 + 0 - 14 = -9$$
$$x_{33}: \quad \bar{c}_{33} = u_3 + v_3 - c_{33} = 5 + 2 - 16 = -9$$

Since x_{31} has the most positive $\bar{c}_{pq}$, it is selected as the entering variable.

Exercise 5.2-1

Show that the values of $\bar{c}_{pq}$ remain unchanged when u_1 is assigned the arbitrary value k (instead of zero) when solving the basic equations.

The equations $u_i + v_j = c_{ij}$, which we use for determining the multipliers, have such a simple structure that it is really unnecessary to write them explicitly. It is usually much simpler to determine the multipliers directly from the transportation tableau by noting that u_i of row i and v_j of column j add up to c_{ij} when row i and column j intersect in a cell containing a *basic* variable x_{ij}. Once u_i and v_j are determined, we can compute $\bar{c}_{pq}$ for all nonbasic x_{pq} by adding u_p of row p and v_q of column q and then subtracting c_{pq} in the cell at the intersection of row p and column q.

Determination of Leaving Variable (Loop Construction)

This step is equivalent to applying the feasibility condition in the simplex method. However, since all the constraint coefficients in the original transportation model are either zero or one, the ratios of the feasibility condition will always have their denominator equal to one. Thus the values of the basic variables will give the associated ratios directly.

For the purpose of determining the minimum ratio, we construct a *closed loop* for the current entering variable (x_{31} in the current iteration). The loop starts and ends at the designated nonbasic variable. It consists of *successive* horizontal and vertical (connected) segments whose end points must be basic variables, except for the end points that are associated with the entering variable. This means that every corner element of the loop must be a cell containing a basic variable. Table 5–9 illustrates a loop for the entering variable x_{31} given the basic solution in Table 5–7. This loop may be defined in terms of the basic variables as $x_{31} \rightarrow x_{11} \rightarrow x_{12} \rightarrow x_{22} \rightarrow x_{24} \rightarrow x_{34} \rightarrow x_{31}$. It is *immaterial whether the loop is traced in a clockwise or counterclockwise direction.* Observe that for a given basic solution only *one unique* loop can be constructed for each nonbasic variable.

We can see from Table 5–9 that if x_{31} (the entering variable) is increased

Table 5-9

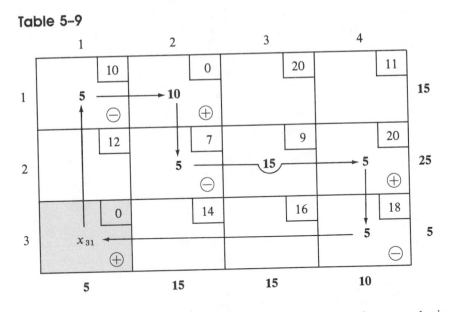

by one unit, then, to maintain the feasibility of the solution, the *corner* basic variables of the x_{31}-loop must be adjusted as follows. Decrease x_{11} by one unit, increase x_{12} by one unit, decrease x_{22} by one unit, increase x_{24} by one unit, and finally decrease x_{34} by one unit. This process is summarized by plus $\oplus$ and minus $\ominus$ signs in the appropriate corners in Table 5-9. The change will keep the supply and demand restrictions satisfied.

The leaving variable is selected from among the corner variables of the loop that will decrease when the entering variable x_{31} increases above zero level. These are indicated in Table 5-9 by the variables in the square labeled by minus signs $\ominus$. From Table 5-9, x_{11}, x_{22}, and x_{34} are the basic variables that will decrease when x_{31} increases. The leaving variable is then selected as the one having the *smallest* value, since it will be the first to reach zero value and any further decrease will cause it to be negative (compare the feasibility condition of the simplex method, where the leaving variable is associated with the minimum ratio). In this example the three $\ominus$-variables x_{11}, x_{22}, and x_{34} have the same value ($= 5$), in which case any *one* of them can be selected as a leaving variable. Suppose that x_{34} is taken as the leaving variable; then the value of x_{31} is increased to 5 and the values of the *corner* (basic) variables are adjusted accordingly (i.e., each is increased or decreased by 5, depending on whether it has $\oplus$ or $\ominus$ associated with it). The new solution is given in Table 5-10. Its new cost is $0 \times 10 + 15 \times 0 + 0 \times 7 + 15 \times 9 + 10 \times 20 + 5 \times 0 = \$335$. This cost differs from the one associated with the starting solution in Table 5-7 by $410 - 335 = \$75$, which is equal to the number of units assigned to x_{31} ($= 5$) multiplied by $\bar{c}_{31}$ ($= \$15$).

The basic solution in Table 5-10 is degenerate, since the basic variables x_{11} and x_{22} are zero. Degeneracy, however, needs no special provisions, and the zero basic variables are treated as any other positive basic variables.

The new basic solution in Table 5-10 is now checked for optimality by computing the *new* multipliers as shown in Table 5-11. The values of $\bar{c}_{pq}$ are

Table 5-10

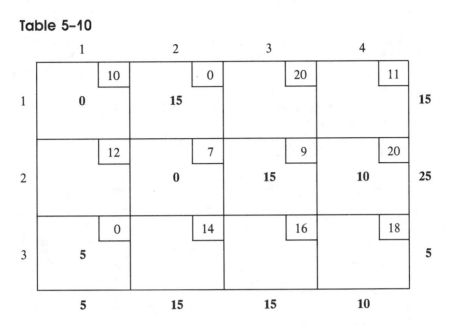

Table 5-11

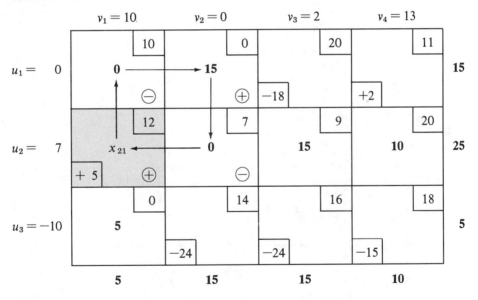

given by the numbers in the *southwest* corner of each nonbasic cell. The nonbasic variable x_{21} with the largest positive $\bar{c}_{pq}$ thus enters the solution. The closed loop associated with x_{21} shows that either x_{11} or x_{22} can be the leaving variable. We arbitrarily select x_{11} to leave the solution.

Exercise 5.2-2

Verify the values of u_i, v_j, and $\bar{c}_{pq}$ in Table 5–11.

Table 5–12 shows the new basic solution that follows from Table 5–11 (x_{21} enters and x_{11} leaves). The new values of u_i, v_j, and $\bar{c}_{pq}$ are computed anew. Table 5–12 gives the entering and leaving variables as x_{14} and x_{24}, respectively. By effecting this change in Table 5–12, we obtain the new solution in Table 5–13. Since all the $\bar{c}_{pq}$ in Table 5–13 are *nonpositive*, the optimum solution has been attained (compare with the minimization optimality condition of the simplex method).

Table 5–12

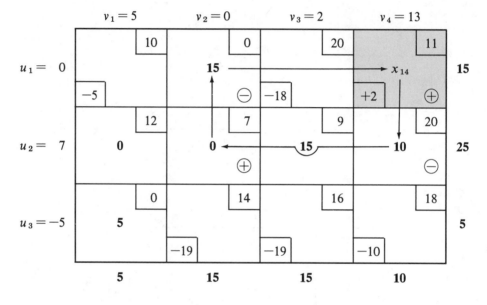

Table 5–13

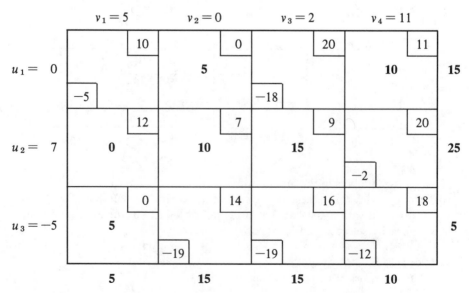

The optimal solution is summarized as follows. Ship 5 units from (source) 1 to (destination) 2 at $5 \times 0 = \$0$, 10 units from 1 to 4 at $10 \times 11 = \$110$, 10 units from 2 to 2 at $10 \times 7 = \$70$, 15 units from 2 to 3 at $15 \times 9 = \$135$, and 5 units from 3 to 1 at $5 \times 0 = \$0$. The total transportation cost of the schedule is $315.

Exercise 5.2-3

Consider the transportation problem discussed earlier.

(a) Compute the improvements (decreases) in the value of the objective function in each of the following cases by using the values of $\bar{c}_{pq}$ directly.

    (1) Solution moves from Table 5–9 to Table 5–10.
        [*Ans.* $5 \times 15 = \$75$.]

    (2) Solution moves from Table 5–11 to Table 5–12.
        [*Ans.* $0 \times 5 = \$0$.]

    (3) Solution moves from Table 5–12 to Table 5–13.
        [*Ans.* $10 \times 2 = \$20$.]

(b) In Table 5–11, determine the change (increase or decrease) in the value of the objective function when each of the following nonbasic variables is forced to enter the basic solution.

    (1) x_{13}. [*Ans.* $+\$270$.]

    (2) x_{14}. [*Ans.* $-\$20$.]

    (3) x_{32}. [*Ans.* $+\$120$.]

    (4) x_{33}. [*Ans.* $+\$120$.]

    (5) x_{34}. [*Ans.* $+\$75$.]

Explanation of the Method of Multipliers as a Simplex Method†

The relationship between the method of multipliers and the simplex method can be established by showing that $\bar{c}_{pq}$, as defined, directly equals the coefficients of the objective equation in the simplex tableau associated with the current iteration. We have seen from the primal–dual computations in Section 4.2 that, given the simplex multipliers of the current iteration, the objective equation coefficients are obtained by taking the difference between the left and right sides of the dual constraints. This relationship will be used to show that the multipliers method is essentially equivalent to the simplex method. Indeed, the multipliers u_i and v_j are nothing but the dual variables (or the simplex multipliers).

To show how the general dual problem is obtained for the transportation model, consider first the special case of $m = 2$ and $n = 3$ given in Table 5–14. Let the dual variables be u_1 and u_2 for the sources constraints and v_1, v_2, and v_3 for the destinations constraints. The dual problem becomes (see Section 4.1)

$$\text{maximize} \quad w = (a_1 u_1 + a_2 u_2) + (b_1 v_1 + b_2 v_2 + b_3 v_3)$$

subject to

---

† The remainder of this section assumes knowledge of duality theory (Chapter 4). It may be skipped without loss of continuity.

$$u_1 + v_1 \qquad\qquad \le c_{11}$$
$$u_1 \qquad + v_2 \qquad \le c_{12}$$
$$u_1 \qquad\qquad + v_3 \le c_{13}$$
$$u_2 + v_1 \qquad\qquad \le c_{21}$$
$$u_2 \qquad + v_2 \qquad \le c_{22}$$
$$u_2 \qquad\qquad + v_3 \le c_{23}$$

u_1, u_2, v_1, v_2, v_3 unrestricted

The special structure of the dual constraints results from the special arrangement of the "1" and "0" elements in the primal problem. Each constraint includes one u-variable and one v-variable only. Also, for each dual constraint, the subscripts of u and v match the double subscripts of the c-element. Thus, in general, if u_i and v_j are the dual variables corresponding to the constraints of the ith source and the jth destination ($i = 1, 2, \ldots, m; j = 1, 2, \ldots, n$) the corresponding dual problem is given by

$$\text{maximize} \quad w = \sum_{i=1}^{m} a_i u_i + \sum_{j=1}^{n} b_j v_j$$

subject to

$$u_i + v_j \le c_{ij}, \qquad \text{for all } i \text{ and } j$$
$$u_i \text{ and } v_j \text{ unrestricted}$$

According to Section 4.2, the objective equation coefficients (and hence the evaluation of the nonbasic variables) are found by substituting the current values of the dual variables (simplex multipliers) in the dual constraints and then taking the difference between its left- and right-hand sides. But whereas in the simplex method the simplex multipliers are immediately available, this is not the case with the transportation tableau. However, the multipliers can be determined indirectly by observing that the dual constraints corresponding to a basic variable must be satisfied as strict equations. This means that

$$u_i + v_j = c_{ij}, \qquad \text{for every } basic \text{ variable } x_{ij}$$

Table 5-14

| | z | Source 1 Variables | | | Source 2 Variables | | | R.H.S. |
		x_{11}	x_{12}	x_{13}	x_{21}	x_{22}	x_{13}	
Objective equation	1	$-c_{11}$	$-c_{12}$	$-c_{13}$	$-c_{21}$	$-c_{22}$	$-c_{23}$	0
Sources constraints $\{$	0	1	1	1				a_1
	0				1	1	1	a_2
Destinations constraints $\{$	0	1			1			b_1
	0		1			1		b_2
	0			1			1	b_3

which gives $m + n - 1$ equations. Therefore, by assuming an arbitrary value for $u_1 (= 0)$, the remaining multipliers can be determined.

The coefficient of nonbasic variable x_{pq} in the objective equation is now given by the difference between the left- and right-hand sides of the corresponding dual constraint, that is, $u_p + v_q - c_{pq}$. Since the transportation problem is a *minimization* problem, the entering variable is the one with the largest *positive* $u_p + v_q - c_{pq}$.

The relationship between the method of multipliers and simplex methods should be clear now. Indeed, at the optimum iteration the multipliers give the *optimal* dual values directly. From Section 4.2, these values should yield the same optimum objective value in the primal and dual. The simplex multipliers associated with the *optimal* solution in Table 5–13 are $u_1 = 0$, $u_2 = 7$, $u_3 = -5$, $v_1 = 5$, $v_2 = 0$, $v_3 = 2$, and $v_4 = 11$. The corresponding value of the dual objective function is

$$\sum_{i=1}^{3} a_i u_i + \sum_{j=1}^{4} b_j v_j = (15 \times 0 + 25 \times 7 + 5 \times -5)$$
$$+ (5 \times 5 + 15 \times 0 + 15 \times 2 + 10 \times 11)$$
$$= 315$$

which is the same as in the primal.

In the foregoing discussion, an arbitrary value is assigned to one of the dual variables (e.g., $u_1 = 0$), which indicates that the simplex multipliers associated with a given basic solution are not unique. This may appear inconsistent with the results in Chapter 4, where the simplex multipliers must be unique. Problem 5–32 resolves this apparent paradox and shows that there is actually no inconsistency.

5.2.2 IMPROVED STARTING SOLUTION

The northwest-corner method presented in Section 5.2.1 does not necessarily produce a "good" starting solution for the transportation model. In this section we present two procedures that determine the starting solution by selecting the "cheap" routes of the model.

The Least-Cost Method

The procedure is as follows. Assign as much as possible to the variable with the smallest *unit* cost in the entire tableau. (Ties are broken arbitrarily.) Cross out the satisfied row or column. (As in the *northwest-corner* method, if both a column and a row are satisfied simultaneously, only one may be crossed out.) After adjusting the supply and demand for all *un* crossed-out rows and columns, repeat the process by assigning as much as possible to the variable with the smallest uncrossed-out unit cost. The procedure is complete when exactly one row *or* one column is left uncrossed out.

The transportation problem in Table 5–6 is used again to illustrate the application of the least-cost method. Table 5–15 gives the resulting starting solution. The steps of the solution are as follows; x_{12} and x_{31} are the variables associated with the smallest unit costs ($c_{12} = c_{31} = 0$). Breaking the tie arbitrarily, select x_{12}. The associated supply and demand units give $x_{12} = 15$, which satisfies both row 1 and column 2. By crossing out column 2, the supply left in row 1 is zero. Next, x_{31} has the smallest uncrossed-out unit cost. Thus $x_{31} = 5$ satisfies both row 3 and column 1. By crossing out row 3, the demand in column 1 is zero. The smallest uncrossed-out element is $c_{23} = 9$. The supply and demand units give $x_{23} = 15$, which crosses out column 3 and leaves 10 units of supply in row 2. The smallest uncrossed-out element is $c_{11} = 10$. Since the remaining supply in row 1 and the remaining demand in column 1 are both zero, $x_{11} = 0$. By crossing out column 1, the supply "left" in row 1 is zero. The remaining basic variables are obtained, respectively, as $x_{14} = 0$ and $x_{24} = 10$. The total cost associated with this solution is $0 \times 10 + 15 \times 0 + 0 \times 11 + 15 \times 9 + 10 \times 20 + 5 \times 0 = \$335$, which is better (smaller) than the one provided by the northwest-corner method.

Exercise 5.2–4

Rework this problem assuming that the least-cost method starts with an assignment to x_{31} instead of x_{12} (both c_{31} and c_{12} equal zero), and compare the resulting starting solution with the one given.
[*Ans.* The positive variables are the same. The only difference occurs in the assignment of the zero variables. This is not a general result, however.]

Vogel's Approximation Method (VAM)

This method is a heuristic and usually provides a better starting solution than the two methods just described. In fact, VAM generally yields an optimum, or close to optimum, starting solution.

The steps of the procedure are as follows.

Step 1: Evaluate a penalty for each row (column) by subtracting the *smallest* cost element in the row (column) from the *next smallest* cost element in the same row (column).

Step 2: Identify the row or column with the largest penalty, breaking ties arbitrarily. Allocate as much as possible to the variable with the least cost in the selected row or column. Adjust the supply and demand and cross out the satisfied row *or* column. If a row and a column are satisfied simultaneously, only one of them is crossed out and the remaining row (column) is assigned a zero supply (demand). *Any row or column with zero supply or demand should not be used in computing future penalties (in step 3).*

Step 3: (a) If exactly one row or one column remains uncrossed out, stop.
     (b) If only one row (column) with *positive* supply (demand) remains uncrossed out, determine the basic variables in the row (column) by the least-cost method.
     (c) If all uncrossed-out rows and columns have (assigned) zero supply

Table 5–15

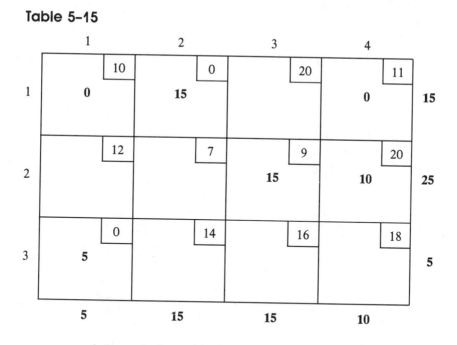

and demand, determine the *zero* basic variables by the least-cost method. Stop.

(d) Otherwise, recompute the penalties for the uncrossed-out rows and columns, then go to step 2. (Notice that the rows and columns with assigned zero supply and demand should not be used in computing these penalties.)

We apply VAM to the problem in Table 5–6. Table 5–16 shows the first set of row and column penalties. Since row 3 has the largest penalty ($= 14$) and since $c_{31} = 0$ is the least unit cost in the same row, the quantity 5 is assigned to x_{31}. Row 3 and column 1 are satisfied simultaneously. Assume that column 1 is crossed out. The remaining supply for row 3 is zero.

Table 5–17 shows the new set of penalties after crossing out column 1 in Table 5–16. (Notice that row 3 with zero supply is not used in computing the penalties.) Row 1 and column 3 have the same penalties. By selecting column 3 arbitrarily, the amount 15 is assigned to x_{23}, which crosses out column 3 and adjusts the supply in row 2 to 10.

Successive applications of VAM yield $x_{22} = 10$ (cross out row 2), $x_{12} = 5$ (cross out column 2), $x_{14} = 10$ (cross out row 1), and $x_{34} = 0$. (Verify.) The cost of the program is \$315, which happens to be optimal.

The given version of VAM breaks ties between penalties arbitrarily. However, breaking of ties may be crucial in rendering a good starting solution. For example, in Table 5–17, if row 1 is selected instead of column 3, a worse starting solution results. (Verify that this solution is $x_{12} = 15$, $x_{23} = 15$, $x_{24} = 10$, $x_{31} = 5$, which will yield a total cost of \$335.) The complete VAM procedure provides details for breaking some of these ties advantageously (see N. Reinfeld and W. Vogel, *Mathematical Programming*, Prentice-Hall, Englewood Cliffs, N.J., 1958).

Table 5-16

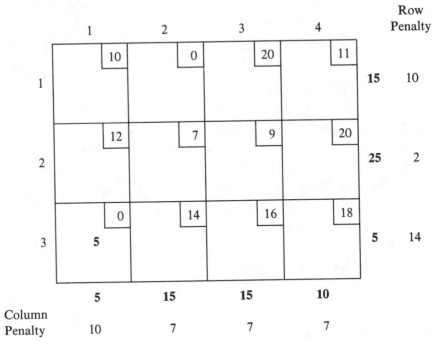

Table 5-17

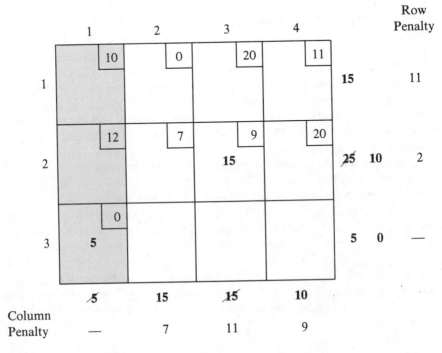

Exercise 5.2-2

Rework the problem by VAM after replacing the value of c_{12} by 2 (instead of zero). For the purpose of this problem, when a column and a row are satisfied simultaneously, always cross out the row.

[*Ans.* The successive assignments are $x_{31} = 5$, $x_{23} = 15$, $x_{22} = 10$, $x_{14} = 10$, $x_{12} = 5$, and $x_{11} = 0$.]

5.3 THE ASSIGNMENT MODEL

Consider the situation of assigning m jobs (or workers) to n machines. A job $i\ (= 1, 2, \ldots, m)$ when assigned to machine $j\ (= 1, 2, \ldots, n)$ incurs a cost c_{ij}. The objective is to assign the jobs to the machines (one job per machine) at the least total cost. The situation is known as the **assignment problem.**

The formulation of this problem may be regarded as a special case of the transportation model. Here jobs represent "sources" and machines represent "destinations." The supply available at each source is 1; that is, $a_i = 1$ for all i. Similarly, the demand required at each destination is 1; that is, $b_j = 1$ for all j. The cost of "transporting" (assigning) job i to machine j is c_{ij}. If a job cannot be assigned to a certain machine, the corresponding c_{ij} is taken equal to M, a very high cost. Table 5–18 gives a general representation of the assignment model.

Table 5-18

	Machine				
	1	2	$\cdots$	n	
Job 1	c_{11}	c_{12}	$\cdots$	c_{1n}	1
2	c_{21}	c_{22}	$\cdots$	c_{2n}	1
$\vdots$	$\vdots$	$\vdots$		$\vdots$	$\vdots$
m	c_{m1}	c_{m2}	$\cdots$	c_{mn}	1
	1	**1**	$\cdots$	**1**	

Before the model can be solved by the transportation technique, it is necessary to balance the problem by adding fictitious jobs or machines, depending on whether $m < n$ or $m > n$. It will thus be assumed that $m = n$ without loss of generality.

The assignment model can be expressed mathematically as follows. Let

$$x_{ij} = \begin{cases} 0, & \text{if the } j\text{th job is } not \text{ assigned to the } i\text{th machine} \\ 1, & \text{if the } j\text{th job is assigned to the } i\text{th machine} \end{cases}$$

The model is thus given by

$$\text{minimize } z = \sum_{i=1}^{n} \sum_{j=1}^{n} c_{ij} x_{ij}$$

subject to

$$\sum_{j=1}^{n} x_{ij} = 1, \qquad i = 1, 2, \ldots, n$$

$$\sum_{i=1}^{n} x_{ij} = 1, \qquad j = 1, 2, \ldots, n$$

$$x_{ij} = 0 \quad \text{or} \quad 1$$

To illustrate the assignment model, consider the problem in Table 5–19 with three jobs and three machines. The initial solution (using the northwest-corner rule) is obviously degenerate. This will always be the case in the assignment model regardless of the method used to obtain the starting basis. In fact, the solution will continue to be degenerate at every iteration.

Table 5–19

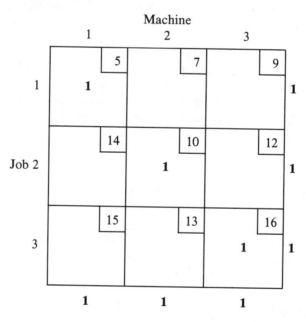

The special structure of the assignment model allows the development of an efficient method of solution. This method will be illustrated by the example just presented.

The optimal solution of the assignment model remains the same if a constant is added or subtracted to any row or column of the cost matrix. This is proved as follows. If p_i and q_j are subtracted from the ith row and the jth column, the new cost elements become $c'_{ij} = c_{ij} - p_i - q_j$. This yields the new objective function

$$z' = \sum_i \sum_j c'_{ij} x_{ij} = \sum_i \sum_j (c_{ij} - p_i - q_j) x_{ij}$$

$$= \sum_i \sum_j c_{ij} x_{ij} - \sum_i p_i \sum_j x_{ij} - \sum_j q_j \sum_i x_{ij}$$

Since $\Sigma_j x_{ij} = \Sigma_i x_{ij} = 1$, we get $z' = z -$ constant. This shows that the minimization of the original objective function z yields the same solution as the minimization of z'.

This idea indicates that if one can create a new c'_{ij}-matrix with zero entries, and if these zero elements or a subset thereof constitute a feasible solution, this feasible solution is optimal, because the cost cannot be negative.

In Table 5–19 the zero elements are created by subtracting the smallest element in each row (column) from the corresponding row (column). If one considers the rows first, the new c'_{ij}-matrix is shown in Table 5–20.

Table 5–20

	1	2	3	
1	0	2	4	$p_1 = 5$
$\|c'_{ij}\| = 2$	4	0	2	$p_2 = 10$
3	2	0	3	$p_3 = 13$

The last matrix can be made to include more zeros by subtracting $q_3 = 2$ from the third column. This yields Table 5–21.

Table 5–21

	1	2	3
1	[0]	2	2
$\|c'_{ij}\| = 2$	4	0	[0]
3	2	[0]	1

The squares in Table 5–21 give the feasible (and hence optimal) assignment (1, 1), (2, 3), and (3, 2), costing $5 + 12 + 13 = 30$. Notice that this cost is equal to $p_1 + p_2 + p_3 + q_3$.

Unfortunately, it is not always possible to obtain a feasible assignment as in the example. Further rules are thus required to find the optimal solution. These rules are illustrated by the example shown in Table 5–22.

Table 5–22

	1	2	3	4
1	1	4	6	3
2	9	7	10	9
3	4	5	11	7
4	8	7	8	5

Now, carrying out the same initial steps as in the previous example, one gets Table 5–23.

Table 5–23

	1	2	3	4
1	0	3	2	2
2	2	0	0	2
3	0	1	4	3
4	3	2	0	0

A feasible assignment to the zero elements is not possible in this case. The procedure then is to draw a *minimum* number of lines through some of the rows and columns such that all the zeros are crossed out. Table 5–24 shows the application of this rule.

Table 5–24

	1	2	3	4
1	0̸	3	2	2
2	~~2~~	~~0~~	~~0~~	~~2~~
3	0̸	1	4	3
4	~~3~~	~~2~~	~~0~~	~~0~~

The next step is to select the *smallest* uncrossed-out element ($= 1$ in Table 5–24). This element is subtracted from every *uncrossed-out* element and added to every element at the intersection of two lines. This yields Table 5–25, which gives the optimal assignment $(1, 1)$, $(2, 3)$, $(3, 2)$, and $(4, 4)$. The corresponding total cost is $1 + 10 + 5 + 5 = 21$.

Table 5–25

	1	2	3	4
1	[0]	2	1	1
2	3	0	[0]	2
3	0	[0]	3	2
4	4	2	0	[0]

It should be noted that if the optimal solution was not obtained in the preceding step, the given procedure of drawing lines should be repeated until a feasible assignment is achieved.

Exercise 5.3–1

Solve the assignment problem in Table 5–20 by the transportation technique and show that it gives the same solution.

5.4 THE TRANSSHIPMENT MODEL

The standard transportation model assumes that the *direct* route between a source and a destination is a *minimum-cost* route. Thus, in Example 5.1–1, the mileage table from the three auto plants to the two distribution centers gives the *shortest routes* between the sources and destinations. This means that preparatory work involving the determination of the shortest routes must be carried out before the unit costs of the standard transportation model can be determined.

In small problems, the determination of the shortest route is a simple matter. As the number of sources and destinations increases, we must resort to the use of a systematic method, called the *shortest-route algorithm* (see Section 6.2), to determine the minimum unit cost for *direct* shipping on a given route.

An alternative method for determining the minimum direct shipping cost is to formulate the problem as a **transshipment model.** The new formulation has the added feature of allowing a (partial or complete) shipment to pass *transiently* through other sources and destinations before it ultimately reaches its designated destination. In essence, the transshipment model automatically seeks the minimum-cost route between a source and a destination without having to determine such a route a priori.

What does transshipment entail? Essentially, the *entire* supply from all sources could potentially pass through any source or destination before it is redistributed again. This means that each node in the transportation network, be it a source or a destination, can be considered both a transient source and a transient destination. Since we do not know a priori which nodes will have this property, we can set up the model so that it would allow each node to be both a source and a destination. Equivalently, this means that the number of sources (destinations) in the transshipment model will equal the *sum* of sources and destinations in the standard model.

To demonstrate this point, consider the MG model (Example 5.1–1). We have three plants and two distribution centers. In terms of the transshipment model, we have *five* sources and five destinations. Figure 5–3 demonstrates the associated network. To allow transient passing of the commodity, an additional buffer stock B is allowed at each source and destination. By definition, the buffer stock *at least* equals the sum of supply (or demand) of the (balanced) standard transportation model. That is,

$$B \geq \sum_{i=1}^{m} a_i = \sum_{j=1}^{n} b_j$$

The unit costs are estimated from the data on the routes connecting the "sources" and "destinations" of the transshipment model. Obviously, the unit costs between the original sources and destinations (see the tinted portion of the network in Figure 5–3) remain as given in Example 5.1–1. Two other remarks concerning unit costs: the shipping cost from a location to itself (e.g., Denver to Denver) is zero and the shipping cost between two locations may differ depending on the direction of travel (e.g., the cost of the Detroit–Denver route

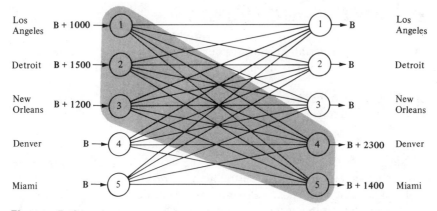

Figure 5-3

may differ from that of Denver–Detroit if different transportation modes are used).

Table 5–26 gives the optimum solution of the transshipment model with the buffer B taken equal to 3700 cars. Note that the cost elements between the original sources (Los Angeles, Detroit, and New Orleans) and the original destinations (Denver and Miami) equal those given in Example 5.1–1. The remaining elements are supposed to have been estimated depending on the mileage and the mode of transportation.

Notice that the diagonal elements in Table 5–26 are the result of using the buffer stock. They do not contribute any information to the final solution. The

Table 5-26

	Los Angeles	Detroit	New Orleans	Denver	Miami	
Los Angeles	0 **3700**	130	90	80 **1000**	215	4700
Detroit	135	0 **3700**	101 **200**	100 **1300**	108	5200
New Orleans	95	105	0 **3500**	102	68 **1400**	4900
Denver	79	99	110	0 **3700**	205	3700
Miami	200	107	72	205	0 **3700**	3700
	3700	3700	3700	6000	5100	

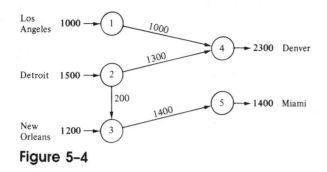

Figure 5–4

off-diagonal elements provide the solution, which is translated graphically in Figure 5–4. Transshipping takes place between the Detroit and the New Orleans plants, where 200 cars are shipped from Detroit to Miami via New Orleans. All other shipments are direct from the plants to the distribution centers.

In addition to the preceding application, there are situations where the commodity actually experiences *transient* shipping. For example, in the MG model (Example 5.1–1), individual dealers rather than distribution centers may represent the final destination of the cars. For simplicity, suppose that only five dealers receive their orders from the Denver and Miami distribution centers. Figure 5–5 shows the distribution centers as temporary storages from which the cars are allocated to their final destination. The respective demands for the five dealers are 800, 500, 750, 1000, and 650 cars. We assume for simplicity that a dealer can fill an order from either distribution center. (Notice that the demands at the distribution centers as given in Example 5.2–1 are no longer assumed to be part of the data of the model.) For the moment, Figure 5–5 shows that transshipping is allowed only through the distribution centers. We subsequently generalize the model to include transshipping within the plants and the distribution centers.

Since the distribution centers (nodes 4 and 5 in Figure 5–5) are the only transshipping locations, they each act both as a destination and as a source. On the other hand, the plants act only as sources and the dealers as destinations. The resulting transshipment model is given in Table 5–27. Note that at the

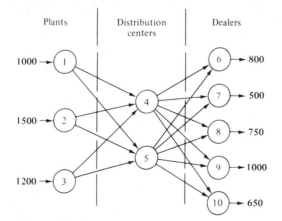

Figure 5–5

distribution centers (nodes 4 or 5) a buffer $B = 3700$ is added to their respective supply and demand quantities. Distinctly, no buffer need be added to the supply of each plant or the demand of each dealer. Remember that *the buffer is added only when a node acts as both a source and a destination,* which can happen when shipments pass transiently through the location. Observe in Figure 5–5 that no direct shipping from plant to dealer is allowed. This is shown in Table 5–27 by blocked (tinted) cells. When solving the model numerically, a blocked route is indicated by assigning to its cell a very high cost, $c_{ij} = M$.

Table 5-27

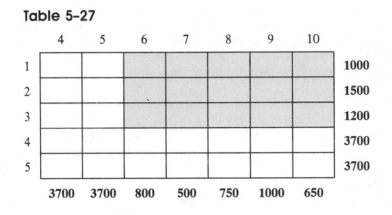

	4	5	6	7	8	9	10	
1								1000
2								1500
3								1200
4								3700
5								3700
	3700	3700	800	500	750	1000	650	

Suppose now that we allow transshipping within and among the plants and distribution centers. Only direct shipping will be allowed from the distribution centers to the dealers. Table 5–28 shows the new model. Notice that each plant and distribution center now acts as both a source and a destination.

Table 5-28

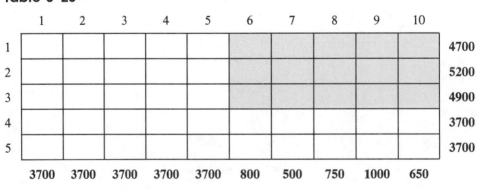

	1	2	3	4	5	6	7	8	9	10	
1											4700
2											5200
3											4900
4											3700
5											3700
	3700	3700	3700	3700	3700	800	500	750	1000	650	

Exercise 5.4-1

Indicate the changes in the model in Table 5–28 when each of the following conditions applies independently.

(a) The Los Angeles plant (node 1) does not receive transient shipments.

[*Ans.* The supply of the first row becomes 1000 and the first column is deleted.]

(b) No transshipping is allowed from the distribution center to the plants.

[*Ans.* Block the routes from sources 4 and 5 to destinations 1, 2, and 3.]

(c) Transshipping is allowed among the dealers.

[*Ans.* Add five sources representing the dealers (nodes 6, 7, 8, 9, and 10), each with a supply of 3700, and block the routes from these sources to destinations 1, 2, 3, 4, and 5. In addition, add 3700 to the demands at each of destinations 6, 7, 8, 9, and 10.]

(d) Distribution center 4 will sell 240 cars directly to customers that otherwise would have been shipped to dealer 9 for sale in the same area.

[*Ans.* Change the demands at destinations 4 and 9 to 3940 and 760 instead of the current values of 3700 and 1000.]

5.5 SUMMARY

Although the transportation model is a linear program, its special structure allows us to modify the details of the simplex algorithm to produce a computationally more efficient technique. In this respect duality theory plays a prominent role in the development of the transportation method.

The classical transportation model deals with shipping (or transshipping) one or more commodities between sources and destinations. The model can be modified to include the **capacitated transportation problem,** which differs from the classical model in that upper limits are set on the capacities of the routes. The model also applies to other situations that do not fall into the classical description, such as the assignment problem and the production–inventory problem.

The transportation model and its variants are but one class of the generalized network models (see Chapter 6). In practice, network models seem to have tremendous successes in solving real-life problems. Indeed, some recent surveys report that perhaps 70% of real-world mathematical programming problems can be treated as networks or network-related problems.

SELECTED REFERENCES

DANTZIG, G. B., *Linear Programming and Extensions,* Princeton University Press, Princeton, N.J., 1963.

ELMAGHRABLY, S., *The Design of Production Systems,* Reinhold, New York, 1966, Chap. 4.

MURTY, KATTA, *Linear Programming,* Wiley, New York, 1983.

REVIEW QUESTIONS

True (T) or False (F)?

1. _____ A transportation problem can always be represented by a balanced model.

2. ____ A transportation model that is initially unbalanced may require the addition of both a dummy source and a dummy destination to effect balancing.

3. ____ In the solution of the transportation model, the amounts shipped from a dummy source to the destinations actually represent shortages at the destinations.

4. ____ The transportation model is restricted to dealing with a single commodity only.

5. ____ A basic requirement for using the transportation technique is that the transportation model be balanced.

6. ____ The transportation technique essentially uses the same steps of the simplex method.

7. ____ If the same starting basic solution is used in both the simplex method and the transportation techniques, the iterations in the two procedures will essentially reproduce one another.

8. ____ In the transportation technique, the closed-loop method for determining the leaving variable is basically different from applying the feasibility condition in the simplex method.

9. ____ The multipliers in the transportation technique are essentially the dual variables of the LP representing the transportation problem.

10. ____ The arbitrary selection of the value of one of the multipliers in a transportation iteration will lead to the determination of different entering variables.

11. ____ If a constant value is added to every cost element c_{ij} in the transportation tableau, the *optimal* values of the variables x_{ij} will change.

12. ____ A balanced transportation model may not have any feasible solution.

13. ____ Every basic solution in the assignment problem is necessarily degenerate.

14. ____ The assignment problem can be solved by the transportation technique.

15. ____ If c_{ij} in the transportation model is the *lowest* unit cost between source i and destination j, both the transportation model and its associated transshipment model will produce the same optimum solution.

16. ____ Determination of the lowest c_{ij} in a transportation model may generally require the use of a special algorithm such as the *shortest-route* technique.

17. ____ The use of transshipping makes it unnecessary to obtain the lowest c_{ij} between source i and destination j for the purpose of determining the minimum-cost transportation schedule between the sources and destinations.

[*Ans.* **1**—T, **2**—F, **3**—T, **4**—F, **5**—T, **6**—T, **7**—T, **8**—F, **9**—T, **10**—F, **11**—F, **12**—F, **13** to **17**—T.]

PROBLEMS

Section	Assigned Problems
5.1	5–1 to 5–9
5.2.1	5–10 to 5–20, 5–32, 5–33
5.2.2	5–21 to 5–23
5.3	5–24 to 5–28
5.4.1	5–29 to 5–31

☐ **5–1** Three refineries with maximum daily capacities of 6 million, 5 million, and 8 million gallons of gasoline supply three distribution areas with daily demands of 4 million, 8 million, and 7 million gallons. Gasoline is transported to the three distribution areas through a network of pipelines. The transportation cost is estimated based on the length of the pipeline at about 1 cent per 100 gallons per mile. The mileage table summarized here shows that refinery 1 is not connected to distribution area 3. Formulate the problem as a transportation model.

		Distribution Area		
		1	2	3
	1	120	180	—
Refinery	2	300	100	80
	3	200	250	120

☐ **5–2** In Problem 5–1, suppose that the capacity of refinery 3 is reduced to 6 million gallons. Also, distribution area 1 must receive all its demand, and any shortage at areas 2 and 3 will result in a penalty of 5 cents per gallon. Formulate the problem as a transportation model.

☐ **5–3** In Problem 5–1, suppose that the daily demand at area 3 drops to 4 million gallons. Any surplus production at refineries 1 and 2 must be diverted to other distribution areas by trucks. The resulting average transportation costs per 100 gallons are $1.50 from refinery 1 and $2.20 from refinery 2. Refinery 3 can divert its surplus gasoline to other chemical processes within the plant. Formulate the problem as a transportation model.

☐ **5–4** Cars are shipped by truck from three distribution centers to five dealers. The shipping cost is based on the mileage between sources and destinations. This cost is independent of whether the truck makes the trip with a partial or a full load. The following table summarizes the mileage between the distribution centers and the dealers as well as the monthly supply and demand figures estimated in *number* of cars. Each truck can carry a maximum of 18 cars. Given that the transportation cost per truck mile is $10, formulate the problem as a transportation model.

		Dealers					Supply
		1	2	3	4	5	
Distribution	1	100	150	200	140	35	400
Centers	2	50	70	60	65	80	200
	3	40	90	100	150	130	150
Demand		100	200	150	160	140	

☐ **5–5** Consider Example 5.1–3, which involves the distribution of four different car models. As the example currently stands, the four models are completely independent, so that it is possible to decompose the problem into four independent transportation models. Suppose that it is possible to substitute a percentage of the demand for one model from the supply for another according to the following table:

Distribution Center	Percentage of Demand	Interchangeable Models
Denver	10	M1, M2
	20	M3, M4
Miami	10	M1, M3
	5	M2, M4

Formulate the problem as a transportation model using the cost information given in Example 5.1–3.
[*Hint:* Add four new destinations corresponding to the new combinations (M1, M2), (M3, M4), (M1, M3), and (M2, M4). The demands at the new destinations are determined from the given percentages.]

☐ **5–6** Consider the problem of assigning four different categories of machines and five types of tasks. The number of machines available in the four categories are 25, 30, 20, and 30. The number of jobs in the five tasks are 20, 20, 30, 10, and 25. Machine category 4 cannot be assigned to task type 4. For the unit costs given, formulate a mathematical model for determining the optimal assignment of machines to tasks.

			Task Type			
		1	2	3	4	5
Machine	1	10	2	3	15	9
	2	5	10	15	2	4
Category	3	15	5	14	7	15
	4	20	15	13	—	8

☐ **5–7 (Caterer Problem).** A caterer who is in charge of serving meals for the next N days must decide on the daily supply of fresh napkins. His needs for the next N days are $b_1, b_2, \ldots, b_N$ and can be satisfied by one of three alternatives,

1. Buy new napkins at p_1 cents each.
2. Send soiled napkins to a 24-hour service laundry at the cost of p_2 cents each.
3. Sent soiled napkins to a 48-hour service laundry at the cost of p_3 cents each.

This problem can be formulated as a transportation model as follows. There are $N + 1$ sources and N destinations. One of the sources represents the new napkins; the remaining N sources represent the soiled napkins at the end of each of the N days. The destinations represent the requirements of the N days.

The unit "transportation" cost is p_1 if napkins are secured from the new napkin source, p_2 if obtained through the 24-hour laundry service, and p_3 if obtained through the 48-hour laundry service.

The total supply for the new napkins source must equal the total requirements for the N days, since it is conceivable that all the demand be satisfied totally by buying new napkins. Also, the supply at the end of day i must equal the demand for the same day.

Formulate the problem as a transportation model.

☐ **5–8** Show that the solution of the transportation model remains unchanged if a constant K is added to all the cost elements of a row or a column of the cost matrix. How will this addition affect the value of the objective function?

☐ **5–9** Rework Problem 5–8 assuming that the constant K is added to every cost element.

☐ **5–10** Apply the northwest-corner rule to find a starting solution for each of the following transportation models. Indicate whether or not the solution is degenerate.

(a) $a_1 = 10, a_2 = 5, \quad a_3 = 4, a_4 = 6$
  $b_1 = 10, b_2 = 5, \quad b_3 = 7, b_4 = 3$

(b) $a_1 = 1, \quad a_2 = 16, a_3 = 7, a_4 = 8$
  $b_1 = 3, \quad b_2 = 4, \quad b_3 = 5, b_4 = 2, b_5 = 8$

(c) $a_1 = 10, a_2 = 3, \quad a_3 = 7$
  $b_1 = 14, b_2 = 3, \quad b_3 = 4, b_4 = 9$

☐ **5–11** Consider the basic solution in Table 5–10.

(a) Determine the loop associated with each nonbasic variable.
(b) If each nonbasic variable is used as an entering variable, determine the associated leaving variable. At what level does each nonbasic variable enter the solution?
(c) For each case in part (b), determine the *total* increase or decrease in the value of the objective function.

☐ **5–12** Solve the following transportation models whose starting solutions are degenerate. Use the northwest-corner method to find the starting solution. (The numbers in the box give c_{ij}.)

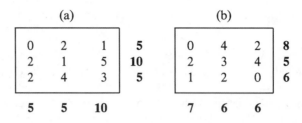

(a)   (b)

□ **5–13** Obtain the optimum solution for each of the following transportation problems.

    (a) Problem 5–1.
    (b) Problem 5–2.
    (c) Problem 5–3.
    (d) Problem 5–4.
    (e) Problem 5–6.
    (f) Problem 5–7. Assume that $p_1 = 50$ cents, $p_2 = 10$ cents, $p_3 = 5$ cents, and $N = 4$ days. The demands for the four days are 110, 210, 190, and 100 napkins, respectively.
    (g) Example 5.1–4.

□ **5–14** In the following transportation problem, the total demand exceeds total supply. Suppose that the penalty costs per unit of unsatisfied demand are 5, 3, and 2 for destinations 1, 2, and 3. Find the optimal solution.

5	1	7	**10**
6	4	6	**80**
3	2	5	**15**

    75   **20**   **50**

□ **5–15** In Problem 5–14, suppose that there are no penalty costs and the demand at destination 3 must be satisfied exactly. Reformulate the problem and find the optimal solution.

□ **5–16** In the unbalanced transportation problem given, if a unit from source i is not shipped out (to one of the destinations), a storage cost must be incurred. Let the storage costs per unit at sources 1, 2, and 3 be 5, 4, and 3. If, in addition, all the supply at source 2 must be shipped out to make room for a new product, find the optimal solution.

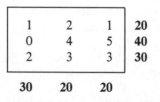

1	2	1	**20**
0	4	5	**40**
2	3	3	**30**

    30   **20**   **20**

□ **5–17** In a (3×3) transportation problem let x_{ij} be the amount shipped from source i to destination j and c_{ij} the corresponding per unit transportation cost. The supplies at sources 1, 2, and 3 are 15, 30, and 85 units, and the demands at destinations 1, 2, and 3 are 20, 30, and 80 units. Assume that the starting solution obtained by the northwest-corner method gives the *optimal* basic solution to the problem. Let the associated values of the multipliers for sources 1, 2, and 3 be −2, 3, and 5, and those for destinations 1, 2, and 3 be 2, 5, and 10.

(a) Find the total optimal transportation cost.

(b) What are the smallest values of c_{ij} for the nonbasic variables that will keep the solution optimal?

☐ **5–18** The transportation problem given here has the indicated *degenerate* basic solution. It is required to minimize the transportation costs. Let the multipliers corresponding to this basic solution be 1, −1 for sources 1 and 2 and −1, 2, −5 for destinations 1, 2, and 3. Let

$$c_{ij} = i + j\theta, \qquad -\infty < \theta < \infty$$

for *all* the *zero* (basic and nonbasic) variables, where c_{ij} is the cost *per unit* shipped from source i to destination j.

(a) If the solution is the optimal, what is the corresponding value of the objective function? (Answer this part in two different ways.)

(b) Under the conditions in part (a), find the single value of θ for which the solution is basic and optimal.

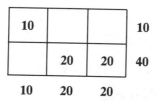

☐ **5–19** Solve this problem by the transportation technique and the simplex method and show that there is a one-to-one correspondence between the iterations of the two methods. Find the starting solution by the northwest-corner method.

1	0	2	4
3	5	4	6
1	2	3	10
3	5	12	

☐ **5–20 (Sensitivity Analysis).** Consider the optimal transportation tableau in Table 5–13. In each of the following cases, determine the change in the amounts of the supply and demand that will keep the current solution feasible.

(a) Source 1 and destination 1.

(b) Source 2 and destination 1.

(c) Source 2 and destination 4.

(d) Source 3 and destination 4.

[*Hint:* Suppose that we change the supply and demand at source i and destination j from a_i and b_j to $a_i + \Delta$ and $b_j + \Delta$, where Δ is unrestricted. If x_{ij} is currently basic, Δ is added directly to the current value of x_{ij} and the solution remains feasible as long as the new value of x_{ij} remains nonnegative. If x_{ij} is nonbasic, construct its loop and from it determine the adjustment in the values of basic variables at the corners of the loop that will keep the solution feasible while netting the value of the nonbasic x_{ij} from Δ to zero.]

□ **5–21** Solve each of the following transportation models by using the north-west-corner method, least-cost method, and Vogel's approximation method to obtain the starting solution. Compare the computations.

(a)   (b)

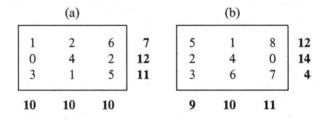

1	2	6	**7**
0	4	2	**12**
3	1	5	**11**
10	**10**	**10**	

5	1	8	**12**
2	4	0	**14**
3	6	7	**4**
9	**10**	**11**	

□ **5–22** Find the starting solution in the following transportation problem by the (a) northwest-corner method, (b) the least-cost method, (c) Vogel's approximation method. Obtain the optimal solution by using the best starting solution.

10	20	5	7	**10**
13	9	12	8	**20**
4	15	7	9	**30**
14	7	1	0	**40**
3	12	5	19	**50**
60	**60**	**20**	**10**	

□ **5–23** Solve the following unbalanced transportation problem using VAM to find the starting solution. The demand at destination 1 must be shipped from source 4.

5	1	0	**20**
3	2	4	**10**
7	5	2	**15**
9	6	0	**15**
5	**10**	**15**	

□ **5–24** Show by the method of multipliers (Section 5.2.1) that the solution in Table 5–25 is optimal.

□ **5–25** Solve the following assignment models.

(a)

3	8	2	10	3
8	7	2	9	7
6	4	2	7	5
8	4	2	3	5
9	10	6	9	10

(b)

3	9	2	3	7
6	1	5	6	6
9	4	7	10	3
2	5	4	2	1
9	6	2	4	6

☐ **5–26** Consider the problem of assigning four operators to four machines. The assignment costs in dollars are given. Operator 1 cannot be assigned to machine 3. Also, operator 3 cannot be assigned to machine 4. Find the optimal assignment.

	Machine 1	2	3	4
Operator 1	5	5	—	2
2	7	4	2	3
3	9	3	5	—
4	7	2	6	7

☐ **5–27** Suppose that in Problem 5–26 a fifth machine is made available. Its respective assignment costs (in dollars) to the four operators are 2, 1, 2, and 8. The new machine replaces an existing one if the replacement can be justified economically. Reformulate the problem as an assignment model and find the optimal solution. In particular, is it economical to replace one of the existing machines? If so, which one?

☐ **5–28** An airline has two-way flights between two cities A and B. The crew based in city A (B) and flying to city B (A) must return to city A (B) on a later flight either on the same day or a following day. An A-based crew can return on an A-destined flight only if there is at least 90 minutes between the arrival time at B and the departure time of the A-destined flight. The objective is to pair the flights so as to minimize the total layover time by all the crews. Solve the problem as an assignment model using the timetable given.
[*Note:* The complete formulation of a similar problem is given in R. Ackoff and M. Sasieni, *Fundamentals of Operations Research,* John Wiley, New York, 1968, pp. 143–145.]

Flight	From A	To B	Flight	From B	To A
1	6:00	8:30	10	7:30	9:30
2	8:15	10:45	20	9:15	11:15
3	13:30	16:00	30	16:30	18:30
4	15:00	17:30	40	20:00	22:00

☐ **5-29** Find the shortest route between nodes 1 and 7 of the network in Figure 5-6 by formulating the problem as a transshipment model. The distances between the different nodes are indicated on the network.

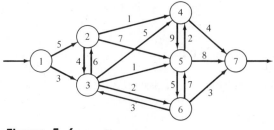

Figure 5-6

☐ **5-30** Consider the transportation problem where two factories are supplying three retail stores with a certain commodity. The number of units available at factories 1 and 2 are 200 and 300; those demanded at stores 1, 2, and 3 are 100, 200, and 50. Rather than ship directly from sources to destinations, it is decided to investigate the possibility of transshipment. Find the optimal shipping schedule. The transportation costs per unit are given.

		Factory		Store		
		1	2	1	2	3
Factory	1	0	6	7	8	9
	2	6	0	5	4	3
Store	1	7	2	0	5	1
	2	1	5	1	0	4
	3	8	9	7	6	0

☐ **5-31** Consider Figure 5-7, which represents an oil pipeline network. The different nodes represent pumping and/or receiving stations. The lengths in miles of the different segments of the network are shown on the respective arcs. Construct the transshipment model for determining the least-cost transportation schedule between pumping stations 1 and 3 and receiving stations 2 and 4. The amounts of supply and demand are shown directly in the network. Assume the transportation cost per gallon pumped through the pipeline to be directly proportional to the length of the pipeline between the source and destination.

☐ **5-32** In the transportation model, one of the dual variables assumes an arbitrary value. Thus for the same basic solution the values of the associated dual variables are not unique. This appears to contradict the theory of linear programming where the simplex multipliers are determined by the product of the vector of the objective coefficients for the basic variables and the associated inverse basic matrix (see Section 4.2.2).

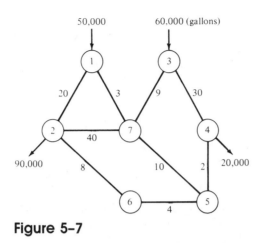

Figure 5-7

Show that for the transportation model although a given inverse matrix is uniquely defined, the vector of *basic* objective coefficients may not be unique for the same problem. Specifically, show that if c_{ij} is changed to $c_{ij} + k$ for all i and j, where k is a constant, the optimal values of x_{ij} remain the same. Thus an arbitrary assignment of a value to a dual variable is implicitly equivalent to assuming that a certain constant k is added to all c_{ij}.

☐ **5–33 (Unbalanced Transportation Problem).** Consider the problem

$$\text{minimize } z = \sum_{i=1}^{m} \sum_{j=1}^{n} c_{ij} x_{ij}$$

subject to

$$\sum_{j=1}^{n} x_{ij} \geq a_i, \qquad i = 1, 2, \ldots, m \tag{1}$$

$$\sum_{i=1}^{m} x_{ij} \geq b_j, \qquad j = 1, 2, \ldots, n \tag{2}$$

$$x_{ij} \geq 0 \qquad \text{for all } i \text{ and } j$$

It may be logical to assume that, at the optimum, inequality (1) *or* (2) will certainly be satisfied in equation form depending on whether $\Sigma \, a_i \geq \Sigma \, b_j$ or $\Sigma \, a_i \leq \Sigma \, b_j$, respectively. A counterexample to this hypothesis is

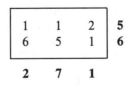

Show that the application of the suggested procedure yields the "optimum" solution $x_{11} = 2$, $x_{12} = 3$, $x_{22} = 4$, and $x_{23} = 2$ with $z = 27$, which is worse than the feasible solution $x_{11} = 2$, $x_{12} = 7$, and $x_{23} = 6$ with $z = 15$. [This

counterexample is due to A. Charnes, F. Glover, and D. Klingman, "A Note on a Distribution Problem," *Operations Research,* Vol. 18, 1970, pp. 1213–1216.]

SHORT CASES

Case 5–1

Ten years ago, a wholesale dealer started a business distributing pharmaceuticals from a central warehouse (*CW*). Orders were delivered to customers by vans. The warehouse has since been expanded to accommodate growing increase in demand. Additionally, two new warehouses (*W*1 and *W*2) have been constructed. The central warehouse, traditionally well stocked, started by occasionally supplying the new warehouses with some of their short items. Currently, the occasional supply of short items has grown into a large-scale operation in which the two new warehouses receive for redistribution about one third of their stock directly from the central warehouse. The following table summarizes the distribution data among the warehouses and the customer locations. A customer location can be thought of as a town that in reality includes several pharmacies:

Route	Number of Orders	Distance (in miles)	Transportation Cost ($/order)
*CW–W*1	2000	5	.08
*W*2	1500	45	.17
*C*1	4800	50	.21
*C*2	3000	30	.13
*C*3	1200	30	.11
*W*1–*C*1	1000	38	.16
*C*3	1100	30	.15
*C*4	1500	8	.06
*C*5	1800	10	.09
*W*2–*C*2	1900	35	.16
*C*5	600	25	.13
*C*6	2200	7	.07

The company's delivery schedule has evolved over the years to its present status. In essence, the schedule was devised in a rather decentralized fashion with each warehouse determining its delivery zone based on certain "self-fulfilling" criteria. Indeed, in some instances, warehouse managers competed for new customers mainly to increase their "sphere of influence" compared to the other warehouses. For instance, the manager of the central warehouse boasts that her delivery zone includes not only regular customers but also the other two warehouses as well. It is not unusual then that supplies to different pharmacies within the same town (customer location) be delivered from more than one warehouse.

Evaluate the present distribution policy of the company.

Case 5–2†

A school district has 10 special education teachers that must be assigned to 10 different schools. A teacher spends the morning period in one school and the evening period in another.

In devising the assignment, teachers would specify their preferences by listing the schools where they wish or do not wish to work. The schools can also specify teachers they prefer for their schools. Such lists are based on past evaluations of teacher's performances. To accommodate the personal conveniences of the teachers, the developed schedule will attempt to assign teachers to schools that are "close" to their place of residence.

Propose a model for this assignment problem specifying the types of data needed to work the problem.

Case 5–3‡

A business executive who is stationed in city A must make six round trips between cities A and B according to the following schedule:

Departure Date from City A	Return Date to City A
Monday, June 3	Friday, June 7
Monday, June 10	Wednesday, June 12
Monday, June 17	Friday, June 21
Tuesday, June 25	Friday, June 28

The basic price of a round-trip air ticket between A and B is $400. A discount of 25% is granted if the dates of arrival and departure span a weekend (Saturday and Sunday). If the stay in B lasts more than 21 days, a 30% discount can be obtained. A one-way ticket from A to B (or B to A) costs $250. How should the executive purchase the tickets?

---

† This case is based on S. M. Lee and M. J. Schniederjuns, "Multicriteria Assignment Problem: A Goal Programming Approach," *Interfaces,* Vol. 13, no. 4, August 1983, pp. 75–81.

‡ This case is based on P. Hansen and R. Wendell, "A Note on Airline Commuting," *Interfaces,* Vol. 11, no. 12, February 1982, pp. 85–87.

Chapter 6

Linear Programming: Networks

In the preceding chapter, we limited our attention to transportation (or distribution) problems that deal with shipping commodities between sources and destinations at minimum transportation costs. The transportation model (and its variants) is but one of the many problems that can be represented and solved as a network. To be specific, consider the following situations:

a. Design of offshore natural-gas pipeline network connecting wellheads in the Gulf of Mexico with an onshore delivery point with the objective of minimizing the cost of constructing the pipeline.

b. Determination of the shortest route joining two cities in an existing network of roads.

c. Determination of the maximum annual capacity in tons of a coal slurry pipeline network joining the coal mines in Wyoming with the power plants in Houston. (Slurry pipelines transport coal by pumping water through suitably designed pipes operating between the coal mines and the desired destination.)

d. Determination of the *minimum-cost* flow schedule from oil fields to refineries and finally to distribution centers. Crude oils and gasoline products can be shipped via tankers, pipelines, and/or trucks. In addition to maximum supply availability at the oil fields and minimum demand requirements at the distribution centers, restrictions on the capacity of the refineries and the modes of transportation must be taken into account.

A study of this representative list reveals that network optimization problems can generally be modeled by one of four models:

1. Network minimization model (situation a).
2. Shortest-route model (situation b).
3. Maximum-flow model (situation c).
4. Minimum-cost capacitated network model (situation d).

The samples cited previously deal with the determination of distances and material flow in a literal sense. There are many applications where the variables of the problem can represent other properties such as inventory or money flow. Examples illustrating these situations will be given in this chapter.

The network models listed can be represented and, in principle, solved as linear programs (see Section 6.4). However, the tremendous number of variables and constraints that normally accompanies a typical network model makes it inadvisable to solve network problems directly by the simplex method. The special structure of these problems allows the development of highly efficient algorithms, which in most cases are based on linear programming theory.

From the practical standpoint, the minimum-cost capacitated network model enjoys a wide variety of application. Indeed, both the shortest-route and maximum-flow problem can be formulated as special cases of the capacitated transportation model. However, the method of solving the capacitated model, called the **out-of-kilter** algorithm, involves considerable computational details that, we feel, should be addressed in specialized books. For this reason, we shall limit the discussion in this chapter on the first three models cited. Our objective is to demonstrate the simplicity of solving network models by specialized algorithms. You may wish to refer to the reference list for more details on the subject.

6.1   NETWORK MINIMIZATION

Network minimization deals with the determination of the branches that can join *all* the nodes of a network (i.e., every pair of nodes is connected by a *chain*) such that the sum of the lengths of the chosen branches is minimized. It is obvious that it is not optimum to include **loops** (or **cycles**) in the solution to the problem. Figure 6–1 illustrates this point. The lengths of the branches connecting nodes 1, 2, and 3 are shown on the respective branches. It is clear that the minimum network occurs when node 3 is connected to both 1 and 2, giving a total minimum length of the branches equal to $4 + 6 = 10$. If nodes

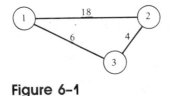

Figure 6-1

1 and 2 are connected, a loop will result and the resulting network is obviously not minimized.

The absence of loops in a minimum network has led to coining the suggestive name **minimal spanning tree.** In any network, the minimal spanning tree is determined iteratively as follows. Start with any node and join it to its *closest* node in the network. The resulting two nodes now form a *connected set,* and the remaining nodes comprise the *unconnected set.* Next, choose a node from the unconnected set that is *closest* (has the shortest distance) to *any* node in the connected sets and add it to the connected set. Redefining the connected and unconnected sets accordingly, the process is repeated until the connected set includes all the nodes of the network. Any tie may be broken arbitrarily. Ties, however, indicate the existence of alternative minimal spanning trees.

Example 6.1-1. The Midwest TV Cable Company is in the process of planning a network for providing cable TV service to five new housing development areas. The cable system network is summarized in Figure 6-2. The numbers associated with each branch represent the miles of cable needed to connect any two locations. Node 1 represents the cable TV station and the remaining nodes (2 through 6) represent the five development areas. A missing branch between two nodes implies that it is prohibitively expensive or physically impossible to connect the associated development areas. It is required to determine the links that will result in the use of minimum cable miles while guaranteeing that all areas are connected (directly or indirectly) to the cable TV station.

The graphical solution is summarized in Figure 6-3 by iterations. The procedure can be started from any node, always ending up with the same optimum solution. In the cable TV example, it is logical to start the computations with

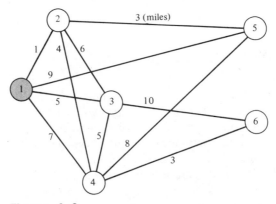

Figure 6-2

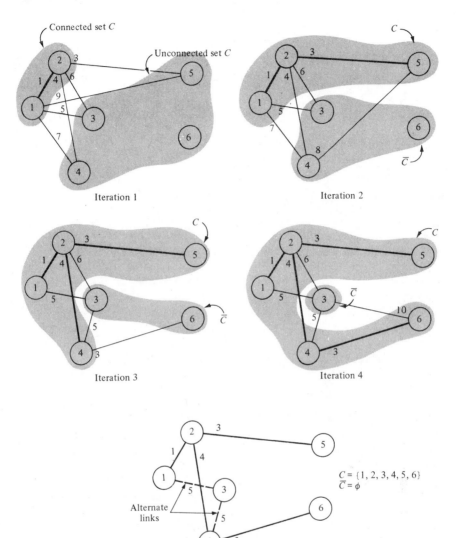

Figure 6-3

node 1. Thus node 1 represents the set of "connected nodes." The set of "uncon-nected nodes" is represented by nodes 2, 3, 4, 5, and 6. Symbolically, we write this as

$$C = \{1\}, \qquad \bar{C} = \{2, 3, 4, 5, 6\}$$

Iteration 1

Node 1 must be connected to node 2, the closest node in $\bar{C} = \{2, 3, 4, 5, 6\}$. Iteration 1 of Figure 6–3 thus shows that

$$C = \{1, 2\}, \qquad \bar{C} = \{3, 4, 5, 6\}$$

Iteration 2

Nodes 1 and 2 (of the set C) are now linked permanently. In iteration 2 we select a node in $\bar{C} = \{3, 4, 5, 6\}$ that is closest to a node in $C = \{1, 2\}$. Since the shortest distance occurs between 2 and 5 (see iteration 1 of Figure 6–3), we have

$$C = \{1, 2, 5\}, \qquad \bar{C} = \{3, 4, 6\}$$

Iteration 3

Iteration 2 of Figure 6–3 gives the distances from the nodes of $C = \{1, 2, 5\}$ to all the nodes in $\bar{C} = \{3, 4, 6\}$. Thus nodes 2 and 4 are connected, which yields

$$C = \{1, 2, 4, 5\}, \qquad \bar{C} = \{3, 6\}$$

Iteration 4

Iteration 3 of Figure 6–3 shows that nodes 4 and 6 must be connected. Thus, we obtain

$$C = \{1, 2, 4, 5, 6\}, \qquad \bar{C} = \{3\}$$

Iteration 5

In iteration 5 we have a tie that may be broken arbitrarily. This means that we can connect 1 and 3 *or* 4 and 3. Both (alternative) solutions lead to

$$C = \{1, 2, 3, 4, 5, 6\}, \qquad \bar{C} = \emptyset$$

Since all the nodes are connected, the procedure is complete. The minimum cable miles used to connect the development areas to the TV station equal $1 + 3 + 4 + 3 + 5 = 16$ miles. ◀

Exercise 6.1–1

Solve Example 6.1–1 using node 4 as the initial connected set; that is, initial $C = \{4\}$. Follow the graphical procedure of Figure 6–3.

[*Ans.* The successive iterations will lead to connecting 4 to 6, 4 to 2, 2 to 1, 2 to 5, and finally, 1 to 3 or 4 to 3. This is the same solution obtained previously, thus demonstrating that the specific choice of the initial set C is arbitrary.]

6.2 SHORTEST-ROUTE PROBLEM

In the obvious sense, the shortest-route problem deals with determining the *connected* roads in a transportation network that collectively comprise the shortest distance between a source and a destination. This section presents other types of applications that can be modeled and solved as a shortest-route problem. The applications are followed by a presentation of the solution algorithms.

6.2.1 EXAMPLES OF THE SHORTEST-ROUTE APPLICATIONS

Example 6.2–1 (Equipment Replacement). A car rental company is developing a replacement plan for its fleet over the next 5 years. A car must be in service for at least 1 year before replacement is considered. Table 6–1 summarizes the replacement cost per car (in thousands of dollars) as a function of time and the number of years in operation. The cost includes purchasing, salvage, operating, and maintenance.

Table 6–1

Year	1	2	3	4	5
1		4.0	5.4	9.8	13.7
2			4.3	6.2	8.1
3				4.8	7.1
4					4.9

This problem can be represented by a network as follows. Each year is represented by a node. The "length" of an arc joining two nodes equals the associated replacement cost given in Table 6–1. Figure 6–4 depicts the network. The problem reduces to finding the shortest "route" from node 1 to node 5.

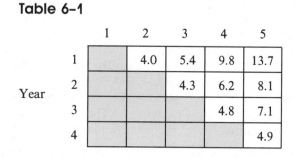

Figure 6–4

The shortest "route" can be determined using the algorithm we will present in Section 6.2.2. The optimal solution will yield the route $1 \rightarrow 2 \rightarrow 5$ with a total cost of $4 + 8.1 = 12.1$ (thousands of dollars). This means that each car should be replaced in year 2 and discarded in year 5.   ◀

Example 6.2–2 (Most Reliable Route).† Ms. I. Q. Smart has to drive daily between her residence and place of work. Having just taken a class in network analysis, she was able to determine the shortest route to work. To her disappointment, she discovered that her shortest route was heavily patrolled by police, who were always stopping her for speeding violations unjustly (or so it seemed, particularly when she was late for work).

_____

† This example paraphrases an industrial application proposed by S. E. Elmaghraby [1970].

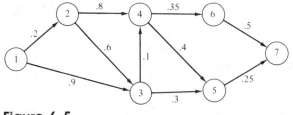

Figure 6-5

   With all the fines she was paying (and the fact that she could not get herself to wake up early enough to reach work on time without exceeding the speed limit), she reached the conclusion that her shortest route is obviously not the most economical. She thus decided to study the problem from a different angle. Ms. Smart would like to choose her route such that the total probability of *not* being stopped by police is maximized. Observing all the feasible road segments between her residence and work, she compiled the probabilities shown on the different arcs (road segments) of Figure 6-5. (The arcs in this case are said to be **directional** or **oriented,** since Ms. Smart must follow certain directions to reach work.)

   Investigating the probability information, she realized that the total probability of *not* being stopped by police on a given route equals the product of the probabilities associated with the road segments comprising the chosen route. For example, the probability associated with the route $1 \to 2 \to 3 \to 5 \to 7$ is $.2 \times .6 \times .3 \times .25 = .009$. Although it is possible to compute all such probabilities (eight different routes in this case), Ms. Smart decided to convert the problem to a shortest-route model by using the following conversion. Letting $P_{1k} = P_1 \times P_2 \times \cdots \times P_k$ be the probability of nosm tabg stopped in the specific route $(1, k)$, then

$$\log P_{1k} = \log P_1 + \log P_2 + \cdots + \log P_k$$

Table 6-2

Road Segment (i, j)	P_{ij}	$\log P_{ij}$	$-\log P_{ij}$
(1, 2)	.2	−.69897	.69897
(1, 3)	.9	−.04576	.04576
(2, 3)	.6	−.22185	.22185
(2, 4)	.8	−.09691	.09691
(3, 4)	.1	−1.0	1.0
(3, 5)	.3	−.52288	.52288
(4, 5)	.4	−.39794	.39794
(4, 6)	.35	−.45593	.45593
(5, 7)	.25	−.60206	.60206
(6, 7)	.5	−.30103	.30103

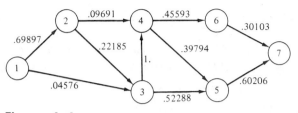

Figure 6-6

A maximization of P_{1k} is algebraically equivalent to maximizing log P_{1k} and, consequently, to maximizing the *sum* of the logarithms of the individual probabilities along the chosen route. Since log $P_j \leq 0$, $j = 1, 2, \ldots, k$, *maximizing the sum of* log P_j is equivalent to *minimizing* the sum of $(-$log $P_j)$. Table 6–2 summarizes the probabilities of Figure 6–5 and their logarithms. Figure 6–6 now expresses Ms. Smart's problem as a shortest-route model.   ◄

Exercise 6.2-1
Suppose that the information in Figure 6–5 represents the probabilities of being stopped by police. Can the same type of analysis be used to choose the route with the *smallest* probability of being stopped?
[*Ans.* No, because the probability of being stopped on a route no longer equals the product of the probabilities of the individual road segments. Specifically, it will equal the complement of the probability of *not* being stopped.]

6.2.2   SHORTEST-ROUTE ALGORITHMS FOR ACYCLIC NETWORKS

In this section we present an algorithm for finding the shortest route in **acyclic** networks. The algorithm is quite simple, but it presents the fundamental concept of **recursive computations,** which is the basis for the dynamic programming calculations we present in Chapter 9.

We first develop the algorithm through a numerical example. The procedure is then explained from the standpoint of recursive computations.

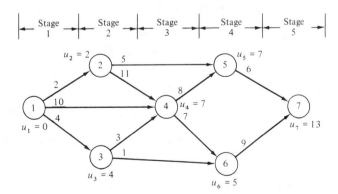

Figure 6-7

Example 6.2–3. Consider the network in Figure 6–7. Node 1 is the starting point (origin) and node 7 is the terminal point (destination). Notice that the network is *acyclic*, since there exist no chains connecting a node to itself.

Before presenting the solution procedure, we need the following definitions:

$$d_{ij} = \text{network distance between } adjacent \text{ nodes } i \text{ and } j$$

$$u_j = shortest \text{ distance from node 1 to node } j, \qquad u_1 = 0$$

The procedure is complete when u_7 is computed. The general formula for computing u_j is

$$u_j = \min_i \left(\begin{array}{c} \text{shortest distance to an } immediately\ preceding \text{ node } i \\ plus \\ \text{distance between present node } j \text{ and its predecessor } i \end{array} \right)$$

$$= \min_i \{u_j + d_{ij}\}$$

The formula implies that the shortest distance u_j to node j can be computed only after we compute the shortest distance to each predecessor node i linked to j by an arc.

Starting the computations at node 1, we see that only u_2 and u_3 can be computed at this point. (Although 4 is linked to 1, its u_4 cannot be computed until u_2 and u_3 are known.) The sequence of computations proceeds in stages as follows:

Stage 1: $u_1 = 0$

Stage 2: $u_2 = u_1 + d_{12} = 0 + 2 = 2$ (from 1)

$\qquad\quad u_3 = u_1 + d_{13} = 0 + 4 = 4$ (from 1)

Stage 3: $u_4 = \min \{u_1 + d_{14},\ u_2 + d_{24},\ u_3 + d_{34}\}$

$\qquad\qquad = \min \{0 + 10,\ 2 + 11,\ 4 + 3\} = 7$ (from 3)

Stage 4: $u_5 = \min \{u_2 + d_{25},\ u_4 + d_{45}\}$

$\qquad\qquad = \min \{2 + 5,\ 7 + 8\} = 7$ (from 2)

$\qquad\quad u_6 = \min \{u_3 + d_{36},\ u_4 + d_{46}\}$

$\qquad\qquad = \min \{4 + 1,\ 7 + 7\} = 5$ (from 3)

Stage 5: $u_7 = \min \{u_5 + d_{57},\ u_6 + d_{67}\}$

$\qquad\qquad = \min \{7 + 6,\ 5 + 9\} = 13$ (from 5)

The minimum distance from 1 to 7 is 13 and follows the route $1 \to 2 \to 5 \to 7$. Notice also that the solution yields the shortest distance from 1 to each of the remaining nodes in the network.

You can actually carry out the computations directly on the network, as shown in Figure 6–7. The value of u_j for node j is computed only after computing u_i for all nodes i immediately preceding j. Thus, starting with $u_1 = 0$, we obtain $u_2 = 2$ and $u_3 = 4$. Next, $u_4 = 7$ can be computed. At this point both $u_5 = 7$ and $u_6 = 5$ are computed. The final step computes $u_7 = 13$.   ◄

Exercise 6.2–2

(a) In Figure 6–7, compute the shortest distance and its designated route to each of the following nodes.

(1) Node 4

[*Ans.* $u_4 = 7$, $1 \rightarrow 3 \rightarrow 4$.]

(2) Node 6

[*Ans.* $u_6 = 5$, $1 \rightarrow 3 \rightarrow 6$.]

(b) Recompute the shortest route in Figure 6–7 when each of the following changes is effected independently.

(1) Node 4 is connected to 7 by an arc of length 5.

[*Ans.* $u_7 = 12$, $1 \rightarrow 3 \rightarrow 4 \rightarrow 7$.]

(2) Node 5 is reached from 6 by an arc of length 2.

[*Ans.* $u_7 = 13$, $1 \rightarrow 2 \rightarrow 5 \rightarrow 7$ or $1 \rightarrow 3 \rightarrow 6 \rightarrow 5 \rightarrow 7$.]

The type of calculations outlined above is interesting because it deals with **recursive computations.** This is characterized by the use of information that summarizes the shortest distances up to the *immediately preceding* node. For example, at node 5, u_5 is computed based on the shortest distances from node 1 to nodes 2 and 4, namely, u_2 and u_4. Note that it is never necessary to know the specific route leading to the shortest distance from 1 to 4. The value of u_4 summarizes all the information we need to know about node 4. This type of summarization is what permits the use of recursive computation. As we shall see in Chapter 9, recursive computation is the basic concept underlying the development of dynamic programming algorithms.

6.2.3 THE SHORTEST-ROUTE PROBLEM VIEWED AS A TRANSSHIPMENT MODEL

We can formulate the shortest-route problem as a transshipment model (see Section 5.4). We can think of the shortest-route network as a transportation model with one source and one destination. The supply at the source is one unit and the demand at the destination is also one unit. The one unit will flow from the source to the destination through the admissable routes of the network. The objective is to minimize the distance traveled by the unit flow from the source to the destination.

To illustrate the construction of the model, we consider the network of Figure 6–8. Unlike the acyclic algorithim, which automatically computes the shortest distances between node 1 and all the other nodes, the transshipment model computes only the shortest distance between two nodes. Thus, assuming that we are interested in determining the shortest distance between nodes 1 and 7, Table 6–3 gives the associated transshipment model of the problem. Note that the buffer stock B (see Section 5.4) equals 1, since at any time during transshipment no more than one unit can pass through any of the nodes of the network. Note also that node 1 does not appear as a transient destination, since it is the (main) source for the network. Similarly, node 7 cannot act as a transient source, since it represents the final destination of the unit flow. The "transportation costs" equal the associated distances. Blocked cells imply that the corre-

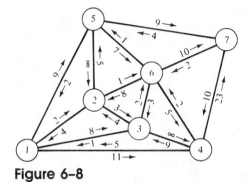

Figure 6-8

sponding route does not exist and must be assigned a very high cost M when solving the model. Finally, the distance from a node to itself is zero.

Table 6–3 also gives the optimum solution, which is obtained by using the transportation technique (Section 5.2.1). The table shows that

$$x_{12} = 1, \quad x_{26} = 1, \quad x_{33} = 1, \quad x_{44} = 1, \quad x_{57} = 1, \quad x_{65} = 1$$

The values of $x_{33} = x_{44} = 1$ do not contribute to the solution since they connect nodes 3 and 4 to themselves. The remaining values can be arranged in the order

$$x_{12} = 1, \quad x_{26} = 1, \quad x_{65} = 1, \quad x_{57} = 1$$

which shows that the optimal route is $1 \to 2 \to 6 \to 5 \to 7$, as obtained previously. (The optimality condition of the transportation will show that an alternate optimum solution exists between nodes 1 and 7. Verify.)

Table 6-3

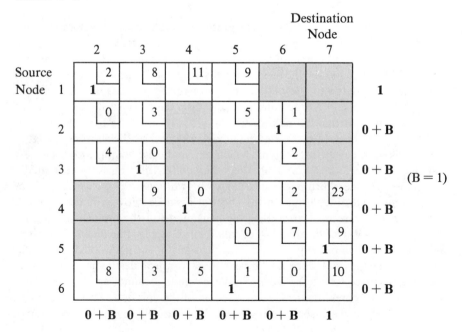

		Destination Node						
		2	3	4	5	6	7	
Source Node 1	**1**	2	8	11	9			1
2		0	3		5	1 **1**		0 + B
3		4	0 **1**			2		0 + B
4			9	0 **1**		2	23	0 + B
5					0	7	9 **1**	0 + B
6		8	3	5	1 **1**	0	10	0 + B
		0 + B	0 + B	0 + B	0 + B	0 + B	1	

(B = 1)

Exercise 6.2–3

Show how you can construct the transshipment model associated with finding the shortest route between nodes 6 and 4.

[*Ans.* The transshipment model will have six sources and six destinations, with no row associated with node 4 (destination node) and no column with node 6 (source node). All nodes will each have one unit of supply or demand.]

If you investigate Table 6–3 closely, you will discover that it has the structure of an assignment model. (Section 5.3). This suggests that it may be possible to formulate the assignment problem as a shortest-route problem. Although this is true, the involved computations are usually more tedious than solving the assignment model directly. The relationship is theoretically interesting, however.

6.3   MAXIMAL-FLOW PROBLEM

This section considers the situation in which a source node and a destination node are linked through a network of unidirectional (or one-way) arcs. Each arc has maximum allowable flow capacity. The objective is to find the maximum amount of flow between the source and the destination. An example of this situation is the case where a number of refineries are connected to distribution terminals through a network of pipelines. Booster and pumping stations are mounted on the pipelines to move the oil products to the distribution terminals. The objective is to maximize the flow between the refineries and the distribution terminals within the capacity limitations of the refineries and the pipelines.

Figure 6–9 illustrates the refinery maximum-flow problem. Here we add a single source that connects to all the refineries and a single sink that receives flow from all the distribution terminals. The nodes between the refineries and the terminal distributions represent the pumping stations. The capacities of the arcs from the single source represent the maximum outputs of the refineries. Each pipeline has a maximum design capacity that determines the maximum flow allowable in the line. The arcs from the distribution terminals to the sink are assumed to have infinite capacities to allow the determination of the maxi-

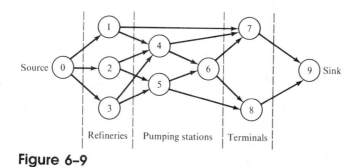

Figure 6–9

mum flow from the refineries. In some cases, we may want to use the demands at the terminals as the capacities of the arcs to the sink.

Example 6.3-1. In this example we introduce the method of computing the maximal flow in a unidirectional network. In Figure 6–10, we need to determine the maximal flow between node 1 and node 5. The minimum and maximum capacities are shown in parentheses on each arc. The algorithm we introduce here is a simple **labeling procedure** in which the computations are carried directly on the network.

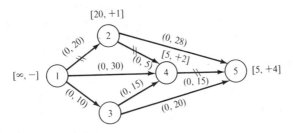

Figure 6-10

The labeling procedure starts from the source node and attempts to find a path that connects it to the sink node. The basic condition on the selection of the path is that it allows a positive flow from the source to the sink. The procedure utilizes labels to commemorate the progress of the computations from one node to the next. A label consists of two elements $[f, n]$, where f is the amount of flow that can be secured from node n across an arc to a succeeding node. The amount f is thus the minimum of flow availability at node n and residue (unused) capacity of the arc leading to the succeeding node.

Applying the proposed idea to the network in Figure 6–10, we start by labeling the source node 1 with $[\infty, -]$, which indicates that the node has no predecessor and hence, theoretically, can release unlimited amount of flow. To find the path of arcs leading to the sink node 5, we choose any of the arcs leaving node 1 (provided the arc has a positive residue capacity). Selecting arc 1–2 arbitrarily, we label node 2 with $[20, +1]$. From node 2, we arbitrarily select arc 2–4. Arc 2–4 has a residue capacity of 5 units that, together with the label of node 2, would label node 4 with $[5, +2]$. Finally, node 5 is labeled with $[5, +4]$. Notice that although arc 4–5 has a capacity of 15 units, node 4 cannot release more than 5 units.

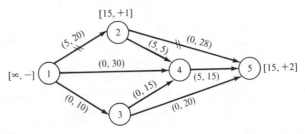

Figure 6-11

The label [5, +4] at node 4 says that the selected path 1–2–4–5 can transport 5 units. We thus update the network to show the used capacity along the indicated path. Figure 6–11 shows the new capacities by raising the lower bound on all the arc of the path 1–2–4–5 by 5 units.

A new labeled path for Figure 6–11 is given by 1–2–5, which changes the minimum flow of the network as shown in Figure 6–12. Given the new flows in Figure 6–12, the path 1–4–5 can be constructed.

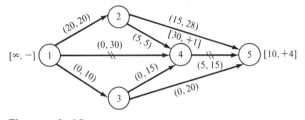

Figure 6–12

Again, Figure 6–13 updates the minimum capacities and specifies a new path, namely, 1–3–5.

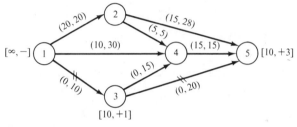

Figure 6–13

In Figure 6–14, we update the network capacities along the path 1–3–5.

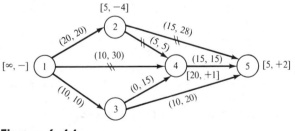

Figure 6–14

The situation in Figure 6–14 differs from those in the preceding networks. Although arc 1–4 still has residue capacity (= 20 units), the only arc out of node 4 is fully utilized. This, however, does not mean that we have reached the end of the calculations. Consider the following possibility: By "shutting off" arc 2–4, we automatically acquire a residue of 5 units in arc 4–5, which we can use to carry a 5-unit flow from node 1 through arc 1–4. Meanwhile, to keep the proper material balance at node 2, we divert the original 5 units of arc 2–4 to arc 2–5 (which has a 13-unit residue capacity). The result of this swapping process is a net flow increase of 5 units between source and sink.

This case can be summarized by introducing the concept of *negative flow* in segments of a selected source–sink path. The path 1–4–2–5 shows that that flow in arc 2–4 occurs in the opposite direction. This special case is indicated by labeling node 2 with [5, −4] with the negative sign commemorating the negative flow from node 4. The opposite flow, in essence, allows us to cancel a previously specified flow in favor of a more advantageous situation, namely, increasing the net flow from source to sink.

The final network update is shown in Figure 6–15. The maximum flow between source and sink is 20 + 15 + 10 = 45 units.

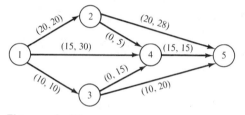

Figure 6–15

A rule of thumb that enhances the labeling procedure is to always move along the arc that has the largest residue capacity. We have purposely not done so in this example to be able to introduce the case of negative labeling (Figure 6–14).   ◀

Exercise 6.2-4

(a) Apply the rule of thumb just mentioned to determine the maximum flow in Figure 6–10.

[*Ans.* Solution is obtained after encountering three labeled paths.]

(b) In Figure 6–15, indicate the arcs that can be eliminated without affecting the flow from source to sink.

[*Ans.* Arcs 2–4 and 3–4. Also, the maximum capacities of arc 2–5 can be reduced by 8 units and that of arc 3–5 by 10 units.]

At this point it is worth introducing the concept of a **minimal cut**. A **cut** in a connected network defines a set of arcs that when set to zero capacity will result in zero flow between 1 and 5. The capacity of a cut is the sum of the capacities of its arcs. In Figure 6–10 some cuts can be identified as follows:

Cut Set	Capacity
(1, 2), (1, 3), (1, 4)	20 + 30 + 10 = 60
(2, 5), (3, 5), (4, 5)	28 + 20 + 15 = 63
(1, 2), (1, 3), (4, 5)	20 + 10 + 15 = 45

Intuitively, we can determine the maximum flow by enumerating *all* the cuts in the network. The cut with the smallest capacity provides the answer. This intuitive result has actually been proved using the important **maximal-flow minimal-cut theorem,** which states that the maximum flow in a network equals the capacity of its minimal cut.

6.4 LINEAR PROGRAMMING REPRESENTATION OF NETWORKS

The shortest route and the maximal flow problems can be formulated as LP models explicitly. We emphasize, however, that the solution of network models by the simplex method is not advisable. On the other hand, a study of LP formulations of networks should assist you in recognizing LP models that may not be networks in the obvious sense, but can be formulated either directly or with modifications as a network. The obvious advantage is that the efficiency of computations may be improved drastically when network formulation is used.

The LP model of a shortest-route problem is constructed as follows:

1. Each variable corresponds to an arc.
2. Each constraint corresponds to a node.

Thus let x_{ij} represent the amount of flow in arc (i, j); the shortest-route model with n nodes is thus given as

$$\text{minimize } z = \sum_{\substack{(i,j) \text{ i8} \\ \text{network}}} d_{ij} x_{ij}$$

subject to

$$\sum_{\substack{(1,j) \text{ in} \\ \text{network}}} x_{1j} = 1 \quad \text{(source)}$$

$$\sum_{\substack{(i,k) \text{ in} \\ \text{network}}} x_{ik} = \sum_{\substack{(k,j) \text{ in} \\ \text{network}}} x_{kj}, \quad \text{for all } k \neq 1 \text{ or } n$$

$$\sum_{\substack{(i,n) \text{ in} \\ \text{network}}} x_{in} = 1 \quad \text{(destination)}$$

$$x_{ij} \geq 0, \quad \text{for all } i \text{ and } j$$

The constraints of the LP model are based on the transshipment formulation of the shortest-route problem introduced in Section 6.2.3. One unit of flow is shipped from node 1 to be received at node n. The first and last constraints say that the total flow (sum of variables) leaving node 1 equals 1, and the total flow received at node n also equals 1. At any intermediate node, the total flow entering the node equals the total flow leaving the same node. The objective function requires that the total distance traveled by the unit flow be minimized.

We must point out that the foregoing formulation would yield a meaningful solution only if $x_{ij} = 0$ or 1; that is, an arc (i,j) falls on the shortest route only if $x_{ij} = 1$. If $x_{ij} = 0$, (i, j) does not fall on the shortest route. Although the requirements $x_{ij} = 0$ or 1 is not spelled out explicitly in the LP model, its special structure always yields an optimal solution in which this condition is satisfied. Indeed, the model possesses the **totally unimodular property**, which guarantees that the LP solution will always yield $x_{ij} = 0$ or 1 (compare with the transshipment model in Section 6.2.3).

Exercise 6.4-1

Consider the shortest-route network in Figure 6–7 (Section 6.2.2). Write the constraint equations associated with each of the following nodes explicitly.
(a) Node 1
  [*Ans.* $x_{12} + x_{13} + x_{14} = 1.$]
(b) Node 4
  [*Ans.* $(x_{14} + x_{24} + x_{34}) - (x_{45} + x_{46}) = 0.$]
(c) Node 7
  [*Ans.* $x_{57} + x_{67} = 1.$]

The maximal-flow problem can be formulated as an LP model in a similar manner. Specifically, let y represent the flow between the source node 1 and terminal node n. Using the same notation x_{ij} to represent the flow in arc (i, j), the LP model becomes:

$$\text{maximize } z = y$$

subject to

$$\sum_{\substack{(1,j) \text{ in} \\ \text{network}}} x_{1j} = y \qquad \text{(source)}$$

$$\sum_{\substack{(i,k) \text{ in} \\ \text{network}}} x_{ik} = \sum_{\substack{(k,j) \text{ in} \\ \text{network}}} x_{kj}, \qquad \text{for all } k \neq 1 \text{ or } n$$

$$\sum_{\substack{(i,n) \text{ in} \\ \text{network}}} x_{in} = y \qquad \text{(destination)}$$

$$0 \leq x_{ij} \leq u_{ij}, \qquad \text{for all } i \text{ and } j$$

where u_{ij} represents the flow capacity in arc (i, j). We note that the construction of the constraints follows the same logic used in developing the shortest-route LP model.

6.5 SUMMARY

In this chapter we explored some applications of network modeling. Problems that are not obviously networks (such as equipment replacement) can be modeled and solved by network models. Although most network models can be modeled as linear programs, special methods exploiting the special structures of network can be developed to solve this type of problem efficiently.

The chapter also introduced the idea of recursive computations for solving the shortest-route problem. Recursive computations form the basis for the development of dynamic programming algorithms which we present in Chapter 9.

Network analysis and applications include more than can be presented in one chapter. Notable among the missing algorithms is the **out-of-kilter method,** which solves the maximal-flow problem. However, rather than maximize the flow itself, the objective is to minimize the *cost* of the flow. The new algorithm is a combined primal–dual method that exploits the special structure of the network in more or less the same manner employed to solve the transportation model (Chapter 5).

Another type of network that is not presented in this chapter is the PERT–CPM technique, which seeks the determination of a feasible schedule of the individual activities of complex projects. The problem is basically equivalent to determining the *longest* route in the project network, which actually gives the minimum time to complete the project. Because of the practical importance of PERT–CPM, Chapter 12 is devoted entirely to the presentation of this technique.

SELECTED REFERENCES

ELMAGHRABY, S. *Some Network Models in Management Science,* Springer-Verlag, New York, 1970.

ELMAGHRABY, S., *Activity Networks,* Wiley, New York, 1977.

FORD, L., and D. FULKERSON, *Flows in Networks,* Princeton University Press, Princeton, N.J., 1962.

JENSEN, P., and J. W. BARNES, *Network Flow Programming,* Wiley, New York, 1980.

REVIEW QUESTIONS

True (T) or False (F)?

1. ____ A network is called *oriented* or *directed* if movement along its arcs is unidirectional.

2. ____ A *chain* linking any two distinct nodes in a network must consist of linked arcs with no "gaps."

3. ____ A *minimal spanning tree* of a network may include *loops.*

4. ____ An *acyclic* network does not include any loops.

5. ____ In a *cyclic* network, the *shortest route* between two nodes may include one or more loops.

6. ____ If the *transshipment* model is used to find the *shortest route* between every two nodes in a network, it will be necessary to solve as many transshipment models as the number of distinct pairs of nodes in the network.

7. ____ A *shortest-route problem* can be treated as a network flow problem with unit flow.

8. ____ In a connected network, if the capacities of the arcs defining a *cut* are set equal to zero, no flow can be expected between the source and terminal nodes.

9. ____ The maximum flow between the two nodes in a network may exceed the capacity of its *minimal cut*.

10. ____ In representing a network optimization problem as a linear program, there will be as many variables as the number of arcs in the network.

11. ____ In the representation of a network optimization problem as a linear program, each node will correspond to a linear programming constraint.

12. ____ A linear programming constraint corresponding to node in a network is simply a *balance equation* that requires the inflow to equal the outflow of that node.

13. ____ The *totally unimodular property* in linear programming representation of network always guarantees an *integer* solution to the network flow problem.

[*Ans.* 1—T, 2—T, 3—F, 4—T, 5—F, 6—T, 7—T, 8—T, 9—F, 10 to 13—T.]

PROBLEMS

Section	Assigned Problems
6.1	6–1 to 6–4
6.2.1	6–5
6.2.2	6–6 to 6–8
6.2.3	6–9, 6–10
6.3	6–11 to 6–14
6.4	6–15

☐ **6–1** Find the minimal spanning tree of the network in Figure 6–2 under each of the following *independent* conditions:
   (a) Nodes 5 and 6 are linked by a 2-mile cable.
   (b) Nodes 2 and 5 cannot be linked.
   (c) Nodes 2 and 6 are linked by a 4-mile cable.
   (d) The cable between nodes 1 and 2 is 8 miles long.
   (e) Nodes 3 and 5 are linked by a 2-mile cable.
   (f) Node 2 cannot be connected directly to nodes 3 and 5.

☐ **6–2** Suppose that it is desired to establish a cable communication network that links the major cities shown in Figure 6–16. Determine how the cities are connected such that the total used cable mileage is minimized.

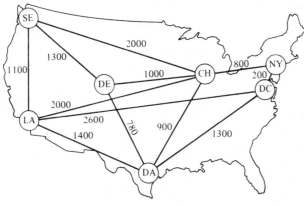

Figure 6–16

☐ **6–3** Figure 6–17 shows the mileage of the feasible links connecting nine offshore natural-gas wellheads with an onshore delivery point. Since the location of wellhead 1 is the closest to shore, it is equipped with sufficient pumping and storage capacity to pump the output of the remaining eight wells to the delivery point. Determine the pipeline network linking all the wellheads to the delivery point that will minimize the total pipeline miles.

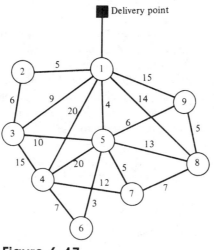

Figure 6–17

☐ **6–4** In Figure 6–17, suppose that the wellheads can be divided into two groups, depending on gas pressure: a high-pressure group, which includes wells 2, 3, 4, and 6, and a low-pressure group, which includes wells 5, 7, 8, and 9. Because of the pressure difference it is not possible to link wellheads from the two groups. However, both groups are connected to the delivery point through wellhead 1. Determine the optimum pipeline network that will connect all wellheads to the delivery point.

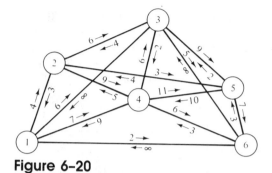

Figure 6-20

(a) Nodes 1 and 5.
(b) Nodes 6 and 3.
(c) Nodes 2 and 6.

☐ **6-11** Three refineries send their gasoline product to two terminals. The capacities of the refineries are estimated at 200,000, 250,000, and 300,000 bb per day. The demands at the terminals are known to be 400,000 and 450,000 bb per day. Any demand that cannot be satisfied from the refineries is acquired from other sources. The gasoline product is transported to the terminals via a network of pipelines that are boosted by three pumping stations. Figure 6-21 summarizes the links of the network together with the capacity of each pipeline. How much flow should be passing through each pumping station?

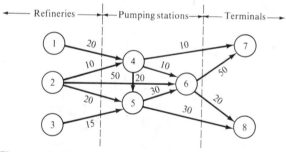

Figure 6-21

☐ **6-12** In Problem 6-11, identify four different cuts. Show that the capacity of these cuts at least equals the maximal flow obtained in Problem 6-11.

☐ **6-13** Consider a version of the transportation problem in which the objective is to maximize the *amounts* transported between m sources and n destinations. Although the sources may have ample supply to satisfy the demand, the limited capacity of the routes connecting the sources and destinations may impede fulfilling the demand completely. Our objective is to determine the shipping schedule that will maximize the *amount* transported (rather than minimize the transportation cost) by expressing the problem as a flow network.

The following table provides the amounts of supply a_i and demand b_j at sources i and destinations j. The maximum capacity c_{ij} of the (i, j) route is given by the (i, j)th element of the tableau. Empty squares indicate that the corresponding route does not exist. Find the schedule that ships the most amounts between the sources and destinations.

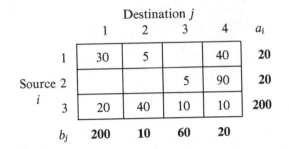

		Destination j				a_i
		1	2	3	4	
	1	30	5		40	**20**
Source	2			5	90	**20**
i	3	20	40	10	10	**200**
b_j		**200**	**10**	**60**	**20**	

☐ **6–14** Solve Problem 6–13 assuming that transshipment is allowed between sources 1, 2, and 3 with a maximum two-way capacity of 50 units each. Also, transshipment is allowed between destinations 1, 2, 3, and 4 with a maximum two-way capacity of 50 units. What is the effect of transshipping on the unsatisfied demands at the different destinations?

☐ **6–15** Express each of the following models as a linear program.
  (a) The equipment-replacement model of Example 6.2–1.
  (b) The most-reliable-route model of Example 6.2–2.
  (c) The maximal-flow network in Figure 6–22.
  (d) The transportation example of Problem 6–13.

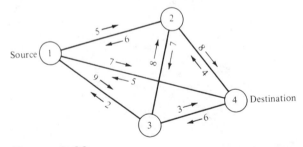

Figure 6-22

SHORT CASE

Case 6–1†
    A number of individuals had set up separate brokerage firms that traded mainly in highly speculative stocks. The brokers operated under a loose financial system that allowed extensive interbrokerage transactions, including buying, selling, borrowing, and lending. For the group of brokers as a whole, the main source of income was the commission they received from sales to outside clients.

---
† This case is based on a true situation that did not take place in the United States.

Eventually, the risky business of trading in speculative stocks became unmanageable, and all brokers declared bankruptcy. At the time the bankruptcy was declared, the financial situation looked as follows. All brokers owed money to outside clients, and furthermore, the interbroker financial entanglements were so complex that almost every broker owed money to every other broker in the group.

The brokers whose personal assets could pay off their debts were declared solvent. The remaining ones were referred to a legal body whose main purpose was to resolve the debt situation in the best interest of outside clients. Since the assets and receivables of the nonsolvent brokers obviously were less than their payables, all debts had to be prorated. The final effect was, in essence, a complete liquidation of all the assets of the nonsolvent brokers.

In resolving the financial entanglements within the group of nonsolvent brokers, it was decided that the transactions would be executed only to satisfy certain legal requirements since, in effect, none of the brokers would be keeping any of the funds owed to him by others. As such, the legal body requested that the number of interbroker transactions be reduced to an absolute minimum. Thus, if A owed B an amount X and B owed A an amount Y, the two "loop" transactions were reduced to one of the amount $|X - Y|$ that went from A to B if $X > Y$ and from B to A if $Y > X$. If $X = Y$, the transactions were completely eliminated. The idea was to be extended to all loop transactions involving any number of brokers.

How would you handle this situation? Specifically, you are required to answer two questions:

1. How should the debts be prorated?

2. How should the number of interbroker transactions be reduced to a minimum?

Chapter 7

Linear Programming: Advanced Topics

In Chapters 3 and 4 we presented the basic computational techniques of linear programming. Several questions were left unanswered for the lack of adequate theory. This chapter presents the basic theory of linear programming. Although an understanding of theory is helpful in appreciating the use and application of linear programming, our primary objective is to utilize the theory to present new solution algorithms that are computationally more efficient than the tableau method of Chapter 3. These algorithms include the revised simplex, decomposition, bounded variables, and parametric programming. We wish to emphasize, however, that all these algorithms utilize the exact steps of the simplex method as presented in Chapter 3. The only difference occurs in the *details* of the computations for identifying the entering and leaving variables.

Since matrix algebra is used throughout this chapter, you are expected to review Appendix A if you have not already studied matrices. You will notice, however, that only elementary knowledge of matrices is needed for understanding the presentation.

7.1 MATRIX DEFINITION OF THE STANDARD LP PROBLEM

The linear programming problem in standard form (all equality constraints with nonnegative right side and all nonnegative variables; see Section 3.1) can be expressed in matrix notation as

$$\begin{bmatrix} \text{maximize} \\ \text{or} \\ \text{minimize} \end{bmatrix} z = \mathbf{CX}$$

subject to

$$(\mathbf{A}, \mathbf{I})\mathbf{X} = \mathbf{b} \qquad (\mathbf{b} \geq \mathbf{0})$$

$$\mathbf{X} \geq \mathbf{0}$$

where $\mathbf{I}$ is the m-identity matrix and

$$\mathbf{X} = (x_1, x_2, \ldots, x_n)^T, \qquad \mathbf{C} = (c_1, c_2, \ldots, c_n)$$

$$\mathbf{A} = \begin{bmatrix} a_{11} & a_{12} & \cdots & a_{1,n\text{-}m} \\ a_{21} & a_{22} & \cdots & a_{2,n\text{-}m} \\ \vdots & \vdots & & \vdots \\ a_{m1} & a_{m2} & \cdots & a_{m,n\text{-}m} \end{bmatrix}, \qquad \mathbf{b} = \begin{bmatrix} b_1 \\ b_2 \\ \vdots \\ b_m \end{bmatrix}$$

The constraints $(\mathbf{A}, \mathbf{I})\mathbf{X} = \mathbf{b}$ also may be written in vector form as

$$\sum_{j=1}^{n} x_j \mathbf{P}_j = \mathbf{b}$$

where $\mathbf{P}_j$ represents the jth column vector of $(\mathbf{A}, \mathbf{I})$.

The identity matrix $\mathbf{I}$ can always be made to appear as shown in the constraint equations by augmenting and arranging the slack and/or artificial variables as necessary (see Chapters 3 and 4). This means that the n elements of the vector $\mathbf{X}$ include any augmented slack or artificial variables, with the rightmost m elements representing the starting solution variables.

To clarify the foregoing definition, we introduce a numerical example.

Example 7.1-1. The linear program

$$\text{maximize } z = 2x_1 + 3x_2 + 4x_3$$

subject to

$$
\begin{aligned}
x_1 + x_2 + x_3 &\geq 5 \\
x_1 + 2x_2 &= 7 \\
5x_1 - 2x_2 + 3x_3 &\leq 9 \\
x_1, x_2, x_3 &\geq 0
\end{aligned}
$$

is expressed in the standard matrix form as

$$\text{maximize } z = (2, 3, 4, 0, -M, -M, 0)\begin{bmatrix} x_1 \\ x_2 \\ x_3 \\ x_4 \\ x_5 \\ x_6 \\ x_7 \end{bmatrix}$$

subject to

$$\begin{bmatrix} 1 & 1 & 1 & -1 & 1 & 0 & 0 \\ 1 & 2 & 0 & 0 & 0 & 1 & 0 \\ 5 & -2 & 3 & 0 & 0 & 0 & 1 \end{bmatrix}\begin{bmatrix} x_1 \\ x_2 \\ x_3 \\ x_4 \\ x_5 \\ x_6 \\ x_7 \end{bmatrix} = \begin{bmatrix} 5 \\ 7 \\ 9 \end{bmatrix}$$

$$x_j \geq 0, \qquad j = 1, 2, \ldots, 7$$

Notice that x_4 is a surplus variable whereas x_7 is a slack. The variables x_5 and x_6 are artificial.

In terms of the matrix definition, we have

$$
\begin{aligned}
\mathbf{X} &= (x_1, x_2, \ldots, x_7)^T \\
\mathbf{C} &= (2, 3, 4, 0, -M, -M, 0) \\
\mathbf{b} &= (5, 7, 9)^T
\end{aligned}
$$

$$\mathbf{A} = \begin{bmatrix} 1 & 1 & 1 & -1 \\ 1 & 2 & 0 & 0 \\ 5 & -2 & 3 & 0 \end{bmatrix}, \qquad \mathbf{I} = \begin{bmatrix} 1 & 0 & 0 \\ 0 & 1 & 0 \\ 0 & 0 & 1 \end{bmatrix}$$

◀

7.2 FOUNDATIONS IN LINEAR PROGRAMMING

This section stresses the important role of basic solutions in solving linear programs. It then utilizes the definition of basic solutions to develop the matrix version of the simplex tableau that will be used for the development of the remainder of the chapter. The matrix simplex tableau is also used to explain directly the role of duality in linear programming computations.

7.2.1 BASIC SOLUTIONS AND BASES

In Chapter 3 we indicated that the fundamental idea underlying the simplex algorithm is that the feasible extreme points of the solution space, which are the candidates for the optimum solution, can be identified completely by the *nonnegative* basic solutions to the constraint equations $(\mathbf{A}, \mathbf{I})\mathbf{X} = \mathbf{b}$. Given $(\mathbf{A}, \mathbf{I})\mathbf{X} = \mathbf{b}$ has m equations and n unknowns, a basic solution is obtained by setting $n - m$ variables at zero and then solving the m equations in the remaining m variables, provided that the solution exists and is unique. Mathematically, a basic solution of the given system is defined as follows. Let

$$(\mathbf{A}, \mathbf{I})\mathbf{X} = \sum_{j=1}^{n} \mathbf{P}_j x_j$$

Then, by setting the variables x_j associated with any $n - m$ vectors equal to zero, the remaining m variables are said to be basic if their associated vectors are **linearly independent.** In other words, the square matrix comprising these m vectors must be **nonsingular.** In this case, the m independent vectors are said to comprise a **basis,** and the resulting solution of the m variables is a basic solution.

For example, in Example 7.1–1, we have $m = 3$ and $n = 7$. Thus we must set $n - m = 4$ variables at zero, and the basis must include three independent vectors. Consider the combination $x_4 = x_5 = x_6 = x_7 = 0$; then the vectors

$$\mathbf{P}_1 = \begin{bmatrix} 1 \\ 1 \\ 5 \end{bmatrix}, \qquad \mathbf{P}_2 = \begin{bmatrix} 1 \\ 2 \\ -2 \end{bmatrix}, \qquad \mathbf{P}_3 = \begin{bmatrix} 1 \\ 0 \\ 3 \end{bmatrix}$$

associated with x_1, x_2, and x_3 form a basis because their matrix,

$$\mathbf{B} = (\mathbf{P}_1, \mathbf{P}_2, \mathbf{P}_3) = \begin{bmatrix} 1 & 1 & 1 \\ 1 & 2 & 0 \\ 5 & -2 & 3 \end{bmatrix}$$

is nonsingular (the determinant of $\mathbf{B} = -9 \neq 0$). In this case the resulting values of x_1, x_2, x_3 obtained from

$$\begin{bmatrix} 1 & 1 & 1 \\ 1 & 2 & 0 \\ 5 & -2 & 3 \end{bmatrix} \begin{bmatrix} x_1 \\ x_2 \\ x_3 \end{bmatrix} = \begin{bmatrix} 5 \\ 7 \\ 9 \end{bmatrix}$$

should yield a basic solution.

We can demonstrate graphically the relationship between vectors and bases by considering the following set of equations in vector form:

$$\begin{bmatrix} 2 \\ 1 \end{bmatrix} x_1 + \begin{bmatrix} 1 \\ 2 \end{bmatrix} x_2 + \begin{bmatrix} 1 \\ 1 \end{bmatrix} x_3 + \begin{bmatrix} 2 \\ -1 \end{bmatrix} x_4 + \begin{bmatrix} 4 \\ 2 \end{bmatrix} x_5 = \begin{bmatrix} 2 \\ 2 \end{bmatrix}$$

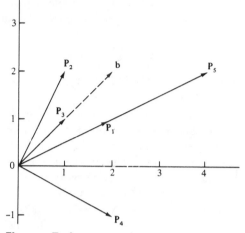

Figure 7–1

or

$$\mathbf{P}_1 x_1 + \mathbf{P}_2 x_2 + \mathbf{P}_3 x_3 + \mathbf{P}_4 x_4 + \mathbf{P}_5 x_5 = \mathbf{b}$$

Figure 7–1 plots the two-dimensional vectors $\mathbf{P}_1$, $\mathbf{P}_2$, $\mathbf{P}_3$, $\mathbf{P}_4$, and $\mathbf{P}_5$. Since we have two equations and five variables, a basis must include exactly $5 - 3 = 2$ independent vectors. We can see in Figure 7–1 that all combinations of two vectors will yield a basis except the combination $(\mathbf{P}_1, \mathbf{P}_5)$ because $\mathbf{P}_1$ and $\mathbf{P}_5$ are *dependent*.

In the simplex method, we deal only with *feasible* basic solutions. A basic solution of the system $(\mathbf{A}, \mathbf{I})\mathbf{X} = \mathbf{b}$ is feasible for the linear program if it also satisfies the nonnegativity restriction $\mathbf{X} \geq \mathbf{0}$.

Exercise 7.2–1

(a) In Figure 7–1, can the combination $(\mathbf{P}_1, \mathbf{P}_4)$ be used as a basis in any of the simplex iterations?

[*Ans.* No, because this basis will result in a *negative* value of x_4, which is not admissible since the simplex method deals with *feasible* basic solutions only]

(b) In the following sets of equations, the first two have unique solutions, the third set has infinity of solutions, and the fourth set has no solution. Use graphical vector representation to show that in each of the first two cases (unique solutions), the left-side vectors are independent and hence form a basis, whereas in the last two cases the left-side vectors are necessarily dependent. In particular, you must get a clear understanding of the meaning of "dependence" and "independence" among vectors.

$$(1) \quad x_1 + 3x_2 = 2 \qquad (2) \quad 2x_1 + 3x_2 = 1$$
$$3x_1 + x_2 = 3 \qquad\qquad 2x_1 - x_2 = 2$$
$$(3) \quad 2x_1 + 6x_2 = 4 \qquad (4) \quad 2x_1 - 4x_2 = 2$$
$$x_1 + 3x_2 = 2 \qquad\qquad -x_1 + 2x_2 = 1$$

[*Ans.* See Figure 7–2.]

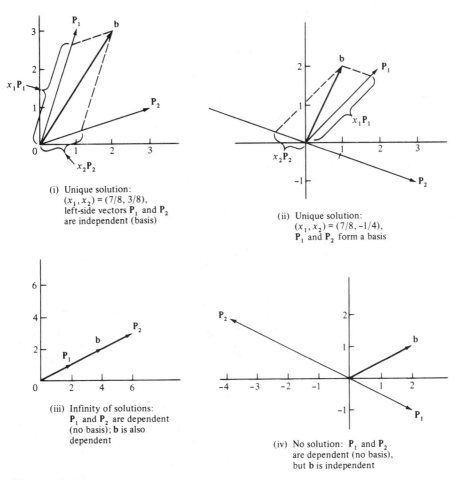

Figure 7-2

(c) In Figure 7–1, how many distinct extreme points are associated with the bases (P_1, P_3), (P_2, P_3), (P_3, P_4), and (P_3, P_5)?

[*Ans.* Exactly one, because P_3 and **b** are dependent. Thus any of these bases will yield $x_3 = 2$ and zero for *all* the remaining variables. The basic solutions are thus *degenerate*. The answer demonstrates that in case of degeneracy, an extreme point may be represented by more than one basic solution. A *nondegenerate* extreme point, however, has exactly one basic solution associated with it.]

7.2.2 MATRIX VERSION OF THE SIMPLEX TABLEAU

We shall use the definition of a basis just given to develop a matrix version of the simplex tableau of Chapter 3. We wish to emphasize that a good understanding of the elements of the simplex tableau will greatly assist you in understanding the material we present later in this chapter. It will also strengthen your knowledge regarding the details of the topics we presented in Chapters 3 and 4.

Consider the matrix definition of the standard linear program presented in Section 7.1. We partition the vector $\mathbf{X}$ into $\mathbf{X_I}$ and $\mathbf{X_{II}}$ such that $\mathbf{X_{II}}$ corresponds to the *starting* basic variables. We also partition the vector $\mathbf{C}$ into $\mathbf{C_I}$ and $\mathbf{C_{II}}$ to correspond to $\mathbf{X_I}$ and $\mathbf{X_{II}}$. For example, in Example 7.1–1,

$$\mathbf{X_I} = (x_1, x_2, x_3, x_4), \qquad \mathbf{X_{II}} = (x_5, x_6, x_7)$$
$$\mathbf{C_I} = (2, 3, 4, 0), \qquad \mathbf{C_{II}} = (-M, -M, 0)$$

Putting the objective equation in the form

$$z - \mathbf{C_I X_I} - \mathbf{C_{II} X_{II}} = 0$$

we can summarize the standard LP problem as

$$\left(\begin{array}{c|c|c} 1 & -\mathbf{C_I} & -\mathbf{C_{II}} \\ \hline 0 & \mathbf{A} & \mathbf{I} \end{array}\right) \begin{pmatrix} z \\ \mathbf{X_I} \\ \mathbf{X_{II}} \end{pmatrix} = \begin{pmatrix} 0 \\ \mathbf{b} \end{pmatrix}$$

At any iteration, suppose that $\mathbf{X_B}$ represents the current basic variables and let $\mathbf{B}$ represent the basis associated with $\mathbf{X_B}$; that is, $\mathbf{B}$ includes the m vectors of $(\mathbf{A}, \mathbf{I})$ associated with the elements of $\mathbf{X_B}$. Since all the remaining nonbasic variables are zero, the current solution is obtained from

$$\mathbf{B X_B} = \mathbf{b} \qquad \text{and} \qquad z = \mathbf{C_B X_B}$$

where $\mathbf{C_B}$ comprises the elements of $\mathbf{C}$ corresponding to $\mathbf{X_B}$.

The current solution equations may be written as

$$\begin{pmatrix} 1 & -\mathbf{C_B} \\ 0 & \mathbf{B} \end{pmatrix} \begin{pmatrix} z \\ \mathbf{X_B} \end{pmatrix} = \begin{pmatrix} 0 \\ \mathbf{b} \end{pmatrix}$$

By using the formulas in Section A.2.7 for the inversion of a partitioned matrix, the current solution is given by

$$\begin{pmatrix} z \\ \mathbf{X_B} \end{pmatrix} = \begin{pmatrix} 1 & \mathbf{C_B B^{-1}} \\ 0 & \mathbf{B^{-1}} \end{pmatrix} \begin{pmatrix} 0 \\ \mathbf{b} \end{pmatrix} = \begin{pmatrix} \mathbf{C_B B^{-1} b} \\ \mathbf{B^{-1} b} \end{pmatrix}$$

The simplex tableau for the current solution can be obtained by considering

$$\begin{pmatrix} 1 & \mathbf{C_B B^{-1}} \\ 0 & \mathbf{B^{-1}} \end{pmatrix} \begin{pmatrix} 1 & -\mathbf{C_I} & -\mathbf{C_{II}} \\ 0 & \mathbf{A} & \mathbf{I} \end{pmatrix} \begin{pmatrix} z \\ \mathbf{X_I} \\ \mathbf{X_{II}} \end{pmatrix} = \begin{pmatrix} 1 & \mathbf{C_B B^{-1}} \\ 0 & \mathbf{B^{-1}} \end{pmatrix} \begin{pmatrix} 0 \\ \mathbf{b} \end{pmatrix}$$

which gives

$$\left(\begin{array}{c|c|c} 1 & \mathbf{C_B B^{-1} A} - \mathbf{C_I} & \mathbf{C_B B^{-1}} - \mathbf{C_{II}} \\ \hline 0 & \mathbf{B^{-1} A} & \mathbf{B^{-1}} \end{array}\right) \begin{pmatrix} z \\ \mathbf{X_I} \\ \mathbf{X_{II}} \end{pmatrix} = \begin{pmatrix} \mathbf{C_B B^{-1} b} \\ \mathbf{B^{-1} b} \end{pmatrix}$$

We can summarize this information in terms of the simplex tableau at the starting and general iterations as follows.

Starting Iteration

Basic	X_I	X_{II}	Solution
z	$C_{II}A - C_I$	0	$C_{II}b$
X_{II}	A	I	b

General Iteration

Basic	X_I	X_{II}	Solution
z	$C_B B^{-1}A - C_I$	$C_B B^{-1} - C_{II}$	$C_B B^{-1}b$
X_B	$B^{-1}A$	B^{-1}	$B^{-1}b$

If you study the starting and general iteration, you will discover that the starting iteration is a special case of the general iteration. Specifically, at the starting iteration we have

$$X_B = X_{II}, \qquad C_B = C_{II}, \qquad B = I, \qquad B^{-1} = I$$

You can verify that a substitution of these special vectors and matrices in the general iteration will yield the starting iteration directly.

Let us elaborate some on the relationship between the matrix version of the tableau given above and the tableau given in Chapter 3. Both are one and the same. Essentially, in Chapter 3 we used the Gauss–Jordan elimination method to secure the basic solution at each iteration, which is computationally equivalent to using matrix inversion.

To illustrate this point, look at the starting iteration in matrix form. If the starting basic variables X_{II} are all slack, then $C_{II} = 0$ and the z-row in the starting iteration will be

Basic	X_I	X_{II}	Solution
z	$-C_I$	0	0

which is exactly what we used in Chapter 3. On the other hand, suppose that X_{II} consists of all artificial variables. Then $C_{II} = (-M, -M, \ldots, -M)$ and the z-row of the starting iteration becomes

Basic	X_I	X_{II}	Solution
z	$(M, M, \ldots, M)A - C_I$	0	0

This result is equivalent to substituting out the artificial variables in the objective equation using the constraint equations as we did in Chapter 3 (see Table 3–1).

If in the general iteration we let $z_j = \mathbf{C}_B \mathbf{B}^{-1}\mathbf{P}_j$, where $\mathbf{P}_j$ is the jth vector of $(\mathbf{A}, \mathbf{I})$, the jth element coefficient of $\mathbf{X}_I$ and $\mathbf{X}_{II}$ in the z-row is thus given by

$$z_j - c_j = \mathbf{C}_B \mathbf{B}^{-1}\mathbf{P}_j - c_j$$

As we showed in Chapter 3, the **optimality condition** of the simplex method requires that at the optimum iteration

$$z_j - c_j \begin{cases} \geq 0, & \text{for all } j \text{ in case of maximization} \\ \leq 0, & \text{for all } j \text{ in case of minimization} \end{cases}$$

A nonbasic variable that does not satisfy this condition becomes an *entering variable*.

The **feasibility condition,** on the other hand, selects the *leaving variable* as the current basic variable associated with

$$\theta = \min_i \left\{ \frac{(\mathbf{B}^{-1}\mathbf{b})_i}{(\mathbf{B}^{-1}\mathbf{P}_k)_i}, \quad (\mathbf{B}^{-1}\mathbf{P}_k)_i > 0 \right\}$$

where $\mathbf{P}_k$ is the entering vector (x_k the entering variable) and the notation $(\cdot)_i$ represents the ith element of the column vector in parentheses.

Exercise 7.2-2

Consider the following linear program:

$$\text{maximize } z = 2x_1 + 3x_2 + 7x_3$$

subject to

$$2x_1 + x_2 + 2x_3 = 4$$
$$3x_1 - x_2 - 2x_3 = 1$$
$$x_1, x_2, x_3 \geq 0$$

Let $\mathbf{P}_1$, $\mathbf{P}_2$, and $\mathbf{P}_3$ be the vectors associated with x_1, x_2, and x_3, and let $\mathbf{b}$ be the right-side vector.
(a) How many vectors are needed to form a basis?
  [*Ans.* Two.]
(b) Can $\mathbf{P}_2$ and $\mathbf{P}_3$ form a basis?
  [*Ans.* No, because they are dependent.]
(c) Suppose that the current basic variables in a simplex iteration are x_1 and x_2; write the associated basis and obtain its inverse.
  $$\left[Ans. \; \mathbf{B} = (\mathbf{P}_1, \mathbf{P}_2) = \begin{pmatrix} 2 & 1 \\ 3 & -1 \end{pmatrix}, \quad \mathbf{B}^{-1} = \begin{pmatrix} 1/5 & 1/5 \\ 3/5 & -2/5 \end{pmatrix}. \right]$$
(d) Find $\mathbf{C}_B$ associated with $\mathbf{X}_B$ in part (c).
  [*Ans.* $\mathbf{C}_B = (2, 3)$.]
(e) Find the current values of z and $\mathbf{X}_B$.
  [*Ans.* $\mathbf{X}_B = (x_1, x_2) = \mathbf{B}^{-1}\mathbf{b} = (1, 2)^T$, $z = \mathbf{C}_B \mathbf{X}_B = 8$.]
(f) Is $\mathbf{X}_B$ in part (c) optimum?
  [*Ans.* No, because $z_3 - c_3 = \mathbf{C}_B \mathbf{B}^{-1}\mathbf{P}_3 - c_3 = -1$. Thus $\mathbf{P}_3$ is the entering vector]
(g) If $\mathbf{P}_3$ enters, which current vector must leave?
  [*Ans.* $\theta = \min_{i=1,2} \{(\mathbf{B}^{-1}\mathbf{b})_i/(\mathbf{B}^{-1}\mathbf{P}_3)_i, \; (\mathbf{B}^{-1}\mathbf{P}_3)_i > 0\} = 1$, corresponding to x_2. Hence $\mathbf{P}_2$ is the leaving vector and the new basis is $\mathbf{B} = (\mathbf{P}_1, \mathbf{P}_3)$.]

In Exercise 7.2–2 we have just gone through a complete simplex iteration by using matrix manipulations (instead of the Gauss–Jordan row operations). Essentially, the computations at a current iteration involve the use of the associated inverse B^{-1} and the *original* data of the problem. As a result, the iterations of the simplex method differ primarily in the definition of the basis B. This observation will be the principal idea for the development of the computational technique called the *revised simplex method,* which we shall present in Section 7.3.

7.2.3 DUALITY AND THE PRIMAL–DUAL RELATIONSHIPS

In this section we consider the dual problem in matrix form. The ultimate objective of this section is to provide the mathematical foundation for the development of the primal–dual relationships that were used in connection with sensitivity analysis in Chapter 4. The information will also be the basis for the development of parametric programming in Section 7.6.

Consider the standard LP problem in the form

$$\text{maximize } z = C_I X_I + C_{II} X_{II}$$

subject to

$$A X_I + I X_{II} = b$$
$$X_I, X_{II} \geq 0$$

By letting $Y = (y_1, y_2, \ldots, y_m)$ represent the dual vector, the rules given in Table 4–2 yield the following dual:

$$\text{minimize } w = Yb$$

subject to

$$YA \geq C_I$$
$$Y \geq C_{II}$$

$$Y \text{ unrestricted vector}$$

Note that the second set of constraints may override the last constraints

We now have the following important result regarding the optimal dual solution.

Optimal Dual Solution. Suppose that B is the optimal *primal* basis and let C_B be its associated objective coefficients; then

$$Y = C_B B^{-1}$$

is the optimal *dual* solution.

To show that this result is correct, we need to verify that

1. $Y = C_B B^{-1}$ is a feasible dual solution.
2. Optimal $w = $ optimal z.

A dual solution Y is feasible if it satisfies the constraints $YA \geq C_I$ and $Y \geq C_{II}$. By the optimality of the primal, we have $z_j - c_j \geq 0$ for all j; that is,

$$C_B \, B^{-1} A_j - C_I \geq 0 \qquad \text{and} \qquad C_B \, B^{-1} - C_{II} \geq 0$$

Thus for $Y = C_B \, B^{-1}$, the condition $z_j - c_j \geq 0$ yields

$$YA - C_I \geq 0 \qquad \text{and} \qquad Y - C_{II} \geq 0$$

which shows that the dual constraints are satisfied.

We can show that optimal $w =$ optimal z directly because

$$w = Yb = C_B \, B^{-1} b$$

and

$$z = C_B \, X_B = C_B \, B^{-1} b$$

This discussion leads to the following conclusion. A pair of primal and dual solutions is optimal if and only if

1. Each solution is feasible for its respective problem.
2. The associated values of z and w are equal.

Exercise 7.2–3
Consider the following linear program:

$$\text{maximize } z = 5x_1 + 12x_2 + 4x_3$$

subject to

$$x_1 + 2x_2 + \ x_3 \leq 5$$
$$2x_1 - \ x_2 + 3x_3 = 2$$
$$x_1, x_2, x_3 \geq 0$$

In each of the following cases, first verify that the basis B is feasible for the primal. Then, using $Y = C_B \, B^{-1}$, compute the associated dual values and verify whether the dual solution is optimal.

(a) $B = (P_{s_1}, P_3)$, where s_1 is the slack in the first constraint.
   [*Ans.* B is feasible. $Y = (0, 4/3)$ is not optimal because the first and second *dual* constraints are not satisfied.]

(b) $B = (P_2, P_3)$
   [*Ans.* B is feasible. $Y = (40/7, -4/7)$ is not optimal because at least the first dual constraint is not satisfied.]

(c) $B = (P_1, P_2)$
   [*Ans.* B is feasible. $Y = (29/5, -2/5)$ is optimal because all dual constraints are satisfied *and* $w = z = 141/5$.]

(d) $B = (P_1, P_3)$
   [*Ans.* B is *not* feasible. Note, however, that the associated $Y = (7, -1)$ satisfies all dual constraints and also the condition $z = w = 31$; yet it is not optimal because B is infeasible.]

We can now explain the origin of the primal–dual computations we introduced in Section 4.2. These computations show that the entire simplex tableau can be generated from the inverse B^{-1} and the original data of the problem. Specifically, we have

1. *Constraint columns computations.* The right-side column of the tableau equals $B^{-1} b$ and the jth left-side column is given by $B^{-1} P_j$.

2. *Objective row computations.* We first compute the dual values by using $\mathbf{Y} = \mathbf{C}_B \mathbf{B}^{-1}$. The objective equation coefficients are then computed as the difference between the left and right sides of the dual constraints; namely,

$$\mathbf{Y}\mathbf{P}_j - c_j = \mathbf{C}_B \mathbf{B}^{-1}\mathbf{P}_j - c_j = z_j - c_j$$

Exercise 7.2-4

Work Exercises 4.2–3 and 4.2–5 (Section 4.2.2) by using the matrix notation.

7.2.4 COMPLEMENTARY SLACKNESS THEOREM

If you investigate the matrix version of the simplex tableau, you will discover the following results regarding the *optimal* primal and dual solutions.

1. If at the optimum a primal variable x_j has $z_j - c_j > 0$, then x_j must be nonbasic and hence at zero level.
2. If at the optimum a dual variable y_i has a positive value, the ith primal constraint $\sum_{j=1}^{n} a_{ij}x_j \leq b_i$ must be satisfied in equation form because its associated slack must be zero.

Let us translate this information mathematically. We note that $z_j - c_j$ represents the difference between the left and right sides of the dual constraint and hence must represent the dual surplus variable. If we assume that v_j and s_i are the surplus and slack variables for the jth dual and ith primal constraints, then

$$v_j = z_j - c_j = \sum_{i=1}^{m} a_{ij}y_i - c_j$$

$$s_i = b_i - \sum_{j=1}^{n} a_{ij}x_j$$

The result given may thus be summarized as follows:

1. When $v_j > 0$, $x_j = 0$.
2. When $y_i > 0$, $s_i = 0$.

We can express both conditions in a compact form as follows. For the optimal primal and dual solutions,

$$v_j x_j = y_i s_i = 0 \qquad \text{for all } i \text{ and } j$$

Equivalently, we have

$$x_j \left(\sum_{i=1}^{m} a_{ij}y_i - c_j \right) = 0, \qquad j = 1, 2, \ldots, n$$

$$y_i \left(\sum_{j=1}^{n} a_{ij}x_j - b_i \right) = 0, \qquad i = 1, 2, \ldots, m$$

These results comprise the **complementary slackness theorem.** The application of the theorem particularly arises in the development of the **primal–dual algo-**

rithm for solving linear programs that start both nonoptimal and infeasible. It is also used for the development of the **out-of-kilter algorithm** in network-flow problems.

7.3 REVISED SIMPLEX METHOD

In the simplex method we developed in Chapter 3, the successive iterations are generated by using the Gauss–Jordan row operations. From the standpoint of automatic computations, this method may result in taxing the computer memory, since the entire tableau must be stored in the machine. The revised simplex method is designed to alleviate this problem. In addition, the new method can result in a reduction in the number of arithmetic operations needed to reach the optimum solution. This point is discussed in Section 7.3.2.

The foregoing observation provides the motivation for the development of the revised simplex method. We emphasize, however, that the revised method utilizes the *exact* steps of the simplex method in Chapter 3. The only difference occurs in the details of the computations.

In the remainder of this section, we first show how the inverse $\mathbf{B}^{-1}$ can be computed for the successive iterations in a manner that is particularly suited for the simplex computations. This information is then used to develop the computational details of the revised simplex method.

7.3.1 PRODUCT FORM OF THE INVERSE

The product-form method is a matrix algebra procedure that computes the inverse of a new basis from the inverse of another basis provided that the two bases differ in exactly one column vector. The procedure is suited particularly for the simplex method computations, since the successive bases differ in exactly one column as a result of interchanging the entering and leaving vectors. In other words, given the current basis $\mathbf{B}$, the next basis $\mathbf{B}_{\text{next}}$ in the immediately succeeding iteration will differ from $\mathbf{B}$ in one column only. The product-form procedure then computes the next inverse $\mathbf{B}_{\text{next}}^{-1}$ by premultiplying the current inverse $\mathbf{B}^{-1}$ by a specially constructed matrix $\mathbf{E}$.

Define the identity matrix $\mathbf{I}_m$ as

$$\mathbf{I}_m = (\mathbf{e}_1, \mathbf{e}_2, \ldots, \mathbf{e}_m)$$

where $\mathbf{e}_i$ is a unit column vector with a one-element in the ith place and zero otherwise. Suppose that we are given $\mathbf{B}$ and $\mathbf{B}^{-1}$ and assume that the vector $\mathbf{P}_r$ in $\mathbf{B}$ is replaced by a new vector $\mathbf{P}_j$ (in terms of the simplex method, $\mathbf{P}_j$ and $\mathbf{P}_r$ are the entering and leaving vectors). For simplicity, define

$$\alpha^j = \mathbf{B}^{-1}\mathbf{P}_j$$

so that α_k^j is the kth element of $\boldsymbol{\alpha}^j$. Then the new inverse $\mathbf{B}_{next}^{-1}$ can be computed as follows:

$$\mathbf{B}_{next}^{-1} = \mathbf{E}\mathbf{B}^{-1}$$

where

$$\mathbf{E} = \quad (\mathbf{e}_1, \ldots, \mathbf{e}_{r-1}, \boldsymbol{\xi}, \mathbf{e}_{r+1}, \ldots, \mathbf{e}_m)$$

and

$$\boldsymbol{\xi} = \begin{pmatrix} -\alpha_1^j/\alpha_r^j \\ -\alpha_2^j/\alpha_r^j \\ \vdots \\ +1/\alpha_r^j \\ \vdots \\ -\alpha_m^j/\alpha_r^j \end{pmatrix} \leftarrow r\text{th place}$$

provided that $\alpha_r^j \neq 0$. If $\alpha_r^j = 0$, $\mathbf{B}_{next}^{-1}$ does not exist.† Note that $\mathbf{E}$ is obtained from $\mathbf{I}_m$ by replacing its rth column $\mathbf{e}_r$ by $\boldsymbol{\xi}$.

To illustrate the procedure, consider the following information:

$$\mathbf{B} = \begin{pmatrix} 2 & 1 & 0 \\ 0 & 2 & 0 \\ 4 & 0 & 1 \end{pmatrix}, \quad \mathbf{B}^{-1} = \begin{pmatrix} 1/2 & -1/4 & 0 \\ 0 & 1/2 & 0 \\ -2 & 1 & 1 \end{pmatrix}$$

If, for example, the third column vector $\mathbf{P}_3 = (0, 0, 1)^T$ of $\mathbf{B}$ is changed to $\mathbf{P}_3^* = (2, 1, 5)^T$, we can find the new inverse as follows:

$$\boldsymbol{\alpha}^3 = \mathbf{B}^{-1}\mathbf{P}_3^* = \begin{pmatrix} 1/2 & -1/4 & 0 \\ 0 & 1/2 & 0 \\ -2 & 1 & 1 \end{pmatrix}\begin{pmatrix} 2 \\ 1 \\ 5 \end{pmatrix} = \begin{pmatrix} 3/4 \\ 1/2 \\ 2 \end{pmatrix} = \begin{pmatrix} \alpha_1^3 \\ \alpha_2^3 \\ \alpha_3^3 \end{pmatrix}$$

$$\boldsymbol{\xi} = \begin{pmatrix} -(3/4)/\ 2 \\ -(1/2)/\ 2 \\ +1/\ 2 \end{pmatrix} = \begin{pmatrix} -3/8 \\ -1/4 \\ 1/2 \end{pmatrix} \leftarrow r = 3$$

$$\mathbf{B}_{next}^{-1} = \begin{pmatrix} 1 & 0 & -3/8 \\ 0 & 1 & -1/4 \\ 0 & 0 & 1/2 \end{pmatrix}\begin{pmatrix} 1/2 & -1/4 & 0 \\ 0 & 1/2 & 0 \\ -2 & 1 & 1 \end{pmatrix}$$

$$= \begin{pmatrix} 5/4 & -5/8 & -3/8 \\ 1/2 & 1/4 & -1/4 \\ -1 & 1/2 & 1/2 \end{pmatrix}$$

---

† The formula $\mathbf{B}_{next}^{-1} = \mathbf{E}\mathbf{B}^{-1}$ can be justified as follows. Define $\mathbf{F} = (\mathbf{e}_1, \ldots, \mathbf{e}_{r-1}, \boldsymbol{\alpha}^j, \mathbf{e}_{r+1}, \ldots, \mathbf{e}_m)$, where $\boldsymbol{\alpha}^j = \mathbf{B}^{-1}\mathbf{P}_j$. When the current basis $\mathbf{B}$ is postmultiplied by $\mathbf{F}$, the result will be the next basis $\mathbf{B}_{next}$; that is, $\mathbf{B}_{next} = \mathbf{B}\mathbf{F}$. $\mathbf{B}_{next}$ is identical to $\mathbf{B}$ except that the rth column of $\mathbf{B}$ is replaced by $\mathbf{P}_j$. Thus

$$\mathbf{B}_{next}^{-1} = (\mathbf{B}\mathbf{F})^{-1} = \mathbf{F}^{-1}\mathbf{B}^{-1}$$

However, $\mathbf{E}$ as defined is nothing but the inverse of $\mathbf{F}$ ($\mathbf{E}\mathbf{F} = \mathbf{I}$). The formula follows directly.

Exercise 7.3-1

(a) Find $\mathbf{B}^{-1}$ for each of the following matrices using the product-form procedure.

$$\mathbf{B}_1 = \begin{pmatrix} 1 & 2 \\ 0 & 4 \end{pmatrix}, \qquad \mathbf{B}_2 = \begin{pmatrix} 1 & 2 \\ 1 & 4 \end{pmatrix}$$

Notice that $\mathbf{B}_1$ differs from $\mathbf{I}$ in only one column and that $\mathbf{B}_2$ differs from $\mathbf{B}_1$ in one column also.

$$\left[Ans. \ \mathbf{B}_1^{-1} = \mathbf{E}_1\mathbf{I} = \begin{pmatrix} 1 & -1/2 \\ 0 & 1/4 \end{pmatrix}, \ \mathbf{B}_2^{-1} = \mathbf{E}_2\mathbf{B}_1^{-1} = \begin{pmatrix} 2 & -1 \\ -1/2 & 1/2 \end{pmatrix}. \right]$$

(b) If the first column of $\mathbf{B}_2$ is $(-1/2, -1)^T$ instead, find $\mathbf{B}_2^{-1}$.

[*Ans.* $\mathbf{B}_2^{-1}$ does not exist because $\alpha_1^1 = 0$. Note the vectors' dependence in the new $\mathbf{B}_2$.]

7.3.2 COMPUTATIONAL DETAILS OF THE REVISED METHOD

The main idea of the revised method is to use the current basis inverse $\mathbf{B}^{-1}$ (and the original data of the problem) to carry out the necessary computations for determining the entering and leaving variables. The use of the *product form* makes it convenient to compute the successive inverses without having to invert any bases directly from raw data. Specifically, as in the simplex method, the starting basis in the revised method is always an identity matrix $\mathbf{I}$ whose inverse is itself. Thus, if $\mathbf{B}_1^{-1}, \mathbf{B}_2^{-1}, \ldots$, and $\mathbf{B}_i^{-1}$ represent the successive inverses up to iteration i and if $\mathbf{E}_1, \mathbf{E}_2, \ldots , \mathbf{E}_i$ are their associated matrices as defined in Section 7.3.1, then

$$\mathbf{B}_1^{-1} = \mathbf{E}_1\mathbf{I}, \ \mathbf{B}_2^{-1} = \mathbf{E}_2\mathbf{B}_1^{-1}, \ldots , \mathbf{B}_i^{-1} = \mathbf{E}_i \mathbf{B}_{i-1}^{-1}$$

Successive substitution then yields

$$\mathbf{B}_i^{-1} = \mathbf{E}_i \mathbf{E}_{i-1} \cdots \mathbf{E}_1$$

We emphasize that the use of the *product form* is not an essential part of the revised method and that any inversion process can be employed at each iteration. What is important from the standpoint of the revised method is that the inverse be computed in a manner that reduces the adverse effect of machine round-off error.

The steps of the revised method are essentially those of the simplex method of Chapter 3. Given the starting basis $\mathbf{I}$, we determine its associated objective coefficients vector $\mathbf{C}_B$ depending on whether the starting basic variables are slack (surplus) and/or artificial.

Step 1: Determination of the Entering Vector $\mathbf{P}_j$. Compute $\mathbf{Y} = \mathbf{C}_B\mathbf{B}^{-1}$. For every nonbasic vector $\mathbf{P}_j$, compute

$$z_j - c_j = \mathbf{Y}\mathbf{P}_j - c_j$$

For maximization (minimization) programs, the entering vector $\mathbf{P}_j$ is selected to have the most negative (positive) $z_j - c_j$ (break ties arbitrarily). Then if *all* $z_j - c_j \geq 0 \ (\leq 0)$, the optimal solution is reached and is given by

$$\mathbf{X}_B = \mathbf{B}^{-1}\mathbf{b} \quad \text{and} \quad z = \mathbf{C}_B\,\mathbf{X}_B$$

Otherwise,

Step 2: Determination of the Leaving Vector $\mathbf{P}_r$. Given the entering vector $\mathbf{P}_j$, compute:

1. The values of the current basic variables, that is,

$$\mathbf{X}_B = \mathbf{B}^{-1}\mathbf{b}$$

2. The constraint coefficients of the entering variables, that is,

$$\boldsymbol{\alpha}^j = \mathbf{B}^{-1}\mathbf{P}_j$$

The leaving vector $\mathbf{P}_r$ (for *both* maximization and minimization) must be associated with

$$\theta = \min_k \left\{ \frac{(\mathbf{B}^{-1}\mathbf{b})_k}{\alpha_k^j},\quad \alpha_k^a > 0 \right\}$$

where $(\mathbf{B}^{-1}\mathbf{b})_k$ and α_k^j are the kth elements of $\mathbf{B}^{-1}\mathbf{b}$ and $\boldsymbol{\alpha}^j$. If all $\alpha_k^j \leq 0$, the problem has no bounded solution.

Step 3: Determination of the Next Basis. Given the current inverse basis $\mathbf{B}^{-1}$, we find that the next inverse basis $\mathbf{B}_{\text{next}}^{-1}$ is given by

$$\mathbf{B}_{\text{next}}^{-1} = \mathbf{E}\mathbf{B}^{-1}$$

Now set $\mathbf{B}^{-1} = \mathbf{B}_{\text{next}}^{-1}$ and go to step 1.

Steps 1 and 2 are exactly equivalent to those of the simplex tableau in Chapter 3, as the following tableau shows:

Basic	x_1	x_2	$\cdots$	x_j	$\cdots$	x_n	Solution
z	$z_1 - c_1$	$z_2 - c_2$	$\cdots$	$z_j - c_j$	$\cdots$	$z_n - c_n$	
$\mathbf{X}_B$				$\mathbf{B}^{-1}\mathbf{P}_j$			$\mathbf{B}^{-1}\mathbf{b}$

Step 1 computes the z-row coefficients and determines the entering variable x_j. Step 2 then determines the leaving variable by computing the right-side elements ($= \mathbf{B}^{-1}\mathbf{b}$) and the constraint coefficients under the entering variable ($= \boldsymbol{\alpha}^j = \mathbf{B}^{-1}\mathbf{P}_j$).

Note: In carrying out the revised simplex computation, it will be helpful initially to summarize the computations of steps 1 and 2 in the tableau form shown.

Example 7.3–1. We shall solve the Reddy Mikks model by the revised method. The same example was solved by the regular simplex method in Section 3.2. You are encouraged to compare the computations of both methods to convince yourself that the two methods are basically equivalent.

The Reddy Mikks model (in standard form) is summarized here. We use x_1 and x_2 in place of x_E and x_I for convenience. Also, the slacks are represented by x_3, x_4, x_5, and x_6.

$$\text{Maximize } z = 3x_1 + 2x_2$$

subject to

$$
\begin{aligned}
x_1 + 2x_2 + x_3 &\qquad\qquad = 6 \\
2x_1 + x_2 \qquad + x_4 &\qquad\qquad = 8 \\
-x_1 + x_2 \qquad\qquad + x_5 &\qquad\quad = 1 \\
x_2 \qquad\qquad\qquad + x_6 &= 2 \\
x_1, x_2, \ldots, x_6 &\geq 0
\end{aligned}
$$

Starting Solution

$$
\begin{aligned}
\mathbf{X}_B &= (x_3, x_4, x_5, x_6)^T \\
\mathbf{C}_B &= (0, 0, 0, 0) \\
\mathbf{B} &= (\mathbf{P}_3, \mathbf{P}_4, \mathbf{P}_5, \mathbf{P}_6) = \mathbf{I} \\
\mathbf{B}^{-1} &= \mathbf{I}
\end{aligned}
$$

First Iteration

Step 1: Computation of $z_j - c_j$ for nonbasic $\mathbf{P}_1$ and $\mathbf{P}_2$.

$$\mathbf{Y} = \mathbf{C}_B \mathbf{B}^{-1} = (0, 0, 0, 0)\mathbf{I} = (0, 0, 0, 0)$$

$$(z_1 - c_1, z_2 - c_2) = \mathbf{Y}(\mathbf{P}_1, \mathbf{P}_2) - (c_1, c_2)$$

$$= (0, 0, 0, 0)\begin{pmatrix} 1 & 2 \\ 2 & 1 \\ -1 & 1 \\ 0 & 1 \end{pmatrix} - (3, 2)$$

$$= (-3, -2)$$

In terms of the simplex tableau of Chapter 3, the computations are represented as

Basic	x_1	x_2	x_3	x_4	x_5	x_6	Solution
z	-3	-2	0	0	0	0	0

(Notice that $z_j - c_j$ automatically equals zero for all basic variables.) Thus $\mathbf{P}_1$ is the entering vector.

Step 2: Determination of the leaving vector given that $\mathbf{P}_1$ enters the basis.

$$\mathbf{X}_B = \mathbf{B}^{-1}\mathbf{b} = \mathbf{I}\mathbf{b} = \mathbf{b} = \begin{pmatrix} 6 \\ 8 \\ 1 \\ 2 \end{pmatrix}$$

$$\alpha^1 = \mathbf{B}^{-1}\mathbf{P}_1 = \mathbf{I}\mathbf{P}_1 = \mathbf{P}_1 = \begin{pmatrix} 1 \\ 2 \\ -1 \\ 0 \end{pmatrix}$$

In terms of the tableau of Chapter 3, the computations for steps 1 and 2 can be summarized as follows:

Basic	x_1	x_2	x_3	x_4	x_5	x_6	Solution
z	-3	-2	0	0	0	0	0
x_3	1						6
x_4	2						8
x_5	-1						1
x_6	0						2

Thus

$$\theta = \min\{6/1, 8/2, -, -,\} = 4, \quad \text{corresponding to } x_4$$

As a result, $\mathbf{P}_4$ is the leaving vector.

Step 3: Determination of the next basis inverse. Since $\mathbf{P}_1$ replaces $\mathbf{P}_4$ and $\boldsymbol{\alpha}^1 = (1, 2, -1, 0)^T$, we have

$$\boldsymbol{\xi} = \begin{pmatrix} -1/2 \\ +1/2 \\ -(-1/2) \\ 0/2 \end{pmatrix} = \begin{pmatrix} -1/2 \\ 1/2 \\ 1/2 \\ 0 \end{pmatrix}$$

and

$$\mathbf{B}_{\text{next}}^{-1} = \mathbf{E}\mathbf{B}^{-1} = \mathbf{E}\mathbf{I} = \mathbf{E} = \begin{pmatrix} 1 & -1/2 & 0 & 0 \\ 0 & 1/2 & 0 & 0 \\ 0 & 1/2 & 1 & 0 \\ 0 & 0 & 0 & 1 \end{pmatrix}$$

The new basis is associated with the basic vector

$$\mathbf{X}_B = (x_3, x_1, x_5, x_6)$$
$$\mathbf{C}_B = (0, 3, 0, 0)$$

Second Iteration

Step 1: Computation of $z_j - c_j$ for nonbasic $\mathbf{P}_2$ and $\mathbf{P}_4$.

$$\mathbf{C}_B \mathbf{B}^{-1} = (0, 3, 0, 0) \begin{pmatrix} 1 & -1/2 & 0 & 0 \\ 0 & 1/2 & 0 & 0 \\ 0 & 1/2 & 1 & 0 \\ 0 & 0 & 0 & 1 \end{pmatrix} = (0, 3/2, 0, 0)$$

$$(z_2 - c_2, z_4 - c_4) = (0, 3/2, 0, 0) \begin{pmatrix} 2 & 0 \\ 1 & 1 \\ 1 & 0 \\ 1 & 0 \end{pmatrix} - (2, 0) = (-1/2, 3/2)$$

Thus $\mathbf{P}_2$ is the entering vector.

Step 2: Determination of the leaving vector given that $\mathbf{P}_2$ enters the basis.

$$\mathbf{X}_B = \mathbf{B}^{-1}\mathbf{b} = \begin{pmatrix} 1 & -1/2 & 0 & 0 \\ 0 & 1/2 & 0 & 0 \\ 0 & 1/2 & 1 & 0 \\ 0 & 0 & 0 & 1 \end{pmatrix} \begin{pmatrix} 6 \\ 8 \\ 1 \\ 2 \end{pmatrix} = \begin{pmatrix} 2 \\ 4 \\ 5 \\ 2 \end{pmatrix}$$

$$\boldsymbol{\alpha}^2 = \mathbf{B}^{-1}\mathbf{P}_2 = \begin{pmatrix} 1 & -1/2 & 0 & 0 \\ 0 & 1/2 & 0 & 0 \\ 0 & 1/2 & 1 & 0 \\ 0 & 0 & 0 & 1 \end{pmatrix} \begin{pmatrix} 2 \\ 1 \\ 1 \\ 1 \end{pmatrix} = \begin{pmatrix} 3/2 \\ 1/2 \\ 3/2 \\ 1 \end{pmatrix}$$

Steps 1 and 2 computations can be summarized in tableau form as follows:

Basis	x_1	x_2	x_3	x_4	x_5	x_6	Solution
z	0	$-1/2$	0	3/2	0	0	
x_3		3/2					2
x_1		1/2					4
x_5		3/2					5
x_6		1					2

Thus

$$\theta = \min\left\{\frac{2}{3/2}, \frac{4}{1/2}, \frac{5}{3/2}, \frac{2}{1}\right\} = 4/3$$

corresponding to x_3. As a result, $\mathbf{P}_3$ is the leaving vector.

Step 3: Determination of the next basis inverse. Since $\mathbf{P}_2$ replaces $\mathbf{P}_3$ and $\boldsymbol{\alpha}^2 = (3/2, 1/2, 3/2, 1)^T$, we have

$$\boldsymbol{\xi} = \begin{pmatrix} +1/(3/2) \\ -(1/2)/(3/2) \\ -(3/2)/(3/2) \\ -1/(3/2) \end{pmatrix} = \begin{pmatrix} 2/3 \\ -1/3 \\ -1 \\ -2/3 \end{pmatrix}$$

$$\mathbf{B}^{-1}_{next} = \begin{pmatrix} 2/3 & 0 & 0 & 0 \\ -1/3 & 1 & 0 & 0 \\ -1 & 0 & 1 & 0 \\ -2/3 & 0 & 0 & 1 \end{pmatrix} \begin{pmatrix} 1 & -1/2 & 0 & 0 \\ 0 & 1/2 & 0 & 0 \\ 0 & 1/2 & 1 & 0 \\ 0 & 0 & 0 & 1 \end{pmatrix}$$

$$= \begin{pmatrix} 2/3 & -1/3 & 0 & 0 \\ -1/3 & 2/3 & 0 & 0 \\ -1 & 1 & 1 & 0 \\ -2/3 & 1/3 & 0 & 1 \end{pmatrix}$$

The new basis is associated with the basic vector

$$\mathbf{X}_B = (x_2, x_1, x_5, x_6)$$
$$\mathbf{C}_B = (2, 3, 0, 0)$$

Third Iteration

Step 1: Computation of $z_j - c_j$ for $\mathbf{P}_3$ and $\mathbf{P}_4$.

$$\mathbf{C}_B\mathbf{B}^{-1} = (2, 3, 0, 0)\begin{pmatrix} 2/3 & -1/3 & 0 & 0 \\ -1/3 & 2/3 & 0 & 0 \\ -1 & 1 & 1 & 0 \\ -2/3 & 1/3 & 0 & 1 \end{pmatrix} = (1/3, 4/3, 0, 0)$$

$$(z_3 - c_3, z_4 - c_4) = (1/3, 4/3, 0, 0)\begin{pmatrix} 1 & 0 \\ 0 & 1 \\ 0 & 0 \\ 0 & 0 \end{pmatrix} - (0, 0) = (1/3, 4/3)$$

Since all $z_j - c_j \geq 0$, the last basis is optimal.

Optimal Solution

$$\begin{pmatrix} x_2 \\ x_1 \\ x_5 \\ x_6 \end{pmatrix} = \mathbf{B}^{-1}\mathbf{b} = \begin{pmatrix} 2/3 & -1/3 & 0 & 0 \\ -1/3 & 2/3 & 0 & 0 \\ -1 & 1 & 1 & 0 \\ -2/3 & 1/3 & 0 & 1 \end{pmatrix}\begin{pmatrix} 6 \\ 8 \\ 1 \\ 2 \end{pmatrix} = \begin{pmatrix} 4/3 \\ 10/3 \\ 3 \\ 2/3 \end{pmatrix}$$

$$z = \mathbf{C}_B\mathbf{X}_B = (2, 3, 0, 0)\begin{pmatrix} 4/3 \\ 10/3 \\ 3 \\ 2/3 \end{pmatrix} = 38/3$$

◄

Exercise 7.3-2

(a) In Example 7.3–1, consider the basis $\mathbf{B}_* = (\mathbf{P}_2, \mathbf{P}_1, \mathbf{P}_5, \mathbf{P}_4)$. Generate $\mathbf{B}_*^{-1}$ from the optimal basis $(\mathbf{P}_2, \mathbf{P}_1, \mathbf{P}_5, \mathbf{P}_6)$ and check if it is feasible and/or optimal.
   [*Ans.* $\mathbf{B}_*$ is feasible and nonoptimal. $(x_2, x_1, x_5, x_4) = (2, 2, 1, 2)$, $z_3 - c_3 = 3$, and $z_6 - c_6 = -4$.]

(b) Show how the revised *dual* simplex computations can be carried out by using matrix manipulations (in place of the Gauss–Jordan row operations as given in Section 4.4).
   [*Ans.* The starting basis equals $\mathbf{I}$. The steps at each iteration are as follows:

Step 1: Compute $\mathbf{X}_B = \mathbf{B}^{-1}\mathbf{b}$, the current values of the basic variables. If $\mathbf{X}_B \geq \mathbf{0}$, the solution is feasible; stop. Otherwise, select the leaving variable x_r as the one having the most negative value among all the elements of $\mathbf{X}_B$.

Step 2:
(a) Compute $z_j - c_j = \mathbf{C}_B\mathbf{B}^{-1}\mathbf{P}_j - c_j$ for all the nonbasic variables x_j.
(b) For all the nonbasic variables x_j, compute the constraint coefficients α_r^j associated with the row of the leaving variable x_r using the formula

$$\alpha_r^j = (\text{row of } \mathbf{B}^{-1} \text{ associated with } x_r) \times \mathbf{P}_j$$

(c) The entering variable is associated with

$$\theta = \min_j\left\{\left|\frac{z_j - c_j}{\alpha_r^j}\right|, \alpha_r^j < 0\right\}$$

(If all $\alpha_r^j \geq 0$, no feasible solution exists.)

Step 3: Obtain the new basis by interchanging the entering and leaving vectors $\mathbf{P}_j$ and $\mathbf{P}_r$ using the familiar formula

$$\mathbf{B}_{next}^{-1} = \mathbf{E}\mathbf{B}^{-1}$$

Set $\mathbf{B}^{-1} = \mathbf{B}_{next}^{-1}$ and go to step 1.

(See Example 7.6–2 for an application of this procedure. Keep in mind that the "new" method is exactly equivalent to the one presented in Section 4.4.)

7.3.3 COMPUTATIONAL ADVANTAGES OF THE REVISED METHOD

The revised simplex method offers two other advantages from the standpoint of the use of the digital computer for solving LP problems:

1. It can reduce the volume of computations.
2. It uses the computer memory efficiently.

The revised method can involve less computations than the tableau method (Chapter 3) if the ratio of the zero to nonzero elements of the matrix $\mathbf{A}$ is high (i.e., $\mathbf{A}$ is *sparse*). The reason is that in the revised method we always deal with the original data. Thus we can take advantage of the zero elements when we carry out the arithmetic operations. In contrast, the tableau method generates each new iteration from the immediately preceding one, a process that will eventually increase the density of the matrix $\mathbf{A}$.

Savings in the use of computer memory can result because the revised method deals primarily with the basis $\mathbf{B}$. In the tableau method, on the other hand, the entire tableau must be stored in the computer memory.

The revised method, unlike that of Chapter 3, has the potential to alleviate the problem of machine round-off error. Specifically, methods (other than the product form) can be used to minimize the effect of round-off error on the accuracy of $\mathbf{B}^{-1}$, and hence all the revised method computations.

7.4 BOUNDED VARIABLES

Applications of linear programming exist where, in addition to the regular constraints, some (or all) variables are bounded from above and below. In this case the problem appears as

$$\text{maximize } z = \mathbf{C}\mathbf{X}$$

subject to

$$(\mathbf{A}, \mathbf{I})\mathbf{X} = \mathbf{b}$$
$$\mathbf{L} \leq \mathbf{X} \leq \mathbf{U}$$

where

$$\mathbf{U} = \begin{pmatrix} u_1 \\ u_2 \\ \vdots \\ u_{n+m} \end{pmatrix} \quad \text{and} \quad \mathbf{L} = \begin{pmatrix} l_1 \\ l_2 \\ \vdots \\ l_{n+m} \end{pmatrix}, \quad \mathbf{U} \geq \mathbf{L} \geq \mathbf{0}$$

The elements of $\mathbf{L}$ and $\mathbf{U}$ for an unbounded variable are 0 and ∞.

The problem can be solved by the regular simplex method, in which case the constraints are put in the form

$$(\mathbf{A}, \mathbf{I})\mathbf{X} = \mathbf{b}$$
$$\mathbf{X} + \mathbf{X}' = \mathbf{U}$$
$$\mathbf{X} - \mathbf{X}'' = \mathbf{L}$$
$$\mathbf{X}, \mathbf{X}', \mathbf{X}'' \geq \mathbf{0}$$

where $\mathbf{X}'$ and $\mathbf{X}''$ are slack and surplus variables. This problem includes $3(m + n)$ variables and $3m + 2n$ constraint equations. However, the size can be reduced considerably through the use of special techniques that will ultimately reduce the constraints to the set

$$(\mathbf{A}, \mathbf{I})\mathbf{X} = \mathbf{b}$$

Consider first the lower-bound constraints. The effect of these constraints can be accounted for by using the substitution

$$\mathbf{X} = \mathbf{L} + \mathbf{X}''$$

to eliminate $\mathbf{X}$ from all the remaining constraints. The new variables of the problem thus become $\mathbf{X}'$ and $\mathbf{X}''$. There is no fear in this case that $\mathbf{X}$ may violate the nonnegativity constraint, since both $\mathbf{L}$ and $\mathbf{X}''$ are nonnegative.

The real problem occurs with the upper-bounded variables. A substitution similar to that of the lower-bound case is illegitimate, since there is no guarantee that $\mathbf{X} = \mathbf{U} - \mathbf{X}'$ will remain nonnegative. This difficulty is overcome by using the following procedure.

The upper-bounded problem may be written as

$$\text{maximize } z = \mathbf{CX}$$

subject to

$$(\mathbf{A}, \mathbf{I})\mathbf{X} = \mathbf{b}$$
$$\mathbf{X} + \mathbf{X}' = \mathbf{U}$$
$$\mathbf{X}, \mathbf{X}' \geq \mathbf{0}$$

It is assumed that $\mathbf{b}$ is a nonnegative vector so that the problem is initially primal-feasible.

Rather than include the constraints

$$\mathbf{X} + \mathbf{X}' = \mathbf{U}$$

in the simplex tableau, one can account for their effect by modifying the feasibility condition of the simplex method. The optimality condition remains the same as in the regular simplex method.

The basic idea for modifying the feasibility condition is that a variable becomes infeasible if it becomes negative or exceeds its upper bound. The nonnegativity condition is treated exactly as in the regular simplex method. The upper-bound condition requires special provisions that will allow a basic variable to become nonbasic at its upper bound. (Compare with the regular simplex method, where all the nonbasic variables are at zero level.) Also, when a nonbasic variable is selected to enter the solution, its entering value should not exceed its upper bound. Thus, in developing the new feasibility condition, two main points must be considered:

1. The nonnegativity and upper-bound constraints for the entering variable.
2. The nonnegativity and upper-bound constraints for those basic variables that may be affected by introducing the entering variable.

To develop these ideas mathematically, consider the linear programming problem without the upper bounds. At every iteration, one guarantees that the solution is feasible as follows. Let x_j be a nonbasic variable at *zero* level that is selected to enter the solution. (Later it is shown that every nonbasic variable can always be put at zero level.) Let $(\mathbf{X}_B)_i = (\mathbf{X}_B^*)_i$ be the ith variable of the current basic solution $\mathbf{X}_B$. Thus introducing x_j into the solution gives

$$(\mathbf{X}_B)_i = (\mathbf{X}_B^*)_i - \alpha_i^j x_j$$

where α_i^j is the ith element of $\boldsymbol{\alpha}^j = \mathbf{B}^{-1}\mathbf{P}_j$ and $\mathbf{P}_j$ is the vector of $(\mathbf{A}, \mathbf{I})$ corresponding to x_j.

Now, according to the guidelines given, x_j remains feasible if

$$0 \leq x_j \leq u_j \tag{i}$$

whereas $(\mathbf{X}_B)_i$ remains feasible if

$$0 \leq (\mathbf{X}_B^*)_i - \alpha_i^j x_j \leq u_i, \qquad i = 1, 2, \ldots, m \tag{ii}$$

Since the introduction of x_j into the solution implies that it must be nonnegative, condition (i) is taken care of by observing the upper bound on x_j. Next, consider condition (ii). From the nonnegativity condition

$$(\mathbf{X}_B)_i = (\mathbf{X}_B^*)_i - \alpha_i^j x_j \geq 0$$

it follows that only $\alpha_i^j > 0$ may cause $(\mathbf{X}_B)_i$ to be negative. Let θ_1 represent the maximum value of x_j resulting from this condition. Thus,

$$\theta_1 = \min_i \left\{ \frac{(\mathbf{X}_B^*)_i}{\alpha_i^j}, \alpha_i^j > 0 \right\}$$

This actually is the same as the feasibility condition of the regular simplex method.

Letting

$$\mathbf{U}_B = \text{upper-bound vector for current basis}$$

then to guarantee that $(\mathbf{X}_B)_i$ will not exceed its upper bound, it is necessary that

$$(\mathbf{X}_B)_i = (\mathbf{X}_B^*)_i + (-\alpha_i^j)x_j \leq (\mathbf{U}_B)_i$$

This condition can be violated if α_i^j is negative. Thus, by letting θ_2 represent the maximum value of x_j resulting from this condition,

$$\theta_2 = \min_i \left\{ \frac{(\mathbf{U}_B)_i - (\mathbf{X}_B^*)_i}{-\alpha_i^j}, \ \alpha_i^j < 0 \right\}$$

Let θ denote the maximum value of x_j that does not violate any of the conditions above. Then

$$\theta = \min \{\theta_1, \theta_2, u_j\}$$

It is noticed that an old basic variable $(\mathbf{X}_B)_i$ can become nonbasic only if the introduction of the entering variable x_j at level θ causes $(\mathbf{X}_B)_i$ to be at zero level or at its upper bound. This means that if $\theta = u_j$, x_j cannot be made basic, since no $(\mathbf{X}_B)_i$ can be dropped from the solution, and thus x_j should remain nonbasic but at its upper bound. (If $\theta = u_j = \theta_1 = \theta_2$, the tie may be broken arbitrarily.)

In the foregoing derivation, the entering variable x_j is assumed to be at zero level before it is introduced into the solution. To maintain the validity of the results, every nonbasic variable x_k at upper bound can be put at zero level by the substitution

$$x_k = u_k - x_k'$$

where $0 \leq x_k' \leq u_k$.

Using these ideas, we can effect the changes in the current basic solution as follows. Let $(\mathbf{X}_B)_r$ be the variable corresponding to $\theta = \min \{\theta_1, \theta_2, u_j\}$; then

1. If $\theta = \theta_1$, $(\mathbf{X}_B)_r$ leaves the solution and x_j enters by using the regular row operations of the simplex method.
2. If $\theta = \theta_2$, $(\mathbf{X}_B)_r$ leaves and x_j enters; then $(\mathbf{X}_B)_r$ being nonbasic at its upper bound must be substituted out by using $(\mathbf{X}_B)_r = u_r - (\mathbf{X}_B)_r'$.
3. If $\theta = u_j$, x_j is substituted at its upper bound $u_j - x_j'$ but remains nonbasic.

Example 7.4-1. Consider the following problem:

$$\text{maximize } z = 3x_1 + 5y + 2x_3$$

subject to

$$x_1 + y + 2x_3 \leq 14$$
$$2x_1 + 4y + 3x_3 \leq 43$$
$$0 \leq x_1 \leq 4, \quad 7 \leq y \leq 10, \quad 0 \leq x_3 \leq 3$$

Since y has a positive lower bound, it must be substituted at its lower bound. Let $y = x_2 + 7$; then $0 \leq x_2 \leq 10 - 7 = 3$, and the starting tableau becomes

Basic	x_1	x_2	x_3	x_4	x_5	Solution
z	-3	-5	-2	0	0	35
x_4	1	1	2	1	0	7
x_5	2	4	3	0	1	15

First Iteration
Select x_2 as the entering variable ($z_2 - c_2 = -5$). Thus,

$$\alpha^2 = \binom{1}{4} > 0$$

and

$$\theta_1 = \min \{7/1, 15/4\} = 3.75$$

Since all $\alpha_i^2 > 0$, it follows that $\theta_2 = \infty$. Thus $\theta = \min \{3.75, \infty, 3\} = 3$.

Because $\theta = u_2$, x_2 is substituted at its upper limit but it remains nonbasic. Thus, putting $x_2 = u_2 - x_2' = 3 - x_2'$, the new tableau becomes

Basic	x_1	x_2'	x_3	x_4	x_5	Solution
z	-3	5	-2	0	0	50
x_4	1	-1	2	1	0	4
x_5	2	-4	3	0	1	3

Second Iteration
Select x_1 as the entering variable ($z_1 - c_1 = -3$). Thus

$$\alpha^1 = \binom{1}{2}$$

$$\theta_1 = \min \{4/1, 3/2\} = 3/2, \quad \text{corresponding to } x_4$$

$$\theta_2 = \infty$$

Hence $\theta = \min \{3/2, \infty, 4\} = 3/2$. Since $\theta = \theta_1$, introduce x_1 and drop x_5. This yields

Basic	x_1	x_2'	x_3	x_4	x_5	Solution
z	0	-1	$5/2$	0	$3/2$	$109/2$
x_4	0	1	$1/2$	1	$-1/2$	$5/2$
x_1	1	-2	$3/2$	0	$1/2$	$3/2$

Third Iteration
Select x_2' as the entering variable. Since

$$\alpha^2 = \binom{1}{-2} \leq 0$$

$$\theta_1 = 5/2$$

$$\theta_2 = \frac{4 - 3/2}{-(-2)} = 5/4, \quad \text{corresponding to } x_1$$

Thus $\theta = \min \{5/2, 5/4, 3\} = 5/4$. Since $\theta = \theta_2$, introduce x_2' into the basis and drop x_1; then substitute x_1 out at its upper bound $(4 - x_1')$. Thus, by removing x_1 and introducing x_2', the tableau becomes

Basic	x_1	x_2'	x_3	x_4	x_5	Solution
z	$-1/2$	0	$7/4$	0	$5/4$	$215/4$
x_4	$1/2$	0	$5/4$	1	$-1/4$	$13/4$
x_2'	$-1/2$	1	$-3/4$	0	$-1/4$	$-3/4$

Now, by substituting for $x_1 = 4 - x_1'$, the final tableau becomes

Basic	x_1'	x_2'	x_3	x_4	x_5	Solution
z	$1/2$	0	$7/4$	0	$5/4$	$223/4$
x_4	$-1/2$	0	$5/4$	1	$-1/4$	$5/4$
x_2'	$1/2$	1	$-3/4$	0	$-1/4$	$5/4$

which is now optimal and feasible.

The optimal solution in terms of the original variables x_1, x_2, and x_3 is found as follows. Since $x_1' = 0$, it follows that $x_1 = 4$. Also, since $x_2' = 5/4$, $x_2 = 3 - 5/4 = 7/4$ and $y = 7 + 7/4 = 35/4$. Finally, x_3 equals 0. These values yield $z = 223/4$, as shown in the optimal tableau. ◄

It might be beneficial at this point to study the effect of the upper bounding technique on the development of the simplex tableau. Specifically, it is required to define $\{z_j - c_j\}$ and the basic solution at every iteration. This will be presented here in matrix notation.

Let $\mathbf{X}_u$ represent the basic *and* nonbasic variables in $\mathbf{X}$ that have been substituted at their upper bound. Also, let $\mathbf{X}_z$ be the remaining basic *and* nonbasic variables. Suppose that the order vectors of $(\mathbf{A}, \mathbf{I})$ corresponding to $\mathbf{X}_z$ and $\mathbf{X}_u$ are given by the matrices $\mathbf{D}_z$ and $\mathbf{D}_u$, and let the vector $\mathbf{C}$ of the objective function be partitioned correspondingly to give $(\mathbf{C}_z, \mathbf{C}_u)$. The equations of the linear programming problem at any iteration then become

$$\begin{pmatrix} 1 & -\mathbf{C}_z & -\mathbf{C}_u \\ 0 & \mathbf{D}_z & \mathbf{D}_u \end{pmatrix} \begin{pmatrix} z \\ \mathbf{X}_z \\ \mathbf{X}_u \end{pmatrix} = \begin{pmatrix} 0 \\ \mathbf{b} \end{pmatrix}$$

Instead of dealing with two types of variables, $\mathbf{X}_z$ and $\mathbf{X}_u$, $\mathbf{X}_u$ is put at zero level by using the substitution

$$\mathbf{X}_u = \mathbf{U}_u - \mathbf{X}_u'$$

where $\mathbf{U}_u$ is a subset of $\mathbf{U}$ representing the upper bounds for the variables in $\mathbf{X}_u$. This gives

$$\begin{pmatrix} 1 & -\mathbf{C}_z & \mathbf{C}_u \\ 0 & \mathbf{D}_z & -\mathbf{D}_u \end{pmatrix} \begin{pmatrix} z \\ \mathbf{X}_z \\ \mathbf{X}_u' \end{pmatrix} = \begin{pmatrix} \mathbf{C}_u \mathbf{U}_u \\ \mathbf{b} - \mathbf{D}_u \mathbf{U}_u \end{pmatrix}$$

The optimality and the feasibility conditions can be developed more easily now since all nonbasic variables are at zero level. However, it is still necessary to check that no basic or nonbasic variable will exceed its upper bound.

Define X_B as the basic variables of the current iteration, and let C_B represent the elements corresponding to X_B in C. Also, let B be the basic matrix corresponding to X_B. The current solution is determined from

$$\begin{pmatrix} 1 & -C_B \\ 0 & B \end{pmatrix}\begin{pmatrix} z \\ X_B \end{pmatrix} = \begin{pmatrix} C_u U_u \\ b - D_u U_u \end{pmatrix}$$

By inverting the partitioned matrix as in Section 7.2.2, the current basic solution is given by

$$\begin{pmatrix} z \\ X_B \end{pmatrix} = \begin{pmatrix} 1 & C_B B^{-1} \\ 0 & B^{-1} \end{pmatrix}\begin{pmatrix} C_u U_u \\ b - D_u U_u \end{pmatrix} = \begin{pmatrix} C_u U_u + C_B B^{-1}(b - D_u U_u) \\ B^{-1}(b - D_u U_u) \end{pmatrix}$$

By using

$$b' = b - D_u U_u$$

the complete simplex tableau corresponding to any iteration is (compare with Section 7.2.2).

Basic	X_z^T	$X_u'^T$	Solution
z	$C_B B^{-1}D_z - C_z$	$-C_B B^{-1}D_u + C_u$	$C_B B^{-1}b' + C_u U_u$
X_B	$B^{-1}D_z$	$-B^{-1}D_u$	$B^{-1}b'$

The arrangement of this tableau is the same as the one presented in Section 7.2.2. For the z-equation, the left-hand side coefficients yield $z_j - c_j$, the optimality indicator for all the nonbasic variables in X_z and X_u', and its right-hand side yields the corresponding value of z. The constraint coefficients $B^{-1}(D_z, -D_u)$ give the corresponding $\{\alpha\}$ for the nonbasic variables. Finally, the right-hand side of the constraint equations $B^{-1}b'$ gives directly the values of X_B.

Exercise 7.4–1
Consider the optimal tableau of Example 7.4–1. Compute the optimal inverse B^{-1}; then show how the entire tableau can be generated from B^{-1} and the original data of the problem by using the matrix representation shown. Can B^{-1} be identified directly from the optimal tableau of Example 7.4–1?

[*Ans.* $B = (P_4, P_{2'})$, $B^{-1} = \begin{pmatrix} 1 & -1/4 \\ 0 & -1/4 \end{pmatrix}$. Yes, B^{-1} is located under the starting solution variables.]

7.5 DECOMPOSITION ALGORITHM

The special structure of certain large linear programs may allow determination of the optimal solution by first decomposing the problem into smaller subproblems and then solving the subproblems almost independently. The procedure has the advantage of making it possible to solve large-scale problems that otherwise may be computationally infeasible.

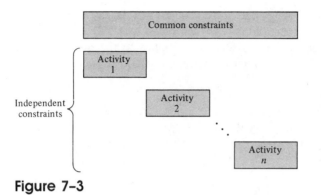

Figure 7-3

A typical situation can arise in the planning of production facilities at the corporate level. Although each facility may have its independent constraints, the different activities are usually tied together at the corporate level by budgetary constraints. The two types of constraints are referred to as *common* and *independent* restrictions.

Figure 7-3 demonstrates a typical structure of an LP model that can be solved by decomposing the problem into smaller subproblems. Note that the independent constraints of the n activities do not overlap. In the absence of the common constraints, the different activities will be completely independent of one another.

Let $\mathbf{D}_j$ $(j = 1, 2, \ldots, n)$ be the technology matrix of the jth activity (e.g., production facility) and let $\mathbf{X}_j$ represent the vector of the corresponding variables. Let the resources of the jth activity be given by the vector $\mathbf{b}_j$. It follows that each set of independent constraints can be written as†

$$\mathbf{D}_j\mathbf{X}_j = \mathbf{b}_j, \qquad j = 1, 2, \ldots, n$$

For the common constraints, let $\mathbf{A}_j$ be the technological matrix of the jth activity and let $\mathbf{b}_0$ be its corresponding resources vector. This gives

$$\mathbf{A}_1\mathbf{X}_1 + \mathbf{A}_2\mathbf{X}_2 + \cdots + \mathbf{A}_n\mathbf{X}_n = \mathbf{b}_0$$

Further, let $\mathbf{C}_j$ represent the vector of the objective function coefficients for the jth activity. Thus the complete problem becomes

$$\text{maximize } z = \mathbf{C}_1\mathbf{X}_1 + \mathbf{C}_2\mathbf{X}_2 + \cdots + \mathbf{C}_n\mathbf{X}_n$$

subject to

$$\begin{aligned}
\mathbf{A}_1\mathbf{X}_1 + \mathbf{A}_2\mathbf{X}_2 + \cdots + \mathbf{A}_n\mathbf{X}_n &= \mathbf{b}_0 \\
\mathbf{D}_1\mathbf{X}_1 \qquad\qquad\qquad\qquad &= \mathbf{b}_1 \\
\mathbf{D}_2\mathbf{X}_2 \qquad\qquad &= \mathbf{b}_2 \\
&\;\;\vdots \\
\mathbf{D}_n\mathbf{X}_n &= \mathbf{b}_n
\end{aligned}$$

$$\mathbf{X}_j \geq \mathbf{0}, \qquad \text{for all } j$$

---

† Slack variables are added as necessary to change inequality constraints into equalities. In this case, the resulting slacks and coefficients are assumed to be part of $\mathbf{X}_j$ and $\mathbf{D}_j$.

If the size of $\mathbf{A}_j$ is $(r_0 \times m_j)$ and that of $\mathbf{D}_j$ is $(r_j \times m_j)$, the problem has $\Sigma_{j=0}^{n} \, r_j$ constraints and $\Sigma_{j=0}^{n} \, m_j$ variables.

The **decomposition principle** as applied to the preceding problem will now be discussed. It is assumed that each of the convex sets

$$\mathbf{D}_j\mathbf{X}_j = \mathbf{b}_j, \qquad \mathbf{X}_j \geq \mathbf{0}, \qquad j = 1, 2, \ldots, n$$

is bounded. Thus, if $\hat{\mathbf{X}}_j^k$, $k = 1, 2, \ldots, K_j$, are the extreme points of the jth set, every point $\mathbf{X}_j$ in this set can be expressed as a convex combination of these extreme points. This means that, for $\beta_j^k \geq 0$ and $\Sigma_{k=1}^{K_j} \, \beta_j^k = 1$,

$$\mathbf{X}_j = \sum_{k=1}^{K_j} \beta_j^k \hat{\mathbf{X}}_j^k, \qquad j = 1, 2, \ldots, n$$

These new equations imply the complete solution space defined by the sets $\mathbf{D}_j\mathbf{X}_j = \mathbf{b}_j$ and $\mathbf{X}_j \geq \mathbf{0}$, $j = 1, 2, \ldots, n$. It is thus possible to eliminate the constraints of the subproblems and reformulate the original problem in the following equivalent form, which we shall call the **master problem:**

$$\text{maximize } z = \sum_{k=1}^{K_1} \mathbf{C}_1\hat{\mathbf{X}}_1^k\beta_1^k + \sum_{k=1}^{K_2} \mathbf{C}_2\hat{\mathbf{X}}_2^k\beta_2^k + \cdots + \sum_{k=1}^{K_n} \mathbf{C}_n\hat{\mathbf{X}}_n^k\beta_n^k$$

subject to

$$\sum_{k=1}^{K_1} \mathbf{A}_1\hat{\mathbf{X}}_1^k\beta_1^k + \sum_{k=1}^{K_2} \mathbf{A}_2\hat{\mathbf{X}}_2^k\beta_2^k + \cdots + \sum_{k=1}^{K_n} \mathbf{A}_n\hat{\mathbf{X}}_n^k\beta_n^k = \mathbf{b}_0$$

$$\sum_{k=1}^{K_1} \beta_1^k \qquad\qquad\qquad\qquad\qquad\qquad = 1$$

$$\sum_{k=1}^{K_2} \beta_2^k \qquad\qquad\qquad = 1$$

$$\vdots$$

$$\sum_{k=1}^{K_n} \beta_n^k = 1$$

$$\beta_j^k \geq 0, \qquad \text{for all } j \text{ and } k$$

Since $\hat{\mathbf{X}}_j^k$ are the "known" extreme points of the set $\mathbf{D}_j\mathbf{X}_j = \mathbf{b}_j$, $\mathbf{X}_j \geq \mathbf{0}$, the new decision variables of the modified problem become β_j^k. Once the optimal values of β_j^k for all j and k are determined, the optimal solution to the original problem is obtained by recognizing that

$$\mathbf{X}_j = \sum_{k=1}^{K_j} \beta_j^k \hat{\mathbf{X}}_j^k$$

Exercise 7.5-1

In each of the following cases, express the solution space as a convex combination of its extreme points.

(a) Solution space in Figure 3–4.

    [*Ans.* $(\bar{x}_1, \bar{x}_2) = \beta_1(0, 0) + \beta_2(0, 2) + \beta_3(4, 0)$, where $\beta_1, \beta_2, \beta_3 \geq 0$ and $\beta_1 + \beta_2 + \beta_3 = 1$.]

(b) Solution space in Figure 3–5.

[*Ans.* $(\bar{x}_1, \bar{x}_2) = \beta_1(0, 0) + \beta_2(2, 0) + \beta_3(3/2, 2) + \beta_4(0, 4)$, where $\beta_1, \beta_2, \beta_3, \beta_4$ ≥ 0 and $\beta_1 + \beta_2 + \beta_3 + \beta_4 = 1$.]

(c) Solution space in Figure 3–8.

[*Ans.* The solution space is unbounded. Hence it is not possible to determine $(\bar{x}_1, \bar{x}_2)$ in terms of the extreme points. However, if we augment the artificial restriction $x_2 \leq M$, where M is very large, then we can represent the solution space as

$$(\bar{x}_1, \bar{x}_2) = \beta_1(0, 0) + \beta_2(1, 0) + \beta_3(4, 6) + \beta_4(4, M) + \beta_5(0, M)$$

where $\beta_1 + \beta_2 + \beta_3 + \beta_4 + \beta_5 = 1$ and $\beta_1, \beta_2, \beta_3, \beta_4, \beta_5 \geq 0$.]

To solve the *master* problem by the (revised) simplex method, we need to determine the entering and leaving vectors. But to do so, it appears, at first thought, that all the extreme points $\hat{\mathbf{X}}_j^k$ must be known in advance. Fortunately, this is not the case. What the decomposition algorithm does is recognize that we need only identify the one extreme point (from among all $\hat{\mathbf{X}}_j^k$) that is associated with the entering variable. Once done, we can determine *numerically* all the elements of the entering vector as well as its objective equation coefficient. The leaving vector can then be determined by using the feasibility condition of the simplex method.

The overall idea of the decomposition algorithm thus involves two principal phases:

1. Convert the original problem into a master problem by implicitly expressing the solution space of each subproblem as a convex combination of its extreme points.

2. Generate the column vector associated with the entering vector and use this information to determine the leaving vector.

The second-phase operation is commonly referred to as **column generation** because it generates the elements of entering vector.

We now show how these ideas can be expressed mathematically. Let $\mathbf{B}$ be the current basis of the *master* problem and $\mathbf{C}_B$ the vector of the corresponding coefficients in the objective function. Thus, according to the revised simplex method, the current solution is optimal if for all nonbasic $\mathbf{P}_j^k$,

$$z_j^k - c_j^k = \mathbf{C}_B \mathbf{B}^{-1} \mathbf{P}_j^k - c_j^k \geq 0$$

where, from the definition of the master problem,

$$c_j^k = \mathbf{C}_j \hat{\mathbf{X}}_j^k \quad \text{and} \quad \mathbf{P}_j^k = n \left\{ \begin{array}{c} r_0 \left\{ \begin{pmatrix} \mathbf{A}_j \hat{\mathbf{X}}_j^k \\ 0 \\ \vdots \\ 1 \\ \vdots \\ 0 \end{pmatrix} \leftarrow (r_0 + j)\text{th place} \right. \end{array} \right.$$

The expression for $z_j^k - c_j^k$ can be simplified as follows. Let

$$\mathbf{B}^{-1} = (\overbrace{\mathbf{R}_0}^{r_0} \mid \overbrace{\mathbf{V}_1, \mathbf{V}_2, \ldots, \mathbf{V}_j, \ldots, \mathbf{V}_n}^{n})$$

where $\mathbf{R}_0$ is the matrix of size $(r_0 + n) \times r_0$ consisting of the first r_0 columns of $\mathbf{B}^{-1}$, and $\mathbf{V}_j$ is the $(r_0 + j)$th column of the same matrix $\mathbf{B}^{-1}$. Thus,

$$z_j^k - c_j^k = (\mathbf{C}_B \mathbf{R}_0 \mathbf{A}_j \hat{\mathbf{X}}_j^k + \mathbf{C}_B \mathbf{V}_j) - \mathbf{C}_j \hat{\mathbf{X}}_j^k$$
$$= (\mathbf{C}_B \mathbf{R}_0 \mathbf{A}_j - \mathbf{C}_j)\hat{\mathbf{X}}_j^k + \mathbf{C}_B \mathbf{V}_j$$

If the current solution is not optimal, the vector $\mathbf{P}_j^k$ having the smallest (most negative) $z_j^k - c_j^k < 0$ is selected to enter the solution. The important point is that $z_j^k - c_j^k$ cannot be evaluated numerically until all the elements of $\mathbf{P}_j^k$ are known. On the other hand, $\mathbf{P}_j^k$ cannot be evaluated numerically until the corresponding extreme point $\hat{\mathbf{X}}_j^k$ is known. This leads to the key point that the evaluation of $z_j^k - c_j^k$ (and hence the selection of the entering variable) depends on the determination of $\hat{\mathbf{X}}_j^k$.

Instead of determining all the extreme points for all the n sets, the problem can be reduced to determining the extreme point $\hat{\mathbf{X}}_j^{k*}$ in every set j that will yield the smallest $z_j^k - c_j^k$. Let

$$\rho_j = \min_k \{z_j^k - c_j^k\} = z_j^{k*} - c_j^{k*}$$

and let $\rho = \min_j \{\rho_j\}$. Consequently, if $\rho < 0$, the variable β_j^{k*} corresponding to ρ is selected as the entering variable. Otherwise, if $\rho \geq 0$, the optimal solution is attained.

The extreme point $\hat{\mathbf{X}}_j^{k*}$ is determined by solving the linear programming problem:

$$\text{minimize } z_j - c_j$$

subject to

$$\mathbf{D}_j \mathbf{X}_j = \mathbf{b}_j$$
$$\mathbf{X}_j \geq 0$$

The superscript k is suppressed for the following reason. Since by assumption the set $\mathbf{D}_j \mathbf{X}_j = \mathbf{b}_j$, $\mathbf{X}_j \geq \mathbf{0}$, is bounded, the minimum value of $z_j - c_j$ is also bounded and must occur at an extreme point of the set. This automatically gives the required extreme point $\hat{\mathbf{X}}_j^{k*}$.

Now, as shown previously,

$$z_j^k - c_j^k = (\mathbf{C}_B \mathbf{R}_0 \mathbf{A}_j - \mathbf{C}_j)\hat{\mathbf{X}}_j^k + \mathbf{C}_B \mathbf{V}_j$$

Since $\mathbf{C}_B \mathbf{V}_j$ is a constant independent of k, the linear programming problem becomes

$$\text{minimize } w_j = (\mathbf{C}_B \mathbf{R}_0 \mathbf{A}_j - \mathbf{C}_j)\mathbf{X}_j$$

subject to

$$\mathbf{D}_j \mathbf{X}_j = \mathbf{b}_j$$
$$\mathbf{X}_j \geq \mathbf{0}$$

It thus follows that

$$\rho_j = w_j^* + \mathbf{C}_B \mathbf{V}_j$$

where w_j^* is the optimum value of w_j. As stated, the variable β_j^k corresponding to $\rho = \min_j \{\rho_j\}$ is then selected to enter the solution. (Notice that the extreme point corresponding to β_j^k is automatically known at this point.)

The leaving variable is determined in the usual manner by using the feasibility condition of the revised simplex method. At the iteration when ρ becomes non-negative, the optimal solution to the original problem is given by

$$\mathbf{X}_j^* = \sum_{k=1}^{K_j} \beta_j^{k*} \hat{\mathbf{X}}_j^{k*}, \qquad j = 1, 2, \ldots, n$$

where β_j^{k*} is the optimal solution to the modified problem and $\hat{\mathbf{X}}_j^{k*}$ is the corresponding extreme point.

The decomposition algorithm is summarized by the following steps:

Step 1: Reduce the original problem to the modified form in terms of the new variables β_j^k.

Step 2: Find an initial basic feasible solution to the modified problem. If such a solution is not immediately obvious, use the artificial variables technique (see Section 3.2.3) to secure an initial basis.

Step 3: For the current iteration, find $\rho_j = w_j^* + \mathbf{C}_B \mathbf{V}_j$ for each subproblem j and then determine $\rho = \min_j \{\rho_j\}$. If $\rho \geq 0$, the current solution is optimal and the process is terminated; otherwise,

Step 4: Introduce the variable β_j^k corresponding to ρ into the basic solution. Determine the leaving variable and then compute the next $\mathbf{B}^{-1}$. Go to step 3.

Example 7.5-1

$$\text{Maximize } z = 3x_1 + 5x_2 + x_3 + x_4$$

subject to

$$
\begin{aligned}
x_1 + x_2 + x_3 + x_4 &\leq 40 \\
5x_1 + x_2 &\leq 12 \\
x_3 + x_4 &\geq 5 \\
x_3 + 5x_4 &\leq 50 \\
x_1, x_2, x_3, x_4 &\geq 0
\end{aligned}
$$

By augmenting the slacks S_1, S_2, S_3, and S_4 to the constraints, the information of the problem can be summarized as follows:

x_1	x_2	S_2	x_3	x_4	S_3	S_4	S_1	
3	5	0	1	1	0	0		
1	1	0	1	1	0	0	1	40
5	1	1						12
			1	1	−1	0		5
			1	5	0	1		50

The problem can be decomposed into two subproblems, $j = 1, 2$. For $j = 1$,

$$\mathbf{X}_1 = (x_1, x_2, S_2)^T, \qquad \mathbf{C}_1 = (3, 5, 0)$$
$$\mathbf{A}_1 = (1, 1, 0), \qquad \mathbf{D}_1 = (5, 1, 1)$$
$$b_1 = 12$$

For $j = 2$,

$$\mathbf{X}_2 = (x_3, x_4, S_3, S_4)^T, \qquad \mathbf{C}_2 = (1, 1, 0, 0)$$
$$\mathbf{A}_2 = (1, 1, 0, 0), \qquad \mathbf{D}_2 = \begin{pmatrix} 1 & 1 & -1 & 0 \\ 1 & 5 & 0 & 0 \end{pmatrix}$$
$$\mathbf{b}_2 = (5, 50)^T$$

The slack variable S_1 does not constitute a subproblem in the sense given above. Thus it is treated separately, as will be shown.

Let R_1 and R_2 be artificial variables. The starting solution for the modified problem is thus given as follows.

β_1^1	β_1^2	$\cdots$	$\beta_1^{K_1}$	β_2^1	β_2^2	$\cdots$	$\beta_2^{K_2}$	S_1	R_1	R_2	
$\mathbf{C}_1\hat{\mathbf{X}}_1^1$	$\mathbf{C}_1\hat{\mathbf{X}}_1^2$	$\cdots$	$\mathbf{C}_1\hat{\mathbf{X}}_1^{K_1}$	$\mathbf{C}_2\hat{\mathbf{X}}_2^1$	$\mathbf{C}_2\hat{\mathbf{X}}_2^2$	$\cdots$	$\mathbf{C}_2\hat{\mathbf{X}}_2^{K_2}$	0	$-M$	$-M$	
$\mathbf{A}_1\hat{\mathbf{X}}_1^1$	$\mathbf{A}_1\hat{\mathbf{X}}_1^2$	$\cdots$	$\mathbf{A}_1\hat{\mathbf{X}}_1^{K_1}$	$\mathbf{A}_2\hat{\mathbf{X}}_2^1$	$\mathbf{A}_2\hat{\mathbf{X}}_2^2$	$\cdots$	$\mathbf{A}_2\hat{\mathbf{X}}_2^{K_2}$	1	0	0	40
1	1	$\cdots$	1	0	0	$\cdots$	0	0	1	0	1
0	0	$\cdots$	0	1	1	$\cdots$	1	0	0	1	1

$\underbrace{\qquad\qquad\qquad\qquad}_{\text{Subproblem 1}}$ $\underbrace{\qquad\qquad\qquad\qquad}_{\text{Subproblem 2}}$ $\underbrace{\qquad\qquad\qquad}_{\text{Starting basic solution}}$

Exercise 7.5–2

If the first (common) constraints of the original problem is $\geq$ instead of $\leq$, indicate how the tableau will be affected.

[*Ans.* Augment an artificial variable to the first equation and change the coefficient of S_1 to -1.]

The information of the starting basic solution is as follows.

$$\mathbf{X}_B = (S_1, R_1, R_2)^T = (40, 1, 1)^T$$

$$\mathbf{B} = \mathbf{B}^{-1} = \begin{pmatrix} 1 & 0 & 0 \\ 0 & 1 & 0 \\ 0 & 0 & 1 \end{pmatrix}, \qquad \mathbf{C}_B = (0, -M, -M)$$

$$\mathbf{R}_0 = \begin{pmatrix} 1 \\ 0 \\ 0 \end{pmatrix}, \qquad \mathbf{V}_1 = \begin{pmatrix} 0 \\ 1 \\ 0 \end{pmatrix}, \qquad \mathbf{V}_2 = \begin{pmatrix} 0 \\ 0 \\ 1 \end{pmatrix}$$

Consequently, $\mathbf{C}_B \mathbf{R}_0 = 0$.

First Iteration

The linear program corresponding to subproblem $j = 1$ is

$$\text{minimize } w_1 = (C_B R_0 A_1 - C_1) X_1$$

subject to

$$D_1 X_1 = b_1, \ X_1 \geq 0$$

Since

$$w_1 = [(0)(1, 1, 0) - (3, 5, 0)] \begin{pmatrix} x_1 \\ x_2 \\ S_2 \end{pmatrix} = -3x_1 - 5x_2$$

the problem becomes

$$\text{minimize } w_1 = -3x_1 - 5x_2$$

subject to

$$(5, 1, 1) \begin{pmatrix} x_1 \\ x_2 \\ S_2 \end{pmatrix} = 12$$

$$x_1, x_2, S_2 \geq 0$$

The optimal solution (obtained by the simplex method) is

$$\hat{X}_1^1 = (0, 12, 0)^T, \qquad w_1^* = -60$$

Thus $\rho_1 = w_1^* + C_B V_I = -60 - M$.

The linear program for $j = 2$ is given as

$$\text{minimize } w_2 = [(0)(1, 1, 0, 0) - (1, 1, 0, 0)] \begin{pmatrix} x_3 \\ x_4 \\ S_3 \\ S_4 \end{pmatrix} = -x_3 - x_4$$

subject to

$$\begin{pmatrix} 1 & 1 & -1 & 0 \\ 1 & 5 & 0 & 1 \end{pmatrix} \begin{pmatrix} x_3 \\ x_4 \\ S_3 \\ S_4 \end{pmatrix} = \begin{pmatrix} 5 \\ 50 \end{pmatrix}$$

$$x_3, x_4, S_3, S_4 \geq 0$$

This has the optimal solution

$$\hat{X}_2^1 = (50, 0, 45, 0)^T, \qquad w_2^* = -50$$

Thus $\rho_2 = w_2^* + C_B V_2 = -50 - M$.

The solution of the two subproblems thus gives $\rho = \min \{\rho_1, \rho_2\} = \rho_1$. Since $\rho = -60 - M < 0$, the variable β_1^1 corresponding to $\hat{X}_1^1$ enters the solution.

The leaving variable is determined by applying the feasibility condition of the revised simplex method. It is noticed that

$$\mathbf{P}_1^1 = \begin{pmatrix} \mathbf{A}_1 \hat{\mathbf{X}}_1^1 \\ 1 \\ 0 \end{pmatrix} = \begin{bmatrix} (1, 1, 0) \begin{pmatrix} 0 \\ 12 \\ 0 \end{pmatrix} \\ 1 \\ 0 \end{bmatrix} = \begin{pmatrix} 12 \\ 1 \\ 0 \end{pmatrix}$$

Thus α for $\mathbf{P}_1^1$ is

$$\alpha = (\alpha_{S_1}, \alpha_{R_1}, \alpha_{R_2})^T = \mathbf{B}^{-1}\mathbf{P}_1^1 = (12, 1, 0)^T$$

Subsequently, $\theta = \min\{40/12, 1/1, -\} = 1$, which corresponds to R_1. Thus R_1 is the leaving variable.

The formula in Section 7.3.1 is used to determine $\mathbf{B}_{\text{next}}^{-1}$; that is,

$$\mathbf{B}_{\text{next}}^{-1} = \mathbf{E}\mathbf{B}^{-1} = \begin{pmatrix} 1 & -12 & 0 \\ 0 & 1 & 0 \\ 0 & 0 & 1 \end{pmatrix}\begin{pmatrix} 1 & 0 & 0 \\ 0 & 1 & 0 \\ 0 & 0 & 1 \end{pmatrix} = \begin{pmatrix} 1 & -12 & 0 \\ 0 & 1 & 0 \\ 0 & 0 & 1 \end{pmatrix}$$

By letting $\mathbf{B}^{-1} = \mathbf{B}_{\text{next}}^{-1}$, the new basic solution is thus given as

$$\mathbf{X}_B = (S_1, \beta_1^1, R_2)^T = \mathbf{B}^{-1}(40, 1, 1)^T = (28, 1, 1)^T$$

The coefficient of β_1^1 in the modified objective function is $c_1^1 = \mathbf{C}_1\hat{\mathbf{X}}_1^1 = 60$. This gives $\mathbf{C}_B = (0, 60, -M)$. Since

$$\mathbf{R}_0 = \begin{pmatrix} 1 \\ 0 \\ 0 \end{pmatrix}, \qquad \mathbf{V}_1 = \begin{pmatrix} -12 \\ 1 \\ 0 \end{pmatrix}, \qquad \mathbf{V}_2 = \begin{pmatrix} 0 \\ 0 \\ 1 \end{pmatrix}$$

it follows that $\mathbf{C}_B\mathbf{R}_0 = 0$.

Second Iteration

$j = 1$: The objective function is $w_1 = -3x_1 - 5x_2$. Thus the solution space $\mathbf{D}_1\mathbf{X}_1 = b_1$, $\mathbf{X}_1 \geq 0$ yields exactly the same optimal solution vector $\mathbf{X}_1^1$ as in the first iteration. Since the corresponding extreme points have been considered previously, the first subproblem yields no new information at this point. [Actually, $\rho_1 = 0$ (verify!) because β_1^1 is a basic variable.]

$j = 2$: The objective function is $w_2 = -x_3 - x_4$. The solution space $\mathbf{D}_2\mathbf{X}_2 = b_2$, $\mathbf{X}_2 \geq 0$, yields the optimal solution

$$\hat{\mathbf{X}}_2^2 = (50, 0, 45, 0)^T, \qquad w_2^* = -50$$
$$\rho_2 = w_2^* + \mathbf{C}_B\mathbf{V}_2 = -50 - M$$

(Notice that $\hat{\mathbf{X}}_2^2$ is the same as $\hat{\mathbf{X}}_2^1$. Yet, unlike $\hat{\mathbf{X}}_1^1$, β_2^1 is not a basic variable. The superscript 2 is used with $\hat{\mathbf{X}}_2^2$ for notational convenience, i.e., to represent the second iteration.) Since $\rho = \rho_2 < 0$, β_2^2 enters the solution.

The leaving variable is now determined. Consider

$$\mathbf{P}_2^2 = \begin{pmatrix} \mathbf{A}_2\hat{\mathbf{X}}_2^2 \\ 0 \\ 1 \end{pmatrix} = \begin{bmatrix} (1, 1, 0, 0)\begin{pmatrix} 50 \\ 0 \\ 45 \\ 0 \end{pmatrix} \\ 0 \\ 1 \end{bmatrix} = \begin{pmatrix} 50 \\ 0 \\ 1 \end{pmatrix}$$

$$\boldsymbol{\alpha} = (\alpha_{S_1}, \alpha_{\beta_1^1}, \alpha_{R_2})^T = \mathbf{B}^{-1}\mathbf{P}_2^2 = (50, 0, 1)^T$$

Subsequently, $\theta = \min\{28/50, \text{—}, 1/1\} = 14/25$. Thus S_1 is the leaving variable. Next, compute $\mathbf{B}_{\text{next}}^{-1}$ and the new basic solution. Thus,

$$\mathbf{B}^{-1} = \begin{pmatrix} 1/50 & 0 & 0 \\ 0 & 1 & 0 \\ -1/50 & 0 & 1 \end{pmatrix}\begin{pmatrix} 1 & -12 & 0 \\ 0 & 1 & 0 \\ 0 & 0 & 1 \end{pmatrix} = \begin{pmatrix} 1/50 & -12/50 & 0 \\ 0 & 1 & 0 \\ -1/50 & 12/50 & 1 \end{pmatrix}$$

$$\mathbf{X}_B = (\beta_2^2, \beta_1^1, R_2)^T = (14/25, 1, 11/25)^T$$

Since $c_2^2 = \mathbf{C}_2\hat{\mathbf{X}}_2^2 = 50$, it follows that $\mathbf{C}_B = (50, 60, -M)$. Also,

$$\mathbf{R}_0 = \begin{pmatrix} 1/50 \\ 0 \\ -1/50 \end{pmatrix}, \quad \mathbf{V}_1 = \begin{pmatrix} -12/50 \\ 1 \\ 12/50 \end{pmatrix}, \quad \mathbf{V}_2 = \begin{pmatrix} 0 \\ 0 \\ 1 \end{pmatrix}$$

Consequently, $\mathbf{C}_B\mathbf{R}_0 = 1 + M/50$.

Third Iteration

$j = 1$: $w_1 = (M/50 - 2)x_1 + (M/50 - 4)x_2$. The associated optimum solution is $\hat{\mathbf{X}}_1^3 = (0, 12, 0)^T$, which is the same as $\hat{\mathbf{X}}_1^1$.

$j = 2$: $w_2 = (M/50)(x_3 + x_4)$. The associated optimum solution is $\hat{\mathbf{X}}_2^3 = (5, 0, 0, 45)^T$ and $w_2^* = M/10$ with $\rho_2 = w_2^* + \mathbf{C}_B\mathbf{V}_2 = -9M/10$.

This iteration differs from the preceding two in that S_1 is now nonbasic and hence must be checked for the possibility of being a candidate for the entering variable. Consider

$$z_{S_1} - c_{S_1} = \mathbf{C}_B\mathbf{B}^{-1}\mathbf{P}_{S_1} - c_{S_1}$$
$$= \left(1 + \frac{M}{50}, 48 - \frac{12M}{50}, -M\right)(1, 0, 0)^T - 0 = 1 + \frac{M}{50}$$

This shows that S_1 cannot improve the solution. Thus $\rho = \rho_2 = -9M/10$ and β_2^3 associated with $\hat{\mathbf{X}}_2^3$ enters the solution.

To determine the leaving variable, consider

$$\mathbf{P}_2^3 = \begin{bmatrix} (1, 1, 0, 0) \begin{pmatrix} 5 \\ 0 \\ 0 \\ 45 \end{pmatrix} \\ 0 \\ 1 \end{bmatrix} = \begin{pmatrix} 5 \\ 0 \\ 1 \end{pmatrix}$$

$$\boldsymbol{\alpha} = (\alpha_{\beta_2^2}, \alpha_{\beta_1^1}, \alpha_{R_2})^T = \mathbf{B}^{-1}\mathbf{P}_2^3 = (1/10, 0, 9/10)^T$$

This gives

$$\theta = \min\left\{\frac{14/25}{1/10}, \text{—}, \frac{11/25}{9/10}\right\} = 22/45$$

which shows that R_2 is the leaving variable. The new solution is thus given by

$$\mathbf{B}^{-1}=\begin{pmatrix}1&0&-1/9\\0&1&0\\0&0&10/9\end{pmatrix}\begin{pmatrix}1/50&-12/50&0\\0&1&0\\-1/50&12/50&0\end{pmatrix}=\begin{pmatrix}1/45&-12/45&-5/45\\0&1&0\\-1/45&12/45&50/45\end{pmatrix}$$

$$\mathbf{X}_B=(\beta_2^2,\beta_1^1,\beta_2^3)^T=(23/45,\ 1,\ 22/45)^T$$

Since $\mathbf{C}_2^3=\mathbf{C}_2\hat{\mathbf{X}}_2^3=5$, it follows that $\mathbf{C}_B=(50,\ 60,\ 5)$. Hence $\mathbf{C}_B\mathbf{R}_0=1$ (verify!).

Fourth Iteration

$j=1$: $w_1=-2x_1-4x_2$. This gives the same solution as in the first iteration.
$j=2$: $w_2=0x_3+0x_4$. Thus $w_2^*=0$ and $\rho_2=48$.
Slack S_1: $z_{S_1}-c_{S_1}=1-0=1$.

This information shows that the last basic is optimal.
The optimal solution to the original problem is

$$\mathbf{X}_1^*=(x_1,x_2,S_2)^T=\beta_1^1\hat{\mathbf{X}}_1^1=(1)(0,\ 12,\ 0)^T=(0,\ 12,\ 0)^T$$
$$\mathbf{X}_2^*=(x_3,x_4,S_3,S_4)^T=\beta_2^2\hat{\mathbf{X}}_2^2+\beta_2^3\hat{\mathbf{X}}_2^3$$
$$=(23/45)(50,\ 0,\ 45,\ 0)^T+(22/45)(5,\ 0,\ 0,\ 45)^T$$
$$=(28,\ 0,\ 23,\ 22)^T$$

All the remaining variables equal zero. ◀

7.6 PARAMETRIC LINEAR PROGRAMMING

Parametric linear programming investigates the changes in the optimum LP solution due to *predetermined continuous* variations in the model's parameters, such as the availability of resources or changes in marginal profits or costs. In the oil industry, for example, where LP has tremendously successful applications, it is normal to investigate changes in the optimal LP solution resulting from changes in the availability and quality of crude oils.

Parametric analysis is based on very much the same ideas employed with sensitivity analysis in Section 4.5. The main difference is that the coefficients of the model vary continuously rather than in a discrete fashion. We must emphasize, however, that the predetermined functions representing changes in the model's coefficients need *not* be linear, for this has nothing to do with *linear* programming. Figure 7-4 depicts functions that can typically represent changes in availability of resources. About the only advantage of utilizing linear function is that the computations become less cumbersome. For this reason, and not to be sidetracked with computational details, the remainder of this section concentrates on the use of linear functions only. Keep in mind, however, that any (single-variable) nonlinear function can be approximated by a piecewise linear function (see Section 19.2.1). Nevertheless, we show throughout the section how nonlinearity can be handled directly and also point to possible computational difficulties.

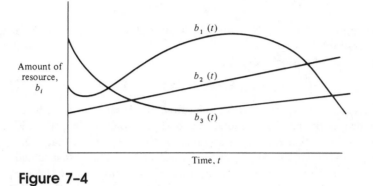

Figure 7-4

As in sensitivity analysis, we consider the following types of variation:

1. Variation in objective coefficients **C**.
2. Variation in resources availability **b**.
3. Variation in constraint coefficients $\mathbf{P}_j$.
4. Simultaneous changes in **C**, **b**, and $\mathbf{P}_j$.

Suppose that t is the parameter with which the different coefficients vary. The general idea of parametric programming is to compute the optimal solution at $t = 0$. Then, using the optimality and feasibility conditions of the simplex method, we find the range of t for which the solution at $t = 0$ remains optimal and feasible. Suppose that this range is given by $(0, t_1)$. This means that any increase in t beyond t_1 will result in an infeasible and/or nonoptimal solution. Thus, at $t = t_1$, we determine a new solution that will remain optimal and feasible from $t = t_1$ to $t = t_2$, where $t_2 > t_1$. The process is then repeated at $t = t_2$ and a new solution is obtained. Eventually, a point on the scale t is reached beyond which the solution either does not change or does not exist. This is where the parametric analysis ends.

We shall now show how the critical values $t_1, t_2, \ldots$ and their associated solutions are determined. We shall consider the different changes separately. Our results will be based primarily on the information in Section 7.2.2 regarding the general simplex iteration expressed in matrix form. You are encouraged to review this material before proceeding with the remainder of this section.

7.6.1 CHANGES IN C

Let $\mathbf{C}(t)$ represent the parameterized objective vector as a function of t. For simplicity, assume that $t \geq 0$. Suppose that $\mathbf{B}_i$ represents the optimal basis at the critical value t_i. We show now how the next critical value t_{i+1} and its optimal basis are determined. Initially, we start at $t = t_0 = 0$ with $\mathbf{B}_0$ as the associated optimal basis.

Let ${}^i\mathbf{X}_B$ be the optimal basic vector at t_i and define $\mathbf{C}_B(t)$ as its associated coefficients of $\mathbf{C}(t)$. As we can see from Section 7.2.2, changes in **C** can affect only the optimality of the current solution. Thus the solution

$$^i\mathbf{X}_B = \mathbf{B}_i^{-1}\mathbf{b}$$

remains optimal for all $t \geq t_i$ for which all $z_j(t) - c_j(t)$ remain nonnegative (maximization).

Mathematically, this is expressed as

$$\mathbf{C}_B(t)\mathbf{B}_i^{-1}\mathbf{P}_j - c_j(t) \geq 0, \qquad \text{for all } j$$

These inequalities are satisfied for the range of t from t_i to t_{i+1}, where t_{i+1} is determined as the largest t beyond which at least one of the given inequalities is violated. Note that *nothing* is specified in the inequalities that requires $\mathbf{C}(t)$ to vary *linearly* with t. Any function is acceptable. The prime difficulty with using nonlinear functions is that the numerical evaluation of the inequalities may be cumbersome. It may even require the use of some type of numerical method to obtain the results.

Example 7.6-1

$$\text{Maximize } z = (3 - 6t)x_1 + (2 - 2t)x_2 + (5 + 5t)x_3$$

subject to

$$x_1 + 2x_2 + x_3 \leq 40$$
$$3x_1 + 2x_3 \leq 60$$
$$x_1 + 4x_2 \leq 30$$
$$x_1, x_2, x_3 \geq 0$$

The parameter t is assumed nonnegative.

According to the problem,

$$\mathbf{C}(t) = (3 - 6t, 2 - 2t, 5 + 5t)$$

We start at $t = t_0 = 0$. The associated optimal tableau follows, where the slacks are represented by x_4, x_5, and x_6.

Optimal Solution at $t = t_0 = 0$

Basic	x_1	x_2	x_3	x_4	x_5	x_6	Solution
z	4	0	0	1	2	0	160
x_2	$-1/4$	1	0	1/2	$-1/4$	0	5
x_3	3/2	0	1	0	1/2	0	30
x_6	2	0	0	-2	1	1	10

Thus

$$^0\mathbf{X}_B = \begin{pmatrix} x_2 \\ x_3 \\ x_6 \end{pmatrix} = \begin{pmatrix} 5 \\ 30 \\ 10 \end{pmatrix} \quad \text{and} \quad \mathbf{B}_0^{-1} = \begin{pmatrix} 1/2 & -1/4 & 0 \\ 0 & 1/2 & 0 \\ -2 & 1 & 1 \end{pmatrix}$$

Now we determine the first critical value $t = t_1$. The parameterized objective coefficients of $^0\mathbf{X}_B$ are

$$\mathbf{C}_B(t) = (2 - 2t, \ 5 + 5t, \ 0)$$

Consequently,

$$\mathbf{C}_B(t)\mathbf{B}_0^{-1} = (1 - t, \ 2 + 3t, \ 0)$$

The values of $z_j(t) - c_j(t)$ for $j = 1, 4$, and 5 (nonbasic x_j) are thus given by

$$\{\mathbf{C}_B(t)\mathbf{B}_0^{-1}\mathbf{P}_j - c_j(t)\}_{j=1,4,5} = (1 - t, \ 2 + 3t, \ 0) \begin{pmatrix} 1 & 1 & 0 \\ 3 & 0 & 1 \\ 1 & 0 & 0 \end{pmatrix} - (3 - 6t, \ 0, \ 0)$$

$$= (4 + 14t, \ 1 - t, \ 2 + 3t)$$

Given $t \geq 0$, the solution $^0\mathbf{X}_B$ (or basis $\mathbf{B}_0^{-1}$) remains optimal as long as the conditions $4 + 14t \geq 0$, $1 - t \geq 0$ and $2 + 3t \geq 0$ are satisfied. The second inequality shows that t must not exceed 1. (All others are satisfied for all $t \geq 0$.) This means that $t_1 = 1$ is the next critical value and $^0\mathbf{X}_B$ remains optimal for the range $(t_0, t_1) = (0, 1)$.

Exercise 7.6-1

Suppose that t is allowed to assume positive, zero, or negative values (instead of $t \geq 0$ only), what is the range of t for which $^0\mathbf{X}_B$ remains optimal?
[*Ans.* $^0\mathbf{X}_B$ remains optimal for $-2/7 \leq t \leq 1$.]

At $t = 1$, we notice that $z_4(t) - c_4(t) = 0$. For $t > 1$, $z_4(t) - c_4(t) < 0$. This means that for $t > 1$, x_4 must enter the basic solution, in which case x_2 must leave (see the optimal table at $t = t_0 = 0$). At $t = 1$, entering x_4 into the basic solution will result in an alternative solution. (Why?) This new solution will remain optimal for the next range (t_1, t_2), where t_2 is the next critical value to be evaluated. We first determine the alternative basis $\mathbf{B}^{-1}$ and then determine the next critical value t_2.

Alternative Optimal Basis at $t = t_1 = 1$

Since $\mathbf{P}_4$ and $\mathbf{P}_2$ are the entering and leaving vectors, we have

$$\alpha^4 = \mathbf{B}_0^{-1}\mathbf{P}_4 = \begin{pmatrix} 1/2 & -1/4 & 0 \\ 0 & 1/2 & 0 \\ -2 & 1 & 1 \end{pmatrix} \begin{pmatrix} 1 \\ 0 \\ 0 \end{pmatrix} = \begin{pmatrix} 1/2 \\ 0 \\ -2 \end{pmatrix}$$

$$\xi = \begin{pmatrix} +1/(1/2) \\ 0/(1/2) \\ -(-2)/(1/2) \end{pmatrix} = \begin{pmatrix} 2 \\ 0 \\ 4 \end{pmatrix}$$

Thus the new basis inverse is given by

$$\mathbf{B}_1^{-1} = \begin{pmatrix} 2 & 0 & 0 \\ 0 & 1 & 0 \\ 4 & 0 & 1 \end{pmatrix} \begin{pmatrix} 1/2 & -1/4 & 0 \\ 0 & 1/2 & 0 \\ -2 & 1 & 1 \end{pmatrix} = \begin{pmatrix} 1 & -1/2 & 0 \\ 0 & 1/2 & 0 \\ 0 & 0 & 1 \end{pmatrix}$$

and

$$\mathbf{X}_B^1 = \begin{pmatrix} x_4 \\ x_3 \\ x_6 \end{pmatrix} = \begin{pmatrix} 1 & -1/2 & 0 \\ 0 & 1/2 & 0 \\ 0 & 0 & 1 \end{pmatrix} \begin{pmatrix} 40 \\ 60 \\ 30 \end{pmatrix} = \begin{pmatrix} 10 \\ 30 \\ 30 \end{pmatrix}$$

We now proceed to compute the next critical value t_2. For $\mathbf{X}_B^1$, we have

$$\mathbf{C}_B(t)\mathbf{B}_1^{-1} = (0, 5 + 5t, 0)\mathbf{B}_1^{-1} = \left(0, \frac{5 + 5t}{2}, 0\right)$$

The values of $z_j(t) - c_j(t)$ for $j = 1, 2,$ and 5 are given by

$$\left(0, \frac{5 + 5t}{2}, 0\right) \begin{pmatrix} 1 & 2 & 0 \\ 3 & 0 & 1 \\ 1 & 4 & 0 \end{pmatrix} - (3 - 6t, 2 - 2t, 0) = \left(\frac{9 + 27t}{2}, -2 + 2t, \frac{5 + 5t}{2}\right)$$

The basis $\mathbf{B}_1$ remains optimal as long as all $z_j(t) - c_j(t) \geq 0$. This condition is satisfied for all $t \geq 1$. Thus $t_2 = \infty$. (Notice that the optimality conditions automatically "remember" that $^1\mathbf{X}_B$ is optimal for a range of t that starts from the last critical value $t_1 = 1$. This will always be the case with parametric computations.)

The optimal solutions for the entire range of t can be summarized as shown in the following table. Notice that the value of z is obtained by direct substitution in the objective function.

t	x_1	x_2	x_3	z
$0 \leq t \leq 1$	0	5	30	$160 + 140t$
$t \geq 1$	0	0	30	$150 + 150t$

◀

Exercise 7.6–2

Consider Example 7.6–1.

(a) Suppose that the objective function is changed to

$$\text{maximize } z = (3 + 18t)x_1 + (2 - 4t)x_2 + (5 + 3t)x_3$$

Find t_1 and indicate which basic variable in $^0\mathbf{X}_B$ leaves the solution and which one enters to obtain the alternative basic solution $^1\mathbf{X}_B$. (Note that at $t = 0$, the optimum tableau is as given for $t = t_0 = 0$.)

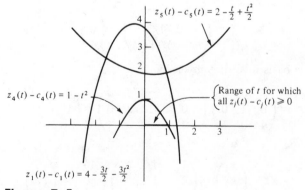

Figure 7–5

[*Ans.* $t_1 = 8/25$; x_1 enters and x_6 leaves.]

(b) Repeat part (a) if the objective function is expressed in terms of the following *nonlinear* parameterization:

$$\text{maximize } z = (3 + 2t^2)x_1 + (2 - 2t^2)x_2 + (5 - t)x_3$$

[*Ans.* $t_1 = 1$. x_4 enters and x_2 leaves. See Figure 7–5 for a graphic representation of $z_j(t) - c_j(t)$. This example is designed to illustrate the computational inconvenience that accompanies the use of nonlinear parameterization.]

7.6.2 CHANGES IN b

Let $b(t)$ represent the parameterized right-side vector as a function of t. As in Section 7.6.1, define $\mathbf{B}_i$ and ${}^i\mathbf{X}_B$ as the basis and the basic vector at the critical value t_i.

Section 7.2.2 shows that changes in the vector $\mathbf{b}$ can affect only the feasibility of the solution. Thus the solution ${}^i\mathbf{X}_B$ remains feasible as long as the condition

$$\mathbf{B}_1^{-1}\mathbf{b}(t) \geq \mathbf{0}$$

is satisfied. The following example shows how the successive critical values are determined.

Example 7.6–2

$$\text{Maximize } z = 3x_1 + 2x_2 + 5x_3$$

subject to

$$
\begin{aligned}
x_1 + 2x_2 + x_3 &\leq 40 - t \\
3x_1 \qquad\; + 2x_3 &\leq 60 + 2t \\
x_1 + 4x_2 \qquad\;\; &\leq 30 - 7t \\
x_1, x_2, x_2 &\geq 0
\end{aligned}
$$

Assume that $t \geq 0$.

At $t = t_0 = 0$, the problem is identical to that of Example 7.6–1 at $t = 0$. Thus

$$
{}^0\mathbf{X}_B = \begin{pmatrix} x_2 \\ x_3 \\ x_6 \end{pmatrix} = \begin{pmatrix} 5 \\ 30 \\ 10 \end{pmatrix} \quad \text{and} \quad \mathbf{B}_0^{-1} = \begin{pmatrix} 1/2 & -1/4 & 0 \\ 0 & 1/2 & 0 \\ -2 & 1 & 1 \end{pmatrix}
$$

The first critical value t_1 is determined by considering

$$\mathbf{B}_0^{-1}\mathbf{b}(t) \geq \mathbf{0}$$

This gives

$$
\begin{pmatrix} x_2 \\ x_3 \\ x_6 \end{pmatrix} = \begin{pmatrix} 1/2 & -1/4 & 0 \\ 0 & 1/2 & 0 \\ -2 & 1 & 1 \end{pmatrix} \begin{pmatrix} 40 - t \\ 60 + 2t \\ 30 - 7t \end{pmatrix} = \begin{pmatrix} 5 - t \\ 30 + t \\ 10 - 3t \end{pmatrix} \geq \begin{pmatrix} 0 \\ 0 \\ 0 \end{pmatrix}
$$

These inequalities are satisfied for $t \leq 10/3$. Thus $t_1 = 10/3$ and the basis $\mathbf{B}_0$ remains feasible for the range $(t_0, t_1) = (0, 10/3)$.

Although the basis $\mathbf{B}_0$ remains unchanged for $0 \leq t \leq 10/3$, the values of the associated basic variables x_2, x_3, and x_6 will be given by $x_2 = 5 - t$, $x_3 = 30 + t$, and $x_6 = 10 - 3t$. The value of z is $2(5 - t) + 5(30 + t) = 160 + 3t$, $0 \leq t \leq 10/3$.

At $t = t_1 = 10/3$, x_6 equals zero. Any increase in t beyond $10/3$ will make x_6 negative. Thus, at the critical value $t_1 = 10/3$, an alternative basis $\mathbf{B}_1$ can be obtained by applying the *dual simplex method* with x_6 as the leaving variable. We use the matrix procedure outlined in Exercise 7.3–2(b) to determine the new basis.

Alternative Basis at $t = t_1 = 10/3$

Since the leaving variable is already known ($= x_6$), we need to determine the entering variable only to compute $\mathbf{B}^{-1}$. Given $^0\mathbf{X}_B = (x_2, x_3, x_6)^T$, $\mathbf{C}_B = (2, 5, 0)$, and $\mathbf{C}_B\mathbf{B}_0^{-1} = (1, 2, 0)$, then $(z_j - c_j)$ for $j = 1$, 4, and 5 are given by

$$\{\mathbf{C}_B\mathbf{B}_0^{-1}\mathbf{P}_j - c_j\} = (1, 2, 0) \begin{pmatrix} 1 & 1 & 0 \\ 3 & 0 & 1 \\ 1 & 0 & 0 \end{pmatrix} - (3, 0, 0) = (4, 1, 2)$$

Next we compute α_r^j for $j = 1$, 4, and 5, and $x_r = x_6$ as

$$(\alpha_6^1, \alpha_6^4, \alpha_6^5) = (\text{row of } \mathbf{B}_0^{-1} \text{ associated with } x_6) \, (\mathbf{P}_1, \mathbf{P}_4, \mathbf{P}_5)$$
$$= (\text{third row of } \mathbf{B}_0^{-1}) \, (\mathbf{P}_1, \mathbf{P}_4, \mathbf{P}_5)$$
$$= (-2, 1, 1) \begin{pmatrix} 1 & 1 & 0 \\ 3 & 0 & 1 \\ 1 & 0 & 0 \end{pmatrix} = (2, -2, 1)$$

Thus, for $j = 1$, 4, 5,

$$\theta = \min \{-, |1/-2|, -\} = 1/2, \quad \text{corresponding to } x_4$$

As a result, $\mathbf{P}_4$ is the leaving vector.

We now determine $\mathbf{B}_1^{-1}$ by interchanging the entering and leaving vectors $\mathbf{P}_4$ and $\mathbf{P}_6$. Thus

$$\alpha^4 = \mathbf{B}_0^{-1}\mathbf{P}_4 = \begin{pmatrix} 1/2 \\ 0 \\ -2 \end{pmatrix}$$

$$\xi = \begin{pmatrix} -(1/2)/(-2) \\ -0/(-2) \\ +1/(-2) \end{pmatrix} = \begin{pmatrix} 1/4 \\ 0 \\ -1/2 \end{pmatrix}$$

$$\mathbf{B}_1^{-1} = \mathbf{E}\mathbf{B}_0^{-1} = \begin{pmatrix} 1 & 0 & 1/4 \\ 0 & 1 & 0 \\ 0 & 0 & -1/2 \end{pmatrix} \begin{pmatrix} 1/2 & -1/4 & 0 \\ 0 & 1/2 & 0 \\ -2 & 1 & 1 \end{pmatrix} = \begin{pmatrix} 0 & 0 & 1/4 \\ 0 & 1/2 & 0 \\ 1 & -1/2 & -1/2 \end{pmatrix}$$

with $^1\mathbf{X}_B = (x_2, x_3, x_4)^T$.

The next critical value t_2 is determined by considering

$$\mathbf{B}_1^{-1}\mathbf{b}(t) \geq \mathbf{0}$$

which yields

$$\begin{pmatrix} x_2 \\ x_3 \\ x_4 \end{pmatrix} = \begin{pmatrix} 0 & 0 & 1/4 \\ 0 & 1/2 & 0 \\ 1 & -1/2 & -1/2 \end{pmatrix} \begin{pmatrix} 40 - t \\ 60 + 2t \\ 30 - 7t \end{pmatrix} = \begin{pmatrix} \dfrac{30 - 7t}{4} \\ 30 + t \\ \dfrac{-10 + 3t}{2} \end{pmatrix} \geq \begin{pmatrix} 0 \\ 0 \\ 0 \end{pmatrix}$$

Thus $\mathbf{B}_1$ remains feasible for $10/3 \leq t \leq 30/7$.

At $t = t_2 = 30/7$, an alternative basis, in which x_2 is the leaving variable, can be obtained by applying the dual simplex method.

Alternative Basis at $t = t_2 = 30/7$

To determine the entering variable (given x_2 is the leaving variable), we compute the ratios of the dual simplex.

Given $^1\mathbf{X}_B = (x_2, x_3, x_4)^T$, $\mathbf{C}_B = (2, 5, 0)$, and $\mathbf{C}_B\mathbf{B}^{-1} = (0, 5/2, 1/2)$, then $(z_j - c_j)$ for $j = 1, 5,$ and 6 are given by

$$\{\mathbf{C}_B\mathbf{B}_1^{-1}\mathbf{P}_j - c_j\} = (0, 5/2, 1/2)\begin{pmatrix} 1 & 0 & 0 \\ 3 & 1 & 0 \\ 1 & 0 & 1 \end{pmatrix} - (3, 0, 0)$$

$$= (5, 5/2, 1/2)$$

Next we compute α_r^j for $j = 1, 5,$ and 6 and $x_r = x_2$ as

$$(\alpha_2^1, \alpha_2^5, \alpha_2^6) = (\text{first row of } \mathbf{B}_1^{-1})\,(\mathbf{P}_1, \mathbf{P}_5, \mathbf{P}_6)$$

$$= (0, 0, 1/4)\begin{pmatrix} 1 & 0 & 0 \\ 3 & 1 & 0 \\ 1 & 0 & 1 \end{pmatrix}$$

$$= (1/4, 0, 1/4)$$

Since all $\alpha_r^j \geq 0$, the problem has no feasible solution for $t > 30/7$ and the parametric analysis ends at $t = t_2 = 30/7$.

The optimal feasible solution for the entire range of t may thus be summarized as follows:

t	x_1	x_2	x_3	z
$0 \leq t \leq 10/3$	0	$5 - t$	$30 + t$	$160 + 3t$
$10/3 \leq t \leq 30/7$	0	$\dfrac{30 - 7t}{4}$	$30 + t$	$165 + \dfrac{3}{2}t$
$t > 30/7$	(no feasible solution)			

◀

Exercise 7.6-3

In Example 6.7-2, find the first critical value t_1 and the vectors of $\mathbf{B}_1$ in each of the following cases.

(a) $\mathbf{b}(t) = (40 + 2t,\ 60 - 3t,\ 30 + 6t)^T$
   [*Ans.* $t_1 = 10,\ \mathbf{B}_1 = (\mathbf{P}_2,\ \mathbf{P}_3,\ \mathbf{P}_4).$]
(b) $\mathbf{b}(t) = 40 - t,\ 60 + 2t,\ 30 - 5t)^T$
   [*Ans.* $t_1 = 5,\ \mathbf{B}_1 = (\mathbf{P}_5,\ \mathbf{P}_3,\ \mathbf{P}_6).$]

7.6.3   CHANGES IN $\mathbf{P}_j$

We shall assume that $\mathbf{P}_j$ is a nonbasic vector in the optimal solution at $t = 0$. If it is basic, the situation would not lend itself neatly to parametric analysis because the basis $\mathbf{B}_0$ will be affected directly.

Let $\mathbf{P}_j(t)$ be the parameterized vector. From Section 7.2.2 we know that variation in a nonbasic vector can affect the optimality of that vector only. The vector $\mathbf{P}_j$ will enter the solution only when its $z_j - c_j$ becomes negative (maximization). Thus a current basis $\mathbf{B}_i$ remains optimal as long as the condition

$$z_j(t) - c_j = \mathbf{C}_B \mathbf{B}_i^{-1} \mathbf{P}_j(t) - c_j \geq 0$$

is satisfied. This inequality thus can be used to determine the next critical value t_{i+1}.

Observe that at $t = t_{i+1}$ an alternative optimal basis $\mathbf{B}_{i+1}$ can be obtained by introducing $\mathbf{P}_j$ into the basis. For $t > t_{i+1}$, $\mathbf{P}_j$ must be the entering vector. At this point it will not be possible to carry out the parametric analysis any further, since $\mathbf{P}_j$ will be in the basis and the resulting situation becomes rather complex. For this reason we limit our attention to determining one critical value t_{i+1} given the basis $\mathbf{B}_i$ at $t = t_i$.

Example 7.6-3. In Example 7.6-1, x_1 is nonbasic at $t = 0$. Suppose that

$$\mathbf{P}_1(t) = \begin{pmatrix} 1 + t \\ 3 - 2t \\ 1 + 3t \end{pmatrix}$$

is the only parameterized vector in the problem. The solution in Example 7.6-1 shows that

$$\mathbf{B}_0^{-1} = \begin{pmatrix} 1/2 & -1/4 & 0 \\ 0 & 1/2 & 0 \\ -2 & 1 & 1 \end{pmatrix}, \qquad {}^0\mathbf{X}_B = \begin{pmatrix} x_2 \\ x_3 \\ x_6 \end{pmatrix}$$

Consequently,

$$z_1(t) - c_1 = \mathbf{C}_B \mathbf{B}_0^{-1} \mathbf{P}_1(t) - c_1$$

$$= (2, 5, 0) \begin{pmatrix} 1/2 & -1/4 & 0 \\ 0 & 1/2 & 0 \\ -2 & 1 & 1 \end{pmatrix} \begin{pmatrix} 1 + t \\ 3 - 2t \\ 1 + 3t \end{pmatrix} - 3$$

$$= 4 - 3t$$

As a result, $\mathbf{B}_0$ will be optimal as long as $\mathbf{P}_1$ remains nonbasic. This will be the case when the condition $4 - 3t \geq 0$ is satisfied. The critical value t_1 is thus equal to $4/3$.

We can obtain the alternative solution at $t = t_1$ by introducing $\mathbf{P}_1$ into the basis and dropping $\mathbf{P}_6$. However, we will be unable to carry out the parametric analysis for $\mathbf{P}_1$ as soon as it enters the basis. ◀

From the practical standpoint, the parametric analysis of nonbasic vectors only does not generally yield useful information. Normally, the parametrization of the different vectors is specified prior to solving the problem for any value of t. Clearly, it will be impossible to carry out any parametric analysis if any of the vectors in the current basis are parameterized. We specifically stress this point here since in practice, managers or decision makers could not care less whether an activity of the model is "basic" or "nonbasic." This is a technical language. All they really want to know is whether parametric analysis can be carried out in such situations, and the answer generally is "no!" This presentation is designed to show you why this answer is justifiable.

Exercise 7.6–4

Find t_1 in Example 7.6–3 assuming that $\mathbf{P}_1(t)$ is given by

$$(1 - 2t, 3 + 3t, 1 - 4t)^T$$

[*Ans.* $\mathbf{B}_0$ remains optimal for all $t \geq 0$. This means that the solution remains unaffected by the parameterization of $\mathbf{P}_1$.]

7.6.4   SIMULTANEOUS CHANGES IN C AND b

In this section we combine the parameterizations of $\mathbf{C}$ and $\mathbf{b}$ so that they are allowed to occur simultaneously. The idea is quite simple. For a given optimal basis $\mathbf{B}_i$, we check the optimality and feasibility separately by applying the procedures given in Sections 7.6.1 and 7.6.2. Let t' and t'' be the next critical values for optimality and feasibility, respectively. We have three cases:

1. If $t' < t''$, $\mathbf{B}_i$ will become nonoptimal first and the next basis $\mathbf{B}_{i+1}$ is obtained at $t_{i+1} = t'$ by using the regular simplex method (see Section 7.6.1).

2. If $t'' < t'$, $\mathbf{B}_i$ will become infeasible first and the next basis $\mathbf{B}_{i+1}$ is obtained at $t_{i+1} = t''$ by using the dual simplex method (see Section 7.6.2).

3. If $t' = t''$, $\mathbf{B}_i$ will become both nonoptimal and infeasible at $t_{i+1} = t' = t''$. In this situation we need to use a *primal–dual simplex* method for determining the next basis $\mathbf{B}_{i+1}$. This case will not be investigated here because the primal–dual algorithm has not been presented. The method is discussed in many specialized LP books. Keep in mind, however, that the primal–dual method is used in much the same way as in the foregoing two cases to secure the new basis.

Example 7.6–4. This example will combine the parameterization of $\mathbf{C}$ and $\mathbf{b}$ as specified in Examples 7.6–1 and 7.6–2. You must carefully note that combining the parameterization of $C(t)$ and $b(t)$ does not generally result in superimposing the critical values obtained by considering each parameterization separately.

$$\text{Maximize } z = (3 - 6t)x_1 + (2 - 2t)x_2 + (5 + 5t)x_3$$

subject to

$$
\begin{aligned}
x_1 + 2x_2 + x_3 &\le 40 - t \\
3x_1 \quad\quad\; + 2x_3 &\le 60 + 2t \\
x_1 + 4x_2 \quad\quad &\le 30 - 7t \\
x_1, x_2, x_3 &\ge 0
\end{aligned}
$$

Optimal Basis at $t = t_0 = 0$

From Example 7.6–1, we have

$$
{}^0X_B = \begin{pmatrix} x_2 \\ x_3 \\ x_6 \end{pmatrix}, \quad
B_0^{-1} = \begin{pmatrix} 1/2 & -1/4 & 0 \\ 0 & 1/2 & 0 \\ -2 & 1 & 1 \end{pmatrix}
$$

Optimality: B_0 remains optimal as long as $z_j(t) - c_j(t)$, $j = 1, 4, 5$, remain nonnegative. That is,

$$\{z_j(t) - c_j(t)\}_{j=1,4,5} = (4 + 14t, \; 1 - t, \; 2 + 3t)$$

$$\ge (0, \, 0, \, 0)$$

(Verify these expressions. They are the same as in Example 7.6–1.) These conditions yield $t' = 1$.

Feasibility: B_0 remains feasible as long as

$$
\begin{pmatrix} x_2 \\ x_3 \\ x_6 \end{pmatrix} = B_0^{-1}b(t) = \begin{pmatrix} 5 - t \\ 30 + t \\ 10 - 3t \end{pmatrix} \ge \begin{pmatrix} 0 \\ 0 \\ 0 \end{pmatrix}
$$

(Verify the computations!) The feasibility is satisfied for $t \le 10/3$ and $t'' = 10/3$.

Optimality and Feasibility: $t_1 = \min \{t', t''\} = t' = 1$, which indicates that B_0 will become nonoptimal first. We thus compute the alternative optimum at $t_1 = 1$ using the regular simplex method.

Alternative Basis at $t = t_1 = 1$

As seen from the optimality conditions, x_4 is the entering variable in the alternative solution. To determine the leaving variable, we carry out the following computations.

$$
\alpha^4 = B_0^{-1}P_4 = \begin{pmatrix} 1/2 & -1/4 & 0 \\ 0 & 1/2 & 0 \\ -2 & 1 & 1 \end{pmatrix} \begin{pmatrix} 1 \\ 0 \\ 0 \end{pmatrix} = \begin{pmatrix} 1/2 \\ 0 \\ -2 \end{pmatrix}
$$

$$
\begin{pmatrix} x_2 \\ x_3 \\ x_6 \end{pmatrix}_{t=1} = B_0^{-1}b(1) = \begin{pmatrix} 5 - 1 \\ 30 + 1 \\ 10 - 3 \times 1 \end{pmatrix} = \begin{pmatrix} 4 \\ 31 \\ 7 \end{pmatrix}
$$

Thus, for x_2, x_3, and x_6,

$$\theta = \min \{\frac{4}{1/2}, -, -\} = 8$$

which means that x_2 is the leaving variable.

The new basis $\mathbf{B}_1$ is thus obtained from $\mathbf{B}_0$ by interchanging $\mathbf{P}_2$ and $\mathbf{P}_4$. Thus

$$\mathbf{X}_B = (x_4, x_3, x_6)^T$$

$$\mathbf{B}_1^{-1} = \begin{pmatrix} 2 & 0 & 0 \\ 0 & 1 & 0 \\ 4 & 0 & 1 \end{pmatrix} \begin{pmatrix} 1/2 & -1/4 & 0 \\ 0 & 1/2 & 0 \\ -2 & 1 & 1 \end{pmatrix}$$

$$= \begin{pmatrix} 1 & -1/2 & 0 \\ 0 & 1/2 & 0 \\ 0 & 0 & 1 \end{pmatrix}$$

Next we compute the new critical value of $t(=t_2)$.

Optimality:

$$\{z_j(t) - c_j(t)\}_{j=1,2,5} = \left(\frac{9+27t}{2}, -2+2t, \frac{5+5t}{2}\right)$$

$$\geq (0, 0, 0)$$

(Verify the computations!) Thus $\mathbf{B}_1$ remains optimal for all $t \geq 1$, which means that $t' = \infty$.

Feasibility:

$$\begin{pmatrix} x_4 \\ x_3 \\ x_6 \end{pmatrix} = \mathbf{B}_1^{-1}\mathbf{b}(t) = \begin{pmatrix} 10-2t \\ 30+t \\ 30-7t \end{pmatrix} \geq \begin{pmatrix} 0 \\ 0 \\ 0 \end{pmatrix}$$

(Verify the computations!) Thus $\mathbf{B}_1$ remains feasible for $t \leq 30/7$ and $t'' = 30/7$.

Optimality and Feasibility: $t_2 = \min \{t', t''\} = 30/7$. Thus $\mathbf{B}_1$ becomes infeasible first.

Alternative Basis at $t = t_2 = 30/7$

The alternative basis is determined by the dual simplex method with x_6 as the leaving variable. To determine the entering variable, we consider the following computations. For $j = 1, 2, 5$, we have

$$\{z_j(t) - c_j(t)\}_{t=30/7} = [9/2 + (27/2)(30/7), -2 + 2(30/7), 5/2 + (5/2)(30/7)]$$

$$= (62.36, 6.57, 13.21)$$

$$(\alpha_6^1, \alpha_6^2, \alpha_6^5) = (0, 0, 1) \begin{pmatrix} 1 & 2 & 0 \\ 3 & 0 & 1 \\ 1 & 4 & 0 \end{pmatrix} = (1, 4, 0)$$

Since all $\alpha_r^j \geq 0$, no feasible solutions exist for $t > 30/7$ and the parametric analysis is complete.

The optimum solution for the entire range of t is summarized as

t	x_1	x_2	x_3	z
$0 \le t \le 1$	0	$5-t$	$30+t$	$7t^2 + 143t + 160$
$1 \le t \le 30/7$	0	0	$30+t$	$5t^2 + 155t + 150$
$t > 30/7$	(No feasible solution)			

You will notice that the critical value $t = 10/3$, which was obtained when $b(t)$ was considered separately (Example 7.6–2), is not encountered when both $b(t)$ and $C(t)$ are considered simultaneously. This is the reason we mentioned earlier that the problem cannot be analyzed by superimposing the critical values obtained when $b(t)$ and $C(t)$ are considered separately. ◀

7.7 SUMMARY

This chapter introduced a number of techniques, including

Revised simplex method
Bounded variables technique
Decomposition principle
Parametric programming

The revised simplex, decomposition, and bounded variables algorithms are designed to reduce the amount of computation and to economize the use of computer memory. Parametric programming investigates changes in the optimum solution resulting from predetermined continuous variations in the model's coefficients. This type of analysis gives LP a dynamic characteristic that allows the solution to respond to possible changes in marginal profits and resources availability.

The LP literature includes additional topics relating to the improvement of computational efficiency and accuracy. Among these are the **primal–dual simplex** method for solving linear programs that start both infeasible and nonoptimal. The **generalized upper bounding** techniques account implicitly for constraints limiting the (partial) sum of the variables by a known upper bound.

SELECTED REFERENCES

DANTZIG, G. B., *Linear Programming and Extensions,* Princeton University Press, Princeton, N.J., 1963.

LASDON, L., *Optimization Theory for Large Systems,* Macmillan, New York, 1970.

SPIVEY, W., and R. M. THRALL, *Linear Optimization,* Holt, Rinehart and Winston, New York, 1970.

WISMER, D., ed., *Optimization Methods for Large-Scale Systems with Applications,* McGraw-Hill, New York, 1971.

REVIEW QUESTIONS

True (T) or False (F)?

1. _____ If a matrix A is nonsingular, its vectors must be independent.

2. _____ If the vectors of a square matrix A are independent, its determinant must equal zero.

3. _____ In the system $AX = b$, the square matrix A forms a basis if all its vectors are independent.

4. _____ The system $AX = b$ will always have a unique solution if the matrix A is nonsingular.

5. _____ If the matrix A in the system $AX = b$ is singular, the system will always have no solution.

6. _____ In the system $AX = b$, if A is singular and b is a dependent vector, the system will have infinity of solutions.

7. _____ A system of m independent linear equations with n unknowns will always have an infinite number of solutions when $m < n$.

8. _____ All the extreme points of an LP solution space are completely identified by the basic solutions of the linear equations representing the solution space.

9. _____ In the simplex method, each two successive bases differ in exactly one vector.

10. _____ All the information in a simplex tableau can be computed from the knowledge of the associated basis and the original linear program.

11. _____ Whereas a primal problem seeks optimality, its dual problem automatically seeks feasibility.

12. _____ If a primal problem is unbounded, its dual will remain infeasible.

13. _____ If a pair of primal and dual solutions are feasible and their corresponding objective values are equal, the two solutions are necessarily optimal.

14. _____ If an optimal dual variable is positive, its associated primal constraint may not be satisfied in equation form.

15. _____ If at the optimal solution, a primal constraint is satisfied as an equation, its associated dual variables will always have a nonzero value.

16. _____ The steps of the revised simplex method differ from those of the tableau simplex method.

17. _____ For a linear program, starting with a given basis, the entering and leaving variables may not be the same in the tableau and revised simplex methods.

18. _____ The revised simplex method can result in computational saving if the original matrix of the constraints is sparse.

19. ____ The revised simplex method is designed to reduce the amount of computation and to economize the use of the computer memory.

20. ____ In the revised method, the round-off is practically controllable once we can control the error in computing the inverses associated with the successive iterations.

21. ____ The product form for computing the inverse is an integral part of the revised simplex method.

22. ____ It is possible to invert any nonsingular matrix by multiplying a series of properly structured matrices (of the type E_i).

23. ____ In using the product form in conjunction with the simplex iteration, it is possible in constructing the vector ξ that the denominator element be equal to zero.

24. ____ The main reason for using the upper-bounding technique is to reduce the computational effort by reducing the explicit number of constraints.

25. ____ Lower bounding constraints can be eliminated by using direct substitution.

26. ____ The main reason for not substituting out upper-bounding constraints is that we have no guarantee that some of the variables will remain nonnegative.

27. ____ In the upper-bounding technique, a variable may be nonbasic at its upper bound.

28. ____ The principal idea of the decomposition algorithm rests in being able to represent the solution spaces of the subproblems in terms of their extreme points.

29. ____ In the decomposition algorithm, the solutions of linear programs associated with the subproblems essentially determine the entering variable for the master problem.

30. ____ In the decomposition algorithm, we must always minimize the subproblems regardless of whether the master problem is maximization or minimization.

31. ____ In parametric linear programming, we can consider only changes in the coefficients that vary *linearly* with a predetermined parameter.

32. ____ The parameterization of the objective function coefficients can affect only the optimality of the solution.

33. ____ The parameterization of the availability of resources can affect only the feasibility of the solution.

34. ____ If we parameterize a nonbasic constraint vector $\mathbf{P}_j$, both optimality and feasibility can be affected.

35. ____ If both the objective function and availability of resources are parameterized, the critical values of the parameter when both types of parame-

terizations are considered simultaneously can be obtained by superimposing the critical values determined by considering each type
separately.

[*Ans.* **1**—T, **2**—F, **3**—T, **4**—T, **5**—F, **6** to **13**—T, **14** to **17**—F, **18**—T, **19**—T, **20**—
T, **21**—F, **22**—T, **23**—F, **24** to **29**—T, **30**—F, **31**—F, **32**—T, **33**—T, **34**—F, **35**—F.]

PROBLEMS

Section	Assigned Problems
7.2.1	7–1 to 7–7
7.2.2	7–8 to 7–15
7.2.3	7–16 to 7–20
7.2.4	7–21
7.3.1	7–22, 7–23
7.3.2	7–24 to 7–27
7.4	7–28 to 7–30
7.5	7–31 to 7–34
7.6	7–35 to 7–49

☐ **7–1** Show graphically whether each of the following matrices forms a basis.

$$B_1 = \begin{pmatrix} 1 & 2 \\ 2 & 3 \end{pmatrix}, \quad B_2 = \begin{pmatrix} 1 & 2 \\ 2 & 1 \end{pmatrix}$$

$$B_3 = \begin{pmatrix} 2 & -4 \\ -1 & 2 \end{pmatrix}, \quad B_4 = \begin{pmatrix} 1 & 5 \\ 2 & 10 \end{pmatrix}$$

☐ **7–2** Solve Problem 7–1 algebraically.

☐ **7–3** Show graphically whether the following systems of equations have a
unique solution, no solution, or an infinity of solutions. For unique solutions,
indicate whether the values of x_1 and x_2 are positive, zero, or negative.

(a) $\begin{pmatrix} 5 & 4 \\ 1 & -3 \end{pmatrix}\begin{pmatrix} x_1 \\ x_2 \end{pmatrix} = \begin{pmatrix} 1 \\ 1 \end{pmatrix}$ (b) $\begin{pmatrix} 2 & -2 \\ 1 & 3 \end{pmatrix}\begin{pmatrix} x_1 \\ x_2 \end{pmatrix} = \begin{pmatrix} 1 \\ 3 \end{pmatrix}$

(c) $\begin{pmatrix} 2 & 4 \\ 1 & 3 \end{pmatrix}\begin{pmatrix} x_1 \\ x_2 \end{pmatrix} = \begin{pmatrix} -2 \\ -1 \end{pmatrix}$ (d) $\begin{pmatrix} 2 & 4 \\ 1 & 2 \end{pmatrix}\begin{pmatrix} x_1 \\ x_2 \end{pmatrix} = \begin{pmatrix} 6 \\ 3 \end{pmatrix}$

(e) $\begin{pmatrix} -2 & 4 \\ 1 & -2 \end{pmatrix}\begin{pmatrix} x_1 \\ x_2 \end{pmatrix} = \begin{pmatrix} 2 \\ 1 \end{pmatrix}$ (f) $\begin{pmatrix} 1 & -2 \\ 0 & 0 \end{pmatrix}\begin{pmatrix} x_1 \\ x_2 \end{pmatrix} = \begin{pmatrix} 1 \\ 1 \end{pmatrix}$

☐ **7–4** Consider the system of equations

$$P_1 x_1 + P_2 x_2 + P_3 x_3 + P_4 x_4 = b$$

where

$$\mathbf{P}_1 = \begin{pmatrix} 1 \\ 2 \\ 3 \end{pmatrix}, \quad \mathbf{P}_2 = \begin{pmatrix} 0 \\ 2 \\ 1 \end{pmatrix}, \quad \mathbf{P}_3 = \begin{pmatrix} 1 \\ 4 \\ 2 \end{pmatrix}, \quad \mathbf{P}_4 = \begin{pmatrix} 2 \\ 0 \\ 0 \end{pmatrix}, \quad \mathbf{b} = \begin{pmatrix} 3 \\ 4 \\ 2 \end{pmatrix}$$

Indicate whether the following vector combinations form a basis.

(a) $(\mathbf{P}_1, \mathbf{P}_2, \mathbf{P}_3)$
(b) $(\mathbf{P}_1, \mathbf{P}_2, \mathbf{P}_4)$
(c) $(\mathbf{P}_2, \mathbf{P}_3, \mathbf{P}_4)$

☐ **7–5** Consider the following system of linear equations in which all $x_j \geq 0$.

$$\begin{pmatrix} 2 & 3 & 1 & 0 \\ 1 & 2 & 0 & 1 \end{pmatrix} \begin{pmatrix} x_1 \\ x_2 \\ x_3 \\ x_4 \end{pmatrix} = \begin{pmatrix} 6 \\ 4 \end{pmatrix}$$

Determine all its *feasible* extreme points by evaluating all its basic feasible solutions. What is the relationship between the numbers of basic solutions and extreme points?

☐ **7–6** Determine all the feasible extreme points and their corresponding basic solutions for the following system of linear equations.

$$\begin{aligned}
3x_1 + 6x_2 + 5x_3 + x_4 &= 12 \\
2x_1 + 4x_2 + x_3 + 2x_5 &= 8 \\
x_1, x_2, x_3, x_4, x_5 &\geq 0
\end{aligned}$$

☐ **7–7** Consider a linear programming problem in which the variable x_k is unrestricted in sign. Prove that by replacing x_k by $x'_k - x''_k$, where x'_k and x''_k are nonnegative variables, then in any of the simplex iterations (including the optimum) it is never possible to have *both* x'_k and x''_k as *basic* variables, nor is it possible that these two variables can replace one another in an *alternative* optimum solution.

☐ **7–8** Given the general linear programming problem with m equations and $(m + n)$ unknowns, what is the maximum number of *adjacent* extreme points that can be reached from a nondegenerate extreme point of the corresponding convex set?

☐ **7–9** In applying the feasibility condition of the simplex method, suppose that $x_r = 0$ is a basic variable and x_j is the entering variable. Why is it necessary to have $(\mathbf{B}^{-1}\mathbf{P}_j)_r > 0$ for x_r to be the leaving variable? What is the fallacy if $(\mathbf{B}^{-1}\mathbf{P}_j)_r \leq 0$?

☐ **7–10** In applying the feasibility condition of the simplex method, what are the conditions for a degenerate solution to appear for the first time in the next iteration? for continuing to obtain a degenerate solution in the next iteration? for removing degeneracy in the next iteration? Express the answer mathematically.

☐ **7-11** What are the relationships between extreme points and basic solutions under each of the following conditions: (a) nondegeneracy, (b) degeneracy? What is the maximum possible number of simplex iterations that can be performed at the same extreme point?

☐ **7-12** Consider the problem, max $z = \mathbf{CX}$ subject to $\mathbf{AX} \le \mathbf{b}$, where $\mathbf{b} \ge 0$ and $\mathbf{X} \ge 0$. Suppose that the entering vector $\mathbf{P}_j$ is such that at least one element of $\mathbf{B}^{-1}\mathbf{P}_j$ is greater than zero. If $\mathbf{P}_j$ is replaced by $\beta\mathbf{P}_j$, where β is a positive scalar, and provided that x_j remains the entering variable, find the relationships between the values of x_j corresponding to $\mathbf{P}_j$ and $\beta\mathbf{P}_j$.

☐ **7-13** Answer Problem 7-12 if, in addition, $\mathbf{b}$ is replaced by $\gamma\mathbf{b}$, where γ is a positive scalar.

☐ **7-14** Prove that for the minimization case, a nonbasic vector $\mathbf{P}_j$ can improve the current solution only if $z_j - c_j$ is greater than zero.

☐ **7-15** Consider the linear programming problem defined in Problem 7-12. After obtaining the optimum solution, it is suggested that a nonbasic variable x_j can be made basic (profitable) by reducing the requirements per unit of x_j for the different resources to $1/\beta$ of their original values, where β is a scalar greater than 1. Since the requirements per unit are reduced, it is expected that the profit per unit of x_j will be reduced to $1/\beta$ of its original value. Will these changes make x_j a profitable variable? What should be recommended for x_j to be an attractive variable?

☐ **7-16** Given the linear programming problem,

$$\text{maximize } z = \mathbf{CX}$$

subject to

$$(\mathbf{A}, \mathbf{I})\mathbf{X} = \mathbf{b}, \qquad \mathbf{X} \ge 0$$

where $\mathbf{X}$ is an $(m + n)$ column vector. Let $\{\mathbf{P}_1, \mathbf{P}_2, \ldots, \mathbf{P}_m\}$ be the vectors corresponding to a *basic* solution, and let $\{c_1, c_2, \ldots, c_m\}$ be the coefficients in the objective function associated with these vectors. If $\{c_1, \ldots, c_m\}$ is changed to $\{d_1, \ldots, d_m\}$, show that $z_j - c_j$ for the *basic* variables will remain equal to zero and interpret the result.

☐ **7-17** For the linear programming problem defined in Problem 7-16, let the given basis be optimal. Prove that for any vector

$$\mathbf{P}_k = \begin{pmatrix} a_{1k} \\ a_{2k} \\ \vdots \\ a_{mk} \end{pmatrix}, \qquad k = 1, 2, \ldots, m+n$$

the following relationship holds.

$$\sum_{j=1}^{m} c_j \alpha_j^k = \sum_{i=1}^{m} y_i^* a_{ik}$$

where $\alpha_j^k = (\mathbf{B}^{-1}\mathbf{P}_k)_j$ and y_i^* is the corresponding optimal dual value.

□ **7-18** Given the primal linear programming problem,

$$\text{maximize } z = \mathbf{CX}$$

subject to

$$\mathbf{DX} = \mathbf{b}, \qquad \mathbf{X} \geq \mathbf{0}$$

(a) Write the complete dual problem.
(b) Find the relationship between the values of the objective function in the primal and dual problems.

□ **7-19** Consider the following linear programming problem:

$$\text{minimize } z = \sum_{j=1}^{n} c_j^0 x_j$$

subject to

$$\sum_{j=1}^{n} a_{ij} x_j = b_i, \quad i = 1, 2, \ldots, m$$

$$x_j \geq 0, \qquad \text{for all } j$$

Let $(x_1^0, \ldots, x_n^0)$ be the optimal solution to the problem, and let $(x_1', \ldots, x_n')$ be its optimal solution if c_j^0 is replaced by c_j', for all j. Prove that

$$\sum_{j=1}^{n} (c_j' - c_j^0)(x_j' - x_j^0) \leq 0$$

If $c_j' < c_j^0$, for $j = k$ and $c_j' = c_j^0$, for $j \neq k$, what can be said about x_k' as compared with x_k^0?

□ **7-20** Consider the following linear programming problem:

$$\text{minimize } z = \mathbf{CX}$$

subject to

$$\mathbf{AX} = \mathbf{b}^0, \ \mathbf{X} \geq \mathbf{0}$$

Let $\mathbf{X}^0$ be the optimal solution, and let $\mathbf{Y}^0$ be its corresponding optimal dual solution. If $\mathbf{X}^*$ is the optimal solution when $\mathbf{b}^0$ is replaced by $\mathbf{b}^*$, prove that

$$\mathbf{C}(\mathbf{X}^0 - \mathbf{X}^*) \leq \mathbf{Y}^0(\mathbf{b}^0 - \mathbf{b}^*)$$

□ **7-21** Write the complementary slackness conditions for the following problem:

$$\text{maximize } z = \mathbf{CX}$$

subject to

$$\mathbf{AX} \leq \mathbf{b}$$
$$\mathbf{L} \leq \mathbf{X} \leq \mathbf{U}$$

☐ **7-22** The matrix $\mathbf{A}$ and its inverse $\mathbf{A}^{-1}$ are given as follows:

$$\mathbf{A} = \begin{pmatrix} 2 & 1 & 0 \\ 0 & 2 & 0 \\ 4 & 0 & 1 \end{pmatrix}, \qquad \mathbf{A}^{-1} = \begin{pmatrix} 1/2 & -1/4 & 0 \\ 0 & 1/2 & 0 \\ -2 & 1 & 1 \end{pmatrix}$$

If the second and third columns of $\mathbf{A}$ are replaced by $(5, -1, 4)^T$ and $(1, 2, 1)^T$, find the new inverse by using the product form introduced in Section 7.3.1.

☐ **7-23** In Problem 7-22, suppose that the third column of $\mathbf{A}$ is replaced by the sum of the first two columns. This change will make $\mathbf{A}$ singular. Show how the product form method discovers that the new matrix is singular.

☐ **7-24** Solve the following problem by the revised simplex method:

$$\text{maximize } z = 6x_1 - 2x_2 + 3x_3$$

subject to

$$2x_1 - x_2 + 2x_3 \leq 2$$
$$x_1 \qquad + 4x_3 \leq 4$$
$$x_1, x_2, x_3 \geq 0$$

☐ **7-25** Solve the following problem by the revised simplex method:

$$\text{maximize } z = 2x_1 + x_2 + 2x_3$$

subject to

$$4x_1 + 3x_2 + 8x_3 \leq 12$$
$$4x_1 + x_2 + 12x_3 \leq 8$$
$$4x_1 - x_2 + 3x_3 \leq 8$$
$$x_1, x_2, x_3 \geq 0$$

☐ **7-26** Solve the following problem by the revised simplex method:

$$\text{minimize } z = 2x_1 + x_2$$

subject to

$$3x_1 + x_2 = 3$$
$$4x_1 + 3x_2 \geq 6$$
$$x_1 + 2x_2 \leq 3$$
$$x_1, x_2 \geq 0$$

☐ **7-27** Solve the following problems by the *revised dual* simplex method outlined in Exercise 7.3-2(b).
  (a) Minimize $z = 2x_1 + 3x_2$
      subject to

$$2x_1 + 3x_2 \leq 30$$
$$x_1 + 2x_2 \geq 10$$
$$x_1, x_2 \geq 0$$

(b) Minimize $z = 5x_1 + 6x_2$
    subject to

$$x_1 + x_2 \geq 2$$
$$4x_1 + x_2 \geq 4$$
$$x_1, x_2 \geq 0$$

☐ **7-28** Solve the following problems by applying the revised simplex to the "two-phase" procedure.
(a) The example in Section 3.2.2.
(b) Problem 3-22.

☐ **7-29** Solve by the upper bounding technique:

$$\text{maximize } z = 6x_1 + 2x_2 + 8x_3 + 4x_4 + 2x_5 + 10x_6$$

subject to

$$8x_1 + x_2 + 8x_3 + 2x_4 + 2x_5 + 4x_6 \leq 13$$
$$0 \leq x_j \leq 1, \quad j = 1, 2, \ldots, 6$$

☐ **7-30** Solve the following problem by the lower- and upper-bounding techniques:

$$\text{maximize } z = 4x_1 + 2x_2 + 6x_3$$

subject to

$$4x_1 - x_2 \qquad \leq 9$$
$$- x_1 + x_2 + 2x_3 \leq 8$$
$$-3x_1 + x_2 + 4x_3 \leq 12$$
$$1 \leq x_1 \leq 3, \quad 0 \leq x_2 \leq 5, \quad 0 \leq x_3 \leq 2$$

☐ **7-31** Apply the decomposition principle to the following problem:

$$\text{maximize } z = 6x_1 + 7x_2 + 3x_3 + 5x_4 + x_5 + x_6$$

subject to

$$x_1 + x_2 + x_3 + x_4 + x_5 + x_6 \leq 50$$
$$x_1 + x_2 \qquad\qquad\qquad \leq 10$$
$$x_2 \qquad\qquad\qquad\qquad \leq 8$$
$$5x_3 + x_4 \qquad\qquad \leq 12$$
$$x_5 + x_6 \geq 5$$
$$x_5 + 5x_6 \leq 50$$
$$x_1, x_2, \ldots, x_6 \geq 0$$

☐ **7-32** Solve the following problem by the decomposition algorithm:

$$\text{maximize } z = x_1 + 3x_2 + 5x_3 + 2x_4$$

subject to

$$
\begin{aligned}
2x_1 + x_2 && \leq 9 \\
5x_1 + 3x_2 + 4x_3 && \geq 10 \\
x_1 + 4x_2 && \leq 8 \\
x_3 - 5x_4 && \leq 4 \\
x_3 + x_4 && \leq 10 \\
x_1, x_2, x_3, x_4 \geq 0 &&
\end{aligned}
$$

☐ **7-33** Indicate the necessary changes in the decomposition algorithm to apply it to minimization problems. Then solve the problem:

$$\text{minimize } z = 5x_1 + 3x_2 + 8x_3 - 5x_4$$

subject to

$$
\begin{aligned}
x_1 + x_2 + x_3 + x_4 &\geq 25 \\
5x_1 + x_2 &\leq 20 \\
5x_1 - x_2 &\geq 5 \\
x_3 + x_4 &= 20 \\
x_1, x_2, x_3, x_4 &\geq 0
\end{aligned}
$$

☐ **7-34** Solve the following problem using the decomposition algorithm:

$$\text{minimize } z = 10y_1 + 2y_2 + 4y_3 + 8y_4 + y_5$$

subject to

$$
\begin{aligned}
y_1 + 4y_2 - y_3 && \geq 8 \\
2y_1 + y_2 + y_3 && \geq 2 \\
3y_1 && + y_4 + y_5 \geq 4 \\
y_1 && + 2y_4 - y_5 \geq 10 \\
y_1, y_2, \ldots, y_5 \geq 0 &&
\end{aligned}
$$

[*Hint:* Consider the dual of Problem 7–33.]

☐ **7-35** Solve Example 7.6–1 assuming that the objective function is given by

$$z = (3 + 3t)x_1 + 2x_2 + (5 - 6t)x_3$$

where t is a nonnegative parameter.

☐ **7-36** Solve Example 7.6–2 assuming that the right-hand side of the constraints is given by

$$\mathbf{b}(t) = \begin{pmatrix} 430 \\ 460 \\ 420 \end{pmatrix} + t \begin{pmatrix} 500 \\ 100 \\ -200 \end{pmatrix}$$

where t is a nonnegative parameter.

[*Hint:* $\mathbf{B}_0^{-1}$ as given in Example 7.6–2 remains feasible for the new $\mathbf{b}(0) = (430, 460, 420)^T$.]

☐ **7–37** Suppose that the parameterization of z and $\mathbf{b}$ given in Problems 7–35 and 7–36 are considered simultaneously. Study the variations in the optimal solution with t.

☐ **7–38** Consider the example in Section 3.2.3. Suppose that the objective function becomes

$$\text{minimize } z = (4 - t)x_1 + (1 - 3t)x_2 + (2 - 2t)x_3$$

where x_5 is an additional variable whose constraint coefficients in the original problem are 2, 2, and 5, respectively. Study the variations in the optimal solution with t. Assume that $t \geq 0$.

☐ **7–39** In Problem 7–38 suppose instead that the right-hand side of the constraints is given by

$$\mathbf{b}(t) = \begin{pmatrix} 3 \\ 6 \\ 4 \end{pmatrix} + t \begin{pmatrix} 3 \\ 2 \\ -1 \end{pmatrix}$$

Study the variation in the optimal solution with t. Assume that $t \geq 0$.

☐ **7–40** Suppose that the parameterization of z and $\mathbf{b}$ as given in Problems 7–38 and 7–39 are considered simultaneously. Study the variation in the optimal solution with t.

☐ **7–41** Consider the following problem:

$$\text{maximize } z = (2 + t)x_1 + (4 - t)x_2 + (4 - 2t)x_3 + (-3 + 3t)x_4$$

subject to

$$
\begin{aligned}
x_1 + x_2 + x_3 \quad\quad &= 4 - t \\
2x_1 + 4x_2 \quad\quad + x_4 &= 8 - t \\
x_1, x_2, x_3, x_4 &\geq 0
\end{aligned}
$$

where t is a nonnegative parameter. Obtain the optimal solution for $t = 0$ using x_3 and x_4 for the starting basic solution. Then study the variation of the optimal solution with t.

☐ **7–42** The linear programming problem

$$\text{maximize } z = 3x_1 + 6x_2$$

subject to

$$
\begin{aligned}
x_1 \quad\quad &\leq 4 \\
3x_1 + 2x_2 &\leq 18 \\
x_1, x_2 &\geq 0
\end{aligned}
$$

has the solution

Basic	x_1	x_2	x_3	x_4	Solution
z	6	0	0	3	54
x_3	1	0	1	0	4
x_2	3/2	1	0	1/2	9

where x_3 and x_4 are slack variables. Let

$$c_2(t) = 6 - 4t$$

$$\mathbf{b}(t) = \binom{4}{18} + t\binom{8}{-24}$$

$$\mathbf{P}_1(t) = \binom{1}{3} + t\binom{2}{-3}$$

while c_1 and $\mathbf{P}_2$ remain as given in the original problem.

If the foregoing parametric functions are introduced *simultaneously*, find the range of t for which the solution remains basic, feasible, and optimal. Assume that $t \geq 0$.

☐ **7–43** In Problem 7–42, suppose that

$$z = (3 + \alpha)x_1 + (6 - \alpha)x_2$$

$$\mathbf{P}_1(\beta) = \binom{1 + \beta}{3 - \beta}$$

while $\mathbf{b}$ remains unparameterized, where α and β are real parameters. Find the relationship between α and β that will always keep the solution in Problem 7–42 optimal.

☐ **7–44** Consider the unparameterized version of Problem 7–42. Suppose that the objective function and the right-hand side vary with the parameter t according to

$$z = (3 + t - t^2)x_1 + (6 - 2t - t^2)x_2$$

$$\mathbf{b}(t) = \binom{4 + t^2}{18 - 2t^2}$$

Study the variation in the optimal solution with the parameter $t(t \geq 0)$. What are the difficulties involved in dealing with the nonlinear functions?

☐ **7–45** Consider the problem

$$\text{maximize } z = (4 - 10t)x_1 + (8 - 4t)x_2$$

subject to

$$x_1 + x_2 \leq 4$$
$$2x_1 + x_2 \leq 3 - t$$
$$x_1, x_2 \geq 0$$

Study the variations in the optimal solution with the parameter t, where $-\infty < t < \infty$. Notice that t may assume negative values in this case.

☐ **7–46** The analysis in this chapter has always assumed that the optimal solution of the problem at $t = 0$ is obtained by the regular simplex method. In some problems, however, it may be more convenient to obtain the optimal solution by the dual simplex method of Section 4.4. Indicate how parametric analysis can be carried out in this case.

☐ **7–47** In the problem introduced in Section 4.4 (dual simplex method), suppose that the objective function is given by

$$z = (2 + t)x_1 + (1 + 4t)x_2$$

Study the variation in the optimal solution with $t \geq 0$.

☐ **7–48** In Problem 7–47, suppose instead that the right-hand side of the constraints is given by

$$\mathbf{b}(t) = \begin{pmatrix} 3 + 2t \\ 6 - t \\ 3 - 4t \end{pmatrix}$$

Study the variation in the optimal solution with t, $t \geq 0$.

☐ **7–49** The parametric programming analysis in this chapter assumes that an optimal basic feasible solution exists at $t = 0$. Show that the selection of $t = 0$ as a datum is arbitrary and that any datum $t = t_0$ at which a basic optimal solution exists can be used to initiate the parametric analysis.

Integer Programming

Integer programming deals with the solution of mathematical programming problems in which some or all the variables can assume nonnegative integer values only. An integer program is called **mixed** or **pure,** depending on whether some or all the variables are restricted to integer values. If in the absence of the integrality conditions the objective and constraint functions are linear, the resulting model is called an **integer linear program.**

Although several finite algorithms have been developed for the integer problem, none of these methods is uniformly efficient from the computational standpoint, particularly as the size of the problem increases. Thus, unlike linear programs, where very large problems have been successfully solved in a reasonable amount of time, the performance of integer algorithms has been erratic.

One of the major difficulties in integer programming computation is the effect of round-off error that results from the inevitable use of the digital computer for solving integer problems. Although algorithms have been developed where, starting with a problem in which all the coefficients are integers, it is never necessary to deal with fractions (hence eliminating machine round-off error), this advantage is acquired only at the expense of (sometimes) extremely slow convergence of the algorithm.

The computational difficulty characterizing integer algorithms has forced some users to think of alternative ways to "solve" the problem. One common approach is to solve the continuous version of the problem and then round the continuous optimum to the closest feasible integers. **Rounding** in this case implies approximation. For example, if the continuous optimum indicates that the "number" of machines required is 10.1, this can be approximated by (rounded to) 10. There is no guarantee, however, that the rounded solution will always satisfy the constraints. This is *always true* if the original problem is linear with some *equality* constraints. From the theory of linear programming, a rounded solution cannot be feasible, since it would imply that the same basis (with all its nonbasic variables at *zero* level) can yield two different solutions.

The infeasibility created by rounding may be tolerated, since, in general, the (estimated) parameters of the problems are not exact. But there are typical *equality* constraints in integer problems where the parameters are exact. The multiple-choice constraint $x_1 + x_2 + \cdots + x_n = 1$, where $x_j = (0, 1)$ for all j, is but one example. Under such conditions, rounding cannot be used, and an exact algorithm becomes essential.

To emphasize further the inadequacy of rounding in general, note that although integer variables are commonly thought of as representing a discrete number of objects (e.g., machines, men, ships), other types represent quantifications of some codes. Thus a decision to finance or not finance a project can be represented by the binary variable $x = 0$ if the project is rejected or $x = 1$ if it is accepted. In this case it is nonsensical to deal with fractional values of x, and the use of rounding as an approximation is logically unacceptable.

To enhance the importance of problems in which "coded" variables are used, the next section presents typical applications in this area. This will also serve to illustrate the importance of integer programming in general.

8.1 SOME APPLICATIONS OF INTEGER PROGRAMMING

In this section a number of integer programming applications are presented. Some of these applications are concerned with the direct formulation of the problem. A more important contribution is the use of integer programming to reformulate "ill-constructed" models into the acceptable format of mathematical programming models. In this case the available techniques can be used to solve problems that otherwise may be difficult to tackle.

8.1.1 FIXED-CHARGE PROBLEM

In a typical production planning problem involving N products, the production cost for product j may consist of a fixed cost (charge) K_j independent of the amount produced and a variable cost c_j per unit. Thus, if x_j is the production level of product j, its production cost function may be written as

$$C_j(x_j) = \begin{cases} K_j + c_j x_j, & x_j > 0 \\ 0, & x_j = 0 \end{cases}$$

The objective criterion then becomes

$$\text{minimize } z = \sum_{j=1}^{N} C_j(x_j)$$

This criterion is nonlinear in x_j because of the discontinuity at the origin. This makes z untractable from the analytic standpoint.

The problem can be made "more" manageable analytically by introducing auxiliary binary variables. Let

$$y_j = \begin{cases} 0, & x_j = 0 \\ 1, & x_j > 0 \end{cases}$$

These conditions can be expressed in the form of a single (linear) constraint as

$$x_j \leq M y_j$$

where $M > 0$ is sufficiently large so that $x_j \leq M$ becomes redundant with respect to any active constraint of the production planning problem. Thus the objective criterion may be written as:

$$\text{minimize } z = \sum_{j=1}^{N} (c_j x_j + K_j y_j)$$

subject to

$$0 \leq x_j \leq M y_j, \qquad \text{all } j$$
$$y_j = 0 \text{ or } 1, \qquad \text{all } j$$

To show that $x_j \leq M y_j$ is a proper constraint, notice that if $x_j > 0$, $y_j = 1$ and the fixed charge K_j is added in the objective function. If $x_j = 0$, y_j is either zero or one, but since $K_j > 0$ and z is minimized, y_j must be at zero level.

It is interesting that the original fixed-charge problem has nothing to do with integer programming. Yet the "transformed" problem becomes a zero–one mixed integer problem. The transformation is introduced only for analytic convenience. Indeed, the added binary variables are "extraneous" in the sense that they do not reveal any new information about the solution. For example, $y_j = 1$ in the optimal solution is already implied by $x_j > 0$.

8.1.2 JOB-SHOP SCHEDULING PROBLEM

Consider the sequencing problem involving the completion of n different operations on a *single* machine in the minimum possible time. Each end product goes through a sequence of different operations whose order must be preserved. Also, each of these end products may have to meet a delivery date.

The problem thus has three types of constraints: (1) sequencing, (2) noninterference, and (3) delivery date. The second type guarantees that no two operations are processed (on one machine) simultaneously.

Consider the first type. Let x_j be the time (beginning from the zero datum) for starting operation j. Let a_j be the processing time required to finish operation j. If operation i is to precede operation j, the resulting sequencing constraint is

$$x_i + a_i \leq x_j$$

Consider next the noninterference constraints. For operations i and j not to occupy the machine simultaneously,

$$\textit{either} \quad x_i - x_j \geq a_j \quad \textit{or} \quad x_j - x_i \geq a_i$$

depending, respectively, on whether j precedes i or i precedes j in the optimal solution.

The presence of the **either–or constraints** poses a problem since the model is no longer in the linear programming format (i.e., the either–or constraint results in a nonconvex solution space). This difficulty is overcome by introducing the binary variable y_{ij} defined by

$$y_{ij} = \begin{cases} 0, & \text{if operation } j \text{ precedes operation } i \\ 1, & \text{if operation } i \text{ precedes operation } j \end{cases}$$

For M sufficiently large, the "either–or" constraints become equivalent to the two *simultaneous* constraints

$$My_{ij} + (x_i - x_j) \geq a_j \quad \textit{and} \quad M(1 - y_{ij}) + (x_j - x_i) \geq a_i$$

The significance of the new transformation is that if in the optimal solution $y_{ij} = 0$, the second constraint becomes redundant. In the meantime the first constraint remains active. Similarly, if $y_{ij} = 1$, the second but not the first constraint becomes active. The introduction of the binary variable y_{ij} has thus reduced these constraints to a form where mixed integer linear programming can be applied.

The delivery dates can be met by adding the following constraints. Suppose that operation j must be completed by time d_j, then

$$x_j + a_j \leq d_j$$

Now, if t is the total time required to finish all n operations the problem becomes

$$\text{minimize } z = t$$

subject to

$$x_j + a_j \leq t, \quad j = 1, 2, \ldots, n$$

together with the sequencing, noninterference, and delivery constraints developed.

8.1.3 DICHOTOMIES

Suppose that in a certain situation, it is required that *any k* out of *m* constraints can be active. However, the specific constraints that must be imposed are not known in advance. This situation can be effected as follows. Let the *m* constraints be of the form

$$g_i(x_1, x_2, \ldots, x_n) \leq b_i, \qquad i = 1, 2, \ldots, m$$

Define

$$y_i = \begin{cases} 0, & \text{if the } i\text{th constraint is active} \\ 1, & \text{if the } i\text{th constraint is inactive} \end{cases}$$

Thus any *k* out of the *m* constraints are guaranteed to be active if, for *M* sufficiently large,

$$g_i(x_1, x_2, \ldots, x_n) \leq b_i + My_i, \qquad i = 1, 2, \ldots, m$$

and

$$y_1 + y_2 + \cdots + y_m = m - k$$

where $y_i = 0$ or 1 for all *i*. This shows for $m - k$ constraints the associated right-hand side will be of the form $b_i + M$, which makes the constraint redundant. It is important to note that the formulation will select the set of *active* constraints that yields the best objective value.

A related situation occurs when the right-hand side of a single constraint is required to assume *one* of several values; that is,

$$g(x_1, x_2, \ldots, x_n) \leq b_1, b_2, \ldots, \text{ or } b_r$$

This can be achieved by transforming the constraint to

$$g(x_1, x_2, \ldots, x_n) \leq \sum_{k=1}^{r} b_k y_k$$

and

$$y_1 + y_2 + \cdots + y_r = 1$$

where

$$y_k = \begin{cases} 1, & \text{if } b_k \text{ is the right-hand side} \\ 0, & \text{if otherwise} \end{cases}$$

Another application of integer programming for the approximation of a non-linear single-variable function is given in Section 19.3.1.

8.2 METHODS OF INTEGER PROGRAMMING

Integer programming methods can be categorized as (1) cutting methods and (2) search methods.

Cutting methods, which are developed primarily for integer *linear* problems, start with the continuous optimum. By systematically adding special "secondary" constraints, which essentially represent necessary conditions for integrality, the continuous solution space is gradually modified until its continuous optimum extreme point satisfies the integer conditions. The name "cutting methods" stems from the fact that the added "secondary" constraints effectively cut (or eliminate) certain parts of the solution space that do not contain feasible integer points.

Search methods originate from the straightforward idea of enumerating all feasible integer points. The basic idea is to develop "clever" tests that consider only a (small) portion of the feasible integers explicitly but automatically account for the remaining points implicitly. The most prominent search method is the branch-and-bound technique. It also starts with the continuous optimum, but systematically "partitions" the solution space into subproblems by deleting parts that contain no feasible integer points.

A special case of the search methods applies when all the integer variables are **binary.** The binary property of the variables simplifies the search procedure greatly.

The algorithms presented in this chapter apply primarily to the linear integer problem. The cutting methods, developed by R. E. Gomory, include the **fractional algorithm,** which applies to the *pure* integer problem, and the **mixed algorithm,** which is designed for the mixed integer problem. The **branch-and-bound** algorithm was originally developed by A. H. Land and A. G. Doig. However, R. J. Dakin's modification offers greater computational advantage ·and his version will be presented here. The third algorithm, called the **additive algorithm,** applies to the *pure* zero–one problem. An extension of the zero–one additive algorithm to *nonlinear* binary problems will also be presented.

8.3 CUTTING-PLANE ALGORITHMS

The concept of the cutting plane will first be illustrated by an example. Consider the integer linear programming problem

$$\text{maximize } z = 7x_1 + 9x_2$$

subject to

$$-x_1 + 3x_2 \le 6$$
$$7x_1 + x_2 \le 35$$

$$x_1, x_2 \text{ nonnegative integers}$$

The optimal continuous solution (ignoring the integrality condition) is shown graphically in Figure 8–1. This is given by $z = 63$, $x_1 = 9/2$, and $x_2 = 7/2$, which is noninteger.

The idea of the cutting-plane algorithm is to change the convex set of the solution space so that the appropriate extreme point becomes all-integer. Such changes in the boundaries of the solution space should result still in a convex set. Also, this change should be made without "slicing off" *any* of the feasible *integer* solutions of the original problem. Figure 8–1 shows how two (arbitrarily

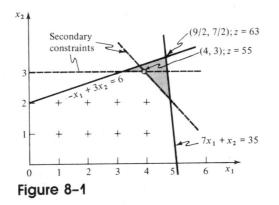

Figure 8–1

selected) secondary constraints are added to the problem with the new extreme point (4, 3) giving the integer optimal solution. Notice that the area sliced off the original solution space (shaded area) does not include any integer values.

The following analysis shows how the secondary constraints are developed systematically for the pure and mixed integer problems.

8.3.1 THE FRACTIONAL (PURE INTEGER) ALGORITHM

A basic requirement for the application of this algorithm is that all the coefficients and the right-hand-side constant of each constraint must be integer. For example, the constraint

$$x_1 + \frac{1}{3} x_2 \leq \frac{13}{2}$$

must be transformed to

$$6x_1 + 2x_2 \leq 39$$

where no fractions are present. The latter is achieved by multiplying both sides of the original constraint by the least common multiple of the denominators.

The foregoing requirement is imposed since, as will be shown later, the pure integer algorithm does not differentiate between the regular and slack variables of the problem in the sense that all variables must be integers. The presence of fractional coefficients in the constraints thus may not allow the slack variables to assume integer values. In this case, the fractional algorithm may indicate that no feasible solution exists, even though the problem may have a feasible integer solution in terms of *non*slack variables. (See Problem 8–7 for an illustration of this case.)

The details of the algorithm will be discussed now. First, the problem is solved as a regular linear programming problem, that is, disregarding the integrality condition. If the optimal solution happens to be integer, there is nothing more to be done. Otherwise, the secondary constraints that will force the solution toward the integer solution are developed as follows. Let the final optimal tableau for the linear program be given by

Basic	x_1	$\cdots$	x_i	$\cdots$	x_m	w_1	$\cdots$	w_j	$\cdots$	w_n	Solution
z	0	$\cdots$	0	$\cdots$	0	$\bar{c}_1$	$\cdots$	$\bar{c}_j$	$\cdots$	$\bar{c}_n$	β_0
x_1	1	$\cdots$	0	$\cdots$	0	α_1^1	$\cdots$	α_1^j	$\cdots$	α_1^n	β_1
x_i	0	$\cdots$	1	$\cdots$	0	α_i^1	$\cdots$	α_i^j	$\cdots$	α_i^n	β_i
x_m	0	$\cdots$	0	$\cdots$	1	α_m^1	$\cdots$	α_m^j	$\cdots$	α_m^n	β_m

The variables x_i ($i = 1, 2, \ldots, m$) represent the basic variables and the variables w_j ($j = 1, 2, \ldots, n$) are the nonbasic variables. These variables have been arranged as such for convenience.

Consider the ith equation where the basic variable x_i assumes a noninteger value.

$$x_i = \beta_i - \sum_{j=1}^{n} \alpha_i^j w_j, \qquad \beta_i \text{ noninteger} \qquad (source\ row)$$

Any such equation will be referred to as a **source row**. Since, in general, the coefficients of the objective function can be made integer, the variable z is also integer and the z-equation may be selected as a source row. Indeed, the convergence proof of the algorithm requires that z be integer.

Let

$$\beta_i = [\beta_i] + f_i$$
$$\alpha_i^j = [\alpha_i^j] + f_{ij}$$

where $N = [a]$ is the largest integer such that $N \le a$. It follows that $0 < f_i < 1$ and $0 \le f_{ij} < 1$; that is, f_i is a strictly positive fraction and f_{ij} is a nonnegative fraction. For example,

a	$[a]$	$f = a - [a]$
$1\frac{1}{2}$	1	$1/2$
$-2\frac{1}{3}$	-3	$2/3$
-1	-1	0
$-2/5$	-1	$3/5$

The source row thus yields

$$f_i - \sum_{j=1}^{n} f_{ij} w_j = x_i - [\beta_i] + \sum_{j=1}^{n} [\alpha_i^j] w_j$$

For *all* the variables x_i and w_j to be integer, the right-hand side of the equation must be integer, which in turn implies that the left-hand side must also be integer. Given $f_{ij} \ge 0$ and $w_j \ge 0$ for all i and j, it follows that $\sum_{j=1}^{n} f_{ij} w_j \ge 0$. Consequently,

$$f_i - \sum_{j=1}^{n} f_{ij} w_j \le f_i$$

This means that $f_i - \sum_{j=1}^n f_{ij} w_j < 1$ because $f_i < 1$. Since the left-hand side must be integer, a *necessary* condition for satisfying integrality becomes

$$f_i - \sum_{j=1}^n f_{ij} w_j \leq 0$$

The last constraint can be put in the form

$$S_i = \sum_{j=1}^n f_{ij} w_j - f_i \qquad \text{(fractional cut)}$$

where S_i is a nonnegative slack variable which by definition must be an integer. This constraint equation defines the **fractional cut.** From the last tableau, $w_j = 0$ and thus $S_i = -f_i$, which is infeasible. This means that the new constraint is not satisfied by the given solution. The dual simplex method (Section 4.4) can then be used to clear this infeasibility, which is equivalent to cutting off the solution space toward the optimal integer solution.

The new tableau after adding the fractional cut will thus become

Basic	x_1	$\cdots$	x_i	$\cdots$	x_m	w_i	$\cdots$	w_j	$\cdots$	w_n	S_i	Solution
z	0	$\cdots$	0	$\cdots$	0	$\bar{c}_1$	$\cdots$	$\bar{c}_j$	$\cdots$	$\bar{c}_n$	0	β_0
x_1	1	$\cdots$	0	$\cdots$	0	α_1^1	$\cdots$	α_1^j	$\cdots$	α_1^n	0	β_1
x_i	0	$\cdots$	1	$\cdots$	0	α_i^1	$\cdots$	α_i^j	$\cdots$	α_i^n	0	β_i
x_m	0	$\cdots$	0	$\cdots$	1	α_m^1	$\cdots$	α_m^j	$\cdots$	α_m^n	0	β_m
S_i	0	$\cdots$	0	$\cdots$	0	$-f_{i1}$	$\cdots$	$-f_{ij}$	$\cdots$	$-f_{in}$	1	$-f_i$

If the new solution (after applying dual simplex method) is integer, the process ends. Otherwise, a new fractional cut is constructed from the *resulting* tableau and the dual simplex method is used again to clear the infeasibility. This procedure is repeated until an integer solution is achieved. However, if at any iteration the dual simplex algorithm indicates that no feasible solution exists, the problem has no feasible *integer* solution.

The algorithm is referred to as the **fractional method** because all the nonzero coefficients of the generated cut are less than one.

The fractional algorithm may indicate, at first thought, that the size of the simplex tableau can become very large as new cuts are augmented to the problem. This is not true. In fact, the total number of constraints in the *augmented* problem cannot exceed the number of variables in the original problem, that is, $(m + n)$. This result follows, since if the augmented problem includes more than $(m + n)$ constraints, one or more of the slack variables S_i associated with the fractional cuts must become basic. In this case the associated equations become redundant and may be dropped from the tableau completely.

The fractional algorithm has two disadvantages:

1. The round-off errors that evolve in automatic calculations will most likely distort the original data, particularly with the increase in problem size.

2. The solution of the problem remains infeasible in the sense that no *integer* solution can be obtained until the optimal integer solution is reached. This means that there will be no "good" integer solution in store if the calculations are stopped prematurely prior to the attainment of the optimal (integer) solution.

The first difficulty is overcome by the development of an **all-integer integer algorithm.** The algorithm starts with an initial all-integer tableau (i.e., all the coefficients are integers) suitable for the application of the dual simplex algorithm. Special cuts are then constructed such that their addition to the tableau will preserve the integrality of all the coefficients. However, the fact that the solution remains infeasible until the integer optimal solution is reached still presents a disadvantage.

The second difficulty was considered by developing cutting-plane algorithms that start integer and feasible but nonoptimal. The iterations continue to be feasible and integral until the optimum solution is reached. In this respect, this algorithm is **primal-feasible** as compared with Gomory's algorithms, which are **dual-feasible.** The primal algorithms do not appear to be computationally promising, however.

Example 8.3-1. Consider the problem that was solved graphically at the beginning of Section 8.3. The optimal continuous solution is given by

Basic	x_1	x_2	x_3	x_4	Solution
z	0	0	28/11	15/11	63
x_2	0	1	7/22	1/22	7/2
x_1	1	0	−1/22	3/22	9/2

Since the solution is noninteger, a fractional cut is added to the tableau. Generally, any of the constraint equations corresponding to a noninteger solution can be selected to generate the cut. However, as a rule of thumb, we usually choose the equation corresponding to $\max_i \{f_i\}$. Since both equations in this problem have the same value of f_i, that is, $f_1 = f_2 = 1/2$, either one may be used. Consider the x_2-equation. This gives

$$x_2 + \frac{7}{22} x_3 + \frac{1}{22} x_4 = 3\tfrac{1}{2}$$

or

$$x_2 + \left(0 + \frac{7}{22}\right) x_3 + \left(0 + \frac{1}{22}\right) x_4 = \left(3 + \frac{1}{2}\right)$$

Hence the corresponding fractional cut is given by

$$S_1 - \frac{7}{22} x_3 - \frac{1}{22} x_4 = -\frac{1}{2}$$

This gives the new tableau

Basic	x_1	x_2	x_3	x_4	S_1	R.H.S.
z	0	0	28/11	15/11	0	63
x_2	0	1	7/22	1/22	0	$3\frac{1}{2}$
x_1	1	0	−1/22	3/22	0	$4\frac{1}{2}$
S_1	0	0	−7/22	−1/22	1	−1/2

The dual simplex method yields

Basic	x_1	x_2	x_3	x_4	S_1	Solution
z	0	0	0	1	8	59
x_2	0	1	0	0	1	3
x_1	1	0	0	1/7	−1/7	$4\frac{4}{7}$
x_3	0	0	1	1/7	−22/7	$1\frac{4}{7}$

Since the solution is still noninteger, a new cut is constructed. The x_1-equation is written as

$$x_1 + \left(0 + \frac{1}{7}\right) x_4 + \left(-1 + \frac{6}{7}\right) S_1 = \left(4 + \frac{4}{7}\right)$$

which gives the cut

$$S_2 - \frac{1}{7} x_4 - \frac{6}{7} S_1 = -\frac{4}{7}$$

Adding this constraint to the last tableau, we get

Basic	x_1	x_2	x_3	x_4	S_1	S_2	R.H.S.
z	0	0	0	1	8	0	59
x_2	0	1	0	0	1	0	3
x_1	1	0	0	1/7	−1/7	0	$4\frac{4}{7}$
x_3	0	0	1	1/7	−22/7	0	$1\frac{4}{7}$
S_2	0	0	0	−1/7	−6/7	1	−4/7

The dual simplex method now yields

Basic	x_1	x_2	x_3	x_4	S_1	S_2	Solution
z	0	0	0	0	2	7	55
x_2	0	1	0	0	1	0	3
x_1	1	0	0	0	-1	1	4
x_3	0	0	1	0	-4	1	1
x_4	0	0	0	1	6	-7	4

which gives the optimal integer solution $z = 55$, $x_1 = 4$, $x_2 = 3$.

The reader can verify graphically that the addition of the developed cuts "cuts" the solution space as desired (see Figure 8–1). The first cut

$$S_1 - \frac{7}{22} x_3 - \frac{1}{22} x_4 = -\frac{1}{2}$$

can be expressed in terms of x_1 and x_2 only by using the appropriate substitution as follows:

$$S_1 - \frac{7}{22}(6 + x_1 - 3x_2) - \frac{1}{22}(35 - 7x_1 - x_2) = -\frac{1}{2}$$

or

$$S_1 + x_2 = 3$$

which is equivalent to

$$x_2 \le 3$$

Similarly, for the second cut,

$$S_2 - \frac{1}{7} x_4 - \frac{6}{7} S_1 = -\frac{4}{7}$$

the equivalent constraint in terms of x_1 and x_2 is

$$x_1 + x_2 \le 7$$

Figure 8–1 shows that the addition of these two constraints will result in the new (optimal) extreme point (4, 3).   ◀

Exercise 8.3–1

Consider the Reddy Mikks model, whose solution is given toward the end of Section 3.2.2. Suppose that all the variables are integers. Determine the cuts associated with the basic variables x_E, x_L, s_3, and s_4 and express them in terms of x_E and x_I only.
[*Ans.* x_I: $2/3s_1 + 2/3s_2 \ge 1/3$, or $2x_E + 2x_I \le 9$.
x_E: Same cut as x_I.
s_3: No cut is possible, since s_3 is already integer.
s_4: $1/3s_1 + 1/3s_2 \ge 2/3$, or $x_E + x_I \le 4$.]

Strength of the Fractional Cut

The foregoing development indicates that the specific inequality defining a cut depends directly on the "source row" from which it is generated. Thus different inequality cuts may be generated from the same simplex tableau. The question naturally arises: Which cut is the "strongest"? Strength could be measured in terms of how deep the inequality cuts into the solution space. This result can be expressed mathematically as follows. Consider the two inequalities

$$\sum_{j=1}^{n} f_{ij} w_j \geq f_i \tag{1}$$

and

$$\sum_{j=1}^{n} f_{kj} w_j \geq f_k \tag{2}$$

Cut (1) is said to be stronger than (2) if $f_i \geq f_k$ and $f_{ij} \leq f_{kj}$, for all j, with the strict inequality holding at least once.

This definition of strength is difficult to implement computationally. Thus empirical rules reflecting this definition are devised. Two such rules call for generating the cut from the source row that has (1) $\max_i \{f_i\}$ or (2) $\max_i \{f_i / \sum_{j=1}^{n} f_{ij}\}$. The second rule is more effective, since it more closely represents the definition of strength given.†

Example 8.3-2. In Example 8.3–1, the optimum continuous solution is $z = 63$, $x_1 = 9/2$, and $x_2 = 7/2$. Since z is already integer, its equation cannot be taken as a source row. According to the empirical rules given since $f_1 = f_2 = 1/2$, the rule is nonconclusive about which source row may be better. But to apply the second rule, it is necessary to develop all the coefficients of the respective fractional cuts from each source row. The cuts from the x_1-row and x_2-row are

$$x_1\text{-row:} \quad \frac{21}{22} x_3 + \frac{3}{22} x_4 \geq \frac{1}{2}$$

$$x_2\text{-row:} \quad \frac{7}{22} x_3 + \frac{1}{22} x_4 \geq \frac{1}{2}$$

Since

$$\frac{1/2}{7/22 + 1/22} > \frac{1/2}{21/22 + 3/22}$$

the x_2-equation is selected as a source row.

The selection of the x_2-equation as a source row in Example 8.3–1 was only accidental. To show that this was a proper choice, the two cuts (from the x_1-row and x_2-row) are compared. In Example 8.3–1, the cut from the x_2-row expressed in terms of x_1 and x_2 is given by

$$x_2 \leq 3$$

---

† Other rules that are based on information in the objective function row may also be found in Taha (1975), pp. 184–185.

By following a similar substitution, the cut from the x_1-row is expressed as

$$x_2 \leq 10/3$$

The first cut is more *restrictive* and hence stronger than the second cut. One must caution, however, that the given rules, being empirical, may not generally yield the strongest cut. ◀

Exercise 8.3–2

In Exercise 8.3–1, determine the strongest of the resulting cuts by using the criterion above; then plot the cuts on the (x_E, x_I)-space to illustrate the concept of cut strength graphically.
[*Ans.* The s_4-cut is the strongest. In the graphical space, it cuts the deepest into the solution space.]

8.3.2 THE MIXED ALGORITHM

Let x_k be an integer variable of the mixed problem. Again, as in the pure integer case, consider the x_k-equation in the optimal continuous solution. This is given by

$$x_k = \beta_k - \sum_{j=1}^{n} \alpha_k^j w_j = [\beta_k] + f_k - \sum_{j=1}^{n} \alpha_k^j w_j \qquad (\textit{source row})$$

or

$$x_k - [\beta_k] = f_k - \sum_{j=1}^{n} \alpha_k^j w_j$$

Because some of the w_j variables may not be restricted to integer values in this case, it is incorrect to use the fractional cut developed in the preceding section. But a new cut can be devised based on the same general idea.

For x_k to be integer, either $x_k \leq [\beta_k]$ or $x_k \geq [\beta_k] + 1$ must be satisfied. From the source row, these conditions are equivalent to

$$\sum_{j=1}^{n} \alpha_k^j w_j \geq f_k \qquad (1)$$

$$\sum_{j=1}^{n} \alpha_k^j w_j \leq f_k - 1 \qquad (2)$$

Let

$$J^+ = \text{set of subscripts } j \text{ for which } \alpha_k^j \geq 0$$
$$J^- = \text{set of subscripts } j \text{ for which } \alpha_k^j < 0$$

Then, from (1) and (2), we get

$$\sum_{j \epsilon J^+} \alpha_k^j w_j \geq f_k \qquad (3)$$

$$\frac{f_k}{f_k - 1} \sum_{j \epsilon J^-} \alpha_k^j w_j \geq f_k \qquad (4)$$

Since (1) and (2), and hence (3) and (4), cannot occur simultaneously, it follows that (3) and (4) can be combined into one constraint of the form

$$S_k - \left\{ \sum_{j \epsilon J^+} \alpha_k^j w_j + \frac{f_k}{f_k - 1} \sum_{j \epsilon J^-} \alpha_k^j w_j \right\} = -f_k \qquad \text{(mixed cut)}$$

where $S_k \geq 0$ is a nonnegative slack variable. The last equation is the required **mixed cut,** and it represents a necessary condition for x_k to be integer. Since all $w_j = 0$ at the current optimal tableau, it follows that the cut is infeasible. The dual simplex method is thus used to clear the infeasibility.

The mixed cut is developed without taking advantage of the fact that some of the w_j variables may be integer. If this is taken into account, the following stronger cut will result:

$$S_k = -f_k + \sum_{j=1}^{n} \lambda_j w_j$$

where

$$\lambda_j = \begin{cases} \alpha_k^j & \text{if } \alpha_k^j \geq 0 \text{ and } w_j \text{ is nonintegral} \\[2mm] \dfrac{f_k}{f_k - 1} \alpha_k^j & \text{if } \alpha_k^j < 0 \text{ and } w_j \text{ is nonintegral} \\[2mm] f_{kj} & \text{if } f_{kj} \leq f_k \text{ and } w_j \text{ is integral} \\[2mm] \dfrac{f_k}{1 - f_k} (1 - f_{kj}) & \text{if } f_{kj} > f_k \text{ and } w_j \text{ is integral} \end{cases}$$

The derivation of this formula is found in Taha [1975, p. 200].

Example 8.3–3. Consider Example 8.3–1. Suppose that x_1 only is restricted to integer values. From the x_1-equation,

$$x_1 - \frac{1}{22} x_3 + \frac{3}{22} x_4 = \left(4 + \frac{1}{2} \right)$$

$$J^- = \{3\}, \qquad J^+ = \{4\}, \qquad f_1 = 1/2$$

Hence the mixed cut is given by

$$S_1 - \left\{ \frac{3}{22} x_4 + \left(\frac{\frac{1}{2}}{\frac{1}{2} - 1} \right) \left(-\frac{1}{22} \right) x_3 \right\} = -\frac{1}{2}$$

or

$$S_1 - \frac{1}{22} x_3 - \frac{3}{22} x_4 = -\frac{1}{2}$$

Adding this to the last tableau gives

Basic	x_1	x_2	x_3	x_4	S_1	R.H.S.
z	0	0	28/11	15/11	0	63
x_2	0	1	7/22	1/22	0	7/2
x_1	1	0	−1/22	3/22	0	9/2
S_1	0	0	−1/22	−3/22	1	−1/2

Now, applying the dual simplex method yields

Basic	x_1	x_2	x_3	x_4	S_1	Solution
z	0	0	23/11	0	10	58
x_2	0	1	10/33	0	−1/3	10/3
x_1	1	0	−1/11	0	1	4
x_4	0	0	1/3	1	−22/3	11/3

which yields the optimal solution $z = 58$, $x_1 = 4$, and $x_2 = 10/3$ with x_1 an integer as required. ◀

Exercise 8.3–3
Consider Example 3.8–3.
(a) Suppose that x_2 is integer also. Develop its mixed cut from the last tableau of the example.
  [*Ans.* $S_2 - 10/33x_3 - 1/6S_1 = -1/3$.]
(b) In the original problem before the x_1-cut is added, if both x_1 and x_2 are integers, then x_3 and x_4 must also be integers. Develop a mixed cut from the x_1-row by using the definition of λ_j given and compare it with the fractional cut for x_1. (Notice that in both cases all the variables x_1, x_2, x_3, and x_4 are integers.)
  [*Ans.* Mixed cut: $-1/2 + 1/22x_3 + 3/22x_4 \geq 0$
          Fractional cut: $-1/2 + 21/22x_3 + 3/22x_4 \geq 0$
  The mixed cut is stronger.]

Computations in Cutting Methods

Although only two types of cuts are presented in this book, several other cuts have been developed, with each new cut alleviating some of the computational difficulties associated with the others. However, no single cut can be considered uniformly superior from the computational standpoint.

Although in some isolated cases of specially structured problems cuts have proven effective, the general consensus among practitioners is that cutting methods cannot be relied on to solve integer problems regardless of size. Experience has shown that some rather small problems could not be solved by the cutting methods. In fact, cases have been reported in which a random change in the order of the constraints has converted a computationally easy problem into a rather formidable one.

Perhaps the general conclusion concerning cutting methods is that they alone cannot be used effectively to solve the general integer problem. However, ideas may be, and indeed have been, borrowed from these methods to enhance the effectiveness of other types of solution techniques [see Taha (1975), pp. 160–161].

8.4 BRANCH-AND-BOUND METHOD

This technique also solves the integer problem by considering its continuous version. But unlike the cutting methods, the branch-and-bound method applies directly to both the *pure* and *mixed* problems.

The general idea of the method is first to solve the problem as a continuous model (linear program). Suppose that x_r is an integer-constrained variable whose optimum continuous value x_r^* is fractional. The range

$$[x_r^*] < x_r < [x_r^*] + 1$$

cannot include any feasible integer solution. Consequently, a feasible integer value of x_r must satisfy *one* of two conditions,

$$x_r \leq [x_r^*] \quad \text{or} \quad x_r \geq [x_r^*] + 1$$

(compare the development of the mixed cut, Section 8.3.2). These two conditions when applied to the continuous model result in two mutually exclusive problems created by imposing the constraints $x_r \leq [x_r^*]$ and $x_r \geq [x_r^*] + 1$ on the original solution space. In this case it is said that the original problem is **branched** (or partitioned) into two subproblems. Actually, the branching process deletes parts of the continuous space that do not include feasible integer points by enforcing *necessary* conditions for integrality.

Now each subproblem may be solved as a linear program (using the same objective function of the original problem). If its *optimum* is feasible with respect to the integer problem, its solution is recorded as the best one so far available. In this case it will be unnecessary to further "branch" this subproblem since this cannot yield a better integer solution. Otherwise, the subproblem must be partitioned into two subproblems by again imposing the integer conditions on one of its integer variables that currently has a fractional optimal value. Naturally, when a *better* integer feasible solution is found for any subproblem, it should replace the one at hand. The process of branching continues, where applicable, until each subproblem terminates either with an integer solution or there is evidence that it cannot yield a better one. In this case the feasible solution at hand, if any, is the optimum.

The efficiency of computations can be enhanced by introducing the concept of **bounding.** This concept indicates that if the *continuous* optimum solution of a subproblem yields a worse objective value than the one associated with the best available integer solution, it does not pay to explore the subproblem any further. In this case the subproblem is said to be **fathomed** and may henceforth be deleted. In other words, once a feasible integer solution is found, its associated objective value can be used as a (upper in case of minimization and lower in case of maximization) **bound** to discard inferior subproblems.

The importance of acquiring a *good* bound at the early stages of the calculations cannot be overemphasized. In terms of the procedure just described, this seems to depend directly on the order in which the different subproblems are *generated* and *scanned*. The specific problems generated depend on the variable selected to effect branching. Unfortunately, there is no definite "best" way for selecting the branching variable or the specific sequence in which the subproblem must be scanned. But empirical rules exist that do enhance the process. These rules are usually implemented in most of the commercial branch-and-bound codes such as UMPIRE (Scientific Control Systems Ltd., London). The details are beyond the scope of this presentation, however [see Taha (1975), pp. 165–171].

Example 8.4-1†

$$\text{Maximize } z = 2x_1 + 3x_2$$

subject to

$$5x_1 + 7x_2 \leq 35$$
$$4x_1 + 9x_2 \leq 36$$
$$x_1, x_2 \geq 0 \text{ and integers}$$

Figure 8–2 summarizes the generated subproblems in the form of a tree. The procedure starts at node 1, where the problem is solved as a (continuous) linear program. The optimum solution at node 1 yields $x_1 = 3\frac{12}{17}$ and $x_2 = 2\frac{6}{17}$. Both variables have fractional values, and either one may be used to start the branching process. If we arbitrarily select x_2 for branching, two subproblems are created by the restrictions $x_2 \leq 2$ and $x_2 \geq 3$ (note that $x_2^* = 2\frac{6}{17}$ so that $[x_2^*] = 2$). These two constraints will result in subproblems 2 and 3 (see the graphical solution space associated with node 1). The two subproblems contain all the *feasible* integer solution of the original problem so that the original feasible integer space remains unchanged by the branching process.

The next step is to select either subproblem 2 or 3 for investigation, and further branching, if necessary. It is important to notice that there are *no definite* rules for making the selection. It is equally important to realize that each choice will result in a different sequence of subproblems and hence in a different number of iterations for reaching the optimum integer solution. We demonstrate this point by using the information in Figure 8–2.

Suppose that we consider node 2 next ($x_2 \leq 2$). The resulting optimum solution is $x_1 = 4\frac{1}{5}$ and $x_2 = 2$. (Algebraically, this solution is determined from the optimum tableau at node 1 by augmenting the constraint $x_2 \leq 2$ and applying the dual simplex method.) Since $x_1 (= 4\frac{1}{5})$ remains noninteger, subproblems 4 and 5 are created by imposing the restrictions $x_1 \leq 4$ and

† This example is adapted from H. Taha, "Integer Programming," *Encyclopedia of Computer Science and Technology,* Vol. 9, Marcel Dekker, New York, 1978, pp. 419–447.

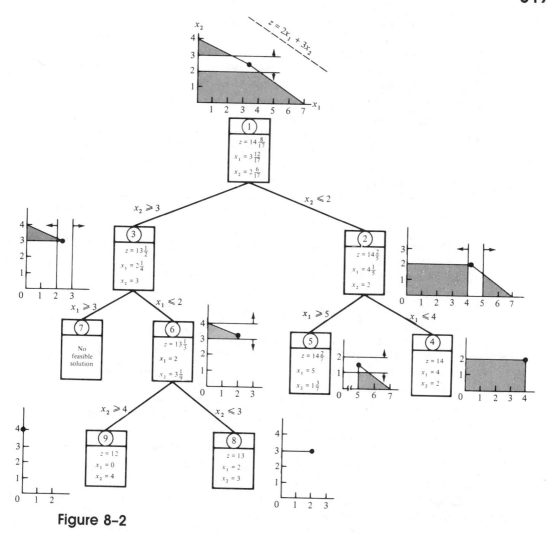

Figure 8-2

$x_1 \geq 5$ on the problem at node 2. Again, suppose that we arbitrarily consider node 4 next (instead of 5 or 3). Solving subproblem 4 yields the integer solution $x_1 = 4$, $x_2 = 2$, and $z = 14$.

The fact that we obtained an integer solution at node 4 does *not* mean that we have encountered *the* optimum. The reason is that subproblems 3 and 5, which have not been solved yet, may yield a better integer solution than $z = 14$. As a result, we say that the integer solution at node 4 yields the **lower bound** $\underline{z} = 14$, which means that any subsequent subproblem that does not have the potential to yield a better optimum value of z must be discarded as nonpromising.

Looking now at subproblem 3, we find that its optimum solution (obtained by imposing $x_2 \geq 3$ on the problem at node 1) yields $z = 13\frac{1}{2}$, which is less than $\underline{z} = 14$. Thus node 3 need not be investigated any further and the search along the branch $x_2 \geq 3$ stops at this point. We are now left with node 5, which must be investigated next. Although optimum z at node 5 is $14\frac{2}{7}$, which is higher than the lower bound $\underline{z} = 14$, any further branching from this node cannot yield a better value than $z = 14$ because the difference between optimal z at node 5 and $\underline{z}$ is less than one *and* all the coefficients of the objective function are integers. Thus further branching of node 5 can *at best* yield another integer solution with $z = 14$. This means that unless we are interested in *alternative* integer solutions, node 5 must be discarded.

This discussion shows that the optimum integer solution is associated with node 4 ($z = 14$, $x_1 = 4$, $x_2 = 2$). This information is obtained by considering the nodes 1, 2, 3, 4, and 5 in the order $1 \rightarrow 2 \rightarrow 4 \rightarrow 3 \rightarrow 5$. (You will reach the same conclusion if node 5 is investigated before node 3. Verify.)

Note that in the sequence of subproblems (namely, $1 \rightarrow 2 \rightarrow 4 \rightarrow 3 \rightarrow 5$), we have no information about the quality of the solution (in terms of the value of z) at nodes 3 and 5 until we *solve* their associated subproblems. This is the reason we cannot tell in advance whether it is advantageous at node 1 to investigate $x_2 \geq 3$ before $x_2 \leq 2$, or vice versa. To illustrate our point, suppose that we consider node 3 first (this means that at this point node 2 will be stored *unsolved* for later scanning). The solution at node 3 is $z = 13\frac{1}{2}$, $x_1 = 2\frac{1}{4}$, and $x_2 = 3$. Since x_1 is not integer, the two branches $x_1 \leq 2$ and $x_1 \geq 3$ are added from node 3, which leads to nodes 6 and 7. Suppose that we consider node 6 next. From its optimum solution, we have $x_2 = 3\frac{1}{9}$, which creates the branches $x_2 \leq 3$ and $x_2 \geq 4$, that is, nodes 8 and 9. If we choose node 8 next, we obtain its optimum (integer) solution $z = 13$, $x_1 = 2$, and $x_2 = 3$. Thus the lower bound is $\underline{z} = 13$. We now have nodes 2, 7, and 9, which we must compare with $\underline{z} = 13$ to see if we can automatically discard them. Node 9 has optimum $z = 12$, which is less than $\underline{z}$ and thus is discarded. Node 7 is discarded because it has no feasible solution. Now, at node 2 we get $z = 14\frac{2}{5}$, which is higher than $\underline{z}$ by at least one. The two branches from node 2 yield nodes 4 and 5. If we consider node 4 next, we obtain the new (better) lower bound $\underline{z} = 14$. This bound will automatically discard node 5, and the search ends with the optimum being given by node 4. The end result is that the optimum is determined by considering the following sequence of nodes, $1 \rightarrow 3 \rightarrow 6 \rightarrow 8 \rightarrow 9 \rightarrow 7 \rightarrow 2 \rightarrow 4 \rightarrow 5$. Thus, by branching to node 3 first, it is necessary to solve nine subproblems as compared with five subproblems only when node 2 is selected. This example illustrates a basic computational disadvantage of the branch-and-bound procedure; namely, we do not know in advance how many subproblems will be solved before the optimum integer is obtained *and verified*. This disadvantage results in a taxation of the computer memory as well as increase in time of computation. ◀

Exercise 8.4–1

(a) Consider Figure 8–2. Suppose that the sequence of nodes starts as $1 \rightarrow 2 \rightarrow 5$. Determine *all* the nodes following node 5 and any lower bound that may result. [*Ans.* See Figure 8–3. The lower bound is $\underline{z} = 14$ if nodes are investigated in the order $5 \rightarrow 11 \rightarrow 13 \rightarrow 14$, or $\underline{z} = 13$ followed by $\underline{z} = 14$ if nodes are investigated in the order $5 \rightarrow 11 \rightarrow 12 \rightarrow 13 \rightarrow 14$.]

(b) In Figure 8–2, verify that the *integer* feasible solutions at nodes 4, 5, 7, 8, and 9 are essentially those of the original problem.

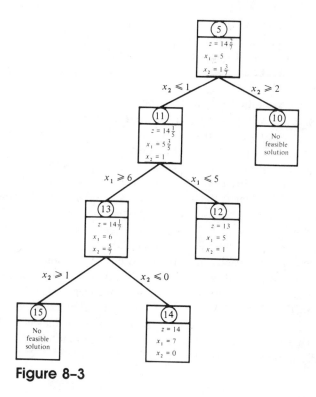

Figure 8–3

(c) In Figure 8–2, suppose that x_2 only is restricted to integer values, indicate the sequence of nodes leading to the optimum solution and find the optimum solution.
[*Ans.* $1 \to 2 \to 3$ or $1 \to 3 \to 2$. The optimum occurs at node 2.]

Computations in Branch-and-Bound Methods

Practical computer codes based on the branch-and-bound technique differ from the given outline mainly in the details of selecting the branching variables at a node and the sequence in which the subproblems are scanned. These rules are based on heuristics developed through experimentation.

A basic disadvantage of the foregoing algorithm is that it is necessary to solve a complete linear program at each node. In large problems, this could be very time consuming, particularly when the only information needed at the node may be its optimum objective value. This point is clarified by realizing that once a "good" bound is obtained, "many" nodes can be discarded from the knowledge of their optimum objective values.

The preceding point led to the development of a procedure whereby it becomes unnecessary to solve all the subproblems of the branch-and-bound tree. The idea is to "estimate" an *upper* bound (assume a maximization problem) on the optimum objective value at each node. Should this upper bound become smaller than the objective value associated with the best available integer solution, the node is discarded. The main advantage is that the upper bounds can be estimated quickly with minimal computations. The general idea is to estimate the **penalties** (i.e., the deterioration in the objective value) resulting from enforcing the conditions $x_k \leq [\beta_k]$ and $x_k \geq [\beta_k] + 1$. This can be achieved by augmenting each of these constraints to the optimum tableau at the node (with which x_k is associated). Then, *under the assumption of no change in basis,* the required penalty can be estimated directly from the objective function coefficients (see Problem 8–27).

Although the penalties are easy to compute, they have the drawback that they are not necessarily proportional to the *true* deterioration in the objective value. In other words, they do not provide a tight bound and hence may not be effective. Various attempts to "strengthen" these penalties have been made. The most interesting of these is the one using information from the cutting methods. Nevertheless, even these "strengthened" penalties appear to be ineffective computationally, particularly when the size of the problem increases. It appears that commercial codes have abandoned the use of simple penalties in favor of heuristics that proved, through experimentation with large and complex problems, to be quite effective [see Taha (1975), pp. 165–171].

In spite of the drawbacks of branch-and-bound methods, it can be stated that, to date, these methods are the most effective in solving integer programs of practical sizes. Indeed, all available commercial codes are based on the branch-and-bound method. This does *not* mean, however, that *every* integer program can be solved by a branch-and-bound method. It only means that when the choice is between a cutting method and a branch-and-bound method, the latter has generally proved superior.

8.5 ZERO–ONE IMPLICIT ENUMERATION

Any integer variable can be equivalently expressed in terms of a number of pure **zero–one (binary)** variables. The simplest way to accomplish this is as follows. Let $0 \leq x \leq n$ be an integer variable where n is an integer upper bound. Then, given that $y_1, y_2, \ldots,$ and y_n are zero–one variables,

$$x = y_1 + y_2 + \cdots + y_n$$

is an exact binary representation of all feasible values of x. Another (more economical) representation in which the number of binary variables is usually less than n is given by

$$x = y_0 + 2y_1 + 2^2 y_2 + \cdots + 2^k y_k$$

where k is the smallest integer satisfying $2^{k+1} - 1 \geq n$.

The fact that every integer problem can be made binary together with the computational simplicity of dealing with zero–one variables (each variable has two values only) has directed attention to exploiting these properties to develop an efficient algorithm.

The algorithm presented next is actually a variation of the more general branch-and-bound method (Section 8.4). It is interesting that the original version of the algorithm is not presented in this context. Perhaps the prime reason is that the original algorithm (and, indeed, its modifications) never require the solution of linear programs. In fact, the only arithmetic operations needed are additions and subtractions. This is why it is sometimes called the **additive algorithm.**

The additive algorithm is presented in the context of the branch-and-bound method. This, perhaps, will make it easier to appreciate the relationship just described.

8.5.1 ADDITIVE ALGORITHM (PURE BINARY PROBLEM)

For the purpose of this algorithm, the continuous version of the zero–one problem must start dual-feasible, that is, optimal but not feasible. Moreover, all the constraints must be of the type ($\leq$), thus ruling out explicit equations. This format can always be achieved as follows. Let the problem be of the minimization type (there is no loss in generality here) and define it as

$$\text{minimize } z = \sum_{j=1}^{n} c_j x_j$$

subject to

$$\sum_{j=1}^{n} a_{ij} x_j + S_i = b_i, \qquad i = 1, 2, \ldots, m$$

$$x_j = 0 \text{ or } 1, \qquad j = 1, 2, \ldots, n$$

$$S_i \geq 0, \qquad i = 1, 2, \ldots, m$$

where S_i is the slack variable associated with the ith constraint.

The continuous version of the foregoing problem is dual-feasible if every $c_j \geq 0$. Any $c_j < 0$ can be converted to the desired format by complementing the variable x_j, that is, by substituting $x_j = 1 - x_j'$, where x_j' is a binary variable, in the objective function and constraints. If, in addition to dual feasibility, the problem is primal-feasible, nothing more need be done since the minimum, in terms of the new variables, is achieved by assigning zero values to all the variables. However, if it is primal infeasible, the additive algorithm is used to find the optimum.

The general idea of the additive algorithm is to enumerate all 2^n possible solutions of the problem. However, it recognizes that some solutions can be discarded automatically without being investigated implicitly. Hence, in the final analysis, only a portion of the 2^n solutions need be investigated explicitly.

In terms of the given zero–one problem, this idea is implemented as follows. Initially, assume that all the variables are at zero level. This is logical, since all $c_j \geq 0$. Since the corresponding solution is not feasible (i.e., some slack variables S_i may be negative), it will be necessary to elevate some variables to level one. The procedure calls for elevating one (or perhaps more) variable at a time, provided there is evidence that this step will be moving the solution toward feasibility, that is, making $S_i \geq 0$ for all i. A number of tests have been developed to ensure the proper selection of the variables to be elevated to level one. These are first presented by means of a numerical example and later formalized mathematically.

Example 8.5-1

$$\text{Maximize } x_0 = 3x_1' + 2x_2' - 5x_3' - 2x_4' + 3x_5'$$

subject to

$$x_1' + x_2' + x_3' + 2x_4' + x_5' \leq 4$$
$$7x_1' + 3x_3' - 4x_4' + 3x_5' \leq 8$$
$$11x_1' - 6x_2' + 3x_4' - 3x_5' \geq 3$$
$$x_j' = 0 \quad \text{or} \quad 1 \quad \text{for all } j$$

This problem is converted into the *minimization* form, in which all variables have nonnegative coefficients in the objective function. First, convert the objective function to minimization form by multiplying x_0 by -1. This leaves x_1', x_2', and x_5' with negative coefficients and x_3' and x_4' with positive coefficients. Thus the substitution

$$x_j' = \begin{cases} 1 - x_j, & j = 1, 2, 5 \\ x_j, & j = 3, 4 \end{cases}$$

converts all the objective function coefficients to nonnegative values as desired. After the third constraint is changed to ($\leq$), the problem is put in the following convenient form (z is the objective value of the *converted* problem).

x_1	x_2	x_3	x_4	x_5	S_1	S_2	S_3	R.H.S.
3	2	5	2	3	0	0	0	z
-1	-1	1	2	-1	1	0	0	1
-7	0	3	-4	-3	0	1	0	-2
11	-6	0	-3	-3	0	0	1	-1

Since initially all $x_j = 0$, the values of slacks are

$$(S_1^0, S_2^0, S_3^0) = (1, -2, -1)$$

[The superscript (0) represents the starting iteration.] The associated objective value is $z^0 = 0$.

It is evident now that the starting solution is not feasible, since S_2^0 and S_3^0 are negative. Thus at least one x_j variable must be elevated to level one. Such a variable must move the solution closer to feasibility, and this can be indicated by the values of the slacks. By investigating the variables at level zero, we see that all the *constraint* coefficients of x_3 corresponding to the negative slacks are *nonnegative*. Thus x_3, if elevated to level one, can only worsen the infeasibility. This means that x_3 must remain at zero level. Although each of x_1, x_2, and x_4 cannot *individually* bring feasibility, a combination of them at level one may lead to feasible values of the slacks. Thus these variables cannot be excluded (as in the case of x_3) at this point. On the other hand, if x_5 is set equal to one, a feasible solution is achieved. Thus the procedure calls for elevating x_5 to level one. This yields

$$(S_1^1, S_2^1, S_3^1) = [S_1^0 - (-1), S_2^0 - (-3), S_3^0 - (-3)] = (2, 1, 2)$$

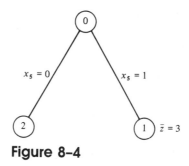

Figure 8–4

with $x_5 = 1$ and $z^1 = 3$. Since this is a feasible solution, it is stored as the best available so far. Thus $\bar{z} = z^1 = 3$ is an upper bound on any future feasible solution. In other words, from here on, we are interested only in feasible solutions with objective values *better* (smaller) than $\bar{z}$.

At this point it is instructive to introduce the foregoing procedure as a branch-and-bound method. Figure 8–4 shows the initial node (0), which represents the problem for which all $x_j = 0$. Two branches emanate from this node. These are associated with $x_5 = 0$ and $x_5 = 1$. By selecting the branch $x_5 = 1$, we obtain the feasible solution at node (1) with $\bar{z} = 3$.

Because all the coefficients of the objective function are positive and since the objective is to minimize z, any branch emanating from node (1) cannot yield a better objective value. In this case the branch $x_5 = 1$ is **fathomed.** Since $x_5 = 0$ is the only remaining branch in the tree, its associated solution must be considered (compare with the branch-and-bound algorithm). This leads to node (2), Figure 8–4, with its solution given by

$$(S_1^2, S_2^2, S_3^2) = (1, -2, -1)$$

with $z^2 = 0$ and all the binary variables equal to zero.

You may wonder about the difference between the solutions at nodes (0) and (2). There is no difference so far as the *values* of the variables and objective function are concerned. But there is the important difference that at node (0) any of the variables x_1, x_2, x_3, x_4, and x_5 is **free** to assume a zero or one value, whereas at node (2) x_5 is fixed at level zero. Thus, in choosing the branching variable at (2), only x_1, x_2, x_3, or x_4 may be considered.

The selection of a variable to be elevated to one at node (2) is done as at node (0). However, there is now the additional information that any variable at level one that leads to an objective value greater than or equal to $\bar{z}$ must remain zero. Thus elevating x_3 to level one is not promising because this will worsen both optimality (it yields $z = 5 > \bar{z}$) and feasibility (it makes the slacks more negative). Also, x_1 may be discarded because $c_1 = 3$ cannot lead to a *better* objective value than $\bar{z}$. Thus a choice must be made between x_2 and x_4. Clearly, neither can individually bring feasibility. In this case a choice is made based on an empirical measure. Define for each *free* variable x_j

$$v_j = \sum_{\text{all } i} \min \{0, S_i - a_{ij}\}$$

This actually may be regarded as a "measure" of the total infeasibility resulting from elevating the free variable x_j to level one. The branching variable selected is the one with the smallest v. Now, for x_2 and x_4,

$$v_2 = 0 + (-2 - 0) + 0 = -2$$
$$v_4 = (1 - 2) + 0 + 0 = \boxed{-1}$$

Hence x_4 is selected as the branching variable, and $x_4 = 1$ leads to node (3), Figure 8–5, where

$$(S_1^3, S_2^3, S_3^3) = (1 - 2, -2 + 4, -1 + 3) = (-1, 2, 2)$$

with $z^3 = 2$.

Node (3) is now defined by $x_5 = 0$ and $x_4 = 1$ so that x_1, x_2, and x_3 are the only free variables at (3). Since $c_1 = 3$, $c_2 = 2$, and $c_3 = 5$, elevating x_1, x_2, or x_3 to level one cannot yield a better objective value than $\bar{z}$, since they yield $z = 2 + 3, 2 + 2$, and $2 + 5$. Hence x_1, x_2, and x_3 are excluded. Since *all* the free variables are nonpromising, no further branching can be effected from node (3) and hence it is fathomed.

The only remaining node is (4). Since it is defined by $x_5 = 0$ and $x_4 = 0$,

$$(S_1^4, S_2^4, S_3^4) = (1, -2, -1)$$

with $z^4 = 0$. Again x_1, x_2, and x_3 are the free variables. The variable x_3 is nonpromising from both optimality and feasibility viewpoints. The remaining variables x_1 and x_2 cannot make the solution at (4) feasible. Thus no branching variables exist at (4) and (4) is fathomed. Since no more unfathomed branches (or nodes) exist in Figure 8–5, the solution is given by node (1) with $z = 3$ and $x_5 = 1$, and all the remaining variables equal zero. This solution can be translated in terms of the original variables to give $x_1' = x_2' = 1$, $x_3' = x_4' = x_5' = 0$, with $x_0 = 5$. ◄

It was stated earlier that the foregoing procedure enumerates (implicitly or explicitly) all 2^n of the problems. This result is illustrated for the examples just presented as follows. The fathoming of node (1) means that all the binary solutions in which $x_5 = 1$ have been accounted for. There are $2^{5-1} = 16$ solutions

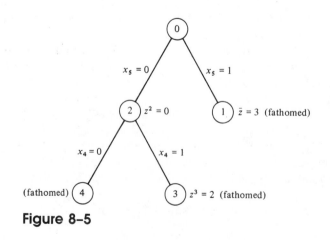

Figure 8–5

in which $x_5 = 1$. Also, node (3) is fathomed. Since it is defined by $x_5 = 0$ and $x_4 = 1$, the number of solutions accounted for at node (3) is $2^{5-2} = 8$. Similarly, node (4) is fathomed, and this accounts for $2^{5-2} = 8$ solutions. Thus, by considering all fathomed nodes together, a total of $16 + 8 + 8 = 2^5$ solutions have been considered, which is the total number of possible binary solutions. One observes that the solutions counted at the different *fathomed* nodes are nonredundant, since the path (branches) leading to each of these nodes is unique.

It is interesting that only 5 (out of 32) solutions are investigated explicitly in the example above. This is why the procedure is called **implicit enumeration.**

The zero–one branch-and-bound tree (e.g., Figure 8–5) can be represented for data manipulation in a very simple manner. To accomplish this, the following definitions are needed.

1. *Free variable.* At any node of the tree, a binary variable is called free if it is not fixed by any of the branches leading to this node. In Figure 8–5, node (3) has x_1, x_2, and x_3 as free variables. A free variable is tentatively at zero level but can be elevated to level one if this can improve the infeasibility of the problem.

2. *Partial solution.* A partial solution provides a specific binary assignment for some of the variables in the sense that it fixes the values of one or more variables at zero or one. A convenient way to summarize this information for the purpose of the (branch-and-bound) algorithm is to express the partial solution as an *ordered* set. Let J_t represent the partial solution at the tth node (or iteration), and let the notation $+j$ $(-j)$ represent $x_j = 1$ $(x_j = 0)$. Thus the elements of J_t consist of the subscripts of the fixed variables with the plus (minus) sign signifying that the variable is one (zero). The set J_t must be *ordered* in the sense that each new element is always augmented *on the right* of the partial solution.

Partial solutions can be used to define the nodes in the branch-and-bound tree, since the branches leading to a node actually represent a partial binary assignment to some variables. In Figure 8–5, the nodes are represented as follows:

$$\text{Node (0):} \quad J_0 = \varnothing$$
$$\text{Node (1):} \quad J_1 = \{5\}$$
$$\text{Node (2):} \quad J_2 = \{-5\}$$
$$\text{Node (3):} \quad J_3 = \{-5, 4\}$$
$$\text{Node (4):} \quad J_4 = \{-5, -4\}$$

Actually, the use of partial solutions eliminates the need for recording generated nodes by means of the branch-and-bound tree. This means that partial solutions can be generated successively from one another. The procedure for generating successive partial solutions (nodes) will be illustrated by the example in Figure 8–5. However, because fathomed partial solutions play a key role in the development of the procedure, the rules for fathoming are summarized here. A partial solution is said to be fathomed if:

1. It cannot lead to a better value of the objective function.
2. It cannot lead to a feasible solution.

A fathomed partial solution means that it is not promising to further branch its associated node, since all the solutions that could be generated from the node are either inferior or infeasible.

The solution of Example 8.5–1 starts with $J_0 = \varnothing$, meaning that all the binary variables are free. Now $x_5 = 1$ leads to $J_1 = \{5\}$, which is feasible. From the special structure of the problem, J_1 is fathomed. At this point the next partial solution is generated by complementing the *rightmost positive* element of J_1. This gives $J_2 = \{-5\}$.† Essentially, J_2 means that the branch $x_5 = 1$ has been considered (fathomed) and hence its complement branch $x_5 = 0$ must now be considered. According to the tests given in Example 8.5–1, J_2 is not fathomed, since $x_4 = 1$ can be augmented. This yields $J_3 = \{-5, 4\}$. Now J_3 is fathomed and J_4 is obtained by complementing the rightmost element of J_3. This yields $J_4 = \{-5, -4\}$. Again, J_4 is fathomed, but since all its elements are negative, the enumeration is complete.

Actually, the general rule for generating the next partial solution from a *fathomed* one is as follows. If *all* the elements of a *fathomed* partial solution are negative, the enumeration is complete. Otherwise, select the rightmost *positive* element, complement it, and *then delete all the (negative) elements to its right*. For example, if $J_t = \{1, 5, 4, -3, 2, 6, -7, -8\}$ is fathomed, $J_{t+1} = \{1, 5, 4, -3, 2, -6\}$. We can see the significance of the negative elements. Since the additive algorithm always adds variables at level one, a negative element means that a *preceding* partial solution (in which this element was positive) was fathomed. Thus, when all the elements of a *fathomed* partial solution are negative, the associated variables have been considered at both zero and one levels. As a result, there are no more branches to consider and the enumeration is complete.

It is important to notice that the *order* of the elements in a partial solution is crucial in properly enumerating all the solutions. Thus $\{1, -3\}$ is not the same as $\{-3, 1\}$, since the first implies that the node reached by $x_1 = 1$ and $x_3 = 1$ has been fathomed while in the second the node reached by $x_3 = 1$ only has been fathomed. This illustrates why a partial solution is defined as an *ordered* set.

The general version of the additive algorithm is now presented by using the concept of partial solutions. The exclusion tests used to fathom partial solutions and augment new variables at level one are also generalized for the zero–one problem.

Consider the following general binary problem:

$$\text{minimize } z = \sum_{j=1}^{n} c_j x_j, \qquad \text{all } c_j \geq 0$$

subject to

$$\sum_{j=1}^{n} a_{ij} x_j + S_i = b_i, \qquad i = 1, 2, \ldots, m$$

$$x_j = 0 \text{ or } 1, \qquad \text{for all } j$$

$$S_i \geq 0, \qquad \text{for all } i$$

---

† The process of complementing the *rightmost positive* element is sometimes called **backtracking**. In the case of J_1, the solution "backtracks" along the branch $x_5 = 1$, and then "moves down" along the branch $x_5 = 0$ to reach J_2 (see Figure 8–5).

Let J_t be the partial solution at node t (initially, $J_0 \equiv \varnothing$, which means that all variables are free) and assume z^t is the associated value of z while $\bar{z}$ is the current best upper bound (initially $\bar{z} = \infty$).

Test 1: For any free variable x_r, if $a_{ir} \geq 0$ for *all* i corresponding to $S_i^t < 0$, then x_r cannot improve the infeasibility of the problem and must be discarded as nonpromising.

Test 2: For any free variable x_r, if

$$c_r + z^t \geq \bar{z}$$

then x_r cannot lead to an improved solution and hence must be discarded.

Test 3: Consider the ith constraint

$$a_{i1}x_1 + a_{i2}x_2 + \cdots + a_{in}x_n + S_i = b_i$$

for which $S_i^t < 0$. Let N_t define the set of *free* variables not discarded by tests 1 and 2. None of the free variables in N_t is promising if for at least one $S_i^t < 0$, the following condition is satisfied:

$$\sum_{j \in N_t} \min \{0, a_{ij}\} > S_i^t$$

This actually says that the set N_t cannot lead to a feasible solution and hence must be discarded altogether. In this case, J_t is said to be fathomed.

Test 4: If $N_t \neq \varnothing$, the branching variable x_k is selected as the one corresponding to

$$v_k^t = \max_{j \in N_t} \{v_j^t\}$$

where

$$v_j^t = \sum_{i=1}^{m} \min \{0, S_i^t - a_{ij}\}$$

If $v_k^t = 0$, $x_k = 1$ together with J_t yields an *improved* feasible solution. In this case, J_{t+1}, which is defined by J_t with $\{k\}$ augmented on the right, is fathomed. Otherwise, the foregoing tests are applied again to J_{t+1} until the enumeration is completed, that is, until *all* the elements of the *fathomed* partial solution are negative.

Example 8.5–2. As a way of summarizing the preceding generalized procedure, Example 8.5–1 is presented by using the new "bookkeeping" method.

Iteration 0
    For $J_0 = \varnothing$, $\bar{z} = \infty$,

$$(S_1^0, S_2^0, S_3^0) = (1, -2, -1), \quad z^0 = 0$$

x_3 is excluded by test 1. By test 3, $N_0 = \{1, 2, 4, 5\}$ cannot be abandoned because

$$S_2: \quad -7 - 4 - 3 = -14 < -2$$
$$S_3: \quad -6 - 3 - 3 = -12 < -1$$

By test 4,

$$v_1^0 = 0 + 0 + (-1 - 11) = -12$$
$$v_2^0 = 0 + (-2 - 0) + 0 = -2$$
$$v_4^0 = (1 - 2) + 0 + 0 = -1$$
$$v_5^0 = 0 + 0 + 0 + 0 = \boxed{0}$$

Hence, $k = 5$.

Iteration 1
For $J_1 = \{5\}$, $\bar{z} = 3$,

$$(S_1^1, S_2^1, S_3^1) = (1 + 1, -2 + 3, -1 + 3) = (2, 1, 2), z^1 = 3$$

Since it is feasible, $\bar{z} = z^1 = 3$. Thus J_1 is fathomed.

Iteration 2
For $J_2 = \{-5\}$, $\bar{z} = 3$

$$(S_1^2, S_2^2, S_3^2) = (1, -2, -1), z^2 = 0$$

Test 1 excludes x_3. Test 2 excludes x_1 and x_3. By test 3, $N_2 = \{2, 4\}$ cannot be abandoned. By test 4, $v_2^2 = -2$ and $v_4^2 = \boxed{-1}$. Hence $k = 4$.

Iteration 3
For $J_3 = \{-5, 4\}$ $\bar{z} = 3$,

$$(S_1^3, S_2^3, S_3^3) = (-1, 2, 2), z^3 = 2$$

Test 1 excludes x_3. Test 2 excludes x_1, x_2, and x_3. Since $N_3 = \emptyset$, J_3 is fathomed.

Iteration 4
For $J_4 = \{-5, -4\}$, $\bar{z} = 3$

$$(S_1^4, S_2^4, S_3^4) = (1, -2, -1), z^4 = 0$$

Test 1 excludes x_3. Test 2 excludes x_1 and x_3. Test 3 indicates $N_4 = \{2\}$ must be abandoned. Thus J_4 is fathomed. Since all the elements of J_4 are negative, the enumeration is complete and J_1 is optimal. ◄

8.5.2 ZERO–ONE POLYNOMIAL PROGRAMMING

Consider the problem

$$\text{maximize } z = f(x_1, \ldots, x_n)$$

subject to

$$g_i(x_1, \ldots, x_n) \leq b_i, \qquad i = 1, 2, \ldots, m$$
$$x_j = 0 \quad \text{or} \quad 1, \qquad j = 1, 2, \ldots, n$$

Assume that f and g_i are polynomials with the kth term generally represented by $d_k \prod_{j=1}^{n_k} x_j^{a_{kj}}$, where a_{kj} is a positive constant exponent and d_k is a constant.

The seemingly highly nonlinear problem shown can be converted into a linear form, which can then be solved as a zero–one linear program. Since x_j is a binary variable, $x_j^{a_{kj}} = x_j$ for any positive exponent a_{kj}. (If $a_{kj} = 0$, obviously the variable x_j will not be present in the kth term.) This means that the kth term can be written as $d_k \prod_{j=1}^{n_k} x_j$.

Let $y_k = \prod_{j=1}^{n_k} x_j$, then y_k is also a binary variable and the kth term of the polynomial reduces to the linear term $d_k y_k$. However, to ensure that $y_k = 1$ when all $x_j = 1$ and zero otherwise, the following constraints must be added[†] for each y_k.

$$\sum_{j=1}^{n_k} x_j - (n_k - 1) \le y_k \tag{1}$$

$$\frac{1}{n_k} \sum_{j=1}^{n_k} x_j \ge y_k \tag{2}$$

If all $x_j = 1$, $\sum_{j=1}^{n_k} x_j = n_k$ and constraint (1) yields $y_k \ge 1$, and constraint (2) gives $y_k \le 1$; that is, $y_k = 1$. On the other hand, if at least one $x_j = 0$, then $\sum_{j=1}^{n_k} x_j < n_k$ and constraints (1) and (2), respectively, yield $y_k \ge - (n_k - 1)$ and $y_k < 1$ with the only feasible value given by $y_k = 0$.

Example 8.5–3. The foregoing linear transformation is illustrated by the following problem:

$$\text{maximize } z = 2x_1 x_2 x_3^2 + x_1^2 x_2$$

subject to

$$5x_1 + 9x_2^2 x_3 \le 15$$

$$x_1, x_2, \text{ and } x_3 \text{ binary}$$

Let $y_1 = x_1 x_2 x_3$, $y_2 = x_1 x_2$, and $y_3 = x_2 x_3$. The problem becomes:

$$\text{maximize } z = 2y_1 + y_2$$

subject to

$$5x_1 + 9y_3 \le 15$$
$$x_1 + x_2 + x_3 - 2 \le y_1$$
$$\tfrac{1}{3}(x_1 + x_2 + x_3) \ge y_1$$
$$x_1 + x_2 - 1 \le y_2$$
$$\tfrac{1}{2}(x_1 + x_2) \ge y_2$$
$$x_2 + x_3 - 1 \le y_3$$
$$\tfrac{1}{2}(x_2 + x_3) \ge y_3$$

where y_1, y_2, y_3, x_1, x_2, and x_3 are binary variables.   ◀

[†] A more efficient procedure that does not require the addition of these constraints and deals directly with the converted linear system of the polynomial problem has been developed by H. Taha, "A Balasian-Based Algorithm for Zero–One Polynomial Programming," *Management Science,* Vol. 18, 1972, pp. B328–B343. This procedure extends the Balas algorithm to the polynomial problem in a straightforward manner.

Computations in Implicit Enumeration

The effectiveness of the additive algorithm is highly dependent on the strength of the (exclusion) tests given in Section 8.5–1. Unfortunately, these tests are not sufficient to produce a computationally efficient algorithm. Successful computer codes are based on much stronger tests. Perhaps the most effective of these is the so-called **surrogate** (or substitute) **constraint.** One notices that the additive algorithm scans the constraints one at a time. The idea then is to develop a constraint that "combines" all the original constraints of the problem into one constraint and does not eliminate any of the original feasible (integer) points of the problem. The new constraint is devised such that it has the potential to reveal information that cannot be conveyed by any of the original constraints considered separately (see Problem 8–28).

Reported computational experiences indicate that the use of the surrogate constraint is effective in improving the computation time. However, because implicit enumeration investigates (implicitly or explicitly) all 2^n binary points, the solution time varies almost exponentially with the number of variables n. This limits the number of variables that can be handled by this method. Although successful cases have been reported for rather large problems (having special structures), it is safe to conclude that, in general, only problems with up to 100 variables can be solved in a reasonable amount of computation time.

One particular observation about implicit enumeration is that the computation time is data-dependent. The specific ordering of the variables and constraints may have a direct effect on the efficiency of the algorithm. For example, the constraints should be ordered with the most restrictive at the top, while the variables could be arranged according to an ascending order of their (nonnegative) objective coefficients. Both conditions are favorable to producing "faster" fathoming of partial solutions.

The general conclusion is that the implicit enumeration method still does not provide the perfect solution to the computational problem. It appears that the branch-and-bound methods (Section 8.4) will continue to dominate, particularly when large practical problems are attempted.

8.6   SUMMARY

As can be inferred from the discussion in this chapter, the most important factor affecting computations in integer programming is the number of variables. This situation is more pronounced in branch-and-bound and implicit enumeration methods. Consequently, in formulating an integer model, it is advantageous to reduce the number of integer variables as much as possible. This may be effected in general by:

1. Approximating integer variables by continuous ones.
2. Restricting the feasible ranges of the integer variables.

3. Eliminating the use of auxiliary binary variables (e.g., as in the fixed-charge problem) by devising more direct solution methods.

4. Avoiding nonlinearity in the model.

These ideas should help in alleviating the computational problem.

Experience with integer codes shows that a user will be disappointed if he expects to feed in input data to the computer and then await the answer at the output end of the machine. In a typical integer code, manual intervention during the computations is almost mandatory. The codes are usually equipped with a number of options, each with a specific advantage in handling the integer problem. The user then plans a strategy for tackling the problem. By monitoring the intermediate information extracted from the machine, he decides whether to continue the same course of action or to select another option. In other words, human judgment during the calculations is needed to ensure that the program is progressing satisfactorily.

The importance of the integer problem in practice is not yet matched by the development of efficient solution methods. This problem is due primarily to the inherent difficulty in dealing with integer computations in general. The current intensive research may eventually lead to a breakthrough. It is more likely, however, that this breakthrough will be achieved by the development of extremely powerful and highly accurate digital computers rather than by the development of more theoretical methods.

SELECTED REFERENCES

GARFINKEL, R., and G. NEMHAUSER, *Integer Programming,* Wiley, New York, 1972.

SALKIN, H., *Integer Programming,* Addison-Wesley, Reading, Mass., 1975.

TAHA, H., *Integer Programming: Theory, Applications and Computations,* Academic Press, New York, 1975.

REVIEW QUESTIONS

True (T) or False (F)?

1. ____ Every integer program, mixed or pure, can be expressed in terms of zero–one variables.

2. ____ It is impossible to obtain a *feasible* integer solution by rounding the continuous optimum of a linear programming problem that originally contains strict equality constraints.

3. ____ The integer programming cuts modify the continuous solution space in a manner that will produce a continuous optimum that satisfies the integer conditions on the variables.

4. ____ The optimum integer solution of a problem can produce a better objective value than its associated continuous optimum.

5. ____ The construction of the fractional cut does not require the slack variables to be integer.

6. ____ In the application of the cutting methods, it is necessary to retain *all* the generated cuts in the simplex tableau until the optimum integer is reached.

7. ____ A cut may eliminate an integer feasible point as long as it is not the optimum integer.

8. ____ If immediately following an application of a cut, the dual simplex does not produce a feasible (continuous or integer) solution, the problem has no integer feasible solution.

9. ____ The mixed cut can eventually produce the optimum integer solution of a problem in which all variables are integers.

10. ____ If a dual cutting-plane algorithm is stopped prematurely, the last available solution can be considered a good feasible solution to the integer problem.

11. ____ In the branch-and-bound procedure, the bounding step sets a lower (upper) limit on the objective value in the case of maximization (minimization) provided a feasible integer solution is encountered.

12. ____ In the branch-and-bound procedure, the branching step effectively removes continuous parts of the feasible space.

13. ____ The dichotomization process used in the branching step of the branch-and-bound procedure is based on the same idea used in the development of the mixed cut of integer programming.

14. ____ The number of subproblems created by the branching step can be reduced drastically if a good bound is discovered at the early stages of computations.

15. ____ A bound obtained by the branch-and-bound procedure may not necessarily be associated with a feasible point of the integer problem.

16. ____ In the branch-and-bound procedure, the available rules for selecting the branching variable at a node guarantee encountering a good bound rapidly.

17. ____ The basic disadvantage of the branch-and-bound procedure is that the number of subproblems created may tax the computer memory.

18. ____ The effect of round-off error in the branch-and-bound procedure is just as bad as in the cutting-plane algorithms.

19. ____ In the branch-and-bound procedure, a branching variable at a current node cannot be branched again at a subsequent node that is generated from the current one.

20. ____ The zero–one implicit enumeration procedure is actually a special case of the general branch-and-bound method.

21. ____ The zero–one implicit enumeration procedure effectively enumerates, either implicitly or explicitly, *all* 2^n binary combinations of an n-variable problem.

22. ____ The application of the branch-and-bound procedure to an n-variable zero–one problem can never produce more than n connected branches from a given node.

[*Ans.* **1**—T, **2**—T, **3**—T, **4 to 7**—F, **8**—T, **9**—T, **10**—F, **11 to 14**—T, **15**—F, **16**—F, **17**—T, **18**—F, **19**—F, **20**—T, **21**—T, **22**—T.]

PROBLEMS

Section	Assigned Problems
8.1	8–1 to 8–6
8.2	None
8.3.1	8–7 to 8–9
8.3.2	8–10 to 8–13
8.4	8–14 to 8–19, 8–27, 8–28
8.5	8–20 to 8–26

☐ **8–1** Consider the problem

$$\text{maximize } z = 20x_1 + 10x_2 + 10x_3$$

subject to

$$2x_1 + 20x_2 + 4x_3 \leq 15$$
$$6x_1 + 20x_2 + 4x_3 = 20$$
$$x_1, x_2, x_3 \text{ nonnegative integers}$$

Solve the problem as a (continuous) linear program; then show that it is impossible to obtain a feasible integer solution by using simple rounding.

☐ **8–2 (Delivery Problem).** Consider the situation in which orders from m different destinations are delivered from a central warehouse. Each destination receives its order in one delivery. Feasible routes are assigned to different carriers, and each carrier may combine at most r orders. Suppose that there are n feasible routes with each route specifying the destinations to which orders are delivered. Assume further that the cost of the jth route is c_j. Overlapping is expected so that the same destination can be reached by more than one carrier. Formulate the problem as an integer model.

☐ **8–3 (Quadratic Assignment).** Consider the assignment of n plants to n different locations. The volume of goods transported between plants i and j is d_{ij}, and the cost of transporting one unit from location p to location q is c_{pq}. Formulate the problem as an integer model so as to minimize the total transportation costs.

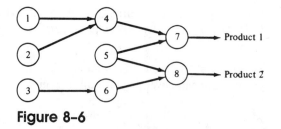

Figure 8-6

☐ **8-4** Consider the job-shop scheduling problem involving eight operations on a single machine with a total of two end products. The sequencing of operations is shown in Figure 8–6. Let b_j be the processing time for the jth operation ($j = 1, 2, \ldots, 8$). Delivery dates for products 1 and 2 are restricted by d_1 and d_2 time units measured from the zero datum. Since each operation requires a special machine setup, it is assumed that any operation once started must be completed before a new operation can be undertaken.

Formulate the problem as a mixed integer programming model to minimize the total processing time on the machine while satisfying all the pertinent constraints.

☐ **8-5** Show how the nonconvex solution spaces (shaded areas in Figure 8–7) can be represented by simultaneous constraints.

Find the optimum solution that maximizes $z = 2x_1 + 3x_2$ subject to the solution space given in Figure 8–7(a).

[*Hint:* Use "either–or" constraints.]

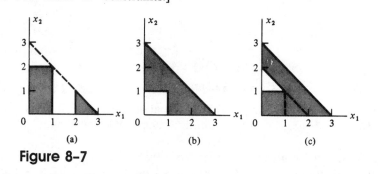

Figure 8-7

☐ **8-6** Show how the solution space indicated by the shaded area in Figure 8–8 can be expressed as simultaneous mixed integer constraints.

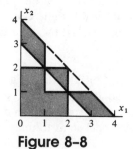

Figure 8-8

☐ **8-7** Consider the problem

$$\text{maximize } z = x_1 + 2x_2$$

subject to

$$x_1 + x_2/2 \leq 13/4$$
$$x_1, x_2 \text{ nonnegative integers}$$

Show that the fractional algorithm does not yield a feasible solution unless the coefficients *and* the right-hand side of the constraint are integers. Then find the optimal solution.

☐ **8-8** Solve by the fractional algorithm

$$\text{maximize } z = 4x_1 + 6x_2 + 2x_3$$

subject to

$$4x_1 - 4x_2 \leq 5$$
$$-x_1 + 6x_2 \leq 5$$
$$-x_1 + x_2 + x_3 \leq 5$$
$$x_1, x_2, x_3 \text{ nonnegative integers}$$

Compare the rounded optimal solution and the integer optimal solution.

☐ **8-9** Solve by the fractional algorithm

$$\text{maximize } z = 3x_1 + x_2 + 3x_3$$

subject to

$$-x_1 + 2x_2 + x_3 \leq 4$$
$$4x_2 - 3x_3 \leq 2$$
$$x_1 - 3x_2 + 2x_3 \leq 3$$
$$x_1, x_2, x_3 \text{ nonnegative integers}$$

Compare the rounded optimal solution and the integer optimal solution.

☐ **8-10** Construct the *first* mixed cut and its stronger version for the following problem and compare the two cuts. By examining the coefficients of the cuts, what is the effect of machine round-off error if the problem is solved by the computer?

$$\text{Maximize } z = 5x_1 + 8x_2 + 6x_3$$

subject to

$$2x_1 + 6.3x_2 + x_3 \leq 11$$
$$9x_1 + 6x_2 + 10x_3 \leq 28$$
$$x_2 \geq 0$$
$$x_1 \text{ and } x_3 \text{ nonnegative integers}$$

□ **8–11** Solve Problem 8–8 by the mixed algorithm assuming that x_1 and x_3 are the only integer variables.

□ **8–12** Solve Problem 8–9 by the mixed algorithm assuming that x_1 and x_3 are the only integer variables.

□ **8–13** Show that the stronger mixed cut (given at the end of Section 8.3.2) when applied to the *pure* integer problem is stronger than the fractional cut (Section 8.3.1) developed from the same source row. Discuss the use of this cut in solving the pure integer problem. In particular, does the pure problem remain pure integer after the first cut is applied?

□ **8–14** Show graphically that the problem

$$\text{maximize } z = 2x_1 + x_2$$

subject to

$$10x_1 + 10x_2 \leq 9$$
$$10x_1 + 5x_2 \geq 1$$

$$x_1, x_2 \text{ nonnegative integers}$$

has no feasible integer solution. Verify the solution algebraically by using
  (a) The fractional algorithm.
  (b) The branch-and-bound algorithm.

□ **8–15** Solve Example 8.4–1 starting at node (1) with the variable x_1 instead of x_2.

□ **8–16** Consider the problem

$$\text{maximize } z = x_1 + x_2$$

subject to

$$2x_1 + 5x_2 \leq 16$$
$$6x_1 + 5x_2 \leq 30$$

$$x_1, x_2 \text{ nonnegative integers}$$

Find the optimal noninteger solution graphically. By using the branch-and-bound algorithm, show graphically the successive parallel changes in the z-value that will lead to the optimal integer solution. Also, define the generated subproblems.

□ **8–17** Solve Problem 8–12 by the branch-and-bound method.

□ **8–18** Consider the following *cargo-loading* problem, where five items are to be loaded on a vessel. The weight w_i and the volume v_i per unit of the different items as well as their corresponding values r_i are tabulated as follows:

Item i	w_i	v_i	r_i
1	5	1	4
2	8	8	7
3	3	6	6
4	2	5	5
5	7	4	4

The maximum cargo weight and volume are given by $W = 112$ and $V = 109$, respectively. It is required to determine the most valuable cargo load in discrete units of each item.

Formulate the problem as an integer programming model and then solve by the branch-and-bound method.

□ **8–19** In Problem 8–9, suppose that all the variables are binary. Find the optimal solution by the additive algorithm.

□ **8–20** In Problem 8–14, assume that x_1 and x_2 are binary variables. Show how the additive algorithm can be used to discover that the problem has no feasible (integer) solution.

□ **8–21** Consider the capital budgeting problem where five projects are being considered for execution over the next 3 years. The expected returns for each project and the yearly expenditures (in thousands of dollars) are shown in the accompanying table. Assume that each approved project will be executed over the 3-year period. The objective is to select the projects that will maximize the total returns.

Project	Expenditures for			Returns
	Year 1	Year 2	Year 3	
1	5	1	8	20
2	4	7	10	40
3	3	9	2	20
4	7	4	1	15
5	8	6	10	30
Maximum available funds	25	25	25	—

Formulate the problem as a zero–one integer programming problem and solve by the additive algorithm.

□ **8–22** Suggest some modifications that will give stronger exclusion tests for the additive algorithm.

☐ **8-23** Solve the following problem

$$\text{maximize } z = x_1 + 2x_2 + 5x_3$$

subject to

$$|-x_1 + 10x_2 - 3x_3| \geq 15$$
$$2x_1 + x_2 + x_3 \leq 10$$
$$x_1, x_2, x_3 \geq 0$$

☐ **8-24** Consider the production planning problem where 2000 units of a certain product are manufactured on three machines. The setup costs, the production costs per unit, and the maximum production capacity for each machine are tabulated below. The objective is to minimize the total production cost of the required lot.

Machine	Setup Cost	Production Cost/Unit	Capacity (units)
1	100	10	600
2	300	2	800
3	200	5	1200

Formulate the problem as an integer programming problem, and find the optimal solution.

☐ **8-25** In an oil-well-drilling problem there are two attractive drilling sites for reaching four targets (or possible oil wells). The preparation costs at each site and the cost of drilling from site i to target j ($i = 1, 2; j = 1, 2, 3, 4$) are given. The objective is to determine the best site for each target so that the total cost is minimized.

Site	Drilling Cost to Target				Preparation Cost
	1	2	3	4	
1	2	1	8	5	5
2	4	6	3	1	6

Formulate the problem as an integer programming model, and suggest a method for obtaining the optimal solution.

☐ **8-26** Solve the following problem assuming that only one of the given constraints holds.

$$\text{Maximize } z = x_1 + 2x_2 - 3x_3$$

subject to

$$20x_1 + 15x_2 - x_3 \le 10$$
$$12x_1 - 3x_2 + 4x_3 \le 20$$
$$x_1, x_2, x_3 \text{ binary}$$

☐ **8-27 (Penalties).** Suppose that the basic variable associated with the current node of a maximization branch-and-bound algorithm is defined by

$$x_k = \beta_k - \sum_{j=1}^{n} \alpha_k^j w_j$$

Branching is effected by the two constraints $x_k - [\beta_k] \le 0$ and $x_k - [\beta_k] - 1 \ge 0$. Let P_d and P_u be the lower bounds on the true degradation in the optimum objective value as a result of activating the first and second constraints. Show (by using the dual simplex algorithm) that, under the assumption of *no change in basis*,

$$P_d = \min_{j \in J^+} \left\{ \frac{(z_j - c_j) f_k}{\alpha_k^j} \right\}$$

$$P_u = \min_{j \in J^-} \left\{ \frac{(z_j - c_j)(f_k - 1)}{\alpha_k^j} \right\}$$

where $(z_j - c_j)$ is the objective coefficient of the jth nonbasic variable at the current node, $f_k = \beta_k - [\beta_k]$, and J^+ (J^-) is the set of nonbasic subscripts for which $\alpha_k^j > 0$ ($\alpha_k^j < 0$). Consequently, the true degradation in the objective value is *at least* equal to $\bar{P} = \min \{P_d, P_u\}$. This means that given z is the optimum objective value at the current node, $z - \bar{P}$ gives an upper bound (assume a maximization problem) on the optimum objective values at the two nodes emanating from the current node. Thus, if $z - \bar{P}$ is less than the best available lower bound, the current node is fathomed.

☐ **8-28 (Surrogate Constraint).** Assume that the set of constraints for the zero-one problem is given in matrix form as $AX \le b$. Let $\mu \ge 0$ be a row vector of nonnegative multipliers. Define the surrogate constraint as

$$\mu(AX - b) \le 0$$

Show that
(a) All the feasible binary solutions of $AX \le b$ are also feasible with respect to the surrogate constraint.
(b) If the surrogate constraint is infeasible, then the original constraints $AX \le b$ are also infeasible.

SHORT CASES

Case 8–1†

A development company owns 90 acres of land in a growing metropolitan area where it intends to construct office buildings and a shopping center. The developed property is rented for seven years, after which time it is sold. The sale price for each building is estimated at 10 times its operating net income in the last year of rental. The company estimates that the project will include a 4.5-million-square-foot shopping center. The master plan calls for constructing three high-rise and four garden office buildings.

The company is faced with a scheduling problem. If a building is completed too early, it may stay vacant; if it is completed too late, potential tenants may be lost to other projects. The demand for office space over the next 7 years estimate based on appropriate market studies is

	Demand (in thousands of square feet)	
Year	High-Rise Space	Garden Space
1	200	100
2	220	110
3	242	121
4	266	133
5	293	146
6	322	161
7	354	177

The following table lists the proposed capacities of the seven buildings:

Garden	Capacity (square feet)	High-Rise Buildings	Capacity (square feet)
1	60,000	1	350,000
2	60,000	2	450,000
3	75,000	3	350,000
4	75,000		

The gross rental income is estimated at $18 per square foot. The operating expenses are $3.75 and $4.75 per square foot for garden and high-rise buildings, respectively. The associated construction costs are $70 and $105 per square foot, respectively. Both construction cost and rental income are estimated to increase at a rate roughly equal to the inflation rate.

How should the company schedule the construction of the seven buildings?

---

† This case is based on R. Perser and S. Andrus, "Phasing of Income-Producing Real Estate," *Interfaces,* Vol. 13, no. 5, 1983, pp. 1–9.

Case 8–2‡

In a National Collegiate Athletic Association (NCAA) women's gymnastic meet consisting of four events (vault, uneven bars, balance beam, and floor exercises), each team may enter the competition with six gymnasts per event. A gymnast is evaluated on a scale of 1 to 10. The total score for a team is determined by summarizing the top five individual scores for each event. An entrant may participate as a specialist in one event or an all-arounder in all four events, but never both. A specialist is allowed to compete in at most three events. Of the six participants allowed for each event, at least four must be all-arounders.

How should a coach select her gymnastic team?

Dynamic (Multistage) Programming

Dynamic programming (DP) is a mathematical procedure designed primarily to improve the computational efficiency of select mathematical programming problems by decomposing them into smaller, and hence computationally simpler, subproblems. Dynamic programming typically solves the problem in **stages,** with each stage involving exactly one optimizing variable. The computations at the different stages are linked through **recursive computations** in a manner that yields a feasible optimal solution to the *entire* problem.

The name *dynamic programming* probably evolved because of its use with applications involving decision making over time (such as inventory problems). However, other situations in which time is not a factor are also solved successfully by DP. For this reason, a more apt name may be **multistage programming,** since the procedure typically determines the solution in stages.

The main unifying theory in DP is the **principle of optimality.** It basically tells us how a properly decomposed problem can be solved in stages (rather than as one entity) through the use of recursive computations.

The subtle concepts used in DP together with the unfamiliar mathematical notations are often a source of confusion, especially to a beginner. However, our experience shows that frequent exposure to DP formulations and solutions will, with some effort, enable a beginner to comprehend these subtle concepts. When this happens, DP becomes amazingly simple and clear.

9.1 ELEMENTS OF THE DP MODEL—THE CAPITAL BUDGETING EXAMPLE

A corporation is entertaining proposals from its three plants for possible expansion of facilities. The corporation is budgeting $5 million for allocation to all three plants. Each plant is requested to submit its proposals giving total cost (c) and total revenue (R) for each proposal. Table 9–1 summarizes the costs and revenues (in millions of dollars). The zero-cost proposals are introduced to allow for the possibility of not allocating funds to individual plants. The goal of the corporation is to maximize the total revenue resulting from the allocation of the $5 million to the three plants.

Table 9–1

	Plant 1		Plant 2		Plant 3	
Proposal	c_1	R_1	c_2	R_2	c_3	R_3
1	0	0	0	0	0	0
2	1	5	2	8	1	3
3	2	6	3	9	—	—
4	—	—	4	12	—	—

A straightforward, and perhaps naïve, way to solve the problem is by exhaustive enumeration. The problem has $3 \times 4 \times 2 = 24$ possible solutions, some of which are infeasible because they require more capital than the $5 million available. The idea of exhaustive enumeration is to compute the total cost for each of the 24 combinations. If it does not exceed the available capital, its total revenue is computed. The optimum solution is the feasible combination yielding the highest total revenue. For example, proposals 2, 3, and 1 for plants 1, 2, and 3 cost $4 million ($<5$) and yield a total revenue of $14 million. On the other hand, the combination comprising proposals 3, 4, and 2 is infeasible because it costs $7 million.

Let us examine the drawbacks of exhaustive enumeration.

1. Each combination defines a decision policy for the *entire* problem, and hence the enumeration of all possible combinations may not be feasible computationally for problems of moderate- and large-size.

2. The infeasible combinations cannot be detected a priori, thus leading to computational inefficiency.

3. Available information regarding previously investigated combinations is not used to eliminate future inferior combinations.

The DP algorithm that we present here is designed to alleviate all the difficulties just noted.

9.1.1   DP Model

In DP, computations are carried out in stages by breaking down the problem into subproblems. Each subproblem is then considered separately with the objective of reducing the volume of computations. However, since the subproblems are interdependent, a procedure must be devised to link the computations in a manner that guarantees that a feasible solution for each stage is also feasible for the entire problem.

A **stage** in DP is defined as the portion of the problem that possesses a set of mutually exclusive alternatives from which the best alternative is to be selected. In terms of the capital budgeting example, each plant defines a stage with the first, second, and third stages having three, four, and two alternatives, respectively. These stages are *interdependent* because all three plants must compete for a *limited* budget. For example, choosing proposal 1 for plant 1 will leave $5 million for plants 2 and 3, whereas choosing proposal 2 for plant 1 will leave $4 million only for plants 2 and 3.

The basic idea of DP is practically to eliminate the effect of interdependence between stages by associating a *state* definition with each stage. A **state** is normally defined to reflect the status of the constraints that bind all the stages together. In the capital budgeting example, we define the states for stages 1, 2, and 3 as follows:

$x_1 =$ amount of capital allocated to stage 1
$x_2 =$ amount of capital allocated to stages 1 and 2
$x_3 =$ amount of capital allocated to stages 1, 2, and 3

We now show how the given definitions of *stages* and *states* are used to decompose the capital budgeting problem into three computationally separate subproblems.

First note that the values of x_1 and x_2 are not known exactly, but must lie somewhere between 0 and 5. In fact, because the costs of the different proposals are discrete, x_1 and x_2 may only assume the values 0, 1, 2, 3, 4, or 5. On the other hand, x_3, which is the total capital allocated to *all* three stages, is equal to 5.

The way we solve the problem is to start with stage (plant) 1. We obtain *conditional* decisions for that stage that answer the following question: Given a specific value of x_1 (= 0, 1, 2, 3, 4, or 5), what would be the best alternative (proposal) for stage 1? The computations for stage 1 are straightforward. Given the value of x_1, we choose the best proposal whose cost does not exceed x_1. The following table summarizes the *conditional* decisions for stage 1.

If Available Capital x_1 Equals	Then, the Resulting Optimal Proposal is	And the Total Revenue of Stage 1 is
0	1	0
1	2	5
2	3	6
3	3	6
4	3	6
5	3	6

So far, we do not know the exact value of x_1. However, by the time we reach stage 3, such information will be available to us, and the problem will then reduce to reading the proper entries in the table.

Exercise 9.1-1

In the preceding table, is it possible that $x_1 > 2$ can be optimal in the final solution? [*Ans.* No, because $x_1 > 2$ represents overspending for stage 1.]

We now consider stage 2 calculations. These calculations also seek a *conditional* optimal solution for stage 2 as a function of the state x_2. However, they differ from those of stage 1 in that the state x_2 now defines the capital to be allocated to stage 1 *and* stage 2. Such a definition will guarantee that a decision made for stage 2 will be automatically feasible for stage 1. The idea now is to choose the alternative in stage 2 given x_2 that yields the best revenue for stages 1 and 2. The following formula summaries the nature of the computations for stage 2:

$$\begin{pmatrix} \text{best revenue} \\ \text{for stages} \\ \text{1 and 2 given} \\ \text{state } x_2 \end{pmatrix} = \begin{matrix} \max \\ \text{all feasible} \\ \text{alternatives} \\ \text{of stage 2} \\ \text{given } x_2 \end{matrix} \left\{ \begin{pmatrix} \text{revenue of} \\ \text{the feasible} \\ \text{alternative} \\ \text{for stage 2} \end{pmatrix} + \begin{pmatrix} \text{best revenue} \\ \text{for stage 1} \\ \text{given its} \\ \text{state } x_1 \end{pmatrix} \right\}$$

where $x_1 = x_2 -$ capital allocated to given alternative of stage 2.

The basic idea of the formula is that a specific choice of an alternative for stage 2 will affect the capital remaining for stage 1, namely, x_1. Thus, by considering *all* the feasible alternatives of stage 2, we are automatically accounting for all the combinations that are possible for stages 1 and 2. Notice that the second term of the right side in the equation is obtained directly from the summary table for stage 1.

We now provide the details for stage 2 computations.

$x_2 = 0$

The only feasible alternative for stage 2 given $x_2 = 0$ is proposal 1 whose cost and revenue are both equal to zero. Thus, the application of the formula yields

$$\begin{pmatrix} \text{best revenue} \\ \text{given } x_2 = 0 \end{pmatrix} = 0 + 0 = 0$$

corresponding to proposal 1.

$x_2 = 1$

For $x_2 = 1$, we only have one feasible alternative for stage 2; namely, proposal 1, which costs zero and yields a revenue of zero. The remaining proposals are infeasible because they cost at least 2. We thus have

$$\binom{\text{best revenue}}{\text{given } x_2 = 1} = 0 + 5 = 5$$

corresponding to proposal 1.

Notice that $x_1 = x_2 -$ cost of proposal $1 = 1 - 0 = 1$. In the summary table of stage 1, we find that the best revenue given $x_1 = 1$ is 5. Notice also that all we need from the calculations in stage 1 is the best revenue associated with given x_1. In other words we do not really care about the *specific proposal* selected at stage 1.

$x_2 = 2$

Here we have two feasible alternatives: proposals 1 and 2 costing 0 and 2 and yielding revenues of 0 and 8, respectively. Thus, the values of x_1 corresponding to proposals 1 and 2 are $2 - 0 = 2$ and $2 - 2 = 0$. The corresponding best revenues from stage 1 given $x_1 = 2$ and $x_1 = 0$ are 6 and zero, respectively. We thus get

$$\binom{\text{best revenue}}{\text{given } x_2 = 2} = \max \{0 + 6, \ 8 + \ 0\} = 8$$

corresponding to proposal 2.

$x_2 = 3$

Feasible alternatives are proposals 1, 2, and 3. The corresponding values of x_1 are $3 - 0 = 3$, $3 - 2 = 1$, and $3 - 3 = 0$, respectively. Thus, we have

$$\binom{\text{best revenue}}{\text{given } x_2 = 3} = \max \{0 + 6, \ 8 + 5, \ 9 + 0\} = 13$$

corresponding to proposal 2.

$x_2 = 4$

Feasible alternatives are proposals 1, 2, 3, and 4. The corresponding values of x_1 are $4 - 0 = 4$, $4 - 2 = 2$, $4 - 3 = 1$, and $4 - 4 = 0$, respectively, which leads to

$$\binom{\text{best revenue}}{\text{given } x_2 = 4} = \max \{0 + 6, \ 8 + 6, \ 9 + 5, \ 12 + 0\} = 14$$

corresponding to proposal 2 or 3

$x_2 = 5$

We have the same feasible alternatives as in $x_2 = 4$. The corresponding values of x_1 are 5, 3, 2, and 1, respectively. Thus,

$$\binom{\text{best revenue}}{\text{given } x_2 = 5} = \max \{0 + 6, \ 8 + 6, \ 9 + 6, \ 12 + 5\} = 17$$

corresponding to proposal 4.

We can summarize stage 2 computation as follows:

If Available Capital x_2 Is	Then, the Resulting Optimal Proposal Is	And the Total Revenue for Stages 1 and 2 Is
0	1	0
1	1	5
2	2	8
3	2	13
4	2 or 3	14
5	4	17

Stage 3 is now considered. The formula for computing the best revenue is similar to that of stage 2 except that x_2 and x_1 are replaced by x_3 and x_2. Similarly, stage 2 and stage 1 are replaced by stage 3 and stage 2. Notice, however, that unlike x_1 or x_2, x_3 now has a single specific value; namely, $x_3 = 5$. Since stage 3 has two proposals whose cost does not exceed 5, both proposals are feasible. The values of x_2 corresponding to proposals 1 and 2 are $5 - 0 = 5$ and $5 - 1 = 4$, respectively. Using the summary table for stage 2 together with x_2, we then obtain

$$\binom{\text{best revenue}}{\text{given } x_3 = 5} = \max \{0 + 17, \ 3 + 14\} = 17$$

corresponding to 1 or 2.

Now that we have completed all the computations, we can *read* the optimal solution directly. Starting from stage 3, we can choose either proposal 1 or 2. If we choose proposal 1, which costs 0, then x_2 for stage 2 will be $5 - 0 = 5$. From the summary table of stage 2, we see that the optimal alternative given $x_2 = 5$ is proposal 4. Since proposal 4 of stage 2 costs 4, we have $x_1 = x_2 - 4 = 5 - 4 = 1$. Again, from the summary table of stage 1, we obtain proposal 2 as the optimal alternative for stage 1.

By combining all the answers for the three stages, an optimal solution calls for selecting proposal 2 for plant 1, proposal 4 for plant 2, and proposal 1 for plant 3. The total cost is 5 and the optimal revenues is 17. Another two solutions can be determined by considering the alternate optimal proposal at stage 3.

Exercise 9.1–2
Identify the remaining two alternate optima for the foregoing example.
[*Ans.* (3, 2, 2) and (2, 3, 2).]

If you study the given procedure carefully, you will find that the computations are actually *recursive*. Thus stage 2 computations are based on stage 1 computations. Similarly, stage 3 computations make use of stage 2 computations only. In other words, the computations at a current stage utilize a summary information from the immediately preceding stage. This summary provides the optimal revenues of *all* stages previously considered. In using this summary, we never concern ourselves about the specific decisions taken in the preceding stages.

Indeed, all *future* decisions are selected optimally without recourse to previously made decisions. This special property constitutes the **principle of optimality,** which is the basis for the validity of DP computations.

To express the recursive equation mathematically, we introduce the following symbols. Let

$R_j(k_j)$ = revenue of alternative k_j at stage j

$f_j(x_j)$ = optimal return of stages 1, 2, . . . , and j given the state x_j

We thus write the recursive equations for the capital budgeting example as

$$f_1(x_1) = \max_{\substack{\text{feasible} \\ \text{proposals } k_1}} \{R_1(k_1)\}$$

$$f_j(x_j) = \max_{\substack{\text{feasible} \\ \text{proposals } k_j}} \{R_j(k_j) + f_{j-1}(x_{j-1})\}, \qquad j = 2, 3$$

There is an important point that we need to clarify regarding the mathematical accuracy of this recursive equation. First, note that $f_j(x_j)$ is a function of the argument x_j only. This requires that the right side of the recursive equation be expressed in terms of x_j rather than x_{j-1}. This is accomplished by recalling that

$$x_{j-1} = x_j - c_j(k_j)$$

where $c_j(k_j)$ is the cost of alternative k_j at stage j.

Another point deals with expressing the feasibility of the proposals mathematically. Specifically, a proposal k_j is feasible if its cost $c_j(k_j)$ does not exceed the state of the system x_j at stage j.

Taking these two points into account, we can write the DP recurrsive equations as

$$f_1(x_1) = \max_{c_1(k_1) \leq x_1} \{R_1(k_1)\}$$

$$f_i(x_j) = \max_{c_j(k_j) \leq x_j} \{R_j(k_j) + f_{j-1}[x_j - c_j(k_j)]\}, \qquad j = 2, 3$$

The implementation of the recursive equations is usually done in a standard tabular form as the following computations illustrate. We must point out, however, that it is always tempting to do the tabular computations in a mechanical fashion without truly understanding *why* they are done. To avoid falling into this trap, we suggest that you always try to relate the tabular computations entries to the corresponding mathematical symbols in the recursive equation.

Stage 1

$$f_1(x_1) = \max_{\substack{c_1(k_1) \leq x_1 \\ k_1 = 1,2,3}} \{R_1(k_1)\}$$

	$R_1(k_1)$			Optimal Solution	
x_1	$k_1=1$	$k_1=2$	$k_1=3$	$f_1(x_1)$	k_1^*
0	**0**	—	—	0	1
1	0	**5**	—	5	2
2	0	5	**6**	6	3
3	0	5	**6**	6	3
4	0	5	**6**	6	3
5	0	5	**6**	6	3

Stage 2

$$f_2(x_2) = \max_{\substack{c_2(k_2) \le x_2 \\ k_2=1,2,3,4}} \{R_2(k_2) + f_1[x_2 - c_2(k_2)]\}$$

	$R_2(k_2) + f_1[x_2 - c_2(k_2)]$				Optimal Solution	
x_2	$k_2=1$	$k_2=2$	$k_2=3$	$k_2=4$	$f_2(x_2)$	k_2^*
0	$0+0=\mathbf{0}$	—	—	—	0	1
1	$0+5=\mathbf{5}$	—	—	—	5	1
2	$0+6=6$	$8+0=\mathbf{8}$	—	—	8	2
3	$0+6=6$	$8+5=\mathbf{13}$	$9+0=9$	—	13	2
4	$0+6=6$	$8+6=\mathbf{14}$	$9+5=\mathbf{14}$	$12+0=12$	14	2 or 3
5	$0+6=6$	$8+6=14$	$9+6=15$	$12+5=\mathbf{17}$	17	4

Stage 3

$$f_3(x_3) = \max_{\substack{c_3(k_3) \le x_3 \\ k_3=1,2}} \{R_3(k_3) + f_2[x_3 - c_3(k_3)]\}$$

	$R_3(k_3) + f_2[x_3 - c_3(k_3)]$		Optimum Solution	
x_3	$k_3=1$	$k_3=2$	$f_3(x_3)$	k_3^*
5	$0+17=\mathbf{17}$	$3+14=\mathbf{17}$	17	1 or 2

The optimum solution can now be read directly from the foregoing tableaus starting with stage 3. For $x_3 = 5$, the optimal proposal is either $k_3^* = 1$ or $k_3^* = 2$. Consider $k_3^* = 1$ first. Since $c_3(1) = 0$, this leaves $x_2 = x_3 - c_3(1) = 5$ for stages 2 and 1. Now, stage 2 shows that $x_2 = 5$ yields $k_2^* = 4$. Since $c_2(4) = 4$, this leaves $x_1 = 5 - 4 = 1$. From stage 1, $x_1 = 1$ gives $k_1^* = 2$. Thus an optimal combination of proposals for stages 1, 2, and 3 is (2, 4, 1). Figure 9–1 shows how all the alternative optima are determined systematically.

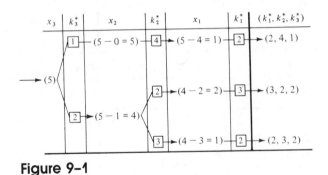

Figure 9–1

Exercise 9.1–3

(a) In each of the following cases, which relate to the capital budgeting example, determine the optimum solution of the problem.

    (1) $x_3 = 2$.

        [*Ans.* $f_3(2) = 8$, and the optimal proposals are (1, 2, 1) or (2, 1, 2).]

    (2) $x_3 = 3$.

        [*Ans.* $f_3(3) = 13$, and the optimal proposals are (2, 2, 1).]

(b) Suppose that the costs of the proposals include fractions of $.1 million rather than being rounded to the closest million dollars as in the example above. How would this change in data affect the tabular computations?

    [*Ans.* x_1 and x_2 must assume discrete values in steps of .1, that is, 0, .1, .2, . . . , 4.9, 5, thus increasing the number of table entries at stages 1 and 2 by approximately 10 times.]

9.1.2 BACKWARD RECURSIVE EQUATION

In Section 9.1.1 the computations are carried out in the order

$$f_1 \rightarrow f_2 \rightarrow f_3$$

This method of computations is known as the **forward procedure** because the computations advance from the first to the last stage. However, when you study most DP literature, you will find out that the recursive equation is set up such that the computations start at the last stage and then "proceed" backward to stage 1. This method is called the **backward procedure.**

    The main difference between the forward and backward methods occurs in the way we define the *state* of the system. To be specific, let us reconsider the capital budgeting example. For the backward procedure, we define the states y_j as

        $y_1 =$ amount of capital allocated to stages 1, 2, and 3

        $y_2 =$ amount of capital allocated to stages 2 and 3

        $y_3 =$ amount of capital allocated to stage 3

To appreciate the difference between the definition of states x_j and y_j in the forward and backward methods, the two definitions are summarized graphically in Figure 9–2.

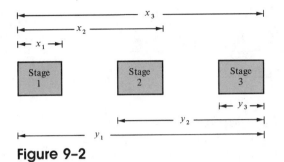

Figure 9-2

Now define

$$f_3(y_3) = \text{optimal revenue for stage 3 given } y_3$$
$$f_2(y_2) = \text{optimal revenue for stages 2 } and \text{ 3 given } y_2$$
$$f_1(y_1) = \text{optimal revenue for stages 1, 2, } and \text{ 3 given } y_1$$

The backward recursive equation is thus written as

$$f_3(y_3) = \max_{\substack{k_3 \\ c_3(k_3) \le y_3}} \{R_3(k_3)\}$$

$$f_j(y_j) = \max_{\substack{k_j \\ c_j(k_j) \le y_j}} \{R_j(k_j) + f_{j+1}[y_j - c_j(k_j)]\}, \qquad j = 1, 2$$

The order of stage computations is thus $f_3 \to f_2 \to f_1$. The computations are now carried out as follows

Stage 3

$$f_3(y_3) = \max_{\substack{c_3(k_3) \le y_3 \\ k_3 = 1,2}} \{R_3(k_3)\}$$

	$R_3(k_3)$		Optimum Solution	
y_3	$k_3 = 1$	$k_3 = 2$	$f_3(y_3)$	k_3^*
0	**0**	—	0	1
1	0	3	3	2
2	0	3	3	2
3	0	3	3	2
4	0	3	3	2
5	0	3	3	2

Stage 2

$$f_2(y_2) = \max_{\substack{c_2(k_2) \le y_2 \\ k_2 = 1,2,3,4}} \{R_2(k_2) + f_3[y_2 - c_2(k_2)]\}$$

y_2	$R_2(k_2)+f_3[y_2-c_2(k_2)]$				Optimum Solution	
	$k_2=1$	$k_2=2$	$k_2=3$	$k_2=4$	$f_2(y_2)$	k_3^*
0	$0+0=0$	—	—	—	0	1
1	$0+3=3$	—	—	—	3	1
2	$0+3=3$	$8+0=8$	—	—	8	2
3	$0+3=3$	$8+3=11$	$9+0=9$	—	11	2
4	$0+3=3$	$8+3=11$	$9+3=12$	$12+0=12$	12	3 or 4
5	$0+3=3$	$8+3=11$	$9+3=12$	$12+3=15$	15	4

Stage 1

$$f_1(y_1)=\max_{\substack{c_1(k_1)\le y_1 \\ k_1=1,2,3}}\{R_1(k_1)+f_2[y_1-c_1(k_1)]\}$$

y_1	$R_1(k_1)+f_2[y_1-c_1(k_1)]$			Optimum Solution	
	$k_1=1$	$k_1=2$	$k_1=3$	$f_1(y_1)$	k_3^*
5	$0+15=15$	$5+12=17$	$6+11=17$	17	2 or 3

The optimal solution is determined by starting with y_1 at stage 1 and proceeding to y_3 at stage 3. Naturally, the solutions are identical with those of the forward method (verify).

Exercise 9.1–4

Compute $f_1(y_1)$ for $y_1=3$ and $y_1=4$ and find the corresponding optimum proposals. [*Ans.* $f_1(3)=13$, optimal proposals for stages 1, 2, and 3 are 2, 2, and 1. $f_1(4)=16$ and the optimal proposals are 2, 2, and 2.]

You may wonder why the backward recursive formulation is needed at all, particularly in that the forward formulation appears more logical and certainly more straightforward. This conclusion is true for the preceding example, since the specific assignment of stages to plants is nonconsequential. In this respect, the forward and backward formulations are in fact computationally equivalent. There are situations, however, where it would make a difference, from the standpoint of computational efficiency, which formulation is used. This is particularly so in problems involving decision making over time, such as inventory and production planning. In this case the stages are designated based on the strict chronological order of the time periods they represent; and the efficiency of computations will depend on whether the forward or backward formulation is used (see Example 9.3–5).

Apparently, experience with DP computations has shown that the backward formulations are generally more efficient. In fact, most of the DP literature is presented in terms of the backward formulation regardless of whether or not it contributes to computational efficiency. Following this tradition, all the presentations in the remainder of this chapter will be based on the backward formulation. The forward formulation will be used only when a comparison is warranted or when the forward formulation offers special advantages (see Section 13.3.4 for an illustrative application in the area of inventory).

9.2 MORE ON THE DEFINITION OF THE STATE

The *state* of the system is perhaps the most important concept in a dynamic programming model. It represents the "link" between (succeeding) stages so that when each stage is optimized *separately,* the resulting decision is automatically feasible for the *entire* problem. Moreover, it allows one to make optimum decisions for the remaining stages without having to check the effect of future decisions on decisions previously made.

The definition of the state is usually the most subtle concept in dynamic programming formulations. There is no easy way to define the state, but clues can usually be found by asking the following two questions:

1. What relationships bind the stages together?
2. What information is needed to make feasible decisions at the current stage without checking the feasibility of decisions made at previous stages?

The following examples are introduced to help you understand the definition of the state.

Example 9.2-1. Consider the capital budgeting problem. As indicated in Section 9.1, each plant represents a stage for which a decision is made. The alternatives are given by the decision variable k_j at stage j, which designates a specific expansion plan. In this case the return function is $R_j(k_j)$.

What defines the state at stage j? Note that the stages are "linked" by the fact that all the plants (stages) are competing for a share of the limited capital C. This suggests that the state should be defined in terms of capital allocation.

Experience has shown that a beginner in dynamic programming will usually define the state at stage j as "the amount of capital allocated to stage j." To see why this definition is *not* correct, consider the manner in which the problem will be solved. The definition of the state should allow one to make a feasible decision for the current stage without checking the decisions made for previous stages. The definition of state given indicates only that the amount allocated to stage j can be as small as zero or as high as the total available capital C. This is not sufficient information to guarantee a feasible decision for the current stage. For example, suppose that it is decided to allocate $.4C$ dollars to the current stage. The feasibility of this decision is not guaranteed without checking the preceding stages to ensure that their total capital allocation did not exceed $C - .4C = .6C$ dollars. This shows that the current stage is not optimized independently, contrary to the basic idea of dynamic programming.

Assume the backward formulation and consider the definition of the state at stage j as "the amount of capital allocated to stages $j, j + 1, \ldots,$ and

N," where N is the total number of plants. This definition is correct, since the difference between the capital allocated to stages $j, j + 1, \ldots, N$—that is, the state of the system at stage j—and the capital allocated to stages $j + 1$, $j + 2, \ldots, N$—that is, the state of the system at stage $(j + 1)$—gives the amount of capital to be allocated to stage j only. This, as we saw in Section 9.1.3, allows us to make a feasible decision for stage j without checking the previous stages.   ◀

The capital budgeting problem represents a typical *allocation* problem in which a resource (or, generally, resources) is distributed (optimally) among a number of activities (stages). The definition of the state of the system for all allocation problems is generally the same: namely, the amount of the resource allocated to a successive number of stages starting from the last stage. Other types of problems do not fall into this category, however, and require a different definition of the state of the system. Two examples are presented here.

Example 9.2-2. A contractor wishes to determine the size of a labor force during each of the next 5 weeks. He knows the minimum number of workers needed for each week. Additional cost is incurred when he hires or fires workers and when workers are idle. The costs of hiring, firing, and idleness per worker are known. The objective is to decide how many workers should be hired or fired each week in order to minimize the total cost.

In looking at this situation from the dynamic programming standpoint, the first element to identify is the stage. Since a decision is to be made for each week, each period (week) represents a stage.

The second element to be defined is the alternatives (decision variables) associated with each stage. In this example the decision variable is the number of workers hired or fired in the period. The return function is represented by the cost of hiring or firing and the cost of idleness.

The third and most important element is the state of the system at a given stage. Unlike allocation problems, no explicit constraint ties the stages together. However, a clue for defining the state can be found by asking the question: What information is needed from all previously considered stages (periods) to make a decision for the current stage without having to examine any of the decisions previously made? With some reflection, we note that the number of workers available at the end of the preceding stage (period) provides sufficient information to decide how many to hire or fire in the current stage. Consequently, the number of workers at the end of preceding stage defines the state of the system at the current stage. In other words, knowing how many were hired or fired in each of the previously considered stages (i.e., previous decisions) is of no consequence in making the decision for the current stage. The only important information is how many workers are on hand before the current decision is made. This information is available from the definition of the state of the system. Stated differently, the state of the system summarizes all the information needed to make a feasible decision for the current stage.   ◀

Example 9.2-3. Consider an equipment replacement situation where at the end of each year a decision is made to keep a machine another year or replace it immediately. If a machine is kept longer, its realized profit declines. On the other hand, replacing a machine incurs the cost of a replacement. The problem is to decide when a machine should be replaced to maximize the total net profit.

In this problem, stage j represents year j. The alternatives at each stage are either to keep the machine or to replace it. Now you should ask: What is the relationship between two successive stages? What information is needed from the preceding stages to make a decision (keep or replace) in the current stage? The answer is: the age of the machine. Thus the state of the system at a stage is defined as the age of the machine at the beginning of the associated period. ◀

9.3   EXAMPLES OF DP MODELS AND COMPUTATIONS

This section presents further examples of dynamic programming models. The first four examples involve both model formulations and computations. The last example introduces a comparison between the forward and backward recursive equations. (Other examples pertaining to DP applications in Markovian processes and inventory are presented in Chapters 13 and 14.)

As you study this section, you will find it helpful to construct each model as a network. As you do this, make sure that you have a clear understanding of the basic elements of the model: (1) stages, (2) states at each stage, and (3) decision alternatives (proposals) at each stage.

As we indicated previously, the concept of *state* is usually the most subtle. Our experience indicates that an understanding of the concept of state is enhanced by trying to "question the validity" of the way it is defined in the book. Try a different definition that may appear "more logical" and use it in the recursive computations. You will eventually discover that the definition given here is not incorrect and, in most cases, may be the only correct definition. In the process you will also gain insight into what the concept of state is all about.

Example 9.3-1 (Cargo-Loading Problem).† Consider loading a vessel with stocks of N items. Each unit of item i has a weight w_i and a value $v_i(i = 1, 2, \ldots, N)$. The maximum cargo weight is W. It is required to determine the most valuable cargo load without exceeding the maximum weight of the vessel. Specifically, consider the following special case of three items and assume that $W = 5$.

i	w_i	v_i
1	2	65
2	3	80
3	1	30

---

† This problem is known also as the **knapsack** or the **flyaway kit** problem.

Note: The optimal solution to this example can be obtained by inspection. A typical problem usually involves a large number of items and hence the solution would not be as obvious. See the discussion immediately following the end of this example.

Consider the general problem of N items first. If k_i is the number of units of item i, the problem becomes

$$\text{maximize } v_1 k_1 + v_2 k_2 + \cdots + v_N k_N$$

subject to

$$w_1 k_1 + w_2 k_2 + \cdots + w_N k_N \leq W$$
$$k_i \text{ nonnegative integer}$$

If k_i is not restricted to integer values, the solution is easily determined by the simplex method. In fact, since there is only one constraint, only one variable will be basic and the problem reduces to selecting the item i for which $v_i W/w_i$ is maximum. Since linear programming is not applicable here, the problem will be attempted by dynamic programming. It must be noted that this problem is also typical of the type that can be solved by integer programming techniques (see Chapter 8).

The DP model is constructed by first considering its three basic elements:

1. *Stage j* is represented by item j, $j = 1, 2, \ldots, N$.
2. *State y_j* at stage j is the total weight assigned to stages $j, j+1, \ldots, N$; $y_1 = W$ and $y_j = 0, 1, \ldots, W$ for $j = 2, 3, \ldots, N$.
3. *Alternative k_j* at stage j is the *number* of units of item j. The value of k_j may be as small as zero or as large as $[W/w_j]$, where $[W/w_j]$ is the largest integer included in (W/w_j).

There is a striking similarity between this problem and the capital budgeting example of Section 9.1, since both are of the resource allocation type. About the only difference is that the alternatives in the cargo-loading model are not given directly as in the capital budgeting model.

Let

$$f_j(y_j) = \text{optimal value of stages } j, j+1, \ldots, N \text{ given the state } y_j$$

The (backward) recursive equation is thus given as

$$f_N(y_N) = \max_{\substack{k_N=0,1,\ldots,[y_N/w_N] \\ y_N=0,1,\ldots,W}} \{v_N k_N\}$$

$$f_j(y_j) = \max_{\substack{k_j=0,1,\ldots,[y_j/w_j] \\ y_j=0,1,\ldots,W}} \{v_j k_j + f_{j+1}(y_j - w_j k_j)\}, \quad j=1,2,\ldots,N-1$$

Note that the maximum *feasible* value of k_j is given by $[y_j/w_j]$. This limit will automatically delete all infeasible alternatives for a given value of the state y_j.

Exercise 9.3-1

Establish the relationship between $R_j(k_j)$ and $c_j(k_j)$ in the capital budgeting model of Section 9.1 and the corresponding elements in the cargo-loading model.
[*Ans.* $R_j(k_j)$ corresponds to $v_j k_j$ and $c_j(k_j)$ is equivalent to $w_j k_j$.]

For the special example given, stage computations are performed as follows.

Stage 3

$$f_3(y_3) = \max_{k_3} \{30 k_3\}, \qquad \max k_3 = [5/1] = 5$$

				30k_3				Optimal Solution	
	$k_3 = 0$	1	2	3	4	5			
y_3	$v_3 k_3 = 0$	30	60	90	120	150	$f_3(y_3)$	k_3^*
0	0	—	—	—	—	—	0	0
1	0	30	—	—	—	—	30	1
2	0	30	60	—	—	—	60	2
3	0	30	60	90	—	—	90	3
4	0	30	60	90	120	—	120	4
5	0	30	60	90	120	150	150	5

Stage 2

$$f_2(y_2) = \max_{k_2} \{80 k_2 + f_3(y_2 - 3k_2)\}, \qquad \max k_2 = [5/3] = 1$$

	80k_2 + f_3(y_2 - 3k_2)		Optimum Solution	
	$k_2 = 0$	1		
y_2	$v_2 k_2 = 0$	80	$f_2(y_2)$	k_2^*
0	$0 + \quad 0 = 0$	—	0	0
1	$0 + \quad 30 = 30$	—	30	0
2	$0 + \quad 60 = 60$	—	60	0
3	$0 + \quad 90 = 90$	$80 + \quad 0 = 80$	90	0
4	$0 + 120 = 120$	$80 + 30 = 110$	120	0
5	$0 + 150 = 150$	$80 + 60 = 140$	150	0

Stage 1

$$f_1(y_1) = \max_{k_1} \{65 k_1 + f_2(y_1 - 2k_1)\}, \qquad \max k_1 = [5/2] = 2$$

		$65k_1 + f_2(y_1 - 2k_1)$		Optimum Solution	
	$k_1 = 0$	1	2		
y_1	$v_1 k_1 = 0$	65	130	$f_1(y_1)$	k_1^*
0	$0 + 0 = 0$	—	—	0	0
1	$0 + 30 = 30$	—	—	30	0
2	$0 + 60 = 60$	$65 + 0 = 65$	—	65	1
3	$0 + 90 = 90$	$65 + 30 = 95$	—	95	1
4	$0 + 120 = 120$	$65 + 60 = 125$	$130 + 0 = 130$	130	2
5	$0 + 150 = 150$	$65 + 90 = 155$	$130 + 30 = 160$	160	2

Given $y_1 = W = 5$, the associated optimum solution is $(k_1^*, k_2^*, k_3^*) = (2, 0, 1)$, with a total value of 160.

Notice that at stage 1, it is sufficient to construct the table for $y_1 = 5$ only. However, by computing the entire table for $y_1 = 0, 1, 2, 3, 4,$ and 5, it is possible to study changes in the optimal solution when the maximum weight allocation is reduced below $W = 5$. This is a form of sensitivity analysis that the DP computations provide automatically. ◄

Exercise 9.3-2

Find the optimal solution to the cargo-loading problem in each of the following cases.

(1) $W = 3$.

[*Ans.* $(k_1^*, k_2^*, k_3^*) = (1, 0, 1)$; total value $= 95$.]

(2) $W = 4$.

[*Ans.* $(k_1^*, k_2^*, k_3^*) = (2, 0, 0)$; total value $= 130$.]

It may appear that the knapsack problem can be solved in general by computing the ratios v_j/w_j for all the variables k_j, and then assigning the largest integer quantities to the variables successively in the order of their ratios until the resource is exhausted. (This procedure actually produces the optimal solution in Example 9.3–1.) Unfortunately, this is not always true, as the following counterexample shows.

$$\text{Maximize } 17k_1 + 72k_2 + 35k_3$$

subject to

$$10k_1 + 41k_2 + 20k_3 \leq 50$$

$$k_1, k_2, k_3 \text{ nonnegative integers}$$

The ratios for k_1, k_2, and k_3 are 1.7, 1.756, and 1.75. Since k_2 has the largest ratio, it is assigned the largest value allowed by the constraint, that is, $k_2 = [50/41] = 1$. The remaining amount of the resource is now $50 - 41 = 9$, which is not sufficient to assign any positive integer values to k_1 or k_3. Thus the trial solution is $k_1 = k_3 = 0$, and $k_2 = 1$ with the objective value equal to 72. This is not optimal, since the feasible solution $(k_1 = 1, k_2 = 0, k_3 = 2)$ yields a better objective value equal to 87.

Example 9.3-2 (Reliability Problem). Consider the design of an electronic device consisting of three main components. The three components are arranged in series so that the failure of one component will cause the failure of the entire device. The reliability (probability of no failure) of the device can be improved by installing standby units in each component. The design calls for using one or two standby units, which means that each main component may include up to three units in parallel. The total capital available for the design of the device is $10,000. The data for the reliability $R_j(k_j)$ and cost $c_j(k_j)$ for the jth component ($j = 1, 2, 3$) given k_j parallel units are summarized next. The objective is to determine the number of parallel units, k_j, in component j that will maximize the reliability of the device without exceeding the allocated capital.

	$j=1$		$j=2$		$j=3$	
k_j	R_1	c_1	R_2	c_2	R_3	c_3
1	.6	1	.7	3	.5	2
2	.8	2	.8	5	.7	4
3	.9	3	.9	6	.9	5

By definition the total reliability R of a device of N series components and k_j parallel units in component j ($j = 1, 2, \ldots, N$) is the *product* of the individual reliabilities. The problem thus becomes:

$$\text{maximize } R = \prod_{j=1}^{N} R_j(k_j)$$

subject to

$$\sum_{j=1}^{N} c_j(k_j) \leq C$$

where C is the total capital available. (Notice that the alternative $k_j = 0$ is meaningless in this problem.)

The reliability problem is similar to the capital budgeting problem in Section 9.1 with the exception that the return function R is the *product*, rather than the sum, of the returns of the individual components. The recursive equation is thus based on **multiplicative** rather than **additive decomposition.** (The reliability problem is also similar to the most-reliable-route problem presented in Example 6.2–2 as a network model.)

The elements of the DP model are defined as follows.

1. *Stage j* represents main component j.
2. *State y_j* is the total capital assigned to components $j, j + 1, \ldots, N$.
3. *Alternative k_j* is the number of parallel units assigned to main component j.

Let $f_j(y_j)$ be the total optimal reliability of components $j, j + 1, \ldots, N$, given the capital y_j. The recursive equations are written as

$$f_N(y_N) = \max_{\substack{k_N \\ c_N(k_N) \le y_N}} \{R_N(k_N)\}$$

$$f_j(y_j) = \max_{\substack{k_j \\ c_j(k_j) \le y_j}} \{R_j(k_j) \cdot f_{j+1}(y_j - c_j(k_j))\}, \qquad j = 1, 2, \ldots, N-1$$

As we have seen previously, the amount of computations at stage j depends directly on the number of values assumed for the state y_j. We show here how we can compute tighter limits on the values of y_j.

Starting with stage 3, since main component 3 must include at least one (parallel) unit, we find that y_3 must at least equal $c_3(1) = 2$. By the same reasoning y_3 cannot exceed $10 - (3 + 1) = 6$; otherwise, the remaining capital will not be sufficient to provide main components 1 and 2 with at least one (parallel) unit each. Following the same reasoning, we see that $y_2 = 5, 6, \ldots$, or 9, and $y_1 = 6, 7, \ldots$, or 10. (Verify!)

Stage 3

$$f_3(y_3) = \max_{k_3=1,2,3} \{R_3(k_3)\}$$

	$R_3(k_3)$			Optimal Solution	
	$k_3 = 1$	$k_3 = 2$	$k_3 = 3$		
y_3	$R = .5, c = 2$	$R = .7, c = 4$	$R = .9, c = 5$	$f_3(y_3)$	k_3^*
2	.5	—	—	.5	1
3	.5	—	—	.5	1
4	.5	.7	—	.7	2
5	.5	.7	.9	.9	3
6	.5	.7	.9	.9	3

Stage 2

$$f_2(y_2) = \max_{k_3=1,2,3} \{R_2(k_2) \cdot f_3[y_2 - c_2(k_2)]\}$$

	$R_2(k_2) \cdot f_3[y_2 - c_2(k_2)]$			Optimal Solution	
	$k_2 = 1$	$k_2 = 2$	$k_2 = 3$		
y_2	$R = .7, c = 3$	$R = .8, c = 5$	$R = .9, c = 6$	$f_2(y_2)$	k_2^*
5	$.7 \times .5 = .35$	—	—	.35	1
6	$.7 \times .5 = .35$	—	—	.35	1
7	$.7 \times .7 = .49$	$.8 \times .5 = .40$	—	.49	1
8	$.7 \times .9 = .63$	$.8 \times .5 = .40$	$.9 \times .5 = .45$	.63	1
9	$.7 \times .9 = .63$	$.8 \times .7 = .56$	$.9 \times .5 = .45$	.63	1

Stage 1

$$f_1(y_1) = \max_{k_1=1,2,3} \{R_1(k_1) \cdot f_2[y_1 - c_1(k_1)]\}$$

y_1	$R_1(k_1) \cdot f_2[y_1 - c_1(k_1)]$			Optimal Solution	
	$R=.6, c=1$	$R=.8, c=2$	$R=.9, c=3$	$f_2(y_2)$	k_2^*
6	$.6 \times .35 = .210$	—	—	.210	1
7	$.6 \times .35 = .210$	$.8 \times .35 = .280$	—	.280	2
8	$.6 \times .49 = .294$	$.8 \times .35 = .280$	$.9 \times .35 = .315$	.315	3
9	$.6 \times .63 = .378$	$.8 \times .49 = .392$	$.9 \times .35 = .315$	.392	2
10	$.6 \times .63 = .378$	$.8 \times .63 = .504$	$.9 \times .49 = .441$	.504	2

The optimal solution given $C = 10$ is $(k_1^*, k_2^*, k_3^*) = (2, 1, 3)$ with $R = .504$. ◀

Exercise 9.3–3

Suppose that a *fourth* main component is added (in series) to the electronic device. The costs and reliabilities of using one, two, or three parallel units in the new component are $R_4(1) = .4$, $c_4(1) = 1$; $R_4(2) = .8$; $c_4(2) = 3$; and $R_4(3) = .95$, $c_4(3) = 7$.

(a) Determine the limits on the values of y_4, y_3, y_2, and y_1.
   [*Ans.* $1 \le y_4 \le 4$, $3 \le y_3 \le 6$, $6 \le y_2 \le 9$, $7 \le y_1 \le 10$.]

(b) Does the definition of states in (a) require recomputing the optimal solutions at all stages?
   [*Ans.* Yes, because stage 4 must be computed first, thus affecting computations at stages 3, 2, and 1.]

(c) Can you find the optimal solution to the entire problem utilizing directly the computations given for the three stages?
   [*Ans.* Yes, but this will require redefining stage 4 as stage 0. Since the order of the components in the device is irrelevant, we can pretend that the new component precedes component 1, numbering it as stage 0. Under this condition, the values of the states are limited by $2 \le y_3 \le 5$, $5 \le y_2 \le 8$, $6 \le y_1 \le 9$, and $7 \le y_0 \le 10$.]

(d) Compute $f_0(y_0)$ and k_0^* as defined in part (c) and find the optimal solution to the four-component problem given $C = 10$.
   [*Ans.* $f_0(7) = .084$, $k_0^* = 1$; $f_0(8) = .112$, $k_0^* = 1$; $f_0(9) = .168$, $k_0^* = 2$; $f_0(10) = .224$, $k_0^* = 2$. The optimal solution given $C = 10$ is $(k_0^*, k_1^*, k_2^*, k_3^*) = (2, 2, 1, 1)$ with $R = .224$.]

Example 9.3–3 (Optimal Subdivision Problem).

Consider the mathematical problem of dividing a quantity q (> 0) into N parts. The objective is to determine the optimum subdivision of q that will maximize the product of the N parts.

Let z_j be the jth portion of q ($j = 1, 2, \ldots, N$). The problem is thus expressed as

$$\text{maximize } p = \prod_{j=1}^{N} z_j$$

subject to

$$\sum_{j=1}^{N} z_j = q, \quad z_j \geq 0, \quad \text{for all } j$$

The DP formulation of this problem is very similar to the reliability model of Example 9.3–2. The main difference occurs in that the variables z_j are continuous, a condition that requires the use of calculus for optimizing each stage's problem.

The elements of the DP model are defined as

1. *Stage j* represents the jth portion of q.
2. *State y_j* is the portion of q allocated to stages $j, j + 1, \ldots, N$.
3. *Alternative z_j* is the portion of q allocated to stage j.

Let $f_j(y_j)$ be the optimum value of the objective function for stages $j, j + 1,$ $\ldots, N$ given the state y_j. The recursive equations are thus given as

$$f_N(y_N) = \max_{z_N \leq y_N} \{z_N\}$$

$$f_j(y_j) = \max_{z_j \leq y_j} \{z_j \cdot f_{j+1}(y_j - z_j)\}, \quad j = 1, 2, \ldots, N-1$$

Stage N

$$f_N(y_N) = \max_{z_N \leq y_N} \{z_N\}$$

Since z_N is a linear function, $\max_{z_N \leq y_N}\{z_N\} = y_N$, which occurs at $z_N^* = y_N$. We can summarize this stage's optimum solution using the form we used in the preceding examples:

	Optimum Solution	
State	$f_N(y_N)$	z_N^*
y_N	y_N	y_N

Stage N − 1

$$f_{N-1}(y_{N-1}) = \max_{z_{N-1} \leq y_{N-1}} \{z_{N-1} \cdot f_N(y_{N-1} - z_{N-1})\}$$

Since $f_N(y_N) = y_N$, we have

$$f_N(y_{N-1} - z_{N-1}) = y_{N-1} - z_{N-1}$$

Thus, by substituting for f_N, the problem for stage $N - 1$ reduces to maximizing $h_{N-1} = z_{N-1}f_N(y_{N-1} - z_{N-1}) = z_{N-1}(y_{N-1} - z_{N-1})$ given $z_{N-1} \leq y_{N-1}$. (This is precisely the same procedure that we follow in the case of tabular computa-

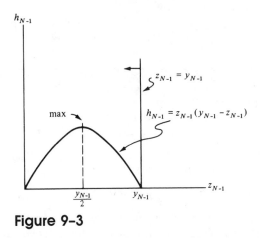

Figure 9-3

tions.) Figure 9–3 shows what the optimization problem entails by plotting the function h_{N-1} in terms of z_{N-1} together with the feasible region $z_{N-1} \leq y_{N-1}$. The optimum feasible solution occurs at $z_{N-1}^* = y_{N-1}/2$. The point $z_{N-1}^* = y_{N-1}/2$ is obtained by differentiating h_{N-1} with respect to z_{N-1}. Since $z_{N-1}^* = y_{N-1}/2$ is feasible, that is, satisfies the condition $z_{N-1} \leq y_{N-1}$, it is the optimum solution.†

The value of $f_{N-1}(y_{N-1})$ is obtained by substituting $z_{N-1} = y_{N-1}/2$ in h_{N-1}. The resulting optimum solution is thus given by

State	Optimum Solution	
	$f_{N-1}(y_{N-1})$	z_{N-1}^*
y_{N-1}	$(y_{N-1}/2)^2$	$(y_{N-1}/2)$

Stage j

$$f_j(y_j) = \max_{z_j \leq y_j} \{ z_j \cdot f_{j+1}(y_j - z_j) \}$$

We can now use *induction* to show that the optimum solution at stage j ($j = 1, 2, \ldots, N$) is summarized as

State	Optimum Solution	
	$f_j(y_j)$	z_j^*
y_j	$\left(\dfrac{y_j}{N-j+1} \right)^{N-j+1}$	$\left(\dfrac{y_j}{N-j+1} \right)$

---

† This simple procedure is applicable here because, among other conditions, h_{N-1} is a concave function. It becomes more involved when these conditions are not satisfied. See Chapter 18 on classical optimization theory for more clarification of this statement.

By inspecting the general solution at stage j, we can obtain the optimum values of z_j given $y_1 = q$ as follows.

$$y_1 = q \to z_1 = \frac{q}{N} \to y_2 = \frac{N-1}{N}q \to \cdots \to y_j = \frac{N-j+1}{N}q \to z_j = \frac{q}{N}$$

Thus the general solution is

$$z_1^* = z_2^* = \cdots = z_j^* = \cdots = z_N^* = \frac{q}{N}$$

and the optimum value of the objective function

$$p = f_1(q) = (q/N)^N$$

This example demonstrates that DP does not give specifics about *how* each stage's problem is optimized. The use of calculus to solve these subproblems has nothing to do with dynamic programming. Note, however, that the decomposition of the "master" problem to smaller subproblems normally simplifies the computations associated with the optimization process. This is the main objective of DP. ◀

Exercise 9.3–4

In Example 9.3–3, suppose that the objective function is $p = \Pi_{j=1}^N a_j z_j$ and the constraint is $\Sigma_{j=1}^N b_j z_j = q$. Write the recursive equation.

[*Ans.* $f_j(y_j) = \max_{z_j \le y_j/b_j} \{a_j z_j \cdot f_{j+1}(y_j - b_j z_j)\}$ for $j = 1, 2, \ldots, N$, where $f_{N+1} \equiv 1$. The state y_j is defined as the portion of q allocated to stages $j, j + 1, \ldots$, and N.]

Example 9.3–4 (Work Force Size).

A contractor needs to decide on the size of his work force over the next 5 weeks. He estimates the *minimum* force size b_i for the 5 weeks to be 5, 7, 8, 4, and 6 workers for $i = 1, 2, 3, 4$, and 5, respectively.

The contractor can maintain the required minimum number of workers by exercising the options of hiring and firing. However, additional hiring cost is incurred every time the work force size of the current week exceeds that of last week. On the other hand, if he maintains a work force for any week that exceeds the minimum requirement, an excess cost is incurred for that week.

Let y_j represent the number of workers for the jth week. Define $C_1(y_j - b_j)$ as the excess cost when y_j exceeds b_j, and $C_2(y_j - y_{j-1})$ as the cost of hiring new workers ($y_j > y_{j-1}$). The contractor's data show that

$$C_1(y_j - b_j) = 3(y_j - b_j), \qquad j = 1, 2, \ldots, 5$$

$$C_2(y_j - y_{j-1}) = \begin{cases} 4 + 2(y_j - y_{j-1}), & y_j > y_{j-1} \\ 0, & y_j \le y_{j-1} \end{cases}$$

Note that the definition of C_2 implies that firing ($y_j \le y_{j-1}$) incurs no additional cost.

If the initial work force y_0 at the beginning of the first week is 5 workers, it is required to determine the optimum sizes of the work force for the 5-week planning horizon.

The definition of *stages* in this example is obvious: Each week represents a stage. However, the definition of the *state* is not as obvious. In all the preceding examples, the definitions of the states are similar because all these examples are of the resource allocation type in which a single resource is distributed optimally among the stages. Our present example is different and we must thus find an appropriate definition of the state.

Recall that the prime objective of the state is to provide sufficient information about all previously considered stages so that future *optimal feasible* decisions can be made without any consideration of how previous decisions were made. In the contractor's problem, the size of the work force at the end of the current week provides sufficient information to make proper feasible decisions for all the remaining weeks. Consequently, the state at stage j is is defined by y_{j-1}.

The only remaining element of the DP model is the definition of alternatives at stage j. This is obviously given by y_j, the work force size at stage j.

To summarize, the elements of the DP model are given as

1. *Stage j* represents the jth week.
2. *State y_{j-1}* at stage j is the number of workers at the end of stage $j - 1$.
3. *Alternative y_j* is the number of workers in week j.

Let $f_j(y_{j-1})$ be the optimal cost for periods (weeks) $j, j + 1, \ldots, 5$, given y_{j-1}. The recursive equations are then written as

$$f_5(y_4) = \min_{y_5 = b_5} \{C_1(y_5 - b_5) + C_2(y_5 - y_4)\}$$

$$f_j(y_{j-1}) = \min_{y_j \geq b_j} \{C_1(y_j - b_j) + C_2(y_j - y_{j-1}) + f_{j+1}(y_j)\}, \qquad j = 1, 2, 3, 4$$

Before we carry out the tabular computations, we need to define the possible values for y_1, y_2, y_3, y_4, and y_5. Since $j = 5$ is the last period and since firing does not incur any cost, y_5 must equal the minimum required number of workers b_5; that is, $y_5 = b_5 = 6$. On the other hand, since $b_4 (= 4) < b_5 (= 6)$, the contractor may maintain $y_4 = 4$, 5, or 6, depending on which level will yield the lowest cost. Following similar reasoning, we can conclude that $y_3 = 8$, $y_2 = 7$ or 8, and $y_1 = 5$, 6, 7, or 8. The initial work force size y_0 is 5, as given by the problem.

Stage 5

$$b_5 = 6$$

y_4	$C_1(y_5 - 6) + C_2(y_5 - y_4)$ $y_5 = 6$	Optimum Solution $f_5(y_4)$	y_5^*
4	$3(0) + 4 + 2(2) = 8$	8	6
5	$3(0) + 4 + 2(1) = 6$	6	6
6	$3(0) + 0 \qquad = 0$	0	6

Stage 4

$$b_4 = 4$$

	$C_1(y_4-4)+C_2(y_4-y_3)+f_5(y_4)$			Optimum Solution	
y_3	$y_4=4$	5	6	$f_4(y_3)$	y_4^*
8	$0+0+8=8$	$3(1)+0+6=9$	$3(2)+0+0=6$	6	6

Stage 3

$$b_3 = 8$$

	$C_1(y_3-8)+C_2(y_3-y_2)+f_4(y_3)$	Optimum Solution	
y_2	$y_3=8$	$f_3(y_2)$	y_3^*
7	$0+4+2(1)+6=12$	12	8
8	$0+0+6 \quad = 6$	6	8

Stage 2

$$b_2 = 7$$

	$C_1(y_2-7)+C_2(y_2-y_1)+f_3(y_2)$		Optimum Solution	
y_1	$y_2=7$	$y_2=8$	$f_2(y_1)$	y_2^*
5	$0+4+2(2)+12=20$	$3(1)+4+2(3)+6=19$	19	8
6	$0+4+2(1)+12=18$	$3(1)+4+2(2)+6=17$	17	8
7	$0+0+12 \quad =12$	$3(1)+4+2(1)+6=15$	12	7
8	$0+0+12 \quad =12$	$3(1)+0+6 \quad =9$	9	8

Stage 1

$$b_1 = 5$$

	$C_1(y_1-5)+C_2(y_1-y_0)+f_2(y_1)$				Optimum Solution	
y_0	$y_1=5$	6	7	8	$f_1(y_0)$	y_1^*
5	$0+0+19=19$	$3(1)+4$ $+2(1)$ $+17=26$	$3(2)+4$ $+2(2)$ $+12=26$	$3(3)+4$ $+2(3)$ $+9=28$	19	5

The optimal solution is obtained as follows:

$$y_0 = 5 \rightarrow y_1^* = 5 \rightarrow y_2^* = 8 \rightarrow y_3^* = 8 \rightarrow y_4^* = 6 \rightarrow y_5^* = 6$$

This solution can be translated to the following plan.

Week j	Minimum Requirement b_j	y_j	Decision
1	5	5	No hiring or firing
2	7	8	Hire 3 workers
3	8	8	No hiring or firing
4	4	6	Fire 2 workers
5	6	6	No hiring or firing

Exercise 9.3–5
Consider the preceding work force size example.
(a) Suppose that the minimum requirements b_j are 6, 5, 3, 6, and 8 for $j = 1, 2, 3, 4$, and 5. Determine all possible values of y_j.
   [*Ans.* $y_5 = 8$, $y_4 = 6$, 7, or 8, $y_3 = 3$, 4, 5, 6, 7, or 8, $y_2 = 5$, 6, 7, or 8, and $y_1 = 6$, 7, or 8.]
(b) If $y_0 = 3$ instead of 5, find the new optimum solution.
   [*Ans.* Same optimum values of y_i except that $f_1(y_0) = 27$.]

Example 9.3–5 (Forward and Backward Recursive Equations).
A farmer owns k sheep. Once every year he decides how many to sell and how many to keep. If he sells, his profit per sheep is p_i in year i. If he keeps, the number of sheep kept in year i will be doubled in year $(i + 1)$. He will sell out completely at the end of n years.

This highly simplified example is designed to illustrate the potential advantages of using backward recursive equations in comparison with the forward method. In general, the forward and backward methods will lead to different computational efficiencies when the stages of the model must be ordered in a specific sequential order. This happens to be the case in this example (also Example 9.3–4), where stage j represents year j. Thus the stages must be considered in the chronological order of the years they represent (compare with Examples 9.3–1 through 9.3–3, where the assignment of stages can be arbitrary).

We first develop the forward and backward recursive equations, and then make a computational comparison between the two methods. The prime difference between the two formulations stems from the definition to state. To facilitate understanding of this point, the problem is summarized graphically in Figure 9–4. For year j, let x_j and y_j represent the number of sheep kept and the number sold, respectively. Define $z_j = x_j + y_j$. Then from the conditions of the problem,

$$z_1 = 2x_0 = 2k$$
$$z_j = 2x_{j-1}, \quad j = 1, 2, \ldots, n$$

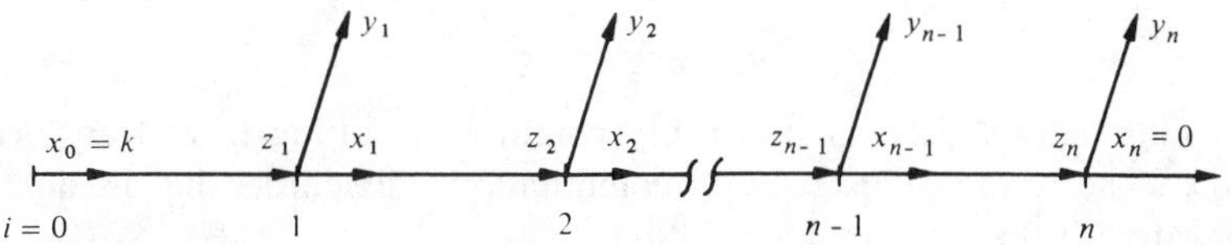

Figure 9–4

The state of the model at stage j may be described by z_j, the number of sheep available at the end of stage j for allocation to stages $j+1, j+2, \ldots, n$; or by x_j, the number of sheep available at the beginning of stage $j+1$ after the decisions at stages $1, 2, \ldots, j$ have been made. The first definition will result in the use of the backward recursive equations, and the second will lead to the use of the forward formulation.

Backward Formulation

Let $f_j(z_j)$ be the optimum profit for stages $j, j+1, \ldots$, and given z_j. The recursive equations are thus given as

$$f_n(z_n) = \max_{y_n = z_n \le 2^n k} \{p_n y_n\}$$

$$f_j(z_j) = \max_{y_j \le z_j \le 2^j k} \{p_j y_j + f_{j+1}(2[z_j - y_j])\}, \quad j = 1, 2, \ldots, n-1$$

Notice that y_j and z_j are nonnegative integers. Also, y_j, the amount sold at the end of period j, must be less than or equal to z_j. The upper limit of z_j is $2^j k$ (where k is the initial size of the flock), which will occur if no sales take place.

Forward Formulation

Let $g_j(x_j)$ be the optimum profit accumulated from stages $1, 2, \ldots, j$ given x_j (where x_j is the size of the flock at the beginning of stage $j+1$). The recursive is thus given as

$$g_1(x_1) = \max_{y_1 = 2k - x_1} \{p_1 y_1\}$$

$$g_j(x_j) = \max_{y_j \le 2^j k - x_j} \left\{p_j y_j + g_{j-1}\left(\frac{y_j + x_j}{2}\right)\right\}, \quad j = 2, 3, \ldots, n$$

$$\left(\frac{x_j + y_j}{2}\right) \text{ integer}$$

A comparison of the two formulations shows that during the course of computation, expressing x_{j-1} in terms of x_j is more difficult than is expressing z_{j+1} in terms of z_j. Namely, $x_{j-1} = (x_j + y_j)/2$ requires that the right side be integer, whereas $z_{j+1} = 2(z_j - y_j)$ does not have such a restriction. Thus, in the case of the forward formulation, the values of y_j and x_j satisfying

$$y_j \le 2^j k - x_j$$

must additionally satisfy an integrality condition resulting from transforming x_{j-1} to x_j. The example illustrates the computational difficulties that are normally associated with the forward formulation. ◀

9.4 PROBLEM OF DIMENSIONALITY IN DYNAMIC PROGRAMMING

In all the dynamic programming problems presented thus far, the states of the system have been described by one variable only. In general, these states may consist of n ($\geq$ 1) variables in which case the dynamic programming model is said to have a multidimensional state vector.

An increase in the state variables signifies an increase in the number of evaluations for the different alternatives at each stage. This is especially true in the case of tabular computations. Since most dynamic programming computations are done on the digital computers, such an increase in the state variables may tax the computer memory and increase the computation time. This problem is known as the **problem of dimensionality** (or the **curse of dimensionality,** as it is called by R. Bellman), and it presents a serious obstacle in solving medium- and large-sized dynamic programming problems.

To illustrate the concept of multidimensional states, consider the following example.

Example 9.4-1. In a house-to-house advertising campaign, D dollars and M man-hours are available for conducting the canvass in N districts. The net return from the jth district is estimated by $R_j(d_j, m_j)$, where d_j is the amount of dollars spent and m_j is the amount of man-hours devoted to the district. The objective is to determine d_j and m_j for each district j in order to maximize the total returns without exceeding available dollars and man-hours.

In the backward recursive equation, the states of the system at any stage j should be described by the amount of capital and man-hours that are allocated to stages j, $j + 1$, . . . , N. This means that the states should be represented by a two-dimensional vector (D_j, M_j), where D_j and M_j represent the capital and man-hours available at stage j for stages j, $j + 1$, . . . , N. Let $f_j(D_j, M_j)$ be the optimal return for stages j through N inclusive given D_j and M_j. The recursive equation is thus given by

$$f_N(D_N, M_N) = \max_{\substack{0 \leq d_N \leq D_N \\ 0 \leq m_N \leq M_N}} \{R_N(d_N, m_N)\}$$

$$f_j(D_j, M_j) = \max_{\substack{0 \leq d_j \leq D_j \\ 0 \leq m_j \leq M_j}} \{R_j(d_j, m_j) + f_{j+1}(D_j - d_j, M_j - m_j)\},$$

$$j = 1, 2, \ldots, N - 1$$

The computations of $f_j(D_j, M_j)$ and (d_j^*, m_j^*) become more difficult in this case, since we have to account for all the feasible combinations of d_j and m_j. This gives rise to the problem of computational infeasibility, especially in tabular computations where the entries of the tables (three dimensional in the example) become too large for available computer storage. In addition, the computation time may become excessively long. For example, suppose that D_j and M_j assume discrete values only in the range (0, 9). An equivalent problem with one state variable will have 10(states) $\times$ 10(alternatives) = 100 entries per table, whereas for the two-dimensional state problem each table will have $100 \times 100 = 10,000$ entries. The storage requirements will thus be increased by approximately 100 times. Also, the computation time will increase roughly by an equivalent amount.

The computer storage requirements and computation time increase rather rapidly with the number of state variables at each stage. Some ramifications and approximation methods have been explored, however, which may partially compensate for the effect of the increase in the number of state variables (see White [1969]). The bulk of the computational difficulties still persists, however, and will probably continue to do so irrespective of the tremendous advancement in the capabilities of modern digital computers.   ◄

9.5 SOLUTION OF LINEAR PROGRAMS BY DYNAMIC PROGRAMMING

The general linear programming problem

$$\text{maximize } z = c_1 x_1 + c_2 x_2 + \cdots + c_n x_n$$

subject to

$$a_{11}x_1 + a_{12}x_2 + \cdots + a_{1n}x_n \le b_1$$
$$a_{21}x_1 + a_{22}x_2 + \cdots + a_{2n}x_n \le b_2$$
$$\vdots \qquad\qquad\qquad \vdots \qquad \vdots$$
$$a_{m1}x_1 + a_{m2}x_2 + \cdots + a_{mn}x_n \le b_m$$
$$x_1, x_2, \ldots, x_n \ge 0$$

can be formulated as a dynamic programming model. Each activity j ($j = 1, 2, \ldots, n$) may be regarded as a stage. The level of activity x_j (≥ 0) represents the alternatives at stage j. Since x_j is continuous, each stage possesses an infinite number of alternatives within the feasible space. For reasons to be stated shortly, it is assumed that all $a_{ij} \ge 0$.

The linear programming problem is an allocation problem. Thus, similar to the examples of Section 9.3, the states may be defined as the amounts of resources to be allocated to the current stage and the succeeding stages. (This will result in a backward recursive equation.) Since there are m resources, the states must be represented by an m-dimensional vector. (In the examples of Section 9.3, each problem has one constraint and hence one state variable.)

Let $(B_{1j}, B_{2j}, \ldots, B_{mj})$ be the states of the system at stage j, that is, the amounts of resources $1, 2, \ldots, m$, allocated to stage $j, j + 1, \ldots, n$. Using the backward recursive equation, we let $f_j(B_{1j}, B_{2j}, \ldots, B_{mj})$ be the optimum value of the objective function for stages (activities) $j, j + 1, \ldots, n$ given the states $B_{1j}, \ldots, B_{mj}$. Thus

$$f_n(B_{1n}, B_{2n}, \ldots, B_{mn}) = \max_{\substack{0 \le a_{in} x_n \le B_{in} \\ i = 1, 2, \ldots, m}} \{c_n x_n\}$$

$$f_j(B_{1j}, B_{2j}, \ldots, B_{mj}) = \max_{\substack{0 \le a_{ij} x_j \le B_{ij} \\ i = 1, \ldots, m}} \{c_j x_j + f_{j+1}(B_{1j} - a_{1j} x_j, \ldots,$$

$$B_{mj} - a_{mj} x_j)\}, \qquad j = 1, 2, \ldots, n-1$$

where $0 \le B_{ij} \le b_i$ for all i and j.

Example 9.5-1. Consider the following linear programming problem:

$$\text{maximize } z = 2x_1 + 5x_2$$

subject to

$$2x_1 + x_2 \le 430$$
$$2x_2 \le 460$$
$$x_1, x_2 \ge 0$$

Because there are two resources, the states of the equivalent dynamic programming model are described by two variables only. Let (v_j, w_j) describe the states at stage j ($j = 1, 2$). Thus

$$f_2(v_2, w_2) = \max_{\substack{0 \le x_2 \le v_2 \\ 0 \le 2x_2 \le w_2}} \{5x_2\}$$

Since $x_2 \le \min \{v_2, w_2/2\}$ and $f_2(x_2 | v_2, w_2) = 5x_2$, then

$$f_2(v_2, w_2) = \max_{x_2} f_2(x_2 | v_2, w_2) = 5 \min \left(v_2, \frac{w_2}{2} \right)$$

and $x_2^* = \min (v_2, w_2/2)$.

Now

$$f_1(v_1, w_1) = \max_{0 \le 2x_1 \le v_1} \{2x_1 + f_2(v_1 - 2x_1, w_1)\}$$

$$= \max_{0 \le 2x_1 \le v_1} \left\{ 2x_1 + 5 \min \left(v_1 - 2x_1, \frac{w_1}{2} \right) \right\}$$

Since this is the last stage, $v_1 = 430$, $w_1 = 460$. Thus $x_1 \le v_1/2 = 215$ and

$$f_1(x_1 | v_1, w_1) = f_1(x_1 | 430, 460)$$

$$= 2x_1 + 5 \min \left(430 - 2x_1, \frac{460}{2} \right)$$

$$= 2x_1 + \begin{cases} 5(230), & 0 \le x_1 \le 100 \\ 5(430 - 2x_1), & 100 \le x_1 \le 215 \end{cases}$$

$$= \begin{cases} 2x_1 + 1150, & 0 \le x_1 \le 100 \\ -8x_1 + 2150, & 100 \le x_1 \le 215 \end{cases}$$

Hence, for the given ranges of x_1,

$$f_1(v_1, w_1) = f_1(430, 460) = \max_{x_1} (2x_1 + 1150, -8x_1 + 2150)$$

$$= 2(100) + 1150 = -8(100) + 2150 = 1350$$

which is achieved at $x_1^* = 100$.

To obtain x_2^*, notice that

$$v_2 = v_1 - 2x_1 = 430 - 200 = 230$$
$$w_2 = w_1 - 0 = 460$$

Hence

$$x_2^* = \min\left(v_2, \frac{w_2}{2}\right) = \min(230, 460/2) = 230$$

Thus the optimal solution is $z = 1350$, $x_1 = 100$, $x_2 = 230$. ◄

In Example 9.5–1 all the constraint coefficients are nonnegative. If some of the coefficients are negative, then for a constraint of the type ($\leq$) it is no longer true that the right-hand side will give the largest value of the state variable. This problem will be more pronounced if the solution happens to be unbounded. The general conclusion then is that dynamic programming is not adequate for solving the general linear programming problem. Perhaps this emphasizes the point that dynamic programming is based on so powerful an optimization principle that it is computationally infeasible for some problems. A case in point is the absence of a general computer code for (even a subclass of) dynamic programming problems.

9.6 SUMMARY

The chapter shows that DP is a procedure designed primarily to enhance the computational efficiency of solving certain mathematical programs by decomposing them into smaller, hence more manageable, problems. We must stress, however, that the principle of optimality provides a well-defined *framework* for solving the problem in stages; but it is "vague" about how each stage should be optimized. In this respect, the principle of optimality is sometimes regarded as being too powerful to be useful in practice; for although a problem can be decomposed properly, a numerical answer still may not be attainable because of the complexity of the optimization process at each stage. We must point out, however, that in spite of this disadvantage, the solution of many problems has been facilitated greatly through the use of DP.

There are several important topics that are not covered in this chapter. Notable among them are methods for reducing dimensionality (or number of state variables), infinite-stage systems, and probabilistic dynamic programming. The last topic is covered in Chapters 13 and 14 as applications of inventory and Markovian decision models. The remaining topics are available in specialized books.

SELECTED REFERENCES

BEIGHTLER, C., D. PHILLIPS, and D. WILDE, *Foundations of Optimization,* 2nd ed., Prentice-Hall, Englewood Cliffs, N.J., 1979.

BELLMAN, R., and S. DREYFUS, *Applied Dynamic Programming,* Princeton University Press, Princeton, N.J., 1962.

DREYFUS, S., and A. LAW, *The Art and Theory of Dynamic Programming,* Academic Press, New York, 1977.

HADLEY, G., *Nonlinear and Dynamic Programming,* Addison-Wesley, Reading, Mass., 1964.

WHITE, D. J., *Dynamic Programming,* Holden-Day, San Francisco, 1969.

REVIEW QUESTIONS

True (T) or False (F)?

1. ____ In DP models, the number of stages equals the number of subproblems.

2. ____ The decision alternatives within a stage must be mutually exclusive in the sense that the optimum solution for each stage can include exactly one alternative.

3. ____ The definition of the state in DP models guarantees that feasible decisions can be made independently at each stage.

4. ____ In tabular DP computations, the amount of computations at each stage depends on the feasible range for state values.

5. ____ In DP it is usually more difficult to define the stages rather than the states.

6. ____ In the network representation of DP models, the nodes represent the state values at each stage and the arcs represent the feasible alternatives.

7. ____ Recursive computations require using the information from each of the previously considered stages in the current stage's calculations.

8. ____ The principle of optimality guarantees that future decisions are made independently of previously made decisions.

9. ____ The forward and backward recursive formulation can result in different optimum solutions to the same problem.

10. ____ Dynamic programming problems can be decomposed either additively or multiplicatively.

11. ____ The use multiplicative or additive decomposition in a single-constraint DP problem is decided by whether the different terms comprising the constraint are multiplied or added to one another.

12. ____ In DP models the assignment of subproblems to successive stages is arbitrary, unless the subproblems are specified by a fixed chronological order.

13. _____ The problem of dimensionality in DP computations arises because of the increase in number of stages.

14. _____ In any DP model, a reduction in the number of constraints that bind all the stages together can lead to computational savings.

15. _____ Dynamic programming provides specific procedures for optimizing the subproblems of each stage.

16. _____ If a linear program has five variables and three constraints, its equivalent DP model will have three stages and the state at each stage will be defined by a five-element vector.

17. _____ It is possible to represent and solve a single-constraint linear program by a network model.

[*Ans.* **1** to **4**—T, **5**—F, **6**—T, **7**—F, **8**—T, **9**—F, **10**—T, **11**—T, **12**—T, **13**—F, **14**—T, **15**—F, **16**—F, **17**—T]

PROBLEMS

Section	Assigned Problems
9.1 and 9.2	9–1 to 9–5
9.3	9–6 to 9–22
9.4	9–23 to 9–25
9.5	9–26, 9–27

☐ **9–1** Consider the capital budgeting problem of Section 9.1. Suppose that plant 3 and plant 1 are renamed as plant 1 and plant 3, respectively, so that the new plant 1 now has the data of old plant 3, and vice versa. Resolve the problem and show that the new designations have no effect on the optimal solution.

☐ **9–2** Consider the capital budgeting example in Section 9.1. Develop the forward DP model associated with cases (a) and (b) and find the optimum solution. Assume that the total available capital is $8 million.

(a)

Proposal	Plant 1		Plant 2		Plant 3	
	c_1	R_1	c_2	R_2	c_3	R_3
1	3	5	3	4	0	0
2	4	6	4	5	2	3
3	—	—	5	8	3	5
4	—	—	—	—	6	9

(b)

Proposal	Plant 1		Plant 2		Plant 3		Plant 4	
	c_1	R_1	c_2	R_2	c_3	R_3	c_4	R_4
1	0	0	1	1.5	0	0	0	0
2	3	5	3	5	1	2.1	2	2.8
3	4	7	4	6	—	—	3	3.6

☐ **9–3** Formulate Problem 9–2 as a DP model using the backward recursive equation and obtain the solution. Compare the computations with those of Problem 9–2.

☐ **9–4** A student must select ten elective courses from four different departments. She must choose at least one course from each department. Her objective is to "allocate" the ten courses to the four departments so as to maximize her "knowledge" in the four fields. She realizes that if she takes over a certain number of courses in one department, her knowledge about the subject will not increase appreciably either because the material becomes too complicated for her comprehension or because the courses repeat themselves. She thus measures her learning ability as a function of the number of courses she takes in each department on a 100-point scale and produces the following chart. (It is assumed that the course groupings satisfy the prerequisites for each department.) Formulate the problem as a dynamic programming model using both the forward and backward recursive equations.

Department	Number of Courses									
	1	2	3	4	5	6	7	8	9	10
I	25	50	60	80	100	100	100	100	100	100
II	20	70	90	100	100	100	100	100	100	100
III	40	60	80	100	100	100	100	100	100	100
IV	10	20	30	40	50	60	70	80	90	100

☐ **9–5** Solve Problem 9–4 by using the forward and backward recursive equations of DP. Define the states and their feasible range of values for each stage.

☐ **9–6** Resolve Example 9.3–1 by using the following data.
  (a) $w_1 = 4$, $v_1 = 70$; $w_2 = 1$, $v_2 = 20$; $w_3 = 2$, $v_3 = 40$; $W = 6$.
  (b) $w_1 = 1$, $v_1 = 30$; $w_2 = 2$, $v_2 = 60$; $w_3 = 3$, $v_3 = 80$; $W = 4$.

☐ **9–7** In Example 9.3–2, suppose that there are four main components with the following data.

m_i	$i = 1$		$i = 2$		$i = 3$		$i = 4$	
	R	c	R	c	R	c	R	c
1	.8	3	.9	3	.6	2	.7	4
2	.82	5	—	—	.8	4	.75	5
3	—	—	—	—	—	—	.85	7

Let $C = 15$. Define the feasible range for the values of the state variables at each of the four stages. Then find the optimal solution.

☐ **9–8** Formulate and solve the forward DP model of the following resource allocation problem:

$$\text{maximize } z = 2x_1 + 3x_2 + 4x_3$$

subject to

$$2x_1 + 2x_2 + 3x_3 \leq 4$$
$$x_1, x_2, x_3 \geq 0 \text{ and integers}$$

☐ **9–9** Solve Problem 9–8 by the DP backward recursive equation.

☐ **9–10** Solve the work force size problem of Example 9.3–4 assuming that the minimum requirements b_j and the initial work force size y_0 are given as follows:
  (a) $b_1 = 6$, $b_2 = 5$, $b_3 = 3$, $b_4 = 6$, $b_5 = 8$, and $y_0 = 5$.
  (b) $b_1 = 8$, $b_2 = 4$, $b_3 = 7$, $b_4 = 8$, $b_5 = 2$, and $y_0 = 6$.

☐ **9–11** (**Shortest-Route Problem**). The network given in Figure 9–5 gives different routes for reaching city B from city A passing through a number of other cities. The lengths of the individual routes are shown on the arrows. It is required to determine the shortest route from A to B. Formulate the problem as a dynamic programming model. Explicitly define the stages, states, and return function; then find the optimal solution.

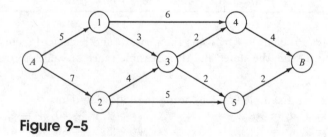

Figure 9-5

☐ **9–12** Formulate the following problem as a dynamic programming model:

$$\text{maximize } z = (x_1 + 2)^2 + x_2 x_3 + (x_4 - 5)^2$$

subject to

$$x_1 + x_2 + x_3 + x_4 \leq 5$$

x_i nonnegative integer

Find the optimum solution. What is the optimum solution if the right-hand side of the constraint is 3 instead of 5?

☐ **9-13** Solve the following problem by dynamic programming:

$$\text{minimize} \sum_{i=1}^{10} y_i^2$$

subject to

$$\prod_{i=1}^{10} y_i = 8, \quad y_i > 0$$

☐ **9-14** An equipment rental business is considering the investment of an initial capital C in buying two types of equipment. If x is the amount of money invested in type I, the corresponding profit at the end of the first year is $g_1(x)$, and the profit from type II is $g_2(C-x)$. The company's policy is to salvage equipment after one year. The salvage values for type I and type II at period t are $p_t x$ and $q_t(C-x)$, where $0 < p_t < 1$ and $0 < q_t < 1$. At the end of each year the company reinvests the returns from salvaging the equipment. This is repeated over the next N years, with the same return functions g_1 and g_2 holding for every year.

Formulate the problem as a dynamic programming model by the backward method, then solve using the following data for $N = 5$.

t:	1	2	3	4	5
p_t	.5	.9	.4	.5	.9
q_t	.6	.1	.5	.7	.5

Assume that $C = \$10,000$, $g_1(z) = .5z$, and $g_2(z) = .7z$.

☐ **9-15** Solve Problem 9-14 if in addition to the return from salvaging equipment 80% of the profit in period t is reinvested in period $t + 1$.

☐ **9-16** Solve Problem 9-14 for $N = 3$, $g_1(z) = .6(z - 1)^2$, $g_2(z) = .5z^2$, and $C = \$10,000$. Use the values of p_t and q_t for the first three periods in Problem 9-14.

☐ **9-17** Formulate the dynamic recursive equation for the problem:

$$\min_{\substack{y_i \\ i=1,2,\dots,n}} \{\max [f(y_1), f(y_2), \dots, f(y_n)]\}$$

subject to

$$\sum_{i=1}^{n} y_i = C, \qquad y_i \geq 0$$

Then solve, assuming that $n = 3$, $C = 10$, and

$$f(y_1) = y_1 + 5$$
$$f(y_2) = 5y_2 + 3$$
$$f(y_3) = y_3 - 2$$

☐ **9–18** Solve Example 9.3–5 by the forward and backward formulations. Assume that $x_0 = 2$, $N = 3$, and $p_1 = 16$, $p_2 = 4$, $p_3 = 4$.

☐ **9–19** A man invests his money in a savings account. At the end of each year he decides how much to spend and how much to reinvest. The interest rate is α ($\alpha > 1$) and the satisfaction he derives from spending an amount y_i in period i is measured by $g(y_i)$. Formulate the problem as a dynamic programming model by the forward and backward formulations. Assuming that the initial capital available is C and $g(y_i) = by_i$, where b is a constant, find the optimal solution to the problem.

☐ **9–20** Solve Problem 9–19 assuming $g(y_i) = b\sqrt{y_i}$.

☐ **9–21** Consider the following equipment replacement problem over N years. New equipment costs C dollars and its salvage value T years hence is $S(T) = N - T$ for $N \geq T$ and zero for $N < T$. The annual profit for T-year-old equipment is $P(T) = N^2 - T^2$, for $N \geq T$, and zero otherwise. Formulate the problem as a dynamic programming model, then solve, assuming that $N = 3$, $C = 10$, and the present equipment is 2 years old.

☐ **9–22** Solve Problem 9–21 assuming that $P(T) = N/(1 + T)$, $C = 6$, $N = 4$, and the equipment is 1 year old.

☐ **9–23** Consider the cargo-loading problem presented in Example 9.3–1. Suppose that in addition to the weight limitation W, there is also the volume limitation V. Formulate the problem as a dynamic programming model given v_i is the volume per unit of item i. The remaining information is the same as in Example 9.3–1.

☐ **9–24** Consider the transportation problem (Chapter 5) with m sources and n destinations. Let a_i be the amount available at source i, $i = 1, 2, \ldots, m$, and let b_j be the amount demanded at destination j, $j = 1, 2, \ldots, n$. If the cost of transporting x_{ij} units from source i to destination j is $h_{ij}(x_{ij})$, formulate the problem as a dynamic programming model.

☐ **9–25** Solve the following linear programming problem by dynamic programming. Assume all the variables to be nonnegative integers.

$$\text{Maximize } z = 8x_1 + 7x_2$$

subject to

$$2x_1 + x_2 \leq 8$$
$$5x_1 + 2x_2 \leq 15$$

x_1 and x_2 nonnegative integers

☐ **9–26** Solve the following linear programming problem using dynamic programming:

$$\text{maximize } z = 4x_1 + 14x_2$$

subject to

$$2x_1 + 7x_2 \leq 21$$
$$7x_1 + 2x_2 \leq 21$$
$$x_1, x_2 \geq 0$$

☐ **9–27** Solve the following nonlinear problem by dynamic programming:

$$\text{maximize } z = 7x_1^2 + 6x_1 + 5x_2^2$$

subject to

$$x_1 + 2x_2 \leq 10$$
$$x_1 - 3x_2 \leq 9$$
$$x_1, x_2 \geq 0$$

Because the second constraint involves a negative coefficient, some computational difficulty is expected if the backward dynamic programming formulation is used. Show that in this example such a difficulty can be eliminated by using the forward formulation (i.e., starting with x_1, whose constraint coefficients are positive).

SHORT CASE

A retail store handles a particular item that has been exhibiting fluctuating purchasing and sales prices. On December 1 of each year, the store manager develops a plan for the upcoming year for the acquisition and sales of the item. The item can be ordered during the month for use at the start of the following month. Although there is no specific limit on the size of the order that can be placed each month, the policy of the store stipulates that the maximum capital tied to the inventory of the item at any one time may never exceed $15,000. The store manager has compiled the following list for the purchasing and sales prices of the item over the next 12 months:

Month	Unit Purchasing Price	Unit Sales Price
1	$30	$33
2	31	33
3	33	37
4	32	35
5	32	34
6	32	31
7	31	31
8	30	31
9	31	32
10	31	34
11	30	35
12	30	34

The stock level as of December 31 is 200 units.

How should the item be handled over the next year?

PART II

PROBABILISTIC MODELS

THIS PART INCLUDES Chapters 10 through 17. Probability theory is reviewed in Chapter 10. Chapter 11 presents decision theory under risk and uncertainty together with game theory. Project scheduling by PERT–CPM is covered in Chapter 12. Chapter 13 develops deterministic and probabilistic inventory models. Chapter 14 covers the Markovian decision process. Queueing theory is presented in Chapter 15. Application of queueing theory is given in Chapter 16. Chapter 17 covers simulation.

The material on decision theory and games (Chapter 11), PERT–CPM (Chapter 12), deterministic inventory models (Sections 13.1 through 13.3), and simulation (Chapter 17) requires minimal knowledge of probability and calculus. The remainder of the material assumes knowledge of probability, stochastic processes, and calculus.

The queueing theory chapter is developed so that it can be presented either at an elementary level or from a higher mathematical standpoint. Section 15.6 assumes knowledge of Markov chain theory, which is given in Section 10.10.

The chapters in this part are developed to support a course in probabilistic models. Their independence should allow flexibility in choosing the content and level of presentation of the course. Material from this part may also be selected to supplement a survey course in operations research.

Chapter 10

Review of Probability Theory

Probability theory is the basis for the development of all the probabilistic decision models presented in the succeeding chapters. The material in this chapter is designed to provide you with a good understanding of the fundamentals of this topic.

10.1   OUTCOMES, SAMPLE SPACES, AND EVENTS

A formal definition of probability requires introducing the concepts of outcomes, sample space, and events. From the viewpoint of probability theory, an **experiment** represents an "activity" whose output is subject to chance (unknown) variation. Such output is usually referred to as the **outcome** of the experiment. The number of outcomes may be finite or infinite, depending on the nature of the experiment. For example, in the experiment of rolling a die, the outcomes are finite and are represented by the six faces of the die, that is, 1, 2, 3, 4, 5, and 6. On the other hand, in the experiment of measuring the time between successive failures of an electronic component, the outcomes are given by the time to failure and may assume any nonnegative real value.

A **sample space** defines the set of *all* possible outcomes of the experiment. For example, in the die-rolling experiment, the sample space is $\{1, 2, 3, 4, 5, 6\}$. Similarly, if t represents the interfailure time of the electronic components, its sample space is $\{0 \leq t < \infty\}$. Again, a sample space may be finite or infinite.

An **event** is a collection of outcomes from the sample space. For example, one may consider the event that a rolled die will turn up a "six" in which case the occurrence of the event is associated with the outcome 6, or the event that the sum of the faces turned up in two consecutive rolls is equal to five, in which case the event is realized when the consecutive outcomes are either (4, 1), (3, 2), (2, 3), or (1, 4). In the case of the electronic component failure, one may consider the event that the interfailure time is less than or equal to some real constant T.

10.2   LAWS OF PROBABILITY

The probability of an event E (usually written as $P\{E\}$) is a *non*negative real number not exceeding one that equals the long-run fraction of trials for which the outcomes successfully describe E. Mathematically, if n is the total number of trials out of which there are m trials describing E, then

$$P\{E\} = \lim_{n \to \infty} \frac{m}{n} \quad \text{and} \quad 0 \leq P\{E\} \leq 1$$

If $P\{E\} = 0$, the event is impossible, while if $P\{E\} = 1$, it is certain. For example, the probability of a rolled die turning up a "seven" is zero (impossible), whereas the probability of a tossed coin turning up a head *or* a tail is one (certain).

The basic probability laws will now be introduced. Consider the two events E and F. The notation $E \cap F$ (or simply EF) means that both E *and* F (or *intersection* of E and F) are realized, and the notation $E \cup F$ (or simply $E + F$) signifies that E or F (or *union* of E and F) occurs. With these notations, the basic probability laws are given by

$$P\{E + F\} = P\{E\} + P\{F\} - P\{EF\} \qquad (1)$$

$$P\{E|F\} = \frac{P\{EF\}}{P\{F\}}, \qquad P\{F\} > 0 \qquad (2)$$

The first law is called the **addition law** and it states that the probability of E *or* F (or both) equals the probability of E plus the probability of F minus the probability of E *and* F. If E and F are **mutually exclusive,** that is, the occurrence of one event signifies the nonoccurrence of the other, then

$$P\{EF\} = 0$$

In this case

$$P\{E + F\} = P\{E\} + P\{F\}$$

In general, if E_1, E_2, . . . , E_n are mutually exclusive,

$$P\{E_1 + E_2 + \cdots + E_n\} = P\{E_1\} + P\{E_2\} + \cdots + P\{E_n\}$$

The second law is called the **conditional probability law.** It computes the probability of event E given F. If event E is contained in event F and if $P\{F\} > 0$,

$$P\{E|F\} = \frac{P\{E\}}{P\{F\}}$$

Two events E and F are said to be **independent** if and only if

$$P\{E|F\} = P\{E\}$$

Hence, from the conditional probability law, E and F are independent if and only if

$$P\{EF\} = P\{E\}P\{F\}$$

In general, events E_1, E_2, . . . , E_n are independent if and only if for all combinations $1 \leq i < j < k < . . . < n$, the following conditions are satisfied simultaneously:

$$P\{E_i E_j\} = P\{E_i\}P\{E_j\}$$
$$P\{E_i E_j E_k\} = P\{E_i\}P\{E_j\}P\{E_k\}$$
$$\vdots$$
$$P\{E_1 E_2 . . . E_n\} = P\{E_1\}P\{E_2\} . . . P\{E_n\}$$

Two mutually exclusive events E and F are independent if and only if

$$P\{EF\} = P\{E\}P\{F\} = 0$$

which can only occur if

$$P\{E\} = 0 \qquad \text{and/or} \qquad P\{F\} = 0$$

10.3 RANDOM VARIABLES AND PROBABILITY DISTRIBUTIONS

The outcomes of an experiment are represented by a random variable if these outcomes are numerical or if real numbers can be assigned to them. For example, in the die-rolling experiment, the corresponding random variable is represented by the set of outcomes {1, 2, 3, 4, 5, 6}; while in the coin-tossing experiment the outcomes, head (H) or tail (T), can be represented as a random variable by assigning 0 to H and 1 to T. In a sense, then, a random variable is a real-valued function that maps the sample space onto the real line.

A random variable may be discrete or continuous. A **discrete** variable takes on specific values at discrete points on the real line, and a **continuous** variable assumes any value over a continuous range of the real line. For example, in the coin-tossing or the die-rolling experiments, the random variables are discrete. In the component-failure experiments, the random variable is continuous.

Associated with the random variable x is a function $f(x)$ that can be used to assign a probability measure to this random variable. This function is called **probability density function** (PDF). If x is a continuous random variable on the range $(-\infty, +\infty)$, its PDF $f(x)$ must satisfy

$$f(x) \geq 0, \qquad -\infty < x < \infty \tag{1}$$

$$\int_{-\infty}^{+\infty} f(x)\, dx = 1 \tag{2}$$

Similarly, if x is a discrete random variable, its PDF $P(x)$, which defines the probability that x assumes a given value, must satisfy

$$P(x) \geq 0 \qquad \text{for all } x \tag{1}$$

$$\sum_{\text{all} x} P(x) = 1 \tag{2}$$

The first condition in both the continuous and discrete distributions indicates that a PDF cannot assume negative values (otherwise, the probability of some events may be negative). The second condition shows that the probability of the entire space must equal 1.0.

Example 10.3-1. To illustrate the properties of the PDF, consider

$$f(x) = \begin{cases} a, & 0 \leq x \leq 10 \\ 0, & \text{otherwise} \end{cases}$$

For $f(x)$ to be a PDF, the condition

$$\int_{-\infty}^{\infty} f(x)\, dx = 1$$

must be satisfied, which is true in this example if

$$\int_{0}^{10} a\, dx = 1 \qquad \text{or} \qquad a = 1/10$$

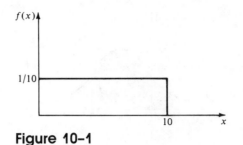

Figure 10-1

Since $a > 0$, it follows that $f(x) \geq 0$. Thus $f(x)$ satisfies the conditions for a PDF. Figure 10–1 illustrates the function graphically.

Similarly, for the die-rolling experiment, the associated probabilities of the discrete random variables are given by

x	1	2	3	4	5	6
$P(x)$	1/6	1/6	1/6	1/6	1/6	1/6

which satisfy the conditions of a PDF. This function is depicted in Figure 10–2. ◀

Another useful probability measure is the **cumulative density function** (CDF). Consider the continuous case first. Let $F(x)$ represent the CDF for the continuous random variable x, $-\infty < x < \infty$. For any real number a, $F(a)$ defines the probability that $x \leq a$, in terms of the PDF $f(x)$ as

$$F(a) = P\{x \leq a\} = \int_{-\infty}^{a} f(x)\,dx$$

Thus $F(a)$ represents the area under $f(x)$ enclosed by the range $-\infty < x < a$.

Notice that for the continuous random variable x,

$$P\{x = a\} = 0$$

since the "enclosed area" is zero. On the other hand, given two real numbers a and b such that $-\infty < a < b < \infty$, the probability of the event $a \leq x \leq b$ (or $a < x < b$) is given by

$$P\{a \leq x \leq b\} = \int_{a}^{b} f(x)\,dx = \int_{-\infty}^{b} f(x)\,dx - \int_{-\infty}^{a} f(x)\,dx = F(b) - F(a)$$

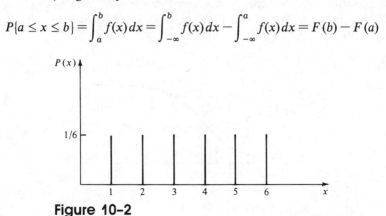

Figure 10-2

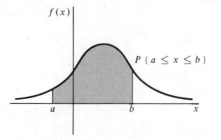

Figure 10-3

This case is illustrated graphically in Figure 10–3. The given probability equals the area under $f(x)$ enclosed in the range $a \leq x \leq b$.

The CDF $F(a)$ has the following properties:

$$\lim_{a \to \infty} F(a) = \lim_{a \to \infty} \int_{-\infty}^{a} f(x)\,dx = 1 \tag{1}$$

$$\lim_{a \to -\infty} F(a) = \lim_{a \to -\infty} \int_{-\infty}^{a} f(x)\,dx = 0 \tag{2}$$

$$F(a) \text{ is monotone nondecreasing in } a \tag{3}$$

A typical CDF is illustrated in Figure 10–4. The vertical scale of $F(x)$ gives directly the probability that x is less than a certain value.

From the relationship between $f(x)$ and $F(x)$, it follows that

$$f(x) = \frac{d}{dx} F(x)$$

Thus the probability law of the random variable x is in general defined completely by either $f(x)$ or $F(x)$.

Similar definitions can be made for the CDF in the discrete case simply by substituting $P(x)$ for $f(x)$ in all the properties shown. This will naturally require changing integration to summation. Differentiation is also replaced by differences. Notice, however, that in the discrete case the probability of a single value is not necessarily equal to zero. Notice also that for a discrete random variable the CDF is in the form of a step function, since the PDF is defined at discrete values only.

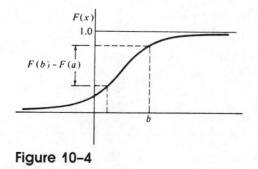

Figure 10-4

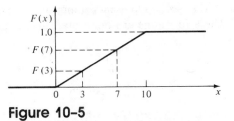

Figure 10-5

Example 10.3-2. Consider again the problems of Example 10.3-1. For the continuous PDF,

$$f(x) = \begin{cases} \dfrac{1}{10}, & 0 \le x \le 10 \\[2mm] 0, & \text{otherwise} \end{cases}$$

the CDF for the range $0 \le x \le 10$ is defined as

$$F(x) = \int_0^x f(u)\,du = \int_0^x \frac{1}{10}\,du = \frac{x}{10}, \qquad 0 \le x \le 10$$

The complete CDF is illustrated graphically in Figure 10–5.

Suppose that we want to compute $P\{3 \le x \le 7\}$. Then

$$P(3 \le x \le 7) = F(7) - F(3) = 7/10 - 3/10 = 4/10$$

as could be seen directly from Figure 10–5.

Again consider the discrete random variable x with the following PDF.

x	1	2	3	4	5	6
$P(x)$	1/6	1/6	1/6	1/6	1/6	1/6

Then

$$F(x) = \sum_{u=1}^{u=x} \frac{1}{6} = \frac{x}{6}, \qquad x = 1, 2, \ldots, 6$$

The complete CDF is shown in Figure 10–6.   ◄

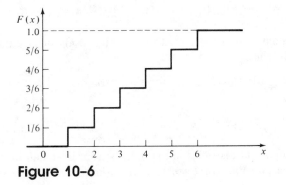

Figure 10-6

The remainder of this section presents important discrete and continuous density functions. These functions are often used in the analysis of inventory and waiting line models.

10.3.1 DISCRETE PROBABILITY DISTRIBUTIONS

A. Independent Bernoulli Trials Distributions

The simplest form of a PDF results from the so-called **independent Bernoulli trials.** A Bernoulli trial is an experiment with only two possible outcomes: "success" (S) and "failure" (F). Examples of such trials occur in the coin-tossing experiment, where the outcome is a head (S) or a tail (F), or in testing a product, where the outcome is defective (F) or nondefective (S).

Let the random variable describing the two outcomes F and S be 0 and 1. The corresponding PDF may be written as

$$P\{x = 0\} = p \quad \text{and} \quad P\{x = 1\} = q = 1 - p$$

where $0 \leq q \leq 1$.

Consider now the case of n independent Bernoulli trials each with the same p. The probability of a *particular* combination of outcomes with k failures and $(n - k)$ successes is

$$p^k q^{n-k}, \quad 0 \leq k \leq n$$

since all n trials are independent. For example, if $n = 5$, the probability that the first trial is F and the remaining four trials are S is equal to pq^4.

Consider next the probability that the *number of failures* in n independent Bernoulli trials equals k, where n is specified in advance. In this case *all* the distinct combinations having k failures (regardless of their order of occurrence in the n trials) should be considered in computing this probability. Under the conditions given, there are $\binom{n}{k} = n!/k!(n - k)!$ distinct combinations. Since the probability of each combination is $p^k q^{n-k}$, the addition law of probability (Section 10.2) gives

$$P\{x = k\} = \binom{n}{k} p^k q^{n-k}, \quad k = 0, 1, 2, \ldots, n$$

This is the **binomial distribution** with parameters n and p. It satisfies the conditions for PDF, since

$$P\{x = k\} \geq 0, \quad \text{for all } k = 0, 1, 2, \ldots, n \tag{1}$$

$$\sum_{k=0}^{n} P\{x = k\} = \sum_{k=0}^{n} \binom{n}{k} p^k q^{n-k} = (p + q)^n = 1 \tag{2}$$

Another important distribution that comes from Bernoulli independent trials is the **negative binomial (or Pascal) distribution.** In the binomial distribution the number of trials n is specified in advance and the random variable is given by the number of failures. In the negative binomial, the random variable will be given by the number of independent trials until a fixed number of failures occur. Let j and c represent the number of trials and the fixed number of failures, respectively. The probability of j trials until c failures occur is thus the product of the following two probabilities:

$$\text{probability of } (c-1) \text{ failures in } (j-1) \text{ trials} = \binom{j-1}{c-1} p^{c-1} q^{j-c}$$

$$\text{probability of a failure on the } j\text{th trial} = p$$

Thus the negative binomial PDF is given by

$$P\{x=j\} = \binom{j-1}{c-1} p^c q^{j-c}, \qquad j = c,\, c+1,\, c+2, \ldots$$

A special case of the negative binomial distribution is the **geometric distribution** that occurs when $c = 1$; that is

$$P\{x=j\} = pq^{j-1}, \qquad j = 1,\, 2,\, 3, \ldots$$

The negative binomial and geometric distributions can be regarded, in a sense, as describing the "time" until a certain number of failures occur. This result can be seen by assuming a fixed time increment with each trial.

B. Poisson Distribution

Consider the random variable x that includes nonnegative integer values only; that is, $k = 0, 1, 2, \ldots$. The PDF

$$P\{x=k\} = \frac{\lambda^k e^{-\lambda}}{k!}, \qquad k = 0, 1, 2, \ldots$$

where $\lambda > 0$, is called the **Poisson distribution.** A typical application of the Poisson distribution occurs in analyzing waiting line problems (Chapter 15).

Under certain conditions, the Poisson distribution approximates the binomial distribution. Suppose that in the binomial distribution (Section 10.3.1–A) $p \rightarrow 0$ and $n \rightarrow \infty$ such that $np \rightarrow \lambda > 0$. Then, under these conditions, the binomial PDF becomes

$$P\{x=k\} = \binom{n}{k} \left(\frac{\lambda}{n}\right)^k \left(1 - \frac{\lambda}{n}\right)^{n-k}$$

It can be shown that, as $n \rightarrow \infty$,

$$P\{x=k\} \rightarrow \frac{\lambda^k e^{-\lambda}}{k!}, \qquad k = 0, 1, 2, \ldots$$

which is the Poisson distribution. Thus, if n is large and p small such that $\lambda = np > 0$, the Poisson distribution approximates the binomial.

10.3.2 CONTINUOUS PROBABILITY DISTRIBUTIONS

A. Normal Distribution

The best known continuous PDF is the **normal distribution** with its density function given by

$$f(x) = \frac{1}{\sqrt{2\pi\sigma^2}}\, e^{-(x-\mu)^2/2\sigma^2}, \qquad -\infty < x < \infty$$

where μ and σ are given parameters. The CDF of the normal distribution is defined by

$$F(x) = \int_{-\infty}^{x} \frac{1}{\sqrt{2\pi\sigma^2}}\, e^{-(y-\mu)^2/2\sigma^2}\, dy$$

Typical PDF and CDF of the normal distribution are shown in Figure 10–7. Notice that $f(x)$ is symmetric around $x = \mu$.

The expression for the CDF $F(x)$ cannot be evaluated in a closed form suitable for computations. For this reason, normal tables that give the values of $F(x)$ as a function of x are available. These tables are based on the following **standard normal** PDF:

$$\phi(z) = \frac{1}{\sqrt{2\pi}}\, e^{-z^2/2}, \qquad -\infty < z < \infty$$

with its parameters given by $\mu = 0$ and $\sigma = 1$. The corresponding CDF is thus given by

$$\Phi(z) = \int_{-\infty}^{z} \frac{1}{\sqrt{2\pi}}\, e^{-y^2/2}\, dy$$

The standard form is obtained from the regular form by making the substitution

$$z = \frac{x - \mu}{\sigma}$$

The normal distribution can approximate the binomial distribution. Specifically, it can be proved for the binomial distribution that, given a fixed p, as $n \to \infty$,

$$\sum_{k=a}^{b} \binom{n}{k} p^k q^{n-k} \to \frac{1}{\sqrt{2\pi}} \int_{(a-\mu-1/2)/\sigma}^{(b-\mu+1/2)/\sigma} e^{-y^2/2}\, dy$$

where $\mu = np$ and $\sigma = \sqrt{npq}$.

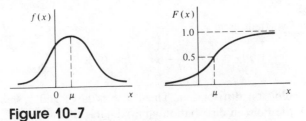

Figure 10–7

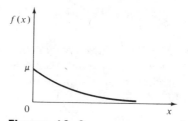

Figure 10-8

B. Exponential Distribution

An important distribution often encountered in waiting line theory is the **exponential distribution** with its PDF given by

$$f(x) = \mu e^{-\mu x}, \qquad x > 0$$

where μ (>0) is a given parameter. The distribution is illustrated in Figure 10-8.

The exponential distribution in the continuous case is analogous to the geometric distribution in the discrete case. For example, if the random variable in the geometric distribution represents the *number* of trials before a failure occurs, its analogous continuous case in the exponential distribution will represent the time to failure. In fact, it can be proved that as $p \to 0$ and the intertrial time $\to 0$, the geometric distribution will in the limit tend to the exponential distribution.

An important relationship exists between the Poisson distribution and the exponential distribution. If the Poisson random variable represents the *number* of failures per unit time, the exponential random variable will represent the *time* between two successive failures. In effect, then, the exponential distribution can be derived from the Poisson distribution (see, e.g., Section 15.2.1).

C. Gamma Distribution

Given n identically distributed and independent exponential random variables, the distribution of the *sum* of these random variables yields a **gamma** or **Erlang distribution** with its PDF given by

$$f(x) = \frac{\mu(\mu x)^{n-1}e^{-\mu x}}{(n-1)!}, \qquad x > 0$$

(See Example 10.9–1 for a proof of this result.) Notice that $n = 1$ reduces the PDF to the exponential density function.

Section 10.3.2B indicates that the exponential distribution is analogous to the geometric distribution. In a similar way, the gamma distribution (sum of exponential random variables) is analogous to the negative binomial (sum of geometric random variables).

10.4 RELATIONSHIPS AMONG PROBABILITY DISTRIBUTIONS

Figure 10–9 summarizes the relationships among common probability distributions. The chart is self-explanatory. It is interesting to see, however, that the independent Bernoulli trials form the basis for all the distributions in the chart.

Some of the relationships in Figure 10–9 have been discussed in previous

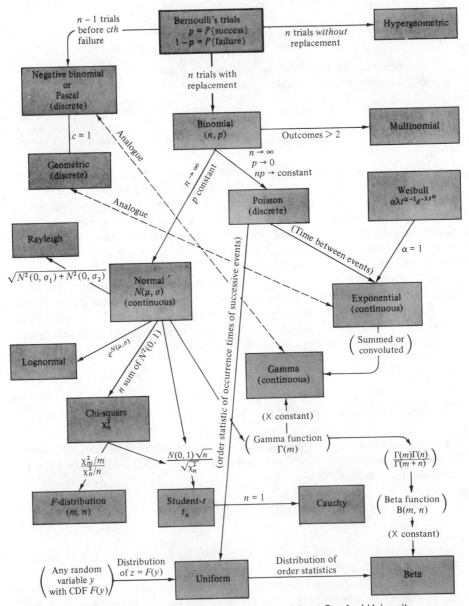

Source: This chart was developed by Roy E. Lave, Jr., while at Stanford University.

Figure 10–9

sections. The remaining relationships are stated without explanation. Also, some of the distributions in the chart are not discussed in this chapter. Any information regarding these distributions can be found in most specialized books in the areas of probability theory and mathematical statistics.

10.5   JOINT PROBABILITY DISTRIBUTIONS

In Section 10.3, distributions with only one random variable have been introduced. In situations where there are two or more distinct types of outcomes, the process may be represented by a corresponding number of random variables.

In general, the notations $f(x_1, x_2, \ldots, x_n)$ and $P(x_1, x_2, \ldots, x_n)$ are used to describe the *joint* PDF of n random variables in the continuous and discrete cases. These functions must satisfy the following conditions.

Continuous Case

$$f(x_1, \ldots, x_n) \geq 0, \qquad -\infty < x_i < \infty, \qquad i = 1, 2, \ldots, n$$

$$\int_{-\infty}^{+\infty} dx_1 \int_{-\infty}^{+\infty} dx_2 \cdots \int_{-\infty}^{+\infty} dx_n \, f(x_1, x_2, \ldots, x_n) = 1$$

Discrete Case

$$P(x_1, x_2, \ldots, x_n) \geq 0 \qquad \text{for all } x_i, \quad i = 1, 2, \ldots, n$$

$$\sum_{\text{all } x_1} \sum_{\text{all } x_2} \cdots \sum_{\text{all } x_n} P(x_1, x_2, \ldots, x_n) = 1$$

The CDF of a joint distribution can be defined as

$$F(x_1, x_2, \ldots, x_n) = \begin{cases} \displaystyle\int_{-\infty}^{x_1} du_1 \int_{-\infty}^{x_2} du_2 \cdots \int_{-\infty}^{x_n} du_n f(u_1, u_2, \ldots, u_n), \\ \qquad\qquad\qquad\qquad\qquad\qquad\qquad\qquad \text{continuous case} \\[2mm] \displaystyle\sum_{u_1 \leq x_1} \sum_{u_2 \leq x_2} \cdots \sum_{u_n \leq x_n} P(u_1, u_2, \ldots, u_n), \\ \qquad\qquad\qquad\qquad\qquad\qquad\qquad\qquad \text{discrete case} \end{cases}$$

This definition indicates that

$$F(-\infty, \ldots, -\infty) = 0 \qquad \text{and} \qquad F(\infty, \ldots, \infty) = 1$$

By a reciprocal operation, the joint PDF for the continuous case is obtained from the joint CDF by using

$$F(x_1, \ldots, x_n) = \frac{\partial^n}{\partial x_1 \cdots \partial x_n} F(x_1, \ldots, x_n)$$

In the discrete case, the difference operation replaces differentiation.

The probability that $a_i \le x_i \le b_i$, $i = 1, 2, \ldots, n$ (where a_i and b_i are given constants defined on the range of the ith random variable) can be determined from

$$P\{a_1 \le x_1 \le b_1 \ldots, a_n \le x_n \le b_n\} = \begin{cases} \int_{a_1}^{b_1} dx_1 \cdots \int_{a_n}^{b_n} dx_n f(x_1, \ldots, x_n), \\ \qquad\qquad\qquad\qquad\qquad \text{continuous case} \\ \sum_{x_1=a_1}^{b_1} \cdots \sum_{x_n=a_n}^{b_n} P(x_1, \ldots, x_n), \\ \qquad\qquad\qquad\qquad\qquad \text{discrete case} \end{cases}$$

Notice that this probability is *not*, in general, equal to $F(b_1, \ldots, b_n) - F(a_1, \ldots, a_n)$.

Example 10.5-1. Consider the joint PDF

$$f(x_1, x_2) = \begin{cases} 2(x_1 + x_2 - 2x_1x_2), & 0 \le x_1 \le 1, \quad 0 \le x_2 \le 1 \\ 0, & \text{otherwise} \end{cases}$$

This is a PDF, since

$$f(x_1, x_2) = 2(x_1 + x_2 - 2x_1x_2) = 2\{x_1(1 - x_2) + x_2(1 - x_1)\}$$

which shows that $f(x_1, x_2)$ is nonnegative for $0 \le x_1, x_2 \le 1$. Also,

$$\int_0^1 \int_0^1 f(x_1, x_2)\, dx_1\, dx_2 = \int_0^1 \int_0^1 2(x_1 + x_2 - 2x_1x_2)\, dx_1\, dx_2 = 1$$

The probability that $0 \le x_1 \le 1/2$ and $0 \le x_2 \le 1/4$ is given by

$$P\{0 \le x_1 \le 1/2, 0 \le x_2 \le 1/4\} = \int_0^{1/2} \int_0^{1/4} 2(x_1 + x_2 - 2x_1x_2)\, dx_1\, dx_2$$
$$= 5/64$$

The CDF is now given by

$$F(x_1, x_2) = \int_0^{x_1} \int_0^{x_2} 2(u_1 + u_2 - 2u_1u_2)\, du_1\, du_2 = x_1^2 x_2 + x_1 x_2^2 - x_1^2 x_2^2$$

Notice that $F(0, x_2) = F(x_1, 0) = 0$ and $F(1, 1) = 1$.   ◀

10.5.1 MARGINAL DISTRIBUTION

Given the continuous joint PDF $f(x_1, \ldots, x_n)$, the **marginal distribution** of any random variable x_i is defined by

$$f_i(x_i) = \int_{-\infty}^{+\infty} dx_i \cdots \int_{-\infty}^{+\infty} dx_{i-1} \int_{-\infty}^{+\infty} dx_{i+1} \cdots \int_{-\infty}^{+\infty} dx_n f(x_1, \ldots, x_n)$$

$$= \frac{\partial}{\partial x_i} F(\infty, \ldots, \infty, x_i, \infty, \ldots, \infty)$$

Similarly, in the discrete case,

$$P_i(x_i) = \sum_{\text{all } x_1} \cdots \sum_{\text{all } x_{i-1}} \sum_{\text{all } x_{i+1}} \cdots \sum_{\text{all } x_n} P(x_1, \ldots, x_n)$$

Example 10.5-2. Consider Example 10.5-1:

$$f_1(x_1) = \int_0^1 2(x_1 + x_2 - 2x_1x_2)\, dx_2 = 1, \qquad 0 \le x_1 \le 1$$

$$f_2(x_2) = \int_0^1 2(x_1 + x_2 - 2x_1x_2)\, dx_1 = 1, \qquad 0 \le x_2 \le 1$$

Both $f_1(x_1)$ and $f_2(x_2)$ satisfy the conditions of a PDF. ◀

10.5.2 CONDITIONAL DISTRIBUTION AND INDEPENDENCE

Consider the joint PDF $f(x_1, x_2)$. The conditional PDF of the random variable x_1 given the random variable x_2 is defined as

$$f(x_1 \mid x_2) = \frac{f(x_1, x_2)}{f_2(x_2)}, \qquad f_2(x_2) > 0$$

where

$$f_2(x_2) = \int_{-\infty}^{\infty} f(x_1, x_2)\, dx_1$$

is the marginal PDF of x_2.

In general, given $f(x_1, \ldots, x_n)$, the conditional PDF of $(x_1, x_2, \ldots, x_{n-1})$ given x_n is

$$f(x_1, x_2, \ldots, x_{n-1} \mid x_n) = \frac{f(x_1, x_2, \ldots, x_n)}{f_n(x_n)}, \qquad f_n(x_n) > 0$$

where

$$f_n(x_n) = \int_{-\infty}^{+\infty} dx_1 \int_{-\infty}^{+\infty} dx_2 \cdots \int_{-\infty}^{+\infty} dx_{n-1} f(x_1, \ldots, x_n)$$

A similar definition can be established for the discrete case.

The conditional PDF must satisfy the two conditions of a PDF:

$$f(x_1, \ldots, x_{n-1} \mid x_n) \ge 0, \qquad -\infty < x_i < \infty, \qquad i = 1, 2, \ldots, n$$

$$\int_{-\infty}^{+\infty} dx_1 \cdots \int_{-\infty}^{+\infty} dx_{n-1} f(x_1, \ldots, x_{n-1} \mid x_n) = 1$$

The conditional CDF can also be defined as

$$F(x_1, \ldots, x_{n-1} \mid x_n) = \int_{-\infty}^{x_1} du_1 \cdots \int_{-\infty}^{x_{n-1}} du_{n-1} f(u_1, \ldots, u_{n-1} \mid x_n)$$

The random variables $(x_1, \ldots, x_n)$ are *independent* if and only if the joint PDF is the *product* of the marginal distributions of the random variables. That is,

$$f(x_1, \ldots, x_n) = f_1(x_1)f_2(x_2) \ldots f_n(x_n) = \prod_{i=1}^{n} f_i(x_i)$$

where $f_i(x_i)$ is the marginal distribution of the random variable x_i.

Now, if x_1 and x_2 are two independent random variables,

$$f(x_1|x_2) = \frac{f(x_1, x_2)}{f_2(x_2)} = \frac{f_1(x_1)f_2(x_2)}{f_2(x_2)} = f_1(x_1)$$

This shows that in the case of independence, the conditional PDF is equal to the marginal PDF.

Example 10.5–3. Consider Example 10.5–1. Example 10.5–2 shows that

$$f_1(x_1) = \begin{cases} 1, & 0 \leq x_1 \leq 1 \\ 0, & \text{otherwise} \end{cases}$$

$$f_2(x_2) = \begin{cases} 1, & 0 \leq x_2 \leq 1 \\ 0, & \text{otherwise} \end{cases}$$

Since $f(x_1, x_2) \neq f_1(x_1) \cdot f_2(x_2)$, x_1 and x_2 are not independent.

Again, the conditional density function of x_1 given x_2 is obtained as

$$g(x_1|x_2 = t) = \frac{f(x_1, t)}{f_2(t)} = \frac{2(x_1 + t - 2tx_1)}{1}$$
$$= 2(x_1 + t - 2tx_1), \qquad 0 \leq x_1 \leq 1$$

Notice that $g(x_1|t) \geq 0$ and that

$$\int_0^1 g(x_1|t)dx_1 = 2 \int_0^1 (x_1 + t - 2tx_1)dx_1$$
$$= 2 \left[\frac{x_1^2}{2} + tx_1 - tx_1^2 \right]_0^1 = 1$$

regardless of the value of $t(0 \leq x_2 = t \leq 1)$. This shows that $g(x_1|t)$ is a PDF.   ◀

10.6 EXPECTATIONS AND MOMENTS OF A RANDOM VARIABLE

The previous sections show how distribution functions are used to represent discrete and continuous random variables. It is often desirable to summarize the characteristics of a probability distribution by meaningful measures from which general conclusions can be drawn about the random variable. These measures are usually specified by the **expected value** of certain functions of the random variable. This section gives the definition and derivation of these measures.

10.6.1 DEFINITION OF EXPECTATION

Let x be a random variable and define $h(x)$ as any function of x. Define $E\{h(x)\}$ as the *expected* value of $h(x)$ with respect to the probability distributions of x. Then

$$E\{h(x)\} = \begin{cases} \int_{-\infty}^{+\infty} h(x)f(x)dx, & x \text{ continuous} \\ \sum_{\text{all} x} h(x)P(x), & x \text{ discrete} \end{cases}$$

The foregoing definition can be extended to the case of a function of n random variables $h(x_1, \ldots, x_n)$ as follows.

$$E\{h(x_1, \ldots, x_n)\} = \begin{cases} \int_{-\infty}^{+\infty} dx_1 \cdots \int_{-\infty}^{+\infty} dx_n h(x_1, \ldots, x_n) f(x_1, \ldots, x_n), \\ \qquad\qquad\qquad\qquad\qquad\qquad\qquad x \text{ continuous} \\ \sum_{\text{all} x_1} \cdots \sum_{\text{all} x_n} h(x_1, \ldots, x_n) P(x_1, \ldots, x_n), \\ \qquad\qquad\qquad\qquad\qquad\qquad\qquad x \text{ discrete} \end{cases}$$

The basic definition of expectation shows that, for any constant b,

$$E\{b\} = b$$
$$E\{bh(x_1, \ldots, x_n)\} = bE\{h(x_1, \ldots, x_n)\}$$
$$E\{b \pm h(x_1, \ldots, x_n)\} = b \pm E\{h(x_1, \ldots, x_n)\}$$

You are encouraged to prove these results.

10.6.2 MEAN, VARIANCE, AND MOMENTS

In summarizing the data of a single random variable x, two measures are commonly used. These are the **mean** $E\{x\}$ and the **variance** var$\{x\}$. The mean is a measure of the central tendency of the distribution, and the variance is a measure of dispersion of the distribution around its mean.

The mean of distribution is defined by making $h(x) = x$ in the expectation formula of Section 10.6.1. Thus

$$E\{x\} = \begin{cases} \int_{-\infty}^{+\infty} xf(x)dx = \int_{-\infty}^{+\infty} xdF(x), & x \text{ continuous} \\ \sum_{\text{all} x} xP(x), & x \text{ discrete} \end{cases}$$

The definition of the mean can be extended to the important case where the random variable x is defined as

$$x = b_1 x_1 + b_2 x_2 + \cdots + b_n x_n$$

where b_i is a constant and x_i is a random variable, $i = 1, 2, \ldots, n$. Then

$$E\{x\} = E\{b_1 x_1 + \cdots + b_n x_n\}$$

$$= \int_{-\infty}^{+\infty} dx_1 \cdots \int_{-\infty}^{+\infty} dx_n (b_1 x_1 + \cdots + b_n x_n) f(x_1, \ldots, x_n)$$

$$= b_1 \int_{-\infty}^{+\infty} x_1 f_1(x_1) dx_1 + \cdots + b_n \int_{-\infty}^{+\infty} x_n f_n(x_n) dx_n$$

$$= b_1 E\{x_1\} + \cdots + b_n E\{x_n\}$$

"E" can be considered as linear operator in this case. For the special case where $b_i = 1$, $i = 1, 2, \ldots, n$, the expected value of the sum of n variables is equal to the sum of their respective expected values. The assumption of independence is not required in this case.

Suppose that x is the product of n random variables; that is,

$$x = x_1 x_2 \cdots x_n$$

Then

$$E\{x\} = \int_{-\infty}^{+\infty} dx_1 \cdots \int_{-\infty}^{+\infty} dx_n (x_1 \cdots x_n) f(x_1, \ldots, x_n)$$

If $x_1, x_2, \ldots, x_n$ are independent,

$$f(x_1, x_2, \ldots, x_n) = f_1(x_1) f_2(x_2) \cdots f_n(x_n)$$

and

$$E\{x_1 x_2 \cdots x_n\} = \int_{-\infty}^{+\infty} x_1 f_1(x_1) dx_1 \cdots \int_{-\infty}^{+\infty} x_n f_n(x_n) dx_n$$

$$= E\{x_1\} E\{x_2\} \cdots E\{x_n\}$$

Thus the expected value of the product of *independent* random variables is the product of their expected values.

The variance of a distribution is obtained by substituting $h(x) = (x - E\{x\})^2$ in the expectation formula given in Section 10.6.1. Thus

$$\text{var } \{x\} = E\{(x - E\{x\})^2\}$$

$$= E\{x^2 - 2xE\{x\} + (E\{x\})^2\}$$

$$= E\{x^2\} - 2E\{x\}E\{x\} + (E\{x\})^2$$

$$= E\{x^2\} - (E\{x\})^2$$

Var $\{x\}$ is thus readily determined from $E\{x^2\}$ and $E\{x\}$.

The quantity $E\{x^2\}$ is called the second moment of the distribution around zero. In general, the nth moment (n is an integer > 0) around zero is defined by putting $h(x) = x^n$ in the expectation formula; that is,

$$E\{x^n\} = \int_{-\infty}^{+\infty} x^n f(x)dx$$

Although these moments can be determined from the given expression, it is usually simpler to use the moment generating function, as explained in the next section.

Before showing how the variance of

$$x = b_1 x_1 + b_2 x_2 + \cdots + b_n x_n$$

is obtained in terms of those of $x_1, x_2, \ldots, x_n$, we must introduce the definition of a **covariance**. Given two random variables x_1 and x_2, the covariance of x_1 and x_2 is defined as

$$\begin{aligned}
\operatorname{cov}\{x_1, x_2\} &= E\{(x_1 - E\{x_1\})(x_2 - E\{x_2\})\} \\
&= E\{(x_1 x_2 - x_1 E\{x_2\} - x_2 E\{x_1\} + E\{x_1\}E\{x_2\})\} \\
&= E\{x_1 x_2\} - E\{x_1\}E\{x_2\}
\end{aligned}$$

If x_1 and x_2 are independent,

$$E\{x_1 x_2\} = E\{x_1\}E\{x_2\} \qquad \text{and} \qquad \operatorname{cov}\{x_1, x_2\} = 0$$

However, the converse is not true in general; that is, if $\operatorname{cov}\{x_1, x_2\} = 0$, this does not mean that x_1 and x_2 are independent.

The covariance of any two random variables x_1 and x_2 can be expressed in terms of their variances as follows:

$$\operatorname{cov}\{x_1, x_2\} = \rho_{12}\sqrt{\operatorname{var}\{x_1\} \operatorname{var}\{x_2\}}$$

or

$$\rho_{12} = \frac{\operatorname{cov}\{x_1, x_2\}}{\sqrt{\operatorname{var}\{x_1\} \operatorname{var}\{x_2\}}}$$

where ρ_{12} is called the **correlation coefficient** of x_1 and x_2. It can be proved that $|\rho_{12}| \leq 1$. When $\rho_{12} = \pm 1$, x_1 and x_2 are linearly related with probability one. When x_1 and x_2 are *independent*, $\rho_{12} = 0$.

It is now possible to obtain the variance of

$$x = b_1 x_1 + b_2 x_1 + \cdots + b_n x_n$$

in terms of the variances of x_i, $i = 1, 2, \ldots, n$. Consider first the case where $x = b_1 x_1 + b_2 x_2$. Then

$$\begin{aligned}
\operatorname{var}\{x\} &= E\{(b_1 x_1 + b_2 x_2 - b_1 E\{x_1\} - b_2 E\{x_2\})^2\} \\
&= b_1^2 E\{x_1 - E\{x_1\})^2\} + b_2^2 E\{(x_2 - E\{x_2\})^2\} \\
&\quad + 2b_1 b_2 E\{(x_1 - E\{x_1\})(x_2 - E\{x_2\})\} \\
&= b_1^2 \operatorname{var}\{x_1\} + b_2^2 \operatorname{var}\{x_2\} + 2b_1 b_2 \operatorname{cov}\{x_1, x_2\}
\end{aligned}$$

If x_1 and x_2 are independent *or* uncorrelated, then

$$\operatorname{var}\{x\} = b_1^2 \operatorname{var}\{x_1\} + b_2^2 \operatorname{var}\{x_2\}$$

In a similar way, it can be proved that for

$$x = b_1 x_1 + \cdots + b_n x_n$$

$$\operatorname{var}\{x\} = \sum_{i=1}^{n} \left[b_i^2 \operatorname{var}\{x_i\} + 2 \sum_{j=i+1}^{n} b_i b_j \operatorname{cov}\{x_i, x_j\} \right]$$

Again, if $x_1, \ldots, x_n$ are independent or *pairwise* uncorrelated, then

$$\operatorname{var}\{x\} = \sum_{i=1}^{n} b_i^2 \operatorname{var}\{x_i\}$$

10.7   MOMENT GENERATING FUNCTION

Evaluation of the moments $E\{x^n\}$ of a probability distribution can be made directly from the definition in Section 10.6.2. The **moment generating function** (MGF) is another transformation method that can be used to generate these moments. Let $M_x(t)$ be the MGF of the random variable x, where t is a parameter greater than zero. Thus, given the PDF $f(x)$,

$$M_x(t) = E\{e^{tx}\} = \begin{cases} \int_{-\infty}^{+\infty} e^{tx} f(x)\, dx, & x \text{ continuous} \\ \sum_{\text{all } x} e^{tx} P(x), & x \text{ discrete} \end{cases}$$

Notice that

$$M_x(0) = \int_{-\infty}^{+\infty} f(x)\, dx = \sum_{\text{all } x} P(x) = 1$$

$$\left[\frac{\partial}{\partial t} M_x(t) \right]_{t=0} = \int_{-\infty}^{\infty} x f(x)\, dx = \sum_{\text{all } x} x P(x) = E\{x\}$$

In general, it can be shown by induction that

$$E\{x^n\} = \left[\frac{\partial^n}{\partial t^n} M_x(t) \right]_{t=0}$$

In the case where all the moments of the distribution exist,† it is possible to express $M_x(t)$ in the form of a power series as follows:

$$M_x(t) = E\{e^{tx}\} = E\left\{ \sum_{n=0}^{\infty} \frac{t^n x^n}{n!} \right\} = \sum_{n=0}^{\infty} \left(\frac{t^n}{n!} \right) E\{x^n\}$$

If the expression for the MGF can be put in this form, a general expression for the nth moment is immediately available.

It can be proved that the MGF for any PDF, when it exists, is unique. Thus, if the MGF's of two random variables are identical, their PDF's are necessarily the same.

Example 10.7-1

For the gamma distribution

$$f(x) = \frac{\mu(\mu x)^{n-1}e^{-\mu x}}{(n-1)!}, \qquad x > 0$$

the MGF is given by

$$M_x(t) = \int_0^\infty e^{tx}\frac{\mu(\mu x)^{n-1}e^{-\mu x}}{(n-1)!}dx$$

$$= \left(\frac{\mu}{\mu-t}\right)^n \int_0^\infty \frac{(\mu-t)[(\mu-t)x]^{n-1}e^{-(\mu-t)x}}{(n-1)!}dx = \left(\frac{\mu}{\mu-t}\right)^n$$

By definition,

$$E\{x\} = \frac{\partial M_x(0)}{\partial t} = \frac{n}{\mu}$$

Also,

$$E\{x^2\} = \frac{\partial M_x(0)}{\partial t^2} = \left(\frac{n}{\mu^2}\right) + \left(\frac{n^2}{\mu^2}\right)$$

Hence

$$\mathrm{var}\,\{x\} = E\{x^2\} - (E\{x\})^2 = \frac{n}{\mu^2} + \frac{n^2}{\mu^2} - \left(\frac{n}{\mu}\right)^2 = \frac{n}{\mu^2}$$

Since the exponential distribution is obtained from the gamma distribution by putting $n = 1$, the MGF for the exponential distribution is

$$M_x(t) = \frac{\mu}{\mu-t}$$

The MGF for the exponential distribution can be put in a power series form as follows:

$$\frac{\mu}{\mu-t} = \frac{1}{1-(t/\mu)} = \sum_{n=0}^\infty \left(\frac{t^n}{n!}\right)\frac{n!}{\mu^n}$$

---

† There are probability distributions for which the moments, and hence the MGF, are not defined. For example, for the **Cauchy distribution,**

$$f(x) = \frac{1}{\pi(1+x^2)}, \qquad -\infty < x < \infty$$

$E\{x^n\}$ does not exist for all $n \geq 1$, and hence its MGF is not defined. Another transformation function can be defined, however, which is proved to exist for any PDF. This is called the **characteristic function** and is defined by

$$\Psi_n(\rho) = E\{e^{i\rho x}\}, \qquad i = \sqrt{-1}, \qquad \rho > 0$$

The characteristic function is used to obtain the moments in much the same way as the moment generating function. Its application is more general, however, since it can be used to derive the PDF.

Thus, from the power series definition of the MGF,

$$E\{x^n\} = \left(\frac{n!}{\mu^n}\right)$$

or

$$E\{x\} = \frac{1}{\mu} \quad \text{and} \quad E\{x^2\} = \frac{2}{\mu^2}$$

Hence

$$\text{var } \{x\} = \frac{2}{\mu^2} - \left(\frac{1}{\mu}\right)^2 = \frac{1}{\mu^2} \qquad \blacktriangleleft$$

The use of the MGF can also be applied to the sum of *independent* random variables. Let

$$y = \sum_{i=1}^{k} x_i$$

be a random variable defined as the sum of k *independent* random variables x_i. The MGF of y as a function of x_i is given by

$$M_y(t) = E\{e^{ty}\} = E\{e^{t(x_1 + x_2 + \cdots + x_k)}\}$$

Since $x_1, x_2, \ldots, x_k$ are independent random variables,

$$M_y(t) = E\{e^{tx_1}\}E\{e^{tx_2}\} \cdots E\{e^{tx_k}\} = \prod_{i=1}^{k} M_{x_i}(t)$$

where $M_{x_i}(t)$ is the MGF of x_i. This means that the MGF of the sum of independent random variables is equal to the product of their MGF's.

Example 10.7–2. Let $x_1, x_2, \ldots, x_n$ be n *independent* and exponentially distributed random variables with parameters $\mu_1, \mu_2, \ldots, \mu_n$, respectively. Let $y = \sum_{i=1}^{n} x_i$. As shown in Example 10.7–1,

$$M_{x_i}(t) = \frac{\mu_i}{\mu_i - t}$$

Hence

$$M_y(t) = \prod_{i=1}^{n} \frac{\mu_i}{\mu_i - t}$$

For the case where all the exponential distributions are identical with the same parameter μ,

$$M_y(t) = \left(\frac{\mu}{\mu - t}\right)^n$$

This is the MGF of a gamma distribution with mean n/μ and variance n/μ^2. (See Example 10.7–1.) Since the MGF is unique, the sum of independent and identically distributed exponential random variables is gamma. $\blacktriangleleft$

10.8 CENTRAL LIMIT THEOREM

Consider the random variable S_n defined as the sum of n independent and identically distributed random variables $x_1, x_2, \ldots, x_n$. That is,

$$S_n = x_1 + x_2 + \cdots + x_n$$

The **central limit theorem** states that, as n becomes large ($n \to \infty$), the distribution of S_n tends toward normality *regardless* of the original distribution of $x_1, \ldots, x_n$. Mathematically, this theorem is expressed as follows. Let the mean and variance of x_i be constant and given by μ and σ^2, respectively. Then $E\{S_n\} = n\mu$ and var $\{S_n\} = n\sigma^2$. Thus the central limit theorem indicates that

$$\lim_{n \to \infty} P\{S_n \le s\} = \lim_{n \to \infty} P\left\{\frac{S_n - n\mu}{\sigma\sqrt{n}} \le \frac{s - n\mu}{\sigma\sqrt{n}}\right\} = \frac{1}{\sqrt{2\pi}} \int_{-\infty}^{z} e^{-y^2/2} \, dy$$

where $z = (s - n\mu)/\sigma\sqrt{n}$. This means that S_n is approximately normal with mean $n\mu$ and variance $n\sigma^2$.

10.9 CONVOLUTIONS

Consider the case of two continuous random variables x and y with a joint PDF $f(x, y)$, where $-\infty < x < \infty$ and $-\infty < y < \infty$. It is required to determine the PDF of $s = x + y$. This will be accomplished by considering the CDF $G(s)$.

$$G(s) = P\{x + y \le s\}$$

$$= \iint_{\{x+y \le s\}} f(x, y) \, dx, \, dy, \qquad -\infty < x < \infty, \qquad -\infty < y < \infty$$

Making the necessary changes in the limits of integration, and noting that $-\infty < x < \infty$ and $-\infty < y < \infty$, we find that

$$G(s) = \int_{-\infty}^{+\infty} \left\{\int_{-\infty}^{s-x} f(x, y) \, dy\right\} dx$$

The PDF $g(s)$ is then obtained by differentiating $G(s)$ with respect to s. Hence†

$$g(s) = \int_{-\infty}^{+\infty} f(x, s - x) \, dx$$

If x and y are independent, $f(x, y) = f_1(x)f_2(y)$ and the resulting PDF of s, denoted by $g_c(s)$, is given by

$$g_c(s) = \int_{-\infty}^{+\infty} f_1(x)f_2(s - x) \, dx$$

The operation of obtaining the density function of the sum of two *independent* random variables is called **convolution,** and the resulting PDF is usually indicated as $g_c = f_1 * f_2$. The convolution operation can also be extended to the discrete case in essentially the same way. Namely,

$$P_c(s) = \sum_{\text{all} x} P_1(x)P_2(s-x)$$

An important case of convolution occurs when the random variables are nonnegative only. In this case the limits of integration in the convolution formula must be nonnegative. This means that if $y = s - x$ is to be nonnegative, x cannot exceed s. Thus

$$G_c(s) = \int_0^s \left\{ \int_0^{s-x} f_2(y)\,dy \right\} f_1(x)\,dx$$

Again, by differentiation, this gives

$$g_c(s) = \int_0^s f_1(x)f_2(s-x)\,dx$$

For the case where it is required to obtain the convolution of three or more random variables, the formulas just given can be applied recursively until the density function of the total sum is obtained. For example, let

$$s = x_1 + x_2 + \cdots + x_n$$

where $x_1, x_2, \ldots, x_n$ are independent random variables. Thus, to obtain the PDF of s, start first by obtaining the PDF of $s_2 = x_1 + x_2$. Next consider $s_3 = s_2 + x_3$ and continue in the same manner until $s = s_{n-1} + x_n$ is obtained.

Example 10.9-1. Consider the exponential distribution

$$f(x) = \mu e^{-\mu x}, \qquad x < 0$$

The n-fold convolution of the exponential random variable is a gamma distribution with mean n/μ and variance n/μ^2.

Let $s_n = x_1 + x_2 + \cdots + x_n$. Consider first $s_2 = x_1 + x_2$. From the convolution formula for positive random variables, we get

$$f_2(s_2) = \int_0^{s_2} \mu e^{-\mu x_2} \mu e^{-\mu(s_2 - x_2)}\,dx_2 = \mu^2 s_2 e^{-\mu s_2}$$

Again, letting $s_3 = s_2 + x_3$, and applying the same formula, we obtain

$$f_3(s_3) = \int_0^{s_3} \mu^2 s_2 e^{-\mu s_2} (\mu e^{-\mu(s_3 - s_2)})\,ds_2 = \frac{\mu^3 s_3^2 e^{-\mu s_3}}{2!}$$

_____

† Given

$$H(y) = \int_{a(y)}^{b(y)} f(x, y)\,dx$$

then

$$\frac{dH(y)}{dy} = \int_{a(y)}^{b(y)} \frac{\partial f(x, y)}{\partial y}\,dx + f[b(y), y]\frac{db(y)}{dy} - f[a(y), y]\frac{da(y)}{dy}$$

Continuing in the same manner, we can show by induction that, for $s_n = s$,

$$f_n(s) = \frac{\mu^n s^{n-1} e^{-\mu s}}{(n-1)!}, \qquad s > 0$$

This is a gamma distribution with mean n/u and variance n/μ^2. This result, obtained by convolution, reaches the same conclusion obtained by the MGF (see Example 10.7–2).   ◄

Example 10.9–2. The discrete density function of the weekly demand of a certain commodity is given by

x	0	1	2
$P(x)$	.1	.4	.5

where x is the number of units demanded per week. If we assume that this distribution is the same for every week (stationary) and that they are statistically independent, the distribution of the demand for 2 weeks can be obtained by applying the convolution formula for the discrete case; namely,

$$P(0) = P_1(0)P_2(0) = .1 \times .1 = .01$$
$$P(1) = P_1(0)P_2(1) + P_1(1)P_2(0)$$
$$= .1 \times .4 + .4 \times .1 = .08$$
$$P(2) = P_1(0)P_2(2) + P_1(1)P_2(1) + P_1(2)P_2(0)$$
$$= .1 \times .5 + .4 \times .4 + .5 \times .1 = .26$$
$$P(3) = P_1(1)P_2(2) + P_1(2)P_2(1)$$
$$= .4 \times .5 + .5 \times .4 = .40$$
$$P(4) = P_1(2)P_2(2) = .5 \times .5 = .25$$

Thus

$x_1 + x_2$	0	1	2	3	4
$P(x_1 + x_2)$	.01	.08	.26	.40	.25

◄

10.10 STOCHASTIC PROCESSES

Consider the discrete points in time $\{t_k\}$ for $k = 1, 2, \ldots$, and let ξ_{t_k} be the random variable that characterizes the state of the system at t_k. The family of random variables $\{\xi_{t_k}\}$ forms a **stochastic process.** The states at time t_k actually represent the (exhaustive and mutually exclusive) outcomes of the system at that time. The number of states may thus be finite or infinite. For example, the Poisson distribution

$$P_n(t) = \frac{e^{-\lambda t}(\lambda t)^n}{n!}, \qquad n = 0, 1, 2, \ldots$$

represents a stochastic process with an infinite number of states. Here the random variable n represents the number of occurrences between 0 and t (assuming that the system starts at time 0). The states of the system at any time t are thus given by $n = 0, 1, 2, \ldots$.

Another example is represented by the coin-tossing game with k trials. Each trial may be viewed as a point in time. The resulting sequence of trials forms a stochastic process. The state of the system at any trial is either a head or a tail.

This section presents a summary of an important class of stochastic systems that includes **Markov processes** and **Markov chains.** A Markov chain is actually a special case of Markov processes. It is used to study the short- and long-run behavior of certain stochastic systems. Markov processes and chains will be especially useful in dealing with queueing (or waiting-line) theory, which is presented in Chapter 15.

10.10.1 MARKOV PROCESSES

A *Markov process* is a stochastic system for which the occurrence of a future state depends on the immediately preceding state and only on it. Thus if $t_0 < t_1 < \cdots t_n$ ($n = 0, 1, 2, \ldots$) represents points in time, the family of random variables $\{\xi_{t_n}\}$ is a Markov process if it possesses the following **Markovian property:**

$$P\{\xi_{t_n} = x_n | \xi_{t_{n-1}} = x_{n-1}, \ldots, \xi_{t_0} = x_0\} = P\{\xi_{t_n} = x_n | \xi_{t_{n-1}} = x_{n-1}\}$$

for all possible values of $\xi_{t_0}, \xi_{t_1}, \ldots, \xi_{t_n}$.

The probability $p_{x_{n-1}x_n} = P\{\xi_{t_n} = x_n | \xi_{t_{n-1}} = x_{n-1}\}$ is called the **transition probability.** It represents the *conditional* probability of the system being in x_n at t_n, given it was in x_{n-1} at t_{n-1}. This probability is also referred to as the **one-step transition probability,** since it describes the system between t_{n-1} and t_n. An m-step transition probability is thus defined by

$$p_{x_n, x_{n+m}} = P\{\xi_{t_{n+m}} = x_{n+m} | \xi_{t_n} = x_n\}$$

10.10.2 MARKOV CHAINS

Let $E_1, E_2, \ldots, E_j$ ($j = 0, 1, 2, \ldots$) represent the exhaustive and mutually exclusive outcomes (states) of a system at any time. Initially, at time t_0, the system may be in any of these states. Let $a_j^{(0)}$ ($j = 0, 1, 2, \ldots$) be the absolute probability that the system is in state E_j at t_0. Assume further that the system is Markovian.

Define

$$p_{ij} = P\{\xi_{t_n} = j | \xi_{t_{n-1}} = i\}$$

as the one-step transition probability of going from state i at t_{n-1} to state j at t_n and assume that these probabilities are stationary (fixed) over time. Thus the transition probabilities from state E_i to state E_j can be more conveniently arranged in a matrix form as follows:

$$\mathbf{P} = \begin{pmatrix} p_{00} & p_{01} & p_{02} & p_{03} & \cdots \\ p_{10} & p_{11} & p_{12} & p_{13} & \cdots \\ p_{20} & p_{21} & p_{22} & p_{23} & \cdots \\ p_{30} & p_{31} & p_{32} & p_{33} & \cdots \\ \vdots & \vdots & \vdots & \vdots & \end{pmatrix}$$

The matrix $\mathbf{P}$ is called a **homogeneous transition** or **stochastic matrix** because all the transition probabilities p_{ij} are fixed and independent of time. The probabilities p_{ij} must satisfy the conditions

$$\sum_j p_{ij} = 1, \qquad \text{for all } i$$

$$p_{ij} \geq 0, \qquad \text{for all } i \text{ and } j$$

The definition of a **Markov chain** is now in order. *A transition matrix $\mathbf{P}$ together with the initial probabilities $\{a_j^{(0)}\}$ associated with the states E_j completely define a Markov chain.* One usually thinks of a Markov chain as describing the transitional behavior of a system over equally spaced intervals of time. Situations exist, however, where the time spacings are dependent on the characteristics of the system and hence may not be equal. This case is referred to as **imbedded Markov chains.**

A. Absolute and Transition Probabilities

Given $\{a_j^{(0)}\}$ and $\mathbf{P}$ of a Markov chain, the absolute probabilities of the system after a specified number of transitions are determined as follows. Let $\{a_j^{(n)}\}$ be the absolute probabilities of the system after n transitions, that is, at t_n. The general expression of $\{a_j^{(n)}\}$ in terms of $\{a_j^{(0)}\}$ and $\mathbf{P}$ can be found as follows.

$$a_j^{(1)} = a_1^{(0)} p_{1j} + a_2^{(0)} p_{2j} + a_3^{(0)} p_{3j} + \cdots = \sum_i a_i^{(0)} p_{ij}$$

Also,

$$a_j^{(2)} = \sum_i a_i^{(1)} p_{ij} = \sum_i \left(\sum_k a_k^{(0)} p_{ki} \right) p_{ij} = \sum_k a_k^{(0)} \left(\sum_i p_{ki} p_{ij} \right) = \sum_k a_k^{(0)} p_{kj}^{(2)}$$

where $p_{kj}^{(2)} = \sum_i p_{ki} p_{ij}$ is the **two-step** or **second-order transition probability,** that is, the probability of going from state k to state j in exactly two transitions.

Similarly, it can be shown by induction that

$$a_j^{(n)} = \sum_i a_i^{(0)} \left(\sum_k p_{ik}^{(n-1)} p_{kj} \right) = \sum_i a_i^{(0)} p_{ij}^{(n)}$$

where $p_{ij}^{(n)}$ is the n-step or n-order transition probability given by the recursive formula

$$p_{ij}^{(n)} = \sum_k p_{ik}^{(n-1)} p_{kj}$$

In general, for all i and j,

$$p_{ij}^{(n)} = \sum_k p_{ik}^{(n-m)} p_{kj}^{(m)}, \qquad 0 < m < n$$

These equations are known as **Chapman–Kolomogorov equations.**

The elements of a higher transition matrix $\|p_{ij}^{(n)}\|$ can be obtained directly by matrix multiplication. Thus

$$\|p_{ij}^{(2)}\| = \|p_{ij}\| \, \|p_{ij}\| = \mathbf{P}^2$$
$$\|p_{ij}^{(3)}\| = \|p_{ij}^{(2)}\| \, \|p_{ij}\| = \mathbf{P}^3$$

and, in general,

$$\|p_{ij}^{(n)}\| = \mathbf{P}^{n-1}\mathbf{P} = \mathbf{P}^n$$

Hence, if the absolute probabilities are defined in vector form as

$$\mathbf{a}^{(n)} = \{a_1^{(n)}, a_2^{(n)}, a_3^{(n)}, \ldots\}$$

then

$$\mathbf{a}^{(n)} = \mathbf{a}^{(0)}\mathbf{P}^n$$

Example 10.10-1. Consider the following Markov chain with two states,

$$\mathbf{P} = \begin{pmatrix} .2 & .8 \\ .6 & .4 \end{pmatrix}$$

with $\mathbf{a}^{(0)} = (.7 \quad .3)$. Determine $\mathbf{a}^{(1)}$, $\mathbf{a}^{(4)}$, and $\mathbf{a}^{(8)}$.

$$\mathbf{P}^2 = \begin{pmatrix} .2 & .8 \\ .6 & .4 \end{pmatrix}\begin{pmatrix} .2 & .8 \\ .6 & .4 \end{pmatrix} = \begin{pmatrix} .52 & .48 \\ .36 & .64 \end{pmatrix}$$

$$\mathbf{P}^4 = \mathbf{P}^2\mathbf{P}^2 = \begin{pmatrix} .52 & .48 \\ .36 & .64 \end{pmatrix}\begin{pmatrix} .52 & .48 \\ .36 & .64 \end{pmatrix} \cong \begin{pmatrix} .443 & .557 \\ .417 & .583 \end{pmatrix}$$

$$\mathbf{P}^8 = \mathbf{P}^4\mathbf{P}^4 = \begin{pmatrix} .443 & .557 \\ .417 & .583 \end{pmatrix}\begin{pmatrix} .443 & .557 \\ .417 & .583 \end{pmatrix} \cong \begin{pmatrix} .4281 & .5719 \\ .4274 & .5726 \end{pmatrix}$$

Thus

$$\mathbf{a}^{(1)} = (.7 \quad .3)\begin{pmatrix} .2 & .8 \\ .6 & .4 \end{pmatrix} = (.32 \quad .68)$$

$$\mathbf{a}^{(4)} = (.7 \quad .3)\begin{pmatrix} .443 & .557 \\ .417 & .583 \end{pmatrix} = (.435 \quad .565)$$

$$\mathbf{a}^{(8)} = (.7 \quad .3)\begin{pmatrix} .4281 & .5719 \\ .4274 & .5726 \end{pmatrix} = (.4279 \quad .5721)$$

The interesting result is that the rows of $\mathbf{P}^8$ tend to be identical. Also, $\mathbf{a}^{(8)}$ tends to be identical with the rows of $\mathbf{P}^8$. This result has to do with the long-run properties of Markov chains, which, as shown in Section 10.10.2, implies that the long-run absolute probabilities are independent of $\mathbf{a}^{(0)}$. In this case the resulting probabilities are known as the **steady-state probabilities.** ◀

B. Classification of States in Markov Chains

In using Markov chain analysis, we may be interested in studying the behavior of the system over a short period of time. In this case the absolute probabilities are computed as shown in the preceding section. A more important study, however, would involve the long-run behavior of the system, that is, when the number of transitions tends to infinity. In such a case the analysis given in the preceding section is inadequate and a systematic procedure that will predict the long-run behavior of the system becomes necessary. This section introduces definitions of the classification of states in Markov chains that will be useful in studying the long-run behavior of the system.

Irreducible Markov Chain

A Markov chain is said to be **irreducible** if every state E_j can be reached from every other state E_i after a finite number of transitions; that is, for $i \neq j$,

$$P_{ij}^{(n)} > 0, \qquad \text{for } 1 \leq n < \infty$$

In this case all the states of the chain **communicate.**

Closed Set and Absorbing States

In a Markov chain, a set C of states is said to be **closed** if the system, once in one of the states of C, will remain in C indefinitely. A special example of a closed set is a single state E_j with transition probability $p_{jj} = 1$. In this case, E_j is called an **absorbing state.** All the states of an irreducible chain must form a closed set and no other subset can be closed. The closed set C must also satisfy all the conditions of a Markov chain and hence may be studied independently.

Example 10.10–2. Consider the following Markov chain:

$$
\mathbf{P} = \begin{array}{c} \\ 0 \\ 1 \\ 2 \\ 3 \end{array}
\begin{pmatrix}
0 & 1 & 2 & 3 \\
1/2 & 1/4 & 1/4 & 0 \\
0 & 0 & 1 & 0 \\
1/3 & 0 & 1/3 & 1/3 \\
0 & 0 & 0 & 1
\end{pmatrix}
$$

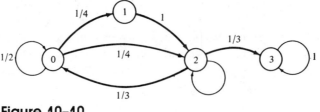

Figure 10–10

This chain is illustrated graphically in Figure 10–10. The figure shows that the four states do *not* constitute an irreducible chain, since states 0, 1, and 2 cannot be reached from state 3. State 3, by itself, forms a closed set and hence it is absorbing. One can also say that state 3 forms an irreducible chain.   ◄

First Return Times

An important definition in Markov chains theory is the **first return time.** Given that the system is initially in state E_j, it may return to E_j *for the first time* at the nth step, $n \geq 1$. The number of steps before the system returns to E_j is called the first return time.

Let $f_{jj}^{(n)}$ denote the probability that the first return to E_j occurs at the nth step. Then given the transition matrix

$$\mathbf{P} = \|p_{ij}\|$$

an expression for $f_{jj}^{(n)}$ can be determined as follows:

$$p_{jj} = f_{jj}^{(1)}$$
$$p_{jj}^{(2)} = f_{jj}^{(2)} + f_{jj}^{(1)} p_{jj}$$

or

$$f_{jj}^{(2)} = p_{jj}^{(2)} - f_{jj}^{(1)} p_{jj}$$

By induction it can be proved in general that

$$p_{jj}^{(n)} = f_{jj}^{(n)} + \sum_{m=1}^{n-1} f_{jj}^{(m)} p_{jj}^{(n-m)}$$

which yields the required expression as

$$f_{jj}^{(n)} = p_{jj}^{(n)} - \sum_{m=1}^{n-1} f_{jj}^{(m)} p_{jj}^{(n-m)}$$

The probability of *at least* one return to state E_j is then given by

$$f_{jj} = \sum_{n=1}^{\infty} f_{jj}^{(n)}$$

Thus the system is certain to return to j if $f_{jj} = 1$. In this case, if μ_{jj} defines the mean return (recurrence) time,

$$\mu_{jj} = \sum_{n=1}^{\infty} n f_{jj}^{(n)}$$

If $f_{jj} < 1$, it is not certain that the system will return to E_j and, consequently, $\mu_{jj} = \infty$.

The states of a Markov chain can thus be classified based on the definition of the first return times as follows:

1. A state is **transient** if $f_{jj} < 1$; that is, $\mu_{jj} = \infty$.
2. A state is **recurrent** (persistent) if $f_{jj} = 1$.
3. A recurrent state is **null** if $\mu_{jj} = \infty$ and **nonnull** if $\mu_{jj} < \infty$ (finite).
4. A state is **periodic** with period t if a return is possible only in t, $2t$, $3t$, ... steps. This means that $p_{jj}^{(n)} = 0$ whenever n is not divisible by t.
5. A recurrent state is **ergodic** if it is nonnull and aperiodic (not periodic).

Ergodic Markov Chains

An *irreducible* Markov chain is ergodic if all its states are ergodic. In this case the absolute probability distribution

$$\mathbf{a}^{(n)} = \mathbf{a}^{(0)}\mathbf{P}^n$$

always converges uniquely to a limiting distribution as $n \rightarrow \infty$, where the limiting distribution is independent of the initial probabilities $\mathbf{a}^{(0)}$.

The following theorem is now in order.

Theorem 10.10–1. *All the states in an irreducible infinite Markov chain may belong to one, and only one, of the following three classes: transient state, recurrent null state, or recurrent nonnull state. In every case all the states communicate and they have the same period. For the special case where the chain has a finite number of states, the chain cannot consist of transient states only nor can it contain any null states.*

C. Limiting Distribution of Irreducible Chains

Example 10.10–1 shows that as the number of transitions increases, the absolute probability becomes independent of the initial distribution. This was referred to as the long-run property of Markov chains. In this section determination of the limiting (long-run) distribution of an *irreducible* chain is presented. The discussion will be restricted to the aperiodic type, since this is the only type needed in this text. In addition, the analysis of the periodic case is rather involved.

The existence of a limiting distribution in an irreducible aperiodic chain depends on the class of its states. Thus, considering the three classes given in Theorem 10.10–1, the following theorem can be stated.

Theorem 10.10–2. *In an irreducible aperiodic Markov chain,*
(a) *If the states are all transient or all null, then $p_{ij}^{(n)} \rightarrow 0$ as $n \rightarrow \infty$ for all i and j and no limiting distribution exists.*

(b) *If all the states are ergodic, then*

$$\lim_{n \to \infty} a_j^{(n)} = \beta_j, \quad j = 0, 1, 2, \ldots$$

where β_j is the limiting (steady-state) distribution. The probabilities β_j exist uniquely and are independent of $a_j^{(0)}$. In this case β_j can be determined from the set of equations†

$$\beta_j = \sum_i \beta_i p_{ij}$$

$$1 = \sum_j \beta_j$$

The mean recurrence time for state j is then given by

$$\mu_{jj} = \frac{1}{\beta_j}$$

Example 10.10-3. Consider Example 10.10–1. To determine its steady-state probability distribution, one has

$$\beta_1 = .2\beta_1 + .6\beta_2$$
$$\beta_2 = .8\beta_1 + .4\beta_2$$
$$1 = \beta_1 + \beta_2$$

(Notice that one of the first two equations is redundant.) The solution yields $\beta_1 = 0.4286$ and $\beta_2 = 0.5714$. These results are very close to the values of $a^{(8)}$ (and the rows of $\mathbf{P}^8$) in Example 10.10–1,

$$\mu_{11} = 1/\beta_1 = 2.3$$
$$\mu_{22} = 1/\beta_2 = 1.75$$

so that the mean recurrent time for the first and second states are 2.3 and 1.75 steps, respectively.   ◀

Example 10.10-4. Consider the following Markov chain with three states:

$$\mathbf{P} = \begin{array}{c} 0 \\ 1 \\ 2 \end{array} \begin{array}{ccc} 0 & 1 & 2 \\ \left(\begin{array}{ccc} 1/2 & 1/4 & 1/4 \\ 1/2 & 1/4 & 1/4 \\ 0 & 1/2 & 1/2 \end{array} \right) \end{array}$$

This is called a **doubly stochastic matrix**, since

$$\sum_{i=1}^{s} p_{ij} = \sum_{j=1}^{s} p_{ij} = 1$$

where s is the number of states. In such cases, the steady-state probabilities are $\beta_j = 1/s$ for all j. Thus, for the matrix given,

$$\beta_0 = \beta_1 = \beta_2 = 1/3$$ ◀

---

† Notice that one of the equations $\beta_j = \sum_i \beta_i p_{ij}$ is redundant.

10.11 THE z-TRANSFORM

An important transformation function used with *nonnegative* discrete random variables is the z-transform. Let n be a discrete random variable defined over a set of nonnegative integers, and let p_n be its PDF. The notation $Z(p_n) \equiv P(z)$ represents the z-transform of p_n, defined by

$$Z(p_n) = P(z) = \sum_{n=0}^{\infty} p_n z^n$$

where z is determined such that $Z(p_n)$ converges. Since $\sum_{n=0}^{\infty} p_n = 1$, $Z(p_n)$ will always converge for $|z| \leq 1$. It should be noted that the z-transform for a given PDF is unique.

The z-transform, like the moment generating function (Section 10.7), can be used to determine the moments of the random variable n. A more important contribution will be the use of the z-transform to determine the PDF of n. The following sections will investigate these points in detail.

10.11.1 CHARACTERISTICS OF THE z-TRANSFORM

Let

$$P^{(n)}(z) = \frac{\partial^n P(z)}{\partial z^n}$$

the following properties can be established.

Property 1

$$P(0) = p_0$$

Property 2

$$P(1) = \sum_{n=0}^{\infty} p_n = 1$$

Property 3

$$p_n = \frac{1}{n!} P^{(n)}(0)$$

Property 3 can be established as follows:

$$P(z) = p_0 + p_1 z + p_2 z^2 + \cdots$$

Hence

$$P^{(1)}(z) = p_1 + 2p_2 z + 3p_3 z^2 + \cdots$$

which yields

$$P^{(1)}(0) = p_1 = 1! p_1$$

Again,

$$P^{(2)}(z) = 2p_2 + 3 \cdot 2p_2 z + 4 \cdot 3p_4 z^2 + \cdots$$

which gives

$$P^{(2)}(0) = 2p_2 = 2!p_2$$

Continuing in the same manner, we can prove property 3 by induction.

Property 4

$$E\{n\} = P^{(1)}(1)$$
$$\text{var}\{n\} = P^{(2)}(1) + P^{(1)}(1) - \{P^{(1)}(1)\}^2$$

This property is derived as follows:

$$P^{(1)}(1) = p_1 + 2p_2 + 3p_3 + \cdots = \sum_{n=1}^{\infty} np_n = E\{n\}$$

Again,

$$P^{(2)}(1) = 2p_2 + 3 \cdot 2p_3 + 4 \cdot 3p_4 + \cdots = \sum_{n=2}^{\infty} n(n-1)p_n$$

$$= \sum_{n=2}^{\infty} n^2 p_n - \sum_{n=2}^{\infty} np_n = E\{n^2\} - E\{n\}$$

Thus

$$\text{var}\{n\} = E\{(n - E\{n\})^2\} = E\{n^2\} - (E\{n\})^2$$
$$= P^{(2)}(1) + P^{(1)}(1) - \{P^{(1)}(1)\}^2$$

Property 3 shows that once the z-transform is known, P_n is completely defined. Property 4 also shows that the mean and variance of n can be derived from the z-transform.

10.11.2 SPECIAL CASES OF THE z-TRANSFORM

This section presents some useful relationships often encountered in the derivations of the z-transform. The z-transform is a general mathematical operation that applies to any nonnegative *sequence* of which a nonnegative random variable represents a special case. Thus, in the following analysis, p_n and q_n will be thought of as general nonnegative sequences.

1. The z-transform of p_{n-1}:

$$Z(p_{n-1}) = \sum_{n=0}^{\infty} p_{n-1}z^n = z \sum_{k=0}^{\infty} P_k z^k = zZ(p_n) = zP(z)$$

2. The z-transform of p_{n+1}:

$$Z(p_{n+1}) = \sum_{n=0}^{\infty} p_{n+1}z^n = \frac{1}{z}\left(\sum_{n=0}^{\infty} p_{n+1}z^{n+1}\right) = \frac{1}{z}\left(\sum_{k=0}^{\infty} p_k z^k - p_0\right)$$

$$= \frac{1}{z}(Z(p_n) - p_0) = \frac{1}{z}(P(z) - p_0)$$

3. The z-transform of $(p_n + q_n)$:

$$Z(p_n + q_n) = \sum_{n=0}^{\infty} (p_n + q_n)z^n = Z(p_n) + Z(q_n) = P(z) + Q(z)$$

4. The z-transform of bp_n, where b is a constant:

$$Z(bp_n) = \sum_{n=0}^{\infty} bp_n z^n = bZ(p_n) = bP(z)$$

5. The z-transform of the *convolution* of p_n and q_n. Let

$$y_n = q_n * p_n = q_0 p_n + q_1 p_{n-1} + \cdots + q_{n-1} p_1 + q_n p_0$$

Then

$$
\begin{aligned}
Z(y_n) &= \sum_{n=0}^{\infty} y_n z^n \\
&= q_0 p_0 + (q_0 p_1 + q_1 p_0)z + (q_0 p_2 + q_1 p_1 + q_2 p_0)z^2 + \cdots \\
&= \{q_0(p_0 + p_1 z + p_2 z^2 + \cdots) \\
&\quad + q_1 z(p_0 + p_1 z + p_2 z^2 + \cdots) \\
&\quad + q_2 z^2(p_0 + p_1 z + p_2 z^2 + \cdots) + \cdots\} \\
&= (q_0 + q_1 z + q_2 z^2 + \cdots)(p_0 + p_1 z + p_2 z^2 + \cdots) \\
&= Z(q_n)Z(p_n) = Q(z)P(z)
\end{aligned}
$$

Thus the z-transform of the convolution of two PDF's is the product of their z-transforms.

The power of the z-transform is realized when dealing with the situations where p_n is expressed implicitly by difference–differential equations such as in queueing theory analysis (see Chapter 15). For such cases, the expression for $P(z)$ is first derived from the given equations, then by using property 3, Section 10.11.1, an expression for p_n can be obtained. There are some cases, however, where the general expression for p_n can be obtained by taking the inverse of $P(z)$; that is, $p_n = Z^{-1}\{P(z)\}$. Tables of the inverse z-transform have been prepared for this purpose. A short list of some important cases that will be used in the next chapters is included in Table 10–1. An illustration of the use of these tables is given in Example 10.11–2.

Example 10.11–1. Consider the z-transform of the following Poisson distribution with parameter λ.

$$p_n = \frac{\lambda^n e^{-\lambda}}{n!}, \qquad n = 0, 1, 2, \ldots$$

The z-transform is given by

$$P(z) = \sum_{n=0}^{\infty} \frac{\lambda^n e^{-\lambda}}{n!} z^n = \frac{e^{-\lambda}}{e^{-\lambda z}} \sum_{n=0}^{\infty} \frac{(\lambda z)^n e^{-\lambda z}}{n!} = e^{\lambda(z-1)}$$

Hence

$$E\{n\} = P^{(1)}(1) = [\lambda e^{\lambda(z-1)}]_{z=1} = \lambda$$

Table 10-1
z-Transform Functions and Their Inverses[a]

Formula	$P(z) = \sum_{n=0}^{\infty} p_n z^n$	$Z^{-1}\{P(z)\} = p_n$
1	$z^k Q(z)$	$p_n = \begin{cases} 0, & n < k \\ q_{n-k}, & n \geq k \end{cases}$
2	$z^{-k} Q(z) - \sum_{j=0}^{k-1} q_j z^{j-k}$	q_{n+k}
3	$Q(a^b z)$	$a^{bn} q_n$
4	$\{(1-z)Q(z) - q_0\}z^{-1}$	$q_{n+1} - q_n$
5	$a/(1-z)$	a
6	$z/(1-z)^2$	n
7	$1/(1-az)$	a^n
8	$1/(1-az)^{k+1}$	$\binom{n+k}{k} a^n$
9	$(a+bz)^m$	$\binom{m}{n} b^n a^{m-n}$
10	$-\ln(1-az)$	a^n/n
11	e^{az}	$a^n/n!$
12	a^z	$(\ln a)^n/n!$

[a] This table is excerpted from C. Beightler, L. Mitten, and G. Nemhauser, "A Short Table of z-Transforms and Generating Functions," *Operations Research*, Vol. 9, 1961, pp. 574–578.

Similarly,

$$P^{(2)}(1) = [\lambda^2 e^{\lambda(z-1)}]_{z=1} = \lambda^2$$

Thus

$$\text{var}\{n\} = P^{(2)}(1) + P^{(1)}(1) - \{P^{(1)}(1)\}^2 = \lambda^2 + \lambda - \lambda^2 = \lambda \qquad \blacktriangleleft$$

Example 10.11-2. Consider the z-transform of the convolution of the following two Poisson distributions:

$$p_n = \frac{\lambda_1^n e^{-\lambda_1}}{n!}, \qquad n = 0, 1, 2, \ldots$$

$$q_n = \frac{\lambda_2^n e^{-\lambda_2}}{n!}, \qquad n = 0, 1, 2, \ldots$$

Let

$$y_n = q_n * p_n$$

Hence, from Section 10.11.2,

$$Z(y_n) = Q(z)P(z)$$

From Example 10.11–1,

$$Q(z) = e^{\lambda_2(z-1)} \qquad \text{and} \qquad P(z) = e^{\lambda_1(z-1)}$$

Consequently,

$$Z(y_n) = e^{\{(\lambda_1 + \lambda_2)(z-1)\}}$$

Formula 11 in Table 10–1 gives

$$Z^{-1}\{e^{az}\} = \frac{a^n}{n!}$$

Thus

$$\begin{aligned}
y_n &= Z^{-1}\{e^{\{(\lambda_1 + \lambda_2)(z-1)\}}\} \\
&= e^{-(\lambda_1 + \lambda_2)} Z^{-1}\{e^{(\lambda_1 + \lambda_2)z}\} \\
&= \frac{e^{-(\lambda_1 + \lambda_2)}(\lambda_1 + \lambda_2)^n}{n!}, \qquad n = 0, 1, 2, \ldots
\end{aligned}$$

This shows that the convolution of two (independent) Poisson distributions with parameters λ_1 and λ_2 is also Poisson with parameter $(\lambda_1 + \lambda_2)$. ◀

SELECTED REFERENCES

FELLER, W., *An Introduction to Probability Theory and Its Applications*, 2nd ed., Vols. I and II, Wiley, New York, 1967.

PARZEN, E., *Modern Probability Theory and Its Applications*, Wiley, New York, 1960.

SPRINGER, M. D., *The Algebra of Random Variables*, Wiley, New York, 1979.

REVIEW QUESTIONS

True (T) or False (F)?

1. ____ The occurrence of an event can imply the occurrence of more than one outcome of an experiment.

2. ____ A sample space includes only a finite number of outcomes.

3. ____ The probability of an event can exceed one to indicate that its occurrence is "very" sure.

4. ____ If two events are mutually exclusive, the probability that one or the other will occur equals the sum of their individual probabilities.

5. ____ Two independent events are always mutually exclusive.

6. ____ Two mutually exclusive events are always independent.

7. ____ There may not be a one-to-one correspondence between the outcomes of an experiment and the values of its associated random variable.

8. ____ The main reason for requiring a density function to be nonnegative is to eliminate the possibility of having an event with negative probability, which is nonsensical.

9. ____ The probability of a fixed value of a continuous random variable is always zero, whereas that of a discrete variable can be positive.

10. ____ If the number of events occurring during a given period t is Poisson, the time between the occurrence of two successive events is exponential.

11. ____ If the number of trials until one failure takes place is geometric, the number of trials until c (> 1) failures take place is negative binomial.

12. ____ A gamma random variable is the sum of identical and independent exponential random variables.

13. ____ Two random variables are independent if their joint density function is the product of their marginal density function.

14. ____ The variance of a random variable may assume negative values.

15. ____ If two random variables are independent, their covariance is zero.

16. ____ If the covariance of two random variables is zero, they must be independent.

17. ____ In a Markovian process, the current state of the system depends only on the immediately preceding state.

18. ____ The steady-state probabilities of a Markov chain can depend on the initial state of the system.

19. ____ Once the system enters an absorbing state in a Markov chain, it remains there indefinitely.

20. ____ If a state of a Markov chain is transient, the steady-state probability of being in that state can be positive.

[*Ans.* **1**—T, **2**—F, **3**—F, **4**—T, **5**—F, **6**—F, **7**—F, **8** to **13**—T, **14**—F, **15**—T, **16**—F, **17**—T, **18**—F, **19**—T, **20**—F]

PROBLEMS

Section	Assigned Problems
10.1	10–1 to 10–3
10.2	10–4 to 10–7
10.3, 10.4	10–8 to 10–12
10.5	10–13 to 10–15
10.6	10–16 to 10–20
10.7	10–21 to 10–24
10.9	10–25 to 10–27
10.10	10–28, 10–29
10.11	10–30, 10–31

☐ **10–1** In each of the following experiments, define the associated sample space in terms of all its outcomes.

(a) Two coins are tossed simultaneously and the outcome is represented by the faces that turn up.

(b) Two dice are thrown simultaneously and the outcome is represented by the sum of the total number of points that turn up.

(c) A coin is tossed followed by the throwing of a die and the outcome is the combination of head or tail and the number of points that turn up on the die.

(d) An experiment to test the useful life of an automobile battery.

☐ **10–2** Consider experiment (b) in Problem 10–1. List the outcomes associated with the following outcomes.

(a) The sum does not exceed 4.

(b) Both numbers that turn up are odd.

(c) The product of both numbers that turn up is even.

(d) Both numbers are not equal.

☐ **10–3** Suppose that a sample space $S = \{1, 2, 3, 4, 5, 6, 7, 8, 9, 10\}$ and let $A = \{1, 2, 3, 4\}$, $B = \{5, 6, 7, 8, 9\}$, and $C = \{2, 4, 6, 8, 10\}$. Suppose that $\bar{A}$, $\bar{B}$, and $\bar{C}$ represent all the elements not in (i.e., complements of) A, B, and C. Determine the elements of the following events.

(a) $A\bar{B}$,   (b) $\bar{A} + B$ (c) $AB + \bar{B}C$ (d) $\bar{A}B + \bar{C}$

(e) $\overline{BCS}$ (f) $A\bar{S}$ (g) $S(A\bar{B})$ (h) $AB\bar{C}$

☐ **10–4** Derive the relationships

(a) $P\{E + F\} = P\{E\} + P\{F\} - P\{EF\}$.

(b) $P\{ABC\} = P\{A\}P\{B|A\}P\{C|AB\}$.

☐ **10–5** Prove that $P\{A + B + C\} \le P\{A\} + P\{B\} + P\{c\}$.

☐ **10–6** Given the events A and B and their complements $\bar{A}$ and $\bar{B}$, show that

$$P\{A\bar{B} + \bar{A}B\} = P\{A\} + P\{B\} - 2P\{AB\}$$

☐ **10–7** In a survey conducted in regional high schools to study the correlation between senior year scores in mathematics and enrolling in engineering colleges, it was found that among the 1000 high school graduates considered, 400 studied mathematics. Out of these 400 students, only 150 enrolled in an engineering college. Determine the following:

(a) The probability that a student studied mathematics but did not study engineering.

(b) The probability that a student studied mathematics and did enroll in engineering.

(c) The probability that a student neither studied mathematics nor enrolled in engineering.

☐ **10–8** Find the value of k so that the following $f(x)$ is a PDF.

$$f(x) = \begin{cases} \dfrac{k}{x^2}, & 10 \leq x \leq 20 \\ 0, & \text{otherwise} \end{cases}$$

☐ **10–9** Given a random variable j that assumes the values c, $c + 1$, $c + 2$, . . . , where c is a positive integer, show that

$$P\{x = j\} = \binom{j-1}{c-1} p^c (1-p)^{j-c}, \qquad 0 < p < 1$$

is a proper density function. (This is the Pascal distribution introduced in Section 10.3.1-A.)

☐ **10–10** A random variable t has the following CDF:

$$F(t) = \begin{cases} 1 - e^{-2t}, & t \geq 0 \\ 0, & t < 0 \end{cases}$$

Find
   (a) The corresponding PDF.
   (b) $P\{5 < t < 10\}$.
   (c) $P\{t = 10\}$.
   (d) $P\{t < 5 \text{ or } t > 10\}$.
   (e) $P\{3 < t < 5 \text{ and } 6 < t < 7\}$.

☐ **10–11** Show that

$$\int_{-\infty}^{+\infty} e^{-x^2/2} \, dx = \sqrt{2\pi}$$

☐ **10–12** Show that the geometric distribution (Section 10.3.1A) approaches the exponential distribution as $p \to 0$ and the intertrial time $\to 0$.

☐ **10–13** Show that

$$f(x, y) = \begin{cases} e^{-x-y}, & x > 0, \ y > 0 \\ 0, & \text{otherwise} \end{cases}$$

is a density function. Then find
   (a) The marginal distributions of x and y.
   (b) The CDF $F(x, y)$.
   (c) $P\{1 < x < 10, \ 5 < y < 7\}$.

☐ **10–14** In Problem 10–13, show that x and y are independent.

☐ **10–15** Given

$$f(x_1, x_2) = \begin{cases} k(1 - x_1 - x_2), & 0 < x_2 < 1 - x_1, \ 0 < x_1 < 1 \\ 0, & \text{otherwise} \end{cases}$$

(a) Find the value of k that will make $f(x_1, x_2)$ a PDF.

(b) Find the conditional PDF $g(x_1|x_2)$, and then show that x_1 and x_2 are not independent.

☐ **10–16** Find the mean and variance for the following uniform PDF.

$$f(x) = \begin{cases} \dfrac{1}{b-a}, & a < x < b \\ 0, & \text{otherwise} \end{cases}$$

☐ **10–17** Find $E\{x_1 \mid x_2\}$ in Problem 10–15.

☐ **10–18** Given that $x_1, x_2 \ldots , x_n$ are identically distributed and independent random variables each with mean μ and variance σ^2, find the expected value of S^2, where

$$S^2 = \sum_{i=1}^{n} \frac{(x_i - \bar{x})^2}{n} \quad \text{and} \quad \bar{x} = \sum_{i=1}^{n} \frac{x_i}{n}$$

☐ **10–19** Given the two random variables x_1 and x_2 with the correlation coefficient ρ_{12}, prove that $|\rho_{12}| \leq 1$.

☐ **10–20** Consider the two random variables x_1 and x_2. If x_1^2 and x_2^2 have a zero correlation coefficient, find var $\{x_1 x_2\}$.

☐ **10–21** The nth moment of a random variable x is given by $E\{x^n\} = n!$. Show that its MGF is that of an exponential distribution with parameter 1.

☐ **10–22** Show by using the MGF that the sum of two independent Poisson distributions with parameters λ_1 and λ_2 is also Poisson with parameter $(\lambda_1 + \lambda_2)$.

☐ **10–23** Find the MGF of a binomial distribution with parameters (p, n). Then find its mean and variance.

☐ **10–24** Find the general expression for the nth moment of a normal distribution with mean zero and variance one by using the MGF.

☐ **10–25** If x_1 and x_2 are independent random variables with uniform distributions on the interval $(0, 1)$, find the PDF of $x_1 + x_2$.

☐ **10–26** Let $x_1, x_2, \ldots , x_n, \ldots$, be independent and identical exponential random variables with mean $1/\mu$. Define m such that

$$x_1 + x_2 + \cdots + x_m \leq t < x_1 + x_2 + \cdots + x_{m+1}$$

show that m is Poisson with mean μt.

□ **10–27** Consider the following PDF of the discrete random variable x that represents the monthly demand of a certain item.

x	0	1	2	3
$p(x)$	0.1	0.2	0.3	0.4

If the monthly demands are independent and identical, find the PDF for a 2-month demand.

□ **10–28** Classify the following Markov chains and find their stationary distributions.

(a)

$$\begin{pmatrix} 1/4 & 1/4 & 1/2 \\ 1/4 & 3/4 & 0 \\ 1/2 & 0 & 1/2 \end{pmatrix}$$

(b)

$$\begin{pmatrix} q & p & 0 & 0 & 0 \\ q & 0 & p & 0 & 0 \\ q & 0 & 0 & p & 0 \\ q & 0 & 0 & 0 & p \\ 1 & 0 & 0 & 0 & 0 \end{pmatrix}, \qquad p+q=1$$

□ **10–29** Find the mean recurrence time for each state of the following Markov chain:

$$\begin{pmatrix} 1/3 & 1/3 & 1/3 \\ 1/2 & 1/4 & 1/4 \\ 1/5 & 3/5 & 1/5 \end{pmatrix}$$

□ **10–30** Find the z-transform of a binomial distribution with parameters (p, n). From this determine its mean and variance.

□ **10–31** Solve for x_n in terms of y_n using the z-transform.

$$x_n = ax_{n-1} + by_n, \qquad n = 2, 3, \ldots$$
$$x_1 = y_1, \qquad x_0 = 0$$

Decision Theory and Games

In Chapters 2 through 9, all decision models are formulated and solved assuming the availability of *perfect* information. This is usually referred to as decision making under **certainty.** For example, in a product-mix problem, the profit per unit, c_j, of the jth product is assumed to be a fixed real value. If x_j is the decision variable representing the level of production for product j, the total profit contribution of the jth product is $c_j x_j$, which again is fixed for a given value of x_j.

The availability of *partial* or *imperfect* information about a problem leads to two new categories of decision-making situations:

1. Decisions under **risk.**
2. Decisions under **uncertainty.**

In the first category, the degree of ignorance about the data is expressed in terms of a probability density function, whereas in the second category no probability density function can be secured. In other words, from the standpoint of data availability, *certainty* and *uncertainty* represent the two extreme cases, and *risk* is the "in-between" situation.

To illustrate risk and uncertainty situations, consider the product-mix example just cited. Under *risk* conditions, the profit c_j will no longer be a fixed value; rather, it is a random variable whose exact numerical value is unknown but can be represented in terms of a probability density function, $f(c_j)$. Thus it does not make sense to talk about c_j without associating some probability statement with it. The profit contribution of the jth variable $c_j x_j$ is also a random variable whose exact value, for a given value of x_j, is unknown.

Under the condition of *uncertainty,* the probability density function $f(c_j)$ is either unknown or cannot be determined. Actually, uncertainty does *not* imply *complete* ignorance about the problem. For example, the decision maker may possess the partial information that c_j is equal to one of three values: c_j', c_j'', and c_j'''. However, as long as he or she cannot associate probabilities with these three values, the situation is regarded as decision making under uncertainty.

The degree of ignorance about data bears directly on how a problem is modeled and solved. For example, suppose that the product-mix problem has a total of n products. Under the assumption of certainty it makes sense to use $z = c_1 x_1 + c_2 x_2 + \cdots + c_n x_n$ as an objective criterion to be maximized subject to proper restrictions. However, under the assumption of risk, the same criterion would be of little value without some type of a probability statement, since z is actually a random variable. The criterion given becomes completely inadequate under the assumption of uncertainty, since the *specific* values of c_j are not known. This simple illustration indicates that insufficient data lead to a more complex decision model and, inevitably, a less satisfactory solution.

Unfortunately, the insufficiency of data has resulted in several, often inconsistent, approaches for quantifying and solving a decision model. There is almost universal acceptance of the criterion of maximizing profit (or minimizing its antithesis, cost) under conditions of certainty. However, several criteria exist in risk and uncertainty situations. For example, under risk the maximization of *expected* profit is sometimes acceptable, but it cannot be applied to every situation. Other criteria range from being completely conservative to being completely permissive. The situation becomes worse under uncertainty.

In most decision models, the problem resolves to selecting a best course(s) of action from a number (possibly infinite) of available options. This was shown in the models in Chapters 2 through 9. However, none of these models assumes that decisions are being made in an environment where the system itself is trying to "defeat" the decision maker. To be specific, suppose one is making a decision that depends on whether it would or would not rain. In this case, the decision maker does not expect nature to be a malevolent opponent.

11.1

2. Combined expected value and variance.
3. Known aspiration level.
4. Most likely occurrence of a future state.

Each of these criteria will now be explained in detail.

11.1.1 EXPECTED VALUE CRITERION

A natural extension of decisions under certainty is the use of expected value criterion, where it is desired to maximize expected profit (or minimize expected cost). This criterion may be expressed in terms of either *actual* money or its **utility.** To illustrate the difference between actual money and its utility, suppose that an investment of $20,000 will result in a gross profit of either zero or $100,000 with equal probabilities. Based on expected value of *money*, the individual's expected net gain is $100,000 \times .5 + 0 \times .5 - 20,000 = \$30,000$. Using this result alone, one would find that the "optimum" decision is to invest the $20,000. However, this decision may not be acceptable to all potential investors. For example, investor A may argue that because of the scarcity of liquid cash, the loss of $20,000 could lead to bankruptcy. Consequently, he may elect not to enter into this arrangement. Investor B, on the other hand, has a surplus of dormant capital that far exceeds any need for liquid money and, consequently, is willing to undertake the venture. What is being illustrated here is the importance of the decision maker's *attitude* toward the worth or utility of money. This point can be dramatized by considering investor A's situation again. Suppose that investor A would in no case be willing to risk the loss of more than $5,000. Suppose further that he has two ventures: Invest $20,000 and obtain a $100,000 gross profit with probability .5 and $0 with probability .5, or invest $5000 and obtain a $23,000 gross profit with probability .5 and $0 with probability .5. This information now shows that investor A has no choice but to select the second alternative even though its *expected* net profit of $6500 is much smaller than the $30,000 expected from the first alternative.

The main result from the illustration is that utility need not be directly proportional to actual money values. Unfortunately, although guidelines for establishing **utility curves** (i.e., actual money versus its utility) have been developed, utility is a rather subtle concept that cannot be quantified easily. In actual practice, the effect of utility may be expressed in terms of additional constraints that reflect the behavior of the decision maker. This point is illustrated by the maximum limit on the dollar loss investor A is willing to accept. In other words, it is not advisable to use expected money value as the only criterion for reaching a decision. Rather, it should serve only as a guide, and the final decision should ultimately be made by considering all the pertinent factors that affect the decision maker's attitude toward the utility of money.

Whether utility or actual money is used in computing expected values, the following drawback is usually cited against the use of the expected value criterion. Expectation implies that the same decision should be repeated a sufficiently large number of times before realizing the net value computed from the expectation formula. Mathematically, this result is expressed as follows. Let z be a random variable with expected value $E\{z\}$ and variance σ^2. If $(z_1, z_2, \ldots, z_n)$ is a random sample of n observed values of z, the sample average $\bar{z} = (z_1 + z_2 + \cdots + z_n)/n$ has a variance σ^2/n. Thus, as $n \to \infty$ (i.e., n becomes very large), $\sigma^2/n \to 0$ and $\bar{z}$ approaches $E\{z\}$. In other words, as the sample size becomes sufficiently large, the difference between the sample average and the expected value tends to zero. Thus, to use the expected value properly in comparing alternatives, one must expect the same decision process to be repeated a sufficiently large number of times. Otherwise, if the process is repeated a small number of times, the sample average $\bar{z}$ may differ considerably from $E\{z\}$. The main conclusion here then is that the use of expectation may be misleading for decisions that are applied only a few number of times.

Example 11.1-1. A preventive maintenance policy requires making decisions about when a machine (or a piece of equipment) should be serviced on a regular basis in order to minimize the cost of sudden breakdown. If the time horizon is specified in terms of equal time periods, the decision entails determining the optimal number of periods between two successive maintenances. If preventive maintenance is applied too frequently, the maintenance cost will increase while the cost of sudden breakdown will decrease. A compromise between the two extreme cases calls for balancing the costs of preventive maintenance and sudden breakdown.

Since we cannot predict in advance when a machine may break down, it is necessary to compute the probability that a machine would break down in a given period t. This is where the element of "risk" enters in the decision process.

The decision situation can be summarized as follows. A machine in a group of n machines is serviced when it breaks down. At the end of T periods, preventive maintenance is performed by servicing all n machines. The decision problem is to determine the optimum T that minimizes the total cost per period of servicing broken machines and applying preventive maintenance.

Let p_t be the probability that a machine would break down in period t, and let n_t be the random variable representing the number of broken machines in the same period. Further, assume that c_1 is the cost of repairing a broken machine and c_2 the preventive maintenance cost per machine.

The application of the expected value criterion to this example is reasonable if one can expect the machines to remain in operation for a large number of periods. The expected cost per period can be written as

$$EC(T) = \frac{c_1 \sum_{t=1}^{T-1} E\{n_t\} + c_2 n}{T}$$

where $E\{n_t\}$ is the expected number of broken machines in period t. Since n_t is a binomial random variable with parameter (n, p_t), $E\{n_t\} = np_t$. Thus

$$EC(T) = \frac{n\left(c_1 \sum_{t=1}^{T-1} p_t + c_2\right)}{T}$$

The necessary conditions for T^* to minimize $EC(T)$ are

$$EC(T^* - 1) \geq EC(T^*) \quad \text{and} \quad EC(T^* + 1) \geq EC(T^*)$$

Thus, by starting with a small value of T, computation of $EC(T)$ is continued until the foregoing conditions are satisfied.

To illustrate this point, suppose that $c_1 = \$100$, $c_2 = \$10$, and $n = 50$. The values of p_t are tabulated below. The table shows that preventive maintenance is applied to all machines every three $(= T^*)$ time periods.

	T	p_T	$\sum_{t=1}^{T-1} p_t$	$EC(T)$	
	1	.05	0	$500	
	2	.07	.05	375	
$T^* \to$	**3**	.10	.12	**366.7**	$\leftarrow EC(T^*)$
	4	.13	.22	400	
	5	.18	.35	450	

Exercise 11.1-1

In Example 11.1–1, suppose that the net worth of production per machine per period is $\$a$ and that it is desired to maximize the expected profit per period, $EP(T)$. Assume that the profit is computed as the difference between the net production worth and the cost of machine breakdown and maintenance.

(a) Write the general expression for $EP(T)$.
   [Ans. $EP(T) = n(a - c_2 - c_1 \sum_{t=1}^{T-1} P_t)/T$.]
(b) Write the necessary condition for determining T^* based on maximizing $EP(T)$.
   [Ans. $EP(T^*) \geq EP(T^* - 1)$ and $EP(T^*) \geq EP(T^* + 1)$.]

11.1.2 EXPECTED VALUE–VARIANCE CRITERION

In Section 11.1.1 we indicated that the expected value criterion is suitable mainly for making "long-run" decisions. The same criterion can be modified to improve its applicability to "short-run" decision problems by considering the following observation. If z is a random variable with variance σ^2, the sample average $\bar{z}$ has a variance σ^2/n, where n is the sample size. Thus, as σ^2 becomes smaller, the variance of $\bar{z}$ also becomes smaller and the probability that $\bar{z}$ approaches $E\{z\}$ becomes larger. This means that it is advantageous to develop a criterion that maximizes expected profit and, simultaneously, minimizes the variance of the profit. This is actually equivalent to considering multiple goals in the same criterion. A possible criterion reflecting this objective is

$$\text{maximize } E\{z\} - K \text{ var } \{z\}$$

where z is a random variable representing profit and K is a prespecified constant.

The constant K is sometimes referred to as **risk aversion factor.** It is actually a weighting factor that indicates the "degree of importance" of var $\{z\}$ relative to $E\{z\}$. For example, a decision maker particularly sensitive to large reductions in profit below $E\{z\}$ may choose K much larger than one. This arrangement will weigh the variance heavily and hence will result in a decision that reduces the chances of having low profit.

It is interesting that the new criterion is compatible with the use of utility in decision making, since the risk aversion factor K is an indicator of the decision maker's attitude toward excessive deviation from the expected values. This intuitive argument has a mathematical basis, and, using Taylor's series expansion, one can show that the first three terms in the expected utility function produce a criterion similar to the one just given (see Problem 11–7).

Example 11.1–2. The expected value–variance criterion is applied to Example 11.1–1. To do so, we need to compute the variance of the cost per period, that is, the variance of

$$C_T = \frac{c_1 \sum_{t=1}^{T-1} n_t + nc_2}{T}$$

C_T is a random variable because n_t $(t = 1, \ldots, T - 1)$ is a random variable. Since n_t is binomial with mean np_t and variance $np_t(1 - p_t)$, it follows that

$$\text{var } \{C_T\} = \left(\frac{c_1}{T}\right)^2 \sum_{t=1}^{T-1} \text{var } \{n_t\}$$

$$= \left(\frac{c_1}{T}\right)^2 \sum_{t=1}^{T-1} np_t(1 - p_t) = n\left(\frac{c_1}{T}\right)^2 \left\{ \sum_{t=1}^{T-1} p_t - \sum_{t=1}^{T-1} p_t^2 \right\}$$

Since $E\{C_T\} = EC(T)$, as given in Example 11.1–1, the criterion becomes

$$\text{minimize } EC(T) + K \text{ var } \{C_T\}$$

The function $EC(T)$ is *added* to K var $\{C_T\}$, since $EC(T)$ is a cost function. With $K = 1$, the criterion becomes

$$\text{minimize } EC(T) + \text{var } \{C_T\} = n\left\{ \left(\frac{c_1}{T} + \frac{c_1^2}{T^2}\right) \sum_{t=1}^{T-1} p_t - \left(\frac{c_1}{T}\right)^2 \sum_{t=1}^{T-1} p_t^2 + \frac{c_2}{T} \right\}$$

Using the same information as that in Example 11.1–1, we can set up the following table to give $T^* = 1$, which indicates that preventive maintenance must be applied every period.

It is interesting that, for the same data, the expected value–variance criterion has resulted in a more conservative decision that applies preventive maintenance every period compared with every third period in Example 11.1–1.

T	p_T	p_T^2	$\sum_{t=1}^{T-1} p_t$	$\sum_{t=1}^{T-1} p_t^2$	$EC(T) + \text{var}\{C_T\}$
1	.05	.0025	0	0	**500.00**
2	.07	.0049	.05	.0025	6312.50
3	.10	.0100	.12	.0074	6622.22
4	.13	.0169	.22	.0174	6731.25
5	.18	.0324	.35	.0343	6764.00

Exercise 11.1–2

Write the expected value–variance criterion associated with Exercise 11.1–1.
[*Ans.* Maximize $EP(T) - nK\,(c_1/T)^2\,\{\sum_{t+1}^{T-1} p_t - \sum_{t=1}^{T-1} p_t^2\}$.]

11.1.3 ASPIRATION-LEVEL CRITERION

The aspiration-level criterion does not yield an optimal decision in the sense of maximizing profit or minimizing cost. Rather, it is a means of determining *acceptable* courses of action. Consider, for example, the situation where a person advertises a used car for sale. On receiving an offer, the seller must decide, within a reasonable time span, whether it is acceptable or not. In this respect, the seller sets a price limit below which the car will not be sold. This is the **aspiration level,** which will allow the seller to accept the first offer that satisfies it. Such a criterion may not yield the optimum, for a later offer may be higher than the one accepted.

In making a decision in the used-car example, there was no mention of a probability distribution. Why then is the aspiration-level criterion classified as a technique for making decisions under risk? It can be argued that in selecting the aspiration level, the owner of the car is aware of the market values of similar cars. This is equivalent to saying that he has a "feeling" of the distribution of used-car prices. Admittedly, this does not provide a formal definition of a probability density function, but there is a basis here for collecting data that can be used to develop such a function. Indeed, one must assume that this is the case, since complete ignorance about the distribution may cause the owner to set the aspiration level too high, in which case no offer would be acceptable, or too low, in which case the owner may not collect a fair value for the car. In any case, one of the advantages of using aspiration-level method is that it may not be necessary to define the probability density function exactly.

This illustration pinpoints the usefulness of the aspiration-level criterion when *all* alternative courses of action are *not* available at the time the decision is made. This need not be the only situation where this criterion is used. Consider, for example, the situation in which a service facility (say, a laundry, a restaurant, or a barbershop) can be operated at different service rates. A high service rate, although it will provide fast, and hence convenient, service for customers, may be too costly for the owner. Conversely, slow service may not be as costly but could result in lost customers and hence in reduced profit. The objective is to determine the "optimum" level at which service may be performed.

In situations such as the foregoing, it is usually possible to determine the probability distributions for the arrivals of customers and their service times. Because such facilities supposedly operate for a long period of time, it appears ideal to determine the optimum service level based on minimizing the *expected* total cost (Section 11.1.1) of the facility per unit time. This includes the expected cost of operating the facility plus the expected cost of customer inconvenience, both being a function of the service level so that the higher is the first, the lower will be the second, and vice versa. However, this criterion becomes impractical because of the difficulty of estimating a cost for customer "inconvenience." Many intangible factors cannot be expressed readily in terms of cost since they depend on the customer's behavior and attitude.

The aspiration-level criterion may apply here. For example, one may decide to select the service level such that the service facility is idle only $\alpha\%$ of the time, while simultaneously requiring that the expected waiting time per customer does not exceed β time units. The parameters α and β are aspiration levels that the decision maker may determine based on a concept of running an "efficient" facility and a knowledge of the customers' behavior. Notice that α and β assume implicit cost values for operating the service facility and for the waiting time of customers, which, if known, could be used in the expected cost model. However, one can see that the determination of α and β is not as demanding as determining cost parameters.

Example 11.1-3. Suppose that the demand x per period on a certain commodity is given by the continuous probability density function $f(x)$. If the amount stocked at the beginning of the period is not sufficient, shortage may occur. If too much is stocked, extra inventory will be held at the end of the period. Both situations are costly. The first reflects loss of potential profit and loss of customers' goodwill; the second reflects an increase in the cost of storing and maintaining the inventory.

Presumably, one would like to balance these two conflicting costs. Since it is generally difficult to estimate the cost of shortage, one may wish to determine the level of stock such that the *expected* shortage quantity does not exceed A_1 units and the *expected* excess quantity does not exceed A_2 units. Mathematically, this is expressed as follows. Let I be the stock level to be determined. Thus

$$\text{expected shortage quantity} = \int_I^\infty (x-I)f(x)\,dx \le A_1$$

$$\text{expected excess quantity} = \int_0^I (I-x)f(x)\,dx \le A_2$$

In general, the selection of the aspiration levels A_1 and A_2 may not yield feasible values for I. In this case it would be necessary to relax one of the two restrictions in order to achieve feasibility.

To illustrate the example numerically, suppose that $f(x)$ is given by the distribution

$$f(x) = \begin{cases} \dfrac{20}{x^2}, & 10 \le x \le 20 \\ 0, & \text{otherwise} \end{cases}$$

It follows that

$$\int_I^{20} (x-I)f(x)\,dx = \int_I^{20} (x-I)\frac{20}{x^2}\,dx = 20\left\{\ln\frac{20}{I} + \frac{I}{20} - 1\right\}$$

$$\int_{10}^I (I-x)f(x)\,dx = \int_{10}^I (I-x)\frac{20}{x^2}\,dx = 20\left\{\ln\frac{10}{I} + \frac{I}{10} - 1\right\}$$

Thus the aspiration level criteria simplify to

$$\ln I - \frac{I}{20} \ge \ln 20 - \frac{A_1}{20} - 1 = 1.996 - \frac{A_1}{20}$$

$$\ln I - \frac{I}{10} \ge \ln 10 - \frac{A_2}{20} - 1 = 1.302 - \frac{A_2}{20}$$

The aspiration levels A_1 and A_2 must be such that the two inequalities can be satisfied simultaneously for at least one value of I.

For example, if $A_1 = 2$ and $A_2 = 4$, the inequalities become

$$\ln I - \frac{I}{20} \ge 1.896$$

$$\ln I - \frac{I}{10} \ge 1.102$$

The value of I must be between 10 and 20, since these are the limits of the demand. The following table shows that the two conditions are satisfied simultaneously for $13 \le I \le 17$. Thus, any of these values provides an answer to the problem.

I	10	11	12	13	14	15	16	17	18	19	20
$\ln I - I/20$	1.8	1.84	1.88	**1.91**	1.94	1.96	1.97	1.98	1.99	1.99	1.99
$\ln I - I/10$	1.3	1.29	1.28	1.26	1.24	1.21	1.17	**1.13**	1.09	1.04	.99 ◀

Exercise 11.1–3
Consider Example 11.1–3.
(a) Indicate if the following combinations of A_1 and A_2 will yield a feasible solution. If so, find the answer.
(1) $A_1 = A_2 = 3$ [*Ans.* $12 \le I \le 16$.]
(2) $A_1 = A_2 = 1$ [*Ans.* No feasible solution.]

(b) Suppose that $f(x) = 1/10$, $0 \leq x \leq 10$. Determine a general expression for determining a feasible range for I given the levels of aspiration A_1 and A_2.

[*Ans.* Max $\{0, 10 - \sqrt{20 A_1}\} \leq I \leq$ min $\{10, \sqrt{20 A_2}\}$, provided that the lower limit does not exceed the upper limit. If it does, no feasible solution exists.]

11.1.4 MOST LIKELY FUTURE CRITERION

This criterion is based on converting the probabilistic situation into a deterministic situation by replacing the random variable with the single value that has the highest probability of occurrence. For example, suppose that the profit per unit of a jth product is c_j, whose (discrete) probability density function is $p_j(c_j)$. Let c_j^* be defined such that $p_j(c_j^*) = $ max $p_j(c_j)$ for all c_j. Then c_j^* is treated as the "deterministic" value representing the per unit profit for the jth product.

This criterion may be thought of as a simplification of the more complex decision under risk. Such a simplification is done not for mere analytic convenience, but primarily for recognizing that, from the practical standpoint, the most probable future provides adequate information for making the decision. For example, there is always a positive probability (small as it may be) that an airplane may crash; yet most passengers fly under the assumption that air travel is always safe.

We must warn against the pitfalls of using the most probable future criterion. Suppose that the random variable under consideration assumes a large number of values each of which has a small probability of occurrence, say, .05, or less. Or consider the case where several values of the random variable occur with the same probability. In both cases, the most likely future criterion becomes inadequate for making a "sound" decision.

11.1.5 EXPERIMENTAL DATA IN DECISIONS UNDER RISK

In developing the criteria for decisions under risk, it is assumed that the probability distributions are known or can be secured. In this respect, these probabilities are referred to as **prior probabilities.**

It is sometimes possible to perform an experiment on the system under study and, depending on the outcomes of the experiment, modify the *prior* probabilities to reflect the availability of new information about the system. The new probabilities are known as the **posterior probabilities.**

Example 11.1–4. A typical situation in which experimentation is used occurs in inspection procedures. Suppose that a manufacturer produces a product in lots of fixed sizes. Because of occasional malfunctions in the production process, bad lots with an unacceptable number of defectives may be produced. Past experience indicates that the probability of producing bad lots is .05, in which case the probability of producing a good lot is .95. These are the prior probabilities. For convenience, let $\theta = \theta_1 (= \theta_2)$ indicate that the lot is good (bad), so that $P\{\theta = \theta_1\} = .95$ and $P\{\theta = \theta_2\} = .05$.

The manufacturer realizes that by shipping out a bad lot he may be penalized. However, based on the prior probabilities, he may decide that the probability of producing a bad lot is "too" small and hence may randomly choose any one of the available lots for shipping (compare the most likely future criterion, Section 11.1.4).

The decision is made without sampling from the shipped lot. For example, if the manufacturer makes his decision *after* testing a sample from the lot, the additional information could affect the final decision. Suppose that he decides to test a sample of two items from the lot. The outcomes of the test may show that

1. Both items are good.
2. One item is good.
3. Both items are defective.

Let z_1, z_2, and z_3 represent these three outcomes, respectively.

Because the sample is drawn either from a good or a bad lot, the conditional probabilities $P\{z_j|\theta_i\}$ are assumed available. The ultimate objective is to use these probabilities together with the prior probabilities to compute the required posterior probabilities which are defined by $P\{\theta_i|z_j\}$, that is, the probability of selecting either a good or a bad lot ($\theta = \theta_1$ or θ_2) given the outcome z_j of the experiment. These probabilities will be the basis for making a decision depending on the outcome of the sample test.

To show how the posterior probabilities $P\{\theta_i|z_j\}$ are computed from the prior probabilities $P\{\theta_i\}$ and the conditional probabilities $P\{z_j|\theta_i\}$, assume a general case in which $\theta = \theta_1, \theta_2, \ldots,$ or θ_m and $z = z_1, z_2, \ldots,$ or z_n. Since

$$P\{z_j\} = \sum_{i=1}^{m} P\{\theta_i, z_j\} = \sum_{i=1}^{m} P\{z_j|\theta_i\}P\{\theta_i\}$$

the posterior probabilities are given by

$$P\{\theta_i \mid z_j\} = \frac{P\{\theta_i, z_j\}}{P\{z_j\}} = \frac{P\{z_j \mid \theta_i\}P\{\theta_i\}}{\sum_{i=1}^{m} P\{z_j \mid \theta_i\}P\{\theta_i\}}$$

These probabilities are also known as **Bayes's probabilities.**

Returning now to the numerical example, we suppose that the percentage of defectives in a good lot is 4%, while a bad lot has 15% defective items. Then based on a binomial distribution and a sample of size 2, the conditional probabilities of an outcome z_j given a lot is good or bad are as follows:

$$P\{z_1 \mid \theta_1\} = C_2^2(.96)^2(.04)^0 = .922$$
$$P\{z_2 \mid \theta_1\} = C_1^2(.96)^1(.04)^1 = .0768$$
$$P\{z_3 \mid \theta_1\} = C_0^2(.96)^0(.04)^2 = .0016$$
$$P\{z_1 \mid \theta_2\} = C_2^2(.85)^2(.15)^0 = .7225$$
$$P\{z_2 \mid \theta_2\} = C_1^2(.85)^1(.15)^1 = .255$$
$$P\{z_3 \mid \theta_2\} = C_0^2(.85)^0(.15)^2 = .0225$$

These probabilities can be summarized conveniently as shown in the following table:

$$P\{z_j \mid \theta_i\} =$$

	z_1	z_2	z_3
θ_1	.922	.0768	.0016
θ_2	.7225	.255	.0225

Given $P\{\theta = \theta_1\} = .95$ and $P\{\theta = \theta_2\} = .05$, the joint probabilities

$$P\{\theta_i, z_j\} = P\{z_j \mid \theta_i\} P\{\theta_i\}$$

can be determined from the foregoing table by multiplying its first row by .95 and its second row by .05. Thus we obtain

$$P\{\theta_i, z_j\} =$$

	z_1	z_2	z_3
θ_1	.8759	.07296	.00152
θ_2	.036125	.01275	.001125

Next, we determine $P\{z_j\}$ by using the formula

$$P\{z_j\} = \sum_{i=1}^{2} P\{\theta_i, z_j\}$$

This is equivalent to summing the columns of the last table. Thus we obtain

$$P\{z_1\} = .912025, \qquad P\{z_2\} = .08571, \qquad P\{z_3\} = .002645$$

Finally, we obtain the posterior probabilities by using the formula

$$P\{\theta_i \mid z_j\} = \frac{P\{\theta_i, z_j\}}{P\{z_j\}}$$

These probabilities are computed by dividing the columns of the last table by the associated $P\{z_j\}$. Thus we obtain the following table:

$$P\{\theta_i \mid z_j\} =$$

	z_1	z_2	z_3
θ_1	.96039	.85124	.57467
θ_2	.03961	.14876	.42533

It is interesting to see how the posterior probabilities can affect the final decision depending on the outcome z_j of the test. If both items tested are good ($z = z_1$), the probability the lot is good is .96039. If both are bad ($z = z_3$), it is almost equally likely that the lot is good or bad. This shows that the final decision can be affected by the outcome z_j. ◀

Exercise 11.1–4

In Example 11.1–4, suppose that the test is applied to a sample of size 1. Specify the outcomes of the test and compute the posterior probabilities using the data of the example.

[*Ans.* There are two outcomes z_1 and z_2, representing whether the tested item is good or bad. $P\{\theta_1 \mid z_1\} = .955474$, $P\{\theta_1 \mid z_2\} = .835165$, $P\{\theta_2 \mid z_1\} = .044526$, and $P\{\theta_2 \mid z_2\} = .164835$.]

We now illustrate how the posterior probabilities are used in decision making.

Example 11.1-5. In Example 11.1–4, suppose that the manufacturer ships lots to two customers, *A* and *B*. The contracts specify that the percentage of defectives for *A* and *B* should not exceed 5 and 8, respectively. A penalty of $100 is incurred per percentage point above the maximum limit. On the other hand, supplying better quality lots will cost the manufacturer $80 per percentage point. Assuming that a sample of size 2 is inspected prior to shipping, how should the manufacturer decide where to ship an inspected lot?

There are two possible actions for this problem, namely,

a_1:   ship the lot to customer *A*
a_2:   ship the lot to customer *B*

Letting θ_1 and θ_2 represent the two types with 4% and 15% defectives, we can develop a cost matrix as follows:

$$C(a, \theta) = \begin{array}{c} \\ a_1 \\ a_2 \end{array} \begin{array}{|c|c|} \hline \theta_1 & \theta_2 \\ \hline \$80 & \$1000 \\ \hline \$320 & \$700 \\ \hline \end{array}$$

Action a_1 specifies that customer *A* will accept lots with 5% defective without penalty. If the lot has 4% defective ($= \theta_1$) it will cost the manufacturer $(5 - 4) \times \$80 = \$80$ for supplying a better quality than needed, but if the lot has 15% defectives ($= \theta_2$), a penalty of $(15 - 5) \times \$100 = \$1000$ will be incurred. A similar reasoning is used to obtain the cost elements of action a_2 (verify!).

We notice now that the decision-making process must be a function of the outcomes z_1, z_2, and z_3 of the sample test. In other words, we must decide which action is preferable (less costly) given that the outcome of the test is two good items, one good item, or two bad items. We shall base our decision on minimization of expected costs. A general formula for computing expected costs is

$$E\{a_k \mid z_j\} = \sum_{\theta_i} C(a_k, \theta_i)P\{\theta_i \mid z_j\}$$

Case 1: The outcome is z_1 (two good items):

$$E\{a_1 \mid z_1\} = 80 \times .96039 + 1000 \times .03961 = \textbf{\$116.44}$$
$$E\{a_2 \mid z_1\} = 320 \times .96039 + 700 \times .03961 = \$335.05$$

Thus, if the outcome is z_1, the decision is to ship the lot to customer *A*, since a_1 yields the lower expected cost.

Case 2: The outcome is z_2 (one good item):

$$E\{a_1 \mid z_2\} = 80 \times .85124 + 1000 \times .14876 = \textbf{\$216.86}$$
$$E\{a_2 \mid z_2\} = 320 \times .85124 + 700 \times .14876 = \$376.53$$

Again, as in case 1, the lot should be shipped to customer A if the outcome of the test indicates one good item.

Case 3: The outcome is z_3 (two defective items):
$$E\{a_1 \mid z_3\} = 80 \times .57467 + 1000 \times .42533 = \textbf{\$471.30}$$
$$E\{a_2 \mid z_3\} = 320 \times .57467 + 700 \times .42533 = \$481.63$$

Thus, in this case also, the lot should be shipped to customer A.

The general decision for the problem, then, is that all lots should be shipped to customer A regardless of the outcome of the test.   ◀

Exercise 11.1–5
Use the data of Example 11.1–5 to determine the optimal decision in Exercise 11.1–4.

[*Ans.* Ship lots to A regardless of the outcome of the test.]

11.2   DECISION TREES

In Section 11.1 we presented decision criteria for evaluating what may be termed as "single-stage" alternatives, in the sense that no future decisions will depend on the the decision taken now. In this section we consider a "multiple-stage" decision process in which dependent decisions are made in tandem. A graphical representation of the decision problem can be made by using a **decision tree.** This representation facilitates the decision-making process. The following example illustrates the basics of the decision tree procedure.

Example 11.2–1.† A company has the options now of building a full-size plant or a small plant that can be expanded later. The decision depends primarily on future demands for the product the plant will manufacture. The construction of a full-size plant can be justified economically if the level of demand is high. Otherwise, it may be advisable to construct a small plant now and then decide in two years whether it should be expanded.

The multistage decision problem arises here because if the company decides to build a small plant now, a future decision must be made in two years regarding expansion. In other words, the decision process involves two stages: a decision now regarding the size of the plant, and a decision two years from now regarding expansion (assuming that it is decided to construct a small plant now).

Figure 11–1 summarizes the problem as a *decision tree.* It is assumed that the demand can be either high or low. The decision tree has two types of nodes: a square (□) represents a *decision point* and a circle (○) stands for a *chance event.* Thus, starting with node 1 (a decision point), we must make a decision regarding the size of the plant. Node 2 is a chance event from which

---

† This example is adapted from J. F. Magee, "Decision Trees for Decision Making," *Harvard Business Review,* July–August 1964, pp. 126–138.

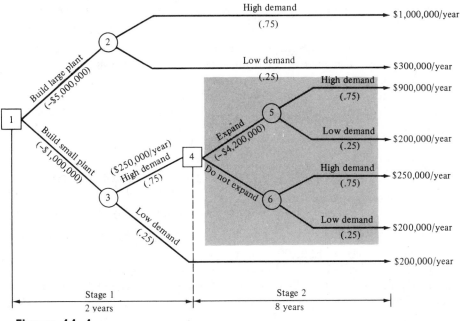

Figure 11-1

two branches representing low and high demand emanate depending on the conditions of the market. These conditions will be represented by associating probabilities with each branch. Node 3 is also a chance event from which two branches representing high and low demands emanate.

Logically, the company will consider possible future expansion of the small plant only if the demand over the first two years turns out to be high. This is the reason node 4 represents a decision point with its two emanating branches representing the "expansion" and "no expansion" decisions. Again, nodes 5 and 6 are chance events, and the branches emanating from each represent high and low demands.

The data for the decision tree must include (1) the probabilities associated with the branches emanating from the chance events and (2) the revenues associated with different alternatives of the problem. Suppose that the company is interested in studying the problem over a 10-year period. A market survey indicates that the probabilities of having high and low demands over the next 10 years are .75 and .25, respectively. The immediate construction of a large plant will cost $5 million and a small plant will cost only $1 million. The expansion of the small plant 2 years from now is estimated to cost $4.2 million. Estimates of annual income for each of the alternatives are given as follows.

1. Full-size plant and high (low) demand will yield $1,000,000 ($300,000) annually.

2. Small plant and low demand will yield $200,000 annually.

3. Small plant and high demand will yield $250,000 for each of the 10 years.

4. Expanded small plant with high (low) demand will yield $900,000 ($200,000) annually.

5. Small plant with no expansion and high demand in the first two years followed by low demand will yield $200,000 in each of the remaining 8 years.

These data are summarized in Figure 11–1. We are now ready to evaluate the alternatives. The final decision must tell us what to do at both decision nodes 1 and 4.

The evaluation of the alternatives is based on the use of the expected value criterion. The computations start at stage 2 and then move backward to stage 1. Thus, for the last 8 years, we can evaluate the two alternatives at node 4 as follows:

$E\{$net profit | expansion$\}$

$$= (900,000 \times .75 + 200,000 \times .25) \times 8 - 4,200,000 = \$1,600,000$$

$E\{$net profit | no expansion$\}$

$$= (250,000 \times .75 + 200,000 \times .25) \times 8 = \textbf{\$1,900,000}$$

Thus, at node 4, the decision calls for no expansion, and the associated expected net profit is $1,900,000.

We can now replace all the branches emanating from node 4 by a single branch with an expected net profit of $1,900,000, representing the net profit for the *last 8 years*. We now make stage 1 computations corresponding to node 1 as follows:

$E\{$net profit | large plant$\}$

$$= (1,000,000 \times .75 + 300,000 \times .25) \times 10 - 5,000,000 = \textbf{\$3,250,000}$$

$E\{$net profit | small plant$\}$

$$= (1,900,000 + 500,000 \times .75 + 2,000,000 \times .25 - 1,000,000 = \$1,300,000$$

Thus the optimal decision at node 1 is to build a full-size plant now. Making this decision now obviously eliminates the need for considering the alternatives at node 4. ◀

Exercise 11.2–1

In Example 11.2–1, suppose that demand during the last 8 years can be high, medium, or low, with probabilities .7, .2, and .1, respectively. The annual incomes are summarized as follows:

1. Expanded small plant with high, medium, and low demands will yield annual income of $900,000, $600,000, and $200,000.
2. Nonexpanded small plant with high, medium, and low demand will yield annual income of $400,000, $280,000, and $150,000.

Determine the optimal decision at node 4.

[*Ans.* $E\{$net profit | expansion$\} = \$1,960,000$ and $E\{$net profit | no expansion$\} = \$2,808,000$. Hence do not expand at node 4.]

11.3 DECISIONS UNDER UNCERTAINTY

This section introduces a number of criteria for making decisions under uncertainty under the assumption that no probability distributions are available. The methods to be presented here include

1. The Laplace criterion.
2. The Minimax criterion.
3. The Savage criterion.
4. The Hurwicz criterion.

The major difference among these criteria is reflected by how conservative the decision maker is in dealing with the prevailing uncertainty conditions. For example, the Laplace criterion is based on more optimistic conditions than the minimax criterion. Also, the Hurwicz criterion can be adjusted to reflect attitudes ranging from the most optimistic to the most pessimistic. In this respect, the criteria, even though they are quantitative in nature, reflect a subjective appraisal of the environment in which the decision is made. Unfortunately, there are no general guides as to which criterion should be implemented since the (changing) mood of the decision maker dictated by the uncertainty of the situation may be an important factor in choosing a suitable criterion.

In the foregoing criteria, it is assumed that the decision maker does not have an *intelligent* opponent. In this case, "nature" is said to be the opponent and there is no reason to believe that "nature" *aims* at inflicting losses on the decision maker.

Situations exist, however, where "nature" is replaced by an intelligent opponent whose interests conflict with those of the decision maker. For example, in a war, opposing armies represent intelligent opponents. The existence of this new element requires special provisions in designing a suitable criterion. **Game theory,** presented in Section 11.4, handles this case.

The information utilized in making decisions under uncertainty is usually summarized in the form of a matrix with its row representing possible **actions** and its columns representing possible future **states** of the system. Consider, for example, the situation where a company is faced with a labor strike. Depending on the length of the strike, a level of inventory for a certain item must be maintained. The future states of the system (columns) are represented by the possible length of the strike, and the actions (rows) are represented by the level of inventory that should be maintained. This means that an action represents a possible decision.

Associated with each action and each future state is an outcome that evaluates the gain (or loss) resulting from taking such action when a given future state occurs. Thus, if a_i represents the ith action ($i = 1, 2, \ldots , m$) and θ_j represents the jth future state ($j = 1, 2, \ldots , n$), then $v(a_i, \theta_j)$ will represent the associated outcome. In general, $v(a_i, \theta_j)$ may be a continuous function of a_i and θ_j. Under discrete conditions, this information is arranged as shown next in the matrix. This representation will be the basis for developing the criteria for decisions under uncertainty.

	θ_1	θ_2	$\cdots$	θ_n
a_1	$v(a_1, \theta_1)$	$v(a_1, \theta_2)$	$\cdots$	$v(a_1, \theta_n)$
a_2	$v(a_2, \theta_1)$	$v(a_2, \theta_2)$	$\cdots$	$v(a_2, \theta_n)$
$\vdots$	$\vdots$	$\vdots$		$\vdots$
a_m	$v(a_m, \theta_1)$	$v(a_m, \theta_2)$	$\cdots$	$v(a_m, \theta_n)$

11.3.1 LAPLACE CRITERION

This criterion is based on what is known as the **principle of insufficient reason.** Since the probabilities associated with the occurrence of $\theta_1, \theta_2, \ldots$, and θ_n are unknown, we do not have enough information to conclude that these probabilities will be different. For if this is not the case, we should be able to determine these probabilities, and the situation will no longer be a decision under uncertainty. Thus, because of insufficient reason to believe otherwise, the states θ_1, $\theta_2, \ldots$, and θ_n are equally likely to occur. When this conclusion is established, the problem is converted into a decision under risk, where one selects the action a_i yielding the largest *expected* gain. That is, select the action a_i^* corresponding to

$$\max_{a_i} \left\{ \frac{1}{n} \sum_{j=1}^{n} v(a_i, \theta_j) \right\}$$

where $1/n$ is the probability that θ_j ($j = 1, 2, \ldots, n$) occurs.

Example 11.3-1. A recreational facility must decide on the level of supplies it must stock to meet the needs of its customers during one of the holidays. The exact number of customers is not known, but it is expected to be in one of four categories: 200, 250, 300, or 350 customers. Four levels of supplies are thus suggested, with level i being ideal (from the viewpoint of incurred costs) if the number of customers falls in category i. Deviation from the ideal levels results in additional costs either because extra supplies are stocked needlessly or because demand cannot be satisfied. The following table provides these costs in thousands of dollars.

	Customer Category			
	θ_1	θ_2	θ_3	θ_4
a_1	5	10	18	25
a_2	8	7	8	23
a_3	21	18	12	21
a_4	30	22	19	15

(Supplies Level at left of a_1–a_4)

The Laplace principle assumes that θ_1, θ_2, θ_3, and θ_4 are equally likely to occur. Thus the associated probabilities are given by $P\{\theta = \theta_j\} = 1/4$, $j = 1$, 2, 3, 4, and the expected costs for the different actions a_1, a_2, a_3, and a_4 are

$$\begin{aligned}
E\{a_1\} &= (1/4)(5 + 10 + 18 + 25) = 14.5 \\
E\{a_2\} &= (1/4)(8 + 7 + 8 + 23) = \mathbf{11.5} \\
E\{a_3\} &= (1/4)(21 + 18 + 12 + 21) = 18.0 \\
E\{a_4\} &= (1/4)(30 + 22 + 19 + 15) = 21.5
\end{aligned}$$

Thus the best level of inventory according to Laplace criterion is specified by a_2. ◄

Exercise 11.3-1

Suppose it is decided that customer category 4 is not a feasible possibility, determine the optimal supply in Example 11.3–1. [*Ans.* a_2.]

11.3.2 MINIMAX (MAXIMIN) CRITERION

This is the most conservative criterion since it is based on making the best out of the worst possible conditions. That is, if the outcome $v(a_i, \theta_j)$ represents loss for the decision maker, then, for a_i, the worst loss regardless of what θ_j may be is $\max_{\theta_j} \{v(a_i, \theta_j)\}$. The **minimax criterion** then selects the action a_i associated with $\min_{a_i} \max_{\theta_j} \{v(a_i, \theta_j)\}$. In a similar manner, if $v(a_i, \theta_j)$ represents gain, the criterion selects the action a_i associated with $\max_{a_i} \min_{\theta_j} \{v(a_i, \theta_j)\}$. This is called the **maximin criterion.**

Example 11.3–2. Consider Example 11.3–1. Since $v(a_i, \theta_j)$ represents cost, the minimax criterion is applicable. The computations are summarized in the matrix. The minimax strategy is a_3.

	θ_1	θ_2	θ_3	θ_4	$\max_{\theta_j}\{v(a_i, \theta_j)\}$
a_1	5	10	18	25	25
a_2	8	7	8	23	23
a_3	21	18	12	21	**21** ← minimax value
a_4	30	22	19	15	30

$v(a_j, \theta_j) =$ for rows above. ◀

Exercise 11.3-2

Apply the minimax criterion to Exercise 11.3–1.
[*Ans.* The minimax value = 8, corresponding to a_2.]

11.3.3 SAVAGE MINIMAX REGRET CRITERION

The minimax criterion of Section 11.3.2 is extremely conservative, to the extent that it may sometimes lead to illogical conclusions. Consider the following *loss* matrix, which is usually quoted as a classic example for justifying the need for the Savage "less conservative" criterion.

	θ_1	θ_2
a_1	\$11,000	\$90
a_2	\$10,000	\$10,000

A minimax criterion applied to this matrix yields a_2. But intuitively we are tempted to choose a_1, since there is a chance that if $\theta = \theta_2$ only \$90 will be lost, whereas it is certain that a_2 will yield a loss of \$10,000 whether $\theta = \theta_1$ or θ_2.

The Savage criterion "rectifies" this point by constructing a new loss matrix in which $v(a_i, \theta_j)$ is replaced by $r(a_i, \theta_j)$, which is defined by

$$r(a_i, \theta_j) = \begin{cases} \max_{a_k} \{v(a_k, \theta_j)\} - v(a_i, \theta_j), & \text{if } v \text{ is profit} \\ v(a_i, \theta_j) - \min_{a_k} \{v(a_k, \theta_j)\}, & \text{if } v \text{ is loss} \end{cases}$$

This means that $r(a_i, \theta_j)$ is the difference between the best choice in column θ_j and the values of $v(a_i, \theta_j)$ in the same column. In essence, $r(a_i, \theta_j)$ is a representation of the "regret" of the decision maker as a result of missing the best choice corresponding to a given future state θ_j. The function $r(a_i, \theta_j)$ is referred to as the **regret matrix**.

To show how the new elements $r(a_i, \theta_j)$ produce a logical conclusion for the foregoing example, consider

$$r(a_i, \theta_j) = \begin{array}{c|cc} & \theta_1 & \theta_2 \\ \hline a_1 & \$1000 & \$0 \\ a_2 & \$0 & \$9900 \end{array}$$

The minimax criterion yields a_1, as is expected.

Notice that whether $v(a_i, \theta_j)$ is a profit or a loss function, $r(a_i, \theta_j)$ is a regret function which, in both cases, represents loss. Thus only the minimax (and not the maximin) criterion can be applied to $r(a_i, \theta_j)$.

Example 11.3–3. Consider Example 11.3–1. The given matrix represents costs. The corresponding regret matrix given here is determined by subtracting 5, 7, 8, and 15 from columns 1, 2, 3, and 4, respectively.

$r(a_i, \theta_j) =$	θ_1	θ_2	θ_3	θ_4	$\max_{\theta_j}\{r(a_i, \theta_j)\}$
a_1	0	3	10	10	10
a_2	3	0	0	8	8 ← minimax value
a_3	16	11	4	6	16
a_4	25 15		11	0	25

Although the same minimax criterion is used to determine the best action (a_2 in this case), the use of $r(a_i, \theta_j)$ has resulted in a different solution from that in Example 11.3–2. ◀

Exercise 11.3–5

Resolve Example 11.3–3 assuming that a_2 is not a possibility.
[*Ans.* The minimax of $r(a_i, \theta_j) = 10$, corresponding to a_1.]

11.3.4 HURWICZ CRITERION

This criterion represents a range of attitudes from the most optimistic to the most pessimistic. Under the most optimistic conditions, one would choose the action yielding $\max_{a_i} \max_{\theta_j} \{v(a_i, \theta_j)\}$. [It is assumed that $v(a_i, \theta_j)$ represents gain or profit.] Similarly, under the most pessimistic conditions, the chosen action corresponds to $\max_{a_i} \min_{\theta_j} \{v(a_i, \theta_j)\}$. The Hurwicz criterion strikes a balance between extreme pessimism and extreme optimism by weighing the above two conditions by the respective weights α and $(1 - \alpha)$, where $0 \leq \alpha \leq 1$. That is, if $v(a_i, \theta_j)$ represents profit, select the action that yields

$$\max_{a_i} \{\alpha \max_{\theta_j} v(a_i, \theta_j) + (1 - \alpha) \min_{\theta_j} v(a_i, \theta_j)\}$$

For the case where $v(a_i, \theta_j)$ represents cost, the criterion selects the action that yields

$$\min_{a_i} \{\alpha \min_{\theta_j} v(a_i, \theta_j) + (1 - \alpha) \max_{\theta_j} v(a_i, \theta_j)\}$$

The parameter α is known as the **index of optimism:** when $\alpha = 1$, the criterion is "too" optimistic; when $\alpha = 0$, it is "too" pessimistic. A value of α between zero and one can be selected depending on whether the decision maker leans toward pessimism or optimism. In the absence of a strong feeling one way or the other, a value of $\alpha = 1/2$ seems to be a reasonable choice.

Example 11.3-4. The Hurwicz principle is applied to Example 11.3-1. It is assumed that $\alpha = 1/2$. The necessary calculations are shown in the table that follows. The optimum solution is given by either a_1 or a_2.

	$\min_{\theta_j} v(a_i, \theta_j)$	$\max_{\theta_j} v(a_i, \theta_j)$	$\alpha \min_{\theta_j} v(a_i, \theta_j) + (1 - \alpha) \max_{\theta_j} v(a_i, \theta_j)$
a_1	5	25	15 $\Big\}$
a_2	7	23	15 $\Big\}$ $\leftarrow \min_{a_i}$
a_3	12	21	16.5
a_4	15	30	22.5 ◄

Exercise 11.3-4
Resolve Example 11.3-4 assuming that $\alpha = .75$.
[*Ans.* Choose a_1 with a value of 10.]

11.4   GAME THEORY

In Section 11.3 the criteria for decisions under uncertainty are developed under the assumption that "nature" is the opponent. In this respect, nature is not malevolent. This section deals with decisions under uncertainty involving two or more *intelligent* opponents in which each opponent aspires to optimize his own decision at the expense of the other opponents. Typical examples include launching advertisement campaigns for competing products and planning war tactics for opposing armies.

In game theory, an opponent is referred to as a **player.** Each player has a number of choices, finite or infinite, called **strategies.** The **outcomes** or **payoffs** of a game are summarized as functions of the different strategies for each player. A game with two players, where a gain of one player *equals* a loss to the other, is known as **two-person zero-sum game.** In such a game it suffices to express the outcomes in terms of the payoff to one player. A matrix similar to the one used in Section 11.3 is usually used to summarize the payoffs to the player whose strategies are given by the rows of the matrix. This section will deal primarily with two-person zero-sum games.

Example 11.4-1. To illustrate the definitions of a *two-person zero-sum* game, consider a coin-matching situation in which each of the two players *A* and *B* selects a head (*H*) or a tail (*T*). If the outcomes match (i.e., *H* and *H*, or *T* and *T*), player *A* wins $1.00 from player *B*. Otherwise, *A* loses $1.00 to *B*.

In this game each player has two strategies (*H* or *T*), which yield the following 2×2 game matrix expressed in terms of the payoff to A:

$$
\begin{array}{cc}
 & \text{Player } B \\
 & \begin{array}{cc} H & \quad T \end{array}
\end{array}
$$

$$
\text{Player } A \; \begin{array}{c} H \\ T \end{array} \left|\begin{array}{cc} 1 & -1 \\ -1 & 1 \end{array}\right.
$$

The "optimal" solution to such a game may require each player to play a **pure strategy** (e.g., either *H* or *T*) or a mixture of pure strategies. The latter case is known as **mixed strategy** selection.   ◀

11.4.1 OPTIMAL SOLUTION OF TWO-PERSON ZERO-SUM GAMES

The selection of a criterion for solving a decision problem depends largely on the available information. Games represent the ultimate case of lack of information in which intelligent opponents are working in a conflicting environment. The result is that a very conservative criterion, called the **minimax-maximin** criterion, is usually proposed for solving two-person zero-sum games. This criterion was introduced in Section 11.3.2. The main difference is that "nature" is not regarded as an active (or malevolent) opponent, whereas in game theory each player is intelligent and hence actively tries to defeat his opponent.

To accommodate the fact that each opponent is working against the other's interest, the minimax criterion selects each player's (mixed or pure) strategy that yields the *best* of the *worst* possible outcomes. An optimal solution is said to be reached if neither player finds it beneficial to alter his strategy. In this case, the game is said to be **stable** or in a state of equilibrium.

Since the game matrix is usually expressed in terms of the payoff to player *A* (whose strategies are represented by the rows), the (conservative) criterion calls for *A* to select the strategy (mixed or pure) that maximizes his minimum gain, the minimum being taken over all the strategies of player *B*. By the same reasoning, player B selects the strategy that minimizes his maximum losses. Again, the maximum is taken over all A's strategies.

The following example illustrates the computations of the minimax and maximin values of a game.

Example 11.4–2. Consider the following payoff matrix, which represents player A's gain. The computations of the minimax and maximin values are shown on the matrix.

		Player B				
		1	2	3	4	Row Minimum
Player A	1	8	2	9	5	2
	2	6	5	7	18	**5** Maximin
	3	7	3	−4	10	−4
Column Maximum		8	**5**	9	18	
			Minimax			

When player A plays his first strategy, he may gain 8, 2, 9, or 5, depending on player B's selected strategy. He can guarantee, however, a gain of at least min $\{8, 2, 9, 5\} = 2$ regardless of B's selected strategy. Similarly, if A plays his second strategy, he is guaranteed an income of at least min $\{6, 5, 7, 18\} = 5$, and if he plays his third strategy, he is guaranteed an income of at least min $\{7, 3, −4, 10\} = −4$. Thus the minimum value in each row represents the minimum gain guaranteed A if he plays his pure strategies. These are indicated in the matrix by "row minimum." Now, player A, by selecting his second strategy, is maximizing his minimum gain. This gain is given by max $\{2, 5, −4\} = 5$. Player A's selection is called the **maximin strategy,** and his corresponding gain is called the **maximin (or lower) value** of the game.

Player B, on the other hand, wants to minimize her losses. She realizes that, if she plays her first pure strategy, she can lose no more than max $\{8, 6, 7\} = 8$ regardless of A's selections. A similar argument can also be applied to the three remaining strategies. The corresponding results are thus indicated in the matrix by "column maximum." Player B will then select the strategy that minimizes her maximum losses. This is given by the second strategy and her corresponding loss is given by min $\{8, 5, 9, 18\} = 5$. Player B's selection is called the **minimax strategy** and her corresponding loss is called the **minimax (or upper) value** of the game.   ◀

From the conditions governing the minimax criterion, the minimax (upper) value is *greater than* or *equal to* the maximum (lower) value (see Problem 11–24). In the case where the equality holds, that is, minimax value = maximin value, the corresponding pure strategies are called "optimal" strategies and the game is said to have a **saddle point.** The value of the game, given by the common entry of the optimal pure strategies, is equal to the maximin and the minimax values. "Optimality" here signifies that neither player is tempted to change his or her strategy, since the opponent can counteract by selecting another strategy yielding less attractive payoff. In general, the value of the game must satisfy the inequality maximin (lower) value $\leq$ value of the game $\leq$ minimax (upper) value

In the preceding example, maximin value = minimax value = 5. This implies that the game has a saddle point which is given by the entry (2, 2) of the matrix. The value of the game is thus equal to 5. Notice that neither player can improve his position by selecting any other strategy.

11.4.2 MIXED STRATEGIES

The preceding section shows that the existence of a saddle point immediately yields the optimal pure strategies for the game. Some games do not have saddle points, however. For example, consider the following zero-sum game:

		\(B\)				
		1	2	3	4	Row Minimum
	1	5	−10	9	0	−10
\(A\)	2	6	7	8	1	1
	3	8	7	15	2	**2** Maximin
	4	3	4	−1	4	−1
Column Maximum		8	7	15	**4**	
					Minimax	

The minimax value (= 4) is greater than the maximin value (= 2). Hence the game does not have a saddle point and the pure maximin-minimax strategies are not optimal. This is true, since each player can improve his payoff by selecting a different strategy. In this case, the game is said to be **unstable.**

The failure of the minimax-maximin (pure) strategies, in general, to give an optimal solution to the game has led to the idea of using mixed strategies. Each player, instead of selecting a pure strategy only, may play all his strategies according to a predetermined set of probabilities. Let $x_1, x_2, \ldots, x_m$ and $y_1, y_2, \ldots, y_n$ be the row and column probabilities by which A and B, respectively, select their pure strategies. Then

$$\sum_{i=1}^{m} x_i = \sum_{j=1}^{n} y_j = 1$$

$$x_i, y_j \geq 0, \quad \text{for all } i \text{ and } j$$

Thus if a_{ij} represents the (i, j)th entry of the game matrix, x_i and y_i will appear as in the following matrix:

		\(B\)			
		y_1	y_2	$\cdots$	y_n
	x_1	a_{11}	a_{12}	$\cdots$	a_{1n}
A	x_2	a_{21}	a_{22}	$\cdots$	a_{2n}
	$\vdots$	$\vdots$	$\vdots$		$\vdots$
	x_m	a_{m1}	a_{m2}	$\cdots$	a_{mn}

The solution of the mixed strategy problem is based also on the minimax criterion given in Section 11.4.1. The only difference is that A selects x_i that maximize the smallest *expected* payoff in a column, whereas B selects y_j that minimize the largest *expected* payoff in a row. Mathematically, the minimax criterion for a mixed strategy case is given as follows. Player A selects x_i ($x_i \geq 0$, $\Sigma_{i=1}^{m} x_i = 1$) that will yield

$$\max_{x_i} \left\{ \min \left(\sum_{i=1}^{m} a_{i1}x_i, \sum_{i=1}^{m} a_{i2}x_i, \ldots , \sum_{i=1}^{m} a_{in}x_i \right) \right\}$$

and player B selects y_j ($y_j \geq 0$, $\Sigma_{j=1}^{n} y_j = 1$) that will yield

$$\min_{y_j} \left\{ \max \left(\sum_{j=1}^{n} a_{1j}y_j, \sum_{j=1}^{n} a_{2j}y_j, \ldots , \sum_{j=1}^{n} a_{mj}y_j \right) \right\}$$

These values are referred to as the maximin and the minimax expected payoffs, respectively.

As in the pure strategies case, the relationship

$$\text{minimax expected payoff} \geq \text{maximin expected payoff}$$

holds. When x_i and y_j correspond to the optimal solution, the equality holds and the resulting values become equal to the (optimal) expected value of the game. This result follows from the **minimax theorem** and is stated here without proof (see Problem 11–31). If x_i^* and y_j^* are the optimal solutions for both players each payoff element a_{ij} will be associated with the probability $(x_i^* y_j^*)$. Thus the optimal expected value of the game is

$$v^* = \sum_{i=1}^{m} \sum_{j=1}^{n} a_{ij} x_i^* y_j^*$$

There are several methods for solving two-person zero-sum games for the optimal values of x_i and y_j. This section presents two methods only. The graphical method for solving $(2 \times n)$ or $(m \times 2)$ games is presented in Section 11.4.3, and the general linear programming method for solving any $(m \times n)$ game is presented in Section 11.4.4.

11.4.3   GRAPHICAL SOLUTION OF $(2 \times N)$ AND $(M \times 2)$ GAMES

Graphical solutions are only applicable to games in which at least one of the players has two strategies only. Consider the following $(2 \times n)$ game.

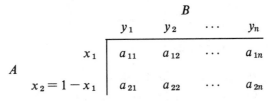

It is assumed that the game does not have a saddle point.

Since A has two strategies, it follows that $x_2 = 1 - x_1$; $x_1 \geq 0$, $x_2 \geq 0$. His expected payoffs corresponding to the *pure* strategies of B are given by

B's Pure Strategy	A's Expected Payoff
1	$(a_{11} - a_{21})x_1 + a_{21}$
2	$(a_{12} - a_{22})x_1 + a_{22}$
$\vdots$	
n	$(a_{1n} - a_{2n})x_1 + a_{2n}$

This shows that A's average payoff varies linearly with x_1.

According to the minimax criterion for mixed strategy games, player A should select the value of x_1 that maximizes his minimum expected payoffs. This may be done by plotting the straight lines as functions of x_1. The following example illustrates the procedure.

Example 11.4-3. Consider the following (2×4) game.

$$
\begin{array}{cc}
 & B \\
 & \begin{array}{cccc} 1 & 2 & 3 & 4 \end{array} \\
A\ \begin{array}{c} 1 \\ 2 \end{array} &
\left|\begin{array}{cccc}
2 & 2 & 3 & -1 \\
4 & 3 & 2 & 6
\end{array}\right.
\end{array}
$$

This games does not have a saddle point. Thus A's expected payoffs corresponding to B's pure strategies are given as follows.

B's Pure Strategies	A's Expected Payoff
1	$-2x_1 + 4$
2	$-x_1 + 3$
3	$x_1 + 2$
4	$-7x_1 + 6$

These four straight lines are then plotted as functions of x_1 as shown in Figure 11–2. The maximin occurs at $x_1^* = 1/2$. This is the point of intersection of *any* two of the lines 2, 3, and 4. Consequently, A's optimal strategy is $(x_1^* = 1/2, x_2^* = 1/2)$, and the value of the game is obtained by substituting for x_1 in the equation of any of the lines passing through the maximin point. This gives

$$
v^* = \begin{cases}
-1/2 + 3 = 5/2 \\
1/2 + 2 = 5/2 \\
-7(1/2) + 6 = 5/2
\end{cases}
$$

To determine B's optimal strategies, it should be noticed that three lines pass through the maximin point. This is an indication that B can mix all three

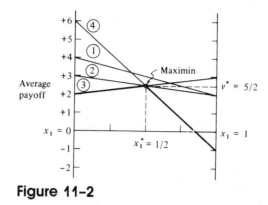

Figure 11-2

strategies. Any two lines having *opposite* signs for their slopes define an alternative optimum solution. Thus, of the three combinations (2, 3), (2, 4), and (3, 4), the combination (2, 4) must be excluded as nonoptimal.

The first combination (2, 3) implies that $y_1^* = y_4^* = 0$. Consequently, $y_3 = 1 - y_2$ and B's average payoffs corresponding to A's pure strategies are given as follows:

A's Pure Strategy	B's Expected Payoff
1	$-y_2 + 3$
2	$y_2 + 2$

Thus y_2^* (corresponding to minimax point) can be determined from

$$-y_2^* + 3 = y_2^* + 2$$

This gives $y_2^* = 1/2$. Notice that by substituting $y_2^* = 1/2$ in B's expected payoffs given, the minimax value is 5/2, which equals the value of the game v^*, as should be expected.

The remaining combination (3, 4) can be treated similarly to obtain an alternative optimal solution. Any weighted average of the combinations (2, 3) and (3, 4) will also yield a new optimal solution that mixes all the three strategies 2, 3, and 4. The treatment of this case is left as an exercise for the reader (see Problem 11–27). ◀

Exercise 11.4-1

Consider Example 11.4–3.
(a) Determine B's pure strategies that can be deleted without affecting the optimal solution.

[*Ans.* Delete B's strategies 1 and 2.]
(b) If B's third pure strategy is eliminated, determine the optimal solution.

[*Ans.* A saddle point optimal solution occurs with each player selecting his second pure strategy.]

Example 11.4-4. Consider the following (4 × 2) game:

$$
\begin{array}{c}
& & B \\
& & \begin{array}{cc} 1 & 2 \end{array} \\
A \begin{array}{c} 1 \\ 2 \\ 3 \\ 4 \end{array} &
\left|
\begin{array}{cc}
2 & 4 \\
2 & 3 \\
3 & 2 \\
-2 & 6
\end{array}
\right.
\end{array}
$$

This game does not have a saddle point. Let y_1 and $y_2 (= 1 - y_1)$ be B's mixed strategies. Thus

A's Pure Strategy	B's Expected Payoff
1	$-2y_1 + 4$
2	$-y_1 + 3$
3	$y_1 + 2$
4	$-8y_1 + 6$

These four lines are plotted in Figure 11-3. In this case the minimax point is determined as the lowest point of the upper envelope. The value of y_1^* is obtained as the point of intersection of lines 1 and 3. This yields $y_1^* = 2/3$ and $v^* = 8/3$.

The lines intersecting at the minimax point correspond to A's pure strategies 1 and 3. This indicates that $x_2^* = x_4^* = 0$. Consequently, $x_1 = 1 - x_3$ and A's average payoffs corresponding to B's pure strategies are

B's Pure Strategy	A's Expected Payoff
1	$-x_1 + 3$
2	$2x_1 + 2$

The point x_1^* is determined by solving

$$-x_1^* + 3 = 2x_1^* + 2$$

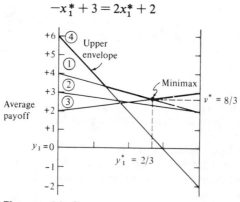

Figure 11-3

This gives $x_1^* = 1/3$. Thus A's optimal strategies are $x_1^* = 1/3$, $x_2^* = 0$, $x_3^* = 2/3$, $x_4^* = 0$. This yields $v^* = 8/3$, as before. ◄

Exercise 11.4-2

(a) In Example 11.4–4, determine A's pure strategies that are strictly dominated by others.

[*Ans.* Strategy 2 is strictly dominated by strategy 1.]

(b) If A's pure strategy 1 is deleted in Example 11.4–4, determine the new optimal solution for A and B.

[*Ans.* Player A plays 2 and 3 with equal probabilities and player B plays his two strategies also with equal probabilities.]

11.4.4 SOLUTION OF ($M \times N$) GAMES BY LINEAR PROGRAMMING

Game theory bears a strong relationship to linear programming, since every finite two-person zero-sum game can be expressed as a linear program and, conversely, every linear program can be represented as a game. In fact, G. Dantzig states [(1963), p. 24] that J. von Neumann, father of game theory, when first introduced to the simplex method of linear programming [1947], immediately recognized this relationship and further pinpointed and stressed the concept of *duality* in linear programming. This section illustrates the solution of game problems by linear programming. It is especially useful for games with large matrices.

Section 11.4.2 shows that A's optimum mixed strategies satisfy

$$\max_{x_i} \left\{ \min \left(\sum_{i=1}^{m} a_{i1}x_i, \sum_{i=1}^{m} a_{i2}x_i, \ldots, \sum_{i=1}^{m} a_{in}x_i \right) \right\}$$

subject to the constraints

$$x_1 + x_2 + \cdots + x_m = 1$$
$$x_i \geq 0, \quad i = 1, 2, \ldots, m$$

This problem can be put in the linear programming form as follows. Let

$$v = \min \left(\sum_{i=1}^{m} a_{i1}x_i, \sum_{i=1}^{m} a_{i2}x_i, \ldots, \sum_{i=1}^{m} a_{in}x_i \right)$$

then the problem becomes (see Example 2.2–5)

$$\text{maximize } z = v$$

subject to

$$\sum_{i=1}^{m} a_{ij}x_i \geq v, \quad j = 1, 2, \ldots, n$$

$$\sum_{i=1}^{m} x_i = 1$$
$$x_i \geq 0, \quad \text{for all } i$$

v represents the value of the game in this case.

The linear programming formulation can be simplified by dividing all $(n + 1)$ constraints by v. This division is correct as long as $v > 0$. Otherwise, if $v < 0$, the direction of the inequality constraints must be reversed. If $v = 0$, the division is illegitimate. This point presents no special problem since a positive constant K can be added to all the entries of the payoff matrix, thus guaranteeing that the value of the game for the "modified" matrix is greater than zero. The *true* value of the game is determined by subtracting K from the *modified* value of the game. In general, if the maximin value of the game is nonnegative, the value of the game is greater than zero (provided that the game has no saddle point).

Thus, assuming that $v > 0$, the constraints of the linear program become

$$a_{11}\frac{x_1}{v} + a_{21}\frac{x_2}{v} + \cdots + a_{m1}\frac{x_m}{v} \geq 1$$

$$a_{12}\frac{x_1}{v} + a_{22}\frac{x_2}{v} + \cdots + a_{m2}\frac{x_m}{v} \geq 1$$

$$\vdots \qquad\qquad\qquad\qquad \vdots$$

$$a_{1n}\frac{x_1}{v} + a_{2n}\frac{x_2}{v} + \cdots + a_{mn}\frac{x_m}{v} \geq 1$$

$$\frac{x_1}{v} + \frac{x_2}{v} + \cdots + \frac{x_m}{v} = \frac{1}{v}$$

Let $X_i = x_i/v$, $i = 1, 2, \ldots, m$. Since

$$\max v \equiv \min \frac{1}{v} = \min\{X_1 + \cdots + X_m\}$$

the problem becomes

$$\text{minimize } z = X_1 + X_2 + \cdots + X_m$$

subject to

$$a_{11}X_1 + a_{21}X_2 + \cdots + a_{m1}X_m \geq 1$$
$$a_{12}X_1 + a_{22}X_2 + \cdots + a_{m2}X_m \geq 1$$
$$\vdots \qquad\qquad\qquad\qquad \vdots$$
$$a_{1n}X_1 + a_{2n}X_2 + \cdots + a_{mn}X_m \geq 1$$
$$X_1, X_2, \ldots, X_m \geq 0$$

Player B's problem is given by

$$\min_{y_j} \left\{ \max \left(\sum_{j=1}^{n} a_{1j}y_j, \sum_{j=1}^{n} a_{2j}y_j, \ldots, \sum_{j=1}^{n} a_{mj}y_j \right) \right\}$$

subject to

$$y_1 + y_2 + \cdots + y_n = 1$$
$$y_j \geq 0, \qquad j = 1, 2, \ldots, n$$

This can also be expressed as a linear program as follows:

$$\text{maximize } w = Y_1 + Y_2 + \cdots + Y_n$$

subject to

$$a_{11}Y_1 + a_{12}Y_2 + \cdots + a_{1n}Y_n \le 1$$
$$a_{21}Y_1 + a_{22}Y_2 + \cdots + a_{2n}Y_n \le 1$$
$$\vdots \qquad\qquad \vdots$$
$$a_{m1}Y_1 + a_{m2}Y_2 + \cdots + a_{mn}Y_n \le 1$$
$$Y_1, Y_2, \ldots, Y_n \ge 0$$

where

$$w = \frac{1}{v}, \qquad Y_i = \frac{y_j}{v}, \qquad j = 1, 2, \ldots, n$$

Notice that B's problem is actually the dual of A's problem. Thus the optimal solution of one problem automatically yields the optimal solution to the other. Player B's problem can be solved by the regular simplex method, and player A's problem is solved by the dual simplex method. The choice of either method will depend on which problem has a smaller number of constraints, which in turn depends on the number of pure strategies for each player.

Example 11.4-5. Consider the following (3×3) game:

		\multicolumn{3}{c}{B}			Row Minimum
		1	2	3		
A	1	3	-1	-3		-3
	2	-3	3	-1		-3
	3	-4	-3	3		-4
Column Maximum		3	3	3		

Since the maximin value is -3, it is possible that the value of the game may be negative or zero. Thus a constant K, which is at *least* equal to the negative of the maximin value, is added to all the elements of the matrix, that is, $K \ge 3$. Let $K = 5$. The preceding matrix becomes

		\multicolumn{3}{c}{B}
		1	2	3
A	1	8	4	2
	2	2	8	4
	3	1	2	8

B's linear programming problem is thus given as

$$\text{maximize } w = Y_1 + Y_2 + Y_3$$

subject to

$$8Y_1 + 4Y_2 + 2Y_3 \leq 1$$
$$2Y_1 + 8Y_2 + 4Y_3 \leq 1$$
$$1Y_1 + 2Y_2 + 8Y_3 \leq 1$$
$$Y_1, Y_2, Y_3 \geq 0$$

The final optimal tableau for this problem is given by

Basic	Y_1	Y_2	Y_3	S_1	S_2	S_3	Solution
w	0	0	0	5/49	11/196	1/14	45/196
Y_1	1	0	0	1/7	−1/14	0	1/14
Y_2	0	1	0	−3/98	31/196	−1/14	11/196
Y_3	0	0	1	−1/98	−3/98	1/7	5/49

Thus, for the original problem,

$$v^* = \frac{1}{w} - K = 196/45 - 5 = -29/45$$

$$y_1^* = \frac{Y_1}{w} = \frac{1/14}{45/196} = 14/45$$

$$y_2^* = \frac{Y_2}{w} = \frac{11/196}{45/196} = 11/45$$

$$y_3^* = \frac{Y_3}{w} = \frac{5/49}{45/196} = 20/45$$

The optimal strategies for A are obtained from the dual solution to the problem above. This is given by

$$z = w = 45/196, \quad X_1 = 5/49, \quad X_2 = 11/196, \quad X_3 = 1/14$$

Hence

$$x_1^* = X_1/z = 20/45, \quad x_2^* = X_2/z = 11/45, \quad x_3^* = X_3/z = 14/45$$

You can verify that these optimal strategies satisfy the minimax theorem. ◀

11.5 SUMMARY

In this chapter a number of decision criteria are discussed for problems with imperfect data. Although some applications have already been presented, Chapters 12 through 17 provide more applications to project scheduling, inventory, Markov processes, queueing, and simulation. You will notice that the decision models in most of these applications are based on the *expected value* criterion. This point requires some explanation.

As stated throughout the chapter, the expected value criterion may not be applicable in certain situations, particularly those where the decision is not repeated a sufficiently large number of times. However, part of the reason for the wide use of expectations is purely traditional. Perhaps also the fact that the expected value criterion is analytically simple makes it particularly appealing to decision makers. For example, the minimax criterion is generally more complex than the expected value criterion.

You should realize that some of the applications in the following chapters may not justify the use of the expected value criterion and, as such, be on the alert to question whether each of these applications is proper. In this respect, different interpretations of the decision problem may give rise to different conclusions about the use of a suitable criterion.

We shall not attempt to compare the applications of the given decision criteria to the different problems we present in Chapter 11, as this may detract from concentrating on the basic elements of the situation for which a decision is made. In general, the use of the expected value criterion should be regarded as an illustration of the application of decision criteria.

SELECTED REFERENCES

DANTZIG, G. B., *Linear Programming and Extensions,* Princeton University Press, Princeton, N.J., 1963.

LUCE, R., and H. RAIFFA, *Games and Decisions,* Wiley, New York, 1957.

MORRIS, W., *The Analysis of Management Decisions,* Irwin, Homewood, Ill., 1964.

WILLIAMS, J., *The Compleat Strategyst,* rev. ed., McGraw-Hill, New York, 1966.

REVIEW QUESTIONS

True (T) and False (F)?

1. _____ The different criteria for making decisions under risk always yield the same optimum choice.

2. _____ The expected value–variance criterion should lead to a more conservative decision, particularly when a high-risk-aversion factor is used.

3. _____ All criteria for decisions under risk inherently assume that the reached decision will be repeated a very large number of times.

4. _____ In decision trees, it is necessary to know in advance all the probabilities of the chance events and all the monetary evaluations of the different alternatives.

5. _____ In decision trees, no more than two alternative courses of action can emanate from a decision point node.

6. _____ In decisions under uncertainty, the Laplace criterion is the least conservative while the minimax criterion is the most conservative.

7. _____ In the Savage criterion, if the payoff matrix represents profit, the optimum selection will be based on the maximin (rather than the minimax) condition.

8. _____ In a two-person zero-sum game, if the optimal solution requires one player to use a pure strategy, the other player *must* do the same.

9. _____ In a $(2 \times n)$ or $(m \times 2)$ two-person zero-sum game with no alternative optima, each player can mix at most two pure strategies.

10. _____ The optimal solution of a two-person zero-sum game always represents a saddle point regardless of whether the players use pure or mixed strategies.

11. _____ If the maximum value of a two-person zero-sum game is negative, the long-run outcome of the game will indicate a net loss to player A.

12. _____ If a two-person zero-sum game is stable, neither player can improve his expected payoff by changing his minimax (maximin) strategies.

13. _____ The addition of a constant to all the elements of a payoff matrix in a two-person zero-sum game can affect only the value of the game, not the optimal mix of strategies.

14. _____ Every two-person zero-sum game can be represented by a pair of primal–dual linear programs.

[*Ans.* 1—F, 2—T, 3—T, 4—T, 5—F, 6—T, 7—T, 8—F, 9 to 14—T.]

PROBLEMS

Section	Assigned Problems
11.1	11–1 to 11–14
11.2	11–15 to 11–18
11.3	11–19 to 11–21
11.4	11–22 to 11–34

☐ **11–1** Solve Example 11.1–1 assuming that $c_1 = 200$, $c_2 = 15$, and $n = 30$. The probabilities are given by

$$P_t = \begin{cases} 0.03, & t = 1 \\ P_{t-1} + .01, & t = 2, 3, \ldots, 10 \\ .13, & t = 10, 11, \ldots \end{cases}$$

☐ **11–2** In a manufacturing process, lots having 8%, 10%, 12%, or 14% defectives are produced according to the respective probabilities .4, .3, .25, and .05. Three customers have contracts to receive lots from the manufacturer. The contracts specify that the percentages of defectives in lots shipped to customers A, B, and C should not exceed 8, 12, and 14, respectively. If a lot has a higher percentage of defectives than stipulated, a penalty of $100 per percentage point is incurred. On the other hand, supplying better quality than required costs the manufacturer $50 per percentage point. If the lots are not inspected prior to shipment, which customer should have the highest priority for receiving the order?

□ **11–3** Daily demand for loaves of bread at a grocery store are given by the following probability distribution:

x	100	150	200	250	300
$p(x)$	.20	.25	.30	.15	.10

If a loaf is not sold the same day, it can be disposed of at 15 cents at the end of the day. Otherwise, the price of a fresh loaf is 49 cents. The cost per loaf to the store is 25 cents. Assuming that the stock level is restricted to one of the demand levels, how many loaves should be stocked daily?

□ **11–4** An automatic machine produces α (thousands of) units of a certain product per day. As α increases, the proportion of defectives p goes up. The probability density function of p in terms of α is given by

$$f(p) = \begin{cases} \alpha p^{\alpha-1}, & 0 \le p \le 1 \\ 0, & \text{otherwise} \end{cases}$$

Each defective item incurs a loss of $50. A good item produces a profit of $5. Determine the value of α that maximizes expected profit.

□ **11–5** The outer diameter d of a cylinder that is processed on an automatic machine has upper and lower tolerance limits of $d + t_U$ and $d - t_L$. If the machine is set at d, the produced diameters can be described by a normal distribution with mean d and standard deviation σ. Cylinders with oversized diameters can be reworked at c_1 dollars per cylinder. Undersized cylinders must be salvaged at a loss of c_2 per cylinder. Determine the best setting of the machine.

□ **11–6** In production processes, maintenance action is periodically applied to cutting tools. If the tool is not sharpened frequently, the percentage of defective items increases. In the meantime, an increase in the frequency of sharpening a tool increases the cost of maintenance. Ideally, a balance between the two extreme costs is desired.

In a typical process, let S_U and S_L represent the upper and lower limits allowed for a measurable dimension machined by the tool. Let $\mu(t)$ be the average of the process at time t after the tool is sharpened, where $\mu(0)$ represents the ideal setting of the machine. Each time the tool is sharpened, a cost c_1 is incurred. A defective item costs c_2 to be reworked. Suppose that the output of the process can be described by a normal distribution with mean $\mu(t)$ and variance σ (σ is independent of time), and that a lot of size Q is to be manufactured at the rate α items per unit time. Determine an expression for the expected cost of sharpening the tool and reworking defectives as a function of the time T that must elapse before maintenance is applied. Show that the optimal value of T is independent of Q and interpret the result. Then determine a numerical value for T by using the data, $c_1 = 10$, $c_2 = 48.85$, $\alpha = 10$, $\mu(t) = \mu(0) + t$, and $\sigma = 1$. [*Hint:* Approximate the number of times a tool is sharpened during the production of Q by $Q/\alpha T$. Also, numerical integration may be needed to obtain a numerical value of T.]

☐ **11–7** Let x be a random variable representing cost and let $f(x)$ be its probability density function. Suppose that $U(x)$ is the utility function of x. Show that the expected value of $U(x)$, $E\{U(x)\}$, can be expanded as a series around the point $E\{x\}$ and that the resulting expansion can be approximated by

$$E\{U(x)\} \cong U(E\{x\}) + K \text{ var } \{x\}$$

Determine the expression for K and show that K assumes a positive value if x represents loss and a negative value if x represents profit. [*Note:* This result is consistent with the derivation of the expected value–variance criterion, Section 11.1.2.]

☐ **11–8** Solve Problem 11–4 by applying the expected value–variance criterion. Compare the optimal solution for the following risk aversion factors: $K = 1$, 2, and 5.

☐ **11–9** The demand for an item is described by the following probability density function:

x	0	1	2	3	4	5
$p(x)$	.1	.15	.4	.15	.1	.1

Determine the stock level so that the probability of running out of stock does not exceed .45. If the average shortage and surplus quantities must not exceed 1 and 2 units, respectively, determine the stock level.

☐ **11–10** In Problem 11–9, suppose that the expected shortage quantity must be strictly less than the expected surplus quantity by at least one unit. Determine the inventory level.

☐ **11–11** In Problem 11–2, suppose that a sample of size $n = 20$ is inspected before each lot is shipped to customers. If four defectives are found in the sample, compute the posterior probabilities of the lot having 8%, 10%, 12%, and 14% defectives. By using the new probabilities, determine which customer has the lowest expected cost.

☐ **11–12** Electronic components are received from two vendors. Vendor A supplies 75% of the components known to include 1% defectives. Vendor B's components include 2% defectives. When a sample of size 5 is inspected, only one defective component is found. By using this information, determine the posterior probability that the components are delivered from vendor A; from vendor B.

☐ **11–13** Consider the following payoff (profit) matrix:

	θ_1	θ_2	θ_3	θ_4
a_1	10	20	−20	13
a_2	12	14	0	15
a_3	7	2	18	9

The a priori probabilities of θ_1, θ_2, θ_3, and θ_4 are .2, .1, .3, and .4. An experiment is conducted and its outcomes z_1 and z_2 are described by the following probabilities:

	θ_1	θ_2	θ_3	θ_4
z_1	.1	.2	.7	.4
z_2	.9	.8	.3	.6

(a) Determine the best action when no data are used.
(b) Determine the best action when the experimental data are used.

☐ **11–14** The probability that it will rain during the rainy season of the year is .7. A fisherman wants to decide whether or not to go fishing tomorrow. The conditional probability that rain is forecast given that it is the rainy season is .85. Find the probability that it will not rain tomorrow given that rain is forecast.

☐ **11–15** The daily demand for loaves of bread in a grocery store can assume one of the following values: 100, 120, or 130 loaves with probabilities .2, .3, and .5. The owner of the store is thus limiting her alternatives to stocking one of the indicated four levels. If she stocks more than she can sell in the same day, she must dispose of the remaining loaves at a discount price of 55 cents/loaf. Assuming that she pays 60 cents per loaf and sells it for $1.05, find the optimum stock level by using a decision tree representation.

☐ **11–16** In Problem 11–15, suppose that the owner wishes to consider her decision problem over a 2-day period. Her alternatives for the second day are determined as follows. If the demand in day 1 is equal to the amount stocked, she will continue to order the same quantity on the second day. Otherwise, if the demand exceeds the amount stocked, she will have the options to order higher levels of stock on the second day. Finally, if day 1's demand is less than the amount stocked, she will have the options to order any of the lower levels of stock for the second day. Express the problem as a decision tree and find the optimum solution using the cost data given in Problem 11–15.

☐ **11–17** Solve Example 11.2–1 assuming that the annual interest rate is 10% and that decisions are made based on the expected value of *discounted* income.

☐ **11–18** In Example 11.2–1, suppose that a third alternative is added which will allow us to expand the small plant to a medium-size plant. This option can be exercised regardless of whether the demand is high or low during the first 2 years. Thus if the 2-year demand is high, the company has three options:

(i) expand the plant fully (cost = $4,200,000), (ii) expand it moderately (cost = $2,800,000), or (iii) do not expand it at all. On the other hand, if the demand is low, the company can expand the plant moderately or elect not to expand it at all. Estimates of annual income for the different alternatives are given as follows:

(i) High demand in the first 2 years and medium-size expansion will yield $700,000 ($250,000) for each of the remaining 8 years if the demand is high (low).
(ii) Low demand in the first 2 years and medium-size expansion will yield $600,000 ($300,000) for each of the remaining 8 years if the demand is high (low).
(iii) Low demand in the first 2 years and no expansion will yield $300,000 ($400,000) for each of the remaining 8 years if the demand is high (low).

The remaining data are as given in Example 11.2–1. Determine the optimal decision based on the optimal value criterion.

☐ 11–19 Consider the following payoff (profit) matrix.

	θ_1	θ_2	θ_3	θ_4	θ_5
a_1	15	10	0	−6	17
a_2	3	14	8	9	2
a_3	1	5	14	20	−3
a_4	7	19	10	2	0

No probabilities are known for the occurrence of the nature states. Compare the solutions obtained by each of the following criteria:
(a) Laplace.
(b) Maximin.
(c) Savage.
(d) Hurwicz (assume that $\alpha = .5$).

☐ 11–20 One of N machines is to be selected for producing a lot whose size Q could assume any value between Q^* and Q^{**} ($Q^* < Q^{**}$). The production cost for machine i is

$$C_i = K_i + c_i Q$$

Solve the problem by each of the following criteria:
(a) Laplace.
(b) Minimax.
(c) Savage.
(d) Hurwicz (assume that $\alpha = .5$).

☐ 11–21 Give numerical answers for Problem 11–20 given $Q^* = 1000$, $Q^{**} = 4000$, and

Machine i	K_i	C_i
1	100	5
2	40	12
3	150	3
4	90	8

☐ **11-22** (a) Find the saddle point and the value of the game for each of the following two games. The payoff is for player A.

$$
A\begin{array}{c} B \\ \begin{array}{|cccc} 8 & 6 & 2 & 8 \\ 8 & 9 & 4 & 5 \\ 7 & 5 & 3 & 5 \end{array} \end{array}
$$
(1)

$$
A\begin{array}{c} B \\ \begin{array}{|cccc} 4 & -4 & -5 & 6 \\ -3 & -4 & -9 & -2 \\ 6 & 7 & -8 & -9 \\ 7 & 3 & -9 & 5 \end{array} \end{array}
$$
(2)

(b) Find the range of values for "p" and "q" that will render the entry (2, 2) a saddle point in the following games.

$$
A\begin{array}{c} B \\ \begin{array}{|ccc} 1 & q & 6 \\ p & 5 & 10 \\ 6 & 2 & 3 \end{array} \end{array}
\qquad
A\begin{array}{c} B \\ \begin{array}{|ccc} 2 & 4 & 5 \\ 10 & 7 & q \\ 4 & p & 6 \end{array} \end{array}
$$

☐ **11-23** Indicate whether the values of the following games are greater than, less than, or equal to zero.

$$
A\begin{array}{c} B \\ \begin{array}{|cccc} 1 & 9 & 6 & 0 \\ 2 & 3 & 8 & 4 \\ -5 & -2 & 10 & -3 \\ 7 & 4 & -2 & -5 \end{array} \end{array}
\qquad
A\begin{array}{c} B \\ \begin{array}{|cccc} 3 & 7 & -1 & 3 \\ 4 & 8 & 0 & -6 \\ 6 & -9 & -2 & 4 \end{array} \end{array}
$$

$$
A\begin{array}{c} B \\ \begin{array}{|cccc} -1 & 9 & 6 & 8 \\ -2 & 10 & 4 & 6 \\ 5 & 3 & 0 & 7 \\ 7 & -2 & 8 & 4 \end{array} \end{array}
\qquad
A\begin{array}{c} B \\ \begin{array}{|ccc} 3 & 6 & 1 \\ 5 & 2 & 3 \\ 4 & 2 & -5 \end{array} \end{array}
$$

□ **11–24** Let a_{ij} be the (i, j)th element of the payoff matrix with m and n strategies, respectively. Prove that

$$\max_i \min_j a_{ij} \leq \min_j \max_i a_{ij}$$

□ **11–25** Two companies A and B are promoting two competing products. Each product currently controls 50% of the market. Because of recent modifications in the two products, the two companies are now preparing to launch a new advertisement campaign. If no advertisement is made by either of the two companies, the present status of the market shares will remain unchanged. However, if either company launches a stronger campaign, the other company will certainly lose a proportional percentage of its customers. A survey of the market indicated that 50% of the potential customers can be reached through television, 30% through newspapers, and the remaining 20% through radio. The objective of each company is to select the appropriate advertisement media.

Formulate the problem as a two-person zero-sum game. Does the problem have a saddle point?

□ **11–26** Consider the game

		B		
		1	2	3
	1	5	50	50
A	2	1	1	0.1
	3	10	1	10

Verify that the strategies (1/6, 0, 5/6) for player A and (49/54, 5/54, 0) for player B are optimal and find the value of the game.

□ **11–27** In Example 11.4–3, show that combination (2, 4) for player B does not yield optimal values for y_j, whereas combination (3, 4) yields the optimal solution. Develop a general expression for all the alternative solutions to the problem.

□ **11–28** Solve the following games graphically.

		B		
A	1	3	−3	7
	2	5	4	−6

		B	
	1	2
A	5	6
	−7	9
	−4	−3
	2	1

		B	
	1	2	5
A	8	4	7
	−1	5	−6

□ **11–29** Consider Colonel Blotto's game, where the Colonel and his enemy are trying to take over two strategic locations. The regiments available for Blotto and his enemy are 2 and 3, respectively. Both sides will distribute their regiments between the two locations. Let n_1 and n_2 be the number of regiments allocated by Colonel Blotto to locations 1 and 2, respectively. Also, let m_1 and m_2 be his enemy's allocations to the respective locations. The payoff of Blotto is computed as follows. If $n_1 > m_1$, he receives $m_{1,} + 1$, and if $n_2 > m_2$, he receives $m_2 + 1$. On the other hand, if $n_1 < m_1$, he loses $n_1 + 1$, and if $n_2 < m_2$, he loses $n_2 + 1$. Finally, if the number of regiments from both sides are the same, each side gets zero. Formulate the problem as a two-person zero-sum game and then solve by linear programming.

☐ **11–30** Verify that B's problem is defined by the linear programming problem given in Section 11.4.4.

☐ **11–31** Prove the minimax theorem by using the relationship between the values of the objective function in the primal and the dual problems of the linear programming problem.

☐ **11–32** Verify that the linear programming solution to Example 11.4–5 satisfies the minimax theorem.

☐ **11–33** Consider the two-finger "Morra" game. Each player shows one or two fingers and simultaneously makes a guess of the number of fingers his opponent has. The player making the correct guess wins an amount equal to the total number of fingers shown by the two players. In all other cases, the game is a draw. Formulate the problem as a two-person zero-sum game and then solve by linear programming.

☐ **11–34** Solve the following games by linear programming:

(a)

	B		
A	−1	1	1
	2	−2	2
	3	3	−3

(b)

	B			
A	1	2	−5	3
	−1	4	7	2
	5	−1	1	9

SHORT CASE

Case 11–1†

In the airline industry, working hours are ruled by agreements between unions and the companies. For example, the maximum length of tour of duty may be limited to 16 hours for Boeing-747 flights and 14 hours for

† This case and its data are based directly on A. Gaballa, "Planning Callout Reserves for Aircraft Delays," *Interfaces*, Vol. 9, no. 2, pt. 2, February 1979, pp. 78–86.

Boeing-707 flights. Whenever these limits are exceeded because of unexpected delays, the crew must be replaced by a fresh crew. The airlines maintain reserve crews for such eventualities. The average annual cost of a reserve crew member is estimated at $30,000. On the other hand, an overnight delay due to unavailability of reserve crew could cost as much $50,000 for each delay. A crew member is on call 4 days a week for 12 consecutive hours. The member may not be called upon during the remaining 3 days of the week. The B-747 crew can be served by two B-707 crews.

The following table summarizes the callout probabilities for reserve crews based on 3-year historical data.

Trip Category	Trip Hours	Probability of Callout of a Reserve Crew	
		B-747	B-707
1	14	.014	.072
2	13	.0	.019
3	$12\frac{1}{2}$	.0	.006
4	12	.016	.006
5	$11\frac{1}{2}$	.003	.003
6	11	.002	.003

As an illustration, the data indicate that for 14 hour-long trips, the probability of a callout is .014 for B-747 and .072 for B-707.

A typical *peak* day schedule is shown here as a function of the time of day.

Time of Day	Aircraft	Trip Category
8:00	707	3
9:00	707	6
	707	2
10:00	707	3
11:00	707	2
	707	4
15:00	747	6
16:00	747	4
19:00	747	1

The present reserve crew policy calls for using two (seven-member) crews between 5:00 and 11:00, four crews between 11:00 and 17:00, and two crews between 17:00 and 23:00.

Evaluate the effectiveness of the present reserve crew policy. Specifically, is the present reserve crew size too large, too small, or just right?

Chapter 12

Project Scheduling by PERT–CPM

A **project** defines a combination of interrelated activities that must be executed in a certain order before the entire task can be completed. The activities are interrelated in a logical sequence in the sense that some activities cannot start until others are completed. An **activity** in a project is usually viewed as a job requiring time and resources for its completion. In general, a project is a one-time effort; that is, the same sequence of activities may not be repeated in the future.

In the past, the scheduling of a project (over time) was done with little planning. The best-known "planning" tool then was the **Gantt bar chart,** which specifies the start and finish times for each activity on a horizontal time scale. Its disadvantage is that the interdependency between the different activities (which mainly controls the progress of the project) cannot be determined from the bar chart. The growing complexities of today's projects have demanded more systematic and more effective planning techniques with the objective of optimizing the efficiency of executing the project. Efficiency here implies effecting the utmost reduction in the time required to complete the project while accounting for the economic feasibility of using available resources.

Project management has evolved as a new field with the development of two analytic techniques for planning, scheduling, and controlling of projects. These are the **critical path method (CPM)** and the **project evaluation and review technique (PERT)**. The two techniques were developed by two different groups almost simultaneously (1956–1958). CPM was first developed by E. I. du Pont de Nemours & Company as an application to construction projects and was later extended to a more advanced status by Mauchly Associates; PERT was developed for the U.S. Navy by a consulting firm for scheduling the research and development activities for the Polaris missile program.

PERT and CPM are basically time-oriented methods in the sense that they both lead to the determination of a time schedule. Although the two methods were developed independently, they are strikingly similar. Perhaps the most important difference is that originally the time estimates for the activities were assumed deterministic in CPM and probabilistic in PERT. Today, PERT and CPM actually comprise one technique and the differences, if any, are only historical. Consequently, both techniques will be referred to as "project scheduling" techniques.

Project scheduling by PERT–CPM consists of three basic phases: **planning, scheduling,** and **controlling.**

The planning phase is initiated by breaking down the project into distinct activities. The time estimates for these activities are then determined, and a network (or arrow) diagram is constructed with each of its arcs (arrows) representing an activity. The entire arrow diagram gives a graphic representation of the interdependencies between the activities of the project. The construction of the arrow diagram as a planning phase has the advantage of studying the different jobs in detail, perhaps suggesting improvements before the project is actually executed. More important will be its use to develop a schedule for the project.

The ultimate objective of the scheduling phase is to construct a time chart showing the start and finish times for each activity as well as its relationship to other activities in the project. In addition, the schedule must pinpoint the critical (in view of time) activities that require special attention if the project is to be completed on time. For the noncritical activities, the schedule must show the amount of slack or float times that can be used advantageously when such activities are delayed or when limited resources are to be used effectively.

The final phase in project management is project control. This includes the use of the arrow diagram and the time chart for making periodic progress reports. The network may thus be updated and analyzed, and, if necessary, a new schedule is determined for the remaining portion of the project.

12.1 ARROW (NETWORK) DIAGRAM REPRESENTATIONS

The arrow diagram represents the interdependencies and precedence relationships among the activities of the project. An **arrow** is commonly used to represent an activity, with its head indicating the direction of progress in the project.

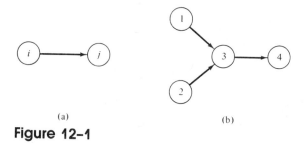

(a)   (b)

Figure 12-1

The precedence relationship between the activities is specified by using events. An **event** represents a point in time that signifies the completion of some activities and the beginning of new ones. The beginning and end points of an activity are thus described by two events usually known as the *tail* and *head* events. Activities originating from a certain event cannot start until the activities terminating at the same event have been completed. In network theory terminology, each activity is represented by a directed arc and each event is represented by a node. The length of the arc need not be proportional to the duration of the activity nor does it have to be drawn as a straight line.

Figure 12-1(a) shows an example of a typical representation of an activity (i, j) with its tail event i and its head event j. Figure 12-1(b) shows another example, where activities $(1, 3)$ and $(2, 3)$ must be completed before activity $(3, 4)$ can start. The direction of progress in each activity is specified by assigning a smaller number to the tail event compared with the number of its head event. This procedure is especially convenient for automatic computations and hence will be adopted throughout this chapter.

The rules for constructing the arrow diagram will be summarized now.

Rule 1. *Each activity is represented by one and only one arrow in the network.* No single activity can be represented twice in the network. This is to be differentiated from the case where one activity is broken down into segments, in which case each segment may be represented by a separate arrow. For example, laying down a pipe may be done in sections rather than as one job.

Rule 2. *No two activities can be identified by the same head and tail events.* A situation like this may arise when two or more activities can be performed concurrently. An example is shown in Figure 12-2(a), where activities A and B have the same end events. The procedure is to introduce a **dummy** activity either between A and one of the end events or between B and one of the end events. The modified representations, after introducing the dummy D, are shown in Figure 12-2(b). As a result of using D, activities A and B can now be identified by unique end events. It must be noted that a dummy activity does not consume time or resources.

Dummy activities are also useful in establishing logic relationships in the arrow diagram that cannot otherwise be represented correctly. Suppose that in a certain project jobs A and B must precede C while job E is preceded by job B only. Figure 12-3(a) shows the incorrect way, since, although the relationship among A, B, and C is correct, the diagram implies that E must be preceded

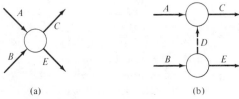

Figure 12-2

by both A and B. The correct representation using the dummy D is shown in Figure 12–3(b). Since D consumes no time (or resources), the precedence relationships indicated are satisfied.

Rule 3. *To ensure the correct precedence relationship in the arrow diagram, the following questions must be answered as every activity is added to the network.*

(a) *What activities must be completed immediately before this activity can start?*

(b) *What activities must follow this activity?*

(c) *What activities must occur concurrently with this activity?*

This rule is self-explanatory. It actually allows for checking (and rechecking) the precedence relationships as one progresses in the development of the network.

Example 12.1–1. Construct the arrow diagram comprising activities A, B, $C, \ldots,$ and L such that the following relationships are satisfied.

1. A, B, and C, the first activities of the project, can start simultaneously.
2. A and B precede D.
3. B precedes E, F, and H.
4. F and C precede G.
5. E and H precede I and J.
6. C, D, F, and J precede K.
7. K precedes L.
8. I, G, and L are the terminal activities of the project.

Figure 12-3

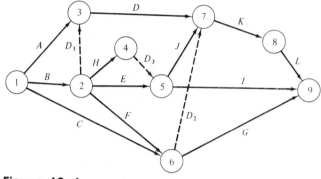

Figure 12-4

The resulting arrow diagram is shown in Figure 12–4. The dummy activities D_1 and D_2 are used to establish correct precedence relationships. D_3 is used to identify activities E and H with unique end events. The events of the project are numbered such that their ascending order indicates the direction of progress in the project.   ◄

Exercise 12.1–1
Consider Example 12.1–1.
(a) Indicate the effect of adding each of the following activities on the precedence relationships in the network. All cases are considered independently.
   (1) Dummy (3, 5).
       [*Ans.* A precedes I and J.]
   (2) Dummy (3, 4).
       [*Ans.* Same as in (1).]
   (3) Dummy (5, 6).
       [*Ans.* E and H precede G.]
   (4) Dummy (3, 6).
       [*Ans.* A precedes G.]
(b) Indicate how each of the following additional relationships can be incorporated in the network.
   (1) Activities A and B precede G.
       [*Ans.* Add dummy (3, 6).]
   (2) Activity D precedes G.
       [*Ans.* Insert dummy between end of D and node 7, then connect end of D and node 6 by a dummy activity.]
   (3) Activity C precedes D.
       [*Ans.* Insert dummy between end of C and node 6, then connect end of C and node 3 by a dummy activity.]

12.2 CRITICAL PATH CALCULATIONS

The application of PERT–CPM should ultimately yield a schedule specifying the start and completion dates of each activity. The arrow diagram represents the first step toward achieving that goal. Because of the interaction among the different activities, the determination of the start and completion times requires special computations. These calculations are performed directly on the arrow diagram using simple arithmetics. The end result is to classify the activities of the project as **critical** or **noncritical.** An activity is said to be critical if a delay in its start will cause a delay in the completion date of the entire project. A noncritical activity is such that the time between its earliest start and its latest completion dates (as allowed by the project) is longer than its actual duration. In this case the noncritical activity is said to have a **slack** or **float** time.

The advantage of pinpointing the critical activities and determining the floats will be discussed in Section 12.3. This section mainly presents the methods for obtaining this information.

12.2.1 DETERMINATION OF THE CRITICAL PATH

A critical path defines a *chain* of critical activities that connects the start and end events of the arrow diagram. In other words, the critical path identifies all the critical activities of the project. The method of determining such a path is illustrated by a numerical example.

Example 12.2–1. Consider the network in Figure 12–5 that starts at node 0 and terminates at node 6. The time required to perform each activity is indicated on the arrows.

The critical path calculations include two phases. The first phase is called the **forward pass,** where calculations begin from the "start" node and move to the "end" node. At each node a number is computed representing the earliest occurrence time of the corresponding event. These numbers are shown in Figure 12–5 in squares □. The second phase, called the **backward pass,** begins calculations from the "end" node and moves to the "start" node. The number computed at each node (shown in triangles △) represents the latest occurrence time of the corresponding event. The forward pass is considered now.

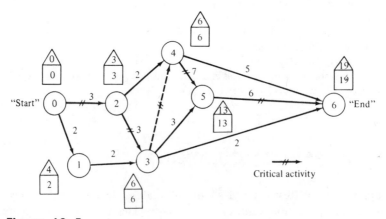

Figure 12–5

Let ES_i be the **earliest start time** of all the activities emanating from event i. Thus ES_i represents the earliest occurrence time of event i. If $i = 0$ is the "start" event, then conventionally, for the critical path calculations, $ES_0 = 0$. Let D_{ij} be the duration of activity (i, j). The forward pass calculations are thus obtained from the formula

$$ES_j = \max_i \{ES_i + D_{ij}\}, \qquad \text{for all } (i, j) \text{ activities defined}$$

where $ES_0 = 0$. Thus, to compute ES_j for event j, ES_i for the tail events of *all* the incoming activities (i, j) must be computed first.

The forward pass calculations applied to Figure 12–5 starts with $ES_0 = 0$, as shown in the square above event 0. Since there is only one incoming activity $(0, 1)$ to event 1 with $D_{01} = 2$,

$$ES_1 = ES_0 + D_{01} = 0 + 2 = 2$$

which is entered in the square associated with event 1. Next, we consider event 2. [Notice that event 3 cannot be considered at this point since ES_2 (event 2) is not yet known.] Thus

$$ES_2 = ES_0 + D_{02} = 0 + 3 = 3$$

which is entered in the square associated with event 2. The next event to be considered is 3. Since there are two incoming activities, $(1, 3)$ and $(2, 3)$, we have

$$ES_3 = \max_{i=1,2} \{ES_i + D_{i\,3}\} = \max \{2 + 2,\ 3 + 3\} = 6$$

which, again, is entered in the square of event 3.

The procedure continues in the same manner until ES_j is computed for all j. Thus

$$ES_4 = \max_{i=2,3} \{ES_i + D_{i\,4}\} = \max \{3 + 2,\ 6 + 0\} = 6$$

$$ES_5 = \max_{i=3,4} \{ES_i + D_{i\,5}\} = \max \{6 + 3,\ 6 + 7\} = 13$$

$$ES_6 = \max_{i=3,4,5} \{ES_i + D_{i\,6}\} = \max \{6 + 2,\ 6 + 5,\ 13 + 6\} = 19$$

These calculations complete the forward pass.

The backward pass starts from the "end" event. The objective of this phase is to compute LC_i, the **latest completion time** for all the activities coming into event i. Thus if $i = n$ is the "end" event, $LC_n = ES_n$ initiates the backward pass. In general, for any node i,

$$LC_i = \min_j \{LC_j - D_{ij}\}, \qquad \text{for all } (i, j) \text{ activities defined}$$

The values of LC (entered in the triangles $\triangle$) are determined as follows.

$$LC_6 = ES_6 = 19$$

$$LC_5 = LC_6 - D_{56} = 19 - 6 = 13$$

$$LC_4 = \min_{j=5,6} \{LC_j - D_{4j}\} = \min \{13 - 7,\ 19 - 5\} = 6$$

$$LC_3 = \min_{j=4,5,6} \{LC_j - D_{3j}\} = \min \{6 - 0, 13 - 3, 19 - 2\} = 6$$

$$LC_2 = \min_{j=3,4} \{LC_j - D_{2j}\} = \min \{6 - 3, 6 - 2\} = 3$$

$$LC_1 = LC_3 - D_{13} = 6 - 2 = 4$$

$$LC_0 = \min_{j=1,2} \{LC_j - D_{0j}\} = \min \{4 - 2, 3 - 3\} = 0$$

The backward pass calculations are now complete.

The critical path activities can now be identified by using the results of the forward and backward passes. An activity (i, j) lies on the **critical path** if it satisfies the following three conditions:

$$ES_i = LC_i \tag{1}$$
$$ES_j = LC_j \tag{2}$$
$$ES_j - ES_i = LC_j - LC_i = D_{ij} \tag{3}$$

These conditions actually indicate that there is no float or slack time between earliest start (completion) and the latest start (completion) of the critical activity. In the arrow diagram these activities are characterized by the numbers in □ and △ being the same at each of the head and the tail events *and* that the difference between the number in □ (or △) at the head event and the number in □ (or △) at the tail event is equal to the duration of the activity.

Activities (0, 2), (2, 3), (3, 4), (4, 5), and (5, 6) define the critical path in Figure 12–5. Actually, the critical path represents the shortest duration needed to complete the project. Notice that activities (2, 4), (3, 5), (3, 6), and (4, 6) satisfy conditions (1) and (2) for critical activities but not condition (3). Hence they are not critical. Notice also that the critical path must form a chain of *connected* activities that span the network from "start" to "end."   ◀

Exercise 12.2-1

For the network in Figure 12–5, determine the critical path(s) for each of the following (independent) cases.

(a) $D_{01} = 4$.

  [*Ans*. (0, 2, 3, 4, 5, 6) and (0, 1, 3, 4, 5, 6).]

(b) $D_{36} = 15$.

  [*Ans*. (0, 2, 3, 6).]

12.2.2 DETERMINATION OF THE FLOATS

Following the determination of the critical path, the floats for the noncritical activities must be computed. Naturally, a critical activity must have zero float. In fact, this is the main reason it is critical.

Before showing how floats are determined, it is necessary to define two new times that are associated with each activity. These are the **latest start** (LS) and the **earliest completion** (EC) times, which are defined for activity (i, j) by

$$LS_{ij} = LC_j - D_{ij}$$

$$EC_{ij} = ES_i + D_{ij}$$

There are two important types of floats: **total float** (TF) and **free float** (FF). The total float TF_{ij} for activity (i, j) is the difference between the maximum time available to perform the activity ($= LC_j - ES_i$) and its duration ($= D_{ij}$); that is,

$$TF_{ij} = LC_j - ES_i - D_{ij} = LC_j - EC_{ij} = LS_{ij} - ES_i$$

The free float is defined by assuming that all the activities start as early as possible. In this case FF_{ij} for activity (i, j) is the excess of available time ($= ES_j - ES_i$) over its duration ($= D_{ij}$); that is,

$$FF_{ij} = ES_j - ES_i - D_{ij}$$

The critical path calculations together with the floats for the noncritical activities can be summarized in the convenient form shown in Table 12–1. Columns (1), (2), (3), and (6) are obtained from the network calculations of Example 12.2–1. The remaining information can be determined from the foregone formulas.

Table 12–1 gives a typical summary of the critical path calculations. It includes all the information necessary to construct the time chart. Notice that a critical activity, and only a critical activity, must have zero *total* float. The free float must also be zero when the total float is zero. The converse is not true, however, in the sense that a *non*critical activity may have zero free float. For example, in Table 12–1, the noncritical activity (0, 1) has zero free float.

Table 12–1

Activity (i, j) (1)	Duration D_{ij} (2)	Earliest Start □ ES_i (3)	Earliest Completion EC_{ij} (4)	Latest Start LS_{ij} (5)	Latest Completion △ LC_j (6)	Total Float TF_{ij} (7)	Free Float FF_{ij} (8)
(0, 1)	2	0	2	2	4	2	0
(0, 2)	3	0	3	0	3	0[a]	0
(1, 3)	2	2	4	4	6	2	2
(2, 3)	3	3	6	3	6	0[a]	0
(2, 4)	2	3	5	4	6	1	1
(3, 4)	0	6	6	6	6	0[a]	0
(3, 5)	3	6	9	10	13	4	4
(3, 6)	2	6	8	17	19	11	11
(4, 5)	7	6	13	6	13	0	0
(4, 6)	5	6	11	14	19	8	8
(5, 6)	6	13	19	13	19	0[a]	0

[a] Critical activity.

Exercise 12.2-2

In Table 12–1, verify the given values of the free and total floats for each of the following activities.

(a) Activity (0, 1).

(b) Activity (3, 4).

(c) Activity (4, 6).

12.3 CONSTRUCTION OF THE TIME CHART AND RESOURCE LEVELING

The end product of network calculations is the construction of the time chart (or schedule). This time chart can be converted easily into a calendar schedule for convenient use in the execution of the project.

The construction of the time chart must be made within the limitations of the available resources, since it may not be possible to execute concurrent activities because of personnel and equipment limitations. This is the point where the total floats for the noncritical activity become useful. By shifting a noncritical activity (back and forth) between its maximum allowable limits, one may be able to lower the maximum resource requirements. In any case, even in the absence of limited resources, it is common practice to use total floats to level resources over the duration of the entire project. In essence this would mean a more steady work force compared to the case where the work force (and equipment) would vary drastically from one day to the next.

The procedure for constructing the time chart will be illustrated by Example 12.3–1. Example 12.3–2 will then show how resource leveling can be effected for the same project.

Example 12.3-1. In this example the time chart for the project given in Example 12.2–1 will be constructed.

The information necessary to construct the time chart is summarized in Table 12–1. The first step is to consider the scheduling of the critical activities. Next, the noncritical activities are considered by indicating their *ES* and *LC* time limits on the chart. The critical activities are shown with solid lines. The time ranges for the noncritical activities are shown by dashed lines, indicating that such activities may be scheduled within those ranges *provided that the precedence relationships are not disturbed*.

Figure 12–6 shows the time chart corresponding to Example 12.2–1. The dummy activity (3, 4) consumes no time and hence is shown by a vertical line. The numbers shown with the noncritical activities represent their durations.

The roles of the *total* and *free* floats in scheduling noncritical activities is explained in terms of two general rules:

1. If the total float *equals* the free float, the noncritical activity can be scheduled *anywhere* between its earliest start and latest completion times (dashed time spans in Figure 12–6).

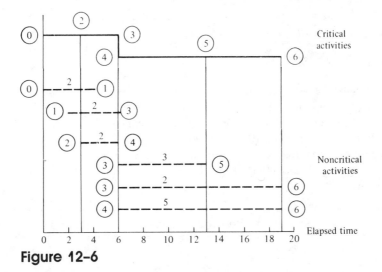

Figure 12-6

2. If the free float is *less than* the total float, the starting of the noncritical activity can be delayed relative to its earliest start time by no more than the amount of its free float without affecting the scheduling of its *immediately* succeeding activities.

In our example, rule 2 applies to activity (0, 1) only, whereas all others are scheduled according to rule 1. The reason is that activity (0, 1) has *zero* free float. Thus, if the starting time for (0, 1) is not delayed beyond its earliest start time ($t = 0$), the immediately succeeding activity (1, 3) can be scheduled anywhere between its earliest start time ($t = 2$) and latest completion time ($t = 6$). On the other hand, if the starting time of (0, 1) is delayed beyond $t = 0$, the earliest start time of (1, 3) must be delayed relative to its earliest start time by at least the same amount. For example, if (0, 1) starts at $t = 1$, it terminates at $t = 3$ and (1, 3) can then be scheduled anywhere between $t = 3$ and $t = 6$. This type of restriction does not apply to any of the remaining noncritical activities because they all have equal total and free floats. We can also see this result in Figure 12–6, since (0, 1) and (1, 2) are the only two tandem activities whose permissible time spans overlap.

In essence, having the free float less than the total float gives us a warning that the scheduling of the activity should not be finalized without first checking its effect on the start times of the *immediately* succeeding activities. This valuable information can be secured only through the use of critical path computations.

◀

Exercise 12.3-1

The following cases represent the total and free floats (*TF* and *FF*) for a noncritical activity. Indicate the maximum delay in the starting time of the activity relative to its earliest start time that will allow all the immediately succeeding activities to be scheduled anywhere between their earliest and latest completion times.

(a) $TF = 10$, $FF = 10$, $D = 4$.
   [*Ans.* Delay = 10.]
(b) $TF = 10$, $FF = 5$, $D = 4$.
   [*Ans.* Delay = 5.]

(c) $TF = 10$, $FF = 0$, $D = 4$.
   [*Ans.* Delay = 0.]
(d) $TF = 10$, $FF = 3$, $D = 4$.
   [*Ans.* Delay = 3.]

Example 12.3–2. In Example 12.3–1 suppose that the following worker requirements are specified for the different activities. It is required to develop a time schedule that will level the worker requirements during the project duration. [Note that activities (0, 1) and (1, 3) require no manual labor, which is indicated by assigning zero number of men to each activity. As a result, the scheduling of (0, 1) and (1, 3) can be made independently of the resource leveling procedure.]

Activity	Number of Workers	Activity	Number of Workers
0, 1	0	3, 5	2
0, 2	5	3, 6	1
1, 3	0	4, 5	2
2, 3	7	4, 6	5
2, 4	3	5, 6	6

Figure 12–7(a) shows the personnel requirements over time if the noncritical activities are scheduled as early as possible, and Figure 12–7(b) shows the same requirements if these activities are scheduled as late as possible. The dashed line shows the requirements for the critical activities that must be satisfied if the project is to be completed on time. [Notice that activities (0, 1) and (1, 3) require no resources.]

The project requires at least 7 workers, as indicated by the requirements of the critical activity (2, 3). The earliest scheduling of the noncritical activities requires a maximum of 10 men, while the latest scheduling of the same activities sets the maximum requirements at 12 workers, which means that the maximum requirements depend on how the floats of the noncritical activities are used. In Figure 12–7, however, regardless of how the floats are allocated, the maximum requirement cannot be fewer than 10 workers, since the range for activity (2, 4) coincides with the time for the critical activity (2, 3). The work force requirement using the earliest scheduling can be improved by rescheduling activity (3, 5) at its latest possible time and activity (3, 6) immediately after activity (4, 6) is completed. This new requirement is shown in Figure 12–8. The new schedule has now resulted in a smoother allocation of resources.

In some projects the objective may be to keep the maximum resource utilization below a certain limit rather than merely leveling the resources. If this objective cannot be accomplished by rescheduling the noncritical activities, it will be necessary to expand the time for some of the critical activities, thus reducing the required daily level of the resource.   ◀

Because of mathematical complexity, no technique has yet been developed that will yield the *optimum* solution to the resource leveling problem, that is,

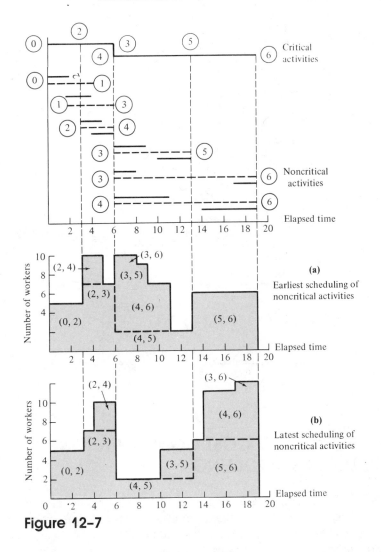

Figure 12-7

minimization of the maximum required resources for the project at any point in time. Rather, heuristic programs similar to the one just outlined are actually used. These programs take advantage of the different floats for the noncritical activities.

Exercise 12.3-2

In Example 12.3–2, suppose that activities (0, 1) and (1, 3) require 8 and 2 workers, respectively. Indicate the changes in Figure 12–7 in each of the following cases.

(a) Both activities are scheduled as early as possible.
   [*Ans.* Changes in the number of workers are 13 for $0 \leq t < 2$, 7 for $2 \leq t < 3$, and 12 for $3 \leq t < 4$.]

(b) Both activities are scheduled as late as possible.
   [*Ans.* Changes in the number of workers are 13 for $2 \leq t < 3$, 15 for $3 \leq t < 4$, and 12 for $4 \leq t < 6$.]

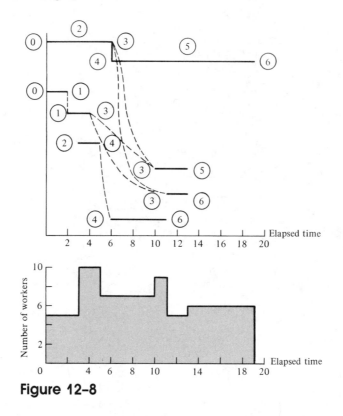

Figure 12-8

12.4 PROBABILITY AND COST CONSIDERATIONS IN PROJECT SCHEDULING

The analysis in Sections 12.1, 12.2, and 12.3 does not take into account the case where time estimates for the different activities are probabilistic. Also, it does not consider explicitly the cost of schedules. This section will thus present both the probability and cost aspects in project scheduling.

12.4.1 PROBABILITY CONSIDERATIONS IN PROJECT SCHEDULING

Probability considerations are incorporated in project scheduling by assuming that the time estimate for each activity is based on three different values:

$a =$ **optimistic time,** which will be required if execution goes extremely well
$b =$ **pessimistic time,** which will be required if everything goes badly
$m =$ **most likely time,** which will be required if execution is normal

The range specified by the optimistic and pessimistic estimates (a and b, respectively) supposedly must enclose every possible estimate of the duration

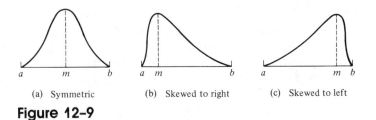

(a) Symmetric   (b) Skewed to right   (c) Skewed to left

Figure 12-9

of the activity. The most likely estimate m need not coincide with the midpoint $(a + b)/2$ and may occur to its left or to its right. Because of these properties it is *intuitively* justified that the duration for each activity may follow a beta distribution with its unimodal point occurring at m and its end points at a and b. Figure 12–9 shows the three cases of the beta distribution, which are (a) symmetric, (b) skewed to the right, and (c) skewed to the left.

The expressions for the mean $\overline{D}$ and variance V of the beta distribution are developed as follows.† The midpoint $(a + b)/2$ is assumed to weight half as much as the most likely point m. Thus $\overline{D}$ is the arithmetic means of $(a + b)/2$ and $2m$; that is,

$$\overline{D} = \frac{(a + b)/2 + 2m}{3} = \frac{a + b + 4m}{6}$$

The range (a, b) is assumed to enclose about 6 standard deviations of the distribution, since about 90% or more of *any* probability density function lies within 3 standard deviations of its mean. Thus

$$V = \left(\frac{b - a}{6}\right)^2$$

The network calculations given in Sections 12.1, 12.2, and 12.3 can now be applied directly, with $\overline{D}$ replacing the single estimate D.

It is now possible to estimate the probability of occurrence of each event in the network. Let μ_i be the earliest occurrence time of event i. Since the times of the activities summing up to i are random variables, μ_i is also a random variable. Assuming that all the activities in the network are statistically independent, one obtains the mean and variance of μ_i as follows. If there is only one path leading from the "start" event to event i, $E\{\mu_i\}$ is given by the sum of the expected durations $\overline{D}$ for the activities along this path, and var $\{\mu_i\}$ is the sum of the variances of the same activities. Complications arise, however, where

---

† The validity of the beta distribution assumption has been challenged. The expressions for $\overline{D}$ and V developed below cannot be satisfied for the beta distribution unless certain restrictive relationships among a, b, and m exist. [See F. Grubbs, "Attempts to Validate Certain PERT Statistics or 'Picking on PERT'," *Operations Research*, Vol. 10, 1962, pp. 912–915.] However, the expressions for $\overline{D}$ and V are based on intuitive arguments regardless of the original beta distribution assumption. Later it will be shown that the network analysis is based on the central limit theorem, which assumes normality regardless of the parent distribution of individual activities. In this respect, whether the real distribution is beta or not seems to be unimportant. The question as to whether $\overline{D}$ and V are the true measures of the parent (unknown) distribution still remains unanswered, however.

more than one path leads to the same event. In this case, if the *exact* $E\{\mu_i\}$ and var $\{\mu_i\}$ are to be computed, one must first develop the statistical distribution for the longest of the different paths (i.e., the distribution of the maximum of several random variables) and then find its expected value and variance. This problem is rather difficult in general and a simplifying assumption is introduced that computes $E\{\mu_i\}$ and var $\{\mu_i\}$ as those of the path to event i having the largest sum of *expected* activity durations. If two or more paths have the same $E\{\mu_i\}$, the one with the largest var $\{\mu_i\}$ is selected, since it reflects greater uncertainty and hence more conservative results. To summarize, $E\{\mu_i\}$ and var $\{\mu_i\}$ are given for the selected path by

$$E\{\mu_i\} = ES_i$$
$$\text{var}\{\mu_i\} = \sum_k V_k$$

where k defines the activities along the longest path leading to i.

The idea is that μ_i is the sum of independent random variables and hence, according to the central limit theorem (see Section 10.8), μ_i is approximately normally distributed with the mean $E\{\mu_i\}$ and variance var $\{\mu_i\}$. Since μ_i represents the earliest occurrence time, event i will meet a certain scheduled time ST_i (specified by the analyst) with probability

$$P\{\mu_i \le ST_i\} = P\left\{\frac{\mu_i - E\{\mu_i\}}{\sqrt{\text{var}\{\mu_i\}}} \le \frac{ST_i - E\{\mu_i\}}{\sqrt{\text{var}\{\mu_i\}}}\right\} = P\{z \le K_i\}$$

where z is the standard normal distribution with mean zero and variance one and

$$K_i = \frac{ST_i - E\{\mu_i\}}{\sqrt{\text{var}\{\mu_i\}}}$$

It is common practice to compute the probability that event i will occur no later than its LC_i. Such probabilities will thus represent the chance that the succeeding events will occur within their (ES_i, LC_i) durations.

Example 12.4-1. Consider the project of Example 12.2–1. To avoid repeating the critical path calculations, the values of a, b, and m shown in Table 12–2 are selected such that $\overline{D}_{ij}$ will have the same value as its corresponding D_{ij} in Example 12.2–1.

Table 12–2

Activity (i, j)	Estimated Times (a, b, m)	Activity (i, j)	Estimated Times (a, b, m)
(0, 1)	(1, 3, 2)	(3, 5)	(1, 7, 2.5)
(0, 2)	(2, 8, 2)	(3, 6)	(1, 3, 2)
(1, 3)	(1, 3, 2)	(4, 5)	(6, 8, 7)
(2, 3)	(1, 11, 1.5)	(4, 6)	(3, 11, 4)
(2, 4)	(.5, 7.5, 1)	(5, 6)	(4, 8, 6)

Table 12-3

Activity	$\bar{D}_{ij}$	V_{ij}	Activity	$\bar{D}_{ij}$	V_{ij}
(0, 1)	2	.33	(3, 5)	3	1.00
(0, 2)	3	1.00	(3, 6)	2	.11
(1, 3)	2	.11	(4, 5)	7	.11
(2, 3)	3	2.78	(4, 6)	5	1.78
(2, 4)	2	1.36	(5, 6)	6	.44

The mean $\bar{D}_{ij}$ and variance V_{ij} for the different activities are given in Table 12–3.

The probabilities are given in Table 12–4. The information in the ST_i column is part of the input data. The values of ST_i can be replaced by LC_i to obtain the probabilities that none of the activities will be delayed beyond its latest occurrence time.

The information under the path column is obtained directly from the network. It defines the *longest* path from event 0 to event i.

After computing $E\{\mu_i\}$ and var $\{\mu_i\}$, the calculations of K_i and $P\{z \leq K_i\}$ are straightforward. The probabilities associated with the realization of each event can then be computed. These probabilities provide information about where resources are needed most to reduce the probability of delays in execution of the project.   ◀

Table 12-4

Event	Path	$E\{\mu_i\}$	var $\{\mu_i\}$	ST_i	K_i	$P\{z \leq K_i\}$
1	(0, 1)	2	.11	4	6.03	1.000
2	(0, 2)	3	1.00	2	−1.000	.159
3	(0, 2, 3)	6	3.78	5	− .514	.304
4	(0, 2, 3, 4)	6	3.78	6	.000	.500
5	(0, 2, 3, 4, 5)	13	3.89	17	2.028	.987
6	(0, 2, 3, 4, 5, 6)	19	4.33	20	.480	.684

12.4.2 COST CONSIDERATIONS IN PROJECT SCHEDULING

The cost aspect is included in project scheduling by defining the cost–duration relationship for each activity in the project. Costs are defined to include direct elements only. Indirect costs such as administrative or supervision cannot be included. Their effect will be included in the final analysis, however. Figure 12–10 shows a typical straight-line relationship used with most projects. The point (D_n, C_n) represents the duration D_n and its associated cost C_n if the activity is executed under **normal** conditions. The duration D_n can be compressed by increasing the allocated resources and hence by increasing the direct costs. There is a limit, called **crash** time, beyond which no further reduction in the duration can be effected. At this point any increase in resources will only increase

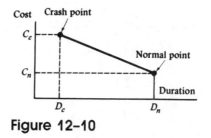

Figure 12–10

the costs without reducing the duration. The crash point is indicated in Figure 12–10 by the point (D_c, C_c).

The straight-line relationship is used mainly for convenience, since it can be determined for each activity from the knowledge of the normal and crash points only, that is, (D_n, C_n) and (D_c, C_c). A nonlinear relationship will complicate the calculations. There is one exceptional case, however, where the nonlinear relationship can be approximated by a *piecewise* linear curve as shown in Figure 12–11. Under such conditions, the activity can be broken down into a number of subactivities each corresponding to one of the line segments. Notice the increasing slopes of the line segments as we move from the normal point to the crash point. If this condition is not satisfied, the approximation is invalid.

After defining the cost–time relationships, the activities of the project are assigned their normal durations. The corresponding critical path is then computed and the associated (direct) costs are recorded. The next step is to consider reducing the duration of the project. Since such a reduction can be effected only if the duration of a critical activity is reduced, attention must be paid to such activities alone. To achieve a reduction in the duration at the least possible cost, one must compress as much as possible the critical activity having the smallest cost–time slope.

The amount by which an activity can be compressed is limited by its crash time. However, other limits must be taken into account before the exact compression amount can be determined. The details of these limits are discussed in Example 12.4–2.

The result of compressing an activity is a new schedule perhaps with a new critical path. The cost of the new schedule must be greater than that of the immediately preceding one. The new schedule must now be considered for compression by selecting the (uncrashed) critical activity with the least slope. The procedure is repeated until all *critical* activities are at their crash times. The final result of these calculations is a cost–time curve for the various schedules

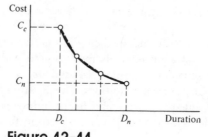

Figure 12–11

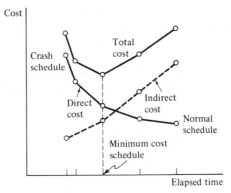

Figure 12-12

and their corresponding costs. A typical curve is shown by a solid line in Figure 12–12. As indicated earlier, it represents the direct costs only.

It is logical to assume that as the duration of the project increases, the *in*direct costs must also increase as shown in Figure 12–12 by a dashed curve. The sum of these two costs (direct + indirect) gives the total cost of the project. The optimum schedule corresponds to the minimum total cost.

Example 12.4–2. Consider the network in Figure 12–13. The normal and crash points for each activity are given in Table 12–5. It is required to compute the different minimum-cost schedules that can occur between normal and crash times.

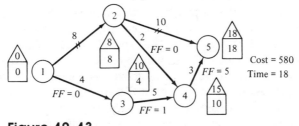

Figure 12-13

Table 12-5

Activity	Normal		Crash	
(i, j)	Duration	Cost	Duration	Cost
(1, 2)	8	100	6	200
(1, 3)	4	150	2	350
(2, 4)	2	50	1	90
(2, 5)	10	100	5	400
(3, 4)	5	100	1	200
(4, 5)	3	80	1	100

The analysis in this problem is dependent mainly on the cost–time slopes for the various activities, which can be computed using the formula

$$\text{slope} = \frac{C_c - C_n}{D_n - D_c}$$

The slopes for the activities of the network are summarized in Table 12–6.

Table 12–6

Activity	Slope
(1, 2)	50
(1, 3)	100
(2, 4)	40
(2, 5)	60
(3, 4)	25
(4, 5)	10

The first step in the calculation procedure is to assume that all activities occur at normal times. The network in Figure 12–13 shows the critical path calculations under normal conditions. Activities (1, 2) and (2, 5) constitute the critical path. The time of the project is 18 and its associated (normal) cost is 580.

The second step is to reduce the time of the project by compressing (as much as possible) the critical activity with the least slope. For the network in Figure 12–13 there are only two critical activities, (1, 2) and (2, 5). Activity (1, 2) is selected for compression because it has the smaller slope. According to the time–cost curve, this activity can be compressed by two time units, a limit that is specified by its crash point (henceforth called **crash limit**). However, compressing a critical activity to its crash point would not necessarily mean that the duration of the entire project will be reduced by an equivalent amount. This result follows, since, as the critical activity is compressed, a *new* critical path may develop. At this point we must discard the old critical activity and pay attention to the activities of the new critical path.

One way of predicting whether a new critical path will develop before reaching crash point is to consider the free floats for the noncritical activities. By definition, these free floats are independent of the start times of the other activities. Thus if during the compression of a critical activity a *positive* free float becomes zero, this critical activity is not to be compressed without further checking because there is a *possibility* that this zero free float activity may become critical. Thus, in addition to the *crash limit,* one must consider the **free float limit.**

To determine the free float limit, we need first to reduce the duration of the critical activity selected for compression by *one* time unit. Then, by recomputing the free floats for all the noncritical activities, we note which of these activities have reduced their *positive* free floats by *one* time unit. The smallest free float (before reduction) of all such activities determines the required free float limit.

By applying this to the network of Figure 12–13, the free floats (FF) are shown on the respective activities. A reduction of activity (1, 2) by one time

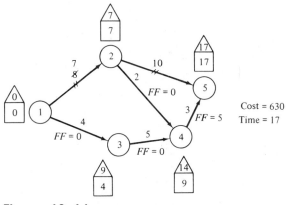

Figure 12–14

unit will drop the free float of activity (3, 4) from one to zero. The free float of activity (4, 5) will remain unchanged at 5. Thus FF-limit = 1. Since the crash limit for (1, 2) is 2, its **compression limit** equals the minimum of its crash and FF-limits, that is, min {2, 1} = 1. The new schedule is shown in Figure 12–14. The corresponding project time is 17 and its associated cost is equal to that of the previous schedule plus the additional cost of the compressed time, that is, $580 + (18 - 17) \times 50 = 630$. Although the free float determines the compression limit, the critical path remains the same. Thus it is *not* always true that a new critical path will arise when the compression limit is specified by the FF-limit.

Since activity (1, 2) is still the best candidate for compression, its corresponding crash and FF-limits are computed. However, since the crash limit for activity (1, 2) is equal to 1, it is not necessary to compute the FF-limit because any positive FF is at least equal to 1. Consequently, activity (1, 2) is compressed by one unit, thus reaching its crash limit. The resulting computations are shown in Figure 12–15, which also shows that the critical path remains unchanged. The time of the project is 16 and its associated cost is $630 + (17 - 16) \times 50 = 680$.

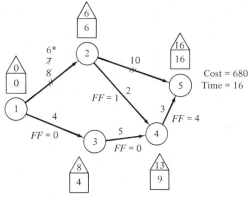

*Signifies that activity has reached its crash limit.

Figure 12–15

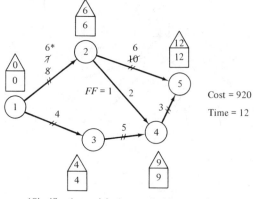

Cost = 920
Time = 12

*Signifies that activity has reached its crash limit.

Figure 12-16

Activity (1, 2) can no longer be compressed. Hence activity (2, 5) is selected for compression. Now

$$\text{crash limit} = 10 - 5 = 5$$

$$\text{FF-limit} = 4, \quad \text{corresponding to activity (4, 5)}$$

$$\text{compression limit} = \min \{5, 4\} = 4$$

The resulting computations are shown in Figure 12–16. There are two critical paths now: (1, 2, 5) and (1, 3, 4, 5). The time for the new project is 12, and its cost is $680 + (16 - 12) \times 60 = 920$.

The appearance of two critical paths indicates that to reduce the time of the project, it will be necessary to reduce the time of the two critical paths simultaneously. The previous rule for selecting the critical activities to be compressed still applies here. For path (1, 2, 5), activity (2, 5) can be compressed by one time unit. For path (1, 3, 4, 5), activity (4, 5) has the least slope and its crash limit is 2. Thus the crash limit for the two paths is equal to min $\{1, 2\} = 1$. The FF-limit is determined for this case by taking the minimum of

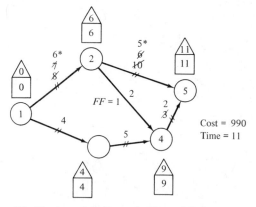

Cost = 990
Time = 11

*Signifies that activity has reached its crash limit.

Figure 12-17

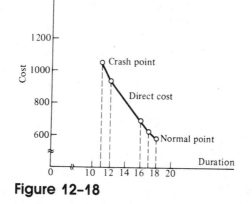

Figure 12-18

the FF-limits obtained by considering each critical path separately. However, since the crash limit is equal to 1, the FF-limit need not be computed.

The new schedule is shown in Figure 12–17. Its time is 11 and its cost is $920 + (12 - 11) \times (10 + 60) = 990$.

The two critical paths of the project remain the same. Since all the activities on the critical path (1, 2, 5) are at crash time, it is no longer possible to reduce the time of the project. The schedule in Figure 12–17 thus gives the crash schedule.

A summary of these computations is given in Figure 12–18, which represents the direct cost of the project. By adding the indirect costs corresponding to each schedule, we can compute the minimum total cost (or optimum) schedule.

◀

Example 12.4–2 summarizes all the rules for compressing activities under the given conditions. There are cases, however, where one may have to expand an already compressed activity before the duration of the entire project can be reduced. Figure 12–19 illustrates a typical case. There are three critical paths: (1, 2, 3, 4), (1, 2, 4), and (1, 3, 4). Activity (2, 3) has been compressed from its normal time 8 to its present time 5. The duration of the project may be reduced by simultaneously reducing one of the activities on each of the critical paths (1, 2, 4) and (1, 3, 4) or by simultaneously compressing activities (1, 2) and (3, 4) and expanding activity (2, 3). The alternative with the smallest *net* sum of slopes is selected. Notice that if activities (1, 2) and (3, 4) are compressed and activity (2, 3) is expanded, the *net* sum of slopes is the sum of the slopes for activities (1, 2) and (3, 4) *minus* the slope for activity (2, 3). In all other

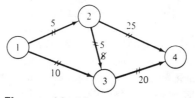

Figure 12-19

cases where there are no expandable activities, the net sum is equal to the sum of the slopes of the compressed activities.

If expansion is necessary, then, in addition to the crash limit and the FF-limit, the expansion limit must also be taken into account. This is equal to the normal time of the activity minus its present compressed time. The compression limit is thus the minimum of the crash limit, the FF-limit, and the expansion limit.

An Alternative Procedure for Detecting New Critical Paths

In Example 12.4–2 the FF-limit was used to detect the possibility of having new critical paths. If the FF-limit is large and equal to the compression limit, we can reduce the duration of the project in large steps. In essence, this procedure has the advantage of minimizing the number of schedules computed between the normal and crash points, which possibly means that the *main* computations of the project are minimized. However, the determination of the FF-limits requires additional computations that increase especially with the increase in the number of critical paths in the project. Consequently, there is no guarantee that the use of the FF-limit method would yield minimum computations.

Another method has thus been developed that completely eliminates the need for the FF-limit.† It is indicated in Example 12.4–2 that if the crash limit is equal to 1, the FF-limit need not be computed, since any positive FF is at least equal to 1. The new procedure thus calls for reducing the duration of the project by one time unit at each cycle of the computations. This is done again by compressing the activity having the least slope. The procedure is repeated on the new schedule—and the new critical path(s), if any—until the crash schedule is attained. Notice that the new method compresses the project duration by one time unit each cycle. Thus, if there are n time units between normal and crash schedules, we should expect a total of n cycles of computations.

There is no conclusive evidence as to which of the methods just described is computationally more efficient. However, for hand computations the non-FF-limit method seems more convenient. You are asked in Problem 12–17 to resolve Example 12.4–2 using the new method and hence compare the amount of computations in each cycle.

12.5   PROJECT CONTROL

There is a tendency among some PERT–CPM users to think that the arrow diagram can be discarded as soon as the time schedule is developed. This is not so. In fact, an important use of the arrow diagram occurs during the execution

---

† There exist other more efficient methods for effecting minimum computations between the normal and crash schedules. The rules for such methods are somewhat complex, however. See, for example, Moder and Phillips [(1970), Chap. 9].

phase of the project. It seldom happens that the planning phase will develop a time schedule that can be followed exactly during the execution phase. Quite often some of the jobs are delayed or expedited, which naturally depends on actual work conditions. As soon as such disturbances occur in the original plan, it becomes necessary to develop a new time schedule for the remaining portion of the project. This section outlines a procedure for monitoring and controlling the project during the execution phase.

It is important to follow the progress of the project on the arrow diagram rather than solely on the time schedule. The time schedule is used principally to check if each activity is on time. The effect of a delay in a certain activity on the remaining portion of the project can best be traced on the arrow diagram.

Suppose that as the project progresses over time, it is discovered that delay in some activities necessitates developing a completely new schedule. How can the new schedule be obtained using the present arrow diagram? The immediate requirement is to update the arrow diagram by assigning zero values to the durations of the completed activities. Partially completed activities are assigned times equivalent to their unfinished portions. Changes in the arrow diagram such as addition or deletion of any future activities must also be made. By repeating the usual computations on the arrow diagram with its new time elements, we can determine the new time schedule and possible changes in the duration of the project. Such information is used until it is again necessary to update the time schedule. In real situations, many revisions of the time schedule are usually required at the early stages of the execution phase. A stable period follows in which little revision of the current schedule may be required.

12.6 SUMMARY

Critical path computations are quite simple, yet they provide valuable information that simplifies the scheduling of complex projects. The result is that PERT–CPM techniques enjoy tremendous popularity among practitioners in the field. The usefulness of the techniques is further enhanced by the availability of specialized computer systems for executing, analyzing, and controlling network projects.

SELECTED REFERENCES

ELMAGHRABY, S., *Activity Networks*, Wiley, New York, 1977.
MODER, J., and C. PHILLIPS, *Project Management with CPM and PERT*, 2nd ed., Van Nostrand Reinhold, New York, 1970.

REVIEW QUESTIONS

True (T) or False (F)?

1. ____ A dummy activity in a project network always has zero duration.

2. ____ In a project network, a sequence of activities may form a loop.

3. ____ Every outgoing activity at a node is necessarily a successor to all the incoming activities at the same node.

4. ____ The critical path of a project network represents the minimum duration needed to complete the network.

5. ____ It is possible to delay the completion of critical activities without delaying the entire project.

6. ____ A critical path need not constitute a connected chain of activities between the start and terminal nodes.

7. ____ A critical activity must have its *total* and *free* floats equal to zero.

8. ____ A network may include more than one critical path.

9. ____ If a network has more than one critical path, the durations of the different paths may not be equal.

10. ____ A noncritical activity cannot have zero total float.

11. ____ A noncritical activity may or may not have zero free float.

12. ____ A noncritical activity can be scheduled anywhere between its earliest start and latest completion.

13. ____ A zero free float for a noncritical activity is a signal that the start time for one or more succeeding activities will depend on when this noncritical activity is completed.

14. ____ The start and completion times for critical activities cannot be changed without necessarily increasing the duration of the entire project.

15. ____ Information about float times is essential for effecting resource leveling.

16. ____ Resource limitations can affect the critical path computations by requiring an expansion of one or more critical activities.

17. ____ The computation of the critical path in PERT is basically different from that in CPM.

18. ____ In a PERT network, there is always a $50:50$ chance that a critical activity will be completed by its earliest (or latest) completion time.

19. ____ The use of the normal distribution to compute the probabilities of events in PERT becomes more accurate with the increase in the number of activities leading from the start point to the event under consideration.

20. ____ It is possible to increase the duration of an activity beyond its normal value without causing an increase in cost.

21. ____ It is possible to reduce an activity's duration below its crash time by allocating more funds and resources.

22. ____ In PERT/cost, the indirect costs can be computed from the network data.

23. _____ The use of crashing in PERT/cost is based on the assumption that the cost–duration trade-off for each activity is convex or linear.

24. _____ The free float plays an important role in making crash-cost analysis in PERT/cost.

25. _____ The use of CPM schedule requires periodic updating to reconcile differences between the schedule and the actual progress in the field.
[*Ans.* 1—T, 2—F, 3—T, 4—T, 5—F, 6—F, 7—T, 8—T, 9—F, 10—T, 11—T, 12—F, 13—T, 14—T, 15—T, 16—T, 17—F, 18—T, 19—T, 20—F, 21—F, 22—F, 23—T, 24—T, 25—T.]

PROBLEMS

Section	Assigned Problems
12.1	12–1 to 12–4
12.2	12–5 to 12–11
12.3	12–12 to 12–14
12.4.1	12–15
12.4.2	12–16 to 12–18

☐ **12–1** Construct the arrow diagram comprising activities $A, B, C, \ldots, P$ that satisfies the following precedence relationships.
  (i) A, B, and C, the first activities of the project, can start simultaneously.
  (ii) Activities D, E, and F start immediately after A is completed.
  (iii) Activities I and G start after both B and D are completed.
  (iv) Activity H starts after both C and G are completed.
  (v) Activities K and L succeed activity I.
  (vi) Activity J succeeds both E and H.
  (vii) Activities M and N succeed F but cannot start until E and H are completed.
  (viii) Activity O succeeds M and I.
  (ix) Activity P succeeds J, L, and O.
  (x) Activities K, N, and P are the terminal jobs of the project.

☐ **12–2** The footings of a building can be completed in four consecutive sections. The activities for each section include digging, placing steel, and pouring concrete. The digging of one section cannot start until the preceding one is completed. The same restriction applies to pouring concrete. Develop a network for the project.

☐ **12–3** Consider Problem 12–2. After digging all sections, plumbing work can be started but only 10% of the job can be completed before *any* concrete is poured. After each section of the footings is completed, an additional 5% of the plumbing can be started provided that the preceding 5% portion is complete. Construct the activity network.

☐ **12–4** An opinion survey involves designing and printing questionnaires, hiring and training personnel, selecting participants, mailing questionnaires, and analyzing the data. Construct a network for this project. Specify all the assumptions made.

☐ **12–5** Table 12–7 provides the data for building a new house. Construct the associated network model and carry out the critical path computations.

☐ **12–6** For the purpose of preparing its next year's budget, a company must gather information from its sales, production, accounting, and treasury departments. Table 12–8 indicates the activities and their durations. Prepare the network model of the problem and carry out the critical path computations.

☐ **12–7** The activities involved in a candlelight choir service are given in Table 12–9. Prepare the network model and carry out the critical path computations.

Table 12-7

Activity	Description	Immediate Predecessor(s)	Duration (days)
A	Clear site	—	1
B	Bring utilities to site	—	2
C	Excavate	A	1
D	Pour foundation	C	2
E	Outside plumbing	B, C	6
F	Frame house	D	10
G	Electric wiring	F	3
H	Lay floor	G	1
I	Lay roof	F	1
J	Inside plumbing	E, H	5
K	Shingling	I	2
L	Outside sheathing insulation	F, J	1
M	Install windows and outside doors	F	2
N	Brick work	L, M	4
O	Insulate walls and ceiling	G, J	2
P	Cover walls and ceiling	O	2
Q	Insulate roof	I, P	1
R	Finish interior	P	7
S	Finish exterior	I, N	7
T	Landscape	S	3

Table 12-8

Activity	Description	Immediate Predecessor(s)	Duration (days)
A	Forecast sales volume	—	10
B	Study competitive market	—	7
C	Design item and facilities	A	5
D	Prepare production schedules	C	3
E	Estimate cost of production	D	2
F	Set sales price	B, E	1
G	Prepare budget	E, F	14

Table 12-9

Activity	Description	Immediate Predecessor(s)	Duration (days)
A	Select music	—	21
B	Learn music	A	14
C	Make copies and buy books	A	14
D	Tryouts	B, C	3
E	Rehearsals	D	70
F	Solo rehearsals	D	70
G	Rent candelabra	D	14
H	Buy candles	G	1
I	Set up and decorate candelabra	H	1
J	Buy decorations	D	1
K	Set up decorations	J	1
L	Order choir robe stoles	D	7
M	Press robes	L	7
N	Check out PA system	D	7
O	Select music tracks	N	14
P	Set up PA system	O	1
Q	Final rehearsal	E, F, P	1
R	Choir party	Q, I, K	1
S	Final program	M, R	1

☐ **12–8** Table 12–10 summarizes the activities for relocating ("reconductoring") 1700 feet of 13.8-kilovolt overhead primary line due to the widening of the road section in which the line is presently installed. Draw the network model and carry out the critical path computations.

☐ **12–9** The activities for buying a new car are summarized in Table 12–11. Draw the network model and carry out the critical path computations.

☐ **12–10** Determine the critical path(s) for projects (a) and (b) (Figure 12–20).

☐ **12–11** In Problem 12–10, compute the total and free floats and summarize the critical path calculations using the format in Table 12–1.

☐ **12–12** In Problem 12–10, using the results of Problem 12–11, construct the corresponding time charts assuming no limits on the resources.

Table 12-10

Activity	Description	Immediate Predecessor(s)	Duration (days)
A	Job review	—	1
B	Advise customers of temporary outage	A	.5
C	Requisition stores	A	1
D	Scout job	A	.5
E	Secure poles and materials	C, D	3
F	Distribute poles	E	3.5
G	Pole location coordination	D	.5
H	Restake	G	.5
I	Dig holes	H	3
J	Frame and set poles	F, I	4
K	Cover old conductors	F, I	1
L	Pull new conductors	J, K	2
M	Install remaining material	L	2
N	Sag conductor	L	2
O	Trim trees	D	2
P	Deenergize and switch lines	B, M, N, O	.1
Q	Energize and phase new line	P	.5
R	Clean up	Q	1
S	Remove old conductor	Q	1
T	Remove old poles	S	2
U	Return material to stores	I	2

Table 12-11

Activity	Description	Immediate Predecessor(s)	Duration (days)
A	Conduct feasibility study	—	3
B	Find potential customer for present car	A	14
C	List possible models	A	1
D	Research all possible models	C	3
E	Conduct interviews with mechanics	C	1
F	Collect dealer propaganda	C	2
G	Compile and organize all pertinent information	D, E, F	1
H	Choose top three models	G	1
I	Test-drive all three choices	H	3
J	Gather warranty and financing information	H	2
K	Choose one car	I, J	2
L	Compare dealers and choose dealer	K	2
M	Search for desired color and options	L	4
N	Test-drive chosen model once again	L	1
O	Purchase new car	B, M, N	3

☐ **12-13** Construct the time schedule for Problem 12–5.

☐ **12-14** Construct the time schedule for Problem 12–6.

☐ **12-15** Suppose that in Problem 12–10 the following personnel requirements are specified for the various activities of projects (a) and (b).

Project (a):

Activity	Number of Workers	Activity	Number of Workers
1, 2	5	3, 6	9
1, 4	4	4, 6	1
1, 5	3	4, 7	10
2, 3	1	5, 6	4
2, 5	2	5, 7	5
2, 6	3	6, 7	2
3, 4	7		

Project (b):

Activity	Number of Workers	Activity	Number of Workers
1, 2	1	3, 7	9
1, 3	2	4, 5	8
1, 4	5	4, 7	7
1, 6	3	5, 6	2
2, 3	1	5, 7	5
2, 5	4	6, 7	3
3, 4	10		

Find the minimum number of workers (as a function of the project time) required during the scheduling of the project. Using resource leveling, estimate the maximum number of workers required.

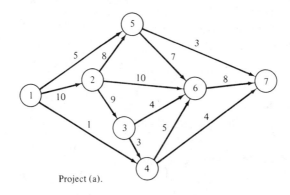

Project (a).

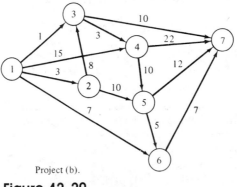

Project (b).

Figure 12–20

☐ **12–16** In Problem 12–10, suppose that the estimates (*a, b, m*) are given as shown in the tables. Find the probabilities that the different events will occur without delay.

Project (a):

Activity	(a, b, m)	Activity	(a, b, m)
1, 2	(5, 8, 6)	3, 6	(3, 5, 4)
1, 4	(1, 4, 3)	4, 6	(4, 10, 8)
1, 5	(2, 5, 4)	4, 7	(5, 8, 6)
2, 3	(4, 6, 5)	5, 6	(9, 15, 10)
2, 5	(7, 10, 8)	5, 7	(4, 8, 6)
2, 6	(8, 13, 9)	6, 7	(3, 5, 4)
3, 4	(5, 10, 9)		

Project (b):

Activity	(a, b, m)	Activity	(a, b, m)
1, 2	(1, 4, 3)	3, 7	(12, 14, 13)
1, 3	(5, 8, 7)	4, 5	(10, 15, 12)
1, 4	(6, 9, 7)	4, 7	(8, 12, 10)
1, 6	(1, 3, 2)	5, 6	(7, 11, 8)
2, 3	(3, 5, 4)	5, 7	(2, 8, 4)
2, 5	(7, 9, 8)	6, 7	(5, 7, 6)
3, 4	(10, 20, 15)		

☐ **12–17** Solve Example 12.4–2 without using the FF-limit method, that is, by compressing the project duration one unit at a time, and compare with the computations in Example 12.4–2.

☐ **12–18** In Problem 12–10, given the following data for the direct costs of the normal and crash durations, find the different minimum cost schedules between the normal and crash points.

Project (a):

Activity (i, j)	Normal		Crash	
	Duration	Cost	Duration	Cost
1, 2	5	100	2	200
1, 4	2	50	1	80
1, 5	2	150	1	180
2, 3	7	200	5	250
2, 5	5	20	2	40
2, 6	4	20	2	40
3, 4	3	60	1	80
3, 6	10	30	6	60
4, 6	5	10	2	20
4, 7	9	70	5	90
5, 6	4	100	1	130
5, 7	3	140	1	160
6, 7	3	200	1	240

Project (b):

Activity (i, j)	Normal		Crash	
	Duration	Cost	Duration	Cost
1, 2	4	100	1	400
1, 3	8	400	5	640
1, 4	9	120	6	180
1, 6	3	20	1	60
2, 3	5	60	3	100
2, 5	9	210	7	270
3, 4	12	400	8	800
3, 7	14	120	12	140
4, 5	15	500	10	750
4, 7	10	200	6	220
5, 6	11	160	8	240
5, 7	8	70	5	110
6, 7	10	100	2	180

Inventory Models

An inventory problem exists when it is necessary to stock physical goods or commodities for the purpose of satisfying demand over a specified time horizon (finite or infinite). Almost every business must stock goods to ensure smooth and efficient running of its operation. Decisions regarding **how much** and **when** to order are typical of every inventory problem. The required demand may be satisfied by stocking once for the entire time horizon or by stocking separately for every time unit of the horizon. The two cases correspond to overstocking (with respect to one time unit) and understocking (with respect to the entire horizon).

An overstock requires higher invested capital per unit time but less frequent occurrences of shortages and placement of orders. An understock, on the other hand, decreases the invested capital per unit time but increases the frequency of ordering as well as the risk of running out of stock. The two extreme situations are costly. Decisions regarding the quantity ordered and the time at which it is ordered may thus be based on the minimization of an appropriate cost function that balances the total costs resulting from overstocking and understocking.

13.1 THE ABC INVENTORY SYSTEM

In most real-life situations, inventory management usually involves a large number of items ranging in price from relatively inexpensive to possibly very expensive units. Since inventory in reality represents idle capital, it is natural that inventory control be exercised on items that are significantly responsible for the increase in capital cost. Thus routine items, such as bolts and nuts, contribute insignificantly to capital cost when compared with items involving expensive spare parts.

Experience has shown that only a relatively small number of inventory items usually incurs a major share of capital cost. Such items are the ones that must be subject to close inventory control.

The ABC system is a simple procedure that can be used to isolate the items that require special attention in terms of inventory control. The procedure calls for plotting percent of total inventory items against the percent of total dollar value of these items for a given time period (usually one year). Figure 13–1 illustrates a typical ABC curve.

The idea of the procedure is to determine the percent of items that contribute 80% of the cumulative dollar value. These items are classified as group A, and they normally constitute about 20% of all the items. Class B items are those corresponding to percent dollar values between 80% and 95%. They normally comprise about 25% of all the items. The remaining items constitute class C.

Class A items represent small quantities of expensive items and must be subject to tight inventory control. Class B items are next in order where a moderate form of inventory control can be applied. Finally class C items should be given the lowest priority in the application of any form of inventory control. Usually, the order size of expensive class A items is expected to be low to reduce the associated capital cost. On the other hand, the order size for class C can be quite large.

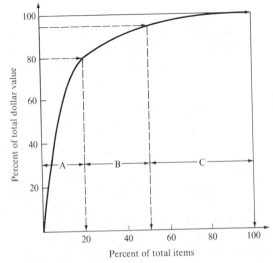

Figure 13–1

The ABC analysis is usually the first step that must be applied in an inventory control situation. Once the important inventory items are identified, models of the types to be presented in the succeeding sections can be used to decide on the ideal way of controlling inventory.

13.2 A GENERALIZED INVENTORY MODEL

The ultimate objective of any inventory model is to answer two questions:

1. How much to order?
2. When to order?

The answer to the first question is expressed in terms of what we call the **order quantity.** It represents the optimum amount that should be ordered every time an order is placed and may vary with time depending on the situation under consideration. The answer to the second question depends on the type of the inventory system. If the system requires **periodic review** at equal time intervals (e.g., every week or month), the time for acquiring a new order usually coincides with the beginning of each time interval. If, on the other hand, the system is of the **continuous review** type, a **reorder point** is usually specified by the *inventory level* at which a new order must be placed.

We can thus express the solution of the general inventory problem as follows:

1. *Periodic review case.* Receive a new order of the amount specified by the *order quantity* at equal intervals of time.

2. *Continuous review case.* When the inventory level reaches the *reorder point,* place a new order whose size equals the *order quantity.*

The order quantity and reorder point are normally determined by minimizing the total inventory cost that can be expressed as a function of these two variables. We can summarize the total cost of a general inventory model as a function of its principal components in the following manner:

$$\begin{pmatrix} \text{total} \\ \text{inventory} \\ \text{cost} \end{pmatrix} = \begin{pmatrix} \text{purchasing} \\ \text{cost} \end{pmatrix} + \begin{pmatrix} \text{setup} \\ \text{cost} \end{pmatrix} + \begin{pmatrix} \text{holding} \\ \text{cost} \end{pmatrix} + \begin{pmatrix} \text{shortage} \\ \text{cost} \end{pmatrix}$$

The **purchasing cost** becomes an important factor when the commodity unit price becomes dependent on the size of the order. This situation is normally expressed in terms of a **quantity discount** or a **price break,** where the unit price of the item decreases with the increase of ordered quantity. The **setup cost** represents the fixed charge incurred when an order is placed. Thus, to satisfy the demand for a given time period, the (more frequent) ordering of smaller quantities will result in a higher setup cost during the period than if the demand is satisfied by placing larger (and hence less frequent) orders. The **holding cost,** which represents the costs of carrying inventory in stock (e.g., interest on invested capital, storage, handling, depreciation, and maintenance), normally increases with the level of inventory. Finally, the **shortage cost** is a penalty incurred when we run out of stock of a needed commodity. It generally includes costs due to loss of customer's goodwill as well as potential loss in income.

Figure 13–2 illustrates the variation of the four cost components of the general inventory model as a function of the inventory level. The optimum inventory level corresponds to the minimum total cost of all four components. Note, however, that an inventory model need not include all four types of costs, either because some of the costs are negligible or will render the total cost function too complex for mathematical analysis. In practice, however, we can delete a cost component only if its effect on the total cost model is negligible. This point should be kept in mind as you study the various models we present in this chapter.

The foregoing general inventory model appears simple enough. Why, then, do we have large varieties of models whose methods of solutions range from the use of simple calculus to the sophisticated applications of dynamic and mathematical programming? The answer lies principally in whether the demand for the item is deterministic (known with certainty) or probabilistic (described by a probability density). Figure 13–3 illustrates the different classifications for demand as they are normally assumed in inventory models. A **deterministic demand** may be **static,** in the sense that the consumption rate remains constant with time, or **dynamic,** where the demand is known with certainty but varies from one time period to the next. The **probabilistic demand** has two similar classifications: the **stationary** case, in which the demand probability density function remains unchanged over time; and the **nonstationary** case, where the probability density function varies with time.

It is rare that a deterministic static demand would occur in real life. We may thus regard this situation as a simplifying case. For example, although demand for staple items, such as bread, may vary from day to day, the variations

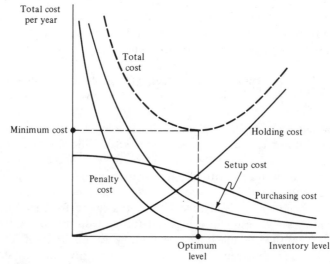

Figure 13–2

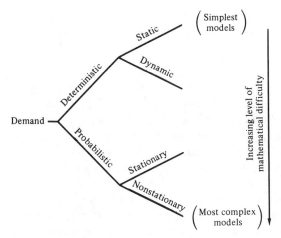

Figure 13-3

may be small and negligible with the result that a static demand assumption may not be too far from reality.

The most accurate representation of demand can perhaps be made by *probabilistic nonstationary* distributions. However, from the mathematical standpoint, the resulting inventory model will be rather complex, especially as the time horizon for the problem increases. Figure 13-3 illustrates this point by showing that the mathematical complexity of the inventory models increases as we move away from the assumption of deterministic static demand to the probabilistic nonstationary demand. Indeed, we can think of the classifications in Figure 13-3 as representing different *levels of abstraction* in demand.

The first level assumes that the probability distribution of demand is stationary over time. That is, the same probability density function is used to represent the demand for all periods over which the study is made. The implication of this assumption is that the effects of seasonal trends in demand, if any, will not be included in the model.

The second level of simplification recognizes the variations in demand between different periods. However, rather than utilize probability distributions, the *average* demand is used to represent the requirements of each period. This simplification has the effect of ignoring the element of risk in the inventory situation. Yet it allows the analyst to consider seasonal trends in demand that for analytic and computational difficulties cannot be included in a probabilistic model. In other words, there appears to be some kind of trade-off between using stationary probability distributions and variable but known demands under the assumption of "assumed certainty."

The third level of simplification eliminates both elements of risk and variability in demand. Thus the demand at any period is assumed equal to the average of the (assumedly) *known* demands for all periods under consideration. The result of this simplification is that demand may be represented as a *constant* rate per unit time.

Although the type of demand is a principal factor in the design of the inventory model, the following factors may also influence the way the model is formulated.

1. *Delivery lags or lead times.* When an order is placed, it may be delivered instantaneously, or it may require some time before delivery is effected. The time between the placement of an order and its receipt is called delivery lag or lead time, which may be deterministic or probabilistic.

2. *Stock replenishment.* Although an inventory system may operate with delivery lags, the actual replenishment of stock may occur instantaneously or uniformly. Instantaneous replenishment can occur when the stock is purchased from outside sources. Uniform replenishment may occur when the product is manufactured locally within the organization. In general, a system may operate with positive delivery lag and also with uniform stock replenishment.

3. *Time horizon.* The time horizon defines the period over which the inventory level will be controlled. This horizon may be finite or infinite, depending on the time period over which demand can be forecast reliably.

4. *Number of supply echelons.* An inventory system may consist of several (rather than one) stocking points. In some cases these stocking points are organized such that one point acts as a supply point for others. This type of operation may be repeated at different levels so that a demand point may again become a new supply point. The situation is usually referred to as a multiechelon system.

5. *Number of items.* An inventory system may involve more than one item (commodity). This case is of interest mainly if some kind of interaction exists between the different items. For example, the items may compete for limited floor space or limited total capital.

13.3 DETERMINISTIC MODELS

It is extremely difficult to develop a general inventory model that accounts for all variations in real systems. Indeed, even if a sufficiently general model can be formulated, it may not be analytically solvable. The models presented in this section are thus meant to be illustrative of some inventory systems. It is unlikely these models will fit a real situation exactly, but the objective of the presentation is to provide different ideas that can be adapted to specific inventory systems.

Five models are discussed in this section. Most of them deal with a single inventory item. Only one treats the effect on the solution of including several competing items. The main difference among these models is whether demand is static or dynamic. The type of cost function is also important in formulating and solving the models. You will notice the diverse methods of solution, which include classical optimization and linear and dynamic programming. These examples underscore the importance of using different optimization techniques in solving inventory models.

13.3.1 SINGLE-ITEM STATIC MODEL

The simplest type of inventory model occurs when demand is constant over time with instantaneous replenishment and no shortages. Typical situations to which this model may apply are

1. The use of light bulbs in a building.
2. The use of clerical supplies, such as paper, pads, and pencils, in a large company.
3. The use of certain industrial supplies, such as bolts and nuts.
4. The consumption of staple food items, such as bread and milk.

Figure 13–4 illustrates the variation of the inventory level. It is assumed that demand occurs at the rate β (per unit time). The highest level of inventory occurs when the order quantity y is delivered. (Delivery lag is assumed a known constant.) The inventory level reaches zero level y/β time units after the order quantity y is received.

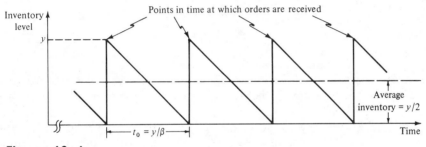

Figure 13–4

The smaller the order quantity y, the more frequent will be the placement of new orders. However, the average level of inventory held in stock will be reduced. On the other hand, larger order quantities indicate larger inventory level but less frequent placement of orders (see Figure 13–5). Because there are costs associated with placing orders and holding inventory in stock, the

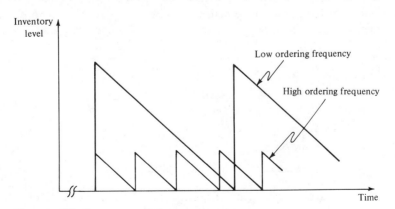

Figure 13–5

quantity y is selected to allow a compromise between the two types of costs. This is the basis for formulating the inventory model.

Let K be the setup cost incurred every time an order is placed and assume that the holding cost per unit inventory *per unit time* is h. Hence the total cost *per unit time TCU* as a function of y may be written as

$$TCU(y) = \text{setup cost/unit time} + \text{holding cost/unit time}$$

$$= \frac{K}{y/\beta} + h\left(\frac{y}{2}\right)$$

As seen from Figure 13–4, the length of each inventory cycle is $t_0 = y/\beta$ and the *average* inventory in stock is $y/2$.

The optimum value of y is obtained by minimizing $TCU(y)$ with respect to y. Thus, assuming that y is a continuous variable, we have

$$\frac{dTCU(y)}{dy} = -\frac{K\beta}{y^2} + \frac{h}{2} = 0$$

which yields the optimum order quantity as

$$y^* = \sqrt{\frac{2K\beta}{h}}$$

[It can be proved that y^* minimizes $TCU(y)$ by showing that the second derivative at y^* is strictly positive.] The order quantity is usually referred to as **Wilson's economic lot size.**

The optimum policy of the model calls for ordering y^* units every $t_0^* = y^*/\beta$ time units. The optimum cost $TCU(y^*)$ obtained by direct substitution is $\sqrt{2K\beta h}$.

Most practical situations usually have a (positive) **lead time** (or time lag) L from the point at which the order is placed until it is actually delivered. The ordering policy of the foregoing model thus must specify the **reorder point.** Figure 13–6 illustrates the situation where reordering occurs L time units before delivery is expected. This information may be translated conveniently for practical implementation by simply specifying the *reorder point,* which is the *level of inventory* at reordering. In practice, the situation is equivalent to observing continuously the level of inventory until the reorder point is reached. Perhaps

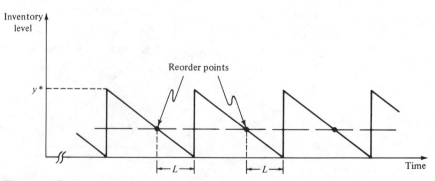

Figure 13–6

this is why the economic lot size model is sometimes classified as a **continuous review model.** Notice that as the system "stabilizes," the lead time L, for the purpose of analysis, may always be taken less than the cycle length t_0^*. The next example illustrates this point.

Example 13.3–1. The daily demand for a commodity is approximately 100 units. Every time an order is placed, a fixed cost of $100 is incurred. The daily holding cost per unit inventory is $.02. If the lead time is 12 days, determine the economic lot size and the reorder point.

From the earlier formulas, the economic lot size is

$$y^* = \sqrt{\frac{2K\beta}{h}} = \sqrt{\frac{2 \times 100 \times 100}{.02}} = 1000 \text{ units}$$

The associated optimum cycle length is thus given as

$$t_0^* = \frac{y^*}{\beta} = \frac{1000}{100} = 10 \text{ days}$$

Since the lead time is 12 days and the cycle length is 10 days, reordering occurs when the level of inventory is sufficient to satisfy the demand for two ($= 12 - 10$) days. Thus the quantity $y^* = 1000$ is ordered when the level of inventory reaches $2 \times 100 = 200$ units.

Notice that the "effective" lead time is taken equal to 2 days rather than 12 days. This result occurs because the lead time is longer than t_0^*. However, after the system stabilizes (it takes two cycles in this example), the situation may be treated as if the lead time is $L - t_0^*$, provided that $L > t_0^*$. Situations such as this exhibit more than one outstanding order at a time. ◀

Exercise 13.3–1

For Example 13.3–1, determine the reorder point in each of the following cases.
(a) Lead time = 15 days. [*Ans.* 500 units.]
(b) Lead time = 23 days. [*Ans.* 300 units.]
(c) Lead time = 8 days. [*Ans.* 800 units.]
(d) Lead time = 10 days. [*Ans.* Zero units.]

The assumptions of the model may not be satisfied for some real situations because demand may be probabilistic. A "crude" procedure has evolved among practitioners that, while retaining the simplicity of applying the economic lot size model, does not completely ignore the effect of probabilistic demand. The idea is quite simple and calls for superimposing a (constant) buffer stock on the inventory level throughout the entire planning horizon. The size of the buffer is determined such that the probability of running out of stock *during lead time L* does not exceed a prespecified value. Suppose that $f(x)$ is the density function of demand *during lead time.* Suppose further that the probability of running out of stock during L must not exceed α. Then the buffer size B is determined from

$$P\{x \geq B + L\beta\} \leq \alpha$$

where $L\beta$ represents the consumption during L. The inventory variation with the buffer is illustrated in Figure 13–7.

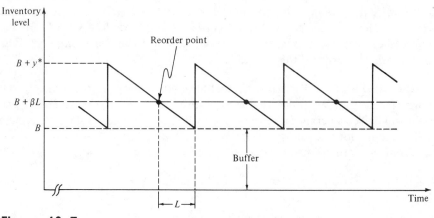

Figure 13–7

Example 13.3–2. Suppose that the demand in Example 13.3–1 is actually an approximation of a probabilistic situation in which the *daily* demand is normal with mean $\mu = 100$ and standard deviation $\sigma = 10$. Determine the size of the buffer stock such that the probability of running out of stock during lead time is at most .05.

From Example 13.3–1, lead time equals 2 days. Because the daily demand is normal, the lead time demand x_L is also normal, with mean $\mu_L = 2 \times 100 = 200$ units and standard deviation $\sigma_L = \sqrt{2 \times 10^2} = 14.14$. Figure 13–8 illustrates the relationship between the distribution of x_L and the buffer size B.

It then follows that

$$P\{x_L \geq \mu_L + B\} \leq \alpha$$

or

$$P\left\{\frac{x_L - \mu_L}{\sigma_L} \geq \frac{B}{\sigma_L}\right\} \leq \alpha$$

or

$$P\left\{\frac{x_L - \mu_L}{\sigma_L} \geq \frac{B}{14.14}\right\} \leq .05$$

From standard tables, this gives $B/14.14 \geq 1.64$ or $B \geq 23.2$. ◀

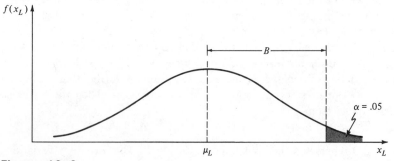

Figure 13–8

Exercise 13.3–2

For Example 13.3–2, determine the buffer stock B in each of the following cases, assuming that the *daily* demand is normal with mean 100 units and variance 30.

(a) Lead time = 15 days. [*Ans.* $\mu_L = 500$, $\sigma_L = 12.25$, $B \geq 20.1$.]

(b) Lead time = 23 days. [*Ans.* $\mu_L = 300$, $\sigma_L = 9.49$, $B \geq 15.6$.]

(c) Lead time = 8 days. [*Ans.* $\mu_L = 800$, $\sigma_L = 15.49$, $B \geq 25.4$.]

(d) Lead time = 10 days. [*Ans.* $\mu_L = \sigma_L = 0$, $B = 0$.]

It is interesting in Example 13.3–2 that the size of B is independent of the mean μ_L. This result is generally expected, since the important factor is the standard deviation. Indeed, if the standard deviation is zero (deterministic case), the buffer size should be zero.

There is no reason to believe that the combined result of superimposing the technique for determining B on the technique for determining the economic lot size is necessarily optimal or near optimal. The fact that some pertinent information is initially ignored, only to be used completely independently at a later stage of the calculations, is sufficient to refute optimality. In fact, the cost of holding the buffer B may merely be considered the "price" for not employing all the information simultaneously in the analysis.

To appreciate the effect of including the probabilistic demand directly in a *continuous review* model, Section 13.4.1 presents a typical case. As should be expected, the new model is necessarily much more complex than the present one. You are encouraged to compare the two models.

Variations of the economic lot size model allow for shortage and uniform (rather than instantaneous) stock replenishment over time. The latter is typical of some production systems where the stock replenishment rate is a function of the production rate. The inventory models in these situations again balance the holding and setup costs. If shortage occurs, a penalty cost is included also in the total cost function. In general, the shortage cost is assumed proportional to the average shortage quantity. Because the analysis of these situations is very similar to the one just given, their details are not presented here. Problems 13–7, 13–10, and 13–13 present the basic results of these models.

13.3.2 SINGLE-ITEM STATIC MODEL WITH PRICE BREAKS

In the models in Section 13.3.1, the purchasing cost per unit time is neglected in the analysis because it is constant and hence should not affect the level of inventory. It often happens, however, that the purchasing price per unit may depend on the size of the quantity purchased. This situation usually occurs in the form of discrete **price breaks** or **quantity discounts.** In such cases, the purchasing price should be considered in the inventory model.

Consider the inventory model with instantaneous stock replenishment and no shortage. Assume that the cost per unit is c_1 for $y < q$ and c_2 for $y \geq q$, where $c_1 > c_2$ and q is the quantity above which a price break is granted. The total cost per cycle will now include the purchasing cost in addition to the setup and holding costs.

The total cost *per unit time* for $y < q$ is

$$TCU_1(y) = \beta c_1 + \frac{K\beta}{y} + \frac{h}{2}y$$

For $y \geq q$ this cost is

$$TCU_2(y) = \beta c_2 + \frac{K\beta}{y} + \frac{h}{2}y$$

These two functions are shown graphically in Figure 13–9(a). Disregarding the effect of price breaks for the moment, we let y_m be the quantity at which the minimum values of TCU_1 and TCU_2 occur. Then

$$y_m = \sqrt{\frac{2K\beta}{h}}$$

The cost functions TCU_1 and TCU_2 in Figure 13–9(a) reveal that the determination of the optimum order quantity y^* depends on where q, the price break point, falls with respect to the three zones I, II, and III shown in the figure. These zones are defined by determining q_1 ($> y_m$) from the equation

$$TCU_1(y_m) = TCU_2(q_1)$$

Since y_m is known ($= \sqrt{2K\beta/h}$), the solution of the equation will yield the value of q_1. In this case the zones are defined as follows.

$$\text{zone I:} \quad 0 \leq q < y_m$$
$$\text{zone II:} \quad y_m \leq q < q_1$$
$$\text{zone III:} \quad q \geq q_1$$

Figure 13–9(b) provides a graphical solution for each case, depending on whether q falls in zone I, II, or III. We thus summarize the optimum order quantity y^* as follows:

$$y^* = \begin{cases} y_m, & \text{if } 0 \leq q < y_m & \text{(zone I)} \\ q, & \text{if } y_m \leq q < q_1 & \text{(zone II)} \\ y_m, & \text{if } q \geq q_1 & \text{(zone III)} \end{cases}$$

The procedure for determining y^* may thus be summarized as follows:

1. Determine $y_m = \sqrt{2K\beta/h}$. If $q < y_m$ (zone I), then $y^* = y_m$, and the procedure ends. Otherwise,

2. Determine q_1 from the equation $TCU_1(y_m) = TCU(q_1)$ and decide whether q falls in zone II or zone III.

    a. If $y_m \leq q < q_1$ (zone II), then $y^* = q$.

    b. If $q \geq q_1$ (zone III), then $y^* = y_m$.

Example 13.3–3. Consider the inventory model with the following information. $K = \$10$, $h = \$1$, $\beta = 5$ units, $c_1 = \$2$, $c_2 = \$1$, and $q = 15$ units. First compute y_m; thus

$$y_m = \sqrt{\frac{2K\beta}{h}} = \sqrt{\frac{2 \times 10 \times 5}{1}} = 10 \text{ units}$$

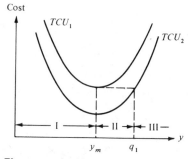

Figure 13-9(a)

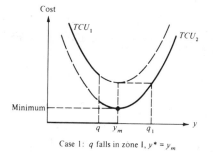

Case 1: q falls in zone I, $y^* = y_m$

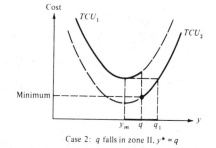

Case 2: q falls in zone II, $y^* = q$

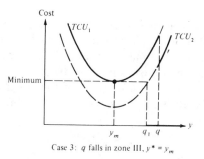

Case 3: q falls in zone III, $y^* = y_m$

Figure 13-9(b)

Since $q > y_m$, it is necessary to check whether q is in zone II or III. The value of q_1 is computed from

$$TCU_1(y_m) = TCU_2(q_1)$$

or

$$c_1\beta + \frac{K\beta}{y_m} + \frac{hy_m}{2} = c_2\beta + \frac{K\beta}{q_1} + \frac{hq_1}{2}$$

Substitution yields

$$2 \times 5 + \frac{10 \times 5}{10} + \frac{1 \times 10}{2} = 1 \times 5 + \frac{10 \times 5}{q_1} + \frac{1 \times q_1}{2}$$

or

$$q_1^2 - 30q_1 + 100 = 0$$

This yields $q_1 = 26.18$ or $q_1 = 3.82$. By definition, q_1 is selected as the larger value. Since $y_m < q < q_1$, q is in zone II. It follows that $y^* = q = 15$ units. The associated total cost per unit time is thus computed as

$$TCU(y^*) = TCU_2(15) = c_2\beta + \frac{K\beta}{15} + \frac{h \times 15}{2}$$

$$= 1 \times 5 + \frac{10 \times 5}{15} + \frac{1 \times 15}{2} = \$15.83/\text{day} \qquad \blacktriangleleft$$

Exercise 13.3–3

In Example 13.3–3, determine y^* and the total cost *per cycle* in each of the following cases.

(a) $q = 30$. [*Ans.* $y^* = 10$, cost/cycle $= \$40$.]
(b) $q = 5$. [*Ans.* $y^* = 10$, cost/cycle $= \$30$.]

13.3.3 MULTIPLE-ITEM STATIC MODEL WITH STORAGE LIMITATION

This model considers the inventory system including n (> 1) items that are competing for a limited storage space. This limitation represents an interaction between the different items and may be included in the model as a constraint.

Let A be the maximum storage area available for the n items and assume that the storage area requirements per unit of the ith item is a_i. If y_i is the amount ordered of the ith item, the storage requirements constraint becomes

$$\sum_{i=1}^{n} a_i y_i \leq A$$

Assume that each item is replenished instantaneously and that there is no quantity discount. Assume further that no shortages are allowed. Let β_i, K_i, and h_i be, respectively, the demand rate per unit time, the setup cost, and the holding cost per unit per unit time corresponding to the ith item. The inventory costs associated with each item should be essentially the same as in the case of an equivalent single-item model. The problem thus becomes

$$\text{minimize } TCU(y_1, \ldots, y_n) = \sum_{i=1}^{n} \left(\frac{K_i \beta_i}{y_i} + \frac{h_i y_i}{2} \right)$$

subject to

$$\sum_{i=1}^{n} a_i y_i \leq A$$

$$y_i > 0, \qquad \text{for all } i$$

The general solution of this problem is obtained by the Lagrange multipliers method.† However, before this is done, it is necessary to check whether the constraint is active by checking whether the unconstrained value

$$y_i^* = \sqrt{\frac{2K_i\beta_i}{h_i}}$$

satisfies the storage constraint. If it does, the constraint is said to be inactive or redundant and may be neglected.

If the constraint is not satisfied by the values of y_i^*, it must be active. In this case, new optimal values of y_i must be found which satisfy the storage constraint in *equality* sense.‡ This result is accomplished by first formulating the Lagrangian function as

$$L(\lambda, y_1, y_2, \ldots, y_n) = TCU(y_1, \ldots, y_n) - \lambda\left(\sum_{i=1}^{n} a_i y_i - A\right)$$

$$= \sum_{i=1}^{n}\left(\frac{K_i\beta_i}{y_i} + \frac{h_i y_i}{2}\right) - \lambda\left(\sum_{i=1}^{n} a_i y_i - A\right)$$

where λ (<0) is the Lagrange multiplier.

The optimum values of y_i and λ can be found by equating the respective first partial derivatives to zero. This gives

$$\frac{\partial L}{\partial y_i} = -\frac{K_i\beta_i}{y_i^2} + \frac{h_i}{2} - \lambda a_i = 0$$

$$\frac{\partial L}{\partial \lambda} = -\sum_{i=1}^{n} a_i y_i + A = 0$$

The second equation implies that y_i^* must satisfy the storage constraint in equality sense.

From the first equation,

$$y_i^* = \sqrt{\frac{2K_i\beta_i}{h_i - 2\lambda^* a_i}}$$

Notice that y_i^* is dependent on λ^*, the optimal value of λ. Also, for $\lambda^* = 0$, y_i^* gives the solution of the unconstrained case.

The value λ^* can be found by systematic trial and error. Since by definition $\lambda < 0$ for the minimization case, by trying successive negative values of λ the value of λ^* should result in simultaneous values of y_i^* that satisfy the given constraint in equality sense. Thus the determination of λ^* automatically yields y_i^*.

---

† See Chapter 18 for a complete analysis of the Lagrangian method.
‡ This procedure happens to yield the correct answer because $TCU(y_1, \ldots, y_n)$ is convex and the problem has a single constraint that is linear (convex solution space). The procedure may not be correct under other conditions or when there is more than one constraint. See Section 18.2.2–A.

Example 13.3–4. Consider the inventory problem with three items ($n = 3$). The parameters of the problem are shown in the table.

Item i	K_i	β_i	h_i	a_i
1	$10	2 units	$.3	1 ft²
2	5	4 units	.1	1 ft²
3	15	4 units	.2	1 ft²

Assume that the total available storage area is given by $A = 25$ ft².
Given the formula

$$y_i^* = \sqrt{\frac{2K_i\beta_i}{h_i - 2\lambda^* a_i}}$$

the following table is constructed:

λ	y_1	y_2	y_3	$\sum_{i=1}^{3} a_i y_i - A$
0	11.5	20.0	24.5	+31
−.05	10.0	14.1	17.3	+16.4
−.10	9.0	11.5	14.9	+10.4
−.15	8.2	10.0	13.4	+ 6.6
−.20	7.6	8.9	12.2	+ 3.7
−.25	7.1	8.2	11.3	+ 1.6
−.30	6.7	7.6	10.6	− 0.1

For $A = 25$ ft², the storage constraint is satisfied in equality sense for some value of λ between −.25 and −.3. This value is equal to λ^* and may be estimated by linear interpolation. The corresponding values of y_i should thus yield y_i^* directly. Since from the table λ^* appears very close to −.3, optimal y_i^* are approximately given by

$$y_1^* = 6.7, \quad y_2^* = 7.6, \quad \text{and} \quad y_3^* = 10.6$$

If $A \geq 52.4$, the unconstrained values of y_i corresponding to $\lambda = 0$ yield y_i^*. In this case the constraint is inactive. ◀

Exercise 13.3–4

Consider Example 13.3–4. Using the second table, determine the range of λ in which λ^* falls assuming that the area A is given as follows.
(a) $A = 45$ ft². [Ans. $0 > \lambda^* > -.05$.]
(b) $A = 30$ ft². [Ans. $-.15 > \lambda^* > -.2$.]
(c) $A = 20$ ft². [Ans. $\lambda^* < -.3$.]

13.3.4   SINGLE-ITEM *N*-PERIOD DYNAMIC MODEL

In this model it is assumed that demand, although known with certainty, may vary from one period to the next. Also, the inventory level is reviewed *periodically* rather than continuously. Although delivery lag (expressed as a fixed number of periods) may be allowed, the model assumes that the stock is replenished instantaneously at the beginning of the period. Finally, no shortages are allowed.

The development of dynamic deterministic models is limited to the study of finite time horizons. The reason is that a numerical solution of these models requires the use of the dynamic programming technique (see Chapter 9), which *in this case* is feasible only for a finite number of periods (stages). This is not a serious limitation, however, since distant future demands usually have little effect on the decisions of the present finite time horizon. In addition, in most situations it is not practical to assume that the item will be held in stock indefinitely.

Define for period i, $i = 1, 2, \ldots, N$,

z_i = amount ordered
D_i = amount demanded
x_i = entering inventory (at the beginning of period i)
h_i = holding cost per unit of inventory carried forward from period i to period $i + 1$
K_i = setup cost
$c_i(z_i)$ = marginal purchasing (production) cost function given z_i

Let

$$C_i(z_i) = \begin{cases} 0, & z_i = 0 \\ K_i + c_i(z_i), & z_i > 0 \end{cases}$$

The function $c_i(z_i)$ is of interest only if the unit purchasing cost varies from one period to the next or if there are price breaks.

Since no shortages are allowed, the objective is to determine the optimal values of z_i that minimize the sum of the setup, purchasing, and holding costs for all N periods. The holding cost is assumed proportional to

$$x_{i+1} = x_i + z_i - D_i$$

which is the amount of inventory carried forward from i to $i + 1$. As a result, the holding cost for period i is $h_i x_{i+1}$. The assumption is introduced only for simplicity, since the model may be readily extended to cover any holding cost function $H_i(x_{i+1})$ by replacing $h_i x_{i+1}$ by $H_i(x_{i+1})$. By the same token, holding cost may be based on x_i or $(x_i + x_{i+1})/2$.

The development of the dynamic programming model is simplified by depicting the problem schematically as shown in Figure 13–10. Each period represents a stage. Using the backward recursive equation, we define the states of the system at stage i as the amount of entering inventory x_i. Let $f_i(x_i)$ be the minimum inventory cost for periods $i, i + 1, \ldots$, and N. The complete recursive equation is given by

$$f_N(x_N) = \min_{\substack{z_N + x_N = D_N \\ z_N \geq 0}} \{C_N(z_N)\}$$

$$f_i(x_i) = \min_{\substack{D_i \leq x_i + z_i \leq D_i + \cdots + D_N \\ z_i \geq 0}} \{C_i(z_i) + h_i(x_i + z_i - D_i) + f_{i+1}(x_i + z_i - D_i)\},$$

$$i = 1, 2, \ldots, N-1$$

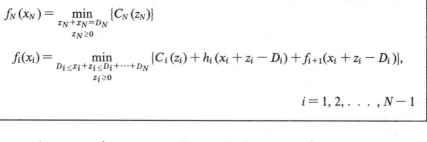

Figure 13–10

The forward recursive equation can be developed by defining the states at stage i as the amount of inventory at the end of period i. From Figure 13–10, these states are given by x_{i+1}. At any stage, the values of x_{i+1} are limited by

$$0 \leq x_{i+1} \leq D_{i+1} + \cdots + D_N$$

Thus, in the extreme case, the amount z_i in period i may be ordered large enough so that the remaining inventory x_{i+1} will satisfy the demand for all the remaining periods.

Let $f_i(x_{i+1})$ be the minimum inventory cost for periods 1, 2, . . . , and i given x_{i+1}, the amount of inventory at the end of period i. The complete recursive equation is then given by

$$f_1(x_2) = \min_{0 \leq z_1 \leq D_1 + x_2} \{C_1(z_1) + h_1 x_2\}$$

$$f_i(x_{i+1}) = \min_{0 \leq z_i \leq D_i + x_{i+1}} \{C_i(z_i) + h_i x_{i+1} + f_{i-1}(x_{i+1} + D_i - z_i)\},$$

$$i = 2, 3, \ldots, N$$

The forward and backward formulations of the model are computationally equivalent. However, the forward algorithm, as indicated later, will prove useful in developing an important special case of the preceding model. The following numerical example is thus used to illustrate the computational procedure of the forward algorithm. The procedure for the backward algorithm is left as an exercise (see Problem 13–22).

Exercise 13.3–5

Suppose that the holding cost in period i is based on the *average* inventory during the period. Write the expressions for the holding cost as they should appear in the forward and backward recursive equations.

[*Ans.* Backward: $h_i\left(x_i + z_i - \dfrac{D_i}{2}\right)$, forward: $h_i\left(x_{i+1} + \dfrac{D_i}{2}\right).$]

Example 13.3-5. Consider a three-period inventory situation with discrete units and dynamic deterministic demand. The data for the problem are as follows:

Period i	Demand D_i	Setup Cost K_i	Holding Cost h_i
1	3 units	$3.00	$1.00
2	2 units	7.00	3.00
3	4 units	6.00	2.00

The entering inventory x_1 to period 1 is 1 unit. Suppose that the marginal purchasing cost is $10 per unit for the first 3 units and $20 for each additional unit. Then

$$c_i(z_i) = \begin{cases} 10z_i, & 0 \le z_i \le 3 \\ 30 + 20(z_i - 3), & z_i \ge 4 \end{cases}$$

The stage calculations for the forward algorithm are as follows.
Stage 1: $D_1 = 3,\ 0 \le x_2 \le 2 + 4 = 6$

									Optimal Solution	
		\multicolumn{8}{c}{$f_1(z_1 \mid x_2) = C_1(z_1) + h_1 x_2$}		
		$z_1 = 2$	3	4	5	6	7	8		
x_2	$h_1 x_2$	$C_1(z_1) = 23$	33	53	73	93	113	133	$f_1(x_2)$	z_1^*
0	0	23							23	2
1	1		34						34	3
2	2			55					55	4
3	3				76				76	5
4	4					97			97	6
5	5						118		118	7
6	6							139	139	8

Since $x_1 = 1$, the smallest value of z_1 is $D_1 - x_1 = 3 - 1 = 2$.
Stage 2: $D_2 = 2,\ 0 \le x_3 \le 4$

									Optimal Solution	
		\multicolumn{7}{c}{$f_2(z_2 \mid x_3) = C_2(z_2) + h_2 x_3 + f_1(x_3 + D_2 - z_2)$}		
		$z_2 = 0$	1	2	3	4	5	6		
x_3	$h_2 x_3$	$C_2(z_2) = 0$	17	27	37	57	77	97	$f_2(x_3)$	z_2^*
0	0	0 + 55 = 55	17 + 34 = 51	27 + 23 = 50					50	2
1	3	3 + 76 = 79	20 + 55 = 75	30 + 34 = 64	40 + 23 = 63				63	3
2	6	6 + 97 = 103	23 + 76 = 99	33 + 55 = 88	43 + 34 = 77	63 + 23 = 86			77	3
3	9	9 + 118 = 127	26 + 97 = 123	36 + 76 = 112	46 + 55 = 101	66 + 34 = 100	86 + 23 = 109		100	4
4	12	12 + 139 = 151	29 + 118 = 147	39 + 97 = 136	49 + 76 = 125	69 + 55 = 124	89 + 34 = 123	109 + 23 = 132	123	5

Stage 3: $D_3 = 4,\ x_4 = 0$

		$f_3(z_3 \mid x_4) = C_3(z_3) + h_3 x_4 + f_2(x_4 + D_3 - z_3)$					Optimal Solution	
		$z_3 = 0$	1	2	3	4		
x_4	$h_3 x_4$	$C_3(z_3) = 0$	16	26	36	56	$f_3(x_4)$	z_3^*
0	0	$0 + 123$ $= 123$	$16 + 100$ $= 116$	$26 + 77$ $= 103$	$36 + 63$ $= 99$	$56 + 50$ $= 106$	99	3

The solution is given as $z_1^* = 2$, $z_2^* = 3$, and $z_3^* = 3$, which costs a total of $99. ◀

Exercise 13.3-6

Consider Example 13.3-5.
(a) Does it make sense to have $x_4 > 0$?
   [*Ans.* No, because it is nonoptimal to terminate the planning horizon with positive inventory.]
(b) In each of the following (independent) cases, determine the feasible ranges for z_1, z_2, z_3, x_2, and x_3. (You will find it helpful to depict each case as in Figure 13–10.)
   (1) $x_1 = 4$ and all other data remain unchanged
      [*Ans.* $0 \le z_1 \le 5$, $0 \le z_2 \le 5$, $0 \le z_3 \le 4$, $1 \le x_2 \le 6$, $0 \le x_3 \le 4$.]
   (2) $x_1 = 0$, $D_1 = 5$, $D_2 = 4$, and $D_3 = 5$
      [*Ans.* $5 \le z_1 \le 14$, $0 \le z_2 \le 9$, $0 \le z_3 \le 5$, $0 \le x_2 \le 9$, $0 \le x_3 \le 5$.]

Special Case with Constant or Decreasing Marginal Costs

The dynamic programming model can be used with *any* cost functions. An important special case of this model occurs when for period i both the purchasing (production) cost *per unit* and the holding cost *per unit* are constant or decreasing functions of z_i and x_{i+1}, respectively. In this case the cost function is said to yield constant or decreasing *marginal* cost. Typical illustrations of such cost functions are shown in Figure 13–11. Mathematically, these functions are concave. Case (a) shows the situation of constant marginal cost. Case (b) is typical of many production (or purchasing) cost functions where, regardless of the amount produced, a setup cost K is charged. A constant marginal cost is then incurred; but if quantity discount or price break is allowed at $z_i = q$, the marginal cost for $z_i > q$ becomes smaller. Finally, case (c) illustrates a general concave function.

Under the conditions just stipulated, it can be proved that:†

1. Given the initial inventory $x_1 = 0$, then at any period i of the N-period

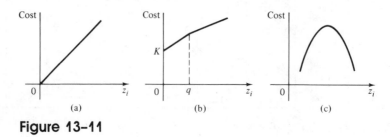

Figure 13–11

model, it is optimal to have a positive quantity z_i^* *or* a positive entering inventory x_i^* but not both; that is, $z_i^* x_i^* = 0$.‡

2. The amount z_i ordered at any period i is optimal only if it is zero *or* satisfies the *exact* demand of one or more succeeding periods. These succeeding periods are such that if the demand in period $i + m$ $(< N)$ is satisfied by z_i^*, then the demands for periods, $i, i + 1, \ldots,$ and $i + m - 1$ must also be satisfied by z_i^*.

The first property (theorem) implies that for any period i, it is not economical to bring in inventory and place an order at the same time. For, suppose that the *least* marginal cost of acquiring and holding *one* additional unit from a previous period i' to the present period i'' $(i' < i'')$ is b', while the marginal cost of ordering one more unit at i'' is b''. If $b'' \leq b'$, the amount ordered at i'' can be increased to cover the exact demand at i'' without an increase in the associated total costs as compared with the case where this demand is satisfied from period i'. This result follows because of the nonincreasing marginal costs. Hence having $x_i'' z_i'' = 0$ will yield a solution that is *at least* as good as any other solution. On the other hand, if $b'' > b'$, it is more economical to increase the order in i' to cover the demand in i' and i'' so that the amount ordered in i'' is equal to zero. This conclusion follows again because of the nonincreasing marginal costs. The implication here then is that the condition $x_i z_i = 0$ will not yield any worse solution provided that the marginal costs are constant or decreasing and the initial inventory is zero. The second property, which calls for ordering the exact amount for one or more periods, follows immediately from the first property.

The properties described, when applicable, will result in a simplified computational procedure, which is still based on the general dynamic programming algorithms presented. This point is explained by using the forward algorithm.

Since by the second property the amount of inventory at the end of period i, that is, x_{i+1}, must satisfy the exact requirements of one or more successive periods, it follows that the number of state values of the system at any period is determined by the number of succeeding *periods* (rather than by the number of *units* demanded in the succeeding periods as in the general model). For example, let $N = 5$ with demands 10, 15, 20, 50, and 70, respectively. Then at the *end* of the third period (stage), the number of state values (x_4) in the general model will be $50 + 70 + 1 = 121$, whereas in the new model it will reduce to three (the remaining number of periods plus one), since x_4 could be 0, 50, or 120 only. A similar argument based on the first property also shows

---

† For details of the proof, see H. Wagner and T. Whitin, "Dynamic Version of the Economic Lot Size Model," *Management Science,* Vol. 5, 1958, pp. 89–96. The original proofs by Wagner and Whitin, however, are developed under the restrictive assumption that the per unit purchasing costs are *constant* and *identical* for all periods. This was later improved by A. F. Veinott, Jr., of Stanford University, to include concave cost functions for each period.

‡ In this special case, the initial inventory x_1 can always be taken equal to zero. If $x_1 > 0$, this amount can be written off from the demands of the successive periods until it is exhausted. Under such conditions, the periods for which the demands have been satisfied are still included in the problem; this time with zero demands. In such a case it is possible to have both z_i and x_i equal to zero.

that the number of alternatives z_i are much smaller in the new model. The result is that the computational effort is reduced tremendously in the new model.

Example 13.3–6. Consider a four-period model with the following data:

Period i	D_i	K_i
1	76	$ 98
2	26	114
3	90	185
4	67	70

The holding cost per unit per period is constant and equal to $1.00. Also the purchasing cost per unit is equal to $2.00 for all the periods. The initial inventory x_1 is 15 units. (The per unit holding and purchasing costs are taken the same over all the periods only for simplicity.)

The solution is obtained by using the same forward algorithm except that the values of states x_{i+1} and the values of alternatives z_i are determined according to the new properties. Since $x_1 = 15$, the demand for the first period is decreased by an equivalent amount and thus equals $76 - 15 = 61$.

Stage 1: $D_1 = 61$

		$f_1(z_1 \mid x_2) = C_1(z_1) + h_1 x_2$				Optimal Solution	
		$z_1 = 61$	87	177	244		
x_2	$h_1 x_2$	$C_1(z_1) = 220$	272	452	586	$f_1(x_2)$	z_1^*
0	0	220	—	—	—	220	61
26	26	—	298	—	—	298	87
116	116	—	—	568	—	568	177
183	183	—	—	—	769	769	244
Order in 1 for:		1	1, 2	1, 2, 3	1, 2, 3, 4		

Stage 2: $D_2 = 26$

		$f_2(z_2 \mid x_3) = C_2(z_2) + h_2 x_3 + f_1(x_3 + D_2 - z_2)$				Optimal Solution	
		$z_2 = 0$	26	116	183		
x_3	$h_2 x_3$	$C_2(z_2) = 0$	166	346	480	$f_2(x_3)$	z_2^*
0	0	$0 + 298$ $= 298$	$166 + 220$ $= 386$	—	—	298	0
90	90	$90 + 568$ $= 586$	—	$436 + 220$ $= 656$	—	656	116
157	157	$157 + 769$ $= 926$	—	—	$637 + 220$ $= 857$	857	183
Order in 2 for:		—	2	2, 3	2, 3, 4		

Stage 3: $D_3 = 90$

		$f_3(z_3 \mid x_4) = C_3(z_3) + h_3 x_4 + f_2(x_4 + D_3 - z_3)$			Optimal Solution	
		$z_3 = 0$	90	157		
x_4	$h_3 x_4$	$C_3(z_3) = 0$	365	499	$f_3(x_4)$	z_3^*
0	0	$0 + 656 = 656$	$365 + 298 = 663$	—	656	0
67	67	$67 + 857 = 924$	—	$566 + 298 = 864$	864	157
Order in 3 for:		—	3	3, 4		

Stage 4: $D_4 = 67$

		$f_4(z_4 \mid x_5) = C_4(z_4) + h_4 x_5 + f_3(x_5 + D_4 - z_4)$		Optimal Solution	
		$z_4 = 0$	67		
x_5	$h_4 x_5$	$C_4(z_4) = 0$	204	$f_4(x_5)$	z_4^*
0	0	$0 + 864 = 864$	$204 + 656 = 860$	860	67
Order in 4 for:		—	4		

The optimal policy is thus given by $z_1^* = 61$, $z_2^* = 116$, $z_3^* = 0$, and $z_4^* = 67$, at a total cost of \$860. ◄

Exercise 13.3–7

In Example 13.3–6, determine the feasible ranges for z_1 in each of the following cases.

(a) $x_1 = 10$ and all the remaining data are unchanged.
   [*Ans.* $z_1 = 66$, 92, 182, or 249.]
(b) $x_1 = 80$ and all the remaining data are unchanged.
   [*Ans.* $z_1 = 0$, 22, 112, or 179.]
(c) $D_1 = 70$, $D_2 = 25$, $D_3 = 80$, $D_4 = 70$, and $x_1 = 10$.
   [*Ans.* $z_1 = 60$, 85, 165, or 235.]

A special case of the *concave* cost model described above occurs when the production cost for a period is defined by the linear function

$$C_i(z_i) = K_i + c_i z_i, \qquad i = 1, 2, \ldots, N$$

provided that $c_{i+1} \leq c_i$ for all i, that is, $c_1 \geq c_2 \geq \cdots \geq c_N$. Under this new condition, the forward algorithm for the concave cost model can be modified such that further savings in computations are possible. To avoid confusion, the names "original" and "modified" algorithms will be used to refer to the forward algorithms associated, respectively, with the concave cost model and the model to be presented shortly.

In the original algorithm each stage i computes the optimal policy by considering ordering in period i for future periods up to and including period j; that is, $i \leq j \leq N$. The modified algorithm defines each stage i such that for period

i, the optimal policy is determined by considering ordering in each of the preceding periods, k, for periods up to and including period i; $1 \le k \le i$. This is expressed mathematically as†

$$f_i = \min \begin{cases} C_1 + h_1(D_2 + \cdots + D_i) + \cdots + h_{i-1}D_i & \text{(order in 1)} \\ C_2 + h_2(D_3 + \cdots + D_i) + \cdots + h_{i-1}D_i + f_i & \text{(order in 2)} \\ \vdots & \\ C_{i-1} + h_{i-1}D_i + f_{i-2} & \text{(order in } i-1) \\ C_i + f_{i-1} & \text{(order in } i) \end{cases}$$

where

f_i = minimum total cost for periods 1 through i, inclusive, $i = 1, 2, \ldots, N$
C_k = total ordering cost (setup + purchasing) for ordering in period k the amount $z_k = D_k + \cdots + D_i$ for periods k through i, $k \le i$

To start with, and without taking advantage of the special feature of the cost function $C_i(z_i)$, the amount of computations in the modified model is less than that in the original one.‡ This result follows because the modified model does not consider explicitly the case where no orders are placed at the different stages. The computations using the modified algorithm may be further reduced by making use of the following theorem.

Planning Horizon Theorem. *In the modified forward algorithm, if for period i^* the minimum cost occurs such that the demand at i^* is satisfied by ordering in a previous period $i^{**} < i^*$, then for all future periods $i > i^*$ it is sufficient to compute the optimal program based on ordering in periods $i^{**}, i^{**} + 1, \ldots, i$ only. In particular, if the optimal policy calls for ordering in i^* for the same period i^* (i.e., $i^* = i^{**}$), then for any future period $i > i^*$ it will always be optimal to order in i^* regardless of future demands. In this case, i^* is said to mark the beginning of a planning horizon.*

This theorem implies two important concepts:

1. During the course of computations, the calculations may be truncated so that the entries for periods $k < i^{**}$ need not be considered. This should lead to computational savings.

2. For the special case where $i^* = i^{**}$, in addition to truncating the computations at i^*, future periods starting with i^* may be considered completely independently of all previous periods. Moreover, it is always optimal to order in i^* regardless of future demands.

When i^{**} is strictly less than i^*, it is not always true that ordering will occur in i^{**}. Indeed, future demands may call for a change in the optimal policy. In this case it will not be possible to break down the problem into

---

† In the modified model, the state of the system x_i is suppressed, since this corresponds directly to the number of the preceding periods, that is, i. The same reasoning could have been used with the original model.

‡ The maximum number of entries are $\{N(N + 1) + (N - 1)N\}/2 = N^2$ in the original table and $N(N + 1)/2$ in the modified one.

independent planning horizons. To avoid confusion, i^{**} will be referred to as the starting period of a *subhorizon* whenever $i^{**} < i^*$.

Example 13.3–7. Consider a six-period inventory model with the following data:

i	D_i	K_i	h_i
1	10	20	1
2	15	17	1
3	7	10	1
4	20	20	3
5	13	5	1
6	25	50	1

The purchasing cost per unit is 2 for all the periods.

The computations for this example are summarized in Table 13–1. These computations are carried out on a row-by-row basis starting with row 1. Each column represents a decision alternative defining the period k in which the demands for periods $k, k + 1, \ldots, i$ are filled, $1 \le k \le i$. Each row represents the limiting period up to which the demand is filled. Thus, for each i, the optimum value f_i, as defined for the modified algorithm, is obtained by considering all feasible decision alternatives k ($\le i$) and then selecting the alternative yielding minimum cumulative cost. For example, given $i = 3$, we have three options: (1) order in 1 for 1, 2, and 3, (2) order in 2 for 2 and 3, and (3) order in 3 for 3. The entries of the table above its main diagonal are infeasible, since no backorders are allowed.

To illustrate the use of planning horizons (and subhorizons), in row 3 f_3 occurs under period 2. This means that it is optimal at this point to order for period 3 (and period 2) in period 2. This is equivalent to saying that $i^{**} = 2$ and $i^* = 3$. According to the theorem, for all $i > 3$ the calculations may go back only to period 2. Period 2 thus marks the beginning of a *sub*horizon. Moving to row 4, we see that f_4 occurs under period 4 signifying it is optimal to order for period 4 in period 4. Thus $i^{**} = i^* = 4$ and $i = 4$ marks the beginning of a planning horizon, which signifies that in the succeeding rows the entries under periods 1, 2, 3 should not be computed. Continuing in this manner, we see in Table 13–1 that another planning horizon commences in period 5, with the result that in the computations in row 6 only the entries under periods 5 and 6 need be computed. Thus periods 1, 4, and 5 mark the beginnings of the three planning horizons of the problem. The advantages of the planning horizon theorem must be clear now, since all the blank entries below the main diagonal of the table represent computational savings.

The optimal solution is obtained by considering the last row in Table 13–1. f_6 indicates that it is optimal to order in 5 the amount $z_5 = 38$ for 5 and 6. Thus from row 4 ($= 5 - 1$), f_4 requires ordering $z_4 = 20$ for period 4 alone. Again, in row 3 ($= 4 - 1$), f_3 calls for ordering in 2 the amount $z_2 = 22$ for 2 and 3. Finally, the amount $z_1 = 10$ is ordered in period 1. The total cost is 274 for the entire problem.   ◀

Table 13-1

Place order in period k for periods up to and including i.

		$k=1$	$k=2$	$k=3$	$k=4$	$k=5$	$k=6$
$i=1$	(1)[a]	20					
	(2)	$10 \times 2 = 20$					
	(3)	0					
	(4)	0					
		—					
		$f_1 \to 40*$					
$i=2$	(1)	20	17				
	(2)	$(10+15) \times 2 = 50$	$15 \times 2 = 30$				
	(3)	15×1	0				
	(4)	0	$f_1 = 40$				
		—	—				
		$f_2 \to 85*$	87				
$i=3$	(1)	20	17	10			
	(2)	$(10+15+7) \times 2 = 64$	$(15+7) \times 2 = 44$	$7 \times 2 = 14$			
	(3)	$22 \times 1 + 7 \times 1 = 29$	$7 \times 1 = 7$	0			
	(4)	0	$f_1 = 40$	$f_2 = 85$			
		—	—	—			
		113	$f_3 \to 108*$	109			

[a] (1) Setup cost; (2) purchasing cost; (3) holding cost; (4) optimum total cost from preceding periods.

$i=4$			
(1)	17	10	20
(2)	$(15+7+20) \times 2 = 84$	$(7+20) \times 2 = 54$	$20 \times 2 = 40$
(3)	$27 \times 1 + 20 \times 1 = 47$	$20 \times 1 = 20$	0
(4)	$f_1 = 40$	$f_2 = 85$	$f_3 = 108$
	—	—	—
	188	169	$f_4 \to 168^*$

$i=5$		
(1)	20	5
(2)	$(20+13) \times 2 = 66$	$13 \times 2 = 26$
(3)	$13 \times 3 = 39$	0
(4)	$f_3 = 108$	$f_4 = 168$
	—	—
	233	$f_5 \to 199^*$

$i=6$		
(1)	5	50
(2)	$(13+25) \times 2 = 76$	$25 \times 2 = 50$
(3)	$25 \times 1 = 25$	0
(4)	$f_4 = 168$	$f_5 = 199$
	—	—
	$f_6 \to 274^*$	299

Exercise 13.3–8

In Example 13.3–7, determine the optimal solution (directly from Table 13–1) assuming that the inventory problem includes the first five periods only.
[*Ans.* Order 13 units in 5 for 5, 20 units in 4 for 4, 22 units in 2 for 2 and 3, and 10 units in 1 for 1.]

13.3.5 *N*-PERIOD PRODUCTION SCHEDULING MODEL

Consider the problem of scheduling production over N successive periods. The demands for the different periods are variable but deterministic. These demands may be met by either fluctuating inventory while keeping production constant or fluctuating production while keeping inventory constant, or a combination of both. Fluctuations in production can be achieved by working overtime, and fluctuation in inventory may be met by holding positive stock on hand or by allowing a backlog of unfilled demand. The objective here is to determine the production schedule for all N periods that minimizes the total relevant costs.

This model assumes zero setup cost in every period. In general, shortages are allowed except that all backlogged demand must be filled by the Nth period. This situation can be represented as a transportation model (see Chapter 5). In particular, by noting the special characteristics of the model for the case where no shortage is allowed, the problem can be solved in an easy way without having to apply the iterative procedure of the transportation technique.

Define the following symbols for period i, $i = 1, \ldots , N$.

c_i = production cost per unit during regular time
d_i = production cost per unit during overtime, $c_i < d_i$
h_i = holding cost per unit forwarded from period i to period $i + 1$
p_i = shortage cost per unit demanded in period i and filled in period $i + 1$
a_{Ri} = production capacity (number of units) during regular time　·
a_{Ti} = production capacity (number of units) during overtime
b_i = demand (number of units)

Notice that c_i, the per unit production cost during regular time, is less than d_i, the per unit production cost during overtime, as shown graphically in Figure

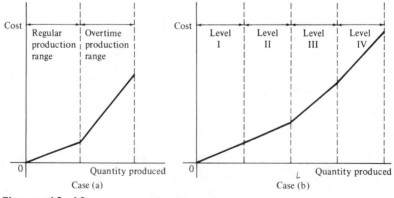

Figure 13–12

13–12(a). The situation may be generalized to the case where there are k levels of production such that the per unit production cost increases with the level of production. A typical illustration is shown in Figure 13–12(b). Under such conditions, the production cost function is said to have increasing marginal costs. Mathematically, the function is said to be convex.

This restriction on the production cost function must be maintained; otherwise, the following model will not be applicable. This point will be justified later after the details of the model have been presented.

No Shortage Model

First, consider the case where no shortage is allowed in the system. According to the terminology of the transportation model (Chapter 5), the *sources* are represented by the regular and overtime productions for the various periods. The *destinations* are given by the demands for the respective periods. The per unit *transportation cost* from any source to any destination is represented by the corresponding per unit production plus holding costs.

The complete cost matrix for the equivalent transportation model (assuming no shortages) is given in Table 13–2.

The surplus column is used to balance the transportation model; that is, $S = \Sigma_i a_i - \Sigma_j b_j$. This is based on the reasonable assumption that the demand is always less than the production capacity of the system. (The cost per unit in the surplus column is equal to zero.) Since no shortage is allowed, a current production period cannot be used to satisfy the demands for its preceding periods. Table 13–2 implements this restriction by shaded squares, which is actually equivalent to assigning a very large per unit cost.

Table 13–2

Demand Period j

	1	2	3	$\cdots$	N	Surplus	
R_1	c_1	c_1+h_1	$c_1+h_1+h_2$		$c_1+h_1+\cdots+h_{N-1}$	0	a_{R1}
T_1	d_1	d_1+h_1	$d_1+h_1+h_2$		$d_1+h_1+\cdots+h_{N-1}$	0	a_{T1}
R_2		c_2	c_2+h_2		$c_2+h_2+\cdots+h_{N-1}$	0	a_{R2}
T_2		d_2	d_2+h_2		$d_2+h_2+\cdots+h_{N-1}$	0	a_{T2}
$\vdots$							$\vdots$
R_N					c_N		a_{RN}
T_N					d_N	0	a_{TN}
	b_1	b_2	b_3	$\cdots$	b_N	S	

Production Period i

Because no backorders are allowed in this model, it is necessary to include the restriction that for every period k, the cumulative amount of demand up to and including that period does not exceed the corresponding cumulative amount of production; that is,

$$\sum_{i=1}^{k} (a_{Ri} + a_{Ti}) \geq \sum_{j=1}^{k} b_j, \qquad \text{for } k = 1, 2, \ldots, N$$

The solution of the problem is greatly simplified by its formulation as a transportation model. Since the demand at period i should be satisfied before those at periods $i + 1$, $i + 2$, . . . , N, and because of the special condition imposed on the production cost function, it will not be necessary to use the regular transportation algorithm in solving the problem. Instead, the demand for period 1 is first satisfied by successively assigning as much amount as possible to the cheapest entries of the first column (period 1). The new values of a_i are then updated to reflect the *remaining* capacities for the different periods. Next, period 2 is considered and its demand is satisfied in the cheapest possible way within the new capacity limitations. The process is continued until the demand for period N is satisfied.†

Because of the increasing marginal costs in the production cost function, the regular production capacity will be exhausted before overtime production can start. If this condition is not satisfied, the transportation model will not be applicable since this might yield meaningless results (such as using overtime production before regular production is exhausted).

Example 13.3–8. Consider a four-period production scheduling problem with the following data:

Period i	Capacity (units)		Demand (units) b_i
	a_{Ri}	a_{Ti}	
1	100	50	120
2	150	80	200
3	100	100	250
4	200	50	200
Totals	550	280	770

The production costs are identical for all the periods; that is, $c_i = 2$ and $d_i = 3$ for all i. The holding cost is also constant for all periods and is given by $h_i = 0.1$ for all i. The cost functions are assumed identical for all periods only for simplicity.

The equivalent transportation model is shown in Table 13–3. The number in the top right-hand corner of each square represents the "transportation" costs; those in the middle of the squares (boldface numbers) represent the solution.

---

† For a proof of the optimality of this procedure, see S. M. Johnson, "Sequential Production Planning over Time at Minimum Cost," *Management Science,* Vol. 3, 1957, pp. 435–437.

Table 13-3

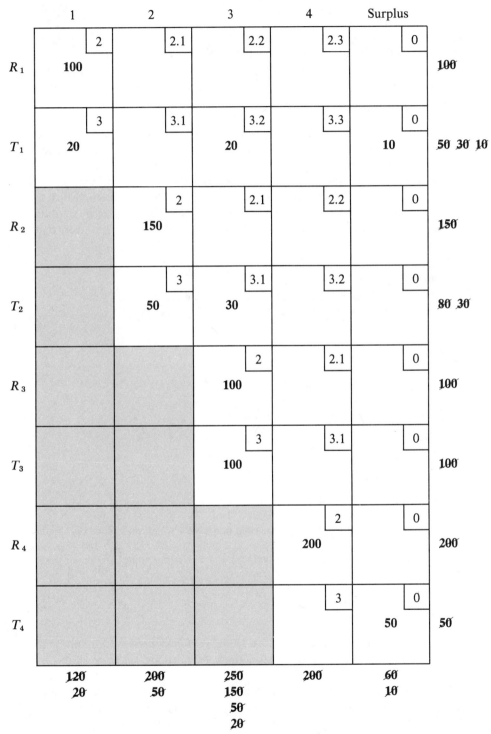

Notice the logic of the solution. For column 1, square $(R_1, 1)$ has the smallest cost per unit $(= 2)$. The maximum amount that can be assigned to this square is 100 units, which exhausts the supply of R_1. The remaining demand units for period 1 can be satisfied by assigning 20 units to square $(T_1, 1)$. This leaves a supply of 30 units for T_1. Next, consider column 2. Square $(R_2, 2)$ has the smallest cost $(= 2)$. A maximum of 150 units can be assigned to it, which will exhaust the R_2-supply. The next smallest cost in column 2 occurs in square $(R_1, 2)$. Since the R_1-supply is zero now, square $(T_2, 2)$, with the next smallest cost element in the same column, must be considered. By assigning 50 units to this square, the demand for period 2 is satisfied. This leaves 30 units in the T_2-supply. The indicated procedure is continued until the demand for period 4 (column 4) is satisfied.

You can verify that this solution is optimal by using the optimality condition of the transportation algorithm (see Section 5.2.1). This is accomplished in the usual manner by computing the simplex multipliers for the present solution and then checking for optimality (see Problem 13–28). Notice, however, that the given "optimal" solution is degenerate. ◀

Exercise 13.3-9

Consider the optimal solution of Example 13.3–8 as given in Table 13–3.
(a) Determine the following amounts.
   (1) Production in period 1 for 1.
       [*Ans.* 120 units.]
   (2) Production in period 1 for 2.
       [*Ans.* None.]
   (3) Production in period 1 for 3.
       [*Ans.* 20 units.]
   (4) Regular and overtime production in period 1.
       [*Ans.* 100 units regular time and 40 units overtime.]
   (5) Inventory carried from period 1 to 3.
       [*Ans.* 20 units.]
   (6) Inventory carried from period 2 to 3.
       [*Ans.* 30 units.]
   (7) Inventory carried from period 3 to 4.
       [*Ans.* None.]
(b) Suppose that 55 additional units are needed in period 4. Determine how they should be produced.
   [*Ans.* Produce 50 overtime units in period 4 and 5 overtime units in period 1.]

Shortage Model

Now consider a generalization of the model in which shortages are allowed. It is assumed that backlogged demand must be filled by the end of the N-period horizon.

Table 13–3 can be modified readily to include the effect of backlogging by introducing the appropriate unit "transportation" costs in the blocked routes. For example, if p_i is the shortage cost per unit demanded in period i and filled in period $i + 1$, the unit transportation costs corresponding to squares

$(R_N, 1)$ and $(T_N, 1)$ are given by $\{c_N + p_1 + p_2 + \cdots + p_{N-1}\}$ and $\{d_N + p_1 + \cdots + p_{N-1}\}$, respectively.

It would seem reasonable that the solution procedure used with the no-shortage case would also apply to the new situation, where shortage is allowed. Unfortunately, this is not true. To justify this claim, the following numerical example is designed to show that the preceding procedure may generally yield an inferior solution.

Example 13.3-9. Consider a three-period model where regular and overtime production are used. The production capacities for the three periods are as follows:

Period	Production Capacity (Units)	
	Regular	Overtime
1	15	10
2	15	0
3	20	15

Table 13-4

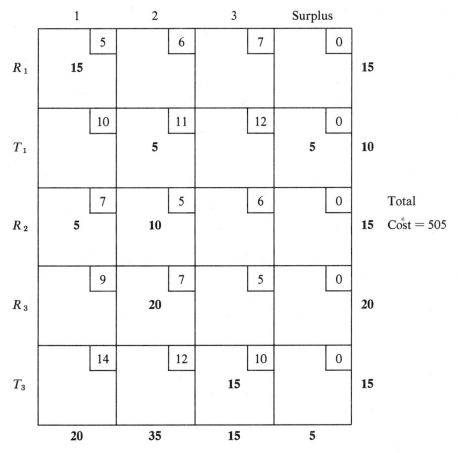

	1	2	3	Surplus		
R_1	5 — **15**	6	7	0	**15**	
T_1	10 — **5**	11	12 — **5**	0	**10**	
R_2	7 — **5**	5 — **10**	6	0	**15**	Total
R_3	9	7 — **20**	5	0	**20**	Cost $= 505$
T_3	14	12	10 — **15**	0	**15**	
	20	**35**	**15**	**5**		

The production cost per unit is 5 for regular production and 10 for overtime production. The holding and shortage costs per unit are given by 1 and 2, respectively. The demand units for the three periods are 20, 35, and 15, respectively.

The equivalent transportation model is given in Table 13–4. Period 2 has no overtime production, since its corresponding capacity is zero.

Table 13–4 also shows the solution of the problem obtained by using the foregoing procedure. Thus, for column 1, 15 units are assigned to $(R_1, 1)$ and 5 units to $(R_2, 1)$. (Notice that the cheapest route is selected from among *all* the entries of the column under consideration.) Next, consider column 2. Assign 10 units to $(R_2, 2)$, 20 units to $(R_3, 2)$, and 5 units to $(T_1, 2)$. Finally, in column 3, assign 15 units to $(T_3, 3)$. The total cost associated with the schedule is $5 \times 15 + 5 \times 7 + 5 \times 11 + 10 \times 5 + 20 \times 7 + 15 \times 10 = 505$.

It can be shown that the solution in Table 13–4 does not satisfy the optimality condition of the transportation algorithm. In fact, Table 13–5 gives the optimal solution to this problem. The associated total cost in this case is $15 \times 5 + 5 \times 10 + 5 \times 11 + 15 \times 5 + 5 \times 7 + 10 \times 12 + 15 \times 5 = 485$. ◄

Table 13–5

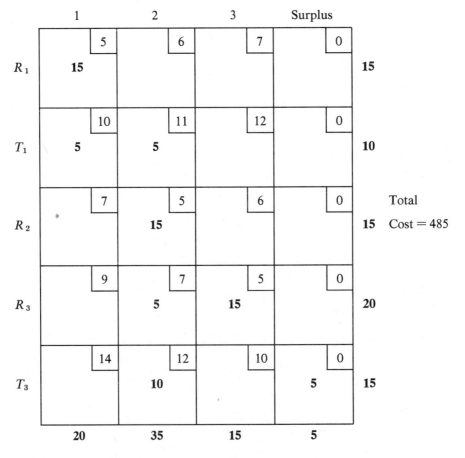

Example 13.2–9 shows that the simple procedure of satisfying the demands for the successive periods does not yield an optimal solution for the shortage model. Consequently, we would have to apply the general transportation algorithm to obtain the optimal solution.

13.4 PROBABILISTIC MODELS

This section presents different (single-item) inventory models with probabilistic demand. The first model extends the deterministic continuous review model (Section 13.3.1) by directly including the probabilistic demand in the formulation. The remaining formulations are categorized under single- and multiple-period models. In the multiple-period models, the distribution of demand is either stationary or nonstationary. Most multiperiod models with stationary demand may be easily extended to the nonstationary case, but the associated computations, especially in the nonstationary case, are almost prohibitive. However, if stationary demand and infinite horizon are assumed, closed-form solutions may usually be obtained for the models.

The basic decision criterion used with probabilistic inventory models in this chapter is the minimization of *expected* costs (or equivalently maximization of expected profit). As mentioned in Chapter 11, other criteria could be used as well. However, since the objective is to concentrate on the development of the inventory problem itself, no other criteria will be discussed here.

13.4.1 A CONTINUOUS REVIEW MODEL

This section introduces a probabilistic model in which the stock is reviewed continuously and an order of size y is placed every time the stock level reaches a certain reorder point R. The objective is to determine the optimum values of y and R that minimize the total expected inventory costs per unit time. In this model, one year represents a unit of time.

The inventory fluctuations corresponding to this situation are depicted in Figure 13–13. A cycle is defined as the time period between two successive arrivals of orders. The assumptions of the model are

1. Lead time between the placement of an order and its receipt is stochastic.
2. Unfilled demand during lead time is backlogged.
3. The distribution of demand during lead time is independent of the time at which it occurs.
4. There is no more than one outstanding order at a time.

Let

$r(x|t) =$ conditional PDF of demand x during lead time t, $\quad x > 0$
$s(t) =$ PDF of lead time t, $\quad t > 0$
$f(x) =$ absolute PDF of demand x during lead time $= \int_0^\infty r(x|t)s(t)\,dt$

$y =$ amount *ordered* per cycle
$D =$ expected total demand per year
$h =$ holding cost per unit per year
$p =$ shortage cost per unit per year

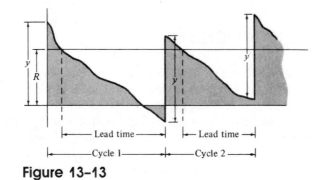

Figure 13–13

The total annual cost for this model includes the average setup cost, the expected holding cost, and the expected shortage cost. The average setup cost is given by (DK/y), where (D/y) is the approximate number of orders per year and K is the setup cost per order.

The expected holding cost is calculated based on the expected net inventory level at the beginning and end of a cycle. The expected stock level at the end of an inventory cycle is equal to $E\{R - x\}$. At the beginning of the cycle (right after an order of size y is received), the expected stock level is $y + E\{R - x\}$. Thus the average inventory per cycle (and hence per year) is given by

$$\bar{H} = \frac{(y + E\{R - x\}) + E\{R - x\}}{2} = \frac{y}{2} + E\{R - x\}$$

Now, given $f(x)$ as defined,

$$E\{R - x\} = \int_0^\infty (R - x)f(x)\,dx = R - E\{x\}$$

Notice that the expression for $\bar{H}$ neglects the case where $R - E\{x\}$ is negative (shortage quantity). This approximation is one of the simplifying assumptions of the model.

Let S be the shortage quantity per cycle. Then

$$S(x) = \begin{cases} 0, & x \leq R \\ x - R, & x > R \end{cases}$$

Consequently, the expected shortage quantity per cycle is

$$\bar{S} = \int_0^\infty S(x)f(x)\,dx = \int_R^\infty (x - R)f(x)\,dx$$

Since there are approximately (D/y) orders per year, the expected annual shortage is then equal to $(D\bar{S}/y)$.

The total annual cost of the system is thus given by

$$TAC(y, R) = \frac{DK}{y} + h\left(\frac{y}{2} + R - E\{x\}\right) + \frac{pD\overline{S}}{y}$$

Notice that the shortage cost $(pD\overline{S}/y)$ is assumed proportional to the shortage quantity only without taking the shortage time into account. This approximation again is another simplifying assumptions in the model, since in the case of backlog, shortage cost is also a function of shortage time.

The solution for optimal y^* and R^* is obtained from

$$\frac{\partial TAC}{\partial y} = -\left(\frac{DK}{y^2}\right) + \frac{h}{2} - \frac{pD\overline{S}}{y^2} = 0$$

$$\frac{\partial TAC}{\partial R} = h - \left(\frac{pD}{y}\right)\int_0^\infty f(x)\, dx = 0$$

From the first equation,

$$y^* = \sqrt{\frac{2D(K + p\overline{S})}{h}} \tag{1}$$

and from the second equation,

$$\int_{R^*}^\infty f(x)\, dx = \frac{hy^*}{pD} \tag{2}$$

An explicit general solution for y^* and R^* is not possible in this case. A convenient numerical method is thus used to solve equations (1) and (2). The following procedure, due to Hadley and Whitin (1963), is proved to converge in a finite number of iterations, provided that a solution exists.

In equation (1), $\overline{S}$ at *least* equals zero, which shows that the *smallest* value of y^* is $\sqrt{2DK/h}$, a result that is achieved when $\overline{S} = 0$ (or $R \to \infty$). Now, at $R = 0$, equation (1) gives

$$y^* = \hat{y} = \sqrt{\frac{2D(K + pE\{x\})}{h}}$$

and equation (2) gives

$$y^* = \tilde{y} = \frac{pD}{h}$$

It can be proved [Hadley and Whitin, (1963), pp. 169–174] that if $\tilde{y} \geq \hat{y}$, the optimal values of y and R exist and are unique. In such a case these values are computed as follows. Compute the first trial value of y^* as $y_1 = \sqrt{2DK/h}$. Next use equation (2) to compute the value of R_1 corresponding to y_1. By using R_1, a new trial value y_2 is obtained from equation (1). Next, R_2 is computed from equation (2) by using y_2. This procedure is repeated until two successive values of R are approximately equal. At this point, the last values computed for y and R will yield y^* and R^*.

Example 13.4–1. Let $K = \$100$, $D = 1000$ units, $p = \$10$, and $h = \$2$ and assume that the demand during lead time follows a uniform distribution over the range 0 to 100.

To check whether the problem has a feasible solution, consider

$$\hat{y} = \sqrt{\frac{2D(K + pE\{x\})}{h}} = \sqrt{\frac{2 \times 1000(100 + 10 \times 50)}{2}}$$

$$= 774.5$$

and

$$\tilde{y} = \frac{pD}{h} = \frac{10 \times 1000}{2} = 5000$$

Since $\tilde{y} > \hat{y}$, a unique solution for y^* and R^* exists.

Now

$$\bar{S} = \int_R^\infty (x - R)f(x)\,dx = \int_R^{100} (x - R)\frac{1}{100}\,dx$$

$$= \frac{R^2}{200} - R + 50 \tag{1}$$

From equation (1),

$$y^* = \sqrt{\frac{2D(K + p\bar{S})}{h}} = \sqrt{\frac{2 \times 1000(100 + 10\bar{S})}{2}}$$

$$= \sqrt{100{,}000 + 10{,}000\bar{S}} \tag{2}$$

where $\bar{S}$ is as given by (1). From equation (2),

$$\int_{R^*}^{100} \frac{1}{100}\,dx = \frac{2y^*}{10 \times 1000}$$

or

$$R^* = 100 - \frac{y^*}{50} \tag{3}$$

Equation (3) is used to compute R_i for a given value of y_i, and equation (2) is used to compute y_{i+1} for a given value of R_i.

Iteration 1

$$y_1 = \sqrt{\frac{2DK}{h}} = \sqrt{\frac{2 \times 1000 \times 100}{2}} = 316$$

$$R_1 = 100 - \frac{316}{50} = 93.68$$

Iteration 2

$$\bar{S} = \frac{R_1^2}{200} - R_1 + 50 = .19971$$

$$y_2 = \sqrt{100{,}000 + 10{,}000 \times .19971} = 319.37$$

Hence

$$R_2 = 100 - \frac{319.37}{50} = 93.612$$

Iteration 3

$$\bar{S} = \frac{R_2^2}{200} - R_2 + 50 = .20403$$

$$y_3 = \sqrt{100{,}000 + 10{,}000 \times .20403} = 319.43$$

Thus

$$R_3 = 100 - \frac{319.43}{50} = 93.611$$

Since R_2 and R_3 are approximately equal, the approximate optimal solution is given by

$$R^* \cong 93.61 \quad \text{and} \quad y^* \cong 319.4 \qquad \blacktriangleleft$$

Exercise 13.4–1

In Example 13.4–1, determine the following values based on the assumptions of the model.
(a) The approximate number of orders per year.
   [*Ans.* Three orders.]
(b) The annual setup cost.
   [*Ans.* $300.]
(c) The expected holding cost per year.
   [*Ans.* $406.62.]
(d) The expected shortage cost per year.
   [*Ans.* $6.39.]
(e) The probability of running out of stock during lead time.
   [*Ans.* .0639.]

13.4.2 SINGLE-PERIOD MODELS

The single-period inventory models occur when an item is ordered once only to satisfy the demand of a specific period. For example, a style item becomes obsolete and hence may not be reordered. In this section the single-period models will be investigated under different conditions, including instantaneous and uniform demand with and without setup cost. It is assumed that stock replenishment occurs instantaneously. The optimal inventory level will be derived based on the minimization of expected inventory costs, which include the ordering (setup + purchasing or production), holding, and shortage. Because the demand is probabilistic, the purchasing (production) cost per unit, although constant, becomes an effective factor in the cost function.

A.   Instantaneous Demand, No Setup Cost

In the models with instantaneous demand, it is assumed that the total demand is filled at the beginning of the period. Thus, depending on the amount demanded D, the inventory position right after demand occurs may be either positive (surplus) or negative (shortage). These two cases are shown in Figure 13–14.

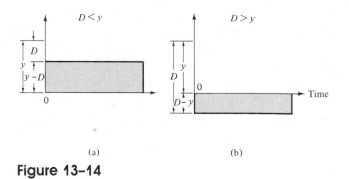

(a)   (b)

Figure 13–14

From Figure 13–14, given y, the amount on hand after an order is received, the holding inventory is generally given by

$$H(y) = \begin{cases} y - D & \text{for } D < y \\ 0, & \text{for } D \geq y \end{cases}$$

The shortage inventory is given by

$$G(y) = \begin{cases} 0, & \text{for } D < y \\ D - y, & \text{for } D \geq y \end{cases}$$

Let x be the amount on hand before an order is placed. Define $f(D)$ as the PDF of demand, and let h and p be the holding and shortage costs per unit per period. Further, let c be the purchasing cost per unit. Assuming that y is continuous and no setup cost is incurred, the expected cost for the period is then given by

$$E\{C(y)\} = \text{purchasing cost} + E\{\text{holding cost}\} + E\{\text{shortage cost}\}$$

$$= c(y - x) + h \int_0^\infty H(y) f(D)\, dD + p \int_0^\infty G(y) f(D)\, dD$$

$$= c(y - x) + h \left\{ \int_0^y (y - D) f(D)\, dD + 0 \right\}$$

$$\qquad\qquad\qquad + p \left\{ 0 + \int_y^\infty (D - y) f(D)\, dD \right\}$$

$$= c(y - x) + h \int_0^y (y - D) f(D)\, dD + p \int_y^\infty (D - y) f(D)\, dD$$

The optimal value of y is obtained by equating the first derivative of $E\{C(y)\}$ to zero.† Thus

$$\frac{\partial E\{C(y)\}}{\partial y} = c + h \int_0^y f(D) \, dD - p \int_y^\infty f(D) \, dD = 0$$

Since

$$\int_y^\infty f(D) \, dD = 1 - \int_0^y f(D) \, dD$$

the equation gives

$$\int_0^{y^*} f(D) \, dD = \frac{p - c}{p + h}$$

The value of y^* is defined only if $p \geq c$. If $p < c$, this is interpreted as discarding the inventory system completely. Now

$$\frac{\partial^2 E\{C(y)\}}{\partial y^2} = (h + p)f(y^*) > 0$$

shows that y^* corresponds to a minimum point. Graphically, the function $E\{C(y)\}$ should appear as shown in Figure 13–15. In such cases $E\{C(y)\}$ is convex. Since y^* is unique, it must give a global minimum. The policy adopted is thus called a **single critical number policy.**

According to the condition given, the value of y^* is selected such that the probability $D \leq y^*$ is equal to

$$q = \frac{p - c}{p + h}, \qquad p > c$$

The optimal ordering policy given x is on hand before an order is placed is given by

$$\begin{aligned} &\text{if } y^* > x, \qquad \text{order } y^* - x \\ &\text{if } y^* \leq x, \qquad \textit{do not order} \end{aligned}$$

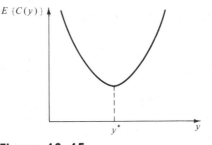

Figure 13–15

---

† See the footnote on page 408.

Example 13.4-2. Consider the one-period model with $h = \$.5$, $p = \$4.5$, and $c = \$.5$. The demand density function is given by

$$f(D) = \begin{cases} 1/10, & 0 \le D \le 10 \\ 0, & D > 10 \end{cases}$$

Thus

$$q = \frac{p-c}{p+h} = \frac{4.5 - .5}{4.5 + .5} = .8$$

and

$$P\{D \le y^*\} = \int_0^{y^*} f(D)\, dD = \int_0^{y^*} \frac{1}{10}\, dD = \frac{y^*}{10}$$

or

$$y^* = 8$$

This solution is illustrated graphically in Figure 13–16. ◀

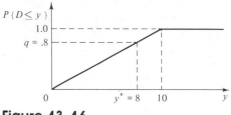

Figure 13–16

Exercise 13.4-2
Consider Example 13.4–2.
(a) Determine the order quantity in each of the following cases.
   (1) Initial inventory = 5 units.
   [*Ans.* Order 3 units.]
   (2) Initial inventory = 10 units.
   [*Ans.* Do not order.]
(b) Determine the probability of not running out of stock during the period.
   [*Ans.* 0.8.]

Suppose now that demand occurs in discrete rather than in continuous units. Then

$$E\{C(y)\} = c(y - x) + h \sum_{D=0}^{y} (y - D)f(D) + p \sum_{D=y+1}^{\infty} (D - y)f(D)$$

In the discrete case, the necessary conditions for a minimum are given by

$$E\{C(y - 1)\} \ge E\{C(y)\} \quad \text{and} \quad E\{C(y + 1)\} \ge E\{C(y)\}$$

Thus

$$E\{C(y - 1)\} = c(y - 1 - x) + h \sum_{D=0}^{y-1} (y - 1 - D)f(D) + p \sum_{D=y}^{\infty} D - y + 1)f(D)$$

Example 13.4-4. Consider Example 13.4–1. Since

$$f(D) = \frac{1}{10}, \qquad 0 \le D \le 10$$

then

$$\int_0^{y^*} \frac{1}{10} \, dD + y^* \int_{y^*}^{10} \frac{1}{10D} \, dD = .8$$

or

$$(1/10)(y^* - y^* \ln y^* + 2.3y^*) = 0.8$$

or

$$3.3y^* - y^* \ln y^* - 8 = 0$$

The solution of this equation is obtained by trial and error and is given by $y^* = 4.5$. Notice the difference between this result and the one given in the case of instantaneous demand.   ◀

C. Instantaneous Demand, Setup Cost—(s-S Policy)

Consider the model in Section 13.4.2A with the exception that the setup cost K will be taken into account. Let $E\{\overline{C}(y)\}$ be the total expected cost of the system inclusive of the setup cost. Thus

$$E\{\overline{C}(y)\} = K + c(y - x) + h \int_0^y (y - D) f(D) \, dD + p \int_y^\infty (D - y) f(D) \, dD$$

$$= K + E\{C(y)\}$$

The minimum value of $E\{C(y)\}$ is shown in Section 13.4.2A to occur at y^*, satisfying

$$\int_0^{y^*} f(D) \, dD = \frac{p - c}{p + h}$$

Since K is constant, the minimum value of $E\{\overline{C}(y)\}$ must also occur at y^*. The curves $E\{C(y)\}$ and $E\{\overline{C}(y)\}$ are shown in Figure 13–18. The new symbols

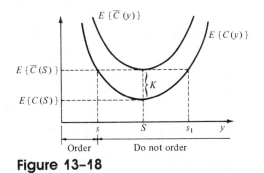

Figure 13–18

s and *S* are defined in the figure for use later in the analysis. The value of *S* is equal to y^*, and the value of *s* is determined from

$$E\{C(s)\} = E\{\overline{C}(S)\} = K + E\{C(S)\}$$

such that $s < S$. (Notice that this equation must yield another value $s_1 > S$, which may be disregarded.)

The question now is: Given *x*, the amount on hand before the order is placed, how much should be ordered, if any? This question is investigated under three conditions:

1. $x < s$.
2. $s \leq x \leq S$.
3. $x > S$.

Case 1: $x < s$. Since *x* is already on hand, its equivalent cost is given by $E\{C(x)\}$. If any additional amount $y - x$ $(y > x)$ is ordered, the corresponding cost given *y* is $E\{\overline{C}(y)\}$, which includes the setup cost *K*. It follows from Figure 13–18 that, for all $x < s$,

$$\min_{y>x} E\{\overline{C}(y)\} = E\{\overline{C}(S)\} < E\{C(x)\}$$

Thus the optimal inventory level must reach $y^* = S$ and the amount ordered must equal to $S - x$.

Case 2: $s \leq x \leq S$. Again, from Figure 13–18,

$$E\{C(x)\} \leq \min_{y>x} E\{\overline{C}(y)\} = E\{\overline{C}(S)\}$$

Thus it is no more costly not to order in this case. Hence $y^* = x$.

Case 3: $x > S$. From Figure 13–18, for $y > x$,

$$E\{C(x)\} < E\{\overline{C}(y)\}$$

which again indicates that it is less costly not to order and hence $y^* = x$.

This policy is called the *s-S* policy and is summarized as follows:

$$\begin{aligned} &\text{if } x < s, \quad &&\text{order } S - x \\ &\text{if } x \geq s, \quad &&\text{do not order} \end{aligned}$$

The optimality of the *s-S* policy follows from the fact that the cost function is convex. In general, when this property is not satisfied, the *s-S* policy will cease to be optimal.

Example 13.4–5. Consider Example 13.4–2. Let $K = \$25$ and assume a zero initial inventory. Since $y^* = 8$, it follows that $S = 8$. To determine the value of *s*, consider

$$E\{C(y)\} = .5(y - x) + .5 \int_0^y \frac{1}{10}(y - D)\, dD + 4.5 \int_y^{10} \frac{1}{10}(D - y)\, dD$$

$$= .5(y - x) + .05 \left[yD - \frac{D^2}{2} \right]_0^y + .45 \left[\frac{D^2}{2} - Dy \right]_y^{10}$$

$$= .25y^2 - 4.0y + 22.5 - .5x$$

The equation

$$E\{C(s)\} = K + E\{C(S)\}$$

thus gives

$$.25s^2 - 4.0s + 22.5 - .5x = 25 + .25S^2 - 4.0S + 22.5 - .5x$$

Setting $S = 8$, we obtain the equation

$$s^2 - 16s - 36 = 0$$

whose solution is

$$s = -2 \quad \text{or} \quad 18$$

The value of $s = 18$ (which is greater than S) should be disregarded. Since the remaining value is negative ($= -2$), s has not feasible value [notice that $E\{C(y)\}$ is defined for nonnegative values of y only]. The optimal solution thus calls for not ordering at all. Clearly, this does not follow the s-S policy, since s is undefined.† The current situation is illustrated graphically in Figure 13–19. This usually occurs when the cost function is flat or when the setup cost K is large compared with the other costs. ◄

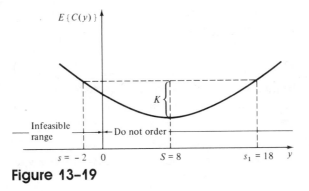

Figure 13–19

Exercise 13.4–4
Find the optimal solution ordering policy of Example 13.4–5 assuming that the setup cost is $5.
[*Ans.* If inventory is below 3.53 units, order up to 8 units.]

13.4.3 MULTIPERIOD MODELS

In this section probabilistic models are considered for the multiperiod (finite or infinite) case under different combinations of the following conditions:

1. Backlogging and no backlogging of demand.
2. Zero and positive delivery lags.

_____

† Conventionally, if $s < 0$, s is set equal to zero and the given s-S policy becomes applicable.

The models are developed mainly for the finite horizon case. Models with infinite periods will be derived from the finite case by taking the limit as the number of periods tends to infinity. It is assumed that no setup cost is incurred in any period. The inclusion of setup costs in the multiperiod case generally leads to difficult computations. As will be shown, all multiperiod models are formulated by dynamic programming.

Although in all previous inventory models the optimal policy is determined by minimizing a cost function, solutions in this section are based on the maximization of a profit function. The objective is to familiarize the reader with the application of the maximization (profit) criterion as an alternative to the minimization (cost) criterion.

Unlike the single-period models, a multiperiod model should take into account the discounted value of money. Thus if α (< 1) is the discount factor per period, an amount of money S after n periods ($n \geq 1$) is equivalent to $\alpha^n S$ now.

The following models are developed under the assumption that the demand distribution is stationary for all periods. In the finite horizon case, stationary models may be extended to cover nonstationary distributions by replacing demand density function $f(D)$ by $f_i(D_i)$, where i designates the period.

A.   Backlog, Zero Delivery Lag

Let the finite planning horizon be limited to N periods. Define

$f_i(x_i) =$ maximum total expected profit for periods $i, i + 1, \ldots, N$, given that x_i is the amount on hand before an order is placed in the ith period

Using the symbols of the preceding section and assuming that r is the revenue per unit, we can formulate the problem as a (backward) dynamic programming model as follows:

$$f_i(x_i) = \max_{y_i \geq x_i} \left(-c(y_i - x_i) + \int_0^y [rD - h(y_i - D)] f(D) \, dD \right.$$

$$+ \int_{y_i}^\infty [ry_i + \alpha r(D - y_i) - p(D - y_i)] f(D) \, dD$$

$$\left. + \alpha \int_0^\infty f_{i+1}(y_i - D) f(D) \, dD \right), \qquad i = 1, 2, \ldots, N$$

with $f_{N+1}(y_N - D) \equiv 0$. Notice that x_i may be negative, since the unfilled demand is backlogged. The quantity $\alpha r(D - y_i)$ in the second integral is included for the following reason. The amount $(D - y_i)$ represents the unfilled demand in the ith period that must be filled in the $(i + 1)$st period. The discounted return is thus $\alpha r(D - y_i)$.

The recursive equation can basically be solved by dynamic programming. However, this procedure is extremely difficult in this case. An important case can be analyzed, however, by considering the infinite period model with its recursive equation given by

$$f(x) = \max_{y \geq x} \left(-c(y-x) + \int_0^y [rD - h(y-D)] f(D)\, dD \right.$$

$$\left. + \int_y^\infty [ry + \alpha r(D-y) - p(D-y)] f(D)\, dD + \alpha \int_0^\infty f(y-D) f(D)\, dD \right)$$

where x and y are the inventory levels for each period before and after an order is received.

The optimal policy for the infinite period case is of the single critical number type. Thus

$$\frac{\partial(\cdot)}{\partial y} = -c - h \int_0^y f(D)\, dD + \int_y^\infty [(1-\alpha)r + p] f(D)\, dD$$

$$+ \alpha \int_0^\infty \frac{\partial f(y-D)}{\partial y} f(D)\, dD = 0$$

The value of

$$\frac{\partial f(y-D)}{\partial y}$$

is determined as follows. If there are δ (> 0) units more on hand at the start of the next period, the profit for the next period will increase by $c\delta$, for this much less has to be ordered. Consequently, the

$$\frac{\partial f(y-D)}{\partial y} = c$$

This equation thus becomes

$$-c - h \int_0^y f(D)\, dD + \left((1-\alpha)r + p \right) \left(1 - \int_0^y f(D)\, dD \right) + \alpha c \int_0^\infty f(D)\, dD = 0$$

which reduces to

$$\int_0^{y^*} f(D)\, dD = \frac{p + (1-\alpha)(r-c)}{p + h + (1-\alpha)r}$$

The optimal policy for each period given its entering inventory x is

$$\text{if } x < y^*, \quad \text{order } y^* - x$$
$$\text{if } x \geq y^*, \quad \text{do not order}$$

It is stated here, without proof, that if in the finite model y_i^* represents the optimal inventory level for period i, the following relationship is always satisfied:

$$y_N^* \leq y_{N-1}^* \leq \cdots \leq y_i^* \leq \cdots \leq y_1^* \leq y^*$$

where y^* is the single critical value in the infinite model. In the finite model the optimal policy calls for ordering less as one comes closer to the end of the horizon. In the meantime, none of the critical values y_i^* can exceed the optimal value y^* in the infinite model.

B. No Backlog, Zero Delivery Lag

This model is similar to the backlog case except when the demand D exceeds the inventory level y_i, in which case the next period will start with $x_{i+1} = 0$. This means that the unfilled demand is lost and hence does not result in any revenue.

The N-period (finite) recursive equation for the no backlog case is thus given by

$$f_i(x_i) = \max_{y_i \geq x_i} \left(-c(y_i - x_i) + \int_0^{y_i} [rD - h(y_i - D)] f(D)\, dD \right.$$

$$+ \int_{y_i}^{\infty} [ry_i - p(D - y_i)] f(D)\, dD + \alpha \left[\int_0^{y_i} f_{i+1}(y_i - D) f(D)\, dD \right.$$

$$\left. \left. + \int_{y_i}^{\infty} f_{i+1}(0) f(D)\, dD \right] \right), \qquad i = 1, 2, \ldots, N$$

with $f_{N+1} \equiv 0$.

It is also difficult to solve this problem by dynamic programming. The corresponding infinite period model is easy to solve, however, since it is of the single critical number type. Thus we get

$$f(x) = \max_{y \geq x} \left(-c(y - x) + \int_0^y [rD - h(y - D)] f(D)\, dD \right.$$

$$+ \int_y^{\infty} [ry - p(D - y)] f(D)\, dD$$

$$\left. + \alpha \left[\int_0^y f(y - D) f(D)\, dD + \int_y^{\infty} f(0) f(D)\, dD \right] \right)$$

Taking the first derivative and equating to zero, we obtain

$$-c - h \int_0^y f(D)\, dD + (r + p) \int_y^{\infty} f(D)\, dD + \alpha \int_0^y \frac{\partial f(y - D)}{\partial y} f(D)\, dD = 0$$

and using the result

$$\frac{\partial f(y - D)}{\partial y} = c$$

we obtain

$$\int_0^{y^*} f(D)\, dD = \frac{r + p - c}{h + r + p - \alpha c}$$

In this model, as in the preceding one, the relationship

$$y_N^* \leq y_{N-1}^* \leq \cdots \leq y_i^* \leq \cdots \leq y_1^* \leq y^*$$

holds, where y_i^* corresponds to the optimal inventory level of period i in the finite model.

C. Backlog, Positive Delivery Lag

In this model it is assumed that an order placed at the beginning of period i will be received k periods later, that is, at period $i + k$, $k \geq 1$. The delivery lag k is assumed constant for all the periods.

Let $z, z_1, \ldots, z_{k-1}$ be the amounts due in (as a result of previous decisions) at the beginning of periods $i, i + 1, \ldots, i + k - 1$ (see Figure 13–20). Let $y = x + z$ be the amount of inventory at the beginning of period i, where x is the entering inventory (possibly negative because of backlogging) in period i. At period i, the decision variable is represented by z_k, the amount ordered now to be received k periods later.

Define $f_i(y, z_1, \ldots, z_{k-1})$ as the present worth of the maximum expected profit for periods $i, i + 1, \ldots, N$ given $y, z_1, \ldots,$ and z_{k-1}. Thus

$$f_i(y, z_1, \ldots, z_{k-1}) = \max_{z_k \geq 0} \left\{ -cz_k + L(y) \right.$$

$$\left. + \alpha \int_0^\infty f_{i+1}(y + z_1 - D, z_2, \ldots, z_k) f(D) dD \right\}, \qquad i = 1, 2, \ldots, N$$

where $f_{N+1} \equiv 0$ and

$$L(y) = \int_0^y [rD - h(y - D)] f D) \, dD + \int_y^\infty [ry + (\alpha r - p)(D - y)] f(D) \, dD$$

$L(y)$ represents the expected revenue minus the holding and penalty costs for period i.

The optimal policy for this model can be expressed in terms of $(y + z_1 + \cdots + z_{k-1})$. This is advantageous computationally, since it reduces the dimensions of the state of the system to one only.

Consider first the special case of a finite horizon consisting of k periods starting with period i. Since z_k is received at the beginning of period $i + k$, it has no effect on the holding and penalty costs or the revenue during the k-

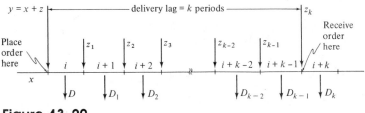

Figure 13–20

period horizon. Let C_k represent the present worth of the expected revenue during the k-period horizon exclusive of the ordering cost cz_k. Thus

$$C_k = L(y) + \alpha E\{L(y + z_1 - D)\} + \alpha^2 E\{L(y + z_1 + z_2 - D - D_1)\}$$

$$+ \cdots + \alpha^{k-1} E\left\{L\left(y + \sum_{j=1}^{k-1} z_j - D - \sum_{j=1}^{k-2} D_j\right)\right\}$$

where D is the demand for period i and D_j is the demand for period $i + j$. The operator "E" is the expectation operator.

Since all the demands are independent and identically distributed, each with PDF $f(D)$, the random variable $s_m = D + D_1 + \cdots + D_{m-1}$, $m = 2, 3, \ldots,$ $k - 1$, is the m-fold convolution of D (see Section 10.10). Let $f_m(s_m)$ be the PDF of s_m. Then

$$E\left\{L\left(y + \sum_{j=1}^{m} z_j - D - \sum_{j=1}^{m-1} D_j\right)\right\} = \int_0^\infty L\left(y + \sum_{j=1}^{m} z_j - s_m\right) f_m(s_m)\, ds_m$$

The expression for C_k is a constant independent of z_k.

To compute the net revenue for period $i + k$, let

$$u = y + (z_1 + \cdots + z_{k-1}) + z_k$$

and

$$v = y + (z_1 + \cdots + z_{k-1}) = u - z_k$$

$s_{k+1} = D + D_1 + \cdots D_k$ represents the demand for periods $i, i + 1, \ldots,$ $i + k$. The holding and shortage inventories for period $i + k$ are thus given by $(u - s_{k+1})$ and $(s_{k+1} - u)$. Consequently, the net revenue (not accounting for the ordering cost cz_k) for period $i + k$ is given by

$$L_{k+1}(u) = \left[\int_0^u \{rs_{k+1} - h(u - s_{k+1})\} f_{k+1}(s_{k+1})\, ds_{k+1}\right.$$

$$\left. + \int_u^\infty \{ru + (\alpha r - p)(s_{k+1} - u)\} f_{k+1}(s_{k+1})\, ds_{k+1} - A\right]$$

where A is a constant representing the expected revenue for periods $i, i + 1,$ $\ldots, i + k - 1$.

Let $g_i(v)$ define the optimal expected profit for periods $i + k, \ldots, N$. Then

$$g_i(v) = \max_{u \geq v} \left\{-c(u - v) + \alpha^k L_{k+1}(u) + \int_0^\infty g_{i+1}(u - D) f(D)\, dD\right\}$$

The expected optimum revenue for periods $i, i + 1, \ldots, N$ is equal to the sum of the expected optimum revenues for periods $i, i + 1, \ldots, i + k - 1$ and for periods $i + k, i + k + 1, \ldots, N; N \geq k$. Since $v = y + z_1 + \cdots + z_{k-1}$, by definition the statement implies that

$$f_i(y, z_1, \ldots, z_{k-1}) = C_k + g_i(y + z_1 + \cdots + z_{k-1})$$

Since C_k is a constant, the optimization problem using f_i must be equivalent to the optimization problem using g_i. The advantage is that the new problem is described by one state $v = y + z_1 + \cdots + z_{k-1}$, which is computationally more attractive. The solution of the modified problem is essentially the same as that of the zero lag case discussed in Section 13.4.3A.

In the infinite horizon case the modified problem becomes

$$g(v) = \max_{u \geq v} \left(-c(u-v) + \alpha^k L_{k+1}(u) + \alpha E\{g(u-D)\} \right)$$

This, as shown in Section 13.4.3A, yields a unique optimal value u^*, which is obtained from

$$\frac{\partial(\cdot)}{\partial u} = -c + \alpha^k L'_{k+1}(u) + \alpha c = 0$$

This gives

$$\int_0^{u^*} f_{k+1}(s_{k+1})\, ds_{k+1} = \frac{p + (1-\alpha)(r - c\alpha^{-k})}{h + p + (1-\alpha)r}$$

The optimal policy at any period i is

if $u^* \geq v$,   order $u^* - v$

if $u^* < v$,   do not order

At period i, the value of v is already known. (See the definition of v.) Also, for $k = 0$, that is, no delivery lag, the result reduces to the same one as that given for the zero delivery lag model (Section 13.4.3A).

D. No Backlog, Positive Delivery Lag

With the same symbols as in Section 13.4.3C, the model for the no backlog case becomes

$$f_i(y, z_1, \ldots, z_{k-1}) = \max_{z_k \geq 0} \left[-c z_k + \int_0^y \{rD - h(y-D)\} f(D)\, dD \right.$$

$$+ \int_y^\infty \{ry - p(D-y)\} f(D)\, dD$$

$$+ \alpha \int_0^y f_{i+1}(y - D + z_1, z_2, \ldots, z_k) f(D)\, dD$$

$$\left. + \alpha \int_y^\infty f_{i+1}(z_1, z_2, \ldots, z_k) f(D)\, dD \right], \qquad i = 1, 2, \ldots, N$$

with $f_{N+1} \equiv 0$.

In general, the dynamic programming solution of this model is very difficult for $k > 1$, since this results in a dimensionality problem that increases with k.

13.5 SUMMARY

An inventory problem is concerned with making optimum decisions regarding *how much* and *when* to order an inventory item. Unfortunately, no unique model can be developed to handle this problem. Instead, a wide variety of models have been designed to represent special cases. An important factor in the design of an inventory model is the nature of demand. The simplest model is associated with static deterministic demand. The models are more complex when the demand is probabilistic.

Most inventory models are based on the optimization of a cost function that includes the setup, purchasing, holding and shortage costs. The shortage cost is usually the most difficult to estimate because it can represent intangible factors such as the loss of customer's goodwill. On the other hand, although it may not be difficult to estimate the setup cost, its inclusion in the cost model often produces a difficult mathematical problem.

In real life it is rare that the available inventory models represent the system accurately. Thus the solutions obtained from an inventory model should be regarded as a guideline rather than a specific recommendation. In some complex cases it may be necessary to model the system by simulation to ensure the reliability of the recommended solution.

SELECTED REFERENCES

HADLEY, G., and T. WHITIN, *Analysis of Inventory Systems,* Prentice-Hall, Englewood Cliffs, N.J., 1963.
LOVE, S., *Inventory Control,* McGraw-Hill, New York, 1979.
PERERSON, R., and E. SILVER, *Decision Systems for Inventory Management and Production Planning,* Wiley, New York, 1979.

REVIEW QUESTIONS

True (T) or False (F)?

1. _____ The setup cost is an inventory situation is independent of the size of the order quantity.

2. _____ In an inventory model, the shortage cost is usually the easiest to estimate.

3. _____ It is permissible to incur both holding and shortage costs in the same inventory model.

4. _____ In periodic review inventory systems, the stock is usually replenished at equal time intervals.

5. _____ In a continuous review inventory system, the reorder point usually occurs after the inventory falls below its maximum level for the cycle.

6. _____ An increase in the setup cost tends to decrease the optimum order quantity.

7. ____ The optimum order quantity decreases with the increase in the shortage cost.

8. ____ As the holding cost increases, the optimum order quantity also increases.

9. ____ In an inventory system, there may be more than one outstanding order at a time.

10. ____ The number of outstanding orders is independent of the length of lead time (or delivery lag).

11. ____ Although deterministic inventory models may be simple mathematically, they are not as accurate as the probabilistic models.

12. ____ A probabilistic demand described by a density function with its standard deviation much smaller than its mean can be approximated by a deterministic demand equal to the mean.

13. ____ If backordering is allowed, the shortage cost can be an important cost component in the inventory model.

14. ____ In the absence of a setup cost, a production cost function with increasing marginal costs is convex.

15. ____ In the presence of a setup cost, a production cost function with decreasing marginal costs is concave.

16. ____ In the absence of a setup cost, a production cost function with decreasing marginal costs is concave.

17. ____ In the presence of a setup cost, a production cost function with increasing marginal costs is concave.

18. ____ In a single or multiperiod model, a single critical number policy will always apply as long as the total cost function is convex.

19. ____ In the (s-S) policy of a single-period model, it may be optimal not to order up to the inventory level S whenever an order is placed.

20. ____ In the (s-S) policy of a single-period model, no order is placed if the initial inventory level is less than s.

[*Ans.* 1—T, 2—F, 3—T, 4—T, 5—T, 6—F, 7—F, 8—F, 9—T, 10—F, 11 to 16-T, 17—F, 18—T, 19—F, 20—F]

PROBLEMS

☐ **13–1** A small manufacturing company keeps 10 items, I1 through I10, in stock. The following table provides the cost per unit and annual usage of each item.

Item	Unit Cost	Annual Usage	Item	Unit Cost	Annual Usage
I1	$.05	2,500	I6	$.35	3,500
I2	.20	1,500	I7	.45	20,000
I3	.10	6,700	I8	.95	8,500
I4	.15	120,000	I9	.10	6,500
I5	.75	50,000	I10	.60	80,000

Apply the ABC analysis to this inventory situation. Which items should be given the tightest inventory control?

☐ **13–2** In each of the following cases, stock is replenished instantaneously and no shortage is allowed. Find the economic lot size, the associated total cost, and the length of time between two orders.
   (a) $K = \$100$, $h = \$.05$, $\beta = 30$ units/day.
   (b) $K = \$50$, $h = \$.05$, $\beta = 30$ units/day.
   (c) $K = \$100$, $h = \$.01$, $\beta = 40$ units/day.
   (d) $K = \$100$, $h = \$.04$, $\beta = 20$ units/day.

☐ **13–3** A company currently replenishes its stock of a certain item by ordering enough supply to cover a 1-month demand. The annual demand of the item is 1500 units. It is estimated that it costs $20 every time an order is placed. The holding cost per unit inventory per month is $2 and no shortage is allowed.
   (a) Determine the optimal order quantity and the time between orders.
   (b) Determine the difference in annual inventory costs between the optimal policy and the current policy of ordering a 1-month supply 12 times a year.

☐ **13–4** A company stocks an item that is consumed at the rate of 50 units per day. It costs the company $25 each time an order is placed. A unit inventory held in stock for 1 week will cost $.70. Determine the optimum number of orders (rounded to the closest integer) that the company has to place each year. Assume that the company has a standing policy of not allowing shortages in demand.

☐ **13–5** In each case in Problem 13–2, determine the reorder point assuming that the lead time is
(1) 14 days.
(2) 40 days.

☐ **13–6** Suppose that the demand distribution per unit time for the four cases in Problem 13–2 is normal with mean $\mu = \beta$ and constant variance $\sigma^2 = 9$. By using the information in Problem 13–5, determine the buffer stock in each case such that the probability of stock out during lead time is at most .02. [*Hint:* For the purpose of estimating the variance of demand during lead time, approximate the cycle length t_0 by its closest integer value.]

☐ **13–7** In the model in Figure 13–4 (Section 13.3.1), suppose that stock is replenished uniformly (rather than instantaneously) at the rate α. Consumption occurs at the uniform rate β at every point in time. If y is the order size and no shortage is allowed, show that
(a) Maximum inventory level at any point in time is $y(1 - \beta/\alpha)$.
(b) Total cost per unit time given y is

$$TCU(y) = \frac{K\beta}{y} + \frac{h}{2}\left(1 - \frac{\beta}{\alpha}\right)y$$

(c) The economic lot size is

$$y^* = \sqrt{\frac{2K\beta}{h(1 - \beta/\alpha)}}, \quad \text{provided } \alpha > \beta$$

(d) Show how the economic lot size of Section 13.3.1 may be derived directly from part (c). [*Hint:* Instantaneous replenishment is equivalent to letting α tend to infinity.]

☐ **13–8** Solve Problem 13–2 assuming that the stock is replenished uniformly at the rate $\alpha = 50$ per unit time.

☐ **13–9** A company can produce an item or buy it from a contractor. If it is produced locally, it will cost $20 each time the machines are set up. The production rate is 100 units per day. If it is bought from a contractor, it will cost $15 each time an order is placed. The cost of maintaining the item in stock, whether bought or produced, is $.02 a day. The company's usage of the item is estimated at 26,000 units annually. Assuming that the company operates with no shortage, should they buy or produce?

□ **13–10** In the model of Figure 13–4 (Section 13.3.1), suppose that shortage is allowed and that shortage cost per unit per unit time is p. If w is the shortage quantity and y is the amount ordered, show that:

(a) Total cost per unit time given y and w is

$$TCU(y,\ w) = \frac{K\beta}{y} + \frac{h(y-w)^2 + pw^2}{2y}$$

(b) $y^* = \sqrt{\dfrac{2K\beta(p+h)}{ph}}$

(c) $w^* = \sqrt{\dfrac{2K\beta h}{p(p+h)}}$

□ **13–11** A stock can be replenished instantaneously upon order. Demand occurs at the constant rate of 50 items per unit time. A fixed cost of $400 is incurred each time an order is placed. Although shortage is allowed, it is the company's policy that the shortage quantity not exceed 20 units. In the meantime, because of budget limitation, no more than 200 units can be ordered at a time. Find the relationship between the implied holding and the shortage cost per unit under optimal conditions.

□ **13–12** Show how the results of the model in Figure 13–4 may be derived from the results in Problem 13–10.

□ **13–13** Generalize the models in Problems 13–6 and 13–10 into a single model in which both uniform replenishment and shortages are present. Show that the following results apply.

(a) $TCU(y,\ w) = \dfrac{K\beta}{y} + \dfrac{h\{y(1-\beta/\alpha) - w\}^2 + pw^2}{2(1-\beta/\alpha)y}$

(b) $y^* = \sqrt{\dfrac{2K\beta(p+h)}{ph(1-\beta/\alpha)}}$

(c) $w^* = \sqrt{\dfrac{2K\beta h(1-\beta/\alpha)}{p(p+h)}}$

□ **13–14** Show that the model in Problem 13–13 may be specialized to produce directly the results of any of the preceding models.

□ **13–15** An item is consumed at the rate of 30 items per day. The holding cost per unit per unit time is $.05 and the setup cost is $100. Suppose that no shortage is allowed and the purchasing cost per unit is $10 for any quantity less than or equal to $q = 300$ and 8 otherwise. Find the economic lot size. What is the answer if $q = 500$ instead?

□ **13–16** An item sells for $4 a unit but a 10% discount is offered for lots of size 150 units or more. A company that consumes this item at the rate of 20 items per day wants to decide whether or not to take advantage of the discount. The setup cost for ordering a lot is $50 and the holding cost per unit per day is $.30. Should the company take advantage of the discount?

☐ **13–17** In Problem 13–16, determine the range on the percentage discount in the price of the item that when offered for lots of size 150 or more will not result in any financial advantage to the company.

☐ **13–18** In the deterministic model with instantaneous stock replenishment, no shortage, and constant demand rate, suppose that the holding cost per unit is given by h_1 for quantities below q and h_2 for quantities above q, $h_1 > h_2$. Find the economic lot size in this case.

☐ **13–19** Four different items are kept in store for continuous use in a manufacturing process. The demand rates are constant for the four items. Shortage is not allowed and stock may be replenished instantaneously upon request. Let D_i be the annual amount demanded for the ith item ($i = 1, 2, 3, 4$). In terms of the regular symbols introduced in the chapter, the data of the problem are given by

Item i	K_i	β_i	h_i	D_i
1	100	10	.1	10,000
2	50	20	.2	5,000
3	90	5	.2	7,500
4	20	10	.1	5,000

Find the economic lot sizes for the four products, assuming that the total number of orders per year (for the four items) cannot exceed 200 orders.

☐ **13–20** Solve Problem 13–19 assuming that there is a limit $C = \$10,000$ on the amount of capital to be invested in inventory at any time. Let c_i be the cost per unit of the ith item, where $c_i = 10, 5, 10,$ and 10, for $i = 1, 2, 3,$ and 4. Disregard the restriction on the number of orders per year.

☐ **13–21** Solve Example 13.3–5 assuming an initial inventory $x_1 = 4$.

☐ **13–22** Solve Example 13.3–5 by the backward recursive equation of dynamic programming.

☐ **13–23** Solve the following four-period deterministic inventory problem.

Period i	Demand D_i	Setup Cost K_i	Holding Cost h_i
1	5	5	1
2	7	7	1
3	11	9	1
4	3	7	1

The purchasing cost per unit is 1 for the first six units and 2 for any additional units.

□ **13–24** Solve Example 13.3–6 assuming an initial inventory $x_1 = 80$ units.

□ **13–25** Solve the following 10-period deterministic inventory problem. Assume an initial inventory of 50 units.

Period i	Demand D_i	Purchasing Cost c_i	Holding Cost h_i	Setup Cost K_i
1	150	6	1	100
2	100	6	1	100
3	20	4	2	100
4	40	4	1	200
5	70	6	2	200
6	90	8	3	200
7	130	4	1	300
8	180	4	4	300
9	140	2	2	300
10	50	6	1	300

□ **13–26** Solve the following five-period deterministic inventory problem by the modified forward algorithm.

Period i	Demand D_i	Holding Cost h_i	Setup Cost K_i
1	50	1	80
2	70	1	70
3	100	1	60
4	30	1	80
5	60	1	60

The ordering cost function specifies a per unit cost of 20 for the first 30 items and 10 for any additional units (quantity discount).

□ **13–27** Solve Problem 13–25 assuming a constant purchasing cost $c_i = 6$ for all the periods. Identify the planning horizons and the subhorizons for the problem.

□ **13–28** Show that the solution of Example 13.3–8 is optimal by showing that the optimality condition of the transportation technique (Section 5.2.3) is satisfied.

□ **13–29** Solve Example 13.3–8 if production costs for the periods are as follows:

Period	R	T
1	2	3
2	3	4
3	3	5
4	1	2

☐ **13–30** An item is manufactured to meet known demand for four periods. The following table summarizes the costs and demand requirements.

Production Range (units)	Production Cost per Unit in			
	Period 1	Period 2	Period 3	Period 4
1–3	1	2	2	3
4–11	1	4	5	4
12–15	2	4	7	5
16–25	5	6	10	7
Holding cost per unit to next period	2	5	3	—
Total demand	11	4	17	29

Find the optimal solution indicating the number of units to be produced in each of the four periods. Suppose that 10 additional units are needed in period 4. In which periods should they be produced?

☐ **13–31** The demand for a product in the next five periods may be filled by regular production, overtime production, and subcontracting. Subcontracting may be used only if overtime capacity is not sufficient. The following data give the supply and demand figures for the five periods.

Period i	Maximum Number of Supply Units			Demand
	Regular Time	Overtime	Subcontracting	
1	100	50	30	153
2	40	60	80	300
3	90	80	70	159
4	60	50	20	134
5	70	50	100	203

The production cost is the same for all periods and is given by 1, 2, and 3 per unit for regular time, overtime, and subcontracting, respectively. The holding cost from period i to period $i + 1$ is .5. A penalty cost of 2 per unit per period is incurred for late delivery. Find the optimal solution. [*Hint:* This problem requires backordering.]

☐ **13–32** Solve Example 13.4–1 assuming that the PDF of demand during lead time is given by

$$f(x) = \begin{cases} 1/50, & 0 \le x \le 50 \\ 0, & \text{otherwise} \end{cases}$$

All the other parameters remain the same as in Example 13.4–1.

☐ **13–33** Find the optimal solution for the continuous review model of Section 13.4–1, assuming that $f(x)$ is normal with mean 100 and variance 4. Assume that $D = 10,000$, $h = 2$, $p = 4$, and $K = 20$.

☐ **13–34** In Problem 13–32, suppose that

$$f(x) = \begin{cases} 1/10, & 20 \le x \le 30 \\ 0, & \text{otherwise} \end{cases}$$

with all the other parameters remaining unchanged. Compare the values of $R*$ and $y*$ in this problem with those of Problem 13–32 and interpret the result. [*Hint:* In both problems $E\{x\}$ is the same, but the variance in this problem is smaller.]

☐ **13–35** The demand for an item during a single period occurs according to an exponential distribution with mean 10. Assuming that the demand occurs instantaneously at the beginning of the period and that the per unit holding and penalty costs for the period are 1 and 3, respectively. The purchasing cost is 2 per unit. Find the optimal order quantity given an initial inventory of 2 units. What is the optimal order quantity if the initial inventory is 5 units?

☐ **13–36** For the discrete case derivation given in Section 13.4.2A, prove that at the optimal solution

$$P\{D \le y\} \ge \frac{p - c}{p + h}$$

☐ **13–37** Solve Problem 13–35 assuming that the demand occurs according to a Poisson distribution with mean 10.

☐ **13–38** The purchasing cost per unit of a product is $10 and its holding cost per unit per period is $1. If the order quantity is 4 units, find the permissible range of p under optimal conditions given the following demand PDF:

D	0	1	2	3	4	5	6	7	8
$f(D)$	.05	.1	.1	.2	.25	.15	.05	.05	.05

☐ **13–39** Suppose that in Problem 13–35 the penalty cost p cannot be estimated easily. Consequently, it is decided to determine the order quantity such that the probability of shortage is at most equal to .1. What is the order quantity in this case? Assuming that all the remaining parameters are as given in Problem 13–35, what is the implied penalty cost under optimal conditions?

☐ **13–40** Consider a one-period inventory model with zero setup cost and zero initial inventory. Let c be the ordering cost per unit and let r and v be the selling price and salvage values per unit ($v < c < r$). The demand D is described by a *discrete* PDF $f(D)$. Find the expression for the total expected *profit* as a function of the order quantity and derive the condition for selecting the optimal value. Assume zero holding and penalty costs.

☐ **13–41** Solve Problem 13–35 assuming that the demand occurs uniformly over the period.

☐ **13–42** Solve Problem 13–35 assuming that the demand occurs uniformly over the period. Derive the general condition for the optimal order quantity given discrete demand units.

☐ **13–43** Solve Problem 13–38 assuming that the demand occurs uniformly over the period.

☐ **13–44** Find the optimal ordering policy for a one-period model with instantaneous demand given that the demand occurs according to the following PDF:

$$f(D) = \begin{cases} 1/5, & 5 \le D \le 10 \\ 0, & \text{otherwise} \end{cases}$$

The cost parameters are $h = 1.0$, $p = 5.0$, and $c = 3.0$; the setup cost is $K = 5.0$. Assume an initial inventory of 10 units. What is the general ordering policy in this case?

☐ **13–45** Repeat Problem 13–44 assuming that

$$f(D) = \begin{cases} e^{-D}, & D > 0 \\ 0, & \text{otherwise} \end{cases}$$

and zero initial inventory.

☐ **13–46** In the single-period model of Section 13.4.2A, suppose instead that profit is to be maximized. Given that r is the selling price per unit and using the information in Section 13.4.2A, develop an expression for the total expected *profit* and find the optimal order quantity.

Suppose that $r = 3$, $c = 2$, $p = 4$, $h = 1$. If a setup cost $K = 10$ is included in the problem, find the optimal ordering policy given that the PDF of demand is uniform for $0 \le D \le 10$.

☐ **13–47** Consider a one-period model where it is desired to maximize the expected profit per period. The demand occurs instantaneously at the *end* of the period. Let r and v be the per unit selling price and salvage value, respectively. Using the notation of the chapter, develop the expression for the expected profit and then find the optimal solution. Assume that the unfilled demand at the end of the period is lost.

□ **13–48** Consider a *two*-period probabilistic inventory model with a backlog and zero delivery lag. Let the demand PDF be given by

$$f(D) = \begin{cases} 1/10, & 0 \le D \le 10 \\ 0, & \text{otherwise} \end{cases}$$

The cost parameters per unit are

selling price $= 2$
purchasing price $= 1$
holding cost $= .1$
penalty cost $= 3$
discount factor $= .8$

Find the optimal ordering policy that will maximize the expected profit over the two periods. Use the dynamic programming formulation.

□ **13–49** By expanding the recursive equation for the infinite horizon model in Section 13.4.3A, show that $f(x)$ is concave. Hence there exists a single critical number y^* for all periods.

□ **13–50** Consider an infinite horizon probabilistic inventory model for which the demand PDF per period is given by

$$f(D) = \begin{cases} .08D & 0 \le D \le 5 \\ 0, & \text{otherwise} \end{cases}$$

The per unit parameters are

selling price $= 10$
purchasing price $= 8$
penalty cost $= 1$
discount factor $= .9$

Find the optimal ordering policy that maximizes the expected profit given that unfilled demand is backlogged and zero delivery lag.

□ **13–51** Solve Problem 13–48 assuming no backlog.

□ **13–52** Solve Problem 13–50 assuming no backlog.

□ **13–53** Consider an infinite horizon inventory model. Rather than developing the optimal policy based on maximization of profit, it is developed based on minimization of expected costs. Using the regular symbols in the chapter, develop an expression for the expected cost and then find the optimal solution. Assume that

holding cost for x units $= hx^2$
penalty cost for x units $= px^2$

Also assume that there is no delivery lag and that all unfilled demand is backlogged.

Show that for the special case where $h = p$, the optimal solution is independent of the specific PDF of demand.

☐ **13–54** Repeat Problem 13–53 assuming no backlog of unfilled demand. In this case, however, when $h = p$, the optimal solution depends on the PDF of demand.

☐ **13–55** Consider a five-period probabilistic inventory model with backlogged demand. Given a delivery lag of three periods and that the PDF of demand per period is exponential with mean one, give a detailed procedure describing how the order quantities for periods 4 and 5 can be determined. Assume, for simplicity, that the receipts for periods 1, 2, and 3 all equal zero and that the initial inventory at the beginning of period 1 equals x_1.

☐ **13–56** Consider an infinite horizon probabilistic inventory model with the following PDF for the demand per period:

$$f(D) = \begin{cases} e^{-D}, & D > 0 \\ 0, & \text{otherwise} \end{cases}$$

If the parameters are $r = 10$, $c = 5$, $p = 15$, $h = 1$, and $\alpha = .9$, find the optimal policy assuming backlog and a two-period delivery lag.

☐ **13–57** Find the expected cost in Problem 13–53 assuming no backlog and a two-period delivery lag.

SHORT CASES

Case 13–1

A company manufactures a final product that requires the use of a single component. The company purchases the component from an outside supplier. The rate of demand for the final product is constant and is estimated at about 20 units per week. Each unit of the final product utilizes 2 units of the purchased component. The following inventory data are available.

	Purchased Component	Final Product
Setup cost per order	$80	$100
Holding cost per unit per week	2	5
Lead time in weeks	2	3

Sales lost due to the unavailability of the final product are estimated to cost the company about $8 per unit lost. On the other hand, any shortage in the purchased component is simply backordered.

How should the company schedule the production of the final product and the purchase of the component?

Case 13–2

A company is faced with the problem of running out of stock during the time interval between the placement and receipt of an order. The following data give a summary of past demands during lead time.

Lead Time Demand	Frequency of Occurrence
≥ 130	10
131–140	30
141–150	90
151–160	150
161–170	101
171–180	50
≥ 181	9

In the past, the reorder level was set at the average demand during lead time. This policy has proven to be inadequate.

Suggest an ordering policy for the company.

Case 13–3

A company deals with a seasonal item for which the monthly demand fluctuates appreciably. The demand data (in number of units) over the past five years are summarized as follows.

Month	Year 1	Year 2	Year 3	Year 4	Year 5
January	10	11	10	12	11
February	50	52	60	50	55
March	8	10	9	15	10
April	99	100	105	110	120
May	120	100	110	115	110
June	100	105	103	90	100
July	130	129	125	130	130
August	70	80	75	75	78
September	50	52	55	54	51
October	120	130	140	160	180
November	210	230	250	280	300
December	40	46	42	41	43

Because of the drastic fluctuations in demand, the inventory control manager has chosen a policy that places orders on a quarterly basis on January 1, April 1, July 1, and October 1. The lead time between placing an order and receiving it is three months. The order size is set to cover the demand for one quarter. Estimates for current year's demand are taken to be equal to last year's actual demand plus an additional 10% safety factor.

A new staff member believes that a better solution can be obtained by using the economic order quantity based on the average monthly demand for the year. She notes that the effect of fluctuations in demand can be "smoothed" out by placing orders to cover demands for consecutive months, with the size of each order approximately equal to the economic lot size. She also believes, unlike the inventory manager, that the estimates for demand should be based on the average of the preceding two years.

The company bases its inventory computations on a holding cost of $.50 per unit inventory per month. A setup cost of $55 is incurred each time a new order is placed.

Analyze the company's inventory control problem.

Chapter 14

Markovian Decision Process

This chapter presents a new application of dynamic programming to the solution of a stochastic decision process that can be described by a finite number of states. The transition probabilities between the states are described by a Markov chain. The reward structure of the process is also described by a matrix whose individual elements represent the revenue (or cost) resulting from moving from one state to another. Both the transition and revenue matrices depend on the decision alternatives available to the decision maker. The objective of the problem is to determine the optimal policy that maximizes the expected revenue of the process over a finite or infinite number of stages.

14.1 SCOPE OF THE MARKOVIAN DECISION PROBLEM—THE GARDENER EXAMPLE

In this section we introduce a simple example that will be used as a vehicle of explanation throughout the chapter. In spite of its simplicity, the example paraphrases a number of important applications in the areas of inventory, replacement, cash flow management, and regulation of water reservoir capacity.

570

An avid gardener attends a plot of land in her backyard. Every year, at the beginning of the gardening season, she uses chemical tests to check the soil's condition. Depending on the outcomes of the tests, she can classify the garden's productivity for the new season as good, fair, or poor.

Over the years, the gardener observed that current year's productivity can be assumed to depend only on last year's soil condition. She is thus able to represent the transition probabilities over a 1-year period from one productivity state to another in terms of the following Markov chain:

$$
\begin{array}{c}
\text{state of} \\
\text{the system} \\
\text{next year} \\
\overbrace{\begin{array}{ccc} 1 & 2 & 3 \end{array}}
\end{array}
$$

$$
\begin{array}{c}
\text{state of} \\
\text{the system} \\
\text{this year}
\end{array}
\begin{array}{c} 1 \\ 2 \\ 3 \end{array}
\left[
\begin{array}{ccc}
.2 & .5 & .3 \\
0 & .5 & .5 \\
0 & 0 & 1
\end{array}
\right] = \mathbf{P}^1
$$

The representation assumes the following correspondence between productivity and the states of the chain:

Productivity (Soil Condition)	State of the System
Good	1
Fair	2
Poor	3

The transition probabilities in $\mathbf{P}^1$ indicate that the productivity for a current year can be no better than last year's. For example, if the soil condition for this year is fair (state 2), next year's productivity may remain fair with probability .5 or become poor (state 3), also with probability .5.

The gardener can alter the transition probabilities $\mathbf{P}^1$ by taking other courses of action available to her. Typically, she may decide to fertilize the garden to boost the soil condition. If she does not, her transition probabilities will remain as given in $\mathbf{P}^1$. But if she does, the following transition matrix $\mathbf{P}^2$ will result:

$$
\mathbf{P}^2 =
\begin{array}{c} 1 \\ 2 \\ 3 \end{array}
\begin{array}{c}
\overset{\begin{array}{ccc} 1 & 2 & 3 \end{array}}{}
\end{array}
\left[
\begin{array}{ccc}
.3 & .6 & .1 \\
.1 & .6 & .3 \\
.05 & .4 & .55
\end{array}
\right]
$$

In the new transition matrix, $\mathbf{P}^2$, it is possible to improve the condition of soil over last year's.

To put the decision problem in perspective, the gardener associates a return function (or a reward structure) with the transition from one state to another. The return function expresses the gain or loss during a 1-year period, depending on the states between which the transition is made. Since the gardener has the options of using or not using fertilizer, her gain and losses are expected to vary depending on the decision she makes. The matrices $\mathbf{R}^1$ and $\mathbf{R}^2$ summarize the return functions in hundreds of dollars associated with the matrices $\mathbf{P}^1$ and $\mathbf{P}^2$, respectively. Thus $\mathbf{R}^1$ applies when no fertilizer is used; otherwise, $\mathbf{R}^2$ is utilized in the representation of the return function.

$$\mathbf{R}^1 = \|r_{ij}^1\| = \begin{array}{c} \\ 1 \\ 2 \\ 3 \end{array} \begin{array}{ccc} 1 & 2 & 3 \\ \left[\begin{array}{ccc} 7 & 6 & 3 \\ 0 & 5 & 1 \\ 0 & 0 & -1 \end{array}\right] \end{array}$$

$$\mathbf{R}^2 = \|r_{ij}^2\| = \begin{array}{c} \\ 1 \\ 2 \\ 3 \end{array} \begin{array}{ccc} 1 & 2 & 3 \\ \left[\begin{array}{ccc} 6 & 5 & -1 \\ 7 & 4 & 0 \\ 6 & 3 & -2 \end{array}\right] \end{array}$$

Notice that the elements r_{ij}^2 of $\mathbf{R}^2$ take into account the cost of applying the fertilizer. For example, if the system were in state 1 and remained in state 1 during next year, its gain would be $r_{11}^2 = 6$ compared to $r_{11}^1 = 7$ when no fertilizer is used.

What kind of a decision problem does the gardener have? First, we must know whether the gardening activity will continue for a limited number of years or, for all practical purposes, indefinitely. These situations are referred to as **finite-stage** and **infinite-stage** decision problems. In both cases, the gardener would need to determine the *best* course of action she should follow (fertilize or do not fertilize) given the outcome of the chemical tests (state of the system). The optimization process will be based on maximization of expected revenue.

The gardener may also be interested in evaluating the expected revenue resulting from following a prespecified course of action whenever a given state of the system occurs. For example, she may decide to fertilize whenever the soil condition is poor (state 3). The decision-making process in this case is said to be represented by a **stationary policy**.

We must note that each stationary policy will be associated with a different transition and return matrices, which, in general, can be constructed from the matrices $\mathbf{P}^1$, $\mathbf{P}^2$, $\mathbf{R}^1$, and $\mathbf{R}^2$. For example, for the stationary policy calling for applying fertilizer only when the soil condition is poor (state 3), the resulting transition and return matrices, $\mathbf{P}$ and $\mathbf{R}$, respectively, are given as

$$\mathbf{P} = \left[\begin{array}{ccc} .2 & .5 & .3 \\ 0 & .5 & .5 \\ .05 & .4 & .55 \end{array}\right], \quad \mathbf{R} = \left[\begin{array}{ccc} 7 & 6 & 3 \\ 0 & 5 & 1 \\ 6 & 3 & -2 \end{array}\right]$$

These matrices differ from $\mathbf{P}^1$ and $\mathbf{R}^1$ in the third rows only, which are taken directly from $\mathbf{P}^2$ and $\mathbf{R}^2$. The reason is that $\mathbf{P}^2$ and $\mathbf{R}^2$ are the matrices that result when fertilizer is applied in *every* state.

Exercise 14.1–1

(a) Identify the matrices $\mathbf{P}$ and $\mathbf{R}$ associated with the stationary policy calling for using fertilizer whenever the soil condition is fair or poor.
[*Ans.*

$$\mathbf{P} = \left[\begin{array}{ccc} .2 & .5 & .3 \\ .1 & .6 & .3 \\ .05 & .4 & .55 \end{array}\right], \quad \mathbf{R} = \left[\begin{array}{ccc} 7 & 6 & 3 \\ 7 & 4 & 0 \\ 6 & 3 & -2 \end{array}\right]$$

(b) Identify all the stationary policies of the gardener's example.
   [*Ans.* The stationary policies call for applying fertilizer whenever the system is in
   (1) state 1, (2) state 2, (3) state 3, (4) state 1 or 2, (5) state 1 or 3, (6) state 2 or
   3, and (7) state 1, 2, or 3.]

Notice that once *all* stationary policies are enumerated, we can apply the
proper analysis to select the *best* policy. This procedure, however, may be impractical even for moderate-size problems, since the number of policies may be
too large. What is needed is a method that determines the best policy systematically without enumerating all the policies in advance. We develop such methods
in the remainder of the chapter, for both finite- and infinite-stage problems.

14.2 FINITE-STAGE DYNAMIC PROGRAMMING MODEL

Suppose that the gardener plans to "retire" from exercising her hobby in N
years. She is thus interested in determining her optimal course of action for
each year (to fertilize or not to fertilize) over a finite planning horizon. Optimality
here is defined such that the gardener will accumulate the highest expected
revenue at the end of N years.

Let $k = 1$ and 2 represent the two courses of action (alternatives) available
to the gardener. The matrices $\mathbf{P}^k$ and $\mathbf{R}^k$ representing the transition probabilities
and reward function for alternative k were given in Section 14.1 and are summarized here for convenience.

$$\mathbf{P}^1 = \|p_{ij}^1\| = \begin{bmatrix} .2 & .5 & .3 \\ 0 & .5 & .5 \\ 0 & 0 & 1 \end{bmatrix}, \quad \mathbf{R}^1 = \|r_{ij}^1\| = \begin{bmatrix} 7 & 6 & 3 \\ 0 & 5 & 1 \\ 0 & 0 & -1 \end{bmatrix}$$

$$\mathbf{P}^2 = \|p_{ij}^2\| = \begin{bmatrix} .3 & .6 & .1 \\ .1 & .6 & .3 \\ .05 & .4 & .55 \end{bmatrix}, \quad \mathbf{R}^2 = \|r_{ij}^2\| = \begin{bmatrix} 6 & 5 & -1 \\ 7 & 4 & 0 \\ 6 & 3 & -2 \end{bmatrix}$$

Recall that the system has three states: good (state 1), fair (state 2), and poor
(state 3).

We can express the gardener's problem as a finite-stage dynamic programming
(DP) model as follows. For the sake of generalization, suppose that the number
of states for each stage (year) is m (= 3 in the gardener's example) and define

$f_n(i)$ = optimal *expected* revenue of stages $n, n + 1, \ldots, N$, given that
the state of the system (soil condition) at the beginning of year n is i

The *backward* recursive equation relating f_n and f_{n+1} can be written as (see
Figure 14–1)

$$f_n(i) = \max_k \left\{ \sum_{j=1}^m p_{ij}^k [r_{ij}^k + f_{n+1}(j)] \right\}, \quad n = 1, 2, \ldots, N$$

where $f_{N+1}(j) \equiv 0$ for all j.

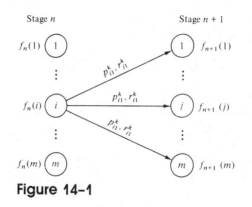

Figure 14-1

A justification for the equation is that the *cumulative* revenue, $r_{ij}^k + f_{n+1}(j)$, resulting from reaching state j at stage $n + 1$ from state i at stage n occurs with probability p_{ij}^k. In fact, if v_i^k represents the expected return resulting from a single transition from state i given alternative k, then v_i^k can be expressed as

$$v_i^k = \sum_{j=1}^{m} p_{ij}^k r_{ij}^k$$

The DP recursive equation can thus be written as

$$f_N(i) = \max_k \{v_i^k\}$$

$$f_n(i) = \max_k \left\{ v_i^k + \sum_{j=1}^{m} p_{ij}^k f_{n+1}(j) \right\}, \qquad n = 1, 2, \ldots, N - 1$$

Before showing how the recursive equation is used to solve the gardener's problem, we illustrate the computation of v_i^k, which is part of the recursive equation. For example, suppose that no fertilizer is used ($k = 1$); then

$$v_1^1 = .2 \times 7 + .5 \times 6 + .3 \times 3 = 5.3$$
$$v_2^1 = 0 \times 0 + .5 \times 5 + .5 \times 1 = 3$$
$$v_3^1 = 0 \times 0 + 0 \times 0 + 1 \times -1 = -1$$

These values show that if the soil condition is found good (state 1) at beginning of the year, a single transition is expected to yield 5.3 for that year. Similarly, if the soil condition is fair (poor), the expected revenue is 3 (-1).

Exercise 14.2-1
Compute the values of v_i^k for all i given that fertilizer is used only if the system is in states 2 or 3 (fair or poor); otherwise, no fertilizer is used.
[*Ans.* $v_1^1 = 5.3$, $v_2^2 = 3.1$, $v_3^2 = .4$.]

The following example will now solve the gardener's problem. We utilize the same format of DP tabular computations introduced in Chapter 9.

Example 14.2-1. In this example, we solve the gardener's problem using the data summarized in the matrices $\mathbf{P}^1$, $\mathbf{P}^2$, $\mathbf{R}^1$, and $\mathbf{R}^2$. It is assumed that the planning horizon comprises 3 years only ($N = 3$).

Since the values of v_i^k will be used repeatedly in the computations, they are summarized here for convenience.

i	v_i^1	v_i^2
1	5.3	4.7
2	3	3.1
3	−1	.4

Stage 3

	v_i^k		Optimal Solution	
i	$k=1$	$k=2$	$f_3(i)$	$k*$
1	5.3	4.7	5.3	1
2	3	3.1	3.1	2
3	−1	.4	.4	2

Stage 2

	$v_i^k + p_{i1}^k f_3(1) + p_{i2}^k f_3(2) + p_{i3}^k f_3(3)$		Optimal Solution	
i	$k=1$	$k=2$	$f_2(i)$	$k*$
1	$5.3 + .2 \times 5.3 + .5 \times 3.1 + .3 \times .4$ $= 8.03$	$4.7 + .3 \times 5.3 + .6 \times 3.1 + .1 \times .4$ $= 8.19$	8.19	2
2	$3 + 0 \times 5.3 + .5 \times 3.1 + .5 \times .4$ $= 4.75$	$3.1 + .1 \times 5.3 + .6 \times 3.1 + .3 \times .4$ $= 5.61$	5.61	2
3	$-1 + 0 \times 5.3 + 0 \times 3.1 + 1 \times .4$ $= -.6$	$.4 + .05 \times 5.3 + .4 \times 3.1 + .55 \times .4$ $\cong 2.13$	2.13	2

Stage 1

	$v_i^k + p_{i1}^k f_2(1) + p_{i2}^k f_2(2) + p_{i3}^k f_2(3)$		Optimal Solution	
i	$k=1$	$k=2$	$f_1(i)$	$k*$
1	$5.3 + .2 \times 8.19 + .5 \times 5.61 + .3 \times 2.13$ $\cong 10.38$	$4.7 + .3 \times 8.19 + .6 \times 5.61 + .1 \times 2.13$ $\cong 10.74$	10.74	2
2	$3 + 0 \times 8.19 + .5 \times 5.61 + .5 \times 2.13$ $= 6.87$	$3.1 + .1 \times 8.19 + .6 \times 5.61 + .3 \times 2.13$ $\cong 7.92$	7.92	2
3	$-1 + 0 \times 8.19 + 0 \times 5.61 + 1 \times 2.13$ $= 1.13$	$.4 + .05 \times 8.19 + .4 \times 5.61 + .55 \times 2.13$ $\cong 4.23$	4.23	2

The optimal solution shows that for years 1 and 2, the gardener should apply fertilizer ($k* = 2$) regardless of the state of the system (soil condition as revealed by the chemical tests). In year 3, however, she should apply fertilizer only if the system is in state 2 or 3 (fair or poor soil condition). The total expected revenues for the three years are $f_1(1) = 10.74$ if the state of the system in year 1 is good, $f_1(2) = 7.92$ if it is fair, and $f_1(3) = 4.23$ if it is poor.   ◀

The DP solution provided above is sometimes referred to as the **value-iteration** approach, because by the very nature of the recursive equation, the values of $f_n(i)$ are determined iteratively.

Exercise 14.2–2

Suppose that the gardener's planning horizon is 4 years; what are her optimal expected revenues and decisions? (*Hint:* You should be able to obtain the solution by adding one iteration only to the DP tableaus.)

[*Ans.* $f_1(1) = 13.097$, $f_1(2) = 10.195$, and $f_1(3) = 6.432$. The optimal decisions call for applying fertilizer in years 1, 2, and 3 regardless of the state of the system. In year 4, fertilizer is used when the state of the system is either fair or poor.]

The (finite horizon) gardener's problem just solved can be generalized in two ways. First, the transition probabilities and their return functions need not be the same for every year. Second, she may apply a discounting factor to the expected revenue of the successive stages so that the values of $f_1(i)$ would represent the *present value* of the expected revenues of all the stages.

The first generalization would require simply that the return values r_{ij}^k and transition probabilities p_{ij}^k be additionally functions of the stage, n. In this case the DP recursive equations appear as

$$
f_N(i) = \max_k \left\{ v_i^{k,N} \right\}
$$

$$
f_n(i) = \max_k \left\{ v_k^{k,n} + \sum_{j=1}^m p_{ij}^{k,n} f_{n+1}(j) \right\}, \qquad n = 1, 2, \ldots , N-1
$$

where

$$
v_i^{k,n} = \sum_{j=1}^m p_{ij}^{k,n} r_{ij}^{k,n}
$$

The second generalization is accomplished as follows. Let $\alpha\ (< 1)$ be the discount factor per year, which is normally computed as $\alpha = 1/(1 + t)$, where t is the annual interest rate. Thus D dollars a year from now are equivalent to αD dollars now. The introduction of the discount factor will modify the original recursive equation as follows:

$$
f_N(i) = \max_k \left\{ v_i^k \right\}
$$

$$
f_n(i) = \max_k \left\{ v_i^k + \alpha \sum_{j=1}^m p_{ij}^k f_{n+1}(j) \right\}, \qquad n = 1, 2, \ldots , N-1
$$

The application of this recursive equation is exactly similar to that given in Example 14.2–1 except that the discount factor α is multiplied by the term that includes $f_{n+1}(j)$. In general, the use of a discount factor may result in a different optimum decision in comparison with when no discount is used.

Exercise 14.2–3

Solve Example 14.2–1 given the discount factor $\alpha = .6$.

[Ans.

$f_3(1) = 5.3, \quad k* = 1; f_3(2) = 3.1, \quad k* = 2; f_3(3) = .4, k* = 2$

$f_2(1) = 6.94, k* = 1; f_2(2) = 4.61, k* = 2; f_2(3) = 1.44, k* = 2$

$f_1(1) = 7.77, k* = 1; f_1(2) = 5.43, k* = 2; f_1(3) = 2.19, k* = 2$

Note that the use of the discount factor α resulted in different optimum decisions: namely, the solution now calls for using no fertilizer in all three years if the state of the system is good.]

The DP recursive equation can be used to evaluate any *stationary policy* for the gardener's problem. Assuming that no discounting is used (i.e., $\alpha = 1$), the recursive equation for evaluating a stationary policy is

$$f_n(i) = v_i + \sum_{j=1}^{m} p_{ij} f_{n+1}(j)$$

where p_{ij} is the (i, j)th element of the transition matrix associated with the policy and v_i is the expected one-step transition revenue of the policy.

To demonstrate the use of the recursive equation above, consider the stationary policy, which calls for applying fertilizer every time the soil condition is poor (state 3). As shown in Section 14.1, we have

$$\mathbf{P} = \begin{bmatrix} .2 & .5 & .3 \\ 0 & .5 & .5 \\ .05 & .4 & .55 \end{bmatrix}, \qquad \mathbf{R} = \begin{bmatrix} 7 & 6 & 3 \\ 0 & 5 & 1 \\ 6 & 3 & -2 \end{bmatrix}$$

Thus we obtain

i	1	2	3
v_i	5.3	3	.4

and the values of $f_n(i)$ are computed as

$$f_3(1) = 5.3, \qquad f_3(2) = 3, \qquad f_3(3) = .4$$
$$f_2(1) = 5.3 + .2 \times 5.3 + .5 \times 3 + .3 \times .4 = 7.98$$
$$f_2(2) = 3 + 0 \times 5.3 + .5 \times 3 + .5 \times .4 = 4.7$$
$$f_2(3) = .4 + .05 \times 5.3 + .4 \times 3 + .55 \times .4 \cong 2.09$$
$$f_1(1) = 5.3 + .2 \times 7.98 + .5 \times 4.7 + .3 \times 2.09 \cong 9.87$$
$$f_1(2) = 3 + 0 \times 7.98 + .5 \times 4.7 + .5 \times 2.09 \cong 6.39$$
$$f_1(3) = .4 + .05 \times 7.98 + .4 \times 4.7 + .55 \times 2.09 \cong 3.83$$

The finite-stage DP model is suitable for decision problems with finite number of periods. However, many decision situations encompass either a very large number of periods or actually continue indefinitely. This makes it awkward, if not impossible, to use the finite model. The following section shows how the problem is tackled by developing an infinite-stage DP model.

14.3 INFINITE-STAGE MODEL

The long-run behavior of a Markovian process is characterized by its independence of the initial state of the system. In this case the system is said to have reached *steady state*. We are thus primarily interested in evaluating policies for which the associated Markov chains allow the existence of a steady-state solution. (Section 10.10 provides the conditions under which a Markov chain can yield steady-state probabilities.)

In this section we are interested in determining the optimum *long-run* policy of a Markovian decision problem. It is logical to base the evaluation of a policy on maximizing (minimizing) the expected revenue (cost) *per transition period*. For example, in the gardener's problem, the selection of the best (infinite-stage) policy is based on the maximum expected revenue per year.

There are two methods for solving the infinite-stage problem. The first method calls for enumerating *all* possible stationary policies of the decision problem. By evaluating each policy, the optimum solution can be determined. This is basically equivalent to an *exhaustive enumeration* process and can be used only if the total number of stationary policies is reasonably small for practical computations.

The second method, called **policy iteration,** alleviates the computational difficulties that could arise in the exhaustive enumeration procedure. The new method is generally efficient in the sense that it determines the optimum policy in a small number of iterations.

Naturally, both methods must lead to the same optimum solution. We demonstrate this point as well as the application of the two methods via the gardener example.

14.3.1 EXHAUSTIVE ENUMERATION METHOD

Suppose that the decision problem has a total of S stationary policies, and assume that $\mathbf{P}^s$ and $\mathbf{R}^s$ are the (one-step) transition and revenue matrices associated with the kth policy, $s = 1, 2, \ldots, S$. The steps of the enumeration method are as follows.

Step 1: Compute v_i^s, the *expected* one-step (one-period) revenue of policy s given state i, $i = 1, 2, \ldots, m$.

Step 2: Compute π_i^s, the long-run stationary probabilities of the transition matrix $\mathbf{P}^s$ associated with policy s. These probabilities, when they exist, are computed from the equations

$$\pi^s P^s = \pi^s$$
$$\pi^s_1 + \pi^s_2 + \cdots + \pi^s_m = 1$$

where $\pi^s = (\pi^s_1, \pi^s_2, \ldots, \pi^s_m)$.

Step 3: Determine E^s, the expected revenue of policy s per transition step (period), by using the formula

$$E^s = \sum_{i=1}^{m} \pi^s_i v^s_i$$

Step 4: The optimum policy $s*$ is determined such that

$$E^{s*} = \max_s \{E^s\}$$

We illustrate the method by solving the gardener problem for an infinite-period planning horizon.

Example 14.3-1. The gardener problem has a total of eight stationary policies, as the following table shows:

Stationary Policy s	Action
1	Do not fertilize at all.
2	Fertilize regardless of the state.
3	Fertilize whenever in state 1.
4	Fertilize whenever in state 2.
5	Fertilize whenever in state 3.
6	Fertilize whenever in state 1 or 2.
7	Fertilize whenever in state 1 or 3.
8	Fertilize whenever in state 2 or 3.

As we explained in Section 14.1, the matrices P^k and R^k for policies 3 through 8 are derived from those of policies 1 and 2. We thus have

$$P^1 = \begin{bmatrix} .2 & .5 & .3 \\ 0 & .5 & .5 \\ 0 & 0 & 1 \end{bmatrix}, \quad R^1 = \begin{bmatrix} 7 & 6 & 3 \\ 0 & 5 & 1 \\ 0 & 0 & -1 \end{bmatrix}$$

$$P^2 = \begin{bmatrix} .3 & .6 & .1 \\ .1 & .6 & .3 \\ .05 & .4 & .55 \end{bmatrix}, \quad R^2 = \begin{bmatrix} 6 & 5 & -1 \\ 7 & 4 & 0 \\ 6 & 3 & -2 \end{bmatrix}$$

$$P^3 = \begin{bmatrix} .3 & .6 & .1 \\ 0 & .5 & .5 \\ 0 & 0 & 1 \end{bmatrix}, \quad R^3 = \begin{bmatrix} 6 & 5 & -1 \\ 0 & 5 & 1 \\ 0 & 0 & -1 \end{bmatrix}$$

$$P^4 = \begin{bmatrix} .2 & .5 & .3 \\ .1 & .6 & .3 \\ 0 & 0 & 1 \end{bmatrix}, \quad R^4 = \begin{bmatrix} 7 & 6 & 3 \\ 7 & 4 & 0 \\ 0 & 0 & -1 \end{bmatrix}$$

$$P^5 = \begin{bmatrix} .2 & .5 & .3 \\ 0 & .5 & .5 \\ .05 & .4 & .55 \end{bmatrix}, \qquad R^5 = \begin{bmatrix} 7 & 6 & 3 \\ 0 & 5 & 1 \\ 6 & 3 & -2 \end{bmatrix}$$

$$P^6 = \begin{bmatrix} .3 & .6 & .1 \\ .1 & .6 & .3 \\ 0 & 0 & 1 \end{bmatrix}, \qquad R^6 = \begin{bmatrix} 6 & 5 & -1 \\ 7 & 4 & 0 \\ 0 & 0 & -1 \end{bmatrix}$$

$$P^7 = \begin{bmatrix} .3 & .6 & .1 \\ 0 & .5 & .5 \\ .05 & .4 & .55 \end{bmatrix}, \qquad R^7 = \begin{bmatrix} 6 & 5 & -1 \\ 0 & 5 & 1 \\ 6 & 3 & -2 \end{bmatrix}$$

$$P^8 = \begin{bmatrix} .2 & .5 & .3 \\ .1 & .6 & .3 \\ .05 & .4 & .55 \end{bmatrix}, \qquad R^8 = \begin{bmatrix} 7 & 6 & 3 \\ 7 & 4 & 0 \\ 6 & 3 & -2 \end{bmatrix}$$

The values of v_i^k can thus be computed as given in the following table.

	v_i^s		
s	$i = 1$	$i = 2$	$i = 3$
1	5.3	3	−1
2	4.7	3.1	.4
3	4.7	3	−1
4	5.3	3.1	−1
5	5.3	3	.4
6	4.7	3.1	−1
7	4.7	3	.4
8	5.3	3.1	.4

The computations of the stationary probabilities are achieved by using the equations

$$\pi^s P^s = \pi^s$$
$$\pi_1 + \pi_2 + \cdots + \pi_m = 1$$

As an illustration, consider $s = 2$. The associated equations are

$$.3\pi_1 + .1\pi_2 + .05\pi_3 = \pi_1$$
$$.6\pi_1 + .6\pi_2 + .4\pi_3 = \pi_2$$
$$.1\pi_1 + .3\pi_2 + .55\pi_3 = \pi_3$$
$$\pi_1 + \pi_2 + \pi_3 = 1$$

(Notice that one of the first three equations is redundant.) The solution yields

$$\pi_1^2 = 6/59, \quad \pi_2^2 = 31/59, \quad \pi_3^2 = 22/59$$

In this case, the expected yearly revenue is

$$E^2 = \sum_{i=1}^{3} \pi_i^2 v_i^2$$

$$= \frac{1}{59}(6 \times 4.7 + 31 \times 3.1 + 22 \times .4) = 2.256$$

The following table summarizes π^k and E^k for all the stationary policies. (Although this will not affect the computations in any way, note that each of policies 1, 3, 4, and 6 has an absorbing state: state 3. This is the reason $\pi_1 = \pi_2 = 0$ and $\pi_3 = 1$ for all these policies.)

s	π_1^s	π_2^s	π_3^s	E^s
1	0	0	1	−1.
2	6/59	31/59	22/59	**2.256**
3	0	0	1	.4
4	0	0	1	−1.
5	5/154	69/154	80/154	1.724
6	0	0	1	−1.
7	5/137	62/137	70/137	1.734
8	12/135	69/135	54/135	2.216

The last table shows that policy 2 yields the largest expected yearly revenue. Consequently, the optimum long-range policy calls for applying fertilizer regardless of the state of the system. ◄

Exercise 14–3–1
Verify the values of π^s and E^s given in the preceding table.

14.3.2 POLICY ITERATION METHOD WITHOUT DISCOUNTING

To gain an appreciation of the difficulty associated with the exhaustive enumeration method, let us assume that the gardener has four courses of action (alternatives) instead of two: do not fertilize, fertilize once during the season, fertilize twice, and fertilize three times. In this case, the gardener would have a total of $4^3 = 256$ stationary policies. Thus, by increasing the number of alternatives from 2 to 4, the number of stationary policies "soars" exponentially from 8 to 256. Not only is it difficult to enumerate all the policies explicitly, but the number of computations involved in the evaluation of these policies may also be prohibitively large.

The policy iteration method is based principally on the following development. For any specific policy, we showed in Section 14.1 that the expected total return at stage n is expressed by the recursive equation

$$f_n(i) = v_i + \sum_{j=1}^{m} p_{ij} f_{n+1}(j), \qquad i = 1, 2, \ldots, m$$

This recursive equation is the basis for the development of the policy iteration method. However, the present form must be modified slightly to allow us to study the asymptotic behavior of the process. In essence, we define η as the number of stages *remaining* for consideration. This is in contrast with n in the equation, which defines the nth stage. The recursive equation is thus written as

$$f_\eta(i) = v_i + \sum_{j-1}^{m} P_{ij} f_{\eta-1}(j), \qquad i = 1, 2, \ldots, m$$

Note that f_η is the cumulative expected revenue given that η is the number of stages *remaining* for consideration. With the new definition, the asymptotic behavior of the process can be studied by letting $\eta \to \infty$.

Given that

$$\boldsymbol{\pi} = (\pi_1, \pi_2, \ldots, \pi_m)$$

is the steady-state probability vector of the transition matrix $\mathbf{P} = \|p_{ij}\|$ and

$$E = \pi_1 v_1 + \pi_2 v_2 + \cdots + \pi_m v_m$$

is the expected revenue per stage as computed in Section 14.3.1, it can be shown that for very large η,

$$f_\eta(i) = \eta E + f(i)$$

where $f(i)$ is a constant term representing the asymptotic intercept of $f_\eta(i)$ given the state i.

Since $f_\eta(i)$ is the cumulative optimum return for η stages given the state i and E is the expected revenue *per stage,* we can see intuitively why $f_\eta(i)$ equals ηE plus a correction factor $f(i)$ that accounts for the specific state i. This result, of course, assumes that η is very large.

Now, using this information, the recursive equation is written as

$$\eta E + f(i) = v_i + \sum_{j=1}^{m} p_{ij} \{(\eta - 1)E + f(j)\}, \qquad i = 1, 2, \ldots, m$$

Simplifying this equation, we get

$$E = v_i + \sum_{j=1}^{m} p_{ij} f(j) - f(i), \qquad i = 1, 2, \ldots, m$$

which results in m equations and $m + 1$ unknowns, the unknowns being $f(1)$, $f(2), \ldots, f(m)$, and E.

As in Section 14.3.1, our ultimate objective is to determine the optimum policy that yields the maximum value of E. Since there are m equations in $m + 1$ unknowns, the optimum value of E cannot be determined in one step. Instead, an iterative approach is utilized which, starting with an arbitrary policy, will then determine a new policy that yields a better value of E. The iterative process ends when two successive policies are identical.

The iterative process consists of two basic components, called the **value determination** step and the **policy improvement** step.

1. *Value determination step.* Choose an arbitrary policy s. Using its associated matrices $\mathbf{P}^s$ and $\mathbf{R}^s$ and arbitrarily assuming $f^s(m) = 0$, solve the equations

$$E^s = v_i^s + \sum_{j=1}^{m} p_{ij}^s f^s(j) - f^s(i), \qquad i = 1, 2, \ldots, m$$

in the unknowns E^s, $f^s(1)$, . . . , and $f^s(m-1)$. Go to the policy improvement step.

2. *Policy improvement step.* For each state i, determine the alternative k that yields

$$\max_{k} \left\{ v_i^k + \sum_{j=1}^{m} p_{ij}^k f^s(j) \right\}, \qquad i = 1, 2, \ldots, m$$

[The values of $f^s(j)$, $j = 1, 2, \ldots, m$, are those determined in the value determination step.] The resulting optimum decisions k for states $1, 2, \ldots, m$ constitute the new policy t. If s and t are identical, stop; t is optimum. Otherwise, set $s = t$ and return to the value determination step.

The optimization problem of the policy improvement step needs clarification. Our objective in this step is to obtain max $\{E\}$. As given,

$$E = v_i + \sum_{j=1}^{m} p_{ij} f(j) - f(i)$$

Since $f(i)$ does not depend on the alternatives k, it follows that the maximization of E over the alternatives k is equivalent to the maximization problem given in the policy improvement step.

Example 14.3–2. We solve the gardener example by the policy iteration method.

Let us start with the arbitrary policy that calls for not applying fertilizer. The associated matrices are

$$\mathbf{P} = \begin{bmatrix} .2 & .5 & .3 \\ 0 & .5 & .5 \\ 0 & 0 & 1 \end{bmatrix}, \qquad \mathbf{R} = \begin{bmatrix} 7 & 6 & 3 \\ 0 & 5 & 1 \\ 0 & 0 & -1 \end{bmatrix}$$

The equations of the value iteration step are

$$\begin{aligned} E + f(1) - .2f(1) - .5f(2) - .3f(3) &= 5.3 \\ E + f(2) \qquad\quad - .5f(2) - .5f(3) &= 3 \\ E + f(3) \qquad\qquad\qquad - f(3) &= -1 \end{aligned}$$

If we arbitrarily let $f(3) = 0$, the equations yield the solution

$$E = -1, \quad f(1) \cong 12.88, \quad f(2) = 8, \quad f(3) = 0$$

Next, we apply the policy improvement step. The associated calculations are shown in the following tableau.

	$v_i^k + p_{i1}^k f(1) + p_{i2}^k f(2) + p_{i3}^k f(3)$		Optimal Solution	
i	$k = 1$	$k = 2$	$f(i)$	$k*$
1	$5.3 + .2 \times 12.88 + .5 \times 8 + .3 \times 0$ $= 11.875$	$4.7 + .3 \times 12.88 + .6 \times 8 + .1 \times 0$ $= 13.36$	13.36	2
2	$3 + 0 \times 12.88 + .5 \times 8 + .5 \times 0$ $= 7$	$3.1 + .1 \times 12.88 + .6 \times 8 + .3 \times 0$ $= 9.19$	9.19	2
3	$-1 + 0 \times 12.88 + 0 \times 8 + 1 \times 0$ $= -1$	$.4 + .05 \times 12.88 + .4 \times 8 + .55 \times 0$ $= 4.24$	4.24	2

The new policy calls for applying fertilizer regardless of the state. Since the new policy differs from the preceding one, the value determination step is entered again. The matrices associated with the new policy are

$$\mathbf{P} = \begin{bmatrix} .3 & .6 & .1 \\ .1 & .6 & .3 \\ .05 & .4 & .55 \end{bmatrix}, \qquad \mathbf{R} = \begin{bmatrix} 6 & 5 & -1 \\ 7 & 4 & 0 \\ 6 & 3 & -2 \end{bmatrix}$$

These matrices yield the following equations:

$$E + f(1) - .3\ f(1) - .6f(2) - .1\ f(3) = 4.7$$
$$E + f(2) - .1\ f(1) - .6f(2) - .3\ f(3) = 3.1$$
$$E + f(3) - .05f(1) - .4f(2) - .55f(3) = \ .4$$

Again, letting $f(3) = 0$, we get the solution

$$E = 2.26, \quad f(1) = 6.75, \quad f(2) = 3.79, \quad f(3) = 0$$

The computations of the policy improvement step are given in the following tableau.

	$v_i^k + p_{i1}^k f(1) + p_{i2}^k f(2) + p_{i3}^k f(3)$		Optimal Solution	
i	$k = 1$	$k = 2$	$f(i)$	$k*$
1	$5.3 + .2 \times 6.75 + .5 \times 3.79 + .3 \times 0$ $= 8.54$	$4.7 + .3 \times 6.75 + .6 \times 3.79 + .1 \times 0$ $= 8.99$	8.99	2
2	$3 + 0 \times 6.75 + .5 \times 3.79 + .5 \times 0$ $= 4.89$	$3.1 + .1 \times 6.75 + .6 \times 3.79 + .3 \times 0$ $= 6.05$	6.05	2
3	$-1 + 0 \times 6.75 + 0 \times 3.79 + 1 \times 0$ $= -1$	$.4 + .05 \times 6.75 + .4 \times 3.79 + .55 \times 0$ $= 2.25$	2.25	2

The new policy, which calls for applying fertilizer regardless of the state, is identical with the preceding one. Thus the last policy is optimal and the iterative process ends. Naturally, this is the same conclusion as that obtained by the exhaustive enumeration method (Section 14.3.1). Note, however, that the policy iteration method converges quickly to the optimum policy, a typical characteristic of the new method. Note that the value of E increased from -1 in the first iteration to 2.26 in the second iteration. The last value equals that obtained for the optimal policy in the exhaustive enumeration method. ◀

Exercise 14.3–2

In Example 14.3–2 we started the iterations by using the arbitrary policy of never applying fertilizer. In general, it may be better, from the viewpoint of convergence, to start with the policy whose individual decisions are those yielding $\max_k \{v_i^k\}$ for each state i. Apply these starting conditions to Example 14.3–2 and obtain the optimal policy by the policy iteration method.

[*Ans.* Starting policy consists of the decisions 1, 2, and 2 for states 1, 2, and 3. Its corresponding **P** and **R** (**P**⁸ and **R**⁸ in Example 14.3–1) used in the value determination step will yield $E = 2.216$, $f(1) = 6.207$, and $f(2) = 3.763$. The solution assumes that $f(3) = 0$. The policy improvement step will yield the decisions (2, 2, 2) with $f(1) = 8.82$, $f(2) = 5.98$, and $f(3) = 2.22$. The next iteration will be identical with that of Example 14.3–2. Notice that $E = 2.216$ in the first iteration compared with $E = -1$ in the first iteration of Example 14.3–2. This exercise demonstrates the possible influence of the starting policy on the convergence of the algorithm.]

14.3.3 POLICY ITERATION METHOD WITH DISCOUNTING

The policy iteration algorithm just described can be extended to include discounting. Specifically, given that $\alpha \, (< 1)$ is the discount factor, the finite-stage recursive equation can be written as (see Section 14.1)

$$f_\eta(i) = \max_k \left\{ v_i^k + \alpha \sum_{j=1}^m p_{ij}^k f_{\eta-1}(j) \right\}$$

(Note that η represents the number of stages *to go*.) It can be proved that as $\eta \to \infty$ (infinite stage model), $f_\eta(i) = f(i)$, where $f(i)$ is the expected present-worth (discounted) revenue given that the system is in state i and operating over an infinite horizon. Thus the long-run behavior of $f_\eta(i)$ as $\eta \to \infty$ is independent of the value of η. This is in contrast with the case of no discounting, where $f_\eta(i) = \eta E + f(i)$, as stated previously. This result should be expected since in the case of discounting the effect of future revenues will asymptotically diminish to zero. Indeed, the present worth $f(i)$ should approach a constant value as $\eta \to \infty$.

Given this information, the steps of the policy iterations are modified as follows.

1. *Value determination step..* For an arbitrary policy s with its matrices **P**ˢ and **R**ˢ, solve the m equations

$$f^s(i) = v_i^s + \alpha \sum_{j=1}^m p_{ij}^s f^s(j), \qquad i = 1, 2, \ldots, m$$

in the m unknowns $f^s(1), f^s(2), \ldots, f^s(m)$. (Note that there are m equations in exactly m unknowns.)

2. *Policy improvement step..* For each state i, determine the alternative k that yields

$$\max_k \left\{ v_i^k + \alpha \sum_{j=1}^m p_{ij}^k f^s(j) \right\}, \qquad i = 1, 2, \ldots, m$$

where $f^s(j)$ are those obtained from the value determination step. If the resulting policy t is the same as s, stop; t is optimum. Otherwise, set $s = t$ and return to the value determination step.

Example 14.3–3. We will solve Example 14.3–2 using a discounting factor $\alpha = .6$.

Starting with the arbitrary policy, $s = \{1, 1, 1\}$. The associated matrices $\mathbf{P}$ and $\mathbf{R}$ ($\mathbf{P}^1$ and $\mathbf{R}^1$ in Example 14.3–1) yield the equations

$$f(1) - .6[.2f(1) + .5f(2) + .3f(3)] = 5.3$$
$$f(2) - .6[\qquad\quad .5f(2) + .5f(3)] = 3$$
$$f(3) - .6[\qquad\qquad\qquad\quad f(3)] = -1$$

The solution of these equations yields

$$f_1 \cong 6.6, \quad f_2 \cong 3.21, \quad f_3 = -2.5$$

A summary of the policy improvement iteration is given in the following tableau.

i	\multicolumn{2}{c}{$v_i^k + p_{i1}^k f(1) + p_{i2}^k f(2) + p_{i3}^k f(3)$}	Optimal Solution	
	$k = 1$	$k = 2$	$f(i)$	$k*$
1	$5.3 + .6[.2 \times 6.6 + .5 \times 3.21 + .3 \times -2.5]$ $= 6.61$	$4.7 + .6[.3 \times 6.6 + .6 \times 3.21 + .1 \times -2.5]$ $= 6.89$	6.89	2
2	$3 + .6[0 \times 6.6 + .5 \times 3.21 + .5 \times -2.5]$ $= 3.21$	$3.1 + .6[.1 \times 6.6 + .6 \times 3.21 + .3 \times -2.5]$ $= 4.2$	4.2	2
3	$-1 + .6[0 \times 6.6 + 0 \times 3.21 + 1 \times -2.5]$ $= -2.5$	$.4 + .6[.05 \times 6.6 + .4 \times 3.21 + .55 \times -2.5]$ $= .54$	.54	2

The value determination step using $\mathbf{P}^2$ and $\mathbf{R}^2$ (Example 14.3–1) yields the following equations:

$$f(1) - .6[.3 \; f(1) + .6f(2) + .1 \; f(3)] = 4.7$$
$$f(2) - .6[.1 \; f(1) + .6f(2) + .3 \; f(3)] = 3.1$$
$$f(3) - .6[.05f(1) + .4f(2) + .55f(3)] = .4$$

The solution of these equations yields

$$f(1) = 8.88, \quad f(2) = 6.62, \quad f(3) = 3.37$$

The policy improvement step yields the following tableau.

i	\multicolumn{2}{c}{$v_i^k + p_{i1}^k f(1) + p_{i2}^k f(2) + p_{i3}^k f(3)$}	Optimal Solution	
	$k = 1$	$k = 2$	$f(i)$	$k*$
1	$5.3 + .6[.2 \times 8.88 + .5 \times 6.62 + .3 \times 3.37]$ $= 8.95$	$4.7 + .6[.3 \times 8.88 + .6 \times 6.62 + .1 \times 3.37]$ $= 8.88$	8.95	1
2	$3 + .6[0 \times 8.88 + .5 \times 6.62 + .5 \times 3.37]$ $= 5.99$	$3.1 + .6[.1 \times 8.88 + .6 \times 6.62 + .3 \times 3.37]$ $= 6.62$	6.62	2
3	$-1 + .6[0 \times 8.88 + 0 \times 6.62 + 1 \times 3.37]$ $= 1.02$	$.4 + .6[.05 \times 8.88 + .4 \times 6.62 + .55 \times 3.37]$ $= 3.37$	3.37	2

Since the new policy $\{1, 2, 2\}$ differs from the preceding one, the value determination step is entered again using $\mathbf{P}^8$ and $\mathbf{R}^8$ (Example 14.3–1). This results in the following equations:

$$f(1) - .6[.2\ f(1) + .5f(2) + .3\ f(3)] = 5.3$$
$$f(2) - .6[.1\ f(1) + .6f(2) + .3\ f(3)] = 3.1$$
$$f(3) - .6[.05f(1) + .4f(2) + .55f(3)] = .4$$

The solution of these equations yields

$$f(1) = 8.98, \quad f(2) = 6.63, \quad f(3) = 3.38$$

The policy improvement step yields the following tableau.

| | $v_i^k + p_{i\,1}^k f(1) + p_{i\,2}^k f(2) + p_{i\,3}^k f(3)$ | | Optimal Solution | |
i	$k = 1$	$k = 2$	$f(i)$	k^*
1	$5.3 + .6[.2 \times 8.98 + .5 \times 6.63 + .3 \times 3.38]$ $= 8.98$	$4.7 + .6[.3 \times 8.98 + .6 \times 6.63 + .1 \times 3.38]$ $= 8.91$	8.98	1
2	$3 + .6[0 \times 8.98 + .5 \times 6.63 + .5 \times 3.38]$ $= 6.00$	$3.1 + .6[1 \times 8.98 + .6 \times 6.63 + .3 \times 3.38]$ $= 6.63$	6.63	2
3	$-1 + .6[0 \times 8.98 + 0 \times 6.63 + 1 \times 3.38]$ $= 1.03$	$.4 + .6[.05 \times 8.98 + .4 \times 6.63 + .55 \times 3.38]$ $= 3.37$	3.37	2

Since the new policy $\{1, 2, 2\}$ is identical with the preceding one, it is optimal. Note that discounting has resulted in a different optimal policy (see Example 14.4–2), which calls for not applying fertilizer if the state of the system is good (state 3). ◀

Exercise 14.3–3
Solve Example 14.3–3 starting with the policy whose decision corresponds to $\max_k \{v_i^k\}$. Compare the resulting iterations with those of Example 14.3–3.
[*Ans.* The starting policy $\{1, 2, 2\}$ is optimal, as is verified after one iteration. The result indicates that it is important to select the starting policy judiciously rather than arbitrarily.]

14.4 LINEAR PROGRAMMING SOLUTION OF THE MARKOVIAN DECISION PROBLEM

The infinite-stage Markovian decision problems, both with discounting and without, can be formulated and solved as linear programs. We consider the no-discounting case first.

In Section 14.3.1 we showed that the infinite-stage Markovian problem with no discounting ultimately reduces to determining the optimal policy $s*$, which corresponds to

$$\max_{s \in S} \left\{ \sum_{i=1}^{m} \pi_i^s v_i^s \mid \pi^s \mathbf{P}^s = \pi^s, \quad \pi_1^s + \pi_2^s + \cdots + \pi_m^s = 1, \right.$$

$$\left. \pi_i^s \geq 0, \quad i = 1, 2, \ldots, m \right\}$$

where S is the collection of all possible policies of the problem. The constraints of the problem ensure that π_i^s, $i = 1, 2, \ldots, m$, represent the steady-state probabilities of the Markov chain $\mathbf{P}^s$.

We solved this problem in Section 14.3.1 by exhaustively enumerating all the elements s of S. Specifically, each policy s is specified by a fixed set of actions (as exemplified by the gardener problem in Example 14.3–1).

The problem is the basis for the development of the LP formulation of the Markovian decision problem. However, we need to modify the unknowns of the problem in such a manner that the optimal solution would *automatically* determine the optimal action (alternative) k when the system is in state i. The collection of all these optimal actions will then define $s*$, the optimal policy.

This objective is achieved as follows. Let

$q_i^k =$ conditional probability of choosing alternative k given that the system is in state i

The problem may thus be expressed as

$$\text{maximize } E = \sum_{i=1}^{m} \pi_i \left(\sum_{k=1}^{K} q_i^k v_i^k \right)$$

subject to

$$\pi_j = \sum_{i=1}^{m} \pi_i p_{ij}, \qquad j = 1, 2, \ldots, m$$

$$\pi_1 + \pi_2 + \cdots + \pi_m = 1$$

$$q_i^1 + q_i^2 + \cdots + q_i^K = 1, \qquad i = 1, 2, \ldots, m$$

$$\pi_i \geq 0, \quad q_i^k \geq 0, \quad \text{all } i \text{ and } k$$

Note that p_{ij} is a function of the policy selected and hence of the specific alternatives k of the policy.

We shall see shortly that the problem can be converted into a linear program by making proper substitutions involving q_i^k. Observe, however, that the formulation is equivalent to the original one in Section 14.3.1 only if $q_i^k = 1$ for exactly *one k for each i,* as this will reduce the sum $\sum_{k=1}^{K} q_i^k v_i^k$ to v_i^{k*}, where $k*$ is the optimal alternative chosen. Fortunately, the linear program we develop here does account for this condition automatically.

Define

$$w_{ik} = \pi_i q_i^k, \qquad \text{for all } i \text{ and } k$$

By definition, w_{ik} represents the *joint* probability of being in state i and making decision k. From probability theory we know that

$$\pi_i = \sum_{k=1}^{K} w_{ik}$$

Hence

$$q_i^k = \frac{w_{ik}}{\sum_{k=1}^{K} w_{ik}}$$

We thus see that the restriction $\sum_{i=1}^{m} \pi_i = 1$ can be written as

$$\sum_{i=1}^{m} \sum_{k=1}^{K} w_{ik} = 1$$

Also, the restriction $\sum_{k=1}^{K} q_i^k = 1$ is automatically implied by the way we defined q_i^k in terms of w_{ik}. (Verify!) Thus the problem can be written as

$$\text{maximize } E = \sum_{i=1}^{m} \sum_{k=1}^{K} v_i^k w_{ik}$$

subject to

$$\sum_{k=1}^{K} w_{jk} - \sum_{i=1}^{m} \sum_{k=1}^{K} p_{ij}^k w_{ik} = 0, \qquad j = 1, 2, \ldots, m$$

$$\sum_{i=1}^{m} \sum_{k=1}^{K} w_{ik} = 1$$

$$w_{ik} \geq 0, \quad i = 1, 2, \ldots, m; \ k = 1, 2, \ldots, K$$

The resulting model is a linear program in w_{ik}. We show now that its optimal solution automatically guarantees that $q_i^k = 1$ for one k for each i. First, note that the linear program has m independent equations (one of the equations associated with $\pi = \pi P$ is redundant). Hence the problem must have m basic variables. However, it can be shown that w_{ik} must be strictly positive for at least one k for each i. From these two results, we conclude that

$$q_i^k = \frac{w_{ik}}{\sum_{k=1}^{K} w_{ik}}$$

can assume a binary value (0 or 1) only, as is desired. (As a matter of fact, the result above also shows that $\pi_i = \sum_{k=1}^{K} w_{ik} = w_{ik*}$, where $k*$ is the alternative corresponding to $w_{ik} > 0$.)

Example 14.4-1. The following is an LP formulation of the gardener problem without discounting:

$$\text{maximize } E = 5.3w_{11} + 4.7w_{12} + 3w_{21} + 3.1w_{22} - w_{31} + .4w_{32}$$

subject to

$$w_{11} + w_{12} - (.2w_{11} + .3w_{12} \qquad\qquad + .1w_{22} \qquad + .05w_{32}) = 0$$
$$w_{21} + w_{22} - (.5w_{11} + .6w_{12} + .5w_{21} + .6w_{22} \qquad + .4w_{32}) = 0$$
$$w_{31} + w_{32} - (.3w_{11} + .1w_{12} + .5w_{21} + .3w_{22} + w_{31} + .55w_{32}) = 0$$
$$w_{11} + \ w_{12} + \ w_{21} + \ w_{22} + w_{31} + \ w_{32} = 1$$
$$w_{ik} \geq 0, \qquad \text{for all } i \text{ and } k$$

The optimal solution is $w_{11} = w_{12} = w_{31} = 0$ and $w_{12} = 6/59$, $w_{22} = 31/59$, and $w_{32} = 22/59$. This result means that $q_1^2 = q_2^2 = q_3^2 = 1$. Thus the optimal policy calls for selecting alternative 2 ($k = 2$) for $i = 1, 2$, and 3. The optimal value of E is $4.7(6/59) + 3.1(31/59) + .4(22/59) = 2.256$. It is interesting that the positive values of w_{ik} exactly equal the values of π_i associated with the optimal policy in the exhaustive enumeration procedure of Example 14.3–1. This observation demonstrates the direct relationship between the two solution methods.   ◄

We next consider the Markovian decision problem with discounting. In Section 14.3.2 the problem is expressed by the recursive equation

$$f(i) = \max_k \left\{ v_i^k + \alpha \sum_{j=1}^{m} p_{ij}^k f(j) \right\}, \qquad i = 1, 2, \ldots, m$$

These equations are equivalent to

$$f(i) \geq \alpha \sum_{j=1}^{m} p_{ij}^k f(j) + v_i^k, \qquad \text{for all } i \text{ and } k$$

provided that $f(i)$ achieves its minimum value for each i. Now consider the objective function

$$\text{minimize } \sum_{i=1}^{m} b_i f(i)$$

where b_i (> 0 for all i) is an arbitrary constant. It can be shown that the optimization of this function subject to the inequalities given will result in the minimum value of $f(i)$, as desired. Thus the problem can be written as

$$\text{minimize } \sum_{i=1}^{m} b_i f(i)$$

subject to

$$f(i) - \alpha \sum_{j=1}^{m} p_{ij}^k f(j) \geq v_i^k, \qquad \text{for all } i \text{ and } k$$

$$f(i) \text{ unrestricted}, \quad i = 1, 2, \ldots, m$$

Now the dual of the problem is

$$\text{maximize } \sum_{i=1}^{m} \sum_{k=1}^{K} v_i^k w_{ik}$$

subject to

$$\sum_{k=1}^{K} w_{jk} - \alpha \sum_{i=1}^{m} \sum_{k=1}^{K} p_{ij}^k w_{ik} = b_j, \qquad j = 1, 2, \ldots, m$$

$$w_{ik} \geq 0, \qquad \text{for } i = 1, 2, \ldots, m; k = 1, 2, \ldots, K$$

Notice that the objective function has the same form as in the case of no discounting, so that w_{ik} can be interpreted similarly. The following example illustrates the application of the model.

Example 14.4–2. Consider the gardener example with the discounting factor $\alpha = .6$. If we let $b_1 = b_2 = b_3 = 1$, the dual LP problem may be written as:

$$\text{maximize } 5.3w_{11} + 4.7w_{12} + 3w_{21} + 3.1w_{22} - w_{31} + .4w_{32}$$

subject to

$$w_{11} + w_{12} - .6[.2w_{11} + .3w_{12} \qquad + .1w_{22} \qquad + .05w_{32}] = 1$$
$$w_{21} + w_{22} - .6[.5w_{11} + .6w_{12} + .5w_{21} + .6w_{22} \qquad + .4w_{32}] = 1$$
$$w_{31} + w_{32} - .6[.3w_{11} + .1w_{12} + .5w_{21} + .3w_{22} + w_{31} + .55w_{32}] = 1$$
$$w_{ik} \geq 0, \qquad \text{for all } i \text{ and } k$$

The optimal solution is $w_{12} = w_{21} = w_{31} = 0$ and $w_{11} = 1.5678$, $w_{22} = 3.3528$, and $w_{32} = 2.8145$. The solution shows that the optimal policy is {1, 2, 2}, as was obtained in Example 14.3–3. ◀

14.5 SUMMARY

This chapter provides models for the solution of the Markovian decision problem. The models developed include the finite-stage models solved directly by the DP recursive equations. In the infinite-stage model, it is shown that exhaustive enumeration is not practical for large problems. The policy iteration algorithm, which is based on the DP recursive equation, is shown to be more efficient computationally than the exhaustive enumeration method, since it normally converges in a small number of iterations. Discounting is shown to result in a possible change of the optimal policy in comparison with the case where no discounting is used. This conclusion applies to both the finite- and infinite-stage models.

The LP formulation is quite interesting but is not as efficient computationally as the policy iteration algorithm. For problems with K decision alternatives and m states, the associated LP model would include $(m + 1)$ constraints and mK variables, which tend to be large for large values of m and K.

Although we presented the simplified gardener example to demonstrate the development of the algorithms, the Markovian decision problem has applications in such areas as inventory, maintenance, replacement, and water resources.

SELECTED REFERENCES

DERMAN, C., *Finite State Markovian Decision Processes,* Academic Press, New York, 1970.

HOWARD, R., *Dynamic Programming and Markov Processes,* MIT Press, Cambridge, Mass., 1960.

PROBLEMS

Section	Assigned Problems
14.2	14–1 to 14–6
14.3.1	14–7 to 14–9
14.3.2	14–10 to 14–12
14.3.3	14–13
14.4	14–14, 14–15

□ **14-1** A company reviews the state of one of its important products on an annual basis and decides whether it is successful (state 1) or unsuccessful (state 2). The company must then decide whether or not to advertize the product to further promote the sales. The matrices P_1 and P_2 given here provide the transition probabilities with and without advertisement during any year. The associated returns are given by the matrices R_1 and R_2. Find the optimal decisions over the next 3 years.

$$P_1 = \begin{bmatrix} .9 & .1 \\ .6 & .4 \end{bmatrix}, \qquad R_1 = \begin{bmatrix} 2 & -1 \\ 1 & -3 \end{bmatrix}$$

$$P_2 = \begin{bmatrix} .7 & .3 \\ .2 & .8 \end{bmatrix}, \qquad R_2 = \begin{bmatrix} 4 & 1 \\ 2 & -1 \end{bmatrix}$$

□ **14-2** A company can use advertisement through one of three media: radio, TV, or newspaper. The weekly costs of advertisement in the three media are estimated at $200, $900, and $300, respectively. The company can classify its sales volume during each week as (1) fair, (2) good, or (3) excellent. A summary of the transition probabilities associated with each advertisement medium follows.

	Radio			TV			Newspaper		
	1	2	3	1	2	3	1	2	3
1	.4	.5	.1	.7	.2	.1	.2	.5	.3
2	.1	.7	.2	.3	.6	.1	0	.7	.3
3	.1	.2	.7	.1	.7	.2	0	.2	.8

The corresponding weekly returns (in thousands of dollars) are

Radio			TV			Newspaper		
400	520	600	1000	1300	1600	400	530	710
300	400	700	800	1000	1700	350	450	800
200	250	500	600	700	1100	250	400	650

Find the optimal advertisement policy over the next 3 weeks.

□ **14-3** A company is introducing a new product into the market. If the sales are high, there is a .5 probability that they will remain so next month. If they are not, the probability that they will become high next month is only .2. The company has the option of launching an advertisement campaign. If it does and the sales are high, the probability that they will remain high next month will increase to .8. On the other hand, an advertising campaign while the sales are low will raise the probability to only .4.

If no advertisement is used and the sales are high, the returns are expected to be 10 if the sales remain high next month and 4 if they do not. The corresponding returns if the product starts with high sales are 7 and −2. Using advertisement will result in returns of 7 if the product starts with high sales and continues to be so and 6 if it does not. If the sales start low, the returns are 3 and −5, depending on whether or not they remain high.

Determine the company's optimal policy over the next 3 months.

☐ **14–4** (Inventory Problem) An appliance store can place orders for refrigerators at the beginning of each month for immediate delivery. A fixed cost of $100 is incurred every time an order is placed. The storage cost per refrigerator per month is $5. The penalty for running out of stock is estimated at $150 per refrigerator per month. The monthly demand is given by the following PDF:

Demand x	0	1	2
$p(x)$	.2	.5	.3

The store's policy is that the maximum stock level should not exceed two refrigerators in any single month.
  (a) Determine the transition probabilities for the different decision alternatives of the problem.
  (b) Determine the expected inventory cost per month as a function of the state of the system and the decision alternative.
  (c) Determine the optimal ordering policy over the next 3 months.

☐ **14–5** Repeat Problem 14–4 assuming that the PDF of demand over the next quarter changes according to the following table.

Demand	Month		
x	1	2	3
0	.1	.3	.2
1	.4	.5	.4
2	.5	.2	.4

☐ **14–6** The market value of a used car is estimated at $2000. The owner believes that he can get more than this amount but is willing to entertain offers from the first three prospective buyers who respond to his advertisement (which means that he must make his decision no later than by the time he receives the third offer). The offers are expected to be $2000, $2200, $2400, and $2600, with equal probabilities. Naturally, once he accepts an offer, all later offers are automatically of no use to him. His objective is to set an acceptance limit that he can use as he receives each of the three offers. These limits may thus be $2000, $2200, $2400, or $2600. Develop an optimal plan for the owner.

☐ **14–7** Solve Problem 14–1 for an infinite number of periods using the exhaustive enumeration method.

☐ **14–8** Solve Problem 14–2 for an infinite planning horizon using the exhaustive enumeration method.

☐ **14–9** Solve Problem 14–3 by the exhaustive enumeration method assuming an infinite horizon.

☐ **14–10** Assume in Problem 14–1 that the planning horizon is infinite. Solve the problem by the policy iteration method.

☐ **14–11** Solve Problem 14–2 by the policy iteration method, assuming an infinite planning horizon. Compare the results with those of Problem 14–8.

☐ **14–12** Solve Problem 14–3 by the policy iteration method assuming an infinite planning horizon, and compare the results with those of Problem 14–9.

☐ **14–13** Repeat the problems listed, assuming a discount factor $\alpha = .9$.
(a) Problem 14–10
(b) Problem 14–11
(c) Problem 14–12

☐ **14–14** Formulate the following problems as linear programs.
(a) Problem 14–10
(b) Problem 14–11
(c) Problem 14–12

☐ **14–15** Formulate the problems in Problem 14–13 as linear programs.

Chapter 15

Queueing Theory (with Miniapplications)

Imagine the following situations:

1. Shoppers waiting in front of checkout stands in a supermarket.
2. Cars waiting at a stoplight.
3. Patients waiting at an outpatient clinic.
4. Planes waiting for takeoff in an airport.
5. Broken machines waiting to be serviced by a repairman.
6. Letters waiting to be typed by a secretary.
7. Programs waiting to be processed by a digital computer.

What these situations have in common is the phenomenon of waiting. It would be most convenient if we could be offered these services, and others like it, without the "nuisance" of having to wait. But like it or not, waiting is part of our daily life, and all we should hope to achieve is to reduce its inconvenience to bearable levels.

The waiting phenomenon is the direct result of *randomness* in the operation of service facilities. In general, the customer's arrival and his or her service time are not known in advance; for otherwise the operation of the facility could be scheduled in a manner that would eliminate waiting completely.

Our objective in studying the operation of a service facility under random conditions is to secure some characteristics that measure the performance of the system under study. For example, a logical measure of performance is how long a customer is expected to wait before being serviced. Another measure is the percentage of time the service facility is not used. The first measure looks at the system from the customer's standpoint, whereas the second measure evaluates the degree of utilization of the facility. We can intuitively see that the larger the customer's waiting time, the smaller is the percentage of time the facility would remain idle, and vice versa. These measures of performance may thus be used to select the level of service (or service rate) that will strike a reasonable balance between the two conflicting situations.

This chapter discusses a number of queueing or waiting line models that account for a variety of service operations. The models developed are basically applications of probability theory and stochastic processes. The ultimate objective of solving these models is to determine the characteristics that measure the performance of the system. We then show in Chapter 16 how this information can be used in seeking an "optimal" design for the service facility.

15.1 BASIC ELEMENTS OF THE QUEUEING MODEL

From the standpoint of a queueing model, a waiting line situation is created in the following manner. As the customer arrives at the facility, he joins a waiting line (or a queue). The server chooses a customer from the waiting line to begin service. Upon the completion of a service, the process of choosing a new (waiting) customer is repeated. It is assumed that no time is lost between the release of a serviced customer from the facility and the admission of a new one from the waiting line.

The principal actors in a queueing situation are the **customer** and the **server.** In queueing models, the interaction between the customer and the server are of interest only in as far as it relates to the *period of time* the customer needs to complete his service. Thus, from the standpoint of customer arrivals, we are interested in the time intervals that separate *successive* arrivals. Also, in the case of service, it is the service time per customer that counts in the analysis.

In queueing models, customer arrivals and service times are summarized in terms of probability distributions normally referred to as **arrivals** and **service time distributions.** These distributions may represent situations where customers arrive and are served *individually* (e.g., banks or supermarkets). In other situations, customers may arrive and/or be served in groups (e.g., restaurants). The latter case is normally referred to as **bulk queues.**

Although the patterns of arrivals and departures are the main factors in the analysis of queues, other factors also figure importantly in the development of the models. The first factor is the manner of choosing customers from the waiting line to start service. This is referred to as the **service discipline.** The most common, and apparently fair, discipline is the FCFS rule (first come, first served). LCFS (last come, first served) and SIRO (service in random order) may also arise in practical situations. We must also add that whereas service discipline regulates the selection of customers from a waiting line, it is also possible that customers arriving at a facility may be put in **priority queues** such that those with a higher priority will receive preference to start service first. The specific selection of customers from each priority queue may, however, follow any service discipline.

The second factor deals with the design of the facility and the execution of service. The facility may include more than one server, thus allowing as many customers as the number of servers to be serviced simultaneously (e.g., bank tellers). In this case, all servers offer the same service and the facility is said to have **parallel servers.** On the other hand, the facility may comprise a number of series stations through which the customer may pass before service is completed (e.g., processing of a product on a sequence of machines). In this case, waiting lines may or may not be allowed between the stations. The resulting situations are normally known as **queues in series** or **tandem queues.** The most general design of a service facility includes both series and parallel processing station. This results in what we call **network queues.**

The third factor concerns admissible **queue size.** In certain situations, only a limited number of customers may be allowed, possibly because of space limitation (e.g., car spaces allowed in a drive-in bank). Once the queue fills to capacity, newly arriving customers are denied service and may not join the queue.

The fourth factor deals with the nature of the source from which calls for service (arrivals of customers) are generated. The **calling source** may be capable of generating a finite number of customers or (theoretically) infinitely many customers. A finite source exists when an arrival affects the rate of arrival of new customers. In a machine shop with a total of M machines, the calling source before any machine breaks down consists of M potential customers.

Once a machine is broken, it becomes a customer and hence incapable of generating new calls until it is repaired. A distinction must be made between the machine shop situation and others where the "cause" for generating calls is limited, yet capable of generating infinity of arrivals. For example, in a typing pool, the number of users is finite, yet each user could generate a limitless number of arrivals, since a user generally need not wait for the completion of previously submitted material before generating new ones.

Queueing models representing situations in which human beings take the roles of customers and/or servers must be designed to account for the effect of **human behavior**. A "human" server may speed up the rate of service when the waiting line builds up in size. A "human" customer may **jockey** from one waiting line to another in hopes of reducing his or her waiting time (the next time you are in a bank or a supermarket you may kill your waiting time by observing this jockeying phenomenon). Some "human" customers also may **balk** from joining a waiting line all together because they *anticipate* a long delay, or they may **renege** *after* being in the queue for a while because their wait has been too long. (Note that in terms of human behavior, a long wait for one person may not be as long as for another.)

Undoubtedly, there are other traits of human behavior that exist in everyday queueing situations. Yet from the standpoint of the queueing *model,* these traits can be accounted for only if they can be quantified in a manner that allows their mathematical inclusion in the model. Also, queueing models cannot account for the *individual* behavior of customers in the sense that all customers in a queue are expected to "behave" equally while in the facility. Thus a "chatty" customer is considered an odd case and his or her behavior is ignored in the design of the system. On the other hand, if the majority of customers happen to be unduly talkative, a *realistic* design of the service facility must be based on the fact that this habit, wasteful as it may be, is an integral part of the operation. A logical way for including the effect of this habit is to increase the service time per customer.

We now see that the basic elements of a queueing model depend on the following factors:

1. Arrivals distribution (single or bulk arrivals).
2. Service-time distributions (single or bulk service).
3. Design of service facility (series, parallel, or network stations).
4. Service discipline (FCFS, LCFS, SIRO) and service priority.
5. Queue size (finite or infinite).
6. Calling source (finite or infinite).
7. Human behavior (jockeying, balking, and reneging).

There are as many queueing models as there are variations on the factors listed. In this chapter we consider a number of models that appear useful in practical applications. The next section shows that the Poisson and exponential distributions play an important role in representing the arrivals and service times in many queueing situations. The succeeding sections then present the selected queueing models and their solutions.

15.2   ROLES OF THE POISSON AND EXPONENTIAL DISTRIBUTIONS

Consider the queueing situation in which the number of arrivals and departures (those served) during an interval of time is controlled by the following conditions.

Condition 1: The probability of an event (arrival or departure) occurring between times t and $t + h$ depends *only* on the length of h, meaning that the probability does not depend on either the number of events that occur up to time t or the specific value of t. (Technically, we say that the probability function has **stationary independent** increments.)

Condition 2: The probability of an event occurring during a very small time interval h is positive but less than 1.

Condition 3: At most one event can occur during a very small time interval h.

The implication of these conditions can be studied by deriving mathematically the probability of n events occurring during a time interval t. Let

$$P_n(t) = \text{probability of } n \text{ events occurring during time } t$$

Condition 1 specifies that $p_n(t)$ has *stationary independent* increments. For $n = 0$, this condition is translated as

$$p_0(t + h) = p_0(t)p_0(h)$$

By condition 2, we have $0 < p_0(h) < 1$ for very small h. It can be shown [see pp. 121–123 of Parzen (1962)] that the solution to the equation is

$$p_0(t) = e^{-\alpha t}, \qquad t \geq 0$$

where α is a *positive* constant.

We show below that α represents the *rate* of arrivals (departures) per unit time when the events represent arrivals (departures). At this point however, we concentrate on showing the significance of the result from the viewpoint of the three conditions we cited.

For $h > 0$ and sufficiently small, we have

$$p_0(h) = e^{-\alpha h}$$
$$= 1 - \alpha h + \frac{(\alpha h)^2}{2!} - \frac{(\alpha h)^3}{3!} + \cdots$$
$$\cong 1 - \alpha h$$

Since condition 3 allows the occurrence of at most one event, it follows that

$$p_1(h) = 1 - p_0(h) \cong \alpha h$$

This result means that the probability of an event occurring during a small interval h is directly proportional to h.

The process described by $p_n(t)$ is completely random in the sense that the time interval remaining until the next event occurs is completely independent of the time that elapsed since the occurrence of the immediately preceding event. Let

$f(t)$ = probability density function (PDF) of the time interval t between the occurrence of *successive events, $t \geq 0$*

$F(t)$ = cumulative density function (CDF) of t

$$= \int_0^t f(x)dx$$

In terms of a probability statement, if T is the time interval since the occurrence of the last event, then we have

$$P\left\{\begin{matrix}\text{interevent time}\\ \text{is not less than } T\end{matrix}\right\} = P\left\{\begin{matrix}\text{no events occur}\\ \text{during } T\end{matrix}\right\}$$

Translating this mathematically, we see that

$$P\{t \geq T\} = p_0(T)$$

Since $f(t)$ is the PDF of t and $p_0(T) = e^{-\alpha T}$, we have

$$\int_T^\infty f(t)dt = e^{-\alpha T}$$

or using the definition of $F(T)$, we have

$$1 - F(T) = e^{-\alpha T}, \qquad T > 0$$

Differentiating both sides with respect to T, we obtain

$$f(T) = \alpha e^{-\alpha T}, \qquad T > 0$$

which is an **exponential distribution.**

The result yields two conclusions:

1. For the process described by the probabilities $p_n(t)$, the time between the occurrence of successive events must follow an exponential distribution.

2. The expected value of the exponential distribution

$$E\{T\} = \frac{1}{\alpha} \text{ time units}$$

represents the average time interval between successive occurrences of events. Thus

$$\frac{1}{E\{T\}} = \alpha \text{ events/unit time}$$

must represent the *rate* (per unit time) at which events are generated. This is the reason we indicated earlier that α represents the arrival (departure) rate when the generated events represent arrivals (departures).

3. The exponential distribution has the unique property that the time until the next event occurs is independent of the time that elapsed since the occurrence of the last event. This result is equivalent to stating that

$$P\{t > T + S \mid t > S\} = P\{t > T\}$$

where t is the random variable describing the interevent time and S is the occurrence time of the last event. To show that this probability is true for the exponential distribution, consider

$$P\{t > T + S \mid t > S\} = \frac{P\{t > T + S,\, t > S\}}{P\{t > S\}} = \frac{P\{t > T + S\}}{P\{t > S\}}$$

$$= \frac{e^{-\alpha(T+S)}}{e^{-\alpha S}} = e^{-\alpha T}$$

$$= P\{t > T\}$$

This property is usually referred to as **forgetfulness** or **lack of memory** of the exponential distribution. The forgetfulness property demonstrates why the process described by $p_n(t)$ is **completely random,** as it shows that the time that has passed since the occurrence of the last event has no effect on the time remaining until the next event occurs.

We defined $p_n(t)$ as the probability of n events occurring during the time interval t and showed that under the three stipulated conditions the *interevent* time $T > 0$ follows the exponential distribution $f(T) = \alpha e^{-\alpha T}$, where α is the *rate* at which the events are generated. We show next how the distribution $p_n(t)$ can be derived when the events represent pure arrivals or pure departures. The derivation will show that the distribution of n during t is Poisson. This point reveals the strong result that whereas interarrival (interdeparture) *time* is exponential, the *number* of arrivals (departures) is Poisson, and vice versa (see also Figure 10–9).

15.2.1 ARRIVALS PROCESS

In terms of the foregoing discussion, events in this situation represent pure arrivals, meaning that customers join the system and never leave. The process is given the suggestive name **pure birth.** Such a process may be illustrated by a *state health department* computerizing all birth records of new babies effective a given date. Birth information for each baby is stored permanently in the computer memory.

Our objective is to derive an expression for the probability $p_n(t)$ of *n arrivals* during time interval t. The derivation is based on the *complete randomness* conditions given earlier. For $h > 0$ and very small, we have, for $n > 0$,

$$p_n(t+h) = P \left\{ \begin{array}{c} n \text{ arrivals during } t \text{ and none during } h \\ or \\ n-1 \text{ arrivals during } t \text{ and one during } h \end{array} \right\}$$

Thus

$$p_n(t+h) = p_n(t)p_0(h) + p_{n-1}(t)p_1(h), \qquad n = 1, 2, \ldots$$
$$p_0(t+h) = p_0(t)p_0(h), \qquad n = 0$$

To use the results developed earlier for $p_0(h)$ and $p_1(h)$, we replace α, the rate of event generation, by the standard notation λ representing the *arrival rate*. Thus, as we indicated previously,

$$p_0(h) = e^{-\lambda h}, \qquad p_1(h) = 1 - p_0(h)$$

Since h is sufficiently small, we have

$$p_0(h) \cong 1 - \lambda h, \qquad p_1(h) \cong \lambda h$$

The preceding equations may thus be written as

$$p_n(t+h) \cong p_n(t)(1 - \lambda h) + p_{n-1}(t)\lambda h, \qquad n > 0$$
$$p_0(t+h) \cong p_0(t)(1 - \lambda h), \qquad n = 0$$

These equations may be written as

$$\frac{p_n(t+h) - p_n(t)}{h} \cong -\lambda p_n(t) + \lambda p_{n-1}(t)$$

$$\frac{p_0(t+h) - p_0(t)}{h} \cong -\lambda p_0(t)$$

Mathematically, as $h \to 0$, the equations reduce to

$$p_n'(t) = -\lambda p_n(t) + \lambda p_{n-1}(t)$$
$$p_0'(t) = -\lambda p_0(t)$$

In this chapter's appendix we show that the solution of these difference–differential equations is given by

$$p_n(t) = \frac{(\lambda t)^n e^{-\lambda t}}{n!}, \qquad n = 0, 1, 2, \ldots$$

The distribution of $p_n(t)$ is thus **Poisson** with mean and variance equal to λt. (Among all commonly known *discrete* distributions, the Poisson probability density function has the unique property of equal mean and variance.)

The conclusion from the development described is as follows: Whereas the *interarrival time* is exponential with mean $1/\lambda$, the *number of arrivals* during time interval t is Poisson with mean λt.

Exercise 15.2-1
Prove that the mean and variance of $p_n(t)$ equal λt.

Example 15.2-1. Consider the *state health department* example. Suppose that births in the state are spaced over time according to an exponential distribution and that the average time between successive births is 2 hours.

To analyze this situation, we realize that since the interbirth (interarrival) time is 2 hours, we have

$$\lambda = \frac{24}{2} = 12 \text{ births per day}$$

Suppose that the department is interested in estimating the size of the permanent computer storage needed annually. This will be equivalent to determining the average number of births per year; that is,

$$\lambda t = 12 \times 365 = 4380 \text{ records per year}$$

Another piece of information that may be of interest is the percentage of time the clerk in charge of storing the received records on the computer remains idle in a day period. This can happen only if no birth records arrive at the computer; that is,

$$p_0(1) = \frac{(12 \times 1)^0 e^{-12 \times 1}}{0!} = e^{-12} = .000006$$

Thus the chance that the clerk will remain completely idle during the day is very negligible.

We may also be interested in the probability of storing 100 records by the end of the third day given that 80 records are already in storage by the end of the second day. We observe that it is equivalent to receiving $100 - 80 = 20$ records during a period of $3 - 2 = 1$ day; that is,

$$p_{20}(1) = \frac{(12 \times 1)^{20} e^{-12 \times 1}}{20!} = .00968$$

The computation of $p_{20}(1)$ is based on the following conditional probability statement, which in turn is based on the complete randomness property of the Poisson stating that $p_n(t)$ has stationary independent increments.

$$P\{n = 100 \text{ in } t = 3 | n = 80 \text{ in } t = 2\} = \frac{P\{n = 100 \text{ in } t = 3 \text{ and } n = 80 \text{ in } t = 2\}}{P\{n = 80 \text{ in } t = 2\}}$$

$$= \frac{P\{n = 80 \text{ in } t = 2\} P\{n = 20 \text{ in } t = 1\}}{P\{n = 80 \text{ in } t = 2\}}$$

$$= P\{n = 20 \text{ in } t = 1\}$$

$$= p_{20}(1) \qquad \blacktriangleleft$$

Exercise 15.2–2†

In Example 15.2–1, suppose that the clerk waits until at least five records accumulate before feeding the information to the computer. What is the probability that he will be feeding a new batch to the computer every 4 hours?
[*Ans.* $p_{n \geq 5}(1/6) = .05265$]

---

† The general computer program in Appendix C is designed to compute the Poisson probabilities as well as the basic results of most Poisson queueing models. You may find it convenient to use this program in solving the exercises and the numerical problems of this chapter.

15.2.2 DEPARTURES PROCESS

The departures process assumes that the system starts with a given number of customers N who leave the facility at the rate μ after being serviced. No new customers are allowed to join the system. The process is given the suggestive name **pure death.**

Inventory situations can be modeled by the pure death process where a stock of N inventory items is made available at the beginning of a planning period. The inventory items are withdrawn from stock at the rate μ units per unit time.

Using the general results developed earlier, we find that the occurrence of events now represents departures at the rate $\alpha = \mu$. Let

$q_n(t) =$ probability of n *departures* during t

As in the case of *pure birth,* under the conditions of complete randomness and for very small $h > 0$, we have

$$q_0(h) = e^{-\mu h} \cong 1 - \mu h$$
$$q_1(h) = 1 - q_0(h) \cong \mu h$$

The equations representing $q_n(t+h)$ may thus be written as

$$q_N(t+h) \cong q_N(t) \cdot 1 + q_{N-1}(t)\mu h, \qquad n = N$$
$$q_n(t+h) \cong q_n(t)(1 - \mu h) + q_{n-1}(t)\mu h, \qquad 1 \le n < N$$
$$q_0(t+h) \cong q_0(t)(1 - \mu h), \qquad n = 0$$

Note in the first equation that if N (or *all*) customers had departed during t, the probability of no departure during the succeeding interval h is certain ($= 1$).

Simplifying the equations and taking limits as $h \to 0$, we get

$$q_N'(t) = \mu q_{N-1}(t), \qquad n = N$$
$$q_n'(t) = -\mu q_n(t) + \mu q_{n-1}(t), \qquad 1 \le n < N$$
$$q_0'(t) = -\mu q_0(t), \qquad n = 0$$

In this chapter's appendix, we show that the solution to these equations is

$$q_n(t) = \frac{(\mu t)^n e^{-\mu t}}{n!}, \qquad n = 0, 1, 2, \ldots, N-1$$

$$q_N(t) = 1 - \sum_{n=0}^{N-1} q_n(t), \qquad n = N$$

Note that $q_n(t)$ represents the probability of n *departures* during t. It is sometimes useful to speak of the probability of n customers *remaining* after a time interval t. To obtain this probability, define

$p_n(t) =$ probability of n customers *remaining* after t

Since the system starts initially with N customers, having n customers remaining after t is equivalent to $N - n$ customers departing during t; that is,

$$p_n(t) = q_{N-n}(t)$$

This yields

$$p_n(t) = \frac{(\mu t)^{N-n} e^{-\mu t}}{(N-n)!}, \qquad n = 1, 2, \ldots, N$$

$$p_0(t) = 1 - \sum_{n=1}^{N} p_n(t)$$

Example 15.2-2. At the beginning of each week, 15 units of an inventory item are stocked for use during the week. Withdrawals from stock occur only during the first 6 days (business is closed on Sundays) and follows a Poisson distribution with mean 3 units/day. When the stock level reaches 5 units, a new order of 15 units is placed for delivery at the beginning of next week. Because of the nature of the item, all units left at the end of the week are discarded.

We can analyze this situation in a number of ways. First, we recognize that the consumption rate is $\mu = 3$ units per day. Suppose that we are interested in computing the probability of having 5 units (the reorder level) on day t; that is,

$$p_5(t) = \frac{(3t)^{15-5} e^{-3t}}{(15-5)!}, \qquad t = 1, 2, \ldots, 6$$

As an illustration of the computations, we have

t (days)	1	2	3	4	5	6
μt	3	6	9	12	15	18
$p_5(t)$	.0008	.0413	.1186	.1048	.0486	.015

Note that $p_5(t)$ represents the probability of reordering *on* day t. This probability peaks at $t = 3$ and then declines as we advance through the week. If we are interested in the probability of reordering *by* day t, we must compute the probability of having 5 units *or* less on day t; that is,

$$
\begin{aligned}
p_{n \leq 5}(t) &= p_0(t) + p_1(t) + \cdots + p_5(t) \\
&= 1 - [p_1(t) + p_2(t) + \cdots + p_{15}(t)] + p_1(t) \\
&\quad + \cdots + p_5(t) \\
&= 1 - [p_6(t) + \cdots + p_{15}(t)] \\
&= 1 - \sum_{n=6}^{15} \frac{(3t)^{15-n} e^{-3t}}{(15-n)!}
\end{aligned}
$$

The formula yields the following table. (You should verify at least one of these computations to develop an appreciation for the use of the formula.)

t (days)	1	2	3	4	5	6
μt	3	6	9	12	15	18
$p_{n \leq 5}(t)$	.0012	.0839	.4126	.7576	.9303	.9847

We can see from the table that the probability of placing the order *by* day t increases monotonically with t.

Another piece of information that is impcrtant in analyzing the situation is determining the average number of inventory units that will be discarded at the end of the week. This is done by computing the expected number of units available on day 6; that is,

$$E\{n|t=6\} = \sum_{n=0}^{15} np_n(6)$$

The following table summarizes the computations given $\mu t = 18$.

n	0	1	2	3	4	5	6	7	8	9	10	11
$p_n(6)$	.792	.0655	.0509	.0368	.0245	.015	.0083	.0042	.0018	.0007	.0002	.0001

and $p_n(6) \cong 0$ for $n = 12, 13, 14,$ and 15. Thus, computing the average, we get

$$E\{n|t=6\} = .5537 \text{ unit}$$

This means that, on the average, less than 1 unit will be discarded at the end of each week. ◀

Exercise 15.2–3

In Example 15.2–2, determine
(a) The probability that the stock is depleted after 3 days.
   [*Ans.* $p_0(3) = .04147.$]
(b) The probability that an inventory unit will be withdrawn by the end of the fourth day given that the last unit was withdrawn at the end of the third day.
   [*Ans.* $P\{\text{time between withdrawals} \le 1\} = .9502.$]
(c) The probability that the time remaining until the next withdrawal is at most 1 day given that the last withdrawal occurs a day earlier.
   [*Ans.* Same as in part (b).]
(d) The average inventory held in stock at the end of the second day.
   [*Ans.* $E\{n|t=2\} = 9.0011$ units.]
(e) The probability that no withdrawals occur during the first day.
   [*Ans.* $q_0(1) = .0498.$]

15.2.3 HOW TO RECOGNIZE A POISSON DISTRIBUTION IN PRACTICE

The Poisson distribution plays an important role in the development of queueing models because it describes many real-life situations. Although we have shown the mathematical conditions under which the Poisson distribution applies, the presentation is still abstract. What we need is to translate these conditions into practical rules that can be used to recognize whether arrivals and/or departures follow a Poisson process.

Naturally, statistical methods exist that are designed to test the hypothesis that a given set of data follows a certain probability distribution. The best known of these is the **chi-square test of goodness of fit** (see Section 16.2). It is based on a comparison between observed and theoretical data, with the theoretical data derived from the theoretical distribution being tested. Although the details of the method will be given in Chapter 16 as part of applications of waiting line models, we wish to present here two *crude* rules that can give us an idea about whether the arrivals or departures of a real-life situation follow the Poisson distribution.

1. If the queueing situation is already in existence, observe the operation for a while. Do successive arrivals (departures) appear to occur *randomly,* or is there a pattern of arrivals (departures)? If they are random, there is a good chance that the process may follow the Poisson distribution.

2. Gather observations about the *number* of arrivals (departures) of customers by recording the number of customers arriving (departing) during appropriate equal time intervals (e.g., hourly). After gathering a "sufficient" amount of data, compute the mean and variance. If the distribution is Poisson, its sample mean and variance will be "approximately" equal (barring, of course, sampling error). This is a unique property of the Poisson among all commonly known *discrete* distributions.

We illustrate the second rule by the following example. Keep in mind that the rule is at best crude and may only be used as a quick way of forming an idea about the nature of the distribution. (The use of the *goodness of fit* test is illustrated in Example 16.2–1.)

Example 15.2–3. Consider data gathered that represent the number of arrivals n per hour as summarized below.

n	0	1	2	3	4	5	6
Frequency f_n	10	31	40	20	10	4	6

The data indicate that during the observation period, 0 arrivals per hour was observed 10 times, 1 arrival 31 times, 2 arrivals 40 times, and so on.

Let $\bar{n}$ and S_n^2 be the mean and variance of n; then given $N = \Sigma_{n=0}^6 f_n$, we have

$$\bar{n} = \left(\sum_{n=0}^{6} n f_n \right) / N$$

$$= \frac{0 \times 10 + 1 \times 31 + 2 \times 40 + 3 \times 20 + 4 \times 10 + 5 \times 4 + 6 \times 6}{10 + 31 + 40 + 20 + 10 + 4 + 6}$$

$$= 2.207 \text{ arrivals per hour}$$

$$S_n^2 = \left(\sum_{n=0}^{6} n^2 f_n - N \bar{n}^2 \right) / (N - 1)$$

$$= \frac{847 - 589.37}{120} = 2.147$$

Since $\bar{n} \cong S_n^2$, there is a "good" chance that the process of arrivals follows a Poisson distribution with mean 2.2 arrivals per hour. The obvious next step is to strengthen this conclusion by carrying out the *goodness of fit* test. ◄

15.3   QUEUES WITH COMBINED ARRIVALS AND DEPARTURES

In this section we study queueing situations in which both arrivals and departures (those serviced) take place simultaneously. We restrict our attention to waiting lines where customers are served by *c parallel* servers so that *c* customers can be serviced simultaneously. All servers offer equal services from the viewpoint of the time it takes to service each customer. Figure 15–1 represents the parallel queueing system schematically. Note that the number of customers in the *system* at any point in time is defined to include those in *queue* and in *service*.

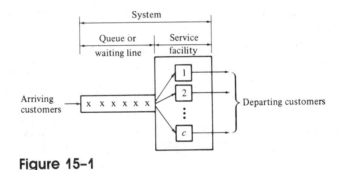

Figure 15–1

A notation that is particularly suited for summarizing the main characteristics of *parallel* queues has been universally standardized in the following format,

$$(a/b/c):(d/e/f)$$

where the symbols *a, b, c, d, e,* and *f* stand for basic elements of the model as follows (see Section 15.1).

    $a \equiv$ arrivals distribution
    $b \equiv$ service time (or departures) distribution
    $c \equiv$ number of parallel servers ($c = 1, 2, \ldots, \infty$)
    $d \equiv$ service discipline (e.g., FCFS, LCFS, SIRO)
    $e \equiv$ maximum number allowed in *system* (in queue + in service)
    $f \equiv$ size of calling source

The standard notation replaces the symbols *a* and *b* for arrivals and departures by the following codes.

    $M \equiv$ Poisson (or Markovian) arrival or departure distribution (or equivalently exponential interarrival or/service-time distribution)
    $D \equiv$ constant or deterministic interarrival or service time

$E_k \equiv$ Erlangian or gamma distribution of interarrival or service time distribution with parameter k

$GI \equiv$ general independent distribution of arrivals (or interarrival time)

$G \equiv$ general distribution of departures (or service time)

To illustrate the notation, consider

$$(M/D/10){:}(GD/N/\infty)$$

Here we have Poisson arrivals, constant service time, and 10 parallel servers in the facility. The service discipline is general (GD) in the sense that it could be FCFS, LCFS, SIRO, or whatever procedure the servers may use to decide on the order in which customers are chosen from the queue to start service. Regardless of how many customers arrive at the facility, the system (queue + service) can hold only a maximum of N customers; all others must seek service elsewhere. Finally, the source generating the arriving customers has an infinite capacity.

The standard notation described was initially devised by D. G. Kendall [1953] in the form $(a/b/c)$ and is known in the literature as the **Kendall notation.** Later, A. M. Lee [1966] added the symbols d and e to the Kendall notation. In this book we find it convenient to augment the Kendall–Lee notation by use of the symbol f, representing the capacity of calling source.

The ultimate objective of analyzing queueing situations is to develop measures of performance for evaluating the real systems. However, since any queueing system operates as a function of time, we must decide in advance whether we are interested in analyzing the system under **transient** or **steady-state** conditions. Transient conditions prevail when the behavior of the system continues to depend on time. Thus the pure birth and death processes (Section 15.2) always operate under transient conditions. On the other hand, queues with combined arrivals and departures start under transient conditions and gradually reach steady state after a *sufficiently large* time has elapsed, provided that the parameters of the system permit reaching steady state (e.g., a queue with arrival rate λ higher than its departure rate μ will never reach steady state regardless of elapsed time, since the queue size will increase with time).

Although the basic equations of the various models that we develop here can be used to study the transient behavior, our analysis will concentrate on the analysis of steady-state results. This conclusion is based on the assumption that most systems are normally designed to stay in operation for a long while. However, we must add also that transient-state analysis is quite complex mathematically and any venture into that area will take us far afield.

Under *steady-state* conditions we shall be interested in determining the following basic measures of performance.

$p_n =$ (steady-state) probability of n customers in *system*

$L_s =$ expected number of customers in *system*

$L_q =$ expected number of customers in *queue*

$W_s =$ expected waiting time in *system* (in queue + in service)

$W_q =$ expected waiting time in *queue*

By definition, we have

$$L_s = \sum_{n=0}^{\infty} n p_n$$

$$L_q = \sum_{n=c}^{\infty} (n-c) p_n$$

A strong relationship exists between L_s and W_s (also L_q and W_q) so that either measure is automatically determined from the other. Specifically, given the arrival rate λ, we have

$$L_s = \lambda W_s$$

$$L_q = \lambda W_q$$

The equations hold under rather general conditions that do not restrict the distribution of arrivals or service time. However, in the special case where customers arrive at the rate λ but not all arrivals can join the system (this can happen, for example, when there is a limit on the maximum number in system), the equations must be modified by redefining λ to include only those customers that actually join the system. Thus letting

$$\lambda_{\text{eff}} = \begin{pmatrix} \text{effective arrival} \\ \text{rate for those who} \\ \text{join the system} \end{pmatrix}$$

we have

$$L_s = \lambda_{\text{eff}} W_s$$

$$L_q = \lambda_{\text{eff}} W_q$$

In general,

$$\lambda_{\text{eff}} = \beta\lambda, \qquad 0 < \beta < 1$$

meaning that only a portion of those arriving can actually join the system. However, we can determine λ_{eff} in terms of L_s and L_q in the following manner. By definition

$$\begin{pmatrix} \text{expected waiting} \\ \text{time in } \textit{system} \end{pmatrix} = \begin{pmatrix} \text{expected waiting} \\ \text{time in } \textit{queue} \end{pmatrix} + \begin{pmatrix} \text{expected } \textit{service} \\ \text{time} \end{pmatrix}$$

Given that μ is the service rate, the expected service time is $1/\mu$ and we get the relationship

$$W_s = W_q + \frac{1}{\mu}$$

Multiplying both sides by λ, we get

$$L_s = L_q + \frac{\lambda}{\mu}$$

The relationship also holds when λ is replaced by λ_{eff} so that we can determine λ_{eff} from knowledge of L_s and L_q as

$$\lambda_{\text{eff}} = \mu(L_s - L_q)$$

In all the queueing models developed here, we concentrate on the derivation of p_n, since with these probabilities we can determine all the basic measures of performance in the following order.

$$p_n \rightarrow L_s = \sum_{n=0}^{\infty} np_n \rightarrow W_s = \frac{L_s}{\lambda} \rightarrow W_q = W_s - \frac{1}{\mu} \rightarrow L_q = \lambda W_q$$

We note that in most queueing models the computation of p_n is usually simple. In contrast, the computation of the waiting-time distribution may be quite complex. Thus the computation of W_s and W_q via L_s and L_q is usually more convenient and straightforward.

Example 15.3–1. Consider the queueing situation with one server in which arrivals occur at the rate $\lambda = 3$ per hour and service is performed at the rate $\mu = 8$ per hour. The probabilities p_n of n customers in the system are computed for the situation as given in the following table.

n	0	1	2	3	4	5	6	7	≥ 8
p_n	.625	.234	.088	.033	.012	.005	.002	.001	0

(As we shall see, these probabilities are normally computed from special formulas developed for the specific queueing model.)

For this situation we can successively obtain the following measures. The expected number in system is

$$L_s = \sum_{n=0}^{\infty} np_n$$
$$= 0 \times .625 + 1 \times .234 + 2 \times .088 + 3 \times .033 + 4 \times .012$$
$$+ 5 \times .005 + 6 \times .002 + 7 \times .001$$
$$= .6 \text{ customers}$$

Since $\lambda = 3$, we get the expected waiting time in the system as

$$W_s = \frac{L_s}{\lambda} = \frac{.6}{3} \cong .2 \text{ hour}$$

Again, with $\mu = 8$, we obtain the expected waiting time in the queue as

$$W_q = W_s - \frac{1}{\mu} = .2 - \frac{1}{8} = .075 \text{ hour}$$

From this we obtain the expected number in the queue as

$$L_q = \lambda W_q = 3 \times .075 = .225 \text{ customer} \qquad \blacktriangleleft$$

Exercise 15.3–1

In Example 15.3–1, compute the following.
(a) The expected number in the queue using p_n directly.
   [*Ans.* $L_q = \sum_{n=2}^{\infty} (n - 1) p_n = .225$, which is the same answer we obtained using the formula $L_q = \lambda W_q$.]
(b) The expected number of customers in the service facility.
   [*Ans.* $L_s - L_q = \lambda/\mu = .375$ customer.]

We present next a number of queueing models with combined arrivals and departures. The characteristics of each model will be summarized in terms of the extended Kendall notation we presented earlier. The basic result of each model is the steady-state probability p_n. However, as we shall see, the derivation of p_n is completely independent of the specific service discipline used to select customers for service. Thus we shall use *GD* (general discipline) in the Kendall notation to indicate that p_n applies for any service discipline. On the other hand, we must state that the service discipline does affect the derivation of the *distribution* of waiting time. Consequently, the service discipline will be specified whenever we derive the waiting-time distribution.

15.3.1 $(M/M/1):(GD/\infty/\infty)$

In this model there is one server, with no limit on the capacity of either the system or the calling source. Arrivals and departures are Poisson with rates λ and μ.

We first derive the difference–differential equations for $p_n(t)$, the probability of having n customers in the system during t. Then, under appropriate conditions, we take limits as $t \to \infty$ to obtain the steady-state probabilities p_n.

The difference–differential equations for $p_n(t)$ are derived using the same ideas advanced with the pure birth and pure death models. For $h > 0$ and very small, we have seen in Section 15.2 that

$$P\{\text{zero arrival in } h\} = e^{-\lambda h} \simeq 1 - \lambda h$$
$$P\{\text{one arrival in } h\} = 1 - e^{-\lambda h} \simeq \lambda h$$
$$P\{\text{zero departure in } h\} = e^{-\mu h} \simeq 1 - \mu h$$
$$P\{\text{one departure in } h\} = 1 - e^{-\mu h} \simeq \mu h$$

The probability $p_n(t + h)$ of n (> 0) in the system at $t + h$ is comprised of the sum of

1. $P\{n$ in the system at t, and no arrivals and no departures during $h\}$.
2. $P\{n - 1$ in the system at t, and one arrival and no departures during $h\}$.
3. $P\{n + 1$ in the system at t, and no arrivals and one departure during $h\}$.

These probabilities are based on the complete randomness property, which admits *at most one* event (arrival or departure) during h. Thus, adding the three probabilities and realizing that terms in h^2 imply the occurrence of two simultaneous events during h and hence will tend to zero as h becomes sufficiently small, we get for $n > 0$,

$$p_n(t + h) \cong p_n(t)(1 - \lambda h)(1 - \mu h) + p_{n-1}(t)(\lambda h)(1 - \mu h) \\ + p_{n-1}(t)(1 - \lambda h)(\mu h)$$

For $n = 0$, we note that the probability of zero departure during h is 1. Thus

$$p_0(t + h) \cong p_0(t)\{(1 - \lambda h) \cdot 1\} + p_1(t)(\mu h)(1 - \lambda h)$$

Now, moving $p_n(t)$ to the left side and dividing both sides by h, we note that the new left side represents a differential, so that when $h \to 0$, we get (verify!)

$$p_n'(t) = \lambda p_{n-1}(t) + \mu p_{n+1}(t) - (\lambda + \mu)p_n(t), \qquad n > 0$$
$$p_0'(t) = -\lambda p_0(t) + \mu p_1(t), \qquad\qquad\qquad n = 0$$

Note carefully that there are *no* approximations in the difference–differential equations. This shows that the approximation introduced in replacing $e^{-\lambda h}$ with $1 - \lambda h$ is inconsequential, since higher-order terms (h^2 or higher) would vanish anyway as $h \to 0$.

Solution of the equations will yield $p_n(t)$, the transient-state probabilities. However, the solution procedure and the solution itself are quite complex. [See Saaty (1961), pp. 88–93, for details.]

The steady-state solution can be proved to exist as $t \to \infty$ when $\lambda < \mu$. Assuming that this restriction holds, one obtains the steady-state equations by recognizing that, as $t \to \infty$, $p_n'(t) \to 0$ and $p_n(t) \to p_n$, for $n = 0, 1, 2, \ldots$. This yields

$$-\lambda p_0 + \mu p_1 = 0, \qquad n = 0$$
$$\lambda p_{n-1} + \mu p_{n+1} - (\lambda + \mu)p_n = 0, \qquad n > 0$$

In this chapter's appendix, we show that the solution is given by

$$\boxed{p_n = (1 - \rho)\rho^n, \quad n = 0, 1, 2, \ldots} \qquad (M/M/1){:}(GD/\infty/\infty)$$

where $\rho = \lambda/\mu < 1$, which is a geometric distribution.

The measure L_s can be derived in the following manner:

$$L_s = \sum_{n=0}^{\infty} n p_n = \sum_{n=0}^{\infty} n(1-\rho)\rho^n$$

$$= (1-\rho)\rho \frac{d}{d\rho} \sum_{n=0}^{\infty} \rho^n$$

$$= (1-\rho)\rho \frac{d}{d\rho} \left(\frac{1}{1-\rho}\right)$$

$$= \frac{\rho}{1-\rho}$$

Note that the convergence of $\sum_{n=0}^{\infty} \rho^n$ is ensured because $\rho < 1$. Using the relationships given earlier, we obtain all the basic measures of performance as

$$L_s = E\{n\} = \frac{\rho}{1-\rho}$$

$$L_q = L_s - \frac{\lambda}{\mu} = \frac{\rho^2}{1-\rho}$$

$$W_s = \frac{L_s}{\lambda} = \frac{1}{\mu(1-\rho)}$$

$$W_q = \frac{L_q}{\lambda} = \frac{\rho}{\mu(1-\rho)}$$

An alternative method for deriving W_s is given following Exercise 15.3–2.

Example 15.3–2. In a car-wash service facility, information gathered indicates that cars arrive for service according to a Poisson distribution with mean 5 per hour. The time for washing and cleaning each car varies but is found to follow an exponential distribution with mean 10 minutes per car. The facility cannot handle more than one car at a time.

To analyze this situation using the results of the $(M/M/1){:}(GD/\infty/\infty)$, we must assume that the calling source is so large that it can be considered infinite. Moreover, there is enough parking space to accommodate all arriving cars.

For the given situation we have $\lambda = 5$ cars per hour and $\mu = 60/10 = 6$ cars per hour. Since $\rho = \lambda/\mu = 5/6$ is less than 1, the system can operate under steady-state conditions. To obtain an idea of how many parking spaces should be available for cars that arrive at the facility, we compute L_q; that is,

$$L_q = \frac{\rho^2}{1-\rho} = \frac{(5/6)^2}{1-(5/6)} = 4.17 \cong 4 \text{ cars}$$

We realize, however, that L_q represents an expected value so that the number of cars waiting at any point in time may be larger or smaller than 4 cars. Thus we may be interested in determining the number of *parking* spaces in a manner that associates a reasonable probability with finding a parking space for arriving cars. For example, suppose that we need enough spaces so that an *arriving* car will be able to park at least 80% of the time. Since n represents the number in the *system,* the requirement is expressed in a probability statement as

$$p_0 + p_1 + p_2 + \cdots + p_s \geq .8$$

where s is the number of parking spaces we seek to determine. (Notice that s does *not* include the extra space inside the facility.) Thus, substituting for the probabilities, we get

$$(1 - \rho) + (1 - \rho)\rho + \cdots + (1 - \rho)\rho^s \geq .8$$

Simplifying the left side, we obtain

$$(1 - \rho)(1 + \rho + \cdots + \rho^s) = (1 - \rho)\frac{1 - \rho^{s+1}}{1 - \rho} = 1 - \rho^{s+1}$$

Consequently,

$$\rho^{s+1} \leq .2$$

Taking the logarithms of both sides, we get

$$(s + 1)\log(5/6) \leq \log(.2)$$

Noting that the logs of values less than 1 are negative, a division by $\log(5/6)$ would reverse the direction of the inequality. This leads to the following inequality for determining the values of s.

$$s \geq \frac{\log(.2)}{\log(5/6)} - 1 = 7.8 \cong 8 \text{ spaces}$$

Thus to accommodate all arriving cars at least 80% of the time, the minimum number of parking spaces must be approximately double that of the expected queue length L_q.

We can obtain more information about the operation of the car-wash facility. For example, the percentage of time the facility is idle equals the probability of having no cars in the facility; that is, $p_0 = 1 - \rho \cong .17$, meaning that the facility is idle 17% of the time. On the other hand, the expected waiting time from the moment the car arrives until it leaves may be useful in determining the convenience of service from the customers' standpoint. Thus we can say that a car is expected to wait

$$W_s = \frac{1}{\mu(1 - \rho)} = \frac{1}{6[1 - (5/6)]} = 1 \text{ hour}$$

This appears quite long and the manager of the facility should think of means to speed up the service rate. ◄

Exercise 15.3–2

In Example 15.3–2:
(a) Determine the probability that an arriving car must wait prior to being washed.
[*Ans.* .8333.]
(b) If there are six parking spaces outside the facility, determine the probability an arriving car will not find a parking space.
[*Ans.* $p_{n \geq 7} = \rho^7 = .279.$]

Waiting-Time Distribution Based on FCFS Service Discipline

In the analysis of the $(M/M/1){:}(GD/\infty/\infty)$, the derivation of p_n is shown to be completely independent of the service discipline. This means that W_s and W_q, the *expected* waiting times in system and in queue, are also completely independent of the service discipline, since they can be determined from $W_s = L_s/\lambda$ and $W_q = L_q/\lambda$.

Although the *expected* waiting time is independent of the service discipline, its probability density function (distribution) does depend on the type of service discipline used. Thus although the distributions may differ depending on the service discipline, their expected values remain unchanged. We illustrate here how the PDF of the waiting time is derived for the queueing model above based on the FCFS discipline. Let τ be the amount of time a person *just arriving* must wait in the *system,* that is, until his service is completed. Based on FCFS service discipline, if an arriving customer finds n persons ahead of him in the system, then

$$\tau = t_1' + t_2 + \cdots + t_{n+1}$$

where t_1' is the time needed for the customer actually in service to complete service and $t_2, t_3, \ldots, t_n$ are the service times for the $n - 1$ customers in queue. The time t_{n+1} thus represents the service time for the arriving customer.

Let $w(\tau \mid n + 1)$ be the conditional PDF of τ given n customers in the system ahead of the arriving customer. Since t_i, for all i, is exponentially distributed, by the forgetfulness property (Section 15.2), t_1' will also have the same exponential distribution as $t_2, t_3, \ldots, t_{n+1}$. Consequently, τ is the sum of $n + 1$ identically distributed and independent exponential distributions. This means that $w(\tau \mid n + 1)$ is gamma-distributed with parameters $(\mu, n + 1)$ (see Example 10.9–1). Thus

$$w(\tau) = \sum_{n=0}^{\infty} w(\tau \mid n + 1)p_n = \sum_{n=0}^{\infty} \frac{\mu(\mu\tau)^n e^{-\mu\tau}}{n!}(1 - \rho)\rho^n$$

$$= (1 - \rho)\mu e^{-\mu\tau} \sum_{n=0}^{\infty} \frac{(\lambda\tau)^n}{n!} = \mu(1 - \rho)e^{-\mu(1-\rho)\tau}, \qquad \tau > 0$$

which is an exponential distribution with mean

$$E\{\tau\} = \frac{1}{\mu(1 - \rho)}$$

The mean $E\{\tau\}$ actually equals W_s, the expected waiting time in the system.

Knowledge of the distribution of waiting time can provide information that is not attainable otherwise. For example, we can obtain an idea about the "reliability" of W_s in indicating the actual time customers wait by computing the probability that customers will wait more than $W_s = 1/\mu(1 - \rho)$; that is,

$$P\{\tau > W_s\} = 1 - \int_0^{W_s} w(\tau)\, dt$$

$$= e^{-\mu(1-\rho)W_s}$$

$$= e^{-1} \cong .368$$

Thus, under FCFS discipline, 36.8% of the customers will wait more than the average waiting time W_s. Naturally, this probability will change with the service discipline and, intuitively, should be higher for the SIRO and LCFS disciplines.

Exercise 15.3-3

In Example 15.3–2, find the following:
(a) The standard deviation of the waiting time in the system.
   [*Ans.* Standard deviation = 1.]
(b) The probability that the waiting time in the system will vary by half a standard deviation around its mean value.
   [*Ans.* Probability = .3834.]

15.3.2 $(M/M/1):(GD/N/\infty)$

The only difference between this model and the $(M/M/1):(GD/\infty/\infty)$ is that the maximum number of customers allowed in the system is N (maximum queue length $= N - 1$). This means that once N customers are in the system, all new arrivals either balk or are not permitted to join the system. The result is that the effective arrival rate λ_{eff} at the facility becomes less than the rate λ at which arrivals are generated from the source.

In terms of the development of the difference–differential equations, the equations for $n = 0$ and $0 < n < N$ remain as given in the $(M/M/1):(GD/\infty/\infty)$. For $n > N$, $p_n(t) = 0$ and for $n = N$, we have

$$p_N(t + h) \cong p_N(t)(1)(1 - \mu h) + p_{N-1}(t)(\lambda h)(1 - \mu h), \qquad n = N$$

Thus, as we found in Section 15.3.1, the steady-state equations representing this situation are

$$-\rho p_0 + p_1 = 0, \qquad n = 0$$
$$-(1 + \rho)p_n + p_{n+1} + \rho p_{n-1} = 0, \qquad 0 < n < N$$
$$-p_N + \rho p_{N-1} = 0, \qquad n = N$$

In this chapter's appendix, we show that the solution is given by

$$p_n = \begin{cases} \left(\dfrac{1-\rho}{1-\rho^{N+1}}\right)\rho^n, & \rho \neq 1 \\[2ex] \dfrac{1}{N+1}, & \rho = 1 \end{cases} \qquad n = 0, 1, 2, \ldots, N \qquad (M/M/1):(GD/N/\infty)$$

Note that $\rho = \lambda/\mu$ need *not* be less than 1 as in the case of the $(M/M/1):(GD/\infty/\infty)$. Intuitively, we see this result because the number allowed in the system is controlled by the queue length $(= N - 1)$, not by the relative rates of arrival and departure, λ and μ.

Using p_n above, we find that the expected number in the system is shown in the chapter's appendix to be given by

$$L_s = \begin{cases} \dfrac{\rho\{1 - (N+1)\rho^N + N\rho^{N+1}\}}{(1-\rho)(1-\rho^{N+1})}, & \rho \neq 1 \\[2ex] \dfrac{N}{2}, & \rho = 1 \end{cases}$$

The measures L_q, W_s, and W_q can be derived from L_s once we determine the effective arrival rate λ_{eff}. Since the probability that a customer does not join the system is equal to p_N, the probability of N in the system, the ratio of customers joining the system must equal $P\{n > N\} = 1 - p_N$. It then follows that

$$\lambda_{\text{eff}} = \lambda(1 - p_N)$$

and

$$W_q = \frac{L_q}{\lambda_{\text{eff}}} = \frac{L_q}{\lambda(1 - p_N)}$$

$$L_s = L_q + \frac{\lambda_{\text{eff}}}{\mu} = L_q + \frac{\lambda(1 - p_N)}{\mu}$$

$$W_s = W_q + \frac{1}{\mu} = \frac{L_s}{\lambda(1 - p_N)}$$

It can be shown that

$$\lambda_{\text{eff}} = \mu(L_s - L_q) = \lambda(1 - p_N)$$

(see Problem 15–58).

Example 15.3–3. Consider the car-wash facility of Example 15.3–2. Suppose that the facility has a total of 5 parking spaces. If the parking lot is full, newly arriving cars balk to seek service elsewhere.

Perhaps the first piece of information that would interest the manager of the facility is to know how many customers he is losing due to the limited parking space. Essentially, this would be equivalent to determining.

$$\lambda - \lambda_{\text{eff}} = \lambda p_N$$

Now $N = 5 + 1 = 6$, $\rho = 5/6$, and

$$p_N = \left[\frac{1 - (5/6)}{1 - (5/6)^7}\right](5/6)^6 = .0774, \qquad N = 6$$

Thus the rate at which cars balk is $5 \times .0774 = .387$ car/hour or, based on a 8-hour day, the facility will lose 3 cars/day on the average. A decision regarding enlarging the parking lot beyond 5 spaces should thus be based on the "worth" of lost business.

Suppose that it is desired to compute the expected waiting time until a car is washed. We first compute L_s from which W_s can be determined.

$$L_s = \frac{(5/6)[1 - 7(5/6)^6 + 6(5/6)^7]}{(1 - 5/6)[1 - (5/6)^7]} = 2.29 \text{ cars}$$

Since $\lambda_{\text{eff}} = \lambda(1 - p_6) = 5(1 - .0774) = 4.613$, it follows that

$$W_s = \frac{L_s}{\lambda_{\text{eff}}} = \frac{2.29}{4.613} = .496 \text{ hour}$$

We note that the expected waiting time has been reduced from 1 hour in the case where all arriving cars are allowed to join the facility (Example 15.3–2) to about half an hour when a limit $N = 6$ is set on the system. This reduction has been achieved at the expense of losing an average of 3 cars a day due to the inavailability of parking spaces. ◄

Exercise 15.3–4

In Example 15.3–3, compute:
(a) The probability that an arriving car will start service immediately upon arrival.
   [*Ans.* .231.]
(b) The expected waiting time until a service starts.
   [*Ans.* .3297 hour.]
(c) The expected number of parking spaces occupied.
   [*Ans.* 1.52 spaces.]

15.3.3 $(M/G/1):(GD/\infty/\infty)$—THE POLLACZEK–KHINTCHINE FORMULA

The Pollaczek–Khintchine (P-K) formula is derived for a single-server situation based on the following three assumptions:

1. Poisson arrivals with arrival rate λ.
2. General service time distribution with mean $E\{t\}$ and variance var$\{t\}$.
3. Steady-state conditions prevail with $\rho = \lambda E\{t\} < 1$.

Assumption 2 changes the situation from the familiar Poisson arrival/Poisson departure analysis. The derivation of an expression for p_n is quite complex and is based on the use of imbedded Markov chains. We thus postpone the derivation of p_n until Section 15.6 and restrict our attention in this section to the development of L_s, from which L_q, W_q, and W_s can be derived.

In this chapter's appendix we provide the details for deriving L_s. The derivation is a bit complex but provides an intriguing exercise in the application of probability theory.

The P-K formula is given by

$$L_s = \lambda E\{t\} + \frac{\lambda^2(E^2\{t\} + \text{var}\{t\})}{2(1 - \lambda E\{t\})}$$

$(M/G/1):(GD/\infty/\infty)$

From this formula we get

$$L_q = L_s - \lambda E\{t\}$$

$$W_q = \frac{L_q}{\lambda}$$

$$W_s = \frac{L_s}{\lambda}$$

Note that the service rate $\mu = 1/E\{t\}$.

Example 15.3–4. Suppose that in the car-wash facility (Example 15.3–2), the washing is done by automatic machines, so that the service time may be considered the same and constant for all cars. The washing machine cycle takes 10 minutes exactly.

To analyze the situation, we note that $\lambda = 5$ per hour (see Example 15.3–2). On the other hand, since the service time is constant, we have $E\{t\} = 10/60 = 1/6$ hour and var $\{t\} = 0$. Thus

$$L_s = 5(1/6) + \frac{5^2[(1/6)^2 + 0]}{2(1 - 5/6)} = 2.917 \text{ cars}$$

$$L_q = 2.917 - (5/6) = 2.083 \text{ cars}$$

$$W_s = \frac{2.917}{5} = .583 \text{ hour}$$

$$W_q = \frac{2.083}{5} = .417 \text{ hour}$$

It is interesting that even though the arrival and departure rates ($\lambda = 5$ and $\mu = 1/E\{t\} = 6$) in this example exactly equal those of Example 15.3–2, where both arrivals and departures are Poisson, the expected waiting times are lower in the current situation where service time is constant. Namely, we have

	Poisson Arrivals and Departures	Poisson Arrivals and Constant Service Time
W_s	1.00 hour	.583 hour
W_q	.83 hour	.417 hour

This conclusion makes sense because a constant service time indicates *more certainty* in the operation of the facility, with the result that the expected waiting time is reduced.   ◀

Exercise 15.3–5

(a) Show that the P-K formula reduces to L_s of the $(M/M/1):(GD/\infty/\infty)$ model when the service-time distribution is exponential with mean $1/\mu$.

(b) In Example 15.3–4, suppose that the automatic machine wash cycle can be adjusted to accommodate large and small cars. Large cars require 12 minutes each, whereas small cars can each be washed in 6 minutes. Although the size of an arriving car is not predictable in advance, it is known that there is a 50:50 chance an arriving car will be either large or small. (In other words, the service time is either 6 minutes with probability .5 or 12 minutes also with probability .5.) Find the expected waiting time until a car starts service.
[*Ans. $W_q = .25$ hour.*]

15.3.4 $(M/M/c):(GD/\infty/\infty)$

In this model arrivals occur at the rate λ and a maximum of c customers can be serviced simultaneously. The average service time per customer is $1/\mu$. Both arrivals and departures occur according to Poisson distributions.

The ultimate effect of using c parallel servers is to "speed up" the rate of service in comparison with the one-server case by allowing a maximum of c customers to be serviced simultaneously. Thus, if the number of customers in the system, n, is *at least* c, the combined rate of service (departure) from the facility is $c\mu$. On the other hand, if n is less than c, the combined service rate is $n\mu$, since no more than n ($< c$) servers will be busy. In essence, the use of a multiple-server model is equivalent to a single-server model in which the service rate varies with n.

To analyze the $(M/M/c)$ model, we shall thus develop a *generalized* single-server model in which both the arrival and service rates depend on n, that is, λ_n and μ_n. The objective would be to derive a general expression for the steady-state probabilities p_n. By putting $\lambda_n = \lambda$ and $\mu_n = n\mu$ for $n < c$ and $\mu_n = c\mu$ for $n \geq c$, we can obtain all the measures of performance for the $(M/M/c)$ model. The general solution of p_n given λ_n and μ_n will also prove useful in deriving the results of other queueing models and, indeed, can be used to derive p_n for the $(M/M/1):(GD/\infty/\infty)$ and $(M/M/1):(GD/N/\infty)$, which we presented in Sections 15.3.1 and 15.3.2.

To distinguish the *generalized* single-server model from the one in Section 15.3.1, we use the notation $(M_n/M_n/1):(GD/\infty/\infty)$ to indicate that arrival and service rates, λ_n and μ_n, are dependent on n. For this model we have

$$P\{\text{zero arrival during } h \text{ given } n \text{ in system}\} \cong 1 - \lambda_n h$$

$$P\{\text{zero departure during } h \text{ given } n \text{ in system}\} \cong 1 - \mu_n h$$

Under the basic condition that *at most* one event (arrival or departure) can occur during h, we can obtain

$$p_n(t+h) \cong p_n(t)(1 - \lambda_n h)(1 - \mu_n h) + p_{n-1}(t)(\lambda_n h)(1 - \mu_n h)$$
$$+ p_{n+1}(t)(1 - \lambda_n h)(\mu_{n+1} h), \qquad n > 0$$
$$p_0(t+h) \cong p_0(t)(1 - \lambda_0 h) \cdot 1 + p_1(t)(1 - \lambda_1 h)\mu_1 h, \qquad n = 0$$

Following the exact procedure advanced in Section 15.3.1, we obtain the following *steady-state* equations:

$$-(\lambda_n + \mu_n)p_n + \mu_{n+1}p_{n+1} + \lambda_{n-1}p_{n-1} = 0, \qquad n > 0$$

$$-\lambda_0 p_0 + \mu_1 p_1 = 0, \qquad n = 0$$

These equations may be written in a convenient form as (verify!)

$$p_1 = \frac{\lambda_0}{\mu_1}p_0$$

$$p_{n+1} = \left(\frac{\lambda_n + \mu_n}{\mu_{n+1}}\right)p_n - \left(\frac{\lambda_{n-1}}{\mu_{n+1}}\right)p_{n-1}, \qquad n > 0$$

Starting with p_1 and successively computing $p_2, p_3, \ldots$, we can show by induction that (verify!)

$$p_n = \frac{\lambda_0 \lambda_1 \cdots \lambda_{n-1}}{\mu_1 \mu_2 \cdots \mu_n}p_0, \qquad n \geq 1$$

$$p_0 = 1 \Big/ \left(1 + \sum_{n=0}^{\infty} \prod_{i=1}^{n} \frac{\lambda_{i-1}}{\mu_i}\right)$$

$$(M_n/M_n/1):(GD/\infty/\infty)$$

The equation for p_0 is obtained from the condition $\sum_{n=0}^{\infty} p_n = 1$.

Exercise 15.3-6

(a) Verify the expression for p_0 given in the $(M_n/M_n/1):(GD/\infty/\infty)$ model.

(b) Obtain p_n for the $(M/M/1):(GD/\infty/\infty)$, Section 15.3.1, using the results of the $(M_n/M_n/1):(GD/\infty/\infty)$.

To derive the results for the $(M/M/c):(GD/\infty/\infty)$ model, we have

$$\lambda_n = \lambda \qquad \text{for all } n \geq 0$$

$$\mu_n = \begin{cases} n\mu, & n \leq c \\ c\mu, & n \geq c \end{cases}$$

Thus, using p_n for the $(M_n/M_n/1):(GD/\infty/\infty)$, we obtain for $n \leq c$,

$$p_n = \frac{\lambda^n}{\mu(2\mu)(3\mu)\cdots(n\mu)}p_0$$

$$= \frac{\lambda^n}{n!\mu^n}p_0$$

and for $n \geq c$, we have

$$p_n = \frac{\lambda^n}{\mu(2\mu)\cdots(c-1)\mu(c\mu)\underbrace{(c\mu)\cdots(c\mu)}}p_0$$

$$(n-c) \text{ times}$$

$$= \frac{\lambda^n}{c!c^{n-c}\mu^n}p_0$$

Letting $\rho = \lambda/\mu$, we obtain

$$p_n = \begin{cases} \left(\dfrac{\rho^n}{n!}\right)p_0, & 0 \le n \le c \\[2ex] \left(\dfrac{\rho^n}{c^{n-c}c!}\right)p_0, & n > c \end{cases}$$

$$p_0 = \left\{\sum_{n=0}^{c-1}\frac{\rho^n}{n!} + \frac{\rho^c}{c!(1-\rho/c)}\right\}^{-1}$$

$$(M/M/c):(GD/\infty/\infty)$$

where

$$\frac{\rho}{c} < 1 \quad \text{or} \quad \frac{\lambda}{\mu c} < 1$$

Also,

$$L_q = \frac{\rho^{c+1}}{(c-1)!(c-\rho)^2}p_0 = \left[\frac{c\rho}{(c-\rho)^2}\right]p_c$$

$$L_s = L_q + \rho$$

$$W_q = \frac{L_q}{\lambda}$$

$$W_s = W_q + \frac{1}{\mu}$$

The expression for p_0 and L_q are derived in this chapter's appendix.

The computations associated with this model may be tedious. Morse [(1958), p. 103] gives two useful approximations for p_0 and L_q. For ρ much smaller than 1,

$$p_0 \cong 1 - \rho \quad \text{and} \quad L_q \cong \frac{\rho^{c+1}}{c^2}$$

and for ρ/c very close to 1,

$$p_0 \cong \frac{(c-\rho)(c-1)!}{c^c} \quad \text{and} \quad L_q \cong \frac{\rho}{c-\rho}$$

Example 15.3–5. A small town is being serviced by two cab companies. Each of the two companies owns two cabs and are known to share the market almost equally. This is evident by the fact that calls arrive at each company's dispatching office at the rate of 10 per hour. The average time per ride is 11.5 minutes. Arrival of calls follows a Poisson distribution, whereas ride times are exponential.

The two companies were recently bought by one of the city's businessmen.

His first action after taking over the two companies was to try to consolidate the two companies into one dispatching office in hope that he would provide faster service for his customers. However, he noticed that the utilization (ratio of hourly arriving calls to rides) for each company is

$$100\frac{\lambda}{c\mu} = \frac{100 \times 10}{2 \times (60/11.5)} = 95.8\%$$

(Note that each cab represents a server.) As a result, the cost of relocating the two companies in one office may not be justifiable because each of the current dispatching offices appears "quite busy," as evidenced by the high utilization factor.

To analyze the new owner's problem, we need in essence to make a comparison between the following two situations:

1. Two independent queues each of the type $(M/M/2){:}(GD/\infty/\infty)$ with $\lambda = 10$ calls per hour and $\mu = 5.217$ rides per hour.
2. One queue of the type $(M/M/4){:}(GD/\infty/\infty)$ with $\lambda = 2 \times 10 = 20$ calls per hour and $\mu = 5.217$ rides per hour.

Note that in both situations, μ represents the number of rides a *single* cab can provide per hour.

The utilization factor in the second situation is

$$\frac{100\lambda}{c\mu} = \frac{100 \times 20}{4 \times 5.217} = 95.8\%$$

which, of course, remains equal to the utilization factors when the two dispatching offices remain unconsolidated. This result seems to confirm the owner's suspicion that the consolidation is unjustified. However, if we consider other measures of performance, the picture will become different. Specifically, let us compute W_q, the expected waiting time by a customer until a cab is dispatched to him or her, both for the separate and consolidated facilities. Thus, for $c = 2$, we have $\rho = \lambda/\mu = 10/5.217 \cong 1.917$ and

$$p_0 = \left[\frac{1.917^0}{0!} + \frac{1.917}{1!} + \frac{1.917^2}{2!(1 - 1.917/2)}\right]^{-1} = .0212$$

Thus

$$W_q = \frac{1}{10}\left[\frac{1.917^3 \times .0212}{1!(2 - 1.917)^2}\right] = 2.16 \text{ hours}$$

On the other hand, for $c = 4$, we have $\lambda/\mu = 20/5.217 = 3.83$ and

$$p_0 = \left[\frac{3.83^0}{0!} + \frac{3.83}{1!} + \frac{3.83^2}{2!} + \frac{3.83^3}{3!} + \frac{3.83^4}{4!(1 - 3.83/4)}\right]^{-1} = .0042$$

Hence

$$W_q = \frac{1}{20}\left[\frac{3.83^5 \times .0042}{3!(4 - 3.83)^2}\right] = 1.05 \text{ hours}$$

The computation indicates that by simply consolidating the two companies, the expected waiting time until a cab is dispatched to the customer is cut by

50%. The conclusion then is that pooling of services should normally result in a more efficient operation in terms of offering speedier service to the customer, even though the utilization factors of individual queues may appear high. (Of course, the owner has more to be concerned about since, even after consolidating the two companies, waiting more than 1 hour for a 10-minute taxi ride is a bit much! Obviously, he needs to increase the size of his fleet.)   ◄

Exercise 15.3–7

In Example 15.3–5, find:
(a) The percentage of time *all* cabs in each of the two companies are "on call."
    [*Ans.* 93.8%.]
(b) The percentage of time *all* cabs in the consolidated operation are "on call."
    [*Ans.* 90.93%. It is interesting to note that the percentage of full occupancy is smaller when the two companies are consolidated into one pool, even though the waiting time W_q is halved.]

Now that you have worked Exercise 15.3–7, you probably have come to realize the inconvenience and inaccuracy of carrying out computations by a hand calculator. For this reason, we include a general FORTRAN program in Appendix C that computes the probabilities and measures of performance of all Poisson queues of the type $(M/M/c)$. The use of the program is quite simple because it uses only five pieces of input data in the same order they appear in the Kendall notation, namely, λ, μ, c, system capacity, and calling source capacity. The program *automatically* selects the appropriate model by using the input data information.

15.3.5 $(M/M/c):(GD/N/\infty)$, $c \leq N$

This queueing situation differs from $(M/M/c):(GD/\infty/\infty)$ in that a limit N is set on the capacity of the system (i.e., maximum queue size $= N - c$). In terms of the $(M_n/M_n/1):(GD/\infty/\infty)$ model, λ_n and μ_n for the current model are given by

$$\lambda_n = \begin{cases} \lambda, & 0 \leq n < N \\ 0, & n \geq N \end{cases}$$

$$\mu_n = \begin{cases} n\mu, & 0 \leq n \leq c \\ c\mu, & c \leq n \leq N \end{cases}$$

Substituting for λ_n and μ_n in the general expression for p_n and noting that $\rho = \lambda/\mu$, we get

$$p_n = \begin{cases} \dfrac{\rho^n}{n!}p_0, & 0 \leq n \leq c \\[2ex] \dfrac{\rho^n}{c!c^{n-c}}p_0, & c \leq n \leq N \end{cases} \qquad (M/M/c):(GD/N/\infty)$$

$$
p_0 = \begin{cases} \left[\displaystyle\sum_{n=0}^{c-1} \frac{\rho^n}{n!} + \frac{\rho^c(1-(\rho/c)^{N-c+1})}{c!(1-\rho/c)} \right]^{-1}, & \rho/c \neq 1 \\[3ex] \left[\displaystyle\sum_{n=0}^{c-1} \frac{\rho^n}{n!} + \frac{\rho^c}{c!}(N-c+1) \right]^{-1}, & \rho/c = 1 \end{cases}
$$

Note that the only difference between p_n in this model and $(M/M/c){:}(GD/\infty/\infty)$ occurs in the expression for p_0. Note also that the *utilization factor* ρ/c need not be less than 1.

We show in this chapter's appendix that

$$
L_q = \begin{cases} p_0 \dfrac{\rho^{c+1}}{(c-1)!(c-\rho)^2} \left\{ 1 - \left(\dfrac{\rho}{c}\right)^{N-c} - (N-c)\left(\dfrac{\rho}{c}\right)^{N-c}\left(1-\dfrac{\rho}{c}\right) \right\}, & \rho/c \neq 1 \\[3ex] p_0 \dfrac{\rho^c(N-c)(N-c+1)}{2c!}, & \rho/c = 1 \end{cases}
$$

$$
L_s = L_q + (c - \bar{c}) = L_q + \frac{\lambda_{\text{eff}}}{\mu}
$$

where

$$
\bar{c} = \text{expected number of idle servers} = \sum_{n=0}^{c}(c-n)p_n
$$

$$
\lambda_{\text{eff}} = \lambda(1-p_N) = \mu(c-\bar{c})
$$

Notice the interpretation of λ_{eff} in this case. Since $(c-\bar{c})$ represents the expected number of busy channels, $\mu(c-\bar{c})$ represents the actual number served per unit time and hence the effective arrival rate.

Example 15.3–6. In the consolidated cab company problem (Example 15.3–5), although the owner realizes that the expected waiting time is excessive, he is unable to obtain funds for the purchase of additional cabs. To alleviate the problem of excessive waiting, however, he instructed the dispatching office to apologize to prospective customers for the unavailability of cabs once the waiting list reaches 16 customers.

To study the effect of the owner's decision on the waiting time, we realize that having a waiting list of 16 customers is equivalent to having $16 + 4 = 20$ customers in the system, since the company has 4 cabs (servers). The queueing model thus reduces to $(M/M/4){:}(GD/20/\infty)$, where $\lambda = 20$ per hour and $\mu = 5.217$ per hour. We can compute W_q as follows:

$$
p_0 = \left\{ 1 + 3.83 + \frac{3.83^2}{2!} + \frac{3.83^3}{3!} + \frac{3.83^4[1-(3.83/4)^{17}]}{4!\,(1-3.83/4)} \right\} = .00753
$$

and

$$L_q = (.00753)\,\frac{3.83^5}{3!(4-3.83)^2}\left[1-\left(\frac{3.83}{4}\right)^{16}-16\left(\frac{3.83}{4}\right)^{16}\left(1-\frac{3.83}{4}\right)\right]$$

$$= 5.85$$

Since

$$p_{20} = \frac{(3.83^{20})(.00753)}{(4!)(4^{16})} = .03433$$

we get

$$\lambda_{\text{eff}} = \lambda(1-p_{20}) = 20(1-.03433) = 19.31$$

As a result,

$$W_q = \frac{L_q}{\lambda_{\text{eff}}} = \frac{5.85}{19.31} = .303 \text{ hour}$$

The expected waiting time W_q before setting a limit on the capacity of the system was 1.05 hour, which is three times higher than the new expected waiting time of .303 hour ($\cong$18 minutes). Note that this remarkable reduction is achieved at the expense of losing about 3.4% of potential customers ($p_{20} = .03433$). Of course, the result does not say how much effect, in the long run, the possible loss of customer's goodwill will have on the operation. ◀

Exercise 15.3–8

In Example 15.3–6 find
(a) The expected number of idle cabs.
    [*Ans.* $\bar{c} = .2986$.]
(b) The probability that a calling customer will be told that no cabs are available.
    [*Ans.* $p_{20} = .03433$.]

15.3.6 $(M/M/\infty){:}(GD/\infty/\infty)$—SELF-SERVICE MODEL

In this model, the number of servers is unlimited because the customer himself or herself is also the server. This normally is the case in self-service facilities. A typical example is taking the written part of a driver's license test. We must caution, however, that situations such as self-service gas stations or 24-hour banks do not fall under this model's category. This conclusion follows because in these situations, the servers really are the gas pump and the bank computer, even though the customer is the one that operates the equipment.

Again, in terms of the $(M_n/M_n/1){:}(GD/\infty/\infty)$ model, we have

$$\lambda_n = \lambda, \qquad \text{for all } n \geq 0$$
$$\mu_n = n\,\mu, \qquad \text{for all } n \geq 0$$

Direct substitution in the expression for p_n in the $(M_n/M_n/1)$ model yields

$$p_n = \frac{\lambda^n}{n!\mu^n}\,p_0 = \frac{\rho^n}{n!}\,p_0$$

Since $\Sigma_{n=0}^{\infty} p_n = 1$, it follows that

$$p_0 = \frac{1}{1 + \rho + \dfrac{\rho^2}{2!} + \cdots} = \frac{1}{e^\rho} = e^{-\rho}$$

As a result,

$$p_n = \frac{e^{-\rho} \rho^n}{n!}, \qquad n = 0, 1, 2, \ldots \qquad\qquad (M/M/\infty){:}(GD/\infty/\infty)$$

which is Poisson with mean $E\{n\} = \rho$. We also have

$$L_s = E\{n\} = \rho$$
$$W_s = \frac{1}{\mu}$$
$$L_q = W_q = 0$$

Note that $W_q = 0$ because each customer services himself or herself. This is the reason W_s is equal to the mean service time $1/\mu$.

It is easy to obtain $p_n(t)$, the transient-state probabilities, for this model. The final results are given as

$$p_n(t) = \frac{e^{-\alpha} \alpha^n}{n!}, \qquad n = 0, 1, 2, \ldots$$

where $\alpha = \rho(1 - e^{-\mu t})$. This is Poisson with mean $E\{n|t\} = \alpha$.

The results of the $(M/M/\infty){:}(GD/\infty/\infty)$ can be used to approximate those of the $(M/M/c){:}(GD/\infty/\infty)$ as c increases "sufficiently." The advantage is obvious, since the computations are much simpler in the $(M/M/\infty)$ model.

We demonstrate the relative accuracy of the approximation by presenting samples of the measures of performance for both models for different values of c and ρ $(= \lambda/\mu)$. Table 15–1 summarizes the results. (The last three elements of the Kendall notation remain unchanged and are deleted for convenience.)

The computations show that as ρ becomes small (i.e., λ much less than μ), the $(M/M/\infty)$ model is a close approximation of the $(M/M/c)$ model even for c as small as 10.

15.3.7 $(M/M/R){:}(GD/K/K)$, $R < K$—MACHINE SERVICING MODEL

This model assumes that R repairmen are available for servicing a total of K machines. Since a broken machine cannot generate new calls while in service, the model is an example of a finite calling source.

Table 15-1

Measure	ρ = .1				ρ = 9			
	M/M/10	M/M/20	M/M/50	M/M/∞	M/M/10	M/M/20	M/M/50	M/M/∞
W_s	.1	.1	.1	.1	1.6	1.0	1.0	1.0
W_q	$.25 \times 10^{-18}$	$.18 \times 10^{-40}$	0	0	.668	.0001	$.6 \times 10^{-22}$	0
L_s	.1	.1	.1	.1	15.02	9.0	9.0	9.0
L_q	$.25 \times 10^{-18}$	$.18 \times 10^{-40}$	0	0	6.02	.0092	$.56 \times 10^{-21}$	0
p_0	.90484	.90484	.90484	.90484	.00007	.00012	.00012	.00012

The model can be treated as a special case of the $(M_n/M_n/1){:}(GD/\infty/\infty)$. If we define λ as the rate of breakdown *per machine*, we have

$$\lambda_n = \begin{cases} (K-n)\lambda, & 0 \le n \le K \\ 0, & n \ge K \end{cases}$$

$$\mu_n = \begin{cases} n\mu, & 0 \le n \le R \\ R\mu, & R \le n \le K \\ 0, & n > K \end{cases}$$

Substituting for λ_n and μ_n in the expression for p_n in the $(M_n/M_n/1)$ model, we get (verify!)

$$p_n = \begin{cases} \dbinom{K}{n} \rho^n p_0, & 0 \le n \le R \\[2mm] \dbinom{K}{n} \dfrac{n! \rho^n}{R! R^{n-R}} p_0, & R \le n \le K \end{cases} \qquad (M/M/R){:}(GD/K/K)$$

$$p_0 = \left\{ \sum_{n=0}^{R} \binom{K}{n} \rho^n + \sum_{n=R+1}^{K} \binom{K}{n} \frac{n! \rho^n}{R! R^{n-R}} \right\}^{-1}$$

The other measures are given by

$$L_q = \sum_{n=R+1}^{K} (n-R) p_n \qquad (R > 1)$$

$$L_s = L_q + (R - \bar{R}) = L_q + \frac{\lambda_{\text{eff}}}{\mu}$$

where

$$\bar{R} = \text{expected number of idle repairmen} = \sum_{n=0}^{R} (R-n) p_n$$

$$\lambda_{\text{eff}} = \mu(R - \bar{R}) = \lambda(K - L_s)$$

The second expression for λ_{eff} is obtained as follows. Since the arrival rate given n machines in the system is $\lambda(K - n)$ (where λ is the rate of breakdown per machine), under steady-state conditions

$$\lambda_{\text{eff}} = E\{\lambda(K - n)\} = \lambda(K - L_s)$$

The results apply to the case of a single repairman simply by putting $R = 1$. In this case it can be shown that

$$L_q = K - \left(1 + \frac{1}{\rho}\right)(1 - p_0) \qquad (R = 1)$$

$$L_s = K - \frac{1 - p_0}{\rho}$$

Example 15.3-7. A company is trying to decide on the number of repairmen to service its 20-machine shop. The rate of breakdown is 10 machines per hour. A repairman takes 3 minutes to repair a broken machine. Machines break down according to a Poisson distribution and the repair time is exponential.

It is almost futile to try to carry out this model's computation by hand calculator. We thus use the computer program in Appendix D to generate the required results. The idea is to increase R in the $(M/M/R):(GD/20/20)$

$$(M/M/\ 4) - (GD/\ 20/\ 20)$$

LAMBDA $= 0.10000E + 02$ LAMBDA EFF $= 0.79919E + 02$
MU $= 0.20000E + 02$ RHO $= 0.50000E + 00$

WS $= 0.15025E = 00$ WQ $= 0.10025E + 00$
LS $= 0.12008E + 02$ LQ $= 0.80121E + 01$

VALUES OF P(N) FOR N $= 0$ TO 20, OTHERWISE P(N) < 0.00001
P(0) $= 0.00001$

0.00015	0.00071	0.00213	0.00452	0.00903	0.01694
0.02964	0.04816	0.07224	0.09933	0.12417	0.13969
0.13969	0.12223	0.09167	0.05729	0.02865	0.01074
0.00269	0.00034				

CUMULATIVE VALUES OF P(N)
P(0) $= 0.00001$

0.00016	0.00087	0.00300	0.00751	0.01655	0.03348
0.06312	0.11128	0.18352	0.28286	0.40702	0.54671
0.68640	0.80863	0.90030	0.95759	0.98624	0.99698
0.99967	1.00000				

$$(M/M/\ 6) - (GD/\ 20/\ 20)$$

LAMBDA $= 0.10000E + 02$ LAMBDA EFF $= 0.11427E + 03$
MU $= 0.20000E + 02$ RHO $= 0.50000E + 00$

WS $= 0.75024E - 01$ WQ $= 0.25024E - 01$
LS $= 0.85729E + 01$ LQ $= 0.28595E + 01$

VALUES OF P(N) FOR N $= 0$ TO 19, OTHERWISE P(N) < 0.00001
P(0) $= 0.00016$

0.00162	0.00772	0.02315	0.04920	0.07871	0.09839
0.11479	0.12435	0.12435	0.11399	0.09499	0.07124
0.04750	0.02771	0.01385	0.00577	0.00192	0.00048
0.00008					

CUMULATIVE VALUES OF P(N)
P(0) $= 0.00016$

0.00179	0.00950	0.03265	0.08185	0.16056	0.25895
0.37374	0.49810	0.62245	0.73644	0.83144	0.90268
0.95018	0.97789	0.99174	0.99751	0.99943	0.99992
1.00000					

Figure 15–2

model and study the resulting values of the corresponding measures of performance until we obtain a specific measure (e.g., W_s) that is acceptable to the company.

The listing in Figure 15–2 is an example of the computer output for the model given $R = 4$ and $R = 6$ ($\lambda = 10$ and $\mu = 20$). It gives all the measures of performance together with $p_0, p_1, \ldots,$ and p_{20}.

We can see that with 4 repairmen, on the average 12 ($L_s = 12$) out of the 20 machines will be either in service or awaiting repair. The time until repair is completed is .15 hours. On the other hand, for 6 repairmen, there are only 8.6 machines down and the average time until repair is completed is cut by half to .075 hour. The specific choice of the number of repairmen depends on what the company regards as acceptable from the viewpoint of the number of broken machines and the time lost during repairs. In Chapter 16 we introduce some cost models for determining the "optimum" number of repairmen. ◀

Exercise 15.3–9

In Example 15.3–7, referring to the computer listing for $R = 6$, find
(a) The expected number of idle repairmen.
   [*Ans.* .286.]
(b) The probability that all repairmen are busy.
   [*Ans.* .839.]

15.4 QUEUES WITH PRIORITIES FOR SERVICE

In queueing models with priority, it is assumed that several parallel queues are formed in front of the facility with each queue, including customers belonging to certain order of priority. If the facility has m queues, we assume that queue 1 has the highest priority for service, and queue m includes customers with lowest priority. Rates of arrival and service may vary for the different priority queues. However, we shall assume that customers within each queue are served on FCFS basis.

Priority service may follow one of two rules:

1. **Preemptive** rule, where the service of a lower-priority customer may be interrupted in favor of an arriving customer with higher priority.

2. **Nonpreemptive** rule, where a customer, once in service, will leave the facility only after his service is completed and regardless of the priority of the arriving customer.

This section will not treat the preemptive case. Two nonpreemptive models applying to single and multiple servers are presented. The single-server model assumes Poisson arrivals and arbitrary service distributions. In the multiple-server case, both arrivals and departures follow the Poisson distribution. The symbol NPRP is used with the Kendall notation to represent the nonpreemptive discipline; M_i and G_i stand for Poisson and arbitrary distributions.

15.4.1 $(Mi/Gi/1):(NPRP/\infty/\infty)$

Let $F_i(t)$ be the CDF of the arbitrary service time distribution for the ith queue ($i = 1, 2, \ldots, m$) and let $E_i\{t\}$ and $\text{var}_i\{t\}$ be the mean and variance, respectively. Let λ_i be the arrival rate at the ith queue per unit time.

Define $L_q^{(k)}$, $L_s^{(k)}$, $W_q^{(k)}$, and $W_s^{(k)}$ in the usual manner except that they now represent the measures of the kth queue. Then the results of this situation are given by

$$
\begin{aligned}
W_q^{(k)} &= \frac{\sum_{i=1}^m \lambda_i (E_i^2\{t\} + \text{var}_i\{t\})}{2(1 - S_{k-1})(1 - S_k)} \\
L_q^{(k)} &= \lambda_k W_q^{(k)} \\
W_s^{(k)} &= W_q^{(k)} + E_k\{t\} \\
L_s^{(k)} &= L_q^{(k)} + \rho_k
\end{aligned}
\qquad (M_i/G_i/1):(NPRP/\infty/\infty)
$$

where

$$
\begin{aligned}
\rho_k &= \lambda_k E_k\{t\} \\
S_k &= \sum_{i=1}^k \rho_i < 1, \qquad k = 1, 2, \ldots, m \\
S_0 &\equiv 0
\end{aligned}
$$

Notice that W_q, the expected waiting time in the queue for *any* customer regardless of his or her priority, is given by

$$
W_q = \sum_{k=1}^m \frac{\lambda_k}{\lambda} W_q^{(k)}
$$

where $\lambda = \sum_{i=1}^m \lambda_i$ and λ_k/λ is the relative weight of $W_q^{(k)}$. A similar result applies to W_s.

Example 15.4–1. Jobs arrive at a production facility in three categories: rush order, regular order, and low-priority order. Although rush jobs are processed prior to any other job and regular jobs take precedence over low-priority orders, any job, once started, must be completed before a new job is taken in. Arrival of orders from the three categories are Poisson with means 4, 3, and 1 per day. The respective service rates are constant and equal to 10, 9, and 5 per day.

In this queueing situation, we have three nonpreemptive priority queues. Assume that queues 1, 2, and 3 represent the three job categories in the order given in the description of the problem. We thus have

$$
\begin{aligned}
\rho_1 &= \lambda_1 E\{t_1\} = 4(1/10) = .4 \\
\rho_2 &= 3(1/9) = .333 \\
\rho_3 &= 1(1/5) = .2
\end{aligned}
$$

We also have

$$S_1 = \rho_1 = .4$$
$$S_2 = \rho_1 + \rho_2 = .733$$
$$S_3 = \rho_1 + \rho_2 + \rho_3 = .933$$

Since $S_3 < 1$, the system can reach steady-state conditions.

We can thus compute the expected waiting time in each queue as follows. Since

$$\sum_{i=1}^{m} \lambda_i(E_i^2\{t\} + \text{var}_i\{t\}$$
$$= 4[(1/10)^2 + 0] + 3[(1/9)^2 + 0] + 1[(1/5)^2 + 0]$$
$$= .117$$

we get

$$W_q^1 = \frac{.117}{2(1-0)(1-.4)} = .0975 \text{ day} \cong 2.34 \text{ hours}$$

$$W_q^2 = \frac{.117}{2(1-.4)(1-.733)} = .365 \text{ day} \cong 8.77 \text{ hours}$$

$$W_q^3 = \frac{.117}{2(1-.733)(1-.933)} = 3.27 \text{ days} \cong 78.5 \text{ hours}$$

The expected overall waiting time for *any* customer regardless of priority is given by

$$W_q = \frac{4 \times 2.34 + 3 \times 8.77 + 1 \times 78.5}{4+3+1} = 14.27 \text{ hours}$$

We can also obtain the average number of jobs awaiting processing in each priority queue.

$$L_q^1 = 4 \times .0975 = .39 \text{ job}$$
$$L_q^2 = 3 \times .365 = 1.095 \text{ jobs}$$
$$L_q^3 = 1 \times 3.27 = 3.27 \text{ jobs} \qquad \blacktriangleleft$$

15.4.2 $(M_i/M/c):(NPRP/\infty/\infty)$

This model assumes that all customers have the same service time distribution regardless of their priorities and that all c channels have identical exponential service distribution with service rate μ. The arrivals at the kth priority queue occur according to a Poisson distribution with an arrival rate λ_k, $k = 1, 2, \ldots, m$. It can be shown for the kth queue that

$$W_q^{(k)} = \frac{E\{\xi_0\}}{(1 - S_{k-1})(1 - S_k)}, \qquad k = 1, 2, \ldots, m$$

where $S_o \equiv 0$ and

$$S_k = \sum_{i=1}^{k} \frac{\lambda_i}{c\mu} < 1, \qquad \text{for all } k$$

$$E\{\xi_0\} = \frac{1}{\mu c \left[\rho^{-c}(c - \rho)(c - 1)! \sum_{n=0}^{c-1} \frac{\rho^n}{n!} + 1 \right]}, \qquad \rho = \frac{\lambda}{\mu}$$

Example 15.4-2. To illustrate the computations in the model, suppose that we have three priority queues with arrival rates $\lambda_1 = 2$, $\lambda_2 = 5$, and $\lambda_3 = 5$ per day. There are two servers and the service rate is 10 per day. Both arrivals and departures follow Poisson distributions.

$$S_1 = \frac{\lambda_1}{c\mu} = \frac{2}{2 \times 10} = .1$$

$$S_2 = S_1 + \frac{\lambda_2}{c\mu} = .1 + \frac{5}{2 \times 10} = .35$$

$$S_3 = S_2 + \frac{\lambda_3}{c\mu} = .35 + \frac{10}{2 \times 10} = .85$$

Since all $S_i < 1$, steady state can be reached.

Now, by definition

$$\rho = \frac{\lambda_1 + \lambda_2 + \lambda_3}{\mu} = \frac{17}{10} = 1.7$$

Hence

$$E\{\xi_0\} = \frac{1}{10 \times 2\{(1.7)^{-2}(2 - 1.7)(1!)(1 + 1.7) + 1\}} = .039$$

Thus

$$W_q^{(1)} = \frac{.039}{(1 - .1)} = .0433$$

$$W_q^{(2)} = \frac{.039}{(1 - .1)(1 - .35)} = .0665$$

$$W_q^{(3)} = \frac{.039}{(1 - 0.35)(1 - .85)} = .4$$

The waiting time in the queue for *any* customer is then given by

$$W_q = \frac{\lambda_1}{\lambda} W_q^{(1)} + \frac{\lambda_2}{\lambda} W_q^{(2)} + \frac{\lambda_3}{\lambda} W_q^{(3)}$$

$$= \frac{2}{17}(.0433) + \frac{5}{17}(.0665) + \frac{10}{17}(.4) = .26$$

Finally, the expected number waiting in the queue for the entire system is given by

$$L_q = \lambda W_q = 17 \times .26 = 4.42 \qquad \blacktriangleleft$$

15.5 TANDEM OR SERIES QUEUES

This section considers Poisson queues with service stations arranged in series so that the customer must pass through all stations before completing his service. We first present the simple case of two series stations where no queues are allowed. Next, we present an important result for serial Poisson queues with no queue limit.

15.5.1 TWO-STATION SERIES MODEL WITH ZERO QUEUE CAPACITY

As an example of the analysis of queues in series, consider a simplified one-channel queueing system consisting of two series stations as shown in Figure 15–3. A customer arriving for service must go through station 1 and station 2. Service times at each station are exponentially distributed with the same service rate μ. Arrivals occur according to a Poisson distribution with rate λ. No queues are allowed in front of station 1 or station 2.

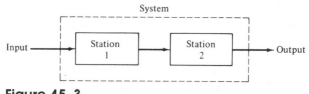

Figure 15–3

Construction of the model requires first that the states of the system at any point in time be identified. This is accomplished as follows. Each station may be either free or busy. Station 1 is said to be blocked if the customer in this station completes his or her service before station 2 becomes free. In this case the customer cannot wait between the stations since this is not allowed. Let the symbols 0, 1, and b represent the free, busy, and blocked states, respectively.

Table 15–2

		\multicolumn{5}{c}{States at $(t+h)$}
		$(0,0)$	$(0,1)$	$(1,0)$	$(1,1)$	$(b,1)$
	$(0,0)$	$1-\lambda h$		λh		
	$(0,1)$	$\mu h(1-\lambda h)$	$1-\mu h-\lambda h$		$\lambda h(1-\mu h)$	
States at t	$(1,0)$		$\mu h(1-\lambda h)$	$1-\mu h$		
	$(1,1)$		$\mu h(1-\lambda h)$	μh	$(1-\mu h)(1-\mu h)$	μh
	$(b,1)$		$\mu h(1-\lambda h)$			$1-\mu h$

Let i and j represent the states of station 1 and station 2. Then the states of the system are given by

$$\{(i,j)\} = \{(0, 0), (1, 0), (0, 1), (1, 1), (b, 1)\}$$

Define $p_{ij}(t)$ as the probability that the system is in state (i,j) at time t. The transition probabilities between times t and $t + h$ (h is a small positive increment in time) are summarized as shown in Table 15–2. The empty squares indicate that transitions between the indicated states at t and $t + h$ are impossible ($= 0$).

The following equations can now be established (neglecting the terms in h^2):

$$p_{00}(t + h) = p_{00}(t)(1 - \lambda h) + p_{01}(t)(\mu h)$$
$$p_{01}(t + h) = p_{01}(t)(1 - \mu h - \lambda h) + p_{10}(t)(\mu h) + p_{b1}(t)(\mu h)$$
$$p_{10}(t + h) = p_{00}(t)(\lambda h) + p_{10}(t)(1 - \mu h) + p_{11}(t)(\mu h)$$
$$p_{11}(t + h) = p_{01}(t)(\lambda h) + p_{11}(t)(1 - 2\mu h)$$
$$p_{b1}(t + h) = p_{11}(t)(\mu h) + p_{b1}(t)(1 - \mu h)$$

By rearranging the terms and taking the appropriate limits, the steady-state equations are given by

$$p_{01} - \rho p_{00} = 0$$
$$p_{10} + p_{b1} - (1 + \rho)p_{01} = 0$$
$$\rho p_{00} + p_{11} - p_{10} = 0$$
$$\rho p_{01} - 2p_{11} = 0$$
$$p_{11} - p_{b1} = 0$$

One of these equations is redundant. Hence adding the condition

$$p_{00} + p_{01} + p_{10} + p_{11} + p_{b1} = 1$$

the solution for p_{ij} is given by

$$p_{00} = \frac{2}{A}$$

$$p_{01} = \frac{2\rho}{A}$$

$$p_{10} = \frac{\rho^2 + 2\rho}{A}$$

$$p_{11} = p_{b1} = \frac{\rho^2}{A}$$

where

$$A = 3\rho^2 + 4\rho + 2$$

Expected number in the system may then be obtained as

$$L_s = 0p_{00} + 1(p_{01} + p_{10}) + 2(p_{11} + p_{b1}) = \frac{5\rho^2 + 4}{A}$$

Example 15.5-1. A two-station subassembly line is operated by a conveyor belt system. The size of the assembled product does not permit storing more than one unit in each station. The product arrives to the subassembly line from another production facility according to a Poisson distribution with mean 10 per hour. The assembly times at stations 1 and 2 are exponential with mean 5 minutes each. All arriving items that cannot enter the assembly line directly are diverted to other subassembly lines.

Since $\lambda = 10$ per hour and $\mu = 60/5 = 12$ per hour, we have $\rho = \lambda/\mu = 10/12 = .833$. We can compute all the probabilities by noting that

$$A = 3(.833)^2 + 4(.833) + 2 = 7.417$$

Thus,

$$p_{00} = \frac{2}{7.417} = .2697$$

$$p_{01} = \frac{2 \times .833}{7.417} = .2247$$

$$p_{10} = \frac{.833^2 + 2 \times .833}{7.417} = .3183$$

$$p_{11} = p_{b1} = \frac{.833^2}{7.417} = .0936$$

The probability that an arriving item will enter station 1 is $p_{00} + p_{01} = .2697 + .2247 = .4944$ so that the effective arrival rate is $\lambda_{\text{eff}} = .4944 \times 10 = 4.944$ jobs per hour. Since

$$L_s = \frac{5 \times .833^2 + 4 \times .833}{7.417} = .917$$

it follows that the waiting time in the system is

$$W_s = \frac{L_s}{\lambda_{\text{eff}}} = \frac{.917}{4.944} = .185 \text{ hour}$$

Observe that W_s represents the expected service time per item since no queues are allowed. We note that an item can be serviced in an average time of $5 + 5 = 10$ minutes or .167 hour provided that station 2 is not blocked. Thus the difference between W_s ($= .185$) and .167 may actually be regarded as the average time an item waits because of the blockage of state 2; that is, $.185 - .167 = .018$ hour or 1.08 minutes. ◀

Exercise 15.5-1

In the two-station series-queues model of Example 15.5–1, determine a general condition relating λ and μ such that the probability of an arriving item joining station 1 directly is higher than the probability of the item being diverted to other subassembly lines.

[*Ans.* $\lambda < .816\mu$.]

15.5.2 *k*-STATION SERIES MODEL WITH INFINITE QUEUE CAPACITY

In this section we state without proof a theorem that is applicable to a *k*-station series with unlimited interqueue capacity [see Saaty (1961), Secs. 12–2 to 12–4, for the proof].

Consider a system with *k* stations in series, as shown in Figure 15–4. Assume that arrivals at station 1 are generated from an infinite population according to a Poisson distribution with mean arrival rate λ. Serviced units will move successively from one station to the next until they are discharged from station *k*. Service time distribution at each station *i* is exponential with mean rate μ_i, $i = 1, 2, \ldots, k$. Further, there is no queue limit at any station.

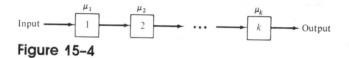

Figure 15–4

Under these conditions, it can be proved that, for all *i*, the output from station *i* (or, equivalently, the input to station *i* = 1) is Poisson with mean rate λ and that each station may be treated *independently* as $(M/M_i/1){:}(GD/\infty/\infty)$. This means that for the *i*th station, the steady-state probabilities p_{n_i} are given by

$$p_{n_i} = (1 - \rho_i)\rho_i^{n_i}, \qquad n_i = 0, 1, 2, \ldots$$

for $i = 1, 2, \ldots, k$, where n_i is the number in the system consisting of station *i* only. Steady-state results will exist only if $\rho_i = \lambda/\mu_i < 1$.

The same result can be extended to the case where station *i* includes c_i parallel servers, each with the same exponential service rate μ_i per unit time (see Figure 15–5). In this case each station may be treated independently as $(M/M_i/c_i){:}(GD/\infty/\infty)$ with mean arrival rate λ. Again, the steady-state results of Section 15.3.4 will prevail only if $\lambda < c_i\mu_i$, for $i = 1, 2, \ldots, k$.

Example 15.5–2. In a production line with five series stations, jobs arrive at station 1 according to a Poisson distribution with mean rate $\lambda = 20$ per

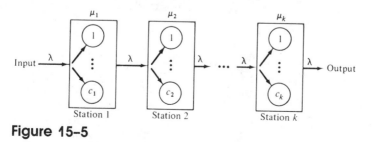

Figure 15–5

hour. The production time in each station is exponential with mean 2 minutes. The output from station k is used as input to station $i = 1$. The portion of good items produced at station i as α_i of the total input to the same station. The remaining portion, $(1 - \alpha_i)$, is defective and is discarded as scrap.

Suppose that we are interested in the size of the storage space between successive stations that will accommodate all incoming items 100β percent of the time. The probability p_{n_i} of n_i items in station i is given by

$$p_{n_i} = (1 - \rho_i)\rho_i^{n_i}$$

where $\rho_i = \lambda_i / \mu_i$. Thus the storage requirement is satisfied if the storage space for station i accommodates $N_i - 1$ items, where N_i is determined from

$$\sum_{n_i=1}^{N_i} p_{n_i} = \sum_{n_i=1}^{N_i} (1 - \rho_i)\rho_i^{n_i} \geq \beta, \qquad i = 1, 2, \ldots, 5$$

Upon simplification, we get (verify)

$$N_i \geq \frac{\ln(1 - \beta)}{\ln \rho_i} - 1, \qquad i = 1, ,2, \ldots, 5$$

Suppose that $\alpha_i = .9$; then

$$\lambda_1 = \lambda = 20$$
$$\lambda_2 = \alpha_1\lambda_1 = 20\alpha_1 = 18$$
$$\lambda_3 = \alpha_2\alpha_1\lambda_1 = 20\alpha_1\alpha_2 = 16.2$$
$$\lambda_4 = 20\alpha_1\alpha_2\alpha_3 = 14.58$$
$$\lambda_5 = 20\alpha_1\alpha_2\alpha_3\alpha_4 = 13.12$$

Since $\mu_i = \mu = 30$ per hour, we have

$$\rho_1 = .67, \quad \rho_2 = .6, \quad \rho_3 = .54, \quad \rho_4 = .486, \quad \rho_5 = .437$$

If we want to establish storage for all incoming items 99% of the time (i.e., $\beta = .99$), the limits on N_i can thus be computed as

$$N_1 \geq 10.499 \,(\cong 11), \quad N_2 \geq 8, \quad N_3 \geq 6.47 \,(\cong 7), \quad N_4 \geq 5.38 \,(\cong 6),$$
$$N_5 \geq 4.57 \,(\cong 5) \qquad \blacktriangleleft$$

Exercise 15.5–2

In Example 15.5–2, find the expected number of defective items that will accumulate each day from all stations.
[*Ans.* Approximately 197 items.]

15.6 ANALYSIS OF QUEUES BY IMBEDDED MARKOV CHAINS†

In this section two general non-Poisson models will be investigated: $(M/G/1)$:$(GD/\infty/\infty)$ and $(GI/M/1)$:$(GD/\infty/\infty)$. Models such as $(E_m/M/1)$, $(D/M/1)$, $(M/E_m/1)$, and $(M/D/1)$ are then treated as special cases of the two general models.

---

† You should review Section 10.10 before proceeding with the material in this section.

The concept of imbedded Markov chains was introduced in 1948 by D. G. Kendall. It implies that $(M/G/1)$ may be studied by observing the state of the system (number in system) at epochs of departure. The transition matrix is determined by the probabilities that certain numbers of arrivals will occur during a service time. Similarly, $(GI/M/1)$ may be studied by observing epochs of arrival with the transition matrix defined by the probabilities of certain numbers of departures occurring during an interarrival time. The indicated epochs of arrival and departure are also known as **regeneration points.**

The advantage of using this analysis is that the problem reduces to a discrete-time Markov chain even though the original system is non-Markovian. It should be noted, however, that the steady-state results obtained for this analysis are slightly different from those obtained by continuous-time analysis. This result follows, since averages are taken over discrete points in time (regeneration points). It suffices to say that for $(GI/M/1)$, if L_s and L_s^d are, respectively, the continuous and discrete time averages of the number in the system, then

$$L_s^d \leq L_s \leq L_s^d + 1$$

In the following models, this point should be kept in mind. No attempt will be made to introduce new symbolism that will reflect the new definitions.

15.6.1 $(M/G/1):(GD/\infty/\infty)$

This model derives the P-K formula of Section 15.3.3. Let $b(t)$ be the PDF of service time. The system is observed only right after a departure takes place. Such instants in time define *regeneration points*. If i customers are in the *system* at observation time, at the next observation time the number in the system j is given by

$$j = \begin{cases} k, & \text{if } i = 0 \\ i - 1 + k, & \text{if } i > 0 \end{cases}$$

where $k \ (= 0, 1, 2, \ldots)$ is the number of arrivals during a service time.

Let q_k be the probability that k persons arrive during a service time. Since arrivals occur according to a Poisson distribution with mean arrival rate λ, then

$$q_k = \int_0^\infty \frac{e^{-\lambda t}(\lambda t)^k}{k!} b(t)\, dt, \qquad k = 0, 1, 2, \ldots$$

The transition matrix between the beginning and end of a service is given by

$$\mathbf{A} = \begin{pmatrix} q_0 & q_1 & q_2 & q_3 & q_4 & \cdots \\ q_0 & q_1 & q_2 & q_3 & q_4 & \cdots \\ 0 & q_0 & q_1 & q_2 & q_3 & \cdots \\ 0 & 0 & q_0 & q_1 & q_2 & \cdots \\ 0 & 0 & 0 & q_0 & q_1 & \cdots \\ \vdots & \vdots & \vdots & \vdots & \vdots & \cdots \end{pmatrix}$$

Exercise 15.6-1

Determine the expression for q_k assuming that the service-time distribution is constant and equal to t_0. [*Ans.* q_k is Poisson with mean λt_0 and parameter k.]

Let p_n $(n = 0, 1, 2, \ldots)$ be the steady-state probabilities that there are n in the system. It can be shown that the states of the matrix $\mathbf{A}$ are ergodic for $\rho = \lambda E\{t\} < 1$. This means that, under the same condition, the corresponding Markov chain is ergodic. Thus the steady-state probabilities are given by

$$(p_0, p_1, p_2, \ldots)\mathbf{A} = (p_0, p_1, p_2, \ldots)$$

or, for $n = 0, 1, 2, \ldots$,

$$p_0 q_n + p_1 q_n + p_2 q_{n-1} + p_3 q_{n-2} + \cdots + p_{n+1} q_0 = p_n$$

These equations can now be put in the general form

$$p_n = p_0 q_n + \sum_{j=1}^{n+1} p_j q_{n+1-j}, \qquad n = 0, 1, 2, \ldots$$

The z-transform gives

$$\sum_{n=0}^{\infty} z^n p_n = \sum_{n=0}^{\infty} z^n p_0 q_n + \sum_{n=0}^{\infty} z^n \left\{ \sum_{j=1}^{n+1} p_j q_{n+1-j} \right\}$$

$$= p_0 \sum_{n=0}^{\infty} z^n q_n + \frac{1}{z} \sum_{m=1}^{\infty} z^m \left\{ \sum_{j=0}^{m} p_j q_{m-j} - p_0 q_m \right\}$$

To simplify this expression, consider

$$r_m = \sum_{j=0}^{m} p_j q_{m-j}$$

Since r_m is the convolution of p_m and q_m, it follows that (see Section 10.11.2)

$$R(z) = Q(z)P(z)$$

Thus, from the transformed equation given,

$$P(z) = p_0 Q(z) + \frac{1}{z}\{R(z) - r_0\} - \frac{p_0}{z}\{Q(z) - q_0\}$$

Since $r_0 = p_0 q_0$ and $R(z) = Q(z)P(z)$,

$$P(z) = p_0 \left\{ \frac{Q(z)}{1 + [1 - Q(z)]/(z - 1)} \right\}$$

To determine p_0, consider

$$P(1) = 1 = \lim_{z \to 1} p_0 \left\{ \frac{Q(z)}{1 + [1 - Q(z)]/(z - 1)} \right\}$$

$$= \lim_{z \to 1} p_0 \{[Q(z)(z - 1)]/[z - Q(z)]\}$$

Applying l'Hôpital's rule, one finally gets

$$p_0 \left\{ \frac{Q(1)}{1 - Q'(1)} \right\} = 1$$

Since $Q(1) = 1$,

$$p_0 = 1 - Q'(1)$$

To compute $Q'(1)$, one needs first to determine the expression for $Q(z)$.

$$Q(z) = \sum_{n=0}^{\infty} q_n z^n = \sum_{n=0}^{\infty} z^n \int_0^{\infty} \frac{(\lambda t)^n e^{-\lambda t}}{n!} b(t) dt$$

$$= \int_0^{\infty} e^{\lambda t(z-1)} b(t) dt = E\{e^{\lambda(z-1)t}\}$$

Thus

$$Q'(1) = \int_0^{\infty} (\lambda t) e^{\lambda t(z-1)} b(t)\, dt \bigg|_{z=1} = \lambda E\{t\}$$

and it follows that

$$p_0 = 1 - \lambda E\{t\}$$

(Notice that $\rho = \lambda E\{t\}$ must be less than one as dictated by the ergodicity of the Markov chain.) The expression for $P(z)$ thus becomes

$$P(z) = \frac{(1 - \lambda E\{t\})(z-1)Q(z)}{z - Q(z)}$$

A general expression is not easily secured for P_n in this case because of the complexity of $P(z)$. Any individual probability can be obtained, however, from the formula

$$P^{(n)}(0) = n! p_n$$

The expected number in the system can be obtained from $P(z)$ by using the condition

$$P^{(1)}(1) = E\{n\} = L_s$$

Thus

$$L_s = \lim_{z \to 1} P^{(1)}(z) = \lambda E\{t\} + \frac{Q''(1)}{2(1 - \lambda E\{t\})}$$

But

$$Q''(1) = \lambda^2 \int_0^{\infty} t^2 e^{\lambda t(z-1)} b(t)\, dt \bigg|_{z=1} = \lambda^2 E\{t^2\} = \lambda^2(\text{var}\{t\} + E^2\{t\})$$

Thus

$$L_s = \lambda E\{t\} + \frac{\lambda^2(\text{var}\{t\} + E^2\{t\})}{2(1 - \lambda E\{t\})}$$

This is the P–K formula obtained in Section 15.3.3.

Exercise 15.6–2
Suppose that the service time is constant and equal to t_0. Determine the probability of having one person in the system for the foregoing model by using the expression for $P(z)$.
[*Ans.* $p_1 = (e^\rho - 1)(1 - \rho)$, where $\rho = \lambda t_0$.]

15.6.2 $(M/E_m/1):(GD/\infty/\infty)$

This model will be treated as a special case of the general model just given. Let the gamma service distribution be given by

$$b(t) = \frac{m\alpha(m\alpha t)^{m-1}e^{-m\alpha t}}{(m-1)!}, \qquad t > 0$$

where

$$E\{t\} = \frac{1}{\alpha} \qquad \text{and} \qquad \text{var}\{t\} = \frac{1}{m\alpha^2}$$

Now, to compute $P(z)$, it is necessary first to compute $Q(z)$. As shown previously,

$$Q(z) = \int_0^\infty e^{\lambda t(z-1)}b(t)dt = \int_0^\infty e^{\lambda t(z-1)}\frac{m\alpha(m\alpha t)^{m-1}e^{-m\alpha t}}{(m-1)!}dt$$

$$= \frac{(m\alpha)^m}{(\lambda(1-z)+m\alpha)^m} = \left(\frac{1-\beta}{1-\beta z}\right)^m$$

where

$$\beta = \frac{\lambda}{\lambda + m\alpha}$$

Hence, letting $\rho = \lambda E\{t\} = \lambda/\alpha$, we obtain

$$P(z) = \frac{(1-\rho)(1-z)(1-\beta)^m}{(1-\beta)^m - z(1-\beta z)^m}$$

The general inverse of $P(z)$ $(= p_n)$ is also difficult to obtain in this case.

For the special case where $m = 1$, this model reduces to $(M/M/1):(GD/\infty/\infty)$. Another result can also be derived as a special case of $(M/E_m/1)$. Since

$$E\{t\} = \frac{1}{\alpha} \qquad \text{and} \qquad \text{var}\{t\} = \frac{1}{m\alpha^2}$$

it is noticed that, as $m \to \infty$, var $\{t\} = 0$. This is the characteristic of constant service time. Thus for $(M/D/1)$, it can be shown that

$$P(z) = \frac{(1-\rho)(1-z)e^{\rho(z-1)}}{e^{\rho(z-1)} - z}, \qquad \rho = \frac{\lambda}{\alpha}$$

15.6.3 $(GI/M/1):(GD/\infty/\infty)$

Let $a(v)$ be the density function of interarrival time and let μ be the service rate. The system is observed only right after an arrival occurs. Thus, if i customers are in the system at observation time, then at the next observation time, the number j in the system is given by

$$j = i + 1 - k, \qquad 0 \le k \le i$$

where k is the number of persons serviced between two successive arrivals.

Let d_k be the probability that k persons are serviced during time v. Since services (departures) occur according to a Poisson distribution with departure rate μ,

$$d_k = \int_0^\infty \frac{(\mu v)^k e^{-\mu v}}{k!} a(v)\, dv$$

The elements of the transition matrix $\mathbf{E} = \|e_{ij}\|$ are given by

$$e_{ij} = d_{i+1-j}, \qquad 1 \le j \le i+1, \qquad i \ge 0$$

$$e_{i0} = 1 - \sum_{j=1}^{i+1} d_{i+1-j} = 1 - \sum_{j=0}^{i} d_{i-j} = 1 - \sum_{j=0}^{i} d_j = h_i$$

Thus

$$\mathbf{E} = \begin{pmatrix} h_0 & d_0 & 0 & 0 & 0 & \cdots \\ h_1 & d_1 & d_0 & 0 & 0 & \cdots \\ h_2 & d_2 & d_1 & d_0 & 0 & \cdots \\ h_3 & d_3 & d_2 & d_1 & d_0 & \cdots \\ \vdots & \vdots & \vdots & \vdots & \vdots & \end{pmatrix}$$

It can be proved that all the states of $\mathbf{E}$ are ergodic for $1/\mu E\{v\} < 1$. Thus the corresponding Markov chain is ergodic under the same condition. The steady-state probabilities p_n are given by

$$(p_0, p_1, p_2, \ldots\,)\mathbf{E} = (p_0, p_1, p_1, \ldots)$$

From the arrangement of the positive elements e_{ij} of $\mathbf{E}$, it is obvious that, for $j \ge 1$,

$$p_j = \sum_{i=j-1}^{\infty} p_i e_{ij} = \sum_{k=0}^{\infty} p_{k+j-1}\, d_k, \qquad j \ge 1$$

Since in computing the steady-state probabilities of a Markov chain, one of the equations is redundant (see Section 10.10.2.C), the preceding set of equations also holds for p_0.

To solve for p_n, consider the trial solution

$$p_n = Bx^n, \qquad n \ge 0$$

where B is a constant ($\neq 0$) to be determined shortly and x is a parameter. This equation holds only for $0 < x < 1$, since, for $x = 0$, $p_n = 0$, and for $x \ge 1$, the trial solution above cannot define a probability distribution because $\sum_{n=0}^{\infty} p_n > 1$. Thus substituting the trial solution in the steady-state equations yields

$$Bx^n = \sum_{k=0}^{\infty} Bx^{k+n-1}\, d_k, \qquad n = 0, 1, 2, \ldots$$

or

$$x = \sum_{k=0}^{\infty} x^k\, d_k \equiv D(x)$$

To solve the equation

$$x = D(x)$$

it is first noticed that $D(x)$ is convex in x. Also, $D(0) = d_0 > 0$ and $D(1) = 1$. Thus plotting $x = D(x)$ as shown in Figure 15-6, it follows that $x = D(x)$ will have a solution in the range $0 < x < 1$ if at $x = 1$,

$$\frac{dD(x)}{dx} > 1$$

that is, if $D'(1) > 1$. Since

$$D(x) = \sum_{j=0}^{\infty} \int_0^{\infty} \frac{e^{-\mu v}(\mu x v)^j}{j!} a(v) \, dv$$

then

$$D'(x) = \int_0^{\infty} e^{-\mu(1-x)v} \mu v a(v) \, dv$$

and

$$D'(1) = \mu \int_0^{\infty} v a(v) \, dv = \mu E\{v\}$$

Thus $D'(1) > 1$ implies that $1/\mu E\{v\} < 1$, which guarantees the ergodicity condition of the Markov chain given.

Under the given condition, let $x = x_0$ $(0 < x_0 < 1)$ be the solution of $x = D(x)$. Now, to determine the constant B, consider

$$\sum_{i=0}^{\infty} p_i = B \sum_{i=0}^{\infty} x_0^i = \frac{B}{1 - x_0}, \qquad 0 < x_0 < 1$$

Hence

$$B = (1 - x_0)$$

and

$$p_n = (1 - x_0) x_0^n, \qquad n = 0, 1, 2, \ldots$$

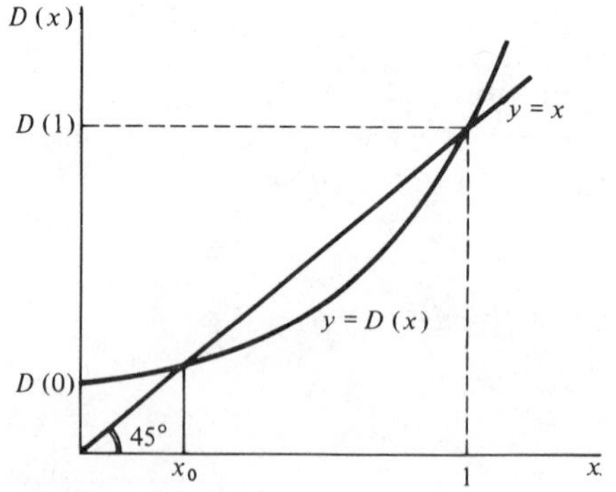

Figure 15-6

It follows that p_n has a geometric distribution that is independent of the arrivals distribution except naturally as it pertains to the determination of x_0. From this, it follows that

$$L_s = E\{n\} = \frac{x_0}{1 - x_0}$$

Distribution of the Waiting Time Based on FCFS Discipline

Since the service-time distribution is exponential, the waiting time in the *system* of an arriving person given there are n persons ahead of him is gamma distributed with parameters $(n + 1)$ and μ. Thus the CDF of the waiting time is given by

$$W(\tau) = \sum_{n=0}^{\infty} W(\tau|n+1)p_n = \sum_{n=0}^{\infty} (1 - x_0)x_0^n \int_0^\tau \frac{\mu(\mu t)^n e^{-\mu t}}{n!}\, dt$$

and the PDF is then given by

$$w(\tau) = \frac{dW(\tau)}{d\tau} = \sum_{n=0}^{\infty} \frac{\mu(\mu\tau)^n e^{-\mu\tau}(1 - x_0)x_0^n}{n!}$$
$$= \mu(1 - x_0)e^{-\mu(1-x_0)\tau}, \qquad \tau > 0$$

which is exponential with mean

$$E\{\tau\} = \frac{1}{\mu(1 - x_0)} = W_s$$

The results for $(E_m/M/1)$ and $(D/M/1)$ may also be derived as special cases of the model by using a procedure similar to the one followed in Section 15.6.1.

15.7 SUMMARY

Queueing theory provides models for analyzing the operation of service facilities in which arrival and/or service of customers occur randomly. The numerical examples introduced throughout the chapter show that queueing analysis yields results that may not be obvious intuitively.

The Poisson and exponential distributions play important roles in queueing analysis. They characterize service facilities in which both arrivals and service are *completely random*. Although other distributions can be implemented in queueing models, the analysis is much more complex than in the Poisson queues. Additionally, the complexity of the analysis does not permit securing as much information as in the Poisson models.

The big question regarding queueing theory is how good it is in practice. The limitations imposed by the mathematical analysis seems to make it difficult to find real applications that fit the model. Nevertheless, many successful queueing applications have been reported over the years. In Chapter 16 we address this point by presenting the use of queueing theory in practice.

15.8 APPENDIX: DERIVATION OF THE QUEUEING RESULTS

Arrivals Process (Section 15.2.1)

The two differential equations of the model can be solved directly by induction (see Problem 15–52). The z-transform introduced in Section 10.11 will be used, however. Thus, the first equation gives

$$\sum_{n=1}^{\infty} p_n'(t)z^n = -\sum_{n=1}^{\infty} \lambda p_n(t)z^n + \sum_{n=1}^{\infty} \lambda p_{n-1}(t)z^n$$

Adding the second equation, we get

$$\sum_{n=0}^{\infty} p_n'(t)z^n = -\sum_{n=0}^{\infty} \lambda p_n(t)z^n + \sum_{n=1}^{\infty} \lambda p_{n-1}(t)z^n$$

Letting

$$P(z,\, t) = \sum_{n=0}^{\infty} p_n(t)z^n$$

we get

$$P'(z,\, t) = \frac{d}{dt}\sum_{n=0}^{\infty} p_n(t)z^n = \sum_{n=0}^{\infty} p_n'(t)z^n$$

The z-transform equation thus becomes

$$P'(z,\, t) = -\lambda P(z,\, t) + \lambda z P(z,\, t)$$

or

$$\frac{dP(z,\, t)}{P(z,\, t)} = \lambda(z-1)\, dt$$

Solving this differential equation in t gives

$$P(z,\, t) = Be^{\lambda(z-1)t}$$

where B is a constant. Since

$$P(z,\, 0) = p_0(0) = 1$$

then $B = 1$. This yields

$$P(z,\, t) = e^{\lambda(z-1)t}$$

Using the inverse z-transform (Table 10–1, formula 11), we find that

$$Z^{-1}\{P(z,\, t)\} = Z^{-1}\{e^{\lambda(z-1)t}\} = e^{-\lambda t}Z^{-1}\{e^{\lambda tz}\}$$

This gives

$$p_n(t) = \frac{e^{-\lambda t}(\lambda t)^n}{n!}, \qquad n = 0, 1, 2, \ldots$$

which is Poisson with mean and variance equal to λt.

Departure Process (Section 15.2.2)

This model will be solved by induction. The equation for $n = 0$,

$$q_0'(t) = -\mu q_0(t)$$

has the solution

$$q_0(t) = Be^{-\mu t}, \qquad B \text{ constant}$$

From the initial condition $q_0(0) = 1$, we get $B = 1$. Thus

$$q_0(t) = e^{-\mu t}$$

Now, from the equations for $0 < n < N$, the equation for $n = 1$ together with $q_0(t) = e^{-\mu t}$ yield

$$q_1'(t) = -q_1(t) + \mu e^{-\mu t}$$

The solution of this differential equation is†

$$q_1(t) = e^{-\mu t}\left(\int \mu e^{-\mu t}e^{\mu t}dt + B\right) = (\mu t + B)e^{-\mu t}$$

Since $q_1(0) = 0$, $B = 0$ and

$$q_1(t) = \mu t e^{-\mu t}$$

It can be proved generally by induction that

$$q_n(t) = \frac{e^{-\mu t}(\mu t)^n}{n!}, \qquad n = 0, 1, 2, \ldots, N-1$$

For $n = N$, the differential equation

$$q_N'(t) = \mu q_{N-1}(t) = \mu \frac{(\mu t)^{N-1}e^{-\mu t}}{(N-1)!}$$

gives the following solution by successive integration by parts:

---

† The general differential equation of the form

$$y' + a(t)y = b(t)$$

has the solution

$$y = e^{-\int a(t)dt}\left\{\int b(t)e^{\int a(t)dt}dt + \text{constant}\right\}$$

$$q_N(t) = 1 - \sum_{n=0}^{N-1} \frac{(\mu t)^n e^{-\mu t}}{n!} = 1 - \sum_{n=0}^{N-1} q_n(t)$$

This result also follows because $\Sigma_{n=0}^{N} q_n(t) = 1$.

(M/M/1):(GD/∞/∞) (Section 15.3.1)

The difference equations just given can be written as

$$(\rho + 1)p_n = \rho p_{n-1} + p_{n+1}, \qquad n > 0$$

$$(\rho + 1)p_0 = p_0 + p_1, \qquad n = 0$$

The z-transform for the two equations is

$$(\rho + 1)P(z) = \rho z P(z) + \frac{1}{z}P(z) + \left(\frac{z-1}{z}\right)p_0$$

This yields

$$P(z) = \frac{z}{(z-1) - \rho z(z-1)}\left(\frac{z-1}{z}\right)p_0 = \left(\frac{1}{1-\rho z}\right)p_0$$

From Table 10–1, the inverse z-transform is

$$Z^{-1}\{P(z)\} = p_0 Z^{-1}\left\{\frac{1}{1-\rho z}\right\} = p_0 \rho^n$$

or

$$p_n = p_0 \rho^n, \qquad n = 0, 1, 2, \ldots$$

The value of p_0 is determined by recognizing that $\Sigma_{n=0}^{\infty} p_n = 1$. Thus

$$p_0 \sum_{n=0}^{\infty} \rho^n = p_0 \frac{1}{1-\rho} = 1$$

or

$$p_0 = 1 - \rho$$

The series $\Sigma_{n=0}^{\infty} \rho^n$ converges only if $\rho < 1$, which agrees with the steady-state requirements. This finally gives

$$p_n = (1-\rho)\rho^n, \qquad n = 0, 1, 2, \ldots$$

The mean $E\{n\} = L_s$ can be obtained from the basic definition of the expected value or from the properties of the z-transform. From Section 10.11.1 (property 4),

$$E\{n\} = P'(1) = p_0\left\{\frac{\rho}{(1-\rho z)^2}\right\}\bigg|_{z=1} = \frac{\rho p_0}{(1-\rho)^2}$$

Since $p_0 = 1 - \rho$,

$$L_s = E\{n\} = \frac{\rho}{1-\rho}$$

$(M/M/1){:}(GD/N/\infty)$ (Section 15.3.2)

The steady-state difference equations can be written as

$$(1 + \rho)p_0 = p_1 + p_0, \qquad n = 0$$
$$(1 + \rho)p_n = p_{n+1} + \rho p_{n-1}, \qquad 0 < n < N$$
$$(1 + \rho)p_N = \rho p_{N-1} + \rho p_N, \qquad n = N$$

Taking the z-transform of these equations, we get

$$(1 + \rho) \sum_{n=0}^{N} z^n p_n = \sum_{n=0}^{N-1} p_{n+1} z^n + \rho \sum_{n=1}^{N} p_{n-1} z^n + p_0 + \rho p_N z^N$$

Since $p_n = 0$ for $n > n$, we get after simplification,

$$(1 + \rho)P(z) = \frac{1}{z}\{P(z) - p_0\} + \rho z\{P(z) - p_N z^N\} + p_0 + \rho p_N z^N$$

or

$$P(z) = p_0 \left(\frac{1}{1 - \rho z} \right) - \rho p_N \left(\frac{z^{N+1}}{1 - \rho z} \right)$$

Taking the inverse z-transform, then from formula 7, Table 10–1, we obtain

$$Z^{-1} \left\{ \frac{p_0}{1 - \rho z} \right\} = p_0 \rho^n, \qquad n = 0, 1, 2, \dots, N$$

and from formula 1 in the same table,

$$Z^{-1} \left\{ \rho p_N \left(\frac{z^{N+1}}{1 - \rho z} \right) \right\} = \begin{cases} 0, & n = 0, 1, 2, \dots, N, \\ p_N \rho^{n-N}, & n = N+1, N+2, \dots \end{cases}$$

Thus, combining the two terms, we get

$$p_n = \begin{cases} p_0 \rho^n, & n = 0, 1, \dots, N \\ p_0 \rho^n - p_N \rho^{n-N}, & n = N+1, N+2, \dots \end{cases}$$

Since

$$p_0 \rho^n - p_N \rho^{n-N} = p_0 \rho^n - p_0 \rho^N \rho^{n-N} = 0$$

then

$$p_n = p_0 \rho^n, \qquad n = 0, 1, \dots, N$$

The value of p_0 can be determined from†

$$p_0 \sum_{n=0}^{N} \rho^n = p_0 \left(\frac{1 - \rho^{N+1}}{1 - \rho} \right) = 1$$

---

† Given $S = 1 + \rho + \rho^2 + \cdots + \rho^N$, then $\rho S = \rho + \rho^2 + \cdots + \rho^{N+1}$. Thus

$$S - S\rho = 1 - \rho^{N+1} \qquad \text{or} \qquad S = \frac{1 - \rho^{N+1}}{1 - \rho}$$

Notice that the derivation does not require ρ to be less than 1.

or

$$p_0 = \frac{1-\rho}{1-\rho^{N+1}}$$

Hence

$$p_n = \left(\frac{1-\rho}{1-\rho^{N+1}}\right)\rho^n, \qquad n = 0, 1, 2, \ldots, N$$

Consequently,

$$
\begin{aligned}
L_s = E\{n\} &= \sum_{n=0}^{N} np_n \\
&= \frac{1-\rho}{1-\rho^{N+1}} \sum_{n=0}^{N} n\rho^n \\
&= \frac{1-\rho}{1-\rho^{N+1}} \rho \frac{d}{d\rho}\left(\frac{1-\rho^{N+1}}{1-\rho}\right) \\
&= \frac{\rho\{1-(N+1)\rho^N + N\rho^{N+1}\}}{(1-\rho)(1-\rho^{N+1})}
\end{aligned}
$$

after simplification. This expression can be derived readily from the z-transform expression (see Problem 15–59).

$(M/G/1){:}(GD/\infty/\infty)$ (Section 15.3.3)

Let

$f(t) =$ service-time distribution with mean $E\{t\}$ and variance $\mathrm{var}\{t\}$
$n =$ number of customers in *the system* right after a customer departs
$t =$ time to service the customer following the one that departed
$k =$ number of new arrivals during t
$n' =$ number of customers left behind the next departing customer

These symbols are illustrated graphically in Figure 15–7, where T represents the time when the jth customer departs and $(T + t)$ represents the time when the next customer, $(j + 1)$st, departs. The notation $j, j + 1, \ldots$ does not necessarily mean that customers are introduced into service on FCFS discipline. Rather, it identifies the different customers departing from the system. Thus, the results of this model are applicable to any of the three service disciplines, FCFS, LCFS, and SIRO.

By steady-state assumption,

$$E\{n\} = E\{n'\} \qquad \text{and} \qquad E\{n^2\} = E\{(n')^2\}$$

Figure 15–7 shows that

$$n' = \begin{cases} k, & \text{if } n = 0 \\ n - 1 + k, & \text{if } n > 0 \end{cases}$$

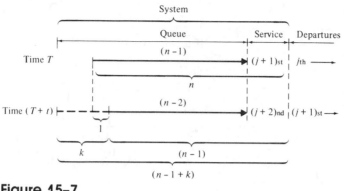

Figure 15-7

Let

$$\delta = \begin{cases} 0, & \text{if } n = 0 \\ 1, & \text{if } n > 0 \end{cases}$$

then

$$n' = n - \delta + k$$

Taking expectation of both sides, we obtain

$$E\{n'\} = E\{n\} - E\{\delta\} + E\{k\}$$

Since $E\{n\} = E\{n'\}$, it follows that

$$E\{\delta\} = E\{k\}$$

Also,

$$(n')^2 = (n + k - \delta)^2 = n^2 + k^2 + 2nk + \delta^2 - 2n\delta - 2k\delta$$

But, by definition, $\delta^2 = \delta$ and $\delta n = n$. Hence

$$(n')^2 = n^2 + k^2 + 2nk + \delta - 2n - 2k\delta$$

Taking expectation and realizing that $E\{(n')^2\} = E\{n^2\}$ gives

$$E\{n\} = \frac{E\{k^2\} - E\{\delta\}(2E\{k\} - 1)}{2(1 - E\{k\})}$$

Substituting for $E\{\delta\} = E\{k\}$ as obtained yields

$$E\{n\} = \frac{E\{k^2\} - E\{k\}(2E\{k\} - 1)}{2(1 - E\{k\})} = \frac{E\{k^2\} + E\{k\} - 2E^2\{k\}}{2(1 - E\{k\})}$$

Now it is necessary to determine $E\{k\}$ and $E\{k^2\}$. Since arrivals occur according to Poisson distribution,

$$E\{k \mid t\} = \lambda t \qquad \text{and} \qquad E\{k^2 \mid t\} = (\lambda t)^2 + \lambda t$$

Hence

$$E\{k\} = \int_0^\infty E\{k \mid t\} f(t) \, dt = \int_0^\infty \lambda t f(t) \, dt = \lambda E\{t\}$$

Also,

$$E\{k^2\} = \int_0^\infty E\{k^2 \mid t\} f(t) \, dt = \int_0^\infty \{(\lambda t)^2 + \lambda t\} f(t) \, dt$$

$$= \lambda^2 \, \mathrm{var}\, \{t\} + \lambda^2 E^2\{t\} + \lambda E\{t\}$$

Thus, after simplification,

$$L_s = E\{n\} = \lambda E\{t\} + \frac{\lambda^2 (E^2\{t\} + \mathrm{var}\, \{t\})}{2(1 - \lambda E\{t\})}$$

Notice that $\lambda E\{t\} < 1$; otherwise, L_s becomes negative and hence undefined.

$(M/M/c)$:$(GD/\infty/\infty)$ (Section 15.3.4)

The value of p_0 is determined from $\Sigma_{n=0}^\infty p_n = 1$, which gives

$$p_0 = \left\{ \sum_{n=0}^{c-1} \frac{\rho^n}{n!} + \frac{\rho^c}{c!} \sum_{n=c}^\infty \frac{\rho^{n-c}}{c^{n-c}} \right\}^{-1} = \left\{ \sum_{n=0}^{c-1} \frac{\rho^n}{n!} + \frac{\rho^c}{c!} \sum_{j=0}^\infty \left(\frac{\rho}{c} \right)^j \right\}^{-1}$$

$$= \left\{ \sum_{n=0}^{c-1} \frac{\rho^n}{n!} + \frac{\rho^c}{c!} \left(\frac{1}{1 - \rho/c} \right) \right\}^{-1}, \qquad \frac{\rho}{c} < 1$$

The expression for L_q is obtained as follows.

$$L_q = \sum_{n=c}^\infty (n - c) p_n = \sum_{k=0}^\infty k p_{k+c} = \sum_{k=0}^\infty \frac{k \rho^{k+c}}{c^k c!} p_0$$

$$= p_0 \frac{\rho^c}{c!} \frac{\rho}{c} \sum_{k=0}^\infty k \left(\frac{\rho}{c} \right)^{k-1} = p_0 \frac{\rho^c}{c!} \frac{\rho}{c} \left[\frac{1}{(1 - \rho/c)^2} \right]$$

$$= \left[\frac{\rho^{c+1}}{(c-1)!(c-\rho)^2} \right] p_0 = \left[\frac{c\rho}{(c-\rho)^2} \right] p_c$$

$(M/M/c)$:$(GD/N/\infty)$ (Section 15.3.5)

$$L_q = \sum_{n=c+1}^N (n - c) p_n = \sum_{j=1}^{N-c} j p_{j+c} = p_0 \frac{\rho^c}{c!} \frac{\rho}{c} \sum_{j=1}^{N-c} j \left(\frac{\rho}{c} \right)^{j-1}$$

$$= p_0 \frac{\rho^{c+1}}{(c-1)!(c-\rho)^2} \left\{ 1 - \left(\frac{\rho}{c} \right)^{N-c} - (N - c) \left(\frac{\rho}{c} \right)^{N-c} \left(1 - \frac{\rho}{c} \right) \right\}$$

To determine λ_{eff}, consider

$$L_q = \sum_{n=c+1}^N (n - c) p_n = \sum_{n=0}^N n p_n - \sum_{n=0}^c n p_n - c \left(1 - \sum_{n=0}^c p_n \right)$$

$$= L_s - \left\{ c - \sum_{n=0}^c (c - n) p_n \right\} = L_s - (c - \bar{c})$$

or

$$L_s = L_q + (c - \bar{c})$$

SELECTED REFERENCES

GROSS, D., and C. HARRIS, *Fundamentals of Queueing Theory,* Wiley, New York, 1974.

LEE, A., *Applied Queueing Theory,* Macmillan, Toronto, Canada, 1966.

MORSE, P., *Queues, Inventories, and Maintenance,* Wiley, New York, 1958.

PARZEN, E., *Stochastic Processes,* Holden-Day, San Francisco, 1962.

SAATY, T., *Elements of Queueing Theory,* McGraw-Hill, New York, 1961.

REVIEW QUESTIONS

True (T) or False (F)?

1. _____ The mean and variance of the Poisson distribution may not be equal.

2. _____ In queueing theory, if the arrivals occur according to a Poisson distribution, the interarrival time is exponential.

3. _____ Under the Poisson assumption, two arrivals can occur during a very small time interval.

4. _____ The arrival rate in the Poisson distribution equals the mean of the exponential interarrival time.

5. _____ If arrivals occur according to a Poisson distribution, the time interval since the occurrence of the last arrival will affect the probability of the remaining time until the next arrival takes place.

6. _____ If the time between successive arrivals is exponential, the time between the occurrences of every third arrival is also exponential.

7. _____ If a waiting customer becomes impatient, he may elect to *renege*.

8. _____ An arriving customer may *balk* if he expects a long waiting time.

9. _____ *Jockeying* is exercised by customers in a multiple-server facility in the hope of reducing their waiting time.

10. _____ The distribution of waiting time is independent of the service discipline used in selecting the waiting customers for service.

11. _____ A facility that limits the maximum number of customers that can be admitted in the system should be expected to yield less average waiting time than a facility that accepts all arriving customers.

12. _____ In a service facility with random arrivals, the average waiting time per customer will be smaller when a random service time is replaced by a constant service time.

13. _____ The effective arrival rate at a facility can never exceed the (uncensored) arrival rate from the calling source.

14. _____ A knowledge of the steady-state probabilities of the number of customers in the system can be used to determine all the basic measures of

performance in a queueing situation, regardless of the types of distributions describing the arrivals and departures.

15. ____ In a single-server queueing model, steady state can be reached after a sufficiently long period only if the arrival rate is less than the service rate, unless the capacity of the queue size is limited.

16. ____ Pooling of identical service facilities cannot lead to a reduction in average waiting time, particularly if the individual facilities have very high utilization factors.

17. ____ A self-service gas station can be modeled as a self-service queueing model.

18. ____ A parking lot can be modeled as a queueing situation in which the number of servers equals the number of available spaces.

19. ____ In preemptive priority queues, a service cannot be interrupted in favor of an arriving customer with higher priority.

20. ____ The output of a Poisson queueing model (number of departures) is also Poisson.

[*Ans.* 1—F, 2—T, 3—F, 4—F, 5—F, 6—F, 7—T, 8—T, 9—T, 10—F, 11 to 15—T, 16—F, 17—F, 18—T, 19—F, 20—T.]

NUMERICAL PROBLEMS

Section	Assigned Problems	Section	Assigned Problems
15.1	15–1, 15–2	15.3.5	15–32 to 15–34
15.2.1	15–3 to 15–9	15.3.6	15–35, 15–36
15.2.2	15–10 to 15–13	15.3.7	15–37 to 15–40
15.3.1	15–14 to 15–20	15.4	15–41 to 15–44
15.3.2	15–21 to 15–23	15.5	15–45 to 15–47
15.3.3	15–24 to 15–26	15.6	15–48 to 15–51
15.3.4	15–27 to 15–31		

☐ **15–1** In each of the following situations, identify the basic elements of the queueing model (i.e., customer, server, design of the facility, service discipline, and limits on the calling source and queue).
   (a) Shoppers in front of checkout stands in a supermarket.
   (b) Cars waiting at a stop light.
   (c) An outpatient clinic.
   (d) Planes taking off in an airport.
   (e) Toll gates on a superhighway.
   (f) A computer center.
   (g) 24-hour bank tellers.

☐ **15–2** Study the following system and identify all the associated queueing situations. For each situation, define the customers, the server(s), the service discipline, the service time, the maximum queue length, and the calling source.

Orders for jobs are received at a workshop for processing. Upon receipt, the supervisor decides whether it is a rush job or a regular job. Some of these orders require use of one type of machine, of which several are available. The remaining orders are processed in a two-stage production line, of which only two are available. In each of the two groups, one facility is especially assigned to handle rush jobs. Jobs arriving at any facility are processed in order of arrival. Completed orders are shipped upon arrival from a terminal shipping zone having a limited capacity.

Sharpened tools for the different machines are supplied from a central tool crib, where operators exchange old tools for new ones. When a machine breaks down, a repairman is called from the service pool to attend it. Machines working on rush orders always receive priorities both in acquiring new tools from the crib and in receiving repair service.

☐ **15–3** During a very small time interval h, at most one arrival can occur. The probability of an arrival occurring is directly proportional to h with the proportionality constant equal to 2. Determine the following:
  (a) The average time between two successive arrivals.
  (b) The probability that no arrivals occurs during a period of .5 time unit.
  (c) The probability that time between two successive arrivals is at least 3 time units.
  (d) The probability that the time between two successive arrivals is at most 2 time units.

☐ **15–4** Customers arrive at a facility according to a Poisson distribution at the rate of two an hour. Find the following:
  (a) The average number of customers arriving in an 8-hour period.
  (b) The probability that there will be at least one customer in a 1-hour period.

☐ **15–5** Customers arrive at a restaurant according to a Poisson distribution at the rate of 20 per hour. The restaurant opens for business at 11:00 A.M. Find the following:
  (a) The probability of having 20 customers in the restaurant at 11:12 A.M. given that there were 18 at 11:07 A.M.
  (b) The probability a new customer will arrive between 11:28 and 11:30 A.M. given that the last customer arrived at 11:25 A.M.

☐ **15–6** Books previously ordered arrive at a university library according to a Poisson distribution at the rate of 25 books per day. Each shelf in the stacks can hold 100 books. Determine the following:
  (a) The expected number of shelves that will be stacked with new books each month.
  (b) The probability that more than 10 book cases will be needed each month given that a book case has five shelves.

☐ **15–7** Two employees, Ann and Jim, of a fast-food restaurant play the following game while waiting for customers to arrive. Jim pays Ann 1 cent if the next customer does not arrive within 1 minute; otherwise, Ann pays Jim 1 cent. Determine Jim's expected gain in an 8-hour period assuming that the customers arrive according to a Poisson distribution with mean rate one per minute.

□ **15–8** Consider Problem 15–7. Suppose the game is such that Jim would pay Ann 1 cent if the next customer arrives after 1.5 minutes, whereas Ann would pay Jim 1 cent if the next customer's arrival is within 1 minute. Determine Jim's expected winnings in an 8-hour period.

□ **15–9** Use the computer program in Appendix D to compute the Poisson probabilities under the following conditions.
   (a) Arrival rate per day is 12.12 customers.
   (b) Same rate as in part (a), but the probabilities are computed for a 2-day period.

□ **15–10** Inventory is withdrawn from a stock of 80 items according to a Poisson distribution at the rate of 5 items a day.
   (a) Find the probability that 10 items are withdrawn during the first 2 days.
   (b) Determine the probability that no items are left at the end of 4 days.
   (c) Determine the *average* number of items withdrawn over a 4-day period.

□ **15–11** Solve Problem 15–10 by using the computer program in Appendix D.

□ **15–12** A machine shop has just stocked 10 pieces of a spare part for the repair of a machine. Stock replenishments of size 10 pieces each take place every 7 days. The Poisson breakdown of the machine occurs three times a week on the average. Determine the probability that the machine will remain broken because of the unavailability of parts for 2 days, for 5 days.

□ **15–13** Demand for an item occurs according to a Poisson distribution with mean 3 a day. The maximum stock level is 25 items, which occurs on each Monday, immediately after a new order is received. The order size thus depends on the number of units left at the end of work week on Saturday (the business is closed on Sundays). Determine the following:
   (a) The *average* weekly size of the orders.
   (b) The probability of incurring shortage in demand after 4 working days.
   (c) The probability that the weekly order size will exceed 5 units.

□ **15–14** A fast-food restaurant has *one* drive-in window. It is estimated that cars arrive according to a Poisson distribution at the rate of 2 every 5 minutes and that there is enough space to accommodate a line of 10 cars. Other arriving cars can wait outside this space, if necessary. It takes 1.5 minutes on the average to fill an order, but the service time actually varies according to an exponential distribution. Determine the following:
   (a) The probability that the facility is idle.
   (b) The expected number of customers waiting but currently not being served.

(c) The expected waiting time until a customer can place his order at the window.

(d) The probability that the waiting line will exceed the capacity of the space leading to the drive-in window.

☐ **15–15** Solve Problem 15–14 by using the computer program in Appendix D.

☐ **15–16** Cars arrive at a toll gate on a freeway according to a Poisson distribution with mean 90 per hour. Average time for passing through the gate is 38 seconds. Drivers complain of the long waiting time. Authorities are willing to decrease the passing time through the gate to 30 seconds by introducing new automatic devices. This can be justified only if under the old system the number of waiting cars exceeds 5. In addition, the percentage of the gate's idle time under the new system should not exceed 10%. Can the new device be justified?

☐ **15–17** Customers arrive at a one-window drive-in bank according to a Poisson distribution with mean 10 per hour. Service time per customer is exponential with mean 5 minutes. There are three spaces in front of the window, including that for the car being serviced. Other arriving cars can wait outside these three spaces.

(a) What is the probability that an arriving customer can enter one of the three spaces in front of the window?

(b) What is the probability that an arriving customer will have to wait outside the three spaces?

(c) How long is an arriving customer expected to wait before starting service?

(d) How many car spaces should be provided in front of the window so that an arriving customer can wait in front of the window at least 20% of the time?

☐ **15–18** Consider Problem 15–14. Determine the probability that the waiting time per customer will exceed the average waiting time in the queue.

☐ **15–19** To attract more business, the owner of the fast-food restaurant in Problem 15–14 decided to give a free drink to each customer that waits more than 5 minutes for service. Normally, a drink costs 50 cents. How much is the owner expected to pay daily for free drinks? Assume that the restaurant is open for 12 hours daily.

☐ **15–20** Solve Problem 15–14 assuming that customers who cannot join the line in front of the service window will normally go elsewhere.

☐ **15–21** Solve Problem 15–20 by using the computer program in Appendix C.

☐ **15–22** A cafeteria can seat a maximum of 50 persons. Customers arrive in a Poisson stream at the rate of 10 per hour. They are serviced at the rate of 12 per hour.

(a) What is the probability that the next customer will not eat in the cafeteria because it is full?

(b) Suppose that three customers (with random arrival times) would like to be seated together. What is the probability that their wish cannot be fulfilled? (Assume that arrangements can be made to seat them together as long as there are three empty seats anywhere in the cafeteria.)

☐ **15–23** Patients arrive at a clinic according to a Poisson distribution at a rate of 30 patients per hour. The waiting room does not accommodate more than 14 patients. Examination time per patient is exponential with mean rate 20 per hour.

(a) Find the effective arrival rate at the clinic.

(b) What is the probability that an arriving patient will not wait? Will find a vacant seat in the room?

(c) What is the expected waiting time until a patient is discharged from the clinic?

☐ **15–24** Solve Example 15.3–4 assuming that the service-time distribution is given as follows:

(a) Uniform from $t = 5$ minutes to $t = 15$ minutes.

(b) Normal with mean 9 minutes and variance 4 minutes^2.

(c) Discrete with values equal to 5, 10, and 15 minutes and probabilities 1/4, 1/2, and 1/4, respectively.

☐ **15–25** A production line consists of two stations. The product must pass through the two stations serially. The time the product spends in the first station is constant and equal to 30 minutes. The second station makes an adjustment (and minor changes) and hence its time will depend on the condition of the item as it is received from station 1. It is estimated that the time in station 2 is uniform between 5 and 10 minutes. The items are received at station 1 in a Poisson stream at the rate of one every 40 minutes. Because of the size of the items, a new unit cannot enter the production line until the one already in the facility clears station 2. Determine the expected number of items waiting in front of station 1.

☐ **15–26** Service is performed in a service facility in three consecutive stages. The service time at each stage is exponential with mean 10 minutes. A new customer must wait until the one already in service passes through stage 3. Customers arrive at stage 1 according to a Poisson distribution with mean rate one per hour. Determine the expected number of customers waiting at stage 1.

☐ **15–27** In Example 15.3–5, use the computer program in Appendix C to compare the performances of using 4, 5, 6, or 7 cars in the consolidated taxi company.

☐ **15–28** In $(M/M/2):(GD/\infty/\infty)$, mean service time is 5 minutes and mean interarrival time is 8 minutes.

(a) What is the probability of a delay?
(b) What is the probability of at least one of the servers being idle?
(c) What is the probability that both servers are idle?

☐ **15–29** A computer center is equipped with three digital computers, all of the same type and capability. The number of users in the center at any time is equal to 10. For *each* user, the time for writing (and key punching) a program is exponential with mean rate .5 per hour. Once a program is completed, it is sent directly to the center for execution. The computer time per program is exponentially distributed with mean rate 2 per hour. Assuming that the center is in operation on a full-time basis, and neglecting the effect of computer downtime, find the following.
   (a) The probability that a program is not executed immediately upon receipt at the center.
   (b) The average time until a program is released from the center.
   (c) The average number of programs awaiting execution.
   (d) The expected number of idle computers.
   (e) The percentage of time the computer center is idle.
   (f) The average percentage of idle time *per computer.*

☐ **15–30** An airport terminal services three types of customers: those arriving from rural areas, those arriving from suburban areas, and the transit customers who are changing planes at the airport. The arrivals distribution for each of the three groups is assumed Poisson with mean arrival rates 10, 5, and 7 per hour, respectively. Assuming that all customers require the same type of service at the terminal and that the service time is exponential with mean rate 10 per hour, how many counters should be provided at the terminal under each of the following conditions?
   (a) The expected waiting time in the system per customer does not exceed 15 minutes.
   (b) The expected number of customers in the system is at most 10.
   (c) The probability that all counters are idle does not exceed .11.

☐ **15–31** In a bank customers arrive in a Poisson stream with mean 36 per hour. The service time per customer is exponential with mean .035 hour. Assuming that the system can accommodate at most 30 customers at a time, how many tellers should be provided under each of the following conditions?
   (a) The probability of having more than 3 customers waiting is less than .20.
   (b) The expected number in the system does not exceed 3.

☐ **15–32** In a parking lot there are 10 parking spaces only. Cars arrive according to a Poisson distribution with mean 10 per hour. The parking time is exponentially distributed with mean 10 minutes. Find the following.
   (a) The expected number of empty parking spaces.
   (b) The probability that an arriving car will not find a parking space.
   (c) The effective arrival rate of the system.

☐ **15-33** For $(M/M/5):(GD/20/\infty)$, the following probabilities are computed based on $\lambda = 1/3$ and $\mu = 1/12$.

n	p_n	n	p_n	n	p_n
0	.013	7	.072	14	.015
1	.053	8	.058	15	.012
2	.105	9	.046	16	.010
3	.141	10	.037	17	.008
4	.141	11	.029	18	.006
5	.112	12	.024	19	.005
6	.090	13	.019	20	.004

(a) Compute L_s and L_q and show that $\lambda_{\text{eff}} = \mu(L_s - L_q) = \mu(c - \bar{c}) = \lambda(1 - p_{20})$, where $c = 5$ and $\bar{c}$ is the expected number of idle channels.
(b) Compute W_s and W_q.

☐ **15-34** Solve Example 15.3–6 by using the computer program in Appendix D.

☐ **15-35** Verify the computations in Table 15–1 by using the computer program in Appendix D.

☐ **15-36** In a self-service facility arrivals occur according to a Poisson distribution with mean 50 per hour. Service time per customer is exponentially distributed with mean 5 minutes.
(a) Find the expected number of customers in service.
(b) What is the percentage of time the facility is idle?

☐ **15-37** Ten machines are being attended by a single overhead crane. When a machine finishes its load, the overhead crane is called to unload the machine and to provide it with a new load from an adjacent storage area. The machine time per load is assumed exponential with mean 30 minutes. The time from the moment the crane moves to service a machine until a new load is installed is also exponential with mean 10 minutes.
(a) Find the percentage of time the crane is idle.
(b) What is the expected number of machines waiting for crane service?

☐ **15-38** Two repairmen are attending five machines in a workshop. Each machine breaks down according to a Poisson distribution with mean 3 per hour. The repair time per machine is exponential with mean 15 minutes.
(a) Find the probability that the two repairmen are idle, that one repairman is idle.
(b) What is the expected number of idle machines not being serviced?

☐ **15-39** Consider the machine servicing models, $(M/M/1):(GD/6/6)$ and $(M/M/3):(GD/20/20)$. The rate of breakdown per machine is one per hour and the service rate is 10 per hour. The probabilities p_n are computed for $(M/M/1):(GD/6/6)$ as

n	p_n	n	p_n	n	p_n
0	.4845	3	.0582	5	.0035
1	.2907	4	.0175	6	.0003
2	.1454				

and for $(M/M/3):(GD/20/20)$ as

n	p_n	n	p_n	n	p_n
0	.13625	5	.04694	10	.00070
1	.27250	6	.02347	11	.00023
2	.25890	7	.01095	12	.00007
3	.15533	8	.00475	13⎫	
4	.08802	9	.00190	⋮ ⎬ .0000	
				20⎭	

Show that although in the first model one repairman is assigned to 6 machines and in the second model each repairman is responsible for $6\frac{2}{3}$ machines, the second model yields a smaller expected waiting time per machine. Justify this conclusion.

☐ **15–40** Solve the following problems by the computer program in Appendix D.
  (a) Problem 15–37.
  (b) Problem 15–38.

☐ **15–41** Job orders arriving at a production facility are divided into three groups. Group 1 will take the highest priority for processing; group 3 will be processed only if there are no waiting orders from groups 1 and 2. It is assumed that a job once admitted to the facility must be completed before any new job is taken in. Orders from groups 1, 2, and 3 occur according to Poisson distributions with means 4, 3, and 2 per day, respectively. The service times for the three groups are *constant* with rates 10, 9, and 10 per day, respectively. Find the following.
  (a) The expected waiting time in the system for each of the three queues.
  (b) The expected waiting time in the system for *any* customer.
  (c) The expected number of waiting jobs in each of the three groups.
  (d) The expected number waiting in the system.

☐ **15–42** Repeat Problem 15–41 given that the service-time distributions are *exponential* with service rates 10, 9, and 10 per day, respectively.

☐ **15–43** Suppose in Problem 15–41 that there are three production facilities in parallel. The service-time distribution for each facility is negative exponential with mean 5 minutes. Find the expected number of waiting jobs in each group. What is the total number of waiting jobs?

☐ **15–44** Repeat Problem 15–43 assuming that there are five production facilities in parallel and compare the results.

☐ **15–45** In a production line suppose that there are k stations in series (Figure 15–8). Assume that jobs arrive at station 1 from an infinite source according to a Poisson distribution with mean rate λ per unit time. The output from station i is used as input to station $i + 1$. Because there are defective items at each station, the percentage of good items from station i is equal to $100\alpha_i$, $0 \leq \alpha_i \leq 1$. The remaining percentage $100(1 - \alpha_i)$ represents the defectives at station i. Assume that service time distribution at station i is exponential with mean rate μ_i per unit time.

  (a) Derive a general expression for determining storage space associated with each station i such that all the arriving (good) items can be accommodated $\beta\%$ of the time.
  (b) Let $\lambda = 20$ items per hour and $\mu_i = 30$ items per hour for all stations. Percentage of defectives at each station may be assumed constant and equal to 10%. Give a numerical answer for part (a) given $k = 5$ and $\beta = 95\%$.
  (c) Using the data in part (b), what is the expected number of defective items from all stations during a period of T hours?

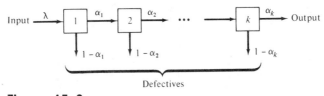

Figure 15–8

☐ **15–46** Suppose in Problem 15–45 that there is one rework station associated with each station in the production line, as shown in Figure 15–9. Defective items are reworked in these "rework" stations and are sent to the station succeeding the one from which they arrive. Assume that for the ith rework station the service-time distribution is exponential with mean rate γ_i and that the percentage of items that can be reworked successfully is equal to $\delta_i\%$.

  (a) Answer part (b) in Problem 15–45 if in addition, $\gamma_i = 4$ items per hour for all i and $\delta_i = 1/(i + 1)$, $i = 1, 2, \ldots, 5$.
  (b) How much space must be provided for each rework station to accommodate all incoming defectives 90% of the time?
  (c) What is the average number of defective items in each rework station (queue + in service)?
  (d) What is the expected waiting time until an arriving item at station 1 is released from station $k = 5$?
  (e) Answer parts (a), (b), (c), and (d) assuming that $\delta_i = 1$ for all i.

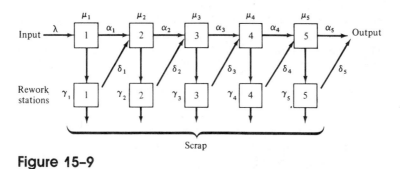

Figure 15-9

☐ **15–47** In Problem 15–46 suppose that all the rework stations are pooled into parallel channels (Figure 15–10). The service-time distribution for each channel is exponential with the same rate γ. Assume that the percentage of reworked items at each channel is equal to δ and the output from the pooled facility is distributed back to the respective production stations according to the same input ratios to the facility.

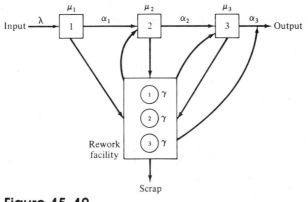

Figure 15-10

Let $k = 3$, $\gamma = 10$, $\mu_i = 15$ for all i, $\alpha_i = 1/2i$, $i = 1, 2, 3$, $\gamma = 10$, and $\delta = 90\%$. Find the following.
(a) The probability of having three items in the production line.
(b) The probability of having two or more items in the rework facility.
(c) The average number of items awaiting processing at each production line station.
(d) The expected waiting time for an arriving item that does not go through a rework facility until it is released from station k.
(e) The expected number in the entire system comprising both production line and rework facility.

☐ **15–48** Repeat Problem 15–47 assuming that $\delta = 75\%$. All the remaining information is unchanged.

☐ **15–49** Arrivals at a single-server queueing system occur according to a Poisson distribution with mean arrival rate of five per hour. If service time (in minutes) follows the uniform distribution,

$$f(x) = \begin{cases} 1/10, & 5 \le x \le 15 \\ 0, & \text{otherwise} \end{cases}$$

find the following.
- (a) The probability that system is busy.
- (b) The expected number of customers in system.
- (c) The expected waiting time in queue.

☐ **15–50** Consider the single-server queueing system, where the server upon commencing on a new customer will decide whether he needs to be serviced or not. The probability that a customer will not need any service is .3. Service time per serviced customer is 10 minutes. If customers arrive according to a Poisson distribution with mean 5 per hour, find the following.
- (a) The probability that there are 10 persons in the system after a service is completed given that there were 4 persons in the system at start of this service.
- (b) The probability that there is no one in the system.
- (c) The expected waiting time in the system.

☐ **15–51** Consider a single-server queueing system with a constant interarrival time of 10 minutes. Departures occur according to a Poisson distribution with mean 10 per hour. Find the following.
- (a) The probability that there are two in the system when an arrival occurs given that there were exactly four in system when the immediately preceding arrival occurred.
- (b) The probability that the system will have two or three customers when an arrival occurs given that the system had exactly three customers when the immediately preceding arrival occurred.
- (c) The expected number in the system.
- (d) The expected waiting time in the system.
- (e) The probability that there is no one in the system; at least two in the system.

THEORETICAL PROBLEMS

Section	Assigned Problems
15.2	15–52, 15–54
15.3	15–55 to 15–69
15.4	none
15.5	15–70
15.6	15–71 to 15–75

☐ **15–52** Solve the difference equations of the model in Section 15.2.1 by using induction.

☐ **15–53** Consider the pure birth process (Section 15.2.1), where the system starts with k customers at $t = 0$. Derive the equations describing the system and then show that

$$P_n(t) = \frac{e^{-\lambda t}(\lambda t)^{n-k}}{(n-k)!}, \qquad n = k, k+1, \dots$$

☐ **15–54** By using the distribution of departures in the pure death model (Section 15.2.2), show that the time between successive departures is exponential.

☐ **15–55** Consider a simplified single-queue model in which the service mechanism allows for only *one* customer in the system. Customers that arrive while the facility is busy leave and never return. Suppose that customers arrive completely randomly (Poisson) with rate λ per unit time and that the service time is exponential with mean value equal to $1/\mu$. Develop the difference equations representing the probabilities of the system and solve the associated steady-state model. Is the condition $\lambda < \mu$ essential for reaching steady state?

☐ **15–56** Show that for $(M/M/1):(FCFS/\infty/\infty)$, the distribution of waiting time in the queue is

$$w_q(T) = \begin{cases} 1 - \rho, & T = 0 \\ \mu\rho(1 - \rho)e^{-(\mu-\lambda)T}, & T > 0 \end{cases}$$

where $\rho = \lambda/\mu$. Then find W_q by using the expression for $w_q(T)$.

☐ **15–57** For $(M/M/1):(GD/\infty/\infty)$ show that
  (a) The expected number in the queue given that the queue is not empty $= 1/(1 - \rho)$.
  (b) The expected waiting time in the queue for those who have to wait $= 1/(\mu - \lambda)$.

☐ **15–58** For $(M/M/1):(GD/N/\infty)$ (Section 15.3.2), show that the two expressions for λ_{eff} are equivalent, namely,

$$\lambda_{\text{eff}} = \lambda(1 - p_N) = \mu(L_s - L_q)$$

☐ **15–59** For $(M/M/1):(GD/N/\infty)$ (Section 15.3.2), find the expression for L_s by using the properties of the z-transform.

☐ **15–60** For the $(M/M/1):(GD/N/\infty)$, prove the formula for p_n and L_s when $\rho = 1$.

☐ **15–61** For $(M_n/M_n/1):(GD/\infty/\infty)$, suppose that

$$\lambda_n = \lambda \qquad \text{and} \qquad \mu_n = n^\alpha \mu$$

where λ, μ, and α are given positive constants. This model represents the case where the server regulates output according to the number of customers n in the system. The constant α is known as the "pressure" coefficient. Find the steady-state difference equations describing the system and then solve, showing that

$$p_0 = \frac{1}{Q}$$

$$p_n = \frac{(\lambda/\mu)^n}{(n!)^\alpha Q}, \qquad n = 1, 2, \ldots$$

where

$$Q = \sum_{n=0}^{\infty} \frac{(\lambda/\mu)^n}{(n!)^\alpha}$$

☐ **15–62** For $(M/M/c){:}(FCFS/\infty/\infty)$, show that the PDF of waiting time in the queue is given by

$$w_q(T) = \begin{cases} 1 - \dfrac{\rho^c p_0}{(c-1)!(c-\rho)}, & T = 0 \\[2ex] \dfrac{\mu \rho^c e^{-\mu(c-\rho)T}}{(c-1)!} p_0, & T > 0 \end{cases}$$

☐ **15–63** Show that for exponential service time with mean $1/\mu$, the P–K formula for L_s (Section 15.3.3) reduces to the corresponding formula for $(M/M/1){:}(GD/\infty/\infty)$.

☐ **15–64** In Problem 15–62, show that

$$P\{T > y\} = P\{T > 0\}e^{-(c\mu - \lambda)y}$$

where $P\{T > 0\}$ is the probability that an arriving customer must wait.

☐ **15–65** For $(M/M/c){:}(FCFS/\infty/\infty)$, show that the PDF of waiting time in the system is given by

$$w(\tau) = \mu e^{-\mu\tau} + \frac{\rho^c \mu e^{-\mu\tau} p_0}{(c-1)!(c-1-\rho)} \left\{ \frac{1}{c-\rho} - e^{-\mu(c-1-\rho)\tau} \right\}$$

for $\tau \geq 0$. [*Hint:* τ is the convolution of the waiting in queue T (Problem 15–62) and the service-time distribution.]

☐ **15–66** (a) In a service facility with c parallel servers, assume that customers arrive according to a Poisson distribution with mean rate λ. Rather than assuming that any customer can join any free server, customers will be assigned to the different servers on a rotational basis so that the first arriving customer is assigned to server 1, the second customer to server 2, and so on. After the cycle is completed for all c servers, the assignment starts again by considering server 1. Find the PDF of interarrival time at each server.

(b) Suppose in part (a) that customers are assigned randomly to the different servers according to the probabilities α_i, where $\alpha_i \geq 0$, and $\alpha_1 + \alpha_2 + \cdots + \alpha_c = 1$. Find the PDF of interarrival time at each server.

☐ **15–67** For $(M/M/c){:}(GD/\infty/\infty)$ show that
(a) The probability that someone is waiting $= [\rho/(c - \rho)]p_c$.
(b) The expected number in the queue if it is given that it is not empty $= c/(c - \rho)$.
(c) The expected waiting time in queue for customers who have to wait $= 1/\mu(c - \rho)$.

☐ **15–68** For $(M/M/c){:}(GD/N/\infty)$, derive the steady-state equations describing the situation for $N = c$; then show that the expression for p_n is given by

$$
p_n =
\begin{cases}
\dfrac{\rho^n}{n!} p_0, & n = 0, 1, 2, \ldots, c \\[2ex]
0, & \text{otherwise}
\end{cases}
$$

where

$$
p_0 = \left\{ \sum_{n=0}^{c} \frac{\rho^n}{n!} \right\}^{-1}
$$

☐ **15–69** For the $(M/M/c){:}(GD/N/\infty)$, prove the formulas for p_0 and L_s when $\rho = c$.

☐ **15–70** Rework the model in Section 15.5.1 assuming three tandem stations. Assume all the remaining conditions to be the same as in the two-station model.

☐ **15–71** Show that for the special case of exponential service time with mean $1/\mu$, the results of the $(M/G/1)$ model reduce to those of the $(M/M/1)$ model (Section 15.3.1).

☐ **15–72** In $(GI/M/1)$, let interarrival time be described by the following gamma distribution:

$$
a(v) = \frac{m\lambda(m\lambda v)^{m-1} e^{-m\lambda v}}{(m - 1)!}, \qquad v > 0
$$

with $E\{v\} = 1/\lambda$ and $\operatorname{var}\{v\} = 1/m\lambda^2$. Show that

$$
D(x_0) = \left(\frac{1 - \sigma}{1 - \sigma x_0} \right)^m
$$

where $\sigma = \mu/(\mu + m\lambda)$, and $D(x_0)$ is as defined in the chapter. Derive $D(x_0)$ for constant interarrival time as a special case of the results for gamma distribution.

☐ **15–73** Show that for exponential interarrival time with mean $1/\lambda$, the results of $(GI/M/1)$ reduce to those of the $(M/M/1)$ model (Section 15.3.1).

☐ **15–74** The proof of the queueing formula $L_s = \lambda W_s$ (Section 15.3) requires that L_s be computed as a steady-state average over all instants of time, whereas only W_s can be calculated at regeneration points. This means that the formula does not apply in general to the results of $(GI/M/1)$, since L_s is computed by averaging over regeneration (imbedded) points. Verify this result by showing that for FCFS service discipline and gamma or constant interarrival time, the values of W_s obtained from the formula and from waiting-time distribution (Section 15.6.3) are not consistent.

☐ **15–75** In Problem 15–74, regardless of the restriction on the derivation of the formula $L_s = \lambda W_s$, show that for exponential interarrival time the two methods for computing W_s will yield the same result. Why is this true only for the case of Poisson arrivals?

Chapter 16

Queueing Theory in Practice

The use of queueing theory in practice involves two major aspects:

1. Selection of the appropriate mathematical model that will represent the real system adequately with the objective of determining the system's measures of performance.

2. Implementation of a decision model based on the system's measures of performance for the purpose of designing the service facility.

Although queueing theory has evolved originally out of successful applications in telephony and communication systems, extension of queueing results to other areas has been somewhat less successful. The apparent reason for this difficulty is that the queueing models available rarely satisfy the conditions under which the real system operates. However, we must keep in mind that this difficulty is typical of all mathematical models. What we need then is a clear recognition of the limitations of available queueing models from the point of view of their application to real-life situations. These limitations should be investigated in a manner that would reveal the degree of sensitivity of approximating a real system by a given mathematical model.

Following the successful selection of a queueing model, the next step is to use the model's results to make decisions regarding the design of the real system. This may entail using the model's measures of performance directly to determine how the system should operate. Alternatively, a cost-optimization model may be implemented to determine the optimal operation of the real system.

16.1 OBSTACLES IN MODELING QUEUEING SYSTEMS[†]

The difficulties in utilizing queueing models in practice can be presented from two principal points of view:

1. Ease of representing the queueing system by a mathematical model.
2. Flexibility of the mathematical model.

The first deals with the degree of applicability of the analytical model to practical systems, and the second looks at using standard models to approximate complex systems.

16.1.1 EASE OF DESCRIBING A QUEUEING SYSTEM MATHEMATICALLY

Standard queueing models with usable results are usually formulated and solved under the assumption that the behavior of the customer and the server can be predicted and quantified (in the form of a probability density function). From this standpoint, real-life queueing systems include three principal types:

1. *Human systems,* where both the server and the customer are human beings, as in the operation of a supermarket.

2. *Semiautomatic systems,* where only the customer or the server is a human being. This system is illustrated by the machine repair situation, where the machine is the customer and the mechanic is the server.

3. *Automatic systems,* in which both the customer and the server are not human beings. For example, in a time-sharing computer facility, the programs represent the customer and the central processing unit plays the role of the server.

The objective of the foregoing classification is to consider the degree of applicability of standard queueing models to real-life problems. *Human* systems should generally be the most difficult to model mathematically, owing to the unpredictability of human behavior. The situation is especially difficult when the interests

---

[†] This section is based on H. A. Taha, "Queueing Theory in Practice," *Interfaces,* February 1981.

of the customer and the server are not mutual. For example, in a supermarket, the customer, whose main interest is to receive fast service, is not directly conscious of the consequences (in terms of costs) of operating the supermarket under such conditions. On the other hand, in a tool crib facility in a factory, no conflict of interest can arise between the customer (machine operator) and the server (tool crib attendant) because they both serve the same organization.

This discussion suggests that the design of queueing systems should be oriented more toward the use of semiautomatic and automatic operation, making the system more amenable to effective analysis by existing queueing models. However, although *human* systems may be the least amenable to mathematical analysis, there is sometimes the possibility that the system can be "doctored" to better "control" the human behavior in the queueing facility. The main objective of doctoring the system is to produce favorable results in terms of its operation and, at the same time, make it conform, as much as possible, to the assumptions of available queueing models. This could entail a redesign of the layout of the facility to "force" the customers to follow a specific service pattern. A good example is the customer service operation in most U.S. post offices and airport check-in facilities. Previously, each server had its separate line of customers. The layout of service area has been changed to a multiple-server waiting line in which *all* arriving customers are forced to form a *single* queue and customers are admitted to service on a strict first-come, first-served basis. The new queueing situation can be analyzed more easily, since it is a nearly perfect example of a multiple-server waiting line model. In the old layout, opportunities for "jockeying" between the queues exist, which should lead to a more complex queueing model.

We cannot prescribe the procedure of doctoring the operation of a system as a routine way for remedying all the ills associated with describing human systems mathematically. Take, for example, the case study reported in Lee [1966] about a queueing problem in a large British airport. To "doctor" the operation of the check-in facility to produce certain favorable results, waiting passengers whose flights are within 5 minutes of boarding time are instructed through a well-placed sign to advance at once to the head of their queue and request priority service. The system has failed because the customers, being mostly British, are "conditioned to very strict queueing behavior" and thus are reluctant to move ahead of others waiting in front of them. Our conclusion, then, is that a human queueing system may be "doctored" only to the extent that the resulting changes would not be met with resistance by either the customer or the server.

16.1.2 MODEL FLEXIBILITY

The analytic complexity of a mathematical queueing model may occur at two levels:

1. Difficulty of formulating and solving the mathematical model even though the distributions of arrivals and departures may be fully known.

2. Difficulty of obtaining *numerical* results from a solved model, owing to the complexity of the mathematical expressions describing the measures of effectiveness of the system.

At the first level, we know from experience that deviations from the Poisson assumptions normally result in complex models. In fact, of all models of the type $(M/G/c)$, analytic solutions exist only for the special cases where the service time is constant or the number of servers equals the number of customers. At the second level, we are seeking numerical evaluation of complex expressions. An example of this case occurs in the transient solution of the $(M/M/1)$ model, which involves numerical evaluation of the Bessel function.

The second type of difficulty appears surmountable, particularly with the availability of high-powered digital computers. What we are interested in exploring is the possibility of approximating complex situations by others whose results are already available. In essence, we would be interested in studying the effect of changing the model's basic assumptions on its measures of performance, such as the expected waiting time and the percentage of time the facility is idle.

There are situations where the use of such approximations is sufficiently evident. For example, the $(M/D/c)$ model can approximate the $(M/G/c)$ model if the standard deviation of the service time is much smaller than its mean value. However, we are really interested in exploring a "bolder" degree of approximation, wherein the basic assumptions of the approximating model are obviously violated. If the results of the model remain relatively insensitive to changes in its assumptions, the model is said to be *flexible* in the sense that it can be used to describe different queueing situations without resulting in excessive bias in the results.

The best example of the most flexible model in queueing theory is the well-known formula $L = \lambda W$; the number of waiting customers equals the rate of arrival times the average waiting time. This formula is general in the sense that its application is independent of the specific distributions of arrivals and departures.

Highly flexible models such as the formula $L = \lambda W$ are not very common in queueing theory. We wish to demonstrate, however, that the idea of using model flexibility to approximate complex systems remains a feasible possibility. Suppose that we use the model $(M/M/1)$ as an approximation of the model $(M/G/1)$; that is, we shall assume that the exponential distribution will always approximate the service-time distribution. Thus if we estimate (through sampling) that the mean service rate is μ, then under the exponential distribution assumption, the mean and variance of service time will be $1/\mu$ and $1/\mu^2$, respectively. Now define α^2 as the ratio of the true to assumed variance of the service-time distribution.

$$\alpha^2 = \frac{\text{true variance}}{\text{assumed variance}}$$

$$= \frac{\text{true variance}}{1/\mu^2}$$

The parameter α represents the degree of error resulting from the use of $1/\mu^2$ as an approximation of the true variance of the service-time distribution.

Suppose that we measure the degree of flexibility of the $(M/M/1)$ in representing the $(M/G/1)$ model by computing the percentage error in the expected number in the system. If L_s' and L_s'' are the expected number in the system given $(M/M/1)$ and $(M/G/1)$, respectively, the percentage error (%E) is defined as

$$\%E = \frac{L_s' - L_s''}{L_s''} \times 100$$

As is known from queueing results,

$$L_s' = \frac{\rho}{1 - \rho}$$

$$L_s'' = \rho + \frac{\rho^2 + \alpha^2\rho^2}{2(1 - \rho)}$$

where

$$\rho = \frac{\lambda}{\mu} = \frac{\text{arrival rate}}{\text{service rate}}$$

Thus

$$\%E = \frac{\rho(1 - \alpha^2)}{2 + \rho(\alpha^2 - 1)} \times 100$$

Figure 16–1 shows the change in %E with α for various values of ρ. The results show that for $\rho = .1$ and $.1 \leq \alpha \leq 1.8$, the error in utilizing L_s' as an approximation of L_s'' remains within 10%. This result is significant because it represents a considerable range of error in the estimation of the true variance, namely, $.01 \leq \alpha^2 \leq 3.24$. Although the flexibility of the $(M/M/1)$ model decreases

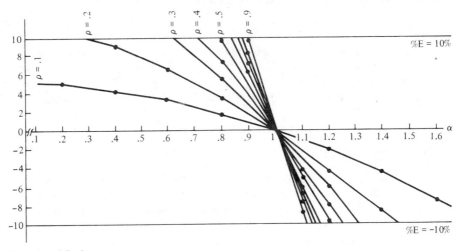

Figure 16–1

with the increase in the utilization factor ρ, Figure 16–1 shows that there is an allowable margin of error in the estimation of the true variance that will maintain %E between ±10%.

This example is not intended to provide a concrete procedure for testing model flexibility because, in general, the measures of effectiveness of the approximated system are not known a priori. It merely demonstrates that opportunities exist in queueing theory in which basic assumptions of the true model can be violated while maintaining the error in computing the system's measures of performance within tolerable limits.

The conclusion from the discussion in Sections 16.1.1 and 16.1.2 is that, on the surface, standard queueing models may not be applicable to many real-life situations. We have demonstrated, however, that a practitioner may overcome this difficulty in two ways:

1. By altering the design and operation of the queueing system in a way that would *logically* produce favorable operational results and in the mean time allow analysis by standard queueing models.

2. By taking advantage of the possibility that certain assumptions of available queueing models can be violated without resulting in considerable error in the system's measures of performance.

The second point appears more promising, since it will result in an increase of the class of problems that can be analyzed by available queueing models. Unfortunately, the theory of queues is almost void of concrete developments in this area. Perhaps it is appropriate to suggest that more attention should be paid to increasing the applicability of the available models rather than concentrating solely on the development of more models of hypothetical situations.

16.2 DATA GATHERING AND TESTING

The selection of a specific method for analyzing a queueing situation, be it analytically or by simulation, is principally determined by the distributions of arrivals and service times. In practice, the determination of these distributions entails observing the queueing system during operation and recording the pertinent data. Two questions normally arise regarding the collection of the necessary data:

1. *When* to observe the system?
2. *How* to collect the data?

Most queueing situations have what are called *busy periods,* during which the system's arrival rate increases in comparison with other times of the day. A typical variation in the arrival rate appears as shown in Figure 16–2. For example, incoming and outgoing traffic on a main highway leading into a city reaches its peak during rush hours around 8:00 A.M. and 5:00 P.M. In situations like these it will be necessary to collect the data during the busy periods. This may be a conservative attitude, but we must keep in mind that congestion in

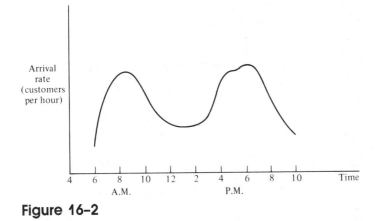

Figure 16-2

queueing systems takes place during busy periods. As such the system must be designed to account for these extreme conditions.

Collecting the data regarding arrivals and departures can be achieved in one of two ways:

1. Measuring the clock time between successive arrivals (departures) to obtain the interarrival (service) times.

2. Counting the number of arrivals (departures) during a selected time unit (e.g., an hour).

The first method is designed to yield the distributions of interarrival or service times, and the second method yields the distributions of the number of arrivals or departures. In most analytic queueing models, we can describe the input and output processes by either the *number* of events (arrivals or departures) or the *time* between events (interarrival or service time).

The mechanism for collecting data may be based on the use of a stopwatch technique or an automatic recording device. An automatic device appears essential when the arrivals occur at a high rate. The use of a manual technique in this case will probably result in distortion of data.

After gathering the data in the manner just outlined, the information must be summarized in a meaningful way that would allow us to determine the associated distribution. This is normally achieved by first summarizing the observations in the form of a frequency histogram. We can then suggest a theoretical distribution that fits the data observed (e.g., Poisson, exponential, normal). A statistical test can then be applied to test the "goodness of fit" of the proposed distribution. The following example demonstrates the determination of the theoretical distributions starting with the raw data.

Example 16.2-1. An automatic device is used to record the volume of traffic at a busy intersection. The device records the time a car arrives at the intersection on a continuous time scale, starting from a zero datum. Table 16–1 demonstrates a typical recording of arrival times (in minutes) for the first 60 cars. (Normally, the data required for determining the arrivals distribution

Table 16-1

Arrival	Arrival Time (minutes)	Arrival	Arrival Time (minutes)	Arrival	Arrival Time (minutes)	Arrival	Arrival Time (minutes)
1	5.2	16	67.6	31	132.7	46	227.8
2	6.7	17	69.3	32	142.3	47	233.5
3	9.1	18	78.6	33	145.2	48	239.8
4	12.5	19	86.6	34	154.3	49	243.6
5	18.9	20	91.3	35	155.6	50	250.5
6	22.6	21	97.2	36	166.2	51	255.8
7	27.4	22	97.9	37	169.2	52	256.5
8	29.9	23	111.5	38	169.5	53	256.9
9	35.4	24	116.7	39	172.4	54	270.3
10	35.7	25	117.3	40	175.3	55	275.1
11	44.4	26	118.2	41	180.1	56	277.1
12	47.1	27	124.1	42	188.8	57	278.1
13	47.5	28	127.4	43	201.2	58	283.6
14	49.7	29	127.6	44	218.4	59	299.8
15	67.1	30	127.8	45	219.9	60	300.0

may be collected over a period of days or weeks. However, we cannot reproduce a full set of data here because of space limitations.)

The data in Table 16-1 can be used to construct the distribution of the *number* of arrivals in the following manner. First, a time unit must be selected. In our example we choose 1 hour as the time unit so that the distribution will represent the number of arrivals per hour. In Table 16-1 there are 14 arrivals during the first hour, 12 during the second, 14 during the third, 8 during the fourth, and 12 during the fifth. This means that for the 5 hours the arrivals per hour are 8 with frequency 1, 12 with frequency 2, and 14 with frequency 2.

Imagine now that we have a full set of data and that a summarization of the number of arrivals per hour, n, is found to have the frequency count f_n, shown in Table 16-2. Our objective is to test whether these data came from a specific theoretical distribution using the *chi-square test of goodness of fit*.

Suppose that we would like to test the hypothesis that the sample in Table 16-2 came from a Poisson distribution. The goodness of fit test compares the observed frequency f_n with the expected frequency that would result if the Poisson distribution is assumed. To compute the expected frequency, we first

Table 16-2

n	0	1	2	3	4	5	6	7	8
f_n	0	0	0	0	0	1	0	3	3

n	9	10	11	12	13	14	15	16	≥ 17
f_n	6	5	9	10	11	8	6	1	0

estimate the mean $\bar{n}$ of the Poisson distribution from the sample. This is given by

$$\bar{n} = \frac{\sum\limits_{n=0}^{16} n f_n}{\sum\limits_{n=0}^{16} f_n} = \frac{734}{63} = 11.65 \text{ cars per hour}$$

The next step is to compute the probabilities p_n for a Poisson distribution with mean 11.65 cars per hour. Table 16–3 summarizes this information. Note that

$$p_n = \frac{(11.65)^n e^{-11.65}}{n!}$$

Since we have a total of 63 observations, the expected frequency can be computed as

$$e_n = \left(\sum_{n=0}^{16} f_n \right) p_n = 63 p_n$$

Following the determination of e_n, we compute the chi-square value as follows:

$$\chi^2\text{-value} = \sum_{n=0}^{\infty} \frac{(f_n - e_n)^2}{e_n}$$

As a rule of thumb, each e_n must at least equal 5. If not, successive values of e_n are combined to satisfy this condition. Thus, in Table 16–2, $n = 0$ through 8 must be combined to yield a theoretical frequency of 11–3. Also, e_n for all n greater than 14 must be combined to yield a theoretical frequency of 12–42. Table 16–4 now shows how the χ^2-value is computed.

We now compare the χ^2-value with the critical value of the χ^2-distribution. To do so, we need to specify the significance level α and the degrees of freedom v. The value of v for the goodness of fit test is given by

$$v = \binom{\text{number of}}{\text{class intervals}} - \binom{\text{number of esti-}}{\text{mated parameters}} - 1$$

In our example, we have eight class intervals (recall that each class interval must include at least five observations). Since we estimated the mean of the

Table 16–3

n	p_n	n	p_n	n	p_n
0	.0000	6	.0303	12	.1138
1	.0001	7	.0504	13	.1020
2	.0006	8	.0734	14	.0848
3	.0023	9	.0950	15	.0659
4	.0067	10	.1106	16	.0479
5	.0156	11	.1172	≥ 17	.0834

Table 16-4

n	f_n	e_n	$\dfrac{(f_n - e_n)^2}{e_n}$
0-4	0 ⎫		
5	1 ⎪		
6	0 ⎬ 7	11.3	1.636
7	3 ⎪		
8	3 ⎭		
9	6	5.99	.000
10	5	6.97	.557
11	9	7.38	.356
12	10	7.17	1.117
13	11	6.43	3.248
14	8	5.34	1.325
15	6 ⎫		
16	1 ⎬ 7	12.42	2.365
≥17	0 ⎭		
Totals	$\overline{63}$	$\overline{63}$	$\overline{10.6} = \chi^2\text{-value}$

Poisson from the sample data, we thus get

$$\nu = 8 - 1 - 1 = 6$$

Using a significance level $\alpha = .05$, the χ^2 tables yield the critical value $\chi^2_6(.05) = 12.592$.

The application of the χ^2-test calls for accepting the hypothesis at the specified level of significance α if χ^2-value $\leq \chi^2_\nu(\alpha)$. Since this condition is satisfied in our example, we thus accept the hypothesis that our sample came from a Poisson distribution with mean 11.65 arrivals per hour. ◀

16.3 QUEUEING DECISION MODELS

16.3.1 COST MODELS

The objective of a queueing cost model is to determine the level of service (either the service rate or the number of servers) that "balances" the following two *conflicting* costs:

1. Cost of offering the service.
2. Cost resulting from delay in offering the service.

The first cost is associated with the operation of the facility, and the second represents the cost of delaying customers. Intuitively, we see that an increase in the level of service should reduce the customer's waiting time, and vice

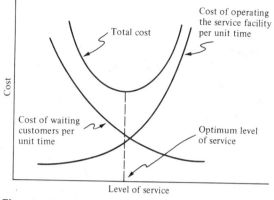

Figure 16–3

versa. This means that as the cost of operating the facility increases (decreases) because of the increase (decrease) of level of service, the cost of waiting should decrease (increase). Figure 16–3 summarizes the two costs as a function of level of service. The optimum level of service is chosen to minimize the sum of the two costs. We note that both costs are defined per unit of time to keep the model dimensionally correct. We shall show how these costs are computed.

Of the two types of costs cited, the cost of waiting is the most difficult to estimate. This is particularly so when the customer is a human being whose interests may not be in harmony with those of the server. For example, a lengthy delay in a grocery store may result in loss of customers' goodwill, for which it may be impossible to assign a monetary value. On the other hand, consider the situation where the customer is represented by a machine awaiting the service of a repairman. In this case the cost of waiting can normally be translated in terms of the cost of lost production, which is relatively easy to estimate.

Our conclusion at this point is that cost models, although computationally simple, may be quite difficult to implement in practice, owing primarily to the difficulty of estimating the cost of waiting. For these difficult cases it may be more practical to use the *aspiration-level* method, which we introduce shortly. Right now we illustrate the use of cost minimization in making queueing decisions by introducing two models. The first deals with the determination of the optimum service rate μ in a single-server model; the second determines the optimal number of servers in a multiple-server queue.

A. Optimum Service Rate μ

Consider the single-channel model with arrival rate λ and service rate μ. It is assumed that the service rate μ is controllable and that it is required to determine its optimum value based on an appropriate cost model. Let

C_1 = cost per unit increase in μ per unit time

C_2 = cost of waiting per unit waiting time per customer

$TC(\mu)$ = expected cost of waiting and service per unit time given μ

Thus

$$TC(\mu) = C_1\mu + C_2 L_s$$

Notice that cost of service per unit time is directly proportional to μ and that cost of waiting per unit time is equal to the expected number of customers in the system, multiplied by the waiting cost per customer per unit time.

Since μ is continuous, its optimum value can be obtained by differentiating $TC(\mu)$ with respect to μ. For example, for the special case of $(M/M/1)$: $(GD/\infty/\infty)$,

$$TC(\mu) = C_1\mu + C_2 \frac{\lambda}{\mu - \lambda}$$

and thus optimum μ is given by

$$\mu = \lambda + \sqrt{\frac{C_2\lambda}{C_1}}$$

This result shows that the optimal value of μ is dependent on the arrival rate λ. This seems logical because if μ is independent of λ, it may cause $\rho \ (= \lambda/\mu)$ to be greater than 1. In the solution, μ is always greater than λ.

For the case where only a maximum of N customers are allowed in the system, that is, $(M/M/1){:}(GD/N/\infty)$, the cost model can be modified to reflect that larger values of N result in a smaller number of lost customers (those who cannot join the system because it is full). In this case N is treated as a decision variable, which, together with μ, is determined by minimizing

$$TC(\mu, N) = C_1\mu + C_2 L_s + C_3 N + C_4 \lambda p_N$$

where

$C_3 =$ cost per unit time per additional accommodation unit

$C_4 =$ cost per lost customer

Notice that λp_N represents the number of lost customers per unit time.

A closed-form solution for this problem is not possible. Consequently, one may have to employ an appropriate numerical technique to obtain the optimal solution.

Example 16.3-1. A computer company owns a central computer that can be accessed by clients through the lease of a card reader and a high-speed printer. One of the clients wishes to determine the optimum speed (in cards per minute) of the card reader he should lease. The client's jobs are generated according to a Poisson stream at the rate of 50 programs per 8-hour day. The average size of a program is 1000 cards. Experience shows that the time for reading programs through the card reader is exponential. The client estimates that the cost of delaying a job one day is $10. The computer company's monthly rate is based on $100 for each incremental 100 cards per minute in the speed of the card reader.

We note that a speed of 100 cards per minute is equivalent to $100 \times 8 \times 60 = 48,000$ cards per day or $48,000/1000 = 48$ jobs per day. Assuming a 22-day work month, the monthly rental rate for the reader is equivalent to

$100/22 = $4.55 a day. Since each unit increase in the speed of the reader is equivalent to processing 48 jobs daily, the cost per additional job per day is $4.55/48 = $.0948. Using the symbols of the cost model, we have $C_1 = $.0948$. Since $C_2 = 10 per job per day and $\lambda = 50$ jobs per day, we obtain the optimum μ as

$$\mu = 50 + \sqrt{\frac{10 \times 50}{.0948}} = 123 \text{ jobs per day}$$

Converting this result to cards per minute, we obtain

$$\text{optimum speed} = \frac{123 \times 1000}{8 \times 60} = 256 \text{ cards per minute} \qquad \blacktriangleleft$$

B. Optimum Number of Servers

Consider the multiple-server model presented in Section 15.3.4. A cost model can be developed in this case for determining the optimal number of servers c. It is assumed that λ and μ are fixed. Following reasoning similar to that of the last model, total cost per unit time is given by

$$TC(c) = cC_1 + C_2 L_s(c)$$

where

$C_1 = $ cost per additional server per unit time

$L_s(c) = $ expected number in the system given c servers

while C_2 is defined as in Section 16.3.1A. (As in the previous model, the effect of allowing a finite number in the system may be similarly included here.)

Since c is discrete, differentiation is not applicable to this case. Although optimum c may be found by direct substitution of successive values of c until the minimum of $TC(c)$ is identified, a more computationally efficient procedure may be developed based on the necessary condition for a minimum of the given function. These are

$$TC(c - 1) \geq TC(c) \qquad \text{and} \qquad TC(c + 1) \geq TC(c)$$

which yield the condition

$$L_s(c) - L_s(c + 1) \leq \frac{C_1}{C_2} \leq L_s(c - 1) - L_s(c)$$

The value of C_1/C_2 now indicates where the search for optimum c should start.

Example 16.3–2. In a tool crib facility, requests for tool exchange occur according to a Poisson distribution with mean 17.5 requests per hour. Each clerk in the facility can handle an average of 10 requests per hour. The cost of adding a new clerk to the facility is estimated at $6 an hour. The cost of lost production per waiting machine per hour is estimated at $30 an hour. How many clerks should staff the facility?

The determination of optimum c is achieved by carrying out the computations as shown in Table 16-5 [note that $L_s(1) = \infty$, since $\lambda > \mu$].

Table 16-5

c	$L_s(c)$	$L_s(c-1) - L_s(c)$	
1	∞	—	
2	7.467	∞	
3	2.217	5.25	
4	1.842	.375	$\leftarrow C_1/C_2 = .2$
5	1.769	.073	
6	1.754	.015	
7	1.75	.004	

Since $C_1/C_2 = 6/30 = .2$, we have

$$L_s(4) - L_s(5) = .073 < .2 < .375 = L_s(3) - L_s(4)$$

Consequently, optimum $C = 4$ clerks.   ◀

Exercise 16.3-1
In Example 16.3-2, find the optimal c given $C_1 = \$10$ and $C_2 = \$20$.
[*Ans. c = 3.*]

16.3.2 ASPIRATION-LEVEL MODEL

The aspiration-level model recognizes the difficulty of estimating cost parameters, and hence it is based on a more straightforward analysis. It makes direct use of the operating characteristics of the system in deciding on the "optimal" values of the design parameters. Optimality here is viewed in the sense of satisfying certain aspiration levels set by the decision maker. These aspiration levels are defined as the upper limits on the values of the conflicting measures that the decision maker wishes to balance.

In the multiple-server model where it is required to determine the optimum value of the number of servers c, the two conflicting measures may be taken as

1. The expected waiting time in the system W_s.
2. The percentage of servers' idle time X.

These two measures reflect the aspirations of customer and server. Let the levels of aspiration (upper limits) for W_s and X be given by α and β. Then the aspiration-level method may be expressed mathematically as follows.

Determine the number of servers c such that

$$W_s \leq \alpha \quad \text{and} \quad X \leq \beta$$

The expression for W_s is known from the analysis of $(M/M/c){:}(GD/\infty/\infty)$. The expression for X is given by

$$X = \frac{100}{c} \sum_{n=0}^{c} (c-n)p_n = 100\left(1 - \frac{\rho}{c}\right)$$

The solution of the problem may be determined more readily by plotting W_s and X against c as shown in Figure 16–4. By locating α and β on the graph, we can immediately determine an acceptable range of c which satisfies both restrictions. Naturally, if these two conditions are not satisfied simultaneously, it would be necessary to relax one or both restrictions before a decision is made.

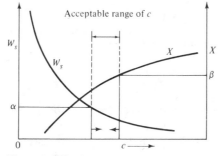

Figure 16–4

Example 16.3–3. In Example 16.3–2, suppose that it is desired to determine the number of clerks such that the expected waiting time until a tool is received remains below 20 minutes. Simultaneously, it is required that the percentage of time that the clerks are idle does not exceed 15%.

Table 16–6 summarizes W_s and X for different values of c. As c increases, W_s decreases and X increases.

Table 16–6

c	1	2	3	4	5	6	7	8
W_s (minutes)	∞	25.6	7.6	6.3	6.1	6.0	6.0	6.0
X (%)	0	12.5	41.7	56.3	65.0	70.8	75.0	78.0

For W_s to stay below 20 minutes, we must have *at least* 3 clerks. On the other hand, keeping the clerks busy 85% of the time requires limiting their number to a *maximum* of 2 clerks. Thus the two aspiration levels cannot be satisfied simultaneously and one of the two conditions must be relaxed if we are to find a feasible solution.

We notice in Table 16–6 that a substantial drop in W_s occurs as c increases from 2 to 3. Any further increase will have little effect on the value of W_s. In terms of X, increasing c from 2 to 3 more than triples the percentage of idle time for the clerks. Thus the choice between $c = 2$ and $c = 3$ should be made in view of whether it is "worthwhile" to reduce the machine's idle time from 25.6 minutes to 7.6 minutes even though the idle time of the clerks will increase from 12.5% to 41.7%. ◄

To assist in making a specific decision in the case of the aspiration-level method, we can compute a range on the cost parameter C_2 resulting from the selection of c for given aspiration levels. We specifically select C_2 instead of C_1 because it is usually more difficult to estimate the cost of waiting in most waiting line models. The procedure we give here thus assumes that C_1, the incremental cost associated with acquiring an additional server, can be estimated without too much difficulty.

From the cost model in Section 16.3.1B, optimum c must satisfy

$$L_s(c) - L_s(c+1) \leq \frac{C_1}{C_2} \leq L_s(c-1) - L_s(c)$$

Thus, under optimal conditions, C_2 falls in the range

$$\frac{C_1}{L_s(c-1) - L_s(c)} \leq C_2 \leq \frac{C_1}{L_s(c) - L_s(c+1)}$$

We illustrate the application of the method by the following example.

Example 16.3-4. In Example 16.3-3 we can estimate ranges on C_2 for c = 2 and c = 3. Using $C_1 = \$6$ as given in Example 16.3-2, we get the following results (see Table 16-5 for the values of L_s).

$c = 2$

$$\frac{6}{\infty} \leq C_2 \leq \frac{6}{5.25}$$

which gives

$$0 \leq C_2 \leq \$1.14$$

$c = 3$

$$\frac{6}{5.25} \leq C_2 \leq \frac{6}{.375}$$

or

$$\$1.14 \leq C_2 \leq \$16$$

Perhaps the ranges on C_2 given c = 2 and c = 3 can help us make a more selective choice between using 2 or 3 clerks. For c = 2, the range on C_2 indicates that the worth of a machine wating for a tool cannot exceed $1.14 in terms of lost production. This estimate regarding the worth of the machine appears very low. Alternatively, for c = 3 an upper limit of $16 on the value of C_2 appears more reasonable. It is thus more logical to use 3 clerks instead of 2. ◄

16.4 SUMMARY

This chapter presents three types of difficulty in applying queueing theory in practice.

1. Difficulty of modeling queueing situations mathematically, especially those in which the customer and/or server are human beings.
2. Difficulty of obtaining usable analytic results for certain mathematical models.
3. Difficulty of estimating cost parameters.

We have suggested some remedies that in certain situations may facilitate the application of queueing theory.

1. The use of computational approximations.
2. The use of model approximation based on testing model flexibility.
3. The use of aspiration levels in situations where it is difficult to estimate costs.

In spite of the apparently great obstacles in using queueing theory in practice, there are situations that lend themselves neatly to this type of analysis. On the other hand, it appears that the use of approximations in applying queueing models can be potentially promising.

SELECTED REFERENCES

ELMAGHRABY, S., *The Design of Production Systems,* Reinhold, New York, 1966, Chap. 5.

LEE, A., *Applied Queueing Theory,* Macmillan, Toronto, Canada, 1966.

PROBLEMS

Section	Assigned Problems
16.1	16–1
16.2	16–2, 16–3
16.3	16–4 to 16–13

☐ **16–1** Classify the queueing situation in Problem 15–1 as *human, semiautomatic,* and *automatic.* In the *human* system, indicate whether the interests of the customer and the server are mutual.

☐ **16–2** Consider the arrival times given in Table 16–2. Carry out a chi-square test of goodness of fit to test the hypothesis that the interarrival time is exponential.

☐ **16–3** The following table summarizes data representing number of arrivals n per hour and the associated frequency of occurrences f_n. Test the hypothesis that the data are generated from a Poisson stream. If the hypothesis is accepted, write down the Poisson density function that you would use in analyzing the queueing situation.

n	0	1	2	3	4	5	6	7	8	≥ 9
f_n	6	15	22	24	15	9	4	3	2	0

☐ **16–4** Jobs arrive at a machine shop according to a Poisson stream at the rate of 10 a day. An automatic machine represents the bottleneck in the shop. It is estimated that a unit increase in the production rate of the machine will cost $100 per week. Delayed jobs normally result in lost business, which is estimated to be $200 per job per week. Determine the optimum speed of the machine in units of the production rate.

☐ **16–5** For $(M/M/1){:}(GD/\infty/\infty)$, find the optimal service rate μ given that the cost per unit increase in μ per unit time is $10 and the cost per unit waiting time per customer is $1. The arrival rate is 20 per unit time.

☐ **16–6** Inventory stock is depleted and replenished according to Poisson distributions. Mean times between depletions and replenishments are equal to $1/\mu$ and $1/\lambda$, respectively. This process may be viewed as an $(M/M/1){:}(GD/\infty/\infty)$ queueing model.

Suppose that every unit of time that inventory is out of stock, a penalty cost C_2 is incurred. Also, every unit of time that n items of inventory are on hand, a holding cost $C_1 n$ is incurred, where $C_2 > C_1$.
  (a) Find an expression for *expected* total cost per unit time.
  (b) What is the optimal value of $\rho = \lambda/\mu$?

☐ **16–7** A company sells two franchized models of restaurant. Model A has a capacity of 80 customers, whereas model B can accommodate 100 customers. The monthly cost of operating model A is $1000 and that of B is $1200. A prospective investor wants to set up a restaurant in his hometown. He estimates that his customers will arrive according to a Poisson distribution at the rate of 30 per hour. Model A will offer service at the rate of 20 customers per hour, and model B will serve 35 customers per hour. Once the restaurant is filled to capacity, new arrivals would normally leave without seeking service. The lost business per customer per day is estimated at about $8.00. A delay in serving those customers waiting inside the restaurant is estimated to cost the owner about $.40 per customer per hour due to the loss in customers' goodwill. Which model should the owner choose? Assume that the restaurant will be open for 10 hours daily.

☐ **16–8** Verify the result

$$L_s(c) - L_s(c+1) \leq \frac{C_1}{C_2} \leq L_s(c-1) - L_s(c)$$

given in Section 16.3.1B for the optimal number of servers c.

☐ **16–9** In Example 16.3–2 suppose that for $(M/M/c):(GD/\infty/\infty)$, $\lambda = 10$ and $\mu = 3$. The costs are $C_1 = 5$ and $C_2 = 25$. Find the number of servers that must be used to minimize total expected costs.

☐ **16–10** Two repairmen are being considered for attending 10 machines in a workshop. The first repairman will be paid at the rate of $3.00 per hour. He can repair machines at the rate of 5 per hour. The second repairman will be paid $5.00 per hour but he can repair machines at the rate of 8 per hour. It is estimated that machine downtime cost is $8.00 per hour.

Assuming that machines break down according to a Poisson distribution with mean 4 per hour and the repair time is exponentially distributed, which repairman should be hired?

☐ **16–11** A particular pipeline booster unit that operates continuously on a 24-hour basis requires service at exponential time intervals with mean time between breakdowns equal to 20 hours. A repairman can service a broken booster on the average in 10 hours with exponential service time. At a station with 10 boosters and two repairmen on duty at all times, each repairman draws a salary of $7.00 an hour. Pipeline schedule losses are estimated to be $15.00 per hour per broken pump. The company is considering hiring an additional repairman.

(a) Determine the cost savings per hour achieved by hiring the additional repairman.
(b) What is the schedule loss in dollars per breakdown with two repairmen on duty?
(c) What is the schedule loss in dollars per breakdown with three repairmen on duty?

☐ **16–12** A company leases a WATS line for telephone service in all states for $2000 a month. The office is open 200 working hours per month. At all other times the WATS line service is used for other purposes and is not available for company business. Access to the WATS line during business hours is extended to 100 executives, each of whom may need the line at any time, but on an average of twice per 8-hour day (assume exponential time between calls). An executive will always wait for the WATS line if it is busy at an estimated inconvenience cost of 1 cent per minute of waiting. It is assumed that no additional needs for calls will arise while the executive waits for a given call. The normal cost of calls (not using the WATS line) averages 50 cents per minute and the duration of each call is exponential with mean 6 minutes. The company is considering leasing (at the same price) a second WATS line to assist in handling the heavy traffic of calls.

(a) Is the single WATS line saving the company money over a "no WATS line" system? How much is the company gaining or losing per month over the "no WATS line" system?
(b) Should the company lease a second WATS line? How much would it gain or lose over the single WATS line system by leasing a second line?

☐ **16–13** A shop utilizes 10 identical machines. The profit per machine is $4.00 per hour of operation. Each machine breaks down on the average once every 7 hours. One person can repair a machine in 4 hours on the average, but the actual repair time varies according to an exponential distribution. The repairman's salary is $6.00 an hour. Determine the following:

(a) The number of repairmen that will minimize the total cost.
(b) The number of repairmen needed so that the expected number of broken machines is less than 4.
(c) The number of repairmen needed so that the expected delay time until a machine is repaired is less than 4 hours.

PROJECTS

The following projects are suggestions for possible areas of queueing theory applications. You should keep in mind that you may not be able to find a mathematical model that fits the system you are studying. This is expected! What we hope to accomplish by attempting to investigate real-life problems is to gain firsthand experience in the use of queueing theory in practice. *Ideally,* the analysis of a queueing system should proceed in the following orderly manner:

1. Gather data.
2. Test the goodness of fit of the empirical distributions.
3. Select a proper mathematical model and determine its measures of performance.
4. Apply an appropriate decision model.

As you work with a real-life situation, you may discover that progress cannot be made, particularly in step 3, where it may be difficult to find a proper mathematical model. When this happens, you should determine the reasons for these difficulties and study the possibility of circumventing them. A clear recognition of such difficulties should help in realizing the limitations of using queueing theory in practice.

Project 1: Select a parking lot around the campus of your college or university. Model the situation as a queueing system for the purpose of determining the number of parking spaces.

Project 2: Study the checkout counters at the university library with the objective of determining the best design of the facility from the standpoint of selecting the service mechanism and the number of clerks.

Project 3: Study the queueing systems of the cashiers in the university cafeteria with the objective of determining the number of cashiers as well as the layout of the facility.

Project 4: Study the mass transit system on your campus at one of the busy bus stops with the purpose of determining how often a bus should pass through the station selected.

Project 5: Study the operation of a drive-in bank with the objective of determining the number of tellers.

SHORT CASES

Case 16–1

A state-run child abuse center operates from 9:00 A.M. to 9:00 P.M. daily. Calls reporting cases of child abuse arrive, as expected, in a completely random fashion. The accompanying table gives the number of calls recorded on an hourly basis over a period seven days.

Starting Hour	Total Number of Calls During Given Hour						
	Day 1	Day 2	Day 3	Day 4	Day 5	Day 6	Day 7
9:00	4	6	8	4	5	3	4
10:00	6	5	5	3	6	4	7
11:00	3	9	6	8	4	7	5
12:00	8	11	10	5	15	12	9
13:00	10	9	8	7	10	16	6
14:00	8	6	10	12	12	11	10
15:00	10	9	12	4	10	6	8
16:00	8	6	9	14	12	10	7
17:00	5	10	10	8	10	10	9
18:00	5	4	6	5	6	7	5
19:00	3	4	6	2	3	4	5
20:00	4	3	2	2	2	3	4
21:00	1	2	1	3	3	5	3

The table does not include lost calls resulting from the caller receiving a busy signal. For those calls that are actually received, each call lasts randomly for up to 12 minutes with an average of 7 minutes.

Past calls show that the center has been experiencing a 15% annual rate of increase in telephone calls.

The center would like to determine the number of telephone lines that must be installed to provide adequate service now and in the future. In particular, special attention is given to reducing the adverse effect of a caller receiving a busy signal.

Case 16–2

A manufacturing company employs three trucks for transporting materials among six departments. Truck users have been demanding that a fourth truck be added to the fleet to alleviate the problem of excessive delays.

The trucks do not have a "home" station from which they can be called. Instead, management considers it more efficient to keep the trucks in (semi-) continuous motion about the factory. A department requesting the use of a truck must await its arrival in the vicinity. If the truck is available, it will respond to the call. Otherwise, the department must await the appearance of another truck.

Data collected regarding the number of calls from all departments are

Number of Calls per Hour	Frequency
0	30
1	90
2	99
3	102
4	120
5	100
6	60
7	47
8	30
9	20
10	12
11	10
12	4

The service time for each department (in minutes) is approximately the same. The following table summarizes a typical service-time histogram for one of the departments.

Service Time t	Frequency
$0 \leq t < 10$	61
$10 \leq t < 20$	34
$20 \leq t < 30$	15
$30 \leq t < 40$	5
$40 \leq t < 50$	8
$50 \leq t < 60$	4
$60 \leq t < 70$	4
$70 \leq t < 80$	3
$80 \leq t < 90$	2
$90 \leq t < 100$	2

What type of recommendation would you make for the management?

Simulation

A simulation model seeks to "duplicate" the behavior of the system under investigation by studying the interactions among its components. The output of a simulation model is normally presented in terms of selected measures that reflect the performance of the system. For example, in the simulation of a drive-in bank operation, we may be interested in estimating the average waiting per customer, the average number of waiting customers, and the percentage of time the facility is idle.

Simulation must be treated as a *statistical experiment.* Unlike the mathematical models presented in the preceding chapters, wherein the output of the model represents a long-run steady-state behavior, the results obtained from running a simulation model are *observations* that are subject to experimental error. This means that any inference regarding the performance of the simulated system must be subject to all the appropriate tests of statistical analysis.

A simulation experiment differs from a regular laboratory experiment in that it can be conducted totally on the computer. By expressing the interactions among the components of the system as mathematical relationships, we are able to gather the necessary information in very much the same way as if we were observing the real system (subject, of course, to the simplifications built into the model). The nature of simulation thus allows greater flexibility in representing complex systems that are normally difficult to analyze by standard mathematical models. We must keep in mind, however, that although simulation is a flexible technique, the development of a simulation model can be both time consuming and costly, particularly when one is trying to *optimize* the simulated system. These points are discussed and stressed throughout the chapter.

17.1 SCOPE OF SIMULATION APPLICATIONS

Simulation has been used to analyze problems of two distinct types:

1. Theoretical problems in basic science areas such as mathematics, physics, and chemistry.
   a. Estimation of the area enclosed by a curve, including the evaluation of multiple integrals.
   b. Matrix inversion.
   c. Estimation of the constant π ($= 3.14159$) in mathematics.
   d. Solution of partial differential equations.
   e. Study of movement of particles in a plane.
   f. Study of particle diffusion.
   g. Solution of simultaneous linear equations.

2. Practical problems in all aspects of real life.
   a. Simulation of industrial problems (e.g., design of chemical processes, inventory control, design of distribution systems, maintenance scheduling, design of queueing systems, job-shop scheduling, design of communication systems).
   b. Simulation of business and economic problems (e.g., operation of total firm, consumer behavior, evaluation of proposed capital expenditures, price determination, market processes, study of national economies under problems of recession and inflation, development plans and balance-of-payments policies in underdeveloped economies, economic forecasting).
   c. Behavioral and social problems (e.g., population dynamics, individual and group behavior).
   d. Simulation of biomedical systems (e.g., fluid balance and electrolyte distribution in the human body, modeling of the brain, blood cell proliferation).
   e. Simulation of war strategies and tactics.

The technique used to solve the theoretical problems cited may be considered a forerunner of simulation in its present-day use. It is called the **Monte Carlo** method and is based on the general idea of using *sampling* to *estimate* a desired result. The sampling process requires describing the problem under study by an appropriate probability distribution from which the samples are drawn. You may find it difficult at this point to establish a link between evaluating an integral, for example, and a probability distribution, since the determination of an integral is a deterministic problem. Keep in mind, however, that Monte Carlo sampling *estimates* rather than determines exactly the value of the integral.

The surge of advances in the use of the Monte Carlo method for solving theoretical problems eventually subsided in the late 1950s and early 1960s. In its place, greater interest was directed to analyzing complex practical problems (such as those just listed) by using **simulation.** Simulation, like the Monte Carlo method, is based on estimating the output of a system through sampling. In this respect, many ideas that were developed in conjunction with Monte Carlo are being used directly in the application of simulation. These ideas include the use of random numbers to obtain samples from a probability distribution as well as techniques for reducing the sample size needed to estimate the desired result reliably. We shall discuss both points in later sections.

The present success of simulation in modeling very complex systems rests squarely on the impressive advances in the capabilities and power of the digital computer. It is unimaginable to think that simulation could have reached any degree of success without the digital computer. For one thing, simulation typically requires very time-consuming computations, albeit simple in nature. To try to do these computations by hand is simply unthinkable.

17.2 TYPES OF SIMULATION

Simulation models are developed to analyze the behavior of systems as a function of time. From that standpoint, there are two types of simulation:

1. Discrete simulation.
2. Continuous simulation.

In discrete simulation, the simulated system is looked at only at selected points in time, whereas in continuous simulations the system is monitored at every point in time. A typical example of discrete simulation is a waiting line system in which customers either join a queue or enter service and then leave the service facility after the service is completed. The continuous case is exemplified by the flow of liquid in a pipeline or by the growth of world population.

In both continuous and discrete simulations, the ultimate objective is to collect pertinent statistics that can be used to describe the behavior of the simulated systems. The manner in which these statistics are collected is the principal factor that decides whether a simulation system is continuous or discrete. Thus, in discrete systems the statistics of the situation can change only when certain **events** take place. For example, in a single-server waiting line model, statistics such as queue length or length of waiting time can change only when a "new" customer arrives for service or when an "old" customer completes its service.

In this respect, we need only observe the system when such events occur. At any other time, the system, from the viewpoint of statistics collection, is at "standstill" and hence need not be monitored. The name "discrete simulation" thus originated because simulation statistics are collected by "jumping" from one (discrete) point to another on the time scale.

In continuous systems, statistics can be collected only by monitoring the situation on a continuous basis. For example, in the study of world population dynamics, some of the variables we would be interested in monitoring as a function of time include

1. Change of population.
2. Change in natural resources.
3. Change in standard of living.

The nature of these variables differs from those of the discrete case in that their change takes place continuously with time.

A wide variety of automatic languages have been developed for both continuous and discrete simulation. The implementation of continuous simulation is basically straightforward, and the role of the language reduces to offering certain aids that facilitate the task of performing the model's tedious computations. In discrete simulation, on the other hand, model construction demands a higher degree of creativity on the part of the user. Three different approaches to discrete modeling have thus evolved, each of which claims specific advantages in facilitating the construction and implementation of discrete models.

In this chapter, we concentrate primarily on discrete simulation. If you are interested in learning more about continuous simulation, you may consult Pritsker [1986] at the end of the chapter.

17.3   DISCRETE SIMULATION APPROACHES

There are three common approaches to discrete simulation:

1. Next-event scheduling.
2. Activity scanning.
3. Process orientation.

It is important to point out that all three approaches are based on the common concept of collecting statistics at the occurrence of key events. For example, in a waiting line model, statistics are collected when a customer arrives and when a service is completed. What makes the approaches appear different is the amount of detailed work (computations and logical decisions) that the approach can perform automatically on behalf of the user. Of the three approaches we listed, process simulation is the most automatic. The other two approaches normally require extensive modeling effort on the part of the user.

For the process-oriented approach to perform most of the logical decisions of the simulation model automatically, it must based on prespecified set of conditions. This restriction makes the approach more rigid when compared with those of next-event scheduling and activity scanning. The trade-off here, then, is between flexibility and the amount of detailed work that the user must provide. Although next-event scheduling and activity scanning are practically limitless in their modeling flexibility, models that can be constructed based on the process-oriented approach are usually compact and easier to develop.

We will explain some of the details of the three approaches by using the simple example of a single-server waiting line model.

17.3.1 NEXT-EVENT SCHEDULING

Consider the situation in which customers arrive at a facility for service. The facility has one server only. When a customer arrives, it must either wait or immediately start service, depending on the status of the server. When a service is completed, the server can draw a waiting customer to start service or, if none is waiting, must remain idle until a new customer arrives. The main objective of simulating this system is to collect statistics such as average queue length, average waiting time per customer, and average idle time per server.

Changes in the systems statistics take place only when the following two events occur:

1. A customer arrives.
2. A customer departs after its service is completed.

The main idea of next-event scheduling is to advance along the time scale until an event is encountered, and, depending on the type of the event, appropriate actions are taken. These actions must be totally exhaustive in the sense that they must account for all the possibilities that may take place when the event occurs. The following is a summary of the actions associated with arrival and departure events.

Arrival Event Actions

1. Check the status of the server (idle or busy).
   (a) If idle, do the following:
      (i) Start the customer in service and generate its departure time by adding its service time to the current simulation time.
      (ii) Indicate that the server is busy and update the idle time statistics of the facility.
   (b) If *busy*, place the arriving customer in the queue and update the queue length statistics.
2. Generate the arrival time of the next customer by adding an interarrival time to the current time of the simulation. (This action is necessary to guarantee a continuation of the simulation.)

Departure Event Actions

1. Check the queue (empty or not empty)
   (a) If empty, revert the facility status to idle,
   (b) if not empty, do the following:
      (i) Choose a waiting customer from the queue by using imposed queue discipline (e.g., FIFO) and update the queue length and the customer waiting time statistics.
      (ii) Start the customer in service and generate its departure time by adding its service time to the current simulation time.

To perform these actions, it is necessary to specify two time elements:

1. The interarrival time for successive customers.
2. The service time for each customer.

In simulation these data elements are either fixed (known with certainty) or, in the majority of cases, are *sampled* from probability distributions that describe the situation. For the time being, we will assume that the interarrival and service time can somehow be secured. In Section 17.4, we provide general procedures for sampling from probability distributions.

The overall implementation of next-event scheduling proceeds as follows. Select the first event on the time scale and perform all its actions including placing any newly generated events in their proper order on the time scale. With the time-scale events now updated, choose the next-in-order event and perform its actions. The procedure is repeated until a desired simulation period is covered.

We now implement the outlined procedure to a specific case. Keep in mind that the values of the interarrival and service times are assumed given throughout the example mainly for simplicity. However, as we stated earlier, these values are usually generated automatically (see Section 17.4).

Suppose that the first arrival event A occurs at $T = 0$ and that the server is initially idle with the queue empty ($Q = 0$). Figure 17–1 summarizes the events lists at $T = 0$.

Since the server is idle, A at $T = 0$ can start service immediately. Supposing that its service time is 17 time units, a departure event D will then occur at $T = 0 + 17 = 17$. Next, to keep the simulation going, we must generate the next arrival. Assuming that the interarrival time is equal to 5 time units, the new A will occur at $T = 5$. Figure 17–2 shows the updated events list.

At $T = 5$, A occurs. Since the server is still busy, A must join the queue whose length now becomes $Q = 1$ at $T = 5$. Assuming that the next arrival occurs after 7 time units, the next A will take place at $T = 5 + 7 = 12$. The new events list is shown in Figure 17–3.

Figure 17–1

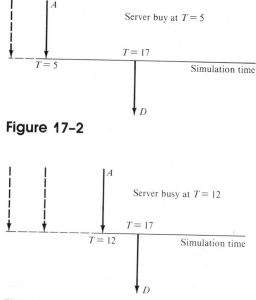

Figure 17–2

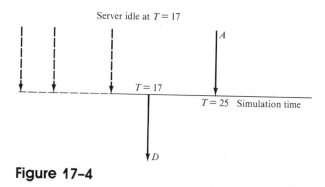

Figure 17–3

Again, at $T = 12$ the server is still busy and A must be put in the queue thus making $Q = 2$ at $T = 12$. Letting the interarrival time equal 13 time units, the next A occurs at $T = 12 + 13 = 25$. Figure 17–4 summarizes the new list of events.

Figure 17–4

At $T = 17$, a departure D occurs, leaving the server idle. Since $Q > 0$, we choose the first waiting customer to start service (FIFO discipline). Since that customer has entered the queue at $T = 5$, its waiting time is computed as $W = 17 - 5 = 12$ time units. Now, assuming that its service time is 5 time units, a D event will take place at $T = 17 + 5 = 22$. The length of the queue at $T = 17$ is $Q = 1$.

This process is repeated as necessary until a desired simulation period is covered. At the end we can compute the needed statistics as follows:

1. Average waiting time per customer

$$= \frac{\Sigma \text{ individual waiting times } W}{\text{number of arrivals}}$$

2. Average server idle time

$$= \frac{\Sigma \text{ idle times}}{\text{number of times server is idle}}$$

3. Average queue length

$$= \frac{\text{area under queue length curve}}{\text{length of simulated time period}}$$

The computing of the first two statistics is straightforward as they represent simple averages. In the case of the queue length, one must imagine that Q is plotted as a function of time as exemplified in Figure 17–5 over the length of the simulated period T. The average is then taken as the area under the curve divided by T. The manner in which these averages are computed gives way to the suggestive names "observation based" in the cases of waiting time and idle time and "time based" in the case of queue length.

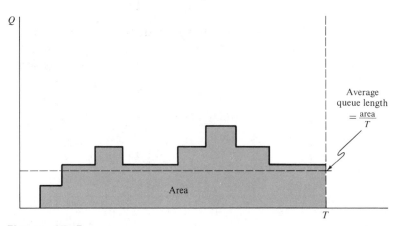

Figure 17–5

Exercise 17.3–1

(a) Continue the simulation in Figure 17–4 until $T = 70$, assuming that the successive interarrival and service times for new customers are as follows:

Interarrival times: 5, 3, 24, 10, 8

Service times: 15, 3, 10, 2, 5, 3

Compute the three statistics defined for this situation.

[*Ans.* Average waiting time = 7.625. Average idle time = 5. Average queue length = .87.]

(b) For a simulated period $T = 70$, is the cumulative waiting time always equal to the area under the queue size curve?

[*Ans.* Yes. Indeed, in part (a), we have $\Sigma W = A = 61$ for $T = 70$.]

Since simulation is normally carried out on the computer, the events list can be thought of as being stored in a (fixed-size) two-dimensional array with each row being associated with an event. The columns should at least carry the event type and its occurrence time. If we keep on storing new events as the sequence of Figures 17–1 to 17–4 suggests, we will eventually exceed the fixed size of the array, no matter how big. For this reason, a more sophisticated method is used in which previously considered events (those shown with dashed lines in Figures 17–1 to 17–4) are dropped and their locations are *reused* for storing new events. The system requires the utilization of pointers to keep track of available locations as well as the chronological order of the events. [See Taha (1987) for details.]

17.3.2 ACTIVITY SCANNING

The activity scanning approach is really very similar to the event scheduling approach. We explain the differences by applying the activity scanning approach to the single-server model. In this situation, the modeler considers three events:

1. An arrival occurs.
2. A service is completed.
3. A service begins.

The actions associated with each of these events are as follows:

1. Arrival
   a. Put the customer in queue.
   b. Generate the next arrival.
2. Service end (departure)
   a. Declare the server idle.
3. Service start
   a. Make the server busy.
   b. Remove customer from queue.
   c. Generate its departure event.

If you compare the actions of the arrival and departure events with those of event scheduling, you will notice that these events *no longer cause movements* of customers. Instead, an arriving customer is automatically put in queue regardless of the status of the server. Also, at the end of a service, the server is always declared idle regardless of the status of the queue.

The "service start" event represents a "new" concept. It differs from the other two in that it cannot really be scheduled in advance. Indeed, its occurrence is conditional upon the satisfaction of two conditions:

1. There are customers waiting.
2. The server is idle.

For this reason, the "service start" event is known as a *conditional* event whose occurrence must be initiated automatically whenever certain conditions of the system are satisfied. Actually, since the conditional event is not a "true" event, it is more appropriate to refer to it as an *activity* that must be scanned every time a true event (arrival or departure) takes place.

You will notice that time advancement in the activity scanning approach is controlled by the arrival and departure events, exactly as in the event scheduling approach. The main difference occurs in the manner in which the actions of the events are executed. In the event scheduling approach, the occurrence of an event results in an *exhaustive* execution of all the conditional and nonconditional actions of the event. In activity scanning, only nonconditional actions (e.g., put customer in queue, generate new arrival, and declare the server idle) are executed. All the conditional actions are moved to activities that must be scanned for possible execution after a real event takes place. For this reason, each activity is defined with a set of conditions that must be satisfied before the activity is executed.

What is the advantage of activity scanning approach? The answer is that it has a simpler structure. If you compare the actions just listed with those of the event scheduling approach, you will get to appreciate this point. The disadvantage, on the other hand, is that all activities must be scanned at the occurrence of each event to check whether or not their conditional actions can be executed. Indeed, the original activity scanning approach calls for listing all the real and conditional events in a single activity list that must be repeatedly scanned from top to bottom until we are sure that none of the conditional events can be executed. With such a procedure, the (real) events need not be kept in a chronological order. This process is obviously computationally inefficient and can be improved at least by placing the conditional and nonconditional events in separate lists.

Apparently, the computational burden of the activity scanning approach outweighs its advantages. Perhaps this is the reason for its lack of popularity among practitioners and researchers alike.

17.3.3   PROCESS-ORIENTED APPROACH

The process-oriented approach differs from the previous two in that it minimizes the effort expended by the user in the development of the model. We demonstrate this point by applying a process-oriented approach to the single-server model. In this case, the modeler may view the system as consisting of three elements:

1. A SOURCE for generating arriving customers.
2. A QUEUE for housing waiting customers.
3. A FACILITY for serving customers.

We can picture the model graphically as shown in Figure 17–6.

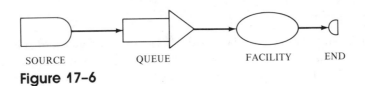

SOURCE   QUEUE   FACILITY   END

Figure 17–6

Figure 17–6 is a complete representation of the simulated system. All the user has to do then is to provide the following basic data for each element:

1. SOURCE: Interarrival time.
2. QUEUE: Queue discipline.
3. FACILITY: Service time.

The model can be embellished by adding such data as limit on the number of customers that can be generated from SOURCE, limit on QUEUE length, and number of parallel servers in FACILITY. However, for the sake of simplicity, we will limit ourselves to the basic data given.

The user normally presents the model in the following format:

SOURCE (interarrival time data)
QUEUE (queue discipline)
FACILITY (service time data)
END

Given this information, the *processor* of the process-oriented approach must do the following:

1. It automatically generates successive arrivals from SOURCE by using the interarrival time data.
2. When a customer arrives, the processor decides whether to put it in QUEUE or send it to FACILITY depending on the status of FACILITY.
3. When a customer starts service, the processor automatically generates its departure time by using the service time data.
4. When FACILITY becomes idle, the processor checks QUEUE for possible waiting customers and then draws one customer as per the discipline of QUEUE.

You will notice that for the processor to be able to carry out these actions automatically, it must have two types of data:

1. The order in which SOURCE, QUEUE, and FACILITY are linked, including possible "skipping" of QUEUE whenever FACILITY is idle.
2. All the data associated with SOURCE, QUEUE, and FACILITY.

What happens inside the processor? Typically, the processor (more or less) emulates the event scheduling approach internally. It keeps track of the arrival events from SOURCE and departure events from FACILITY in a chronological order and then performs their actions as if the situation were carried out by the event scheduling approach. In this respect, the main driving force of the processor is still the arrival and departure events, exactly as in event scheduling.

The obvious advantage of process-oriented simulation is its simplicity. The user provides a "capsule" summary of the model's data and is completely relieved of all the details of how the actions are executed. The disadvantage is that the developed processor, being based on preset conditions, may be rigid at times and hence unfit to model certain complex situations. Clever development of the processor, however, can be made to account for most of the situation that can arise in real life. It is no wonder, then, that process-oriented simulation is the most popular among the three available approaches.

17.4 ROLE OF RANDOM NUMBERS

In Section 17.3, we indicated that a typical simulation usually requires sampling from probability distributions. For example, the interarrival and service times used in the single-server model computations are actually samples taken from associated probability distributions. In this section, we show how samples are computed in simulation. The presentation takes into account the fact that the digital computer is an essential tool for carrying out simulation computations.

In simulation models, sampling from *any* probability distribution is based on the use of [0, 1] random numbers. Before explaining how sampling takes place, we first introduce the statistical conditions that must be satisfied by the [0, 1] random numbers:

1. All (continuous) values in the interval [0, 1] are equally likely to occur; that is, they are *uniformly* distributed.
2. Successive values are generated in the interval [0, 1] in a completely random fashion, meaning that they are independent and uncorrelated.

Arithmetic methods, particularly those suited for computer operation, are available for generating [0, 1] random numbers. The most common method is the **multiplicative congruential** process, which produces random numbers by using a properly constructed recursive equation. The method is shown by appropriate statistical tests to produce numbers that are random and uniform on the [0, 1] interval. Also, the parameters of the recursive equation can be selected in a manner that will produce enough random numbers to carry out a complete simulation run before the numbers generated repeat themselves. Table 17–1 lists 180 random numbers generated by this method. The leftmost decimal point is deleted for convenience. (It is argued that random numbers generated by arithmetic operation cannot be truly random, even though they satisfy the required statistical tests. The reason is that all such numbers are known *in advance* once the **seed** initializing the recursive computations is given. As a result, these computed random numbers are sometimes referred to as **pseudorandom numbers** as compared to truly random numbers that can be generated by special electronic devices.)

One of the principal advantages of being able to generate random numbers arithmetically is the ability to produce the same sequence of random numbers, whenever desired. Thus if one is comparing two alternative designs, by using the same sequence of random numbers one is assured that the difference in the output measures of the experiment is due basically to differences in the alternative designs, not to experimental error.

Let us now illustrate how [0, 1] random numbers are used in simulation sampling by a simple example.

Example 17.4–1 (Coin-Tossing Game). In a coin-tossing game, player *A* wins $10 from player *B* if a head (H) turns up and loses $10 to *B* if the outcome is a tail (T). We assume that the coin is fair so that there is a 50:50 chance of turning up H or T. In other words, the outcomes H and T of the game occur with the following probabilities:

$$p\{H\} = .5, \qquad p\{T\} = .5$$

Since by definition all random numbers are uniformly distributed in the [0, 1] interval, we can devise the following rules for determining the outcome of the game. Let R represent the generated random number; then;

    1. If $0 \leq R \leq .5$, the outcome is H.
    2. If $.5 < R \leq 1$, the outcome is T.

This partitioning of R in the interval [0, 1] is exactly equivalent to stipulating that H or T have equal probabilities of occurrence.

To illustrate how the coin-tossing game is simulated, suppose that players A and B want to repeat the tossing 10 times. This would be equivalent to drawing ten [0, 1] random numbers. Suppose now that we use the first 10

Table 17-1

058962	352943	586999	345500	790012	630566
673284	364609	128099	487110	769774	234646
479909	767638	286650	811154	287072	422037
948578	893129	821570	891254	953397	699089
613960	391962	826125	429090	139412	974665
593277	787674	386551	230243	902490	342756
934123	519930	712470	595449	160463	603734
178239	635823	210783	542293	356707	259606
347270	747163	357549	420822	306993	054556
564395	895364	292630	697501	551336	030504
220998	051455	319740	455344	854407	028341
480379	627205	439814	994034	005875	088946
480798	084273	178448	312230	267349	794021
357985	001720	788457	715256	195421	735221
652531	298202	916428	814746	640618	510998
300419	203535	517436	272799	979865	424002
725225	535325	684924	291621	585405	887838
058382	359749	633051	560561	665907	950391
709176	701533	826616	645896	435829	801902
888953	116597	699005	144655	576880	159382
764371	151787	031383	822207	650795	504908
172286	489546	386699	914280	005383	803777
774217	411304	499874	297506	286166	039436
661127	611829	720835	818541	423735	175532
239580	857687	989902	220226	412235	491375
238133	006422	895334	314207	827234	135543
368147	988993	620635	822874	351526	703292
056012	006447	534569	149393	085234	166865
234088	902734	309612	733068	611896	073756
935472	949026	274906	108197	175026	076380

random numbers in column 1 of Table 17–1 to represent the 10 tossings. The outcomes of the game will thus successively be H, T, H, T, T, T, T, H, H, and T. The net result of 10 tossings is that A will lose $60 - 40 = \$20$ to B. Naturally, as the number of tossings increases sufficiently, we expect a "draw" game, with neither player claiming gain or suffering loss. ◀

Exercise 17.4–1

(a) Suppose that we replace coin tossing by die throwing. If the die is fair, show how the [0, 1] random numbers R are partitioned to identify the outcomes 1, 2, 3, 4, 5, and 6.
   [*Ans.* 1: $0 \leq R \leq 1/6$; 2: $1/6 < R \leq 1/3$; 3: $1/3 < R \leq 1/2$; 4: $1/2 < R \leq 2/3$; 5: $2/3 < R \leq 5/6$; 6: $5/6 < R \leq 1$.]
(b) By using the first 10 random numbers in column 1 of Table 17–1, identify the sequence of outcomes in the die-throwing game.
   [*Ans.* 1, 5, 3, 6, 4, 4, 6, 2, 3, 4.]

It may appear that the use of [0, 1] random numbers is limited to sampling from distributions in which all the outcomes occur with equal probabilities, such as the coin-tossing and die-throwing examples. This is not so. Indeed, we show below that the [0, 1] random numbers can be used to generate outcomes from any probability distribution.

Before generalizing the use of [0, 1] random numbers to sampling from any distributions, we first consider the die-throwing game from a slightly different angle. Let x represent the outcome (random variable) and define $p(x)$ and $F(x)$ as its PDF and CDF, respectively. Then we have

x:	1	2	3	4	5	6
$p(x)$	1/6	1/6	1/6	1/6	1/6	1/6
$F(x)$	1/6	1/3	1/2	2/3	5/6	1

Figure 17–7 plots x versus $F(x)$. Examining the $F(x)$ scale, we notice that the partitioning of the [0, 1] random numbers among the different values of x is automatically given as

$$x = 1: \quad 0 \leq F(x) \leq 1/6, \qquad x = 4: \quad 1/2 < F(x) \leq 2/3$$
$$x = 2: \quad 1/6 < F(x) \leq 1/3, \qquad x = 5: \quad 2/3 < F(x) \leq 5/6$$
$$x = 3: \quad 1/3 < F(x) \leq 1/2, \qquad x = 6: \quad 5/6 < F(x) \leq 1$$

We can see from Figure 17–1 that by generating a [0, 1] random number R and assigning it to $F(x)$, we can obtain the outcome x by simply inverting $F(x)$. For example, for $R = .4$, $F(x) = .4$, which yields $x = 3$, as can be seen both graphically in Figure 17–7 and also by the ranges we tabulated. This procedure is precisely equivalent to inverting $F(x)$ and, indeed, is referred to as the **method of inversion.**

The inversion method can be applied to *any* probability distribution provided that we can show that $F(x)$ is uniformly distributed in the interval $0 \leq F(x) \leq 1$. This result is easily proved for any $F(x)$ as follows. Let

$$y = F(x), \qquad 0 \leq y \leq 1$$

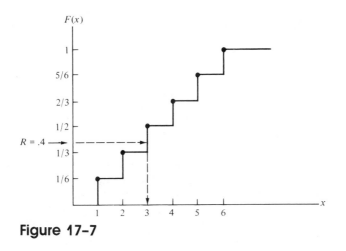

Figure 17-7

and assume that $G(Y)$ is the CDF of $y = F(x)$. Then

$$G(Y) = P\{y \le Y\} = P\{F(x) \le Y\}$$
$$= P\{x \le F^{-1}(Y)\}$$
$$= F\{F^{-1}(Y)\}$$
$$= Y, \qquad 0 \le Y \le 1$$

This property can be satisfied only if $y = F(x)$ is *uniformly* distributed for $0 \le F(x) \le 1$. The result is true for both discrete and continuous distributions.

The application of the inversion method to any discrete distribution should now be obvious, since we follow the same procedure employed with the die-throwing example. We illustrate the application of the method to continuous distributions by the following example.

Example 17.4-2 (Sampling the Exponential Distribution). Suppose that in a service facility the service time t is exponential with service rate μ per unit time. Then the PDF of t is

$$f(t) = \mu e^{-\mu t}, \qquad t > 0$$

Thus we have

$$F(t) = \int_0^t \mu e^{-\mu x}\, dx = 1 - e^{-\mu t}$$

If R is a [0, 1] random number, then putting $F(t) = R$, we get

$$R = 1 - e^{-\mu t}$$

or

$$t = -\frac{1}{\mu} \ln(1 - R) = -\frac{1}{\mu} \ln R$$

The last step is justified because if R is [0, 1] random number, then so is $(1 - R)$ and we may replace $(1 - R)$ by R for convenience. ◀

Exercise 17.4-2

(a) Suppose that the service-time distribution in the service facility problem is exponential with mean 10 minutes. Determine the service times for the first five customers using the first five random numbers in column 1 of Table 17–1.
[*Ans.* 28.3, 3.96, 7.34, .53, 4.88 minutes.]

(b) Suppose that a gambler's loss or gain in a roulette game is uniformly distributed such that he can never lose more than $60 or gain more than $40. What is the net income (loss or gain) if he plays the game five times? Assume that the outcome of the game is continuous and use the first five [0, 1] random numbers in column 1 of Table 17–1.
[*Ans.* The gambler's outcome is uniformly distributed between −$60 and $40. His outcome in each play is $-60 + 100R$, and his net income in five plays is $(-54.1 + 7.3 - 12 + 34.9 + 1.4) = -\$22.5$, which is a loss.]

(c) Generalize the result to compute the game's outcome given that it is uniformly distributed on the continuous scale from A to B.
[*Ans.* Outcome $= A + R(B - A)$.]

The inversion method cannot be used with *continuous* distributions for which the CDF $F(x)$ cannot be determined analytically. Typical examples are the normal, gamma, and Poisson distributions. In situations such as these, we may use one of two methods for the purpose of sampling:

1. Approximate the continuous $F(x)$ by a discrete $F(x)$ and then sample from the approximate CDF.

2. Use appropriate relationships in the theory of statistics to obtain the required information from other distributions that are easy to manipulate analytically.

The discrete approximation approach is straightforward and convenient. In fact, some automatic simulation languages, such as GPSS, always express the CDF as a discrete approximation. Let us illustrate the application of this approach by the following example.

Example 17.4-3 (Discretization of Standard Normal Distribution). To approximate the standard normal by a discrete distribution, we first divide the range of the random variable x into appropriate intervals. Naturally, the smaller the intervals, the better the approximation. In this example we use large division intervals of .5 each primarily for economy of space. From the standard normal tables, we obtain the information in Table 17–2.

Table 17-2

x	$F(x)$	x	$F(x)$
−3	.00135	.5	.6915
−2.5	.00621	1.	.8413
−2.	.0228	1.5	.9332
−1.5	.0668	2.	.9772
−1.	.1587	2.5	.99379
− .5	.3085	3.	.99865
0	.5	∞	1.

To approximate the standard normal by a discrete distribution using intervals of .5 would be equivalent to replacing the continuous scale $-\infty < x < \infty$ with the discrete values $-3.25, -2.75, -2.25, \ldots, 2.25, 2.75,$ and 3.25. In other words, we have replaced each interval by its midpoint. The approximation takes into account that, for all practical purposes, the standard normal distribution is totally included in the interval $x = -3.5$ to $x = 3.5$. Table 17–3 now associates the discrete outcomes with the appropriate ranges of the [0, 1] random number R. As we mentioned previously, the approximation can be improved by choosing smaller intervals. ◄

Table 17-3

x	Limits on R	x	Limits on R
-3.25	$0 \leq R \leq .00135$	$.25$	$.5 < R \leq .6915$
-2.75	$.00135 < R \leq .00621$	$.75$	$.6915 < R \leq .8413$
-2.25	$.00621 < R \leq .0228$	1.25	$.8413 < R \leq .9332$
-1.75	$.0228 < R \leq .0668$	1.75	$.9332 < R \leq .9772$
-1.25	$.0668 < R \leq .1587$	2.25	$.9772 < R \leq .99379$
$-.75$	$.1587 < R \leq .3085$	2.75	$.99379 < R \leq .99865$
$-.25$	$.3085 < R \leq .5$	3.25	$.99865 < R \leq 1.$

Exercise 17.4–3

Show how the discrete approximation in Table 17–3 can be used to sample from any normal distribution with mean μ and standard deviation σ.
[*Ans.* Simply replace x in Table 17–3 by $\mu + \sigma x$.]

The second method for sampling from analytically complex distribution is based on the use of the relationships between statistical distributions. The main idea of the approach is to take advantage of the fact that certain distributions that can be sampled easily have direct statistical relationships to other more complex distributions (see Figure 10–9). For example, the gamma random variable is the sum of exponential random variable and the time between arrivals of a Poisson process is also exponential. The normal distribution can also be related to the uniform distribution by the central limit theorem. Let us consider these cases in detail.

Example 17.4-4 (Sampling the Gamma Distribution by Using the Exponential).

The random variable of a gamma distribution with parameters n and μ is the sum (convolution) of n independent and identically distributed exponential random variables each with parameter μ (see Section 10.9). Thus, if T represents a gamma random variable, then

$$T = t_1 + t_2 + \cdots + t_n$$

where t_i is the ith exponential random variable with parameter μ. As given in Example 17.4–2,

$$t_i = \frac{-1}{\mu} \ln R_i$$

Consequently,

$$T = \left(\frac{-1}{\mu} \ln R_1\right) + \left(\frac{-1}{\mu} \ln R_2\right) + \cdots + \left(\frac{-1}{\mu} \ln R_n\right)$$

$$= \frac{-1}{\mu} \ln (R_1 R_2 \cdots R_n)$$

To illustrate the use of this formula, suppose $n = 3$ and $1/\mu = 10$ minutes. Thus, to generate the first gamma sample, we need three [0, 1] random numbers. Suppose that $R_1 = .09656$, $R_2 = .96657$, and $R_3 = .64842$, then

$$T = -10 \ln (R_1 R_2 R_3) = 28.05 \text{ minutes.} \qquad \blacktriangleleft$$

Example 17.4–5 (Sampling the Poisson Distribution by Using the Exponential). We know from probability theory (see Section 10.3.2B) that if the number of events can be described by a Poisson stream, then the time between the occurrence of events must be exponential. In the Poisson distribution, the outcome is expressed as the number of events n occurring in a time period t. Thus, to sample a Poisson distribution with mean λt, all we have to do is to sample the exponential distribution with mean $1/\lambda$ as many times as necessary until the sum of the exponential random variables generated exceeds t *for the first time*. In this case the sampled Poisson value n is taken equal to the number of times we sampled the exponential distribution *less one*.

We illustrate the process numerically as follows. Suppose that we wish to sample a Poisson distribution with mean rate $\lambda = 3$ events per hour during a period of 1.4 hours ($t = 1.4$). Thus we sample from an exponential distribution with mean $1/3$ hour and record the number of samples needed to exceed $t = 1.4$ hours for the first time. For the exponential distribution we have

$$t_i = -\frac{1}{\lambda} \ln R_i = \frac{-1}{3} \ln R_i$$

Suppose that the random numbers from column 1 of Table 17–1 are used. We then obtain the following results:

n:	1	2	3	4	5
R_n	.058962	.673284	.479909	.948578	.61396
t_n	.9436	.1318	.2447	.0176	.1626
$\sum_{i=1}^{n} t_i$	.9436	1.0754	1.3201	1.3377	1.5003

We can now see that $\sum_{i=1}^{n} t_i$ becomes larger than $t = 1.4$ hours for the first time when $n = 5$. This means that the sampled Poisson value is $n = 5 - 1 = 4$. $\qquad \blacktriangleleft$

Exercise 17.4–4

In Example 17.4–5, find the Poisson value given the following values of t. (Use the sequential random numbers in column 1 of Table 17–1 as necessary.)

(a) $t = 1$.
   [*Ans.* $n = 1$.]
(b) $t = 1.75$.
   [*Ans.* $n = 7$.]

We can actually simplify the computations of the procedure as follows. First, note that n is defined such that

$$\sum_{i=1}^{n} t_i \leq t < \sum_{i=1}^{n+1} t_i$$

Substituting for $t_i = -(1/\lambda) \ln R_i$, we see that

$$-\sum_{i=1}^{n} \ln R_i \leq \lambda t < -\sum_{i=1}^{n+1} \ln R_i$$

Multiplying throughout by -1 (which reverses the direction of the inequalities) and noting that

$$\sum_{i=1}^{n} \ln R_i = \ln \prod_{i=1}^{n} R_i$$

we get

$$\ln \prod_{i=1}^{n} R_i \geq -\lambda t > \ln \prod_{i=1}^{n+1} R_i$$

or, equivalently,

$$\prod_{i=1}^{n} R_i \geq e^{-\lambda t} > \prod_{i=1}^{n+1} R_i$$

This formula now simplifies the computations considerably. For example, in the case of Example 17.4–5, $\lambda t = 4.2$. Using the same sequence of random numbers employed earlier, we note that since $e^{-\lambda t} = .01499$ lies between the two values

$$\prod_{i=1}^{4} R_i = .01807 \quad \text{and} \quad \prod_{i=1}^{5} R_i = .01109$$

the sampled Poisson value is $n = 4$.

Exercise 17.4–5
Repeat Exercise 17.4–4 using the product of random variables. You must obtain the same answer, of course.

Example 17.4–6 (Sampling the Normal Distribution by Using the Central Limit Theorem).
From Section 10.8 we show that the sum of n identically distributed and independent random variables tends to be normally distributed as n becomes sufficiently large. This means that we can sample normal distributions by using uniform [0, 1] random numbers. The idea is as follows. Using the random numbers $R_1, R_2, \ldots, R_n$, we define

$$T = R_1 + R_2 + \cdots + R_n$$

Then

$$E\{T\} = \frac{n}{2}$$

$$\text{var}\{T\} = \sum_{i=1}^{n} \text{var}\{R_i\} = \frac{n}{12}$$

According to the central limit theorem, T is asymptotically normal with mean $n/2$ and variance $n/12$. Now let

$$z = \frac{T - n/2}{\sqrt{n/12}}$$

which is normal with zero mean and unit variance.

For any normal distribution with mean μ and variance σ^2, the random deviate y corresponding to the n random numbers is obtained by considering

$$\frac{y - \mu}{\sigma} = z = \frac{T - n/2}{\sqrt{n/12}}$$

Hence

$$y = \mu + \frac{\sigma}{\sqrt{n/12}}\left(T - \frac{n}{2}\right) = \mu + \frac{\sigma}{\sqrt{n/12}}\left(\sum_{i=1}^{n} R_i - \frac{n}{2}\right)$$

By the central limit theorem, normality is approached rapidly even for relatively small values of n. In practice, n is usually chosen equal to 12. In this case, the formula simplifies to

$$y = \mu + \sigma\left(\sum_{i=1}^{12} R_i - 6\right) \qquad \blacktriangleleft$$

This method requires 12 random numbers for each normal sample, which obviously is inefficient. A method was thus developed to alleviate this difficulty. It can be proved that for the pair of [0, 1] random numbers R_1 and R_2, the random variables x_1 and x_2 defined as

$$x_1 = \sqrt{-2 \ln R_1} \cos (2\pi R_2)$$
$$x_2 = \sqrt{-2 \ln R_2} \sin (2\pi R_1)$$

are standard normal with mean 0 and variance 1. By using these two formulas, we can generate two standard normal samples. For example, given $R_1 = .09656$ and $R_2 = .96657$, then

$$x_1 = \sqrt{-2 \ln .09656} \cos (2\pi .96657) = 4.5724$$
$$x_2 = \sqrt{-2 \ln .96657} \sin (2\pi .09656) = .1487$$

The two samples can be converted to any $N(\mu, \sigma)$ by using the formula

$$y = \mu + \sigma x$$

From the standpoint of computations, the foregoing formulas appear efficient. However, the computation of the trigonometric function on the computer requires the use of series approximation, which is time consuming. For this reason, the method was modified to a recursive formulation that eliminates the use of the sine and cosine functions. The new procedure is based on what is known as the *acceptance–rejection* method. The details of this method can be found in Fishman [1978].

17.4.1 TRANSIENT AND STEADY STATES IN SIMULATION

Simulation, being a *statistical experiment,* typically exhibits variations in output. These variations occur as a function of time. For example, typical variations with time of the cumulative average queue length in a single-server model is shown in Figure 17–8 for two different simulation runs. The two runs differ primarily in the sequences of random numbers that are used to sample the interarrival and service times. You will notice that the average queue length in the two runs differs initially, but eventually tends to converge toward a fixed value as the length of the simulation run increases.

The behavior of simulation output exhibited in Figure 17–8 is typical of all simulation models. The output normally comprises two regions: the *transient* state and the *steady* state. The transient state is characterized by the output being a function of the simulation time. In the steady state, the output becomes independent of time.

The design of most systems is usually based on steady-state results. Consequently, it is essential that the output of the simulation be measured after the steady state has been reached. Unfortunately, in a typical simulation model, the length of transient state cannot be predicted in advance. Indeed, a number of factors can affect the length of the transient state. Most prominent among these is how close are the initial conditions (at the start of the simulation run) to those representing the steady state. Another factor deals with the design parameters of the system. For example, in the single-server model, the steady state will be reached faster as the interarrival time increases relative to the

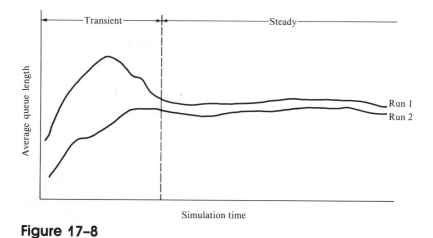

Figure 17–8

service time. Indeed, steady state will never be attained if the service time is higher than the interarrival time.

Unfortunately, there are no practical procedures for deciding when the transient state ends and the steady state begins. All available procedures are difficult to implement in the context of a simulation model [see Fishman (1978)]. Perhaps the only general guideline that can be offered in this respect is what the very definition of steady state stands for, namely, longer simulation runs are more likely to push the system toward the steady state.

17.5 GATHERING OBSERVATIONS IN SIMULATION

Having taken care of the mechanics of constructing and executing simulation models, we now turn our attention to the important question of how to gather observations in simulation. Since simulation is basically an experiment, observations gathered must be statistically independent and identically distributed to be able to make proper statistical inferences about the simulated system.

In any physical experiment, an estimate of an output variable is normally based on the *average* value of n independent observations. The value of n is determined such that a certain *confidence level* is acquired. In simulation, an estimate of the simulated system's measure of performance must also be based on n observations. Nevertheless, gathering independent observations in simulation is far more difficult than it is in a regular laboratory experiment. We have seen in Figure 17.8 that early output of the simulation experiment is unstable (transient state) and that stability (steady state) is usually reached after the simulation run becomes "sufficiently" long. As a result, care must be taken that observations are not gathered during the early stages of the simulation run, because the information obtained is subject to large variation and hence may not be representative of the true behavior of the system. In essence, we are interested in gathering observations after steady state prevails because they will yield smaller sampling error (as measured by the standard deviation) and hence more accurate results.

Since our main objective is to gather observations with the sampling error as small as possible, ideally we can achieve this result by

1. Making "very" long simulation runs to improve the chances of reaching steady state.
2. Replicating the simulation runs with different sequences of random numbers and using each run to represent an observation.

By using different sequences of random numbers, the resulting observations are necessarily independent, as desired. Although the sampling error is reduced by gathering the observations under steady-state conditions, it can be reduced further by taking the *average* of these observations. This result follows, since the standard deviation of the average of n observations is $1/\sqrt{n}$ of the standard deviation of the individual observations.

In spite of the fact that the procedure just described will produce a small sampling error, at stake in this process is the cost of gathering the observations. In other words, although it is important to reduce sampling error, such an objective should not be acquired at any cost. Clearly, making simulation runs very long in order to bypass the transient state is costly, as it may involve long nonproductive use of computer time.

In practice, simulation observations are gathered with two points in mind:

1. The cost of simulation can be controlled essentially by reducing the length of simulation runs.

2. The sampling error can be reduced by utilizing improved sampling techniques specially designed to reduce statistical error.

Naturally, we cannot have something for nothing. As we show shortly, a reduction in the length of simulation runs can be achieved either at the expense of sampling the system while it is still in transient state or, to reach steady state, gathering observations that may be correlated. The techniques for reducing sampling error, known as **variance reduction** methods, are useful, but their implementation may be demanding in terms of the design of the simulation model. We discuss both points.

17.5.1 PROCEDURES FOR GATHERING OBSERVATIONS

In this section we present the following three procedures for gathering observations in simulation:

1. The replication method.
2. The subintervals method.
3. The cycles method.

Other methods are available also [see Pritsker (1986)], but the three procedures just listed seem to be the most useful in practice.

In all methods used to gather observations, it is important to allow an initial "warm-up" period so that the simulation can reach steady state. Naturally, the warm-up period differs with the type of simulation model as well as its initial conditions. There are methods, however, that can determine within experimental error whether or not steady state has been reached. Such methods are called **truncation procedures** because they specify the length of the initial simulation period that should be truncated before one can start gathering observations. A summary of these procedures has been given by Pritsker [1986].

Replication Method

In this method each observation is represented by a separate run, with each run started with the same initial conditions but with a different sequence of random numbers. The *advantage* of this method is that observations are statistically independent, which is a basic underlying assumption for applying any statistical test. The *disadvantage* is that each observation may be heavily biased because of the effect of the initial conditions (transient state). As mentioned previously, we cannot alleviate this problem by making long runs, as this may be costly from the standpoint of using the computer.

Suppose that $x_1, x_2, \ldots$, and x_n are n observations obtained by the replication method for a certain measure of the system. Then the measure's best estimate is given by the average

$$\bar{x} = \frac{\sum\limits_{i=1}^{n} x_i}{n}$$

and the 100 $(1 - \alpha)$ confidence interval $(0 \le \alpha \le 1)$ on the exact mean μ is given by

$$\bar{x} - \frac{s}{\sqrt{n}} t_{\alpha/2, n-1} \le \mu \le \bar{x} + \frac{s}{\sqrt{n}} t_{\alpha/2, n-1}$$

where

$$s^2 = \frac{\sum\limits_{i=1}^{n} (x_i - \bar{x})^2}{n - 1}$$

and $t_{\alpha/2, n-1}$ is the t-statistic with $n - 1$ degrees of freedom.

Subintervals Method

The subintervals method is designed to reduce the effect of the transient conditions associated with the replication method. It calls for dividing each simulation run into equal time intervals. The beginning of each interval is the starting point for recording the information of a new observation. The *advantage* of the method is that the effect of transient condition should "fade away" with time, which means that the observations will be more representative of the real-life conditions. The *disadvantage* is that the assumption of independence is destroyed because the values at the beginning of an interval obviously depend on the end conditions of the immediately preceding interval. This means that autocovariance exists between successive intervals. The effect of autocorrelation can be reduced by (1) increasing the number of observations n, and (2) increasing the interval size associated with each observation. We note, however, that both requirements will result in increasing the computer time and hence the cost.

Example 17.5-1. In this example we demonstrate the application of the subinterval method. Figure 17–9 demonstrates the change with time of queue size Q of a single run in a typical single-server facility. We have used simple numbers and have divided the area under the curve into squares to facilitate area computations for the purpose of computing the mean values.

In the subinterval method we first choose the subinterval size that must be large enough to allow gleaning important information about the variable under consideration. In Figure 17–9 we choose the interval size equal to 5 time units.

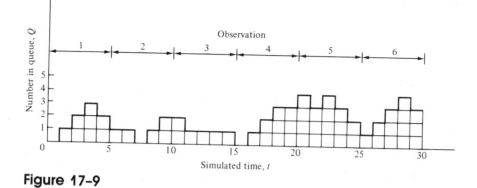

Figure 17-9

Thus the successive observations are collected for the intervals $0 \le t \le 5$, $5 \le t \le 10$, $10 \le t \le 15, \ldots$, where t represents the simulated time. Our objective in this example is to gather observations about the average queue size. Let Q_i represent the ith observation; then, as explained in Figure 17–5,

$$Q_i = \frac{\text{area under } Q \text{ in interval } i}{\text{size of interval } i} = \frac{A_i}{b_i - a_i}$$

where $a_i \le t \le b_i$ represent the ith interval. For example, for the first observation, $i = 1$, we have $0 \le t \le 5$ and $A_1 = 8$ so that

$$Q_1 = \frac{8}{5} = 1.6 \text{ customers}$$

Table 17–4 summarizes the computations of all six observations as specified in Figure 17–9.

Table 17-4

Observation i	A_i	Interval Size	Q_i
1	8	5	1.6
2	5	5	1.
3	6	5	1.2
4	9	5	1.8
5	16	5	3.2
6	13	5	2.6

Based on these observations, a best estimate of the queue size is

$$\overline{Q} = \frac{Q_1 + Q_2 + \cdots + Q_6}{6} = \frac{11.4}{6} = 1.9 \text{ customers}$$

and the variance of Q_i is

$$s^2 = \sum_{i=1}^{6} \frac{(Q_i - \overline{Q})^2}{5} = .716$$

so that a 95% confidence interval on the expected queue length is given by

$$\bar{Q} \pm \frac{s}{\sqrt{6}} t_{.025,5} = 1.9 \pm .888 \qquad \blacktriangleleft$$

Exercise 17.5-1

Repeat the computations in Table 17–6 using an interval size of 6 time units in Figure 17–8; then compute the resulting confidence interval.
[*Ans.* Observations Q_i are 1.5, 1.17, 1, 3.33, 2.5. $\bar{Q} = 1.9$ and $s^2 = .977$. The 95% confidence interval is 1.9 ± 1.227.]

A comparison between the computations in Table 17–4 and those in Exercise 17.5–1 reveals that $\bar{Q}$ remains that same ($= 1.9$) regardless of the number of observations provided that we use the same length T of the simulation run ($= 30$), whereas the sample variance s^2 tends to decrease with the increase in the number of observations. We must caution, however, that the computation of s^2 by using the common formula given in Example 17.5–1 is only an approximation of the true variance estimate. The reason is that the effect of autocorrelation is ignored. Indeed, the formula for estimating the variance of the autocorrelated sample $x_1, x_2, \ldots, x_n$ is approximated by

$$s^2 = v_0 + 2 \sum_{j=1}^{m} \left(1 - \frac{j}{m}\right) v_j$$

where

$$v_j = \frac{1}{n-j} \sum_{k=1}^{n-j} (x_k - \bar{x})(x_{k+j} - \bar{x}), \qquad j = 0, 1, 2, \ldots$$

is the sample covariance between x_k and x_{k+j}. Note that v_0 is the sample variance used in Example 17.5–1 with the denominator $n - 1$ replaced by n. The approximation requires the value of m to be much smaller than n under the assumption that the value of v_j becomes negligible for $j > m$. This condition can normally be accounted for by increasing the interval size associated with each observation. Also, as n becomes large, the effect of autocorrelation becomes negligible, and one can then use v_0 to estimate the sample variance.

Cycles Method

The cycles method attempts to reduce the effect of autocorrelation that is present in the subintervals method by starting each interval at time points with similar conditions. For example, in Figure 17–8 we can consider the intervals $0 \le t \le 7$ and $7 \le t \le 15$ to represent the first two intervals in the cycles method. This follows because each of the two intervals start with the same condition regarding the variable under consideration: namely, queue size $Q = 0$. It is now obvious why we use the name *cycles method* because each observation is associated with a time cycle that starts with identical conditions (from the standpoint of the variable under consideration).

Although the cycles method may reduce the effect of autocorrelation, it has the disadvantage of yielding a smaller number of observations compared with the subintervals method. This follows because we cannot tell in advance when each new cycle will begin or how long each cycle will be. Under steady-state conditions, however, we should expect the occurrence of cycles in a more or less regular fashion.

Since both the cycle length and the (time integrated) area enclosed by the cycle are random variables, the estimation of the value of each observation is not as simple because it is now represented by the quotient of two random variables (compare the subintervals method). Let

t_i = length of cycle i (time units)

z_i = value of variable of interest in cycle i

Then the ith observation x_i is given by

$$x_i = \frac{z_i}{t_i}$$

A *crude* way of analyzing x_i is to apply the regular mean and variance formulas directly to x_i. However, because x_i is the ratio of two random variables, these formulas normally result in biased estimates.

An unbiased estimate of the sample average based on the cycles method can be shown to be

$$\bar{y} = \frac{\sum\limits_{i=1}^{n} y_i}{n}$$

where

$$y_i = \frac{n\bar{z}}{\bar{t}} - \frac{(n-1)(n\bar{z} - z_i)}{n\bar{t} - t_i}, \qquad i = 1, 2, \ldots, n$$

$$\bar{z} = \frac{\sum\limits_{i=1}^{n} z_i}{n}$$

$$\bar{t} = \frac{\sum\limits_{i=1}^{n} t_i}{n}$$

In this case a confidence interval can be set on the true mean by using $\bar{y}$ and the variance s_y^2 of y_i; that is, a $100(1 - \alpha)$ confidence interval on the true mean is given by

$$\bar{y} \pm \frac{s_y}{\sqrt{n}} t_{\alpha/2, n-1}$$

Example 17.5–2. Figure 17–10 represents the number in system (queue + in service) of a single-server facility. It is required to estimate the percentage of time the facility is idle by using the cycles method.

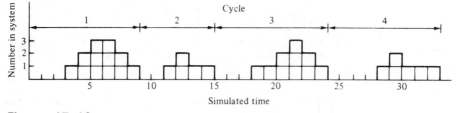

Figure 17-10

The simulated period T (= 32 time units) in Figure 17-10 has four cycles. The percentage of idle time is represented by the proportion of time when no customers are in the system during a cycle. For example, cycle 1 is 9 time units, during which the system was empty 3 time units. Thus the proportion of idle time is (3/9) × 100 or 33.3%. Letting z_i represent the idle period during cycle i, we can compile the information required for all four cycles and compute the unbiased estimators y_i as shown in Table 17-5. The order of computations requires calculating $\bar{z}$ and $\bar{t}$ first, from which y_i can be computed. Thus, with $\bar{z} = 3$ and $\bar{t} = 8.25$, we have

$$y_i = \frac{4 \times 3}{8.25} - \frac{3(4 \times 3 - z_i)}{4 \times 8.25 - t_i} = 1.4545 - \frac{3(12 - z_i)}{33 - t_i}$$

Table 17-5

Cycle i	z_i	t_i	y_i
1	3	9	.3295
2	2	6	.3434
3	3	9	.3295
4	4	9	.4545
	$\bar{z} = 3$	$\bar{t} = 8.25$	$\bar{y} = .3642$

We then have

$$s_y^2 = \frac{\sum\limits_{i=1}^{4} (y_i - \bar{y})^2}{3} = .00366$$

A 95% confidence interval in the true mean of y_i is thus given by

$$\bar{y} \pm \frac{s_y}{\sqrt{4}} t_{.025,3} = .3642 \pm .0963 \qquad \blacktriangleleft$$

Exercise 17.5-2

Estimate the average number in system using the cycles in Figure 17-9 and establish a confidence interval on the true value.
[*Ans.* 1.0035 ± .5107.]

17.5.2 VARIANCE REDUCTION TECHNIQUE

As we mentioned previously, a straightforward method for reducing sampling error is to use a large sample size and a long simulation run. This procedure may not be economical, however. For this reason, other methods have been developed which, for the same sample size, will result in a smaller sampling error. These methods are called variance reduction techniques.

The most practical variance reduction technique in simulation is the **antithetic method.** It is based on the observation that when we generate a random number R, its complement $1 - R$ is also a random number. The idea of the antithetic technique is to make one simulation run with the sequence $R_1, R_2, \ldots, R_k$ and then another with the *complement* sequence $1 - R_1, 1 - R_2, \ldots, 1 - R_k$. Let $x_1, x_2, \ldots, x_k$ and $y_1, y_2, \ldots, y_k$ represent the outcomes of the two runs. The two sequences $(x_1, x_2, \ldots, x_n)$ and $(y_1, y_2, \ldots, y_n)$ are negatively correlated because when a low (random) value of x_i is drawn, a high (random) value of y_i is generated, and vice versa. Additionally, both sequences must have the same mean μ. In this case an estimate of μ is given by

$$\bar{z} = \frac{1}{n} \sum_{i=1}^{n} \frac{x_i + y_i}{2}$$

We observe that estimating μ by using $\bar{z}$ as defined is equivalent to taking a sample of size $2n$.

The advantage of estimating the mean in this manner is that $\bar{z}$ has a smaller variance than that obtained from a sample of size $2n$ in which the observations are not correlated. To prove this point, we note that the proposed procedure is equivalent to generating observations y_i satisfying

$$z_i = \frac{x_i}{2} + \frac{y_i}{2}$$

so that

$$\text{var}\{z\} = E\left\{\left(\frac{x}{2} + \frac{y}{2}\right)^2\right\} - \left(E\left\{\frac{x}{2} + \frac{y}{2}\right\}\right)^2$$

$$= \frac{1}{4}\left(E\{x^2\} - E^2\{x\}\right) + \frac{1}{4}\left(E\{y^2\} - E^2\{y\}\right)$$

$$+ \frac{1}{2}\left(E\{xy\} - E\{x\}E\{y\}\right)$$

$$= \frac{1}{4}\left(\text{var}\{x\} + \text{var}\{y\}\right) + \frac{1}{2}\text{cov}\{x, y\}$$

Since x and y are negatively correlated, it follows that $\text{cov}(x, y) < 0$ and

$$\text{var}\{z\} < \frac{1}{4}\left(\text{var}\{x\} + \text{var}\{y\}\right)$$

We can see from this inequality that, under the worst conditions, var $\{z\}$ is less than one half of the larger of var $\{x\}$ and var $\{y\}$.

Example 17.5-3. In this example we show the superiority of the antithetic technique in reducing the sampling error. It is required to estimate the mean value of the exponential distribution

$$f(x) = .5e^{-.5x}, \qquad x > 0$$

whose exact mean value is $1/\mu = 2$. We do so by sampling the exponential distribution using a sequence $\{R_i\}$ and its complementary sequence $\{1 - R_i\}$. If $\{x_i\}$ and $\{y_i\}$ are the exponential random variables given $\{R_i\}$ and $\{1 - R_i\}$, respectively, then

$$x_i = -2 \ln R_i$$
$$y_i = -2 \ln (1 - R_i)$$

The process is repeated for $n = 10$ random numbers. The antithetic technique is applied by computing

$$z_i = \frac{x_i + y_i}{2}$$

Table 17–6 provides the necessary computations. The antithetic computations are based on $n = 5$ (instead of $n = 10$) to allow a meaningful comparison, because from the standpoint of the number of computations, $n = 5$ in the antithetic method is actually equivalent to a sample of size $n = 10$.

Table 17-6

i	R_i	x_i	$1 - R_i$	y_i	$z_i = (x_i + y_i)/2$
1	.05896	2.830	.94104	.122	1.476
2	.673284	.791	.326716	2.237	1.514
3	.479909	1.468	.520091	1.308	1.388
4	.948578	.106	.051422	5.935	3.021
5	.613960	.976	.38604	1.904	1.440
6	.593277	1.044	.406723	1.799	
7	.934123	.136	.065877	5.440	
8	.178239	3.449	.821761	.393	$\bar{z} = 1.7678$
9	.347270	2.115	.65273	.853	$s_z^2 = .4929$
10	.564395	1.144	.435605	1.662	
	$\bar{x} = 1.406,$	$s_x^2 = 1.197$	$\bar{y} = 2.1653,$	$s_y^2 = 3.9076$	

We can see from the results in Table 17–6 that the antithetic technique yields the smallest variance ($s_z^2 = .4929$). Note that whereas the sequence y_i yields a closer estimate of the mean (exact $\mu = 2$), its variance is much larger. Naturally, if we increase the sample size of z_i to 10, we should expect the estimate of $\bar{z}$ to be closer to the true value ($1/\mu = 2$). Additionally, s_z^2 should also be smaller than that based on five observations only (see Exercise 17.5–3). ◄

Exercise 17.5-3
Compute the value of z and s_z^2 in Table 17–8 based on $n = 10$.
[*Ans.* $\bar{z} = 1.7857$, $s_z^2 = .3738$.]

In an actual simulation model, the variable x_i would represent a measure of performance of the system under study. This means that each x_i will not normally be generated by a single random number as Example 17.5-3 might suggest. Indeed, the gathering of each observation x_i or its antithesis y_i should be based on one of the methods we presented in Section 17.5.1 (i.e., replication, subintervals, and cycles methods). Thus the negative correlation property resulting from having two runs based on the sequences $\{R_i\}$ and $\{1 - R_i\}$ may not hold in general because we do not know how the desired measure of the system would change when the sequence $\{R_i\}$ is replaced by $\{1 - R_i\}$. This difficulty may not be as pronounced when we use the *replication* or *subinterval methods* because in both techniques we can base the two observations x_i and y_i on *equal* time intervals, with the result that negative correlation between x_i and y_i may be maintained. In the *cycles method,* however, the use of the sequences $\{R_i\}$ and $\{1 - R_i\}$ will probably result in different cycle lengths for the kth pair x_k and y_k. This means that the sequence of random numbers used in the run yielding x_k may not correlate negatively with that used to generate y_k, primarily because the beginnings and ends of the kth cycle may not coincide (timewise) in the two runs. This is a very important point that should be taken into account when applying a variance reduction technique.

17.6 OPTIMIZATION IN SIMULATION— AN INVENTORY EXAMPLE

A distinct difference exists between the optimization of well-defined mathematical models and simulation models. In a mathematical model, the optimization problem is expressed in terms of explicit mathematical functions of the decision variables. The optimization problem is then *solved* to yield the values of the decision variables that optimize (maximize or minimize) the model's objective function. In this respect, the optimum values of the decision variables (together with the optimum objective value) represent the *output* of the mathematical model.

Simulation models, on the other hand, usually are *not* constructed in the framework of an optimization process. A simulation model merely measures the output of the system for *predetermined* values of the decision variables. This means that the values of the decision variables are considered part of the *input* data.

We can thus see that the implementation of an optimization process within the context of simulation can generally be achieved by systematically changing the values of the decision variables and then measuring the output by making proper simulation runs. If the model includes only one decision variable, we may use a systematic search technique (similar to the one described in Section

19.2) to locate the "optimal" values of the decision variables. The convergence of this technique depends on the properties of the function that measures the output of the model (response surface). When the number of decision variables exceeds one, the optimization process becomes much more complex, owing to the lack of an efficient procedure for locating the optimum. Although heuristics exist for optimizing multivariable decision models, their effectiveness remains doubtful, particularly in that the specific mathematical properties of the response surface (measure of output) are usually unknown in simulation models.

The optimization of simulation models becomes all the more difficult because of sampling error. Thus variations in the output measure of the model can be the result of sampling error rather than changes in the (input) values of the decision variables. The inability to separate the two sources of variation would make it difficult to compare different combinations of the decision variables for the purpose of identifying a better solution point. Although we cannot eliminate sampling error completely, we may be able to reduce its effect through replication. By using n runs (observations) with each combination of input values for the decision variables and basing the output measure on the average value of the n observations, we can reduce the sampling error by a multiple of $1/\sqrt{n}$. We can see, however, that a comparison of the average output measures of any two combinations of the decision variables must be based on proper statistical inference tests.

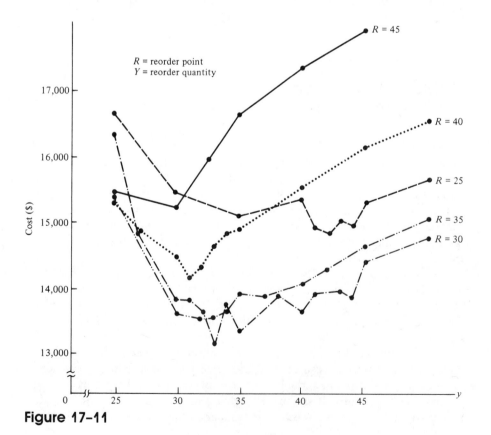

Figure 17-11

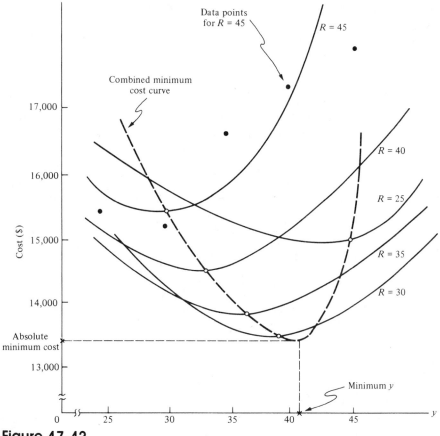

Figure 17-12

To illustrate the difficulties of optimizing a simulation model, we consider an inventory situation in which demand and lead time are both probabilistic. The lead time represents the time period between placement and receipt of an order. A new order of size y is placed whenever the inventory reaches the reorder level R. The objective of simulating the inventory system is to determine the optimum values of y and R that minimize the total inventory cost, including setup, holding, and shortage costs.

A similar model is investigated mathematically in Section 13.4.1. The mathematical model is useful only if the PDF of demand during lead time can be expressed mathematically. We will apply the simulation model, however, to a situation where the distributions of demand and lead time are approximated by discrete PDF's.

Because our objective is to demonstrate the difficulty of optimizing by simulation, we do not present the details of the inventory model. Rather, we simply show how the results are obtained after the model has been constructed. To carry out the optimization, we run the simulation for various values of y and a fixed value of R. We then repeat the process for different fixed values of R.

Figure 17-11 shows the total inventory cost versus the order quantity y for $R = 25$, 30, 35, 40, and 45 units. Each point on the curves represents one run (one observation), so that no replication is effected. We did this intentionally to show the effect of sampling error. We can see a general trend in the different curves representing R in that each tends to be convex, meaning that for a fixed value of R, there is a value of y that minimizes the total cost.

The resulting cost curves will become smoother (less erratic change in cost) if we increase the number of observations associated with the determination of each combination (y, R) and then average the resulting cost values. With a sufficiently large number of observations, the resulting cost curves associated with the different R should approach the smooth curves shown in Figure 17–12. By identifying the minimum-cost point associated with each R, a *combined* cost curve can be constructed by joining the minimum points as shown in Figure 17–12. Each point on the combined (dashed) curve corresponds to a fixed value of R. The minimum point on the dashed curve thus yields optimum y and R. Thus, in Figure 17–12, optimum $y = 41$ units, whereas optimum R appears to be 28 units. A more accurate estimate of optimum y and R would require a more refined search in the region around $y = 41$ and $R = 28$.

This example demonstrates the computational difficulty of optimizing by simulation. Although other systematic methods have been developed to locate the "optimum" without the use of exhaustive enumeration, the methods are merely heuristics and their effectiveness remains unproved. It can thus be concluded that optimization by simulation is generally unreliable.

17.7   SIMULATION LANGUAGES

In a typical simulation model, we view the system as comprising *entities* or *transactions* that, at any point in time during the simulation, may either be in service or waiting for service. To process waiting entities (or transactions) in a desired order, a simulation language must provide automatic means for storing and retrieving these entities whenever needed by utilizing files or ordered lists. These files can be thought of as occupying a two-dimensional array, with each row being associated with a single entity. The columns of the array represent the *attributes* describing the entity. For example, an attribute may represent the entity's priority for service compared with the other entities waiting in a queue. These attributes remain "attached" to the entity as it "travels" throughout the model and can be modified whenever desired.

Available discrete simulation languages can be categorized into four groups:

1. Event scheduling.
2. Activity scanning.
3. Process.
4. Combined process and event scheduling.

The most flexible simulation languages are those based on event scheduling and activity scanning. Flexibility here implies that the language can readily model any complex situation. This flexibility, however, implies that the user must expend additional effort in the development of the model. Process-oriented languages, on the other hand, are typically more compact and more user-friendly than are those of event scheduling and activity scanning. Again, such an advantage is achieved at the expense of losing flexibility.

The combined process–event scheduling languages aim at reaching a compromise between flexibility and ease of use. In particular, the process modeling capabilities of these languages significantly reduce the effort expended in the development of the model.

A. Event Scheduling Languages (SIMSCRIPT, GASP, SLAM)

The two most prominent event scheduling languages are SIMSCRIPT and GASP IV and, more recently, SLAM. SLAM is a sequel to GASP IV, which also offers both network and continuous simulation capabilities. The network portion of SLAM is discussed later in this section. SIMSCRIPT has its own compiler, whereas GASP IV and SLAM are coded totally in standard FORTRAN. The design of the languages is based (more or less) on the following general framework.

1. A *main* program that is used to set up the model's files (including the definition of the attributes) and to initialize the first event that will be executed (e.g., the arrival of a customer) as well as set the length of the simulation run.

2. A "subroutine" for each of the model's events that describes (totally and exhaustively) all the actions that must be taken at the time the event occurs.

Both SIMSCRIPT and SLAM (GASP IV) automatically process, store, and retrieve the model's events in proper chronological order. They also provide appropriate coding facilities that allow the user to access and manipulate internal data (e.g., current queue length) during the course of the simulation. The output in SIMSCRIPT is completely user specified both in format and content. In SLAM, the system provides a standard output summary. Additionally, the user is given the option to write a user-specified output subroutine.

One advantage of SIMSCRIPT is its English-like syntax that tends to make the developed model readable and, to an extent, self-documenting. However, for modelers who are accustomed to concise and compact coding, the verbosity of the language could be distracting. In SLAM, the model is coded entirely in FORTRAN. In essence, all SLAM actually offers is a number of general FORTRAN subroutines whose basic function is to relieve the user from the burden of coding the tedious and repetitive details of simulation. The obvious disadvantage of SLAM is that the cryptic FORTRAN coding renders models that are practically undecipherable to outsiders.

B. Activity Scanning Languages (ECSL)

Activity scanning languages are more popular in Europe, and particularly in England. The best known activity scanning language is ECSL (Extended Control and Simulation Language). It is coded in FORTRAN and hence can be run on practically any machine. The language also allows the user to enter the simulation model interactively. Such a procedure works well for small models, but it tends to be slow for complex cases.

C. Process-Oriented Language

In process-oriented languages, there are three types of model representation:

1. Blocks.
2. Network with nodes and branches.
3. Statements.

In block languages, each block is designed to perform a certain function in simulation. For example, a GENERATE block introduces entities or transactions into the system, whereas as an ADVANCE block will effect a delay of the transaction. Each block possesses a number of operands that determine the exact manner in which the block is executed. For example, the ADVANCE block must specify the length of delay for the transaction.

Network representation is based on the assumption that most simulation models are typically queueing situations. Network languages define nodes whose function is very much like that of blocks in block-based languages. The main difference occurs in the manner in which the nodes are linked. Here, network-based languages utilize branches emanating from each node. These branches may be deterministic, probabilistic, or conditional. Thus, copies of a transaction leaving a node may be sent to other nodes by using proper branching. Although the same result can be effected in block-based languages, it usually requires combining more than one block.

Statement-based languages provide the user with powerful statements that can be used in devising the simulation model. Such languages, however, are usually not as user-friendly as are block-based or network-based languages.

Block-Based Process Languages (GPSS, SIMAN)

The best-known block-based language is GPSS, which is written in assembly language. As an illustration of the use of the language, suppose that customers arrive at a single-server facility named FAC with the interarrival time being uniformly distributed between 4 and 8 time units. The service time at the facility is also uniform between 3 and 9 time units. The graphical block model and the model statements are then given in GPSS as shown in Figure 17–13.

Notice that the model in Figure 17–13 describes the entire life of the transaction from "birth" at GENERATE to "death" at TERMINATE. The GENERATE block automatically creates a new transaction using an interarrival time that is uniformly distributed between 4 and 8 (6 ± 2). A transaction leaving GENERATE will attempt to SEIZE FAC. If FAC is free, the transaction

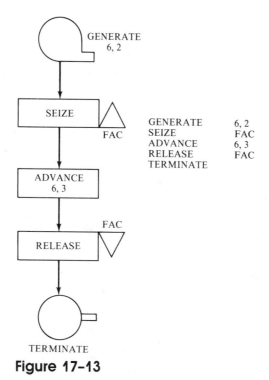

GENERATE	6, 2
SEIZE	FAC
ADVANCE	6, 3
RELEASE	FAC
TERMINATE	

Figure 17-13

will pass to ADVANCE where it will be delayed by the equivalent of its service time (6 ± 3). If FAC is not available, the transaction will wait in SEIZE until FAC becomes free. After a specified service time delay in ADVANCE, the transaction will enter RELEASE, at which point FAC becomes free and hence available to process the first transaction waiting in SEIZE, if any. The "death" of the transaction occurs when it passes through TERMINATE.

One of the drawbacks of GPSS is that the simulation clock must be advanced in integer value only. Another drawback is that sampling from any probability distribution other than the uniform must be done by approximating the distribution by a discrete function. A third drawback is that arithmetic computations are done on a rather limited and restricted scale.

Newer versions of GPSS allow the user to use external functions written in FORTRAN to obtain random samples. They also allow the user to carry out difficult computations by writing special FORTRAN subroutine that can then be linked to the GPSS model.

A recently developed block-based language promises to alleviate some of the drawbacks associated with GPSS. The new language, named SIMAN, is written entirely in standard FORTRAN. The basic idea of blocks in SIMAN is almost identical to that of GPSS. Special blocks are added, however, to handle the simulation of materials handling and manufacturing systems. SIMAN also allows the use of FORTRAN inserts for modeling complex cases that cannot be handled by the language's standard blocks. As in SLAM, SIMAN provides capabilities for continuous simulation.

Network-Based Languages (Q-GERT, SLAM, SIMNET)

The first network-based language to be developed is Q-GERT, which is written in standard FORTRAN. The basic structure of a Q-GERT simulation model is a network comprised of nodes and activities (branches). Figure 17–14 shows the Q-GERT simulation of a single-server model together with the data statements that translate the network into a computer code.

In the network of Figure 17–14, transactions leave SOURCE (node 1) at exponential intervals (EX, 1)† with the first transaction being created at time 0. Each time the branch (EX, 1) loops back into SOURCE, it will create a single transaction. The code M in the SOURCE node indicates that the transaction will carry its creation time with it as it flows through the network.

Transactions leaving SOURCE will enter a QUEUE (node 2), which is followed by an ACTIVITY that plays the role of a facility with one or more parallel servers. If a server is available, service is started for a length of (EX, 2); otherwise, the transaction must wait in QUEUE. After the service is completed, SINK (node 3), which is released by a single transaction, will automatically compute the time interval between SOURCE (where M code is recorded) and SINK, that is, the time the transaction stays in the system. This requirement is indicated by the code I in the SINK node.

Q-GERT offers more coding facilities than can be demonstrated by the preceding simple example. In that respect, the language is quite powerful. However, the language is definitely not user-friendly, mainly because the user will, more often than not, find it necessary to write special FORTRAN subroutines or functions to accommodate certain situations that cannot be coded directly by the available facilities of Q-GERT. All in all, the language employs some 60 special functions and subroutines, and because FORTRAN restricts the subprogram names to six characters, keeping track of all those cryptic names can be somewhat frustrating.

A sequel to Q-GERT is the new language called SLAM, which provides both network and discrete event capabilities (as well as continuous simulation). SLAM network is similar to that of Q-GERT. However, as seen by the single-server model of Figure 17–15, the computer code representing the network is more user-friendly than in Q-GERT.

In general, the capabilities of SLAM's network approach are not powerful enough to handle complex situations (e.g., job-shop scheduling problems). For this reason, SLAM provides the event scheduling approach that is basically equal to the one adopted in GASP. The event scheduling approach can be used either independently or in conjunction with the network segment of the language. In dealing with complex cases, SLAM's event scheduling should be regarded as the language's principal coding facility. The use of concurrent network does enhance the capabilities of the language, however.

A new network-based language, called SIMNET, is designed to allow modeling of complex situations directly. The language is written entirely in FOR-

---

† (EX, 1) does not mean that the exponential has mean 1. Rather, it means that the parameters are stored in special parameter set 1.

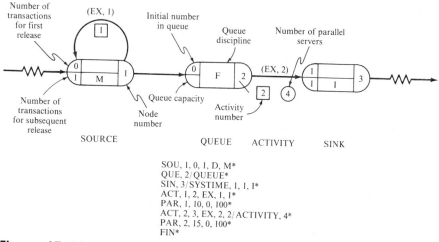

Figure 17-14

TRAN. The single-server model (which is simulated in Q-GERT and SLAM, Figures 17–14 and 17–15) is again simulated in Figure 17–16 by using SIMNET.

SIMNET utilizes exactly four nodes: a *source* that creates transactions, a *queue* where transactions waits, a *facility* where transactions are served, and an *auxiliary* that serves to increase the modeling flexibility of the language. Each node is assigned an arbitrary name (e.g., SOU, QUE, and FAC in Figure 17–16) that is followed by the symbol *S, *Q, *F, or *A to identify the name as being that of a Source, Queue, Facility, or Auxiliary. Nodes are linked by seven types of branches (identified by the reserved symbol *B) that control the flow of transactions in the network (in SLAM, activities act as both facilities

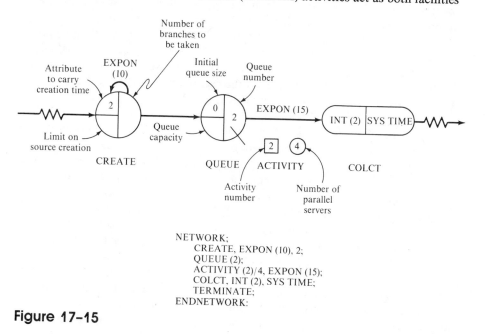

Figure 17-15

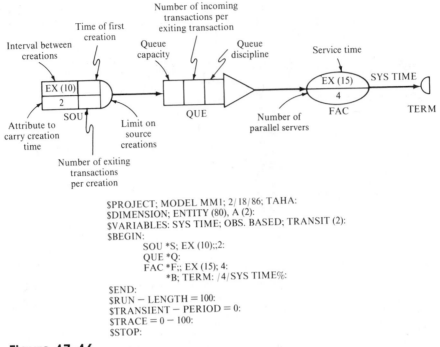

```
$PROJECT; MODEL MM1; 2/18/86; TAHA:
$DIMENSION; ENTITY (80), A (2):
$VARIABLES: SYS TIME; OBS. BASED; TRANSIT (2):
$BEGIN:
        SOU *S; EX (10);;2:
        QUE *Q:
        FAC *F;; EX (15); 4:
            *B; TERM: /4/SYS TIME%:
$END:
$RUN − LENGTH = 100:
$TRANSIENT − PERIOD = 0:
$TRACE = 0 − 100:
$STOP:
```

Figure 17–16

and branches). Each branch operates independently of all other branches emanating from the same node. As transactions transverse a branch, they can execute arithmetic computations; collect variable statistics; suspend or resume creations from a source; effect premature termination of a service in a busy facility; and cause deletion, insertion, reordering, and/or movement of transactions in and out of a queue. Passing transactions may also modify their own attributes, retrieve old ones, and copy or replace attributes of transactions residing in queues or facilities. In brief, SIMNET allows the user to exercise complete control of transaction flow anywhere in the network without the need for writing special FORTRAN inserts (as in Q-GERT, SLAM, and SIMAN).

The design of SIMNET's nodes is based on an *input-output* concept that defines all the data needed to effect the flow of transactions in and out of (as well as within) each node. This concept proved effective, not only in eliminating the need for introducing special-purpose nodes, but also in modeling complex situations (e.g., job shop scheduling) that, in other languages, may require the design of special-purpose nodes or the use of external FORTRAN inserts.

SIMNET offers a detailed trace report that automatically accounts for *all* the changes that take place during the simulation. Figure 17–17 illustrates the trace capability of SIMNET by applying it to the single-server model of Figure 17–16. All the user has to do to produce this report is to include the command $TRACE=T1-T2 as part of the model, where T1 and T2 are the trace time limits. The power of SIMNET's trace report as a debugging tool becomes more evident in complex models in which transaction flow is controlled from a "remote" point in the network. In such cases, the report automatically provides

```
TIME                    ACTION
----                    ------

0.0000E+00    EXIT 'SOU            '
                   ATR(S): 0.0000E+00    0.0000E+00
                   FILE NEXT 'SOU  ' CREATION AT T = 0.3530E+02
                   ATR(S): 0.0000E+00    0.3530E+02
                   SKIP 'QUE  ' -- CUR.LEN =    0
                   ATR(S): 0.0000E+00    0.0000E+00
                   ENTER 'FAC  ' -- CUR.UTILIZ =    1
                   ATR(S): 0.0000E+00    0.0000E+00
                   FILE DEPARTURE FROM 'FAC  ' AT T = 0.3723E+02
                   ATR(S): 0.0000E+00    0.0000E+00

0.3530E+02    EXIT 'SOU            '
                   ATR(S): 0.0000E+00    0.3530E+02
                   FILE NEXT 'SOU  ' CREATION AT T = 0.3959E+02
                   ATR(S): 0.0000E+00    0.3959E+02
                   ENTER 'QUE  ' -- CUR.LEN =    1
                   ATR(S): 0.0000E+00    0.3530E+02

0.3723E+02    EXIT 'FAC            '
                   ATR(S): 0.0000E+00    0.0000E+00
                   'SYS TIME  ' VALUE =0.3723E+02
                   TERMINATE TRANSACTION
                   MAKE SERVER IN 'FAC  ' IDLE -- CUR.UTILIZ =    0
                   LEAVE 'QUE  ' -- CUR.LEN =    0
                   ATR(S): 0.0000E+00    0.3530E+02
                   ENTER 'FAC  ' -- CUR.UTILIZ =    1
                   ATR(S): 0.0000E+00    0.3530E+02
                   FILE DEPARTURE FROM 'FAC  ' AT T = 0.4626E+02
                   ATR(S): 0.0000E+00    0.3530E+02
```

Figure 17–17

all the details of the flow in a standard format with no information left out.

SIMNET's output capabilities include providing means for estimating the length of the transient period by using a graphical output, executing runs with different initial data in the same simulation session, obtaining a global statistical summary based on either the *subinterval* or *replication* methods for collecting data, and obtaining a complete count of the transaction flow at each node of the network. All of this information is obtained by including simple commands (similar to $TRACE) directly in the model. The standard output report will then provide the desired data at the termination of the simulation session.

17.8 SUMMARY

Simulation is a highly flexible tool that can be used effectively for analyzing complex systems. However, simulation has the drawback that its output is subject to statistical error and hence must be interpreted by the appropriate statistical test. The special nature of the simulation experiment makes it difficult to gather observations that are both independent and representative of steady-state conditions.

Simulation languages are of the event-scheduling, activity-scanning, or process-oriented type. Event scheduling and activity scanning offer the most flexibility in modeling, whereas process-oriented languages are the most compact and possibly the easiest to use and implement. A number of languages offer the combined capabilities of event scheduling and process orientation.

SELECTED REFERENCES

FISHMAN, G., *Principles of Discrete Event Simulation,* Wiley, New York, 1978.
GORDON, G., *System Simulation,* Prentice-Hall, Englewood Cliffs, N.J., 1978.
PRITSKER, A., *Introduction to Simulation and SLAM II,* 3rd ed., Halsted Press–Wiley, New York, 1986.
TAHA, H. A., *Simulation Modeling and SIMNET,* Prentice-Hall, Englewood Cliffs, N.J., 1987.

REVIEW QUESTIONS

True (T) or False (F)?

1. _____ A [0, 1] random number can be converted to a random number in the interval [0, 100] through multiplication by the constant 100.

2. _____ A [0, 1] random number can be converted to a random number in the interval [50, 150] through multiplication by the constant 100.

3. _____ The [0, 1] random numbers can be used to generate samples (values of a random variable) only if the sampled distribution is uniform.

4. _____ The principal reason we are able to use [0, 1] random numbers to sample from *any* PDF is that the CDF of any random variable is itself uniformly distributed in the [0, 1] interval.

5. _____ The output of a simulation model is independent of the length of the simulation run.

6. _____ As the length of the simulation run becomes sufficiently large, the output of the model becomes independent of the specific sequence of random numbers used in the simulation.

7. _____ Any simulation model represents a statistical experiment whose output is subject to statistical error.

8. _____ A better estimate of the output of a simulation model is obtained by averaging observations.

9. _____ The precision of the output of a simulation model (as measured by the standard deviation) is increased with an increase in the length of the simulation run.

10. _____ In a simulation experiment, there is always a clear-cut indication of when the transient state ends and the steady state begins.

11. _____ The computer time needed to execute a simulation model is directly proportional to the actual length of the simulation time period.

12. _____ In executing a simulation model, the gathering of pertinent information can occur at points in time when no key events occur.

13. ____ In executing a simulation model, it is necessary to generate in advance *all* the events in the entire simulated period *T*.

14. ____ In executing a simulation model, all events that precede the current one can be deleted without affecting future computations.

15. ____ The basic advantage of the *replication* method is that the observations gathered are statistically independent.

16. ____ In the subintervals method, although the effect of transient conditions is minimized with the increase the length of the simulation run, the successive observations are correlated.

17. ____ The cycles method has the advantage of eliminating the correlation between the successive observations.

18. ____ The *antithetic* method for reducing the variance may not be effective when each observation and its antithetic value are not generated by the exact antithetic sequences of [0, 1] random numbers.

19. ____ The optimization process in simulation is complicated by the existence of the sampling error inherent in any simulation experiment.

20. ____ An optimum solution obtained by simulation cannot be expressed by single fixed values of the decision variables. Rather, it must be expressed in terms of confidence intervals.

[*Ans.* **1**—T, **2**—F, **3**—F, **4**—T, **5**—F, **6** to **9**—T, **10** to **13**—F, **14** to **20**—T.]

PROBLEMS

Section	Assigned Problems
17.3	17–1 to 17–5
17.4	17–6 to 17–17
17.5	17–18 to 17–20
17.6	17–21
17.7	17–22 to 17–24

☐ **17–1** Write a computer simulation for a single-server facility using any of the familiar programming languages (e.g., FORTRAN). The time between arrivals of customers at the facility is uniformly distributed between 1 and 5 minutes. The service time at the facility is constant and equal to 2 minutes. Carry out the simulation for an 8-hour period. Compute the average utilization of the facility, its average idle time, and the average queue length.

☐ **17–2** In Problem 17–1, suppose that the service time is increased to 5 minutes. From the theoretical standpoint, the length of the queue should continue to increase with time because the average service time exceeds the average interarrival time. Devise a simulation experiment that can be used to detect this unstable situation.

□ **17–3** In Problem 17–1, suppose that there are two types of customers. Type I has priority for service, so that type II cannot be serviced until all waiting type I customers have completed their service. However, a type II customer already in service cannot be preempted. Define all the events necessary to simulate the problem. What input data are needed to carry out the simulation?

□ **17–4** In Problem 17–3, suppose that the service times for types I and II are 3 and 2 minutes per customer and the interarrival times are exponential with means 1.5 and 3 minutes, respectively. Simulate the system by hand computations for the first 20 arrivals. Compute the average number of waiting customers for each type, as well as the combined average. Assume that the system starts empty.

□ **17–5** The following table represents the variation in the number of waiting customers with simulation time. Determine:
   (a) The percentage of time the queue is empty.
   (b) The average waiting time per customer assuming a total of 30 arrivals.
   (c) The average number of waiting customers.

Simulation Time t (hours)	Number of Waiting Customers
$0 \leq t \leq 3$	0
$3 \leq t \leq 4$	1
$4 \leq t \leq 6$	2
$6 \leq t \leq 7$	1
$7 \leq t \leq 10$	0
$10 \leq t \leq 12$	2
$12 \leq t \leq 18$	4
$18 \leq t \leq 20$	1
$20 \leq t \leq 25$	0

□ **17–6** A person plays a game in which a fair die is thrown. If the outcomes are 1 or 6, he wins; otherwise, he loses.
   (a) Specify the distribution describing the situation, then show how it can be sampled for the purpose of simulating the game.
   (b) Using the [0, 1] random numbers in column 1 of Table 17–1, indicate whether the player would win or lose in each of the first 10 throws.

□ **17–7** The time between successive arrivals at a facility follows an exponential distribution with mean 10 minutes. By using the random numbers in column 1 of Table 17–1, determine the arrival times of the first five customers.

□ **17–8** The demand per day for an item occurs according to the following discrete PDF:

Demand	0	1	2	3
Probability	.2	.3	.4	.1

By using the random numbers in column 1 of Table 17–1, determine the demands in each of the first 5 days.

☐ **17–9** A gambler's loss or gain in a game is normally distributed with mean $5 and variance 16. By using the discrete approximation in Table 17–3 (Example 17.4–3) and the first column of random numbers in Table 17–1, determine the gambler's net gain or loss after playing the game five times.

☐ **17–10** Customers arrive at a facility according to a Poisson distribution at the rate of 10 per customers per hour. Because the facility cannot handle all customers, the owner has decided to admit every other arriving customer. By using the random numbers in column 1 of Table 17–1, determine the times of arrival of the first three customers who are *admitted* into the facility.

☐ **17–11** In a service facility, the service rate in the next hour is adjusted depending on the number of customers who arrived in the immediately preceding hour. Customers normally arrive according to a Poisson distribution with mean rate 5 per hour. By using the random numbers in column 1 of Table 17–1, determine the *number* of arrivals in the first and second hours of operation.

☐ **17–12** Show how the geometric distribution may be sampled in simulation. From this information, show how the negative binomial samples can be determined.
[*Hint:* The negative binomial is the convolution of identically distributed and independent geometric distributions.]

☐ **17–13** Repeat Problem 17–9 using the central limit procedure of Example 17.4–6.

☐ **17–14** The lead time for receiving an order can be 1 or 2 days with equal probabilities. The demand per day follows the distribution

Demand	0	1	2
Probability	.2	.5	.3

Use the random numbers in columns 1 and 2 of Table 17–1 to generate an estimate of the joint distribution of the demand and lead time. From the joint distribution, estimate the absolute PDF of demand during lead time.
[*Hint:* The demand during lead time may assume one of the values 0, 1, 2, 3, and 4.]

☐ **17–15** Use the first 30 random numbers in column 1 of Table 17–1 to estimate the area of an equilateral triangle where the length of each side is equal to 2 inches (use the first random number to compute the *y*-coordinate.)

☐ **17–16** Estimate the value of the following integral by using the first 30 random numbers in column 1 of Table 17–1.

$$\int_0^1 x^2\, dx$$

[*Hint:* The integral represents the area under x^2 enclosed in the interval $0 \le x \le 1$.]

☐ **17–17** Repeat Problem 17–16 assuming that the limits on the integral are from 3 to 6 (use the first random number to compute the y-coordinate).

☐ **17–18** Use the information in Problem 17–5 to determine five subinterval observations, each over a simulated period of 5 time units. From these observations determine the expected number in the queue and a 90% confidence interval.

☐ **17–19** In Problem 17–18, determine the observations by using the cycles method, assuming that each cycle starts with zero waiting customers.

☐ **17–20** Use the antithetic technique to estimate the mean of a uniform distribution in the interval 5 to 12. Use the first 10 random numbers in column 1 of Table 17–1.

☐ **17–21** The inventory problem in Section 17.6 is based on the following information. The demand size for an item can consist of one, two, or three units, with probabilities .7, .2, and .1, respectively. The time between demands has the following discrete approximation:

Days	1	2	3	4	5
Probability	.6	.343	.05	.005	.002

Lead time (time between order and receipt of replenishment stock) is normally distributed with mean 30 days and standard deviation 5 days. If the system is unable to fill a demand due to a "stockout," the unfilled portion of the demand is lost, at a cost of $5.00 per unit. Interest, storage, insurance, risk, waste, and other inventory carrying costs can be estimated at $3.75 per day per 100 units in storage. Upon each demand, the level of inventory is reviewed, and a decision must be made whether to order replenishment stock and, if so, how much to order. The cost of placing an order is estimated at $15.00. It is the management's policy not to place a second order while awaiting receipt of a shipment.

You are to decide at what level of inventory an order should be placed (reorder point) and how much to order (order quantity).

Simulate the problem by using an automatic language (such as FORTRAN, GPSS, or SLAM) assuming an initial inventory level of 50 units.

☐ **17–22** A bank operates two drive-in lanes. Each lane can accommodate a maximum of four cars, including the cars being served. Cars arrive at the bank in a Poisson stream with an average interarrival time of 1 minute. If the two lanes are full, arriving cars drive away without seeking service. Preference is always given to the shorter lane. However, when both lanes are equal, customers tend to join the right lane. Service times at the two tellers are exponential with mean 1.5 minutes. Jockeying between the lanes occurs whenever one lane is shorter than the other by at least two cars. In this case, the last car in the longer lane moves to join the other lane. Simulate the system to determine the following statistics:

(a) The average time a car spends in the system.
(b) The average time between arrivals of cars that cannot join because the lanes are full.
(c) The average utilization of the letters.
(d) The average length of each of the two queues.

☐ **17–23** A machine utilizes a certain tool that is subject to breakdown. When the tool fails, a new tool is brought from stock that is then installed on the machine by an operator. The operator also repairs the broken tool. However, the repair process may be interrupted in favor of installing a new tool, provided that the tool is available in stock.

The stock includes a total of three good tools. The time to install the tool is uniformly distributed between 1 and 3 minutes. Repair of broken tool is also uniformly distributed between 40 and 120 minutes. A tool will function properly before breakdown for an exponential time period with mean 90 minutes. It takes half a minute to return a repaired tool back to stock.

Develop a simulation model for the purpose of collecting the following statistics:

(a) The net utilization of the repair facility, excluding interruption due to new part installation.
(b) The average number of tools awaiting repair.
(c) The average utilization of the operator.
(d) The average number of tools in stock.

☐ **17–24** The initial stock of an inventory item includes 20 units. Demand for the item arrive according to a Poisson stream with an average interarrival time is equal to .4 day. The number of units per order is Poisson distributed with mean 3 units. An order may be partially filled with the unsatisfied portion of the order being backordered. Priority is always given to backordered units. The reorder point is 15 units, at which point the stock is replenished to reach 20 units. There is a delay of 1.5 days for receiving a placed order. There can be no more than one outstanding order at a time.

Simulate the inventory system and determine the following:

(a) Average stock level.
(b) Average order size.
(c) Average number of order in the system.
(d) Average number of backorders.

PROJECTS

Each of the following situations suggests a possible application of simulation. The implementation of the simulation requires gathering the pertinent data particularly the probability distributions that describe segments of the system.

It is your responsibility to obtain these data. Keep in mind, however, that probability distributions in simulation models, unlike in queueing, need not be fitted into known PDF's. Any relative frequency histogram can be used in the simulator.

You may write the simulation model in any automatic language. What is more important is how you gather the data in a manner that does not violate the assumptions of statistical analysis. Special attention must be paid to the transient versus steady-state conditions and the independence of the observations. You should state clearly how these requirements are met in the simulator you develop.

1. Develop simulation models for the queueing projects listed at the end of Chapter 16.

2. Develop a simulation model for the game room in the student's recreation center (The Union). The purpose of the model is to determine how many of each game machine should be provided to meet the demand.

3. Develop a simulation model for the (interdepartmental) campus mail delivery at your college or university. The purpose of the model is to assign routes to the carriers in a manner that ensures efficiency of operation and reliability of the delivery service. (It is up to you to define measures for "efficiency" of operation and "reliability" of service.)

4. Simulate the operation of remote computer terminals rooms that serve the students in your college. The purpose of the simulator is to determine the number and types of terminals needed to serve the users.

5. Simulate the operation of the student's infirmary on your campus. The purpose of the simulator is to determine the number of doctors, nurses, and hospital beds needed in the clinic.

6. Imagine a bay of a heavy industry factory in which the machines are serviced by one or more overhead cranes. Simulate the operation of the cranes and the machines for the purpose of determining the "economical" number of cranes.

7. Copper coils are manufactured by feeding straight copper tubing to an automatic coiling machine. The machine releases coils at a steady rate of 40 coils per hour. Coils are then fed into a revolving tray that transports the coils into a furnace for heat treating. The capacity of the tray is 50 coils. Whenever this capacity is reached, the coiling machine is automatically stopped. Heat treatment last between 4 and 6 minutes, uniformly distributed. However, the furnace can house up to 15 coils at a time. Treated coils are packed 4 in a carton, and the packing time is approximately 1 minute. Packed cartons are carried by a gravity conveyor to a storage area, where they are labeled for shipping. When the capacity of the storage area becomes full, cartons will fill up the gravity conveyor, which automatically shuts off delivery from the packing machine to the conveyor and from the revolving tray to the furnace. All the units inside the furnace will be burned and msut be discarded as they come out in cartons after the system is reactivated.

Simulate the system for the purpose of studying the bottlenecks.

PART III

NONLINEAR PROGRAMMING

THIS PART INCLUDES Chapters 18 and 19. The material is designed to provide basic foundations in nonlinear programming. Further information about the subject can be found in specialized books.

Chapter 18 introduces classical optimization theory, including unconstrained optima, constrained (Jacobian and Lagrangean) methods, wherein all constraints are equations, and the Kuhn–Tucker conditions for constrained nonlinear problems. Chapter 19 concentrates on the *computational* aspects of optimizing unconstrained and constrained functions.

The material in this part assumes higher ability in mathematics than that in Parts I and II. One reason is that this type of material cannot be presented meaningfully at an elementary level. Another reason is that one of the objectives of the book is to increase your level of mathematical sophistication as you successively complete Parts I, II, and III.

To assist in establishing the prerequisites for Part III, Appendix B summarizes fundamental theorems with which you must be familiar. This is equivalent to completing a course in advanced calculus. Although matrix algebra is used in Part III, it is not a mandatory prerequisite and the material in Appendix A should be sufficient for this purpose.

Chapter 18

Classical Optimization Theory

Classical optimization theory develops the use of differential calculus to determine points of maxima and minima (extrema) for unconstrained and constrained functions. The methods developed may not be suitable for efficient numerical computations. However, the underlying theory provides the basis for devising most nonlinear programming algorithms (see Chapter 19).

This chapter develops necessary and sufficient conditions for determining unconstrained extrema, the *Jacobian* and *Lagrangean* methods for problems with equality constraints, and the *Kuhn–Tucker* conditions for problems with inequality constraints.

18.1 UNCONSTRAINED EXTREMAL PROBLEMS

An extreme point of a function $f(\mathbf{X})$ defines either a maximum or a minimum of the function. Mathematically, a point $\mathbf{X}_0 = (x_1, \ldots, x_j, \ldots, x_n)$ is a maximum if

$$f(\mathbf{X}_0 + \mathbf{h}) \leq f(\mathbf{X}_0)$$

for all $\mathbf{h} = (h_1, \ldots, h_j, \ldots, h_n)$ such that $|h_j|$ is sufficiently small for all j. In other words, $\mathbf{X}_0$ is a maximum if the value of f at every point in the neighborhood of $\mathbf{X}_0$ does not exceed $f(\mathbf{X}_0)$. In a similar manner, $\mathbf{X}_0$ is a minimum if for $\mathbf{h}$ as defined

$$f(\mathbf{X}_0 + \mathbf{h}) \geq f(\mathbf{X}_0)$$

Figure 18–1 illustrates the maxima and minima of a single-variable function $f(x)$ over the interval $[a, b]$. [The interval $a \leq x \leq b$ is not meant to represent restrictions on $f(x)$.] The points $x_1, x_2, x_3, x_4,$ and x_6 are all extrema of $f(x)$. These include $x_1, x_3,$ and x_6 as maxima and x_2 and x_4 as minima. Since

$$f(x_6) = \max \{f(x_1), f(x_3), f(x_6)\}$$

$f(x_6)$ is called **global** or **absolute** maximum, and $f(x_1)$ and $f(x_3)$ are **local** or **relative** maxima. Similarly, $f(x_4)$ is a local minimum and $f(x_2)$ is a global minimum.

Although x_1 (in Figure 18–1) is a maximum point, it differs from remaining local maxima in that the value of f corresponding to at least one point in the neighborhood of x_1 is equal to $f(x_1)$. In this respect, x_1 is called a **weak maximum** compared with x_3, for example, where $f(x_3)$ defines a **strong maximum**. A weak maximum thus implies (an infinite number of) alternative maxima. Similar results may be developed for the weak minimum at x_4. In general, $\mathbf{X}_0$ is a weak maximum if $f(\mathbf{X}_0 + \mathbf{h}) \leq f(\mathbf{X}_0)$ and a strong maximum if $f(\mathbf{X}_0 + \mathbf{h}) < f(\mathbf{X}_0)$, where $\mathbf{h}$ is as defined earlier.

An interesting observation about the extrema in Figure 18–1 is that the first derivative (slope) of f vanishes at these points. However, this property is not unique to extrema. For example, the slope of $f(x)$ at x_5 is zero.

Because a vanishing first derivative (generally, gradient) plays an important role in identifying maxima and minima (see the next section), it is essential to define points such as x_5 separately. These points are known as **inflection** (or, in special cases, **saddle**) points. If a point with zero slope (gradient) is not an

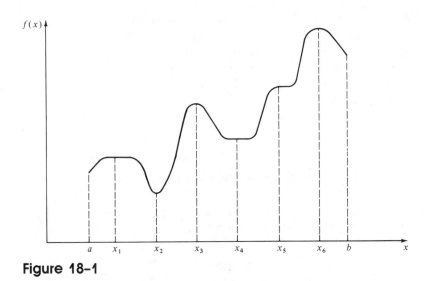

Figure 18–1

extremum (maximum or minimum), then it must automatically be an inflection point.

18.1.1 NECESSARY AND SUFFICIENT CONDITIONS FOR EXTREMA

This section develops theorems for establishing necessary and sufficient conditions for an n-variable function $f(\mathbf{X})$ to have extrema. It is assumed throughout that the first and second partial derivatives of $f(\mathbf{X})$ are continuous at every $\mathbf{X}$.

Theorem 18.1–1. *A necessary condition for* $\mathbf{X}_0$ *to be an extreme point of* $f(\mathbf{X})$ *is that*

$$\nabla f(\mathbf{X}_0) = \mathbf{0}$$

PROOF. By Taylor's theorem (Section B.5.1), for $0 < \theta < 1$,

$$f(\mathbf{X}_0 + \mathbf{h}) - f(\mathbf{X}_0) = \nabla f(\mathbf{X}_0)\mathbf{h} + (1/2)\mathbf{h}^T \mathbf{H} \mathbf{h} \Big|_{\mathbf{X}_0 + \theta \mathbf{h}}$$

where $\mathbf{h}$ is as defined.

For sufficiently small $|h_j|$, the remainder term $(1/2)(\mathbf{h}^T \mathbf{H} \mathbf{h})$ is of the order h_j^2, and hence the expansion may be approximated as

$$f(\mathbf{X}_0 + \mathbf{h}) - f(\mathbf{X}_0) = \nabla f(\mathbf{X}_0)\mathbf{h} + 0(h_j^2) \cong \nabla f(\mathbf{X}_0)\mathbf{h}$$

Suppose now that $\mathbf{X}_0$ is a minimum point; it is shown by contradiction that $\nabla f(\mathbf{X}_0)$ must vanish. Suppose it does not; then for a specific j, either

$$\frac{\partial f(\mathbf{X}_0)}{\partial x_j} < 0 \qquad \text{or} \qquad \frac{\partial f(\mathbf{X}_0)}{\partial x_j} > 0$$

By selecting h_j with appropriate sign, it is always possible to have

$$h_j \frac{\partial f(\mathbf{X}_0)}{\partial x_j} < 0$$

By setting all other h_j equal to zero, Taylor's expansion yields

$$f(\mathbf{X}_0 + h) < f(\mathbf{X}_0)$$

This result contradicts the assumption $\mathbf{X}_0$ is a minimum point. Consequently, $\nabla f(\mathbf{X}_0)$ must vanish. A similar proof can be established for the maximization case.

The conclusion from Theorem 18.1–1 is that, at any extreme point, the condition

$$\nabla f(\mathbf{X}_0) = \mathbf{0}$$

must be satisfied; that is, the gradient vector must be null.

For functions with one variable only (say, y), the condition reduces to

$$f'(y_0) = 0$$

As stated previously, the condition is also satisfied for inflection and saddle points. Consequently, these conditions are necessary but not sufficient for identifying extreme points. It is thus more appropriate to refer to the points obtained from the solution of

$$\nabla f(\mathbf{X}_0) = \mathbf{0}$$

as **stationary** points. The next theorem establishes the sufficiency conditions for $\mathbf{X}_0$ to be an extreme point.

Theorem 18.1-2. *A sufficient condition for a stationary point $\mathbf{X}_0$ to be extremum is that the Hessian matrix $\mathbf{H}$ evaluated at $\mathbf{X}_0$ is*
  (i) *Positive definite when $\mathbf{X}_0$ is a minimum point.*
  (ii) *Negative definite when $\mathbf{X}_0$ is a maximum point.*

PROOF. By Taylor's theorem, for $0 < \theta < 1$,

$$f(\mathbf{X}_0 + \mathbf{h}) - f(\mathbf{X}_0) = \nabla f(\mathbf{X}_0)\mathbf{h} + (1/2)\mathbf{h}^T \mathbf{H}\mathbf{h}\Big|_{\mathbf{X}_0 + \theta \mathbf{h}}$$

Since $\mathbf{X}_0$ is a stationary point, by Theorem 18.1–1, $\nabla f(\mathbf{X}_0) = \mathbf{0}$. Thus,

$$f(\mathbf{X}_0 + \mathbf{h}) - f(\mathbf{X}_0) = (1/2)\mathbf{h}^T \mathbf{H}\mathbf{h}\Big|_{\mathbf{X}_0 + \theta \mathbf{h}}$$

Let $\mathbf{X}_0$ be a minimum point; then, by definition,

$$f(\mathbf{X}_0 + \mathbf{h}) > f(\mathbf{X}_0)$$

for all nonnull $\mathbf{h}$. This means that for $\mathbf{X}_0$ to be a minimum, it must be true that

$$(1/2)\mathbf{h}^T \mathbf{H}\mathbf{h}\Big|_{\mathbf{X}_0 + \theta \mathbf{h}} > 0$$

However, continuity of the second partial derivative guarantees that the expression $(1/2)\mathbf{h}^T \mathbf{H}\mathbf{h}$ must yield the same sign when evaluated at both $\mathbf{X}_0$ and $\mathbf{X}_0 + \theta \mathbf{h}$. Since $\mathbf{h}^T \mathbf{H}\mathbf{h}|_{\mathbf{X}_0}$ defines a quadratic form (see Section A.3), this expression (and hence $\mathbf{h}^T \mathbf{H}\mathbf{h}|_{\mathbf{X}_0 + \theta \mathbf{h}}$) is positive if and only if $\mathbf{H}|_{\mathbf{X}_0}$ is positive-definite. This means that a sufficient condition for the stationary point $\mathbf{X}_0$ to be a minimum is that the Hessian matrix evaluated at the same point is positive-definite. A similar proof can be established for the maximization case to show that the corresponding Hessian matrix is negative-definite.

Example 18.1-1. Consider the function

$$f(x_1, x_2, x_3) = x_1 + 2x_3 + x_2 x_3 - x_1^2 - x_2^2 - x_3^2$$

The necessary condition

$$\nabla f(\mathbf{X}_0) = \mathbf{0}$$

gives

$$\frac{\partial f}{\partial x_1} = 1 - 2x_1 = 0$$

$$\frac{\partial f}{\partial x_2} = x_3 - 2x_2 = 0$$

$$\frac{\partial f}{\partial x_3} = 2 + x_2 - 2x_3 = 0$$

The solution of these simultaneous equations is given by

$$X_0 = (1/2, 2/3, 4/3)$$

To establish sufficiency, consider

$$H\Big|_{X_0} = \begin{pmatrix} \dfrac{\partial^2 f}{\partial x_1^2} & \dfrac{\partial^2 f}{\partial x_1 \partial x_2} & \dfrac{\partial^2 f}{\partial x_1 \partial x_3} \\[2mm] \dfrac{\partial^2 f}{\partial x_2 \partial x_1} & \dfrac{\partial^2 f}{\partial x_2^2} & \dfrac{\partial^2 f}{\partial x_2 \partial x_3} \\[2mm] \dfrac{\partial^2 f}{\partial x_3 \partial x_1} & \dfrac{\partial^2 f}{\partial x_3 \partial x_2} & \dfrac{\partial^2 f}{\partial x_3^2} \end{pmatrix}_{X_0} = \begin{pmatrix} -2 & 0 & 0 \\ 0 & -2 & 1 \\ 0 & 1 & -2 \end{pmatrix}$$

The principal minor determinants of $H|_{X_0}$ have the values -2, 4, and -6, respectively. Thus as indicated in Section A.3, $H|_{X_0}$ is negative-definite and $X_0 = (1/2, 2/3, 4/3)$ represents a maximum point.   ◀

Exercise 18.1–1

Resolve Example 18.1–1 assuming that $f(x_1, x_2)$ is replaced by $-f(x_1, x_2)$.
[*Ans.* $X_0 = (1/2, 2/3, 4/3)$ is minimum point because the associated Hessian matrix is positive-definite.]

In general, if $H|_{X_0}$ is indefinite, X_0 must be a saddle point. However, for the case where it is nonconclusive, X_0 may or may not be an extremum and the development of a sufficiency condition becomes rather involved since it would be necessary to consider higher-order terms in Taylor's expansion. (See Theorem 18.1–3 for an illustration of this point to single-variable functions.) However, in some cases such complex procedures may not be necessary since the diagonalization of H may lead to more conclusive information. The following example illustrates this point.

Example 18.1–2. Consider the function

$$f(x_1, x_2) = 8x_1 x_2 + 3x_2^2$$

Thus,

$$\nabla f(x_1, x_2) = (8x_2, 8x_1 + 6x_2) = (0, 0)$$

This gives the stationary point $X_0 = (0, 0)$. The Hessian matrix at X_0 is

$$H = \begin{pmatrix} 0 & 8 \\ 8 & 6 \end{pmatrix}$$

which is nonconclusive. By using one of the diagonalization methods [see, e.g., Hadley (1961), Sec. 7–10], the transformed Hessian matrix becomes

$$\mathbf{H}_t = \begin{pmatrix} -64/6 & 0 \\ 0 & 6 \end{pmatrix}$$

By the principal minor determinants test, $\mathbf{H}_t$ (and hence $\mathbf{H}$) is indefinite. This concludes that $\mathbf{X}_0$ is a saddle point.   ◄

The sufficiency condition established by Theorem 18.1–2 reduces readily to single-variable functions. Given y_0 is a stationary point, then

(i) $f''(y_0) < 0$ is a sufficient condition for y_0 to be maximum.
(ii) $f''(y_0) > 0$ is a sufficient condition for y_0 to be minimum.

These conditions are directly determined by considering the Hessian matrix with one element.

If in the single-variable case $f''(y_0)$ vanishes, higher-order derivatives must be investigated as shown by the following theorem.

Theorem 18.1–3. *If at a stationary point y_0 of $f(y)$, the first $(n-1)$ derivatives vanish and $f^{(n)}(y) \neq 0$, then at $y = y_0, f(y)$ has*
(i) *An inflection point if n is odd.*
(ii) *An extreme point if n is even. This extreme point will be a maximum if $f^{(n)}(y_0) < 0$ and a minimum if $f^{(n)}(y_0) > 0$.*
The proof of this theorem is left as an exercise.

Example 18.1–3. Consider the two functions

$$f(y) = y^4 \qquad \text{and} \qquad g(y) = y^3$$

For $f(y) = y^4$,

$$f'(y) = 4y^3 = 0$$

which yields the stationary point $y_0 = 0$. Now

$$f'(0) = f''(0) = f^{(3)}(0) = 0$$

But $f^{(4)}(0) = 24 > 0$, hence $y_0 = 0$ is a minimum point (see Figure 18–2).

For $g(y) = y^3$,

$$g'(y) = 3y^2 = 0$$

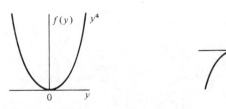

Figure 18–2

This yields $y_0 = 0$ as a stationary point. Since $g^{(n)}(0)$ is not zero at $n = 3$, $y_0 = 0$ is an inflection point. ◀

Exercise 18.1–2

Find the maxima and minima of $f(x) = x^3 + x^4$.
[*Ans.* $x_0 = -3/4$ is inflection point and $x_0 = 0$ is minimum.]

18.1.2 THE NEWTON–RAPHSON METHOD

A drawback of using the necessary condition $\nabla f(\mathbf{X}) = \mathbf{0}$ to determine stationary points is the difficulty of solving the resulting simultaneous equations numerically. The Newton–Raphson method is an iterative procedure for solving simultaneous nonlinear equations. Although the method is presented here in this context, it is actually part of the **gradient methods** for optimizing unconstrained functions numerically (see Section 19.1.2).

Consider the simultaneous equations

$$f_i(\mathbf{X}) = 0, \qquad i = 1, 2, \ldots, m$$

Let $\mathbf{X}^k$ be a given point. Then by Taylor's expansion

$$f_i(\mathbf{X}) \cong f_i(\mathbf{X}^k) + \nabla f_i(\mathbf{X}^k)(\mathbf{X} - \mathbf{X}^k), \qquad i = 1, 2, \ldots, m$$

Thus the original equations may be approximated by

$$f_i(\mathbf{X}^k) + \nabla f_i(\mathbf{X}^k)(\mathbf{X} - \mathbf{X}^k) = 0, \qquad i = 1, 2, \ldots, m$$

These equations may be written in matrix notation as

$$\mathbf{A}_k + \mathbf{B}_k(\mathbf{X} - \mathbf{X}^k) = \mathbf{0}$$

Under the assumption that all $f_i(\mathbf{X})$ are independent, $\mathbf{B}_k$ is necessarily nonsingular. Thus the preceding equation gives

$$\mathbf{X} = \mathbf{X}^k - \mathbf{B}_k^{-1}\mathbf{A}_k$$

The idea of the method is to start from an initial point $\mathbf{X}^0$. By using the foregoing equation, a new point $\mathbf{X}^{k+1}$ can always be determined from $\mathbf{X}^k$. The procedure is terminated with $\mathbf{X}^m$ as the solution when $\mathbf{X}^m \cong \mathbf{X}^{m-1}$.

Exercise 18.1–3

Given the necessary conditions $\nabla f(\mathbf{X}) = \mathbf{0}$, develop the associated Newton–Raphson equation for determining $\mathbf{X}^{k+1}$ given a current trial point $\mathbf{X}^k$.
[*Ans.* $\mathbf{X}^{k+1} = \mathbf{X}^k - \mathbf{H}_k^{-1}(\nabla f(\mathbf{X}^k))^T$, where $\mathbf{H}_k$ is the Hessian matrix evaluated at $\mathbf{X}^k$.]

A geometric interpretation of the method is illustrated by a single-variable function in Figure 18–3. The relationship between x^k and x^{k+1} for a single-variable function $f(x)$ reduces to

$$x^{k+1} = x^k - \frac{f(x^k)}{f'(x^k)}$$

or

$$f'(x^k) = \frac{f(x^k)}{x^k - x^{k+1}}$$

An investigation of Figure 18–3 shows that x^{k+1} is determined from the slope of $f(x)$ at x^k, where $\tan \theta = f'(x^k)$.

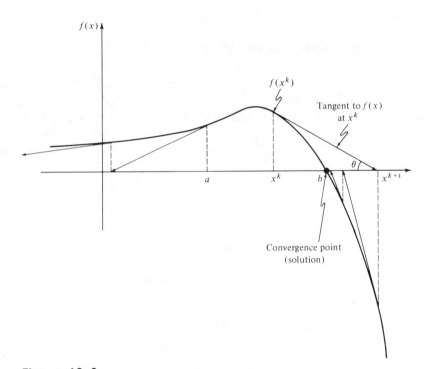

Figure 18–3

One difficulty with the method is that convergence is not always guaranteed unless the function f is well behaved. In Figure 18–3, if the initial point x_0 is a, the method will diverge. There is no easy way for locating a "good" initial x_0. Perhaps a remedy to this difficulty is to use trial and error.

18.2 CONSTRAINED EXTREMAL PROBLEMS

This section deals with the optimization of continuous functions subject to side conditions or constraints. Such constraints may be in the form of equation or inequation. Section 18.2.1 introduces the case with equality constraints and Section 18.2.2 introduces the other case with inequality constraints. The presentation in Section 18.2.1 is covered for the most part in Beightler and Wilde [1979, pp. 45–55].

18.2.1 EQUALITY CONSTRAINTS

This section presents two methods for optimizing functions subject to equality constraints. The first is the **Jacobian** method. This method may be considered a generalization of the simplex method for linear programming. Indeed, the simplex method conditions can be derived by the Jacobian method. The second method is the **Lagrangean** procedure, which is shown to be closely related to, and indeed may be developed logically from, the Jacobian method. This relationship allows an interesting economic interpretation of the Lagrangean method.

A. Constrained Derivatives (Jacobian) Method

Consider the problem

$$\text{minimize } z = f(\mathbf{X})$$

subject to

$$\mathbf{g}(\mathbf{X}) = \mathbf{0}$$

where

$$\mathbf{X} = (x_1, x_2, \ldots, x_n)$$
$$\mathbf{g} = (g_1, g_2, \ldots, g_m)^T$$

The functions $f(\mathbf{X})$ and $g_i(\mathbf{X})$, $i = 1, 2, \ldots, m$, are assumed twice continuously differentiable.

The idea of using constrained derivatives for solving the problem is to find a closed-form expression for the first partial derivatives of $f(\mathbf{X})$ at all points that satisfy the constraints $\mathbf{g}(\mathbf{X}) = \mathbf{0}$. The corresponding stationary points are thus identified as the points at which these partial derivatives vanish. The sufficiency conditions introduced in Section 18.1 can then be used to check the identity of stationary points.

To clarify this concept, consider $f(x_1, x_2)$ illustrated in Figure 18–4. This function is to be minimized subject to the constraint

$$g_1(x_1, x_2) = x_2 - b = 0$$

where b is a constant. From Figure 18–4, the curve designated by the three points A, B, and C represents the values of $f(x_1, x_2)$ for which the given constraint is always satisfied. The constrained derivatives method then defines the gradient of $f(x_1, x_2)$ at any point on the curve ABC. The point at which the constrained derivatives vanish represents a stationary point for the constrained problem. In Figure 18–4 this is given by B. The figure also shows an example of the incremental *constrained* value $\partial_c f$ of f.

The method is now developed mathematically. By Taylor's theorem, for the points $\mathbf{X} + \Delta\mathbf{X}$ in the feasible neighborhood of $\mathbf{X}$, it follows that

$$f(\mathbf{X} + \Delta\mathbf{X}) - f(\mathbf{X}) = \nabla f(\mathbf{X})\Delta\mathbf{X} + 0(\Delta x_j^2)$$

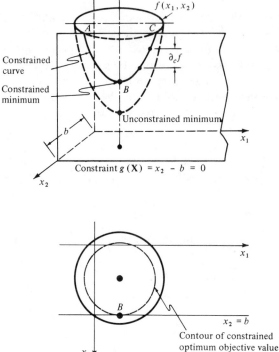

Figure 18–4

and

$$g(X + \Delta X) - g(X) = \nabla g(X)\Delta X + 0(\Delta x_j^2)$$

As $\Delta x_j \to 0$, the equations can be shown to reduce to

$$\partial f(X) = \nabla f(X) \, \partial X$$

and

$$\partial g(X) = \nabla g(X) \, \partial X$$

Since $g(X) = 0$, $\partial g(X) = 0$ for feasibility, and it follows that

$$\partial f(X) - \nabla f(X) \, \partial X = 0$$
$$\nabla g(X) \, \partial X = 0$$

This reduces to $(m + 1)$ equations in $(n + 1)$ unknowns, the unknowns being given by $\partial f(X)$ and ∂X. The unknown $\partial f(X)$ is determined, however, as soon as ∂X is known. This means that there are, in effect, m equations in n unknowns.

If $m > n$, at least $(m - n)$ equations are redundant. After eliminating this redundancy, the system reduces to an effective number of independent equations such that $m \le n$. For the case where $m = n$ the solution is $\partial X = 0$. This shows that X has no feasible neighborhood and hence the solution space consists of one point only. Such a case is of no interest. The remaining case, where $m < n$, will be considered in detail.

Let

$$\mathbf{X} = (\mathbf{Y}, \mathbf{Z})$$

where

$$\mathbf{Y} = (y_1, y_2, \ldots, y_m) \quad \text{and} \quad \mathbf{Z} = (z_1, z_2, \ldots, z_{n-m})$$

are the *dependent* and *independent* variables, respectively, corresponding to the vector $\mathbf{X}$. Rewriting the gradient vectors of f and $\mathbf{g}$ in terms of $\mathbf{Y}$ and $\mathbf{Z}$, we find that

$$\nabla f(\mathbf{Y}, \mathbf{Z}) = (\nabla_\mathbf{Y} f, \nabla_\mathbf{Z} f)$$
$$\nabla \mathbf{g}(\mathbf{Y}, \mathbf{Z}) = (\nabla_\mathbf{Y} \mathbf{g}, \nabla_\mathbf{Z} \mathbf{g})$$

Define

$$\mathbf{J} = \nabla_\mathbf{Y} \mathbf{g} = \begin{pmatrix} \nabla_\mathbf{Y} g_1 \\ \vdots \\ \nabla_\mathbf{Y} g_m \end{pmatrix}$$

$$\mathbf{C} = \nabla_\mathbf{Z} \mathbf{g} = \begin{pmatrix} \nabla_\mathbf{Z} g_1 \\ \vdots \\ \nabla_\mathbf{Z} g_m \end{pmatrix}$$

$\mathbf{J}_{m \times m}$ is called the **Jacobian matrix** and $\mathbf{C}_{m \times n-m}$ the **control matrix.** The Jacobian $\mathbf{J}$ is assumed nonsingular. This is always possible, since the given m equations are independent by definition. The components of the vector $\mathbf{Y}$ can thus be selected from those of $\mathbf{X}$ such that $\mathbf{J}$ is nonsingular.

Using the definitions given, we write the original set of equations in $\partial f(\mathbf{X})$ and $\partial \mathbf{X}$ as

$$\partial f(\mathbf{Y}, \mathbf{Z}) = \nabla_\mathbf{Y} f \, \partial \mathbf{Y} + \nabla_\mathbf{Z} f \, \partial \mathbf{Z}$$

and

$$\mathbf{J} \, \partial \mathbf{Y} = -\mathbf{C} \, \partial \mathbf{Z}$$

Since $\mathbf{J}$ is nonsingular, its inverse $\mathbf{J}^{-1}$ exists. Hence

$$\partial \mathbf{Y} = -\mathbf{J}^{-1} \mathbf{C} \, \partial \mathbf{Z}$$

This set of equations relates the effect of variation in $\partial \mathbf{Z}$ ($\mathbf{Z}$ being the independent vector) on $\partial \mathbf{Y}$. Substituting for $\partial \mathbf{Y}$ in the equation for $\partial f(\mathbf{Y}, \mathbf{Z})$ gives ∂f as a function of $\partial \mathbf{Z}$. That is,

$$\partial f(\mathbf{Y}, \mathbf{Z}) = (\nabla_\mathbf{Z} f - \nabla_\mathbf{Y} f \mathbf{J}^{-1} \mathbf{C}) \, \partial \mathbf{Z}$$

From this equation, the constrained derivative with respect to the independent vector $\mathbf{Z}$ is given by

$$\nabla_c f = \frac{\partial_c f(\mathbf{Y}, \mathbf{Z})}{\partial_c \mathbf{Z}} = \nabla_\mathbf{Z} f - \nabla_\mathbf{Y} f \mathbf{J}^{-1} \mathbf{C}$$

where $\nabla_c f$ represents the **constrained gradient** vector of f with respect to $\mathbf{Z}$. Thus $\nabla_c f(\mathbf{Y}, \mathbf{Z})$ must be null at stationary points.

The sufficiency conditions are similar to those developed in Section 18.1. In this case, however, the Hessian matrix will correspond to the independent vector $\mathbf{Z}$. In the meantime, the elements of the Hessian matrix must be the *constrained* second derivatives. To show how this is obtained, let

$$\nabla_c f = \nabla_z f - \mathbf{WC}$$

It thus follows the ith row of the (constrained) Hessian matrix is $\partial \nabla_c f / \partial z_i$. Notice that $\mathbf{W}$ is a function of $\mathbf{Y}$ and $\mathbf{Y}$ is a function of $\mathbf{Z}$; that is,

$$\partial \mathbf{Y} = -\mathbf{J}^{-1} \mathbf{C}\, \partial \mathbf{Z}$$

Thus, in taking the partial derivation of $\nabla_c f$ with respect to z_i, the chain rule must apply to $\mathbf{W}$. This means that

$$\frac{\partial w_j}{\partial z_i} = \frac{\partial w_j}{\partial y_j} \frac{\partial y_j}{\partial z_i}$$

Example 18.2-1. In this example it will be shown how $\partial_c f$ can be estimated at a given point using the formulas given. Example 18.2-2 will then illustrate the application of the constrained derivative.

Consider the problem in which

$$f(\mathbf{X}) = x_1^2 + 3x_2^2 + 5x_1 x_3^2$$
$$g_1(\mathbf{X}) = x_1 x_3 + 2x_2 + x_2^2 - 11 = 0$$
$$g_2(\mathbf{X}) = x_1^2 + 2x_1 x_2 + x_3^2 - 14 = 0$$

Given the feasible point $\mathbf{X}^0 = (1, 2, 3)$, it is required to study the variation in $f(= \partial_c f)$ in the feasible neighborhood of $\mathbf{X}^0$.

Let

$$\mathbf{Y} = (x_1, x_3) \qquad \text{and} \qquad \mathbf{Z} = x_2$$

Thus

$$\nabla_{\mathbf{Y}} f = \left(\frac{\partial f}{\partial x_1}, \frac{\partial f}{\partial x_3} \right) = (2x_1 + 5x_3^2, 10x_1 x_3)$$

$$\nabla_{\mathbf{Z}} f = \frac{\partial f}{\partial x_2} = 6x_2$$

$$\mathbf{J} = \begin{pmatrix} \dfrac{\partial g_1}{\partial x_1} & \dfrac{\partial g_1}{\partial x_3} \\[2mm] \dfrac{\partial g_2}{\partial x_1} & \dfrac{\partial g_2}{\partial x_3} \end{pmatrix} = \begin{pmatrix} x_3 & x_1 \\[2mm] 2x_1 + 2x_2 & 2x_3 \end{pmatrix}$$

$$\mathbf{C} = \begin{pmatrix} \dfrac{\partial g_1}{\partial x_2} \\[2mm] \dfrac{\partial g_2}{\partial x_2} \end{pmatrix} = \begin{pmatrix} 2x_2 + 2 \\[2mm] 2x_1 \end{pmatrix}$$

An estimate of $\partial_c f$ in the feasible neighborhood of the feasible point $\mathbf{X}^0 = (1, 2, 3)$ resulting from a small change $\partial x_2 = .01$ is obtained as follows:

$$\mathbf{J}^{-1}\mathbf{C} = \begin{pmatrix} 3 & 1 \\ 6 & 6 \end{pmatrix}^{-1} \begin{pmatrix} 6 \\ 2 \end{pmatrix} = \begin{pmatrix} 6/12 & -1/12 \\ -6/12 & 3/12 \end{pmatrix} \begin{pmatrix} 6 \\ 2 \end{pmatrix} \cong \begin{pmatrix} 2.83 \\ -2.50 \end{pmatrix}$$

Hence,

$$\partial_c f = (\nabla_\mathbf{Z} f - \nabla_\mathbf{Y} f \mathbf{J}^{-1}\mathbf{C})\, \partial\mathbf{Z} = \left(6(2) - (47, 30) \begin{bmatrix} 2.83 \\ -2.5 \end{bmatrix} \right) \partial x_2$$

$$\cong -46 \partial x_2 = -.46$$

By specifying the value of ∂x_2 for the *independent* variable x_2, feasible values of ∂x_1 and ∂x_2 are automatically determined for the *dependent* variables x_1 and x_3 from the formula

$$\partial\mathbf{Y} = -\mathbf{J}^{-1}\mathbf{C}\, \partial\mathbf{Z}$$

This gives for $\partial x_2 = .01$,

$$\begin{pmatrix} \partial x_1 \\ \partial x_3 \end{pmatrix} = -\mathbf{J}^{-1}\mathbf{C}\, \partial x_2 = \begin{pmatrix} -.0283 \\ .0250 \end{pmatrix}$$

To check the value of $\partial_c f$ obtained, we can compute the value of f at $\mathbf{X}^0$ and $\mathbf{X}^0 + \partial\mathbf{X}$. Thus

$$\mathbf{X}^0 + \partial\mathbf{X} = (1 - .0283, 2 + .01, 3 + .025) = (.9717, 2.01, 3.025)$$

This yields

$$f(\mathbf{X}^0) = 58 \qquad \text{and} \qquad f(\mathbf{X}^0 + \partial\mathbf{X}) = 57.523$$

or

$$\partial_c f = f(\mathbf{X}^0 + \partial\mathbf{X}) - f(\mathbf{X}^0) = -.477$$

This indicates a decrease in the value of f as obtained by the formula for $\partial_c f$. The difference between the two answers ($-.477$ and $-.46$) is the result of the linear approximation at $\mathbf{X}^0$. The formula is good only for very small variations around $\mathbf{X}^0$.   ◀

Exercise 18.2–1

Consider Example 18.2–1.

(a) Compute $\partial_c f$ by the two methods presented in the example, using $\partial x_2 = .001$ instead of $\partial x_2 = .01$. Does the effect of linear approximation become more negligible with the decrease in the value of ∂x_2?
[*Ans.* Yes; $\partial_c f = -.046$ and $-.04618$.]

(b) Specify a relationship among ∂x_1, ∂x_2, and ∂x_3 at the feasible point $\mathbf{X}^0 = (1, 2, 3)$ that will keep the point $(x_1^0 + \partial x_1, x_2^0 + \partial x_2, x_3^0 + \partial x_3)$ feasible.
[*Ans.* $\partial x_1 = -.283 \partial x_2$, $\partial x_3 = .25 \partial x_2$.]

(c) If $\mathbf{Y} = (x_2, x_3)$ and $\mathbf{Z} = x_1$, what is the value of ∂x_1 that will produce the same value of $\partial_c f$ given in the example?
[*Ans.* $\partial x_1 = -.0283$.]

(d) Verify that the result in part (c) will yield $\partial_c f = -.46$.

Example 18.2-2. This example illustrates the use of constrained derivatives. Consider the problem

$$\text{minimize} f(X) = x_1^2 + x_2^2 + x_3^2$$

subject to

$$g_1(X) = x_1 + x_2 + 3x_3 - 2 = 0$$
$$g_2(X) = 5x_1 + 2x_2 + x_3 - 5 = 0$$

We determine the constrained extreme points as follows. Let

$$Y = (x_1, x_2) \quad \text{and} \quad Z = x_3$$

Thus

$$\nabla_Y f = \left(\frac{\partial f}{\partial x_1}, \frac{\partial f}{\partial x_2}\right) = (2x_1, 2x_2), \quad \nabla_Z f = \frac{\partial f}{\partial x_3} = 2x_3$$

$$J = \begin{pmatrix} 1 & 1 \\ 5 & 2 \end{pmatrix}, \quad J^{-1} = \begin{pmatrix} -2/3 & 1/3 \\ 5/3 & -1/3 \end{pmatrix}, \quad C = \begin{pmatrix} 3 \\ 1 \end{pmatrix}$$

Hence

$$\nabla_c f = \frac{\partial_c f}{\partial_c x_3} = 2x_3 - (2x_1, 2x_2)\begin{pmatrix} -2/3 & 1/3 \\ 5/3 & -1/3 \end{pmatrix}\begin{pmatrix} 3 \\ 1 \end{pmatrix}$$

$$= \frac{10}{3}x_1 - \frac{28}{3}x_2 + 2x_3$$

At a stationary point, $\nabla_c f = 0$, which together with $g_1(X) = 0$ and $g_2(X) = 0$ give the required stationary point(s). That is, the equations

$$\begin{pmatrix} 10 & -28 & 6 \\ 1 & 1 & 3 \\ 5 & 2 & 1 \end{pmatrix}\begin{pmatrix} x_1 \\ x_2 \\ x_3 \end{pmatrix} = \begin{pmatrix} 0 \\ 2 \\ 5 \end{pmatrix}$$

give the solution

$$X^0 \cong (.81, .35, .28)$$

The identity of this stationary point is now checked by considering the sufficiency condition. Given the independent variable x_3, it follows from $\nabla_c f$ that

$$\frac{\partial_c^2 f}{\partial_c x_3^2} = \frac{10}{3}\left(\frac{dx_1}{dx_3}\right) - \frac{28}{3}\left(\frac{dx_2}{dx_3}\right) + 2 = \left(\frac{10}{3}, -\frac{28}{3}\right)\begin{pmatrix} \frac{dx_1}{dx_3} \\ \frac{dx_2}{dx_3} \end{pmatrix} + 2$$

From the development of the Jacobian method,

$$\begin{pmatrix} \frac{dx_1}{dx_3} \\ \frac{dx_2}{dx_3} \end{pmatrix} = -J^{-1}C = \begin{pmatrix} 5/3 \\ -14/3 \end{pmatrix}$$

Substitution gives $\partial_c^2 f/\partial_c x_3^2 = 460/9 > 0$. Hence X^0 is the minimum point. ◄

Exercise 18.2-2

Suppose that Example 18.2–2 is solved in the following manner. First, solve the constraints expressing x_1 and x_2 in terms of x_3; then use the resulting equations to express the objective function in terms of x_3 only. By taking the derivative of the new objective function with respect to x_3, we can determine the points of maxima and minima.

(a) Would the derivative of the new objective function (expressed in terms of x_3) be different from that obtained by the Jacobian method?
   [*Ans.* No, the necessary and sufficient conditions are exactly the same in both methods.]

(b) What is the prime difference between the procedure outlined and the Jacobian method?
   [*Ans.* The Jacobian method computes the *constrained* gradient of the objective function directly, whereas the proposed method computes the equation of the constrained objective function from which we can compute the constrained gradient.]

The use of the Jacobian method as presented is hindered, in general, by the difficulty of obtaining $\mathbf{J}^{-1}$ for a large number of constraints. This difficulty can be overcome by applying Cramer's rule to solve for ∂f in terms of $\partial \mathbf{Z}$. Thus, if z_j represents the jth element of $\mathbf{Z}$ and y_i represents the ith element of $\mathbf{Y}$, it can be shown that

$$\frac{\partial_c f}{\partial_c z_j} = \frac{\partial(f, g_1, \ldots, g_m)/\partial(z_j, y_1, \ldots, y_m)}{\partial(g_1, \ldots, g_m)/\partial(y_1, \ldots, y_m)}$$

where

$$\frac{\partial(f, g_1, \ldots, g_m)}{\partial(z_j, y_1, \ldots, y_m)} \equiv \begin{vmatrix} \dfrac{\partial f}{\partial z_j} & \dfrac{\partial f}{\partial y_1} & \cdots & \dfrac{\partial f}{\partial y_m} \\ \dfrac{\partial g_1}{\partial z_j} & \dfrac{\partial g_1}{\partial y_1} & \cdots & \dfrac{\partial g_1}{\partial y_m} \\ \vdots & \vdots & & \vdots \\ \dfrac{\partial g_m}{\partial z_j} & \dfrac{\partial g_m}{\partial y_1} & \cdots & \dfrac{\partial g_m}{\partial y_m} \end{vmatrix}$$

and

$$\frac{\partial(g_1, \ldots, g_m)}{\partial(y_1, \ldots, y_m)} \equiv \begin{vmatrix} \dfrac{\partial g_1}{\partial y_1} & \cdots & \dfrac{\partial g_1}{\partial y_m} \\ \vdots & & \vdots \\ \dfrac{\partial g_m}{\partial y_1} & \cdots & \dfrac{\partial g_m}{\partial y_m} \end{vmatrix} = |\mathbf{J}|$$

Thus the necessary conditions become

$$\frac{\partial_c f}{\partial_c z_j} = 0, \qquad j = 1, 2, \ldots, n - m$$

Similarly, in the matrix expression

$$\frac{\partial \mathbf{Y}}{\partial \mathbf{Z}} = -\mathbf{J}^{-1}\mathbf{C}$$

the (i, j)th element is given by

$$\frac{\partial y_i}{\partial z_j} = -\frac{\partial(g_1, \ldots, g_m)/\partial(y_1, \ldots, y_{i-1}, z_j, y_{i+1}, \ldots, y_m)}{\partial(g_1, \ldots, g_m)/\partial(y_1, \ldots, y_m)}$$

which represents the rate of variation of the dependent variable y_i with respect to the independent variable z_j.

Finally, to obtain the sufficiency condition given previously, determinant expressions for the elements of $\mathbf{W} \equiv \nabla_Y f \mathbf{J}^{-1}$ must be given. Thus the ith element of $\mathbf{W}$ is given by

$$w_i = \frac{\partial(g_1, \ldots, g_{i-1}, f, g_{i+1}, \ldots, g_m)/\partial(y_1, \ldots, y_m)}{\partial(g_1, \ldots, g_m)/\partial(y_1, \ldots, y_m)}$$

To illustrate the application of the method just described, consider the determination of the necessary condition for Example 18.2–2. Thus,

$$\frac{\partial_c f}{\partial_c x_3} = \frac{\begin{vmatrix} 2x_3 & 2x_1 & 2x_2 \\ 3 & 1 & 1 \\ 1 & 5 & 2 \end{vmatrix}}{\begin{vmatrix} 1 & 1 \\ 5 & 2 \end{vmatrix}} = \frac{10}{3}x_1 - \frac{28}{3}x_2 + 2x_3$$

Sensitivity Analysis in the Jacobian Method

The Jacobian method can be used to study the sensitivity of the optimal value of f due to small changes in the right-hand sides of the constraints. For example, suppose that the right-hand side of the ith constraint $g_i(\mathbf{X}) = 0$ is changed to ∂g_i instead of zero. What effect will this have on the optimum value of f? This type of investigation is called **sensitivity analysis** and, in some sense, is similar to that carried out in linear programming (see Chapter 4). However, sensitivity analysis in nonlinear programming is valid only in the small neighborhood of the extreme point due to the absence of linearity. Nevertheless, the development will be helpful in studying the Lagrangean method (see the next section).

It is shown that

$$\partial f(\mathbf{Y}, \mathbf{Z}) = \nabla_Y f \, \partial \mathbf{Y} + \nabla_Z f \, \partial \mathbf{Z}$$
$$\partial \mathbf{g} = \mathbf{J} \, \partial \mathbf{Y} + \mathbf{C} \, \partial \mathbf{Z}$$

Suppose that $\partial \mathbf{g} \neq \mathbf{0}$; then

$$\partial \mathbf{Y} = \mathbf{J}^{-1} \, \partial \mathbf{g} - \mathbf{J}^{-1} \mathbf{C} \, \partial \mathbf{Z}$$

Substituting in the equation for $\partial f(\mathbf{Y}, \mathbf{Z})$ gives

$$\partial f(\mathbf{Y}, \mathbf{Z}) = \nabla_Y f \mathbf{J}^{-1} \, \partial \mathbf{g} + \nabla_c f \, \partial \mathbf{Z}$$

where

$$\nabla_c f = \nabla_Z f - \nabla_Y f \mathbf{J}^{-1} \mathbf{C}$$

as defined previously. The expression for $\partial f(\mathbf{Y}, \mathbf{Z})$ can be used to study variation in f in the feasible neighborhood of a feasible point $\mathbf{X}^0$ due to small changes $\partial \mathbf{g}$ and $\partial \mathbf{Z}$.

Now, at the extreme (indeed, any stationary) point $\mathbf{X}_0 = (\mathbf{Y}_0, \mathbf{Z}_0)$, the constrained gradient $\nabla_c f$ must vanish. Thus,

$$\partial f(\mathbf{Y}_0, \mathbf{Z}_0) = \nabla_{\mathbf{Y}_0} f \, \mathbf{J}^{-1} \, \partial \mathbf{g}(\mathbf{Y}_0, \mathbf{Z}_0)$$

or

$$\frac{\partial f}{\partial \mathbf{g}} = \nabla_{\mathbf{Y}_0} f \mathbf{J}^{-1}$$

evaluated at $\mathbf{X}_0$. Consequently, the effect of small variations in $\mathbf{g}(=\partial \mathbf{g})$ on the *optimum* value of f can be studied by evaluating the rate of change of f with respect to $\mathbf{g}$. These rates are usually referred to as **sensitivity coefficients.**

In general, at the optimum point, $\partial f / \partial \mathbf{g}$ is independent of the specific choice of the variables in the vector $\mathbf{Y}$. This follows, since the expression for the sensitivity coefficients does not include $\mathbf{Z}$. Hence the partitioning of $\mathbf{X}$ between $\mathbf{Y}$ and $\mathbf{Z}$ is not an effective factor in this case. The given coefficients are thus constant regardless of the specific choice of $\mathbf{Y}$.

Example 18.2–3. Consider the same problem of Example 18.2–2. The optimum point is given by $\mathbf{X}_0 = (x_1^0, x_2^0, x_3^0) = (.81, .35, .28)$. Since $\mathbf{Y}_0 = (x_1^0, x_2^0)$, then

$$\nabla_{\mathbf{Y}_0} f = \left(\frac{\partial f}{\partial x_1}, \frac{\partial f}{\partial x_2} \right) = (2x_1^0, 2x_2^0) = (1.62, .70)$$

Consequently,

$$\left(\frac{\partial f}{\partial g_1}, \frac{\partial f}{\partial g_2} \right) = \nabla_{\mathbf{Y}_0} f \mathbf{J}^{-1} = (1.62, .7) \begin{pmatrix} -2/3 & 1/3 \\ 5/3 & -1/3 \end{pmatrix} = (.0876, .3067)$$

This implies that if $\partial g_1 = 1$, f will increase *approximately* by .0867. Similarly, if $\partial g_2 = 1$, f will increase *approximately* by .3067. ◀

Example of Application of the Jacobian Method to a Linear Programming Problem

Consider the linear programming problem

$$\text{maximize } z = 2x_1 + 3x_2$$

subject to

$$x_1 + x_2 + x_3 \qquad = 5$$
$$x_1 - x_2 \qquad + x_4 = 3$$
$$x_1, x_2, x_3, x_4 \geq 0$$

Consider the nonnegativity constraints $x_j \geq 0$. Let w_j^2 be the corresponding (nonnegative) slack variable. Thus $x_j - w_j^2 = 0$, or $x_j = w_j^2$. With this substitution, the nonnegativity conditions become implicit and the original problem becomes

$$\text{maximize } z = 2w_1^2 + 3w_2^2$$

subject to

$$w_1^2 + w_2^2 + w_3^2 = 5$$
$$w_1^2 - w_2^2 + w_4^2 = 3$$

To apply the Jacobian method, let

$$\mathbf{Y} = (w_1, w_2) \quad \text{and} \quad \mathbf{Z} = (w_3, w_4)$$

(Notice that in the terminology of linear programming, $\mathbf{Y}$ and $\mathbf{Z}$ correspond to the basic and nonbasic variables, respectively.) Thus

$$\mathbf{J} = \begin{pmatrix} 2w_1 & 2w_2 \\ 2w_1 & -2w_2 \end{pmatrix}, \quad \mathbf{J}^{-1} = \begin{pmatrix} \dfrac{1}{4w_1} & \dfrac{1}{4w_1} \\ \dfrac{1}{4w_2} & \dfrac{-1}{4w_2} \end{pmatrix}, \quad w_1 \text{ and } w_2 \neq 0$$

$$\mathbf{C} = \begin{pmatrix} 2w_3 & 0 \\ 0 & 2w_4 \end{pmatrix}, \quad \nabla_{\mathbf{Y}}f = (4w_1, 6w_2), \quad \nabla_{\mathbf{Z}}f = (0, 0)$$

so that

$$\nabla_c f = (0, 0) - (4w_1, 6w_2) \begin{pmatrix} \dfrac{1}{4w_1} & \dfrac{1}{4w_1} \\ \dfrac{1}{4w_2} & \dfrac{-1}{4w_2} \end{pmatrix} \begin{pmatrix} 2w_3 & 0 \\ 0 & 2w_4 \end{pmatrix} = (-5w_3, w_4)$$

Solution of $\nabla_c f = \mathbf{0}$ together with the constraints of the problem yield the stationary point ($w_1 = 2$, $w_2 = 1$, $w_3 = 0$, $w_4 = 0$). The Hessian is given by

$$\mathbf{H}_c = \begin{pmatrix} \dfrac{\partial_c^2 f}{\partial_c w_3^2} & \dfrac{\partial_c^2 f}{\partial_c w_3 \, \partial_c w_4} \\ \dfrac{\partial_c^2 f}{\partial_c w_3 \, \partial_c w_4} & \dfrac{\partial_c^2 f}{\partial_c w_4^2} \end{pmatrix} = \begin{pmatrix} -5 & 0 \\ 0 & 1 \end{pmatrix}$$

Since $\mathbf{H}_c$ is indefinite, the stationary point does not yield a maximum.

Actually, the result is not surprising, since the (nonbasic) variables w_3 and w_4 (and hence x_3 and x_4) equal zero, as contemplated by the theory of linear programming. This means that, depending on the specific choice of $\mathbf{Y}$ and $\mathbf{Z}$, the Jacobian method solution determines the corresponding extreme point of the solution space. This may or may not be the optimal solution. The Jacobian method has the power, however, to identify the optimum point through the use of the sufficiency conditions.

$$\nabla_c f = (4w_1, 0) - (6w_2, 0) \begin{pmatrix} \dfrac{1}{2w_2} & 0 \\ \dfrac{1}{2w_4} & \dfrac{1}{2w_4} \end{pmatrix} \begin{pmatrix} 2w_1 & 2w_3 \\ 2w_1 & 0 \end{pmatrix} = (-2w_1, -6w_3)$$

The corresponding stationary point is given by $w_1 = 0$, $w_2 = \sqrt{5}$, $w_3 = 0$, $w_4 = \sqrt{8}$. Now

$$\mathbf{H}_c = \begin{pmatrix} -2 & 0 \\ 0 & -6 \end{pmatrix}$$

is negative-definite. Thus the given solution corresponds to a maximum point.

The result is verified graphically in Figure 18–5. The first solution ($x_1 = 4$, $x_2 = 1$) is not optimal, whereas the second ($x_1 = 0$, $x_2 = 5$) gives the optimal solution. You can verify that the remaining two extreme points of the solution space do not yield maximum points. In fact, the extreme point ($x_1 = 0$, $x_2 = 0$) can be shown by the sufficiency condition to yield a minimum point.

It is interesting that when applied to linear programming, the sensitivity coefficients $\nabla_{Y_0} f \, \mathbf{J}^{-1}$ introduced previously will actually yield its dual values. To illustrate this point for the given numerical example, let u_1 and u_2 be corresponding dual variables. At the optimum point ($w_1 = 0$, $w_2 = \sqrt{5}$, $w_3 = 0$, $w_4 = \sqrt{8}$), these dual variables are given by

$$(u_1, u_2) = \nabla_{Y_0} \mathbf{J}^{-1} = (6w_2, 0) \begin{pmatrix} \dfrac{1}{2w_2} & 0 \\ \dfrac{1}{2w_2} & \dfrac{1}{2w_4} \end{pmatrix} = (3, 0)$$

The corresponding dual objective value is equal to $5u_1 + 3u_2 = 15$, which is the same as the optimal primal objective value. The given solution also satisfies the dual constraints and hence is optimal and feasible. This shows that the sensitivity coefficients are the same as the dual variables. In fact, one notices that both have the same interpretation.

It is now possible to draw some general conclusions from the application of the Jacobian method to the linear programming problem. From the numerical

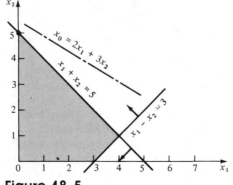

Figure 18–5

example, the necessary conditions require the independent variables to equal zero. Also, the sufficiency conditions indicate that the Hessian matrix is a diagonal matrix. Thus all its diagonal elements must be positive for a minimum and negative for a maximum.

The observations suggest that the necessary condition is equivalent to specifying that only "basic" (feasible) solutions are needed to locate the optimum solution. In this case the independent variables are equivalent to the nonbasic variables in the linear programming problem. Also, the sufficiency condition suggests that there may be a strong relationship between the diagonal elements of the Hessian matrix and the optimality indicator $z_j - c_j$ (see Section 7.2.2) in the simplex method.†

B.   Lagrangean Method

Section 8.2.1A shows that the sensitivity coefficients

$$\frac{\partial f}{\partial \mathbf{g}} = \nabla_{\mathbf{Y}_0}\mathbf{J}^{-1}$$

can be used to study the effect of small variations in the constraints on the *optimum* value of *f.* It is also indicated that these coefficients are constant. These properties can be used to solve the constrained problems with equality constraints.

Let

$$\boldsymbol{\lambda} = \nabla_{\mathbf{Y}_0}\mathbf{J}^{-1} = \frac{\partial f}{\partial \mathbf{g}}$$

Thus

$$\partial f - \boldsymbol{\lambda}\,\partial \mathbf{g} = 0$$

This equation satisfies the *necessary* conditions for stationary points since the expression for $\partial f/\partial \mathbf{g}$ is computed such that $\nabla_c f = \mathbf{0}$. A more convenient form for presenting these equations, however, is obtained by taking their partial derivatives with respect to all x_j. This yields

$$\frac{\partial}{\partial x_j}(f - \boldsymbol{\lambda}\mathbf{g}) = 0, \qquad j = 1, 2, \ldots, n$$

The resulting equations together with the constraint equations $\mathbf{g} = \mathbf{0}$ yield the feasible values of $\mathbf{X}$ and $\boldsymbol{\lambda}$ that satisfy the *necessary* conditions for stationary points.

---

† For a formal proof of the validity of these results for the general linear programming problem, see H. Taha and G. Curry, "Classical Derivation of the Necessary and Sufficient Conditions for Optimal Linear Programs," *Operations Research,* Vol. 19, 1971, pp. 1045–1049. The paper shows that all the key ideas of the simplex method can be derived by the Jacobian method.

The procedure described defines the so-called *Lagrangean method* for identifying the stationary points of optimization problems with *equality* constraints. This procedure can be developed formally as follows. Let

$$L(\mathbf{X}, \lambda) = f(\mathbf{X}) - \lambda \mathbf{g}(\mathbf{X})$$

The function L is called the **Lagrangean function** and the parameters λ the **Lagrange multipliers.** By definition, these multipliers have the same interpretation as the sensitivity coefficients introduced in Section 18.2.1–A.

The equations

$$\frac{\partial L}{\partial \lambda} = 0 \quad \text{and} \quad \frac{\partial L}{\partial \mathbf{X}} = 0$$

yield the same necessary conditions given above and hence the Lagrangean function can be used directly to generate the necessary conditions. This means that optimization of $f(\mathbf{X})$ subject to $\mathbf{g}(\mathbf{X}) = 0$ is equivalent to optimization of the Lagrangean function $L(\mathbf{X}, \lambda)$.

The sufficiency conditions for the Lagrangean method will be stated without proof. Define

$$\mathbf{H}^B = \left(\begin{array}{c|c} \mathbf{0} & \mathbf{P} \\ \hline \mathbf{P}^T & \mathbf{Q} \end{array} \right)_{(m+n) \times (m+n)}$$

where

$$\mathbf{P} = \begin{pmatrix} \nabla g_1(\mathbf{X}) \\ \vdots \\ \nabla g_m(\mathbf{X}) \end{pmatrix}_{m \times n} \quad \text{and} \quad \mathbf{Q} = \left\| \frac{\partial^2 L(\mathbf{X}, \lambda)}{\partial x_i\, \partial x_j} \right\|_{n \times n}, \quad \text{for all } i \text{ and } j$$

The matrix $\mathbf{H}^B$ is called the **bordered Hessian matrix.**

Given the stationary point $(\mathbf{X}_0, \lambda_0)$ for the Lagrangean function $L(\mathbf{X}, \lambda)$ and the bordered Hessian matrix $\mathbf{H}^B$ evaluated at $(\mathbf{X}_0, \lambda_0)$, then $\mathbf{X}_0$ is

1. A maximum point if, starting with the principal major determinant of order $(2m + 1)$, the *last* $(n - m)$ principal minor determinants of $\mathbf{H}^B$ form an alternating sign pattern starting with $(-1)^{m+1}$.

2. A minimum point if, starting with the principal minor determinant of order $(2m + 1)$, the *last* $(n - m)$ principal minor determinants of $\mathbf{H}^B$ have the sign of $(-1)^m$.

These conditions are sufficient for identifying an extreme point, but not necessary. In other words, a stationary point may be an extreme point without satisfying these conditions.

Other conditions exist that are both necessary and sufficient for identifying extreme points. The disadvantage here is that this procedure is computationally infeasible for most practical purposes. Define the matrix

$$\Delta = \left(\begin{array}{c|c} \mathbf{0} & \mathbf{P} \\ \hline \mathbf{P}^T & \mathbf{Q} - \mu\mathbf{I} \end{array} \right)$$

evaluated at the stationary point $(\mathbf{X}_0, \boldsymbol{\lambda}_0)$, where $\mathbf{P}$ and $\mathbf{Q}$ are as defined and μ is an unknown parameter. Consider the determinant $|\Delta|$; then each of the real $(n - m)$ roots u_i of the polynomial

$$|\Delta| = 0$$

must be

1. Negative if $\mathbf{X}_0$ is a maximum point.
2. Positive if $\mathbf{X}_0$ is a minimum point.

Example 18.2–4. Consider the same problem of Example 18.2–2. The Lagrangean function is

$$L(\mathbf{X}, \boldsymbol{\lambda}) = x_1^2 + x_2^2 + x_3^2 - \lambda_1(x_1 + x_2 + 3x_3 - 2) - \lambda_2(5x_1 + 2x_2 + x_3 - 5)$$

This yields the following necessary conditions:

$$\frac{\partial L}{\partial x_1} = 2x_1 - \lambda_1 - 5\lambda_2 = 0$$

$$\frac{\partial L}{\partial x_2} = 2x_2 - \lambda_1 - 2\lambda_2 = 0$$

$$\frac{\partial L}{\partial x_3} = 2x_3 - 3\lambda_1 - \lambda_2 = 0$$

$$\frac{\partial L}{\partial \lambda_1} = -(x_1 + x_2 + 3x_3 - 2) = 0$$

$$\frac{\partial L}{\partial \lambda_2} = -(5x_1 + 2x_2 + x_3 - 5) = 0$$

The solution to these simultaneous equations yields

$$\mathbf{X}_0 = (x_1, x_2, x_3) = (.81, .35, .28)$$
$$\boldsymbol{\lambda} = (\lambda_1, \lambda_2) = (.0867, .3067)$$

This solution combines the results of Examples 18.2–2 and 18.2–3. The values of the Lagrange multipliers $\boldsymbol{\lambda}$ are the same as the sensitivity coefficients obtained in Example 18.2–3. This shows that these coefficients are independent of the choice of the dependent vector $\mathbf{Y}$ in the Jacobian method.

To show that the given point is a minimum, consider

$$\mathbf{H}^B = \left(\begin{array}{cc|ccc} 0 & 0 & 1 & 1 & 3 \\ 0 & 0 & 5 & 2 & 1 \\ \hline 1 & 5 & 2 & 0 & 0 \\ 1 & 2 & 0 & 2 & 0 \\ 3 & 1 & 0 & 0 & 2 \end{array} \right)$$

Since $n = 3$ and $m = 2$, it follows that $n - m = 1$. Thus we need to check the determinant of $\mathbf{H}^B$ only, which must have the sign of $(-1)^2$ at a minimum. Since det $\mathbf{H}^B = 460 > 0$, $\mathbf{X}_0$ is a minimum point. ◄

A method that is sometimes convenient for solving equations resulting from the necessary conditions is to select successive numerical values of $\boldsymbol{\lambda}$ and then solve the given equations for X. This is repeated until for some values of λ, the resulting X satisfies all the active constraints in equation form. This method was illustrated in Chapter 13 as an application to the single-constraint inventory problem (see Example 13.3–4). This procedure becomes very tedious computationally, however, as the number of constraints increases. In this case one may resort to an appropriate numerical technique, such as the Newton–Raphson method (Section 18.1.2) to solve the resulting equations.

Example 18.2–5. Consider the problem

$$\text{minimize } z = x_1^2 + x_2^2 + x_3^2$$

subject to

$$4x_1 + x_2^2 + 2x_3 - 14 = 0$$

The Lagrangean function is

$$L(X, \lambda) = x_1^2 + x_2^2 + x_3^2 - \lambda(4x_1 + x_2^2 + 2x_3 - 14)$$

This yields the following necessary conditions:

$$\frac{\partial L}{\partial x_1} = 2x_1 - 4\lambda = 0$$

$$\frac{\partial L}{\partial x_2} = 2x_2 - 2\lambda x_2 = 0$$

$$\frac{\partial L}{\partial x_3} = 2x_3 - 2\lambda = 0$$

$$\frac{\partial L}{\partial \lambda} = -(4x_1 + x_2^2 + 2x_3 - 14) = 0$$

whose solution is

$$(X_0, \lambda_0)_1 = (2, 2, 1, 1)$$
$$(X_0, \lambda_0)_2 = (2, -2, 1, 1)$$
$$(X_0, \lambda_0)_3 = (2.8, 0, 1.4, 1.4)$$

Applying the sufficiency conditions yields

$$H^B = \begin{pmatrix} 0 & 4 & 2x_2 & 2 \\ 4 & 2 & 0 & 0 \\ 2x_2 & 0 & 2 - 2\lambda & 0 \\ 2 & 0 & 0 & 2 \end{pmatrix}$$

Since $m = 1$ and $n = 3$, for a stationary point to be a minimum, the sign of the last $(3 - 1) = 2$ principal minor determinants must be that of $(-1)^m = -1$. Thus for $(X_0, \lambda_0)_1 = (2, 2, 1, 1)$,

$$\begin{vmatrix} 0 & 4 & 4 \\ 4 & 2 & 0 \\ 4 & 0 & 0 \end{vmatrix} = -32 < 0 \quad \text{and} \quad \begin{vmatrix} 0 & 4 & 4 & 2 \\ 4 & 2 & 0 & 0 \\ 4 & 0 & 0 & 0 \\ 2 & 0 & 0 & 2 \end{vmatrix} = -64 < 0$$

For $(X_0, \lambda_0)_2 = (2, -2, 1, 1)$,

$$\begin{vmatrix} 0 & 4 & -4 \\ 4 & 2 & 0 \\ -4 & 0 & 0 \end{vmatrix} = -32 < 0 \quad \text{and} \quad \begin{vmatrix} 0 & 4 & -4 & 2 \\ 4 & 2 & 0 & 0 \\ -4 & 0 & 0 & 0 \\ 2 & 0 & 0 & 2 \end{vmatrix} = -64 < 0$$

Finally, for $(\mathbf{X}_0, \boldsymbol{\lambda}_0)_3 = (2.8, 0, 1.4, 1.4)$,

$$\begin{vmatrix} 0 & 4 & 0 \\ 4 & 2 & 0 \\ 0 & 0 & -.8 \end{vmatrix} = 12.8 > 0 \quad \text{and} \quad \begin{vmatrix} 0 & 4 & 0 & 2 \\ 4 & 2 & 0 & 0 \\ 0 & 0 & -.8 & 0 \\ 2 & 0 & 0 & 2 \end{vmatrix} = 32 > 0$$

This shows that $(\mathbf{X}_0)_1$ and $(\mathbf{X}_0)_2$ are minimum points. The fact that $(\mathbf{X}_0)_3$ does not satisfy the sufficiency conditions of either a maximum or a minimum does not necessarily mean that it is not an extreme point. This, as explained earlier, follows since the given conditions, although sufficient, may not be satisfied for every extreme point. In such a case it is necessary to use the other sufficiency condition.

To illustrate the use of the other sufficiency condition that employs the roots of polynomial, consider

$$\Delta = \begin{pmatrix} 0 & 4 & 2x_2 & 2 \\ 4 & 2-\mu & 0 & 0 \\ 2x_2 & 0 & 2-2\lambda-\mu & 0 \\ 2 & 0 & 0 & 2-\mu \end{pmatrix}$$

Now, for $(\mathbf{X}_0, \boldsymbol{\lambda}_0)_1 = (2, 2, 1, 1)$,

$$|\Delta| = 9\mu^2 - 26\mu + 16 = 0$$

This gives $\mu = 2$ or $8/9$. Since all $\mu > 0$, $(\mathbf{X}_0)_1 = (2, 2, 1)$ is a minimum point. Again, for $(\mathbf{X}_0, \boldsymbol{\lambda}_0)_2 = (2, -2, 1, 1)$,

$$|\Delta| = 9\mu^2 - 26\mu + 16 = 0$$

which is the same as in the previous case. Hence $(\mathbf{X}_0)_2 = (2, -2, 1)$ is a minimum point. Finally, for $(\mathbf{X}_0, \boldsymbol{\lambda}_0)_3 = (2.8, 0, 1.4, 1.4)$,

$$|\Delta| = 5\mu^2 - 6\mu - 8 = 0$$

This gives $\mu = 2$ and $-.8$, which means that $(\mathbf{X}_0)_3 = (2.8, 0, 1.4)$ is not an extreme point. ◀

18.2.2 INEQUALITY CONSTRAINTS

This section shows how the Lagrangean method may, in a restricted manner, be extended to handle inequality constraints. The main contribution of the section is the development of the Kuhn–Tucker conditions, which provide the basic theory for nonlinear programming.

A. Extension of the Lagrangean Method

Suppose that the problem is given by

$$\text{maximize } z = f(\mathbf{X})$$

subject to

$$g_i(\mathbf{X}) \leq 0, \qquad i = 1, 2, \ldots, m$$

Nonnegativity constraints $\mathbf{X} \geq 0$, if any, are assumed included in the m constraints.

The general idea of extending the Lagrangean procedure is that if the *un* constrained optimum of $f(\mathbf{X})$ does not satisfy all constraints, the constrained optimum must occur at a boundary point of the solution space. This means that one, or more, of the m constraints must be satisfied in equation form. The procedure thus involves the following steps.

Step 1: Solve the unconstrained problem

$$\text{maximize } z = f(\mathbf{X})$$

If the resulting optimum satisfies all the constraints, there is nothing more to be done since all constraints are redundant. Otherwise, set $k = 1$ and go to step 2.

Step 2: Activate any k constraints (i.e., convert them into equalities) and optimize $f(\mathbf{X})$ subject to the k active constraints by the Lagrangean method. If the resulting solution is feasible with respect to the remaining constraints, stop; it is a *local* optimum.† Otherwise, activate another set of k constraints and repeat the step. If *all* sets of active constraints taken k at a time are considered without encountering a feasible solution, go to step 3.

Step 3: If $k = m$, stop; no feasible solution exists. Otherwise, set $k = k + 1$ and go to step 2.

An important point often neglected in presenting the procedure described above is that, as should be expected, it does *not* guarantee global optimality even when the problem is well behaved (possess a *unique* optimum). Another important point is the implicit misconception that, for $p < q$, the optimum of $f(\mathbf{X})$ subject to p equality constraints is always better than its optimum subject to q equality constraints. Unfortunately, this is true, in general, only if the q constraints form a subset of the p constraints. The following example is designed to illustrate these points.

Example 18.2–6

$$\text{Maximize } z = -(2x_1 - 5)^2 - (2x_2 - 1)^2$$

---

† A *local* optimum is defined from among all the optima resulting from optimizing $f(\mathbf{X})$ subject to *all* combinations of k *equality* constraints, $k = 1, 2, \ldots, m$.

subject to

$$x_1 + 2x_2 \leq 2$$
$$x_1, x_2 \geq 0$$

The graphical representation in Figure 18–6 should assist in understanding the analytic procedure. Observe that the problem is well behaved (concave objective function subject to a convex solution space) with the result that a reasonably well-defined algorithm would guarantee global optimality. Yet, as will be shown, the extended Lagrangean method produces a local maximum only.

The unconstrained optimum is obtained by solving

$$\frac{\partial z}{\partial x_1} = -4(2x_1 - 5) = 0$$

$$\frac{\partial z}{\partial x_2} = -4(2x_2 - 1) = 0$$

This gives $(x_1, x_2) = (5/2, 1/2)$. Since this solution violates $x_1 + 2x_2 \leq 2$, the constraints are activated one at a time. Consider $x_1 = 0$. The Lagrangean function is

$$L(x_1, x_2, \lambda) = -(2x_1 - 5)^2 - (2x_2 - 1)^2 - \lambda x_1$$

Thus

$$\frac{\partial L}{\partial x_1} = -4(2x_1 - 5) - \lambda = 0$$

$$\frac{\partial L}{\partial x_2} = -4(2x_2 - 1) \qquad = 0$$

$$\frac{\partial L}{\partial \lambda} = -x_1 \qquad\qquad = 0$$

This gives the solution point $(x_1, x_2) = (0, 1/2)$, which can be shown by the sufficiency condition to be a maximum. Since this point satisfies all other constraints, the procedure terminates with $(x_1, x_2) = (0, 1/2)$ as a local optimal solution to the problem. (Notice that the remaining constraints $x_2 \geq 0$ and

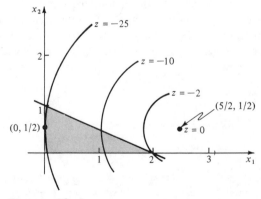

Figure 18–6

$x_1 = 2x_2 \leq 2$, activated one at a time, yield infeasible solutions.) The objective value is $z = -25$.

However, observe in Figure 18–6 that the feasible solution $(x_1, x_2) = (2, 0)$, which is the point of intersection of the *two* constraints $x_2 = 0$ and $x_1 + 2x_2 = 2$, yields the objective value $z = -2$. This value is better than the one obtained with one active constraint.   ◀

The procedure just described illustrates that the best to be hoped for in using the extended Lagrangean method is a (possibly) good feasible solution to the problem. This is particularly so if the objective function is not unimodal. Of course, if the functions of the problem are well behaved (e.g., the problem possesses a unique constrained optimum as in Example 18.2–6), the procedure can be rectified to locate the global optimum. Specifically, consider the unconstrained optimum and the constrained optima subject to *all* sets of one active constraint at a time, then two active constraints at a time, and so on, until all m constraints are activated. The best of *all* such *feasible* optima would then be the global optimum.

If this procedure is followed in Example 18.2–6, it will be necessary to solve seven problems before global optimality is verified. This indicates the limited use of the method in solving problems of any practical size.

B. The Kuhn–Tucker Conditions

This section develops the Kuhn–Tucker *necessary* conditions for identifying stationary points of a nonlinear constrained problem subject to inequality constraints. The development is based on the Lagrangean method. These conditions are also sufficient under certain limitations, which will be stated later.

Consider the problem

$$\text{maximize } z = f(\mathbf{X})$$

subject to

$$\mathbf{g}(\mathbf{X}) \leq 0$$

The inequality constraints may be converted into equations by adding the appropriate *nonnegative* slack variables. Thus, to satisfy the nonnegativity conditions, let $S_i^2 (\geq 0)$ be the slack quantity added to the ith constraint $g_i(\mathbf{X}) \leq 0$. Define

$$\mathbf{S} = (S_1, S_2, \ldots, S_m)^T \quad \text{and} \quad \mathbf{S}^2 = (S_1^2, S_2^2, \ldots, S_m^2)^T$$

where m is the total number of inequality constraints. The Lagrangean function is thus given by

$$L(\mathbf{X}, \mathbf{S}, \lambda) = f(\mathbf{X}) - \lambda[\mathbf{g}(\mathbf{X}) + \mathbf{S}^2]$$

Given the constraints

$$\mathbf{g}(\mathbf{X}) \leq 0$$

a necessary condition for optimality is that λ be nonnegative (nonpositive) for maximization (minimization) problems. This is justified as follows. Consider the maximization case. Since λ measures the rate of variation of f with respect to $\mathbf{g}$, that is,

$$\lambda = \frac{\partial f}{\partial \mathbf{g}}$$

as the right-hand side of the constraint $\mathbf{g} \leq \mathbf{0}$ increases above zero, the solution space becomes less constrained and hence f cannot decrease. This means that $\lambda \geq 0$. Similarly, for minimization as a resource increases, f cannot increase, which implies that $\lambda \leq \mathbf{0}$. If the constraints are equalities, that is, $\mathbf{g}(\mathbf{X}) = \mathbf{0}$, then λ becomes unrestricted in sign (see Problem 18–17).

The restrictions on λ given must hold as part of the Kuhn–Tucker necessary conditions. The remaining conditions will now be derived.

Taking the partial derivatives of L with respect to $\mathbf{X}$, $\mathbf{S}$, and λ, we obtain

$$\frac{\partial L}{\partial \mathbf{X}} = \nabla f(\mathbf{X}) - \lambda \nabla \mathbf{g}(\mathbf{X}) = \mathbf{0}$$

$$\frac{\partial L}{\partial S_i} = -2\lambda_i S_i = 0, \qquad i = 1, 2, \ldots, m$$

$$\frac{\partial L}{\partial \lambda} = -(\mathbf{g}(\mathbf{X}) + \mathbf{S}^2) = \mathbf{0}$$

The second set of equations reveals the following results.

1. If λ_i is greater than zero, $S_i^2 = 0$. This means that the corresponding resource is scarce, and consequently it is exhausted completely (equality constraint).

2. If $S_i^2 > 0$, $\lambda_i = 0$. This means the ith resource is not scarce and, consequently, it does not affect the value of f ($\lambda_i = \partial f/\partial g_i = 0$).

From the second and third sets of equations it follows that

$$\lambda_i g_i(\mathbf{X}) = 0, \qquad i = 1, 2, \ldots, m$$

This new condition essentially repeats the foregoing argument, since if $\lambda_i > 0$, $g_i(\mathbf{X}) = 0$ or $S_i^2 = 0$. Similarly, if $g_i(\mathbf{X}) < 0$, that is, $S_i^2 > 0$, then $\lambda_i = 0$.

The Kuhn–Tucker conditions necessary for $\mathbf{X}$ and λ to be a stationary point of the maximization problem can now be summarized as follows:

$$\lambda \geq \mathbf{0},$$
$$\nabla f(\mathbf{X}) - \lambda \nabla \mathbf{g}(\mathbf{X}) = \mathbf{0}$$
$$\lambda_i g_i(\mathbf{X}) = 0, \qquad i = 1, 2, \ldots, m$$
$$\mathbf{g}(\mathbf{X}) \leq \mathbf{0}$$

You can verify that these conditions apply to the minimization case as well, with the exception that λ must be nonpositive, as shown previously. In both maximization and minimization, the Lagrange multipliers corresponding to *equality* constraints must be unrestricted in sign.

Exercise 18.2-3

Consider the problem just defined:

$$\text{maximize } f(\mathbf{X})$$

subject to

$$g(\mathbf{X}) \leq 0$$

Suppose that the Lagrangean function is formulated as

$$L(\mathbf{X}, \lambda, \mathbf{S}) = f(\mathbf{X}) + \lambda[g(\mathbf{X}) + \mathbf{S}^2]$$

How would this change affect the Kuhn–Tucker conditions?

[*Ans.* λ will be nonpositive instead of nonnegative because λ is defined equal to $-\partial f/\partial g$ instead of $+\partial f/\partial g$.]

Sufficiency of the Kuhn–Tucker Conditions

The Kuhn–Tucker necessary conditions are also sufficient if the objective function and the solution space satisfy certain conditions regarding convexity and concavity. These conditions are summarized in Table 18–1.

Table 18-1

Sense of Optimization	Required Conditions	
	Objective Function	Solution Space
Maximization	Concave	Convex set
Minimization	Convex	Convex set

It is simpler to verify that a function is convex or concave than it is to prove that a solution space is a convex set. For this reason, we provide a list of conditions that are easier to apply in practice in the sense that the convexity of the solution space can be established by checking directly the convexity or concavity of the constraint functions. To provide these conditions, we define the generalized nonlinear problems as

$$\begin{pmatrix} \text{maximize} \\ \text{or} \\ \text{minimize} \end{pmatrix} z = f(\mathbf{X})$$

subject to

$$g_i(\mathbf{X}) \leq 0, \qquad i = 1, 2, \ldots, r$$
$$g_i(\mathbf{X}) \geq 0, \qquad i = r+1, \ldots, p$$
$$g_i(\mathbf{X}) = 0, \qquad i = p+1, \ldots, m$$

$$L(\mathbf{X}, \mathbf{S}, \lambda) = f(\mathbf{X}) - \sum_{i=1}^{r} \lambda_i [g_i(\mathbf{X}) + \mathbf{S}_i^2] - \sum_{i=r+1}^{p} \lambda_i [g_i(\mathbf{X}) - \mathbf{S}_i^2] - \sum_{i=p+1}^{m} \lambda_i g_i(\mathbf{X})$$

where λ_i is the Lagrangean multiplier associated with constraint i. The conditions for establishing the sufficiency of the Kuhn–Tucker conditions can thus be summarized as shown in Table 18–2.

Table 18-2

Sense of Optimization	Conditions Required			
	$f(X)$	$g_i(X)$	λ_i	
Maximization	Concave	Convex	≥ 0	$(1 \leq i \leq r)$
		Concave	≤ 0	$(r+1 \leq i \leq p)$
		Linear	Unrestricted	$(p+1 \leq i \leq m)$
Minimization	Convex	Convex	≤ 0	$(1 \leq i \leq r)$
		Concave	≥ 0	$(r+1 \leq i \leq p)$
		Linear	Unrestricted	$(p+1 \leq i \leq m)$

We must remark that the conditions in Table 18–2 represent only a subset of the conditions in Table 18–1. The reason is that a solution space may be convex without satisfying the conditions stipulated in Table 18–2 on the functions $g_i(X)$.

The validity of Table 18–2 rests on the fact that the given conditions yield a concave Lagrangean function $L(X, S, \lambda)$ in case of maximization and a convex $L(X, S, \lambda)$ in case of minimization. This result can be verified directly by noticing that if $g_i(x)$ is convex, then $\lambda_i g_i(x)$ is convex if $\lambda_i \geq 0$ and concave if $\lambda_i \leq 0$. Similar interpretations can be established for all the remaining conditions. We must indicate, however, that a linear function, by definition, is both convex and concave. Notice also that if a function f is concave, then $-f$ is convex, and vice versa.

Exercise 18.2-4

Develop the conditions similar to those in Tables 18–1 and 18–2 assuming the Lagrangean function is expressed as

$$L(X, \lambda, S) = f(X) + \sum_{i=1}^{r} \lambda_i [g_i(X) + S_i^2] + \sum_{i=r+1}^{p} \lambda_i [g_i(X) - S_i^2] + \sum_{i=p+1}^{m} \lambda_i g_i(X)$$

[*Ans.* Table 18–1 remains unchanged. In Table 18–2 the signs of *restricted* λ_i must be reversed while the rest of the conditions remain unchanged.]

Example 18.2-7.

Consider the following *minimization* problem:

$$\text{minimize } f(X) = x_1^2 + x_2^2 + x_3^2$$

subject to

$$g_1(X) = 2x_1 + x_2 - 5 \leq 0$$
$$g_2(X) = x_1 + x_3 - 2 \leq 0$$
$$g_3(X) = 1 - x_1 \leq 0$$
$$g_4(X) = 2 - x_2 \leq 0$$
$$g_5(X) = -x_3 \leq 0$$

Since this is a minimization problem, it follows that $\lambda \leq 0$. The Kuhn–Tucker conditions are thus given as follows.

$$(\lambda_1, \lambda_2, \lambda_3, \lambda_4, \lambda_5) \leq 0$$

$$(2x_1, 2x_2, 2x_3) - (\lambda_1, \lambda_2, \lambda_3, \lambda_4, \lambda_5)\begin{pmatrix} 2 & 1 & 0 \\ 1 & 0 & 1 \\ -1 & 0 & 0 \\ 0 & -1 & 0 \\ 0 & 0 & -1 \end{pmatrix} = 0$$

$$\lambda_1 g_1 = \lambda_2 g_2 = \cdots = \lambda_5 g_5 = 0$$

$$g(\mathbf{X}) \leq 0$$

These conditions simplify to the following:

$$\lambda_1, \lambda_2, \lambda_3, \lambda_4, \lambda_5 \leq 0$$
$$2x_1 - 2\lambda_1 - \lambda_2 + \lambda_3 = 0$$
$$2x_2 - \lambda_1 + \lambda_4 = 0$$
$$2x_3 - \lambda_2 + \lambda_5 = 0$$
$$\lambda_1(2x_1 + x_2 - 5) = 0$$
$$\lambda_2(x_1 + x_3 - 2) = 0$$
$$\lambda_3(1 - x_1) = 0$$
$$\lambda_4(2 - x_2) = 0$$
$$\lambda_5 x_3 = 0$$
$$2x_1 + x_2 \leq 5$$
$$x_1 + x_3 \leq 2$$
$$x_1 \geq 1, \quad x_2 \geq 2, \quad x_3 \geq 0$$

The solution is $x_1 = 1$, $x_2 = 2$, $x_3 = 0$; $\lambda_1 = \lambda_2 = \lambda_5 = 0$, $\lambda_3 = -2$, $\lambda_4 = -4$. Since the function $f(\mathbf{X})$ is convex and the solution space $g(\mathbf{X}) \leq 0$ is also convex, $L(\mathbf{X}, \mathbf{S}, \boldsymbol{\lambda})$ must be convex and the resulting stationary point yields a global constrained minimum. The example given shows, however, that it is difficult in general to solve the resulting conditions explicitly. Consequently, the procedure is not suitable for numerical computations. The importance of the Kuhn–Tucker conditions will be clear in developing the nonlinear programming algorithms in Chapter 19.   ◀

18.3   SUMMARY

In this chapter we provided the classical theory for locating the points of maxima and minima of constrained nonlinear problems. We observe that the theory presented is generally not suitable for computational purposes. Few exceptions exist, however, where the Kuhn–Tucker theory is the basis for the development of efficient computational algorithms. *Quadratic programming*, which we present in the next chapter, is an excellent example of the use of the Kuhn–Tucker necessary conditions.

We must emphasize that no sufficiency conditions (similar to those of unconstrained problems and problems with *equality* constraints) can be established for nonlinear programs with inequality constraints. Thus, unless the conditions given in Table 18–1 or 18–2 can be established *in advance,* there is no way of verifying whether the convergence of a nonlinear programming algorithm leads to a local or a global optimum.

SELECTED REFERENCES

BAZARAA, M., and C. SHETTY, *Nonlinear Programming Theory and Algorithms,* Wiley, New York, 1979.

BEIGHTLER, C., D. PHILLIPS, and D. WILDE, *Foundations of Optimization,* 2nd ed., Prentice-Hall, Englewood Cliffs, N.J., 1979.

COURANT, R., and D. HILBERT, *Methods of Mathematical Physics,* Vol. I, Interscience, New York, 1953.

HADLEY, G., *Matrix Algebra,* Addison-Wesley, Reading, Mass., 1961.

REVIEW QUESTIONS

True (T) or False (F)?

1. ____ If the gradient vector of a function at a given point is null, the point can only be a maximum or a minimum.

2. ____ In a single-variable function, the second derivative at a maximum (minimum) point must be negative (positive).

3. ____ If a single-variable function has two local minima, it must have at least one local maximum.

4. ____ If the Hessian matrix at a stationary point is positive-definite (negative-definite), the point is a minimum (maximum).

5. ____ If the Hessian matrix at a stationary point is indefinite, the point is an inflection (or saddle) point.

6. ____ If the Hessian matrix at a stationary point is semidefinite, the identity of the point cannot be determined without further testing.

7. ____ Convergence of the Newton–Raphson method is assured regardless of the choice of the starting point.

8. ____ The Jacobian and Lagrangean methods are designed for nonlinear programs with equality constraints only.

9. ____ The Jacobian and Lagrangean methods automatically account for the nonnegativity restrictions on the variables.

10. ____ The Jacobian matrix in the Jacobian method can be singular.

11. ____ The basic idea of the Jacobian method is to determine the gradient of the objective function at every feasible point satisfying the equality constraints.

12. ____ The Lagrangean multiplier associated with a constraint is defined as the rate of change in the objective function with respect to a small increase in the right side of the constraint.

13. ____ The Kuhn–Tucker conditions are also sufficient as long as the solution space of the problem is convex and the objective function is concave in the case of maximization and convex in the case of minimization.

14. ____ The solution space of a nonlinear problem may be convex even though the functions of the constraints may be neither convex nor concave.

15. ____ A linear function is either convex or concave.

16. ____ A convex (concave) function multiplied by a positive constant is also convex (concave).

17. ____ The sum of two convex (concave) functions is also a convex (concave) function.

18. ____ A concave function minus a convex function is a concave function.

19. ____ A convex function minus a concave function is a convex function.

20. ____ If $f(x)$ is convex, then $1/f(x)$ is concave.

[*Ans.* **1**—F, **2** to **6**—T, **7**—F, **8**—T, **9**—F, **10**—F, **11** to **20**—T.]

PROBLEMS

Section	Assigned Problems
18.1.1	18–1 to 18–5
18.1.2	18–6
18.2.1A	18–7 to 18–12
18.2.1B	18–13 to 18–16
18.2.2A	None
18.2.2B	18–17 to 18–20

☐ **18–1** Examine the following functions for extreme points.
(a) $f(x) = x^3 + x$.
(b) $f(x) = x^4 + x^2$.
(c) $f(x) = 4x^4 - x^2 + 5$.
(d) $f(x) = (3x - 2)^2(2x - 3)^2$.
(e) $f(x) = 6x^5 - 4x^3 + 10$.

☐ **18–2** Examine the following functions for extreme points.
(a) $f(\mathbf{X}) = x_1^3 + x_2^3 - 3x_1x_2$.
(b) $f(\mathbf{X}) = 2x_1^2 + x_2^2 + x_3^2 + 6(x_1 + x_2 + x_3) + 2x_1x_2x_3$.

☐ **18–3** Verify that the function

$$f(x_1, x_2, x_3) = 2x_1x_2x_3 - 4x_1x_3 - 2x_2x_3 + x_1^2 + x_2^2 + x_3^2 - 2x_1 - 4x_2 + 4x_3$$

has the stationary points (0, 3, 1), (0, 1, −1), (1, 2, 0), (2, 1, 1), and (2, 3, −1). Use the sufficiency condition to find the extreme points.

☐ **18–4** Solve the following simultaneous equations by converting the system to a nonlinear objective function with no constraints.

$$x_2 - x_1^2 = 0$$
$$x_2 - x_1 = 2$$

[*Hint:* min $f^2(x_1, x_2)$ occurs at $f(x_1, x_2) = 0$.]

☐ **18–5** Apply the Newton–Raphson method to Problem 18–1(c) and Problem 18–2(b).

☐ **18–6** Prove Theorem 18.1–3.

☐ **18–7** Apply the Jacobian method to Example 18.2–1 by selecting $Y = (x_2, x_3)$ and $Z = (x_1)$.

☐ **18–8** Solve by the Jacobian method:

$$\text{minimize } f(X) = \sum_{i=1}^{n} x_i^2$$

subject to

$$\prod_{i=1}^{n} x_i = C$$

where C is a positive constant. Suppose that the right-hand side of the constraint is changed to $C + \delta$, where δ is a small positive quantity. Find the corresponding change in the optimal value of f.

☐ **18–9** Solve by the Jacobian method

$$\text{minimize } f(X) = 5x_1^2 + x_2^2 + 2x_1 x_2$$

subject to

$$g(X) = x_1 x_2 - 10 = 0$$

(a) Find the change in the optimal value of $f(X)$ if the constraint is replaced by $x_1 x_2 - 9.99 = 0$.

(b) Find the change in value of $f(X)$ in the neighborhood of the feasible point (2, 5) given that $x_1 x_2 = 9.99$ and $\partial x_1 = .01$.

☐ **18–10** Consider the problem:

$$\text{maximize } f(X) = x_1^2 + 2x_2^2 + 10x_3^2 + 5x_1 x_2$$

subject to

$$g_1(X) = x_1 + x_2^2 + 3x_2 x_3 - 5 = 0$$
$$g_2(X) = x_1^2 + 5x_1 x_2 + x_3^2 - 7 = 0$$

Apply the Jacobian method to find $\partial f(\mathbf{X})$ in the feasible neighborhood of the feasible point $(1, 1, 1)$. Assume that this feasible neighborhood is specified by $\partial g_1 = -.01$, $\partial g_2 = .02$, and $\partial x_1 = .01$.

☐ **18–11** Consider the problem

$$\text{minimize } f(\mathbf{X}) = x_1^2 + x_2^2 + x_3^2 + x_4^2$$

subject to

$$g_1(\mathbf{X}) = x_1 + 2x_2 + 3x_3 + 5x_4 - 10 = 0$$
$$g_2(\mathbf{X}) = x_1 + 2x_2 + 5x_3 + 6x_4 - 15 = 0$$

Show that by selecting x_3 and x_4 as the independent variables, the Jacobian method fails to give the solution. Then solve the problem using x_1 and x_3 as the independent variables and apply the sufficiency condition to examine the resulting stationary point. Find the sensitivity coefficients of the problem.

☐ **18–12** Consider the linear programming problem.

$$\text{maximize } f(\mathbf{X}) = \sum_{j=1}^{n} c_j x_j$$

subject to

$$g_i(\mathbf{X}) = \sum_{j=1}^{n} a_{ij} x_j - b_i = 0, \qquad i = 1, 2, \ldots, m$$

$$x_j \geq 0, \qquad j = 1, 2, \ldots, n$$

Neglecting the nonnegativity constraint, show that the constrained derivatives $\nabla_c f(\mathbf{X})$ for this problem yield the same expression for $\{z_j - c_j\}$ defined by the optimality condition of the linear programming problem (Section 7.2.2). That is,

$$\{z_j - c_j\} = \{\mathbf{C}_B \mathbf{B}^{-1} \mathbf{P}_j - c_j\}, \qquad \text{for all } j$$

Can the constrained-derivative method be applied directly to the linear programming problem? Why or why not?

☐ **18–13** Solve the following linear programming problem by both the Jacobian and the Lagrangean methods:

$$\text{maximize } f(\mathbf{X}) = 5x_1 + 3x_2$$

subject to

$$g_1(\mathbf{X}) = x_1 + 2x_2 + x_3 - 6 = 0$$
$$g_2(\mathbf{X}) = 3x_1 + x_2 + x_4 - 9 = 0$$
$$x_1, x_2, x_3, x_4 \geq 0$$

☐ **18–14** Find the optimal solution to the problem

$$\text{minimize } f(\mathbf{X}) = x_1^2 + 2x_2^2 + 10x_3^2$$

subject to

$$g_1(X) = x_1 + x_2^2 + x_3 - 5 = 0$$
$$g_2(X) = x_1 + 5x_2 + x_3 - 7 = 0$$

Suppose that $g_1(X) = .01$ and $g_2(X) = .02$. Find the corresponding change in the optimal value of $f(X)$.

□ **18–15** Solve Problem 18–11 by the Lagrangean method and verify that the value of the Lagrange multipliers are the same as the sensitivity coefficients obtained in Problem 18–11.

□ **18–16** Show that the Kuhn–Tucker conditions for the problem:

$$\text{maximize } f(X)$$

subject to

$$g(X) \geq 0$$

are the same as in Section 18.2.2B except that the Lagrange multipliers λ are nonpositive.

□ **18–17** Show that the Kuhn–Tucker conditions for the problem:

$$\text{maximize } f(X)$$

subject to

$$g(X) = 0$$

are

$$\lambda \text{ unrestricted in sign}$$
$$\nabla f(X) - \lambda \nabla g(X) = 0$$
$$g(X) = 0$$

□ **18–18** Write the Kuhn–Tucker necessary conditions for the following problems.

(a)
$$\text{Maximize } f(X) = x_1^3 - x_2^2 + x_1 x_3^2$$

subject to

$$x_1 + x_2^2 + x_3 = 5$$
$$5x_1^2 - x_2^2 - x_3 \geq 0$$
$$x_1, x_2, x_3 \geq 0$$

(b)
$$\text{Minimize } f(X) = x_1^4 + x_2^2 + 5x_1 x_2 x_3$$

subject to

$$x_1^2 - x_2^2 + x_3^3 \leq 10$$
$$x_1^3 + x_2^2 + 4x_3^2 \geq 20$$

☐ **18–19** Consider the problem:

$$\text{maximize } f(\mathbf{X})$$

subject to

$$\mathbf{g}(\mathbf{X}) = \mathbf{0}$$

Given $f(\mathbf{X})$ is concave and $g_i(\mathbf{X})$ ($i = 1, 2, \ldots, m$) is a *linear* function, show that the Kuhn–Tucker necessary conditions are also sufficient. Is this result true if $g_i(\mathbf{X})$ is a convex *non*linear function for all i? Why?

☐ **18–20** Consider the problem

$$\text{maximize } f(\mathbf{X})$$

subject to

$$g_1(\mathbf{X}) \geq 0, \qquad g_2(\mathbf{X}) = 0, \qquad g_3(\mathbf{X}) \leq 0$$

Develop the Kuhn–Tucker conditions for the problem; then give the stipulations under which the conditions are sufficient.

Chapter 19

Nonlinear Programming Algorithms

Chapter 18 has presented the theory for optimizing unconstrained and constrained nonlinear functions. However, the techniques developed are not suitable for computational purposes. This chapter develops working algorithms for both the unconstrained and constrained problems. Because of space limitations, the material presented here is meant to include only a selected sample of nonlinear programming algorithms.

19.1 UNCONSTRAINED NONLINEAR ALGORITHMS

This section presents two algorithms for the unconstrained problem: the *direct search* algorithm and the *gradient* algorithm. As evident from the names, the first algorithm locates the optimum by direct search over a specified region, and the second utilizes the gradient of the function to find the optimum.

780

19.1.1 DIRECT SEARCH METHOD

Direct search methods have been developed primarily for single-variable functions. Although this may appear trivial from the practical standpoint, it is shown in Section 19.1.2 that optimization of single-variable functions may evolve as part of the algorithms for multivariable functions.

The general idea of direct search methods is rather simple. First, an interval (called **interval of uncertainty**) that is known to include the optimum is identified. The size of the interval is then systematically reduced in a manner guaranteeing that the optimum is not missed. The procedure does not determine the exact optimum but rather minimizes the length of the interval that includes the optimum point. Theoretically, the length of the interval including the optimum can be made as small as desired.

One of the limitations of search methods is that the optimized function is assumed unimodal over the search interval. This guarantees only one local optimum. In addition, no finite intervals exist in which the slope of the function is zero. With this additional assumption, the optimized function may be referred to as *strictly unimodal*.

This section presents a method called **dichotomous search**. Suppose that the initial interval in which a local optimum occurs is defined by $a \leq x \leq b$. Suppose for convenience that the function $f(x)$ is maximized. Define the two points x_1 and x_2 symmetrically with respect to a and b such that the intervals $a \leq x \leq x_2$ and $x_1 \leq x \leq b$ overlap by a finite amount Δ (see Figure 19–1). Now evaluate $f(x_1)$ and $f(x_2)$. Three cases will result:

1. If $f(x_1) > f(x_2)$, x^* (optimum x) must lie between a and x_2.
2. If $f(x_1) < f(x_2)$, $x_1 < x^* < b$.
3. If $f(x_1) = f(x_2)$, $x_1 < x^* < x_2$.

These results follow directly from strict unimodality of $f(x)$. In each of these cases, the interval(s) not including x^* is discarded in future iterations.

The result of the search is that the maximum of $f(x)$ is now confined to a smaller interval. The new interval may thus be dichotomized into two (overlapping) intervals in the same manner followed for the interval $a \leq x \leq b$. Continuing in this manner, one can narrow (in the limit) the interval in which the local maximum lies to the length Δ. This means that Δ should be chosen reasonably small.

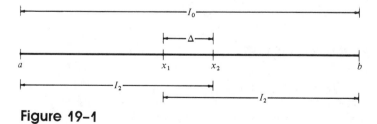

Figure 19–1

Example 19.1-1

$$\text{Maximize } f(x) = \begin{cases} 3x, & 0 \le x \le 2 \\ -\dfrac{x}{3} + \dfrac{20}{3}, & 2 \le x \le 3 \end{cases}$$

Obviously, max $f(x)$ occurs at $x = 2$. The dichotomous search method is now used to solve the problem. Let x_L and x_R define the (left and right) boundaries of the *current* interval. Initially, $x_L = 0$ and $x_R = 3$. Define x_1 and x_2 as the points dichotomizing an interval so that the associated overlapping intervals are $x_L \le x \le x_2$ and $x_1 \le x \le x_R$, where $x_1 - x_L = x_R - x_2$ and $\Delta = x_2 - x_1$. This means that

$$x_1 = x_L + \frac{x_R - x_L - \Delta}{2}$$

$$x_2 = x_L + \frac{x_R - x_L + \Delta}{2}$$

Table 19–1 summarizes the computations given $\Delta = .001$.

Table 19-1
Computations for Dichotomous Search Method[a]

x_L	x_R	x_1	x_2	$f(x_1)$	$f(x_2)$
0	3	1.4995^L	1.5005	4.4985	4.5015[b]
1.4995	3	2.24925	2.25025^R	5.91692[b]	5.91658
1.4995	2.25025	1.87437^L	1.87537	5.62312	5.62612[b]
1.87437	2.25025	2.06181	2.06281^R	5.97939[b]	5.97906
1.87437	2.06281	1.96809^L	1.96909	5.90427	5.90727[b]
1.96809	2.06281	2.01495	2.01595^R	5.99502[b]	5.99447
1.96809	2.01595	1.99152^L	1.99252	5.97456	5.97756[b]
1.99152	2.01595	2.00323	2.00423^R	5.99892[b]	5.99859
1.99152	2.00423	1.99737^L	1.99837	5.99213	5.99511[b]
1.99737	2.00423				

[a] $L(R)$ indicates that $x_L(x_R)$ is set equal to $x_1(x_2)$ in the next step.
[b] max $\{f(x_1), f(x_2)\}$.

The last step in Table 19–1 gives $x_L = 1.99737$ and $x_R = 2.00423$. This means that max $f(x)$ occurs at x^*, satisfying $1.99737 \le x^* \le 2.00423$. If the midpoint is used, this will give $x = 2.0008$, which is very close to the exact optimum $x^* = 2.0$. ◀

19.1.2 GRADIENT METHOD

This section develops a method for optimizing functions that are twice continuously differentiable. The general idea is to generate successive points, starting from a given initial point, in the direction of the fastest increase (maximization) of the function. The technique is known as the *gradient method* because the gradient of the function at a point is indicative of the fastest rate of increase.

A gradient method, the Newton–Raphson method, was presented in Section 18.1.2. The method is based on solving the simultaneous equations representing the necessary condition for optimality, namely, $\nabla f(\mathbf{X}) = \mathbf{0}$. This section presents another technique, called the **steepest ascent** method.

Termination of the gradient method is effected at the point where the gradient vector becomes null. This is only a necessary condition for optimality. It is thus emphasized that optimality cannot be verified unless it is known a priori that $f(\mathbf{X})$ is concave or convex.

Suppose that $f(\mathbf{X})$ is maximized. Let $\mathbf{X}^0$ be the initial point from which the procedure starts and define $\nabla f(\mathbf{X}^k)$ as the gradient of f at the kth point $\mathbf{X}^k$. The idea of the method is to determine a particular path p along which df/dp is maximized at a given point. This result is achieved if successive points $\mathbf{X}^k$ and $\mathbf{X}^{k+1}$ are selected such that

$$\mathbf{X}^{k+1} = \mathbf{X}^k + r^k \nabla f(\mathbf{X}^k)$$

where r^k is a parameter called the optimal **step size.**

The parameter r^k is determined such that $\mathbf{X}^{k+1}$ results in the largest improvement in f. In other words, if a function $h(r)$ is defined such that

$$h(r) = f[\mathbf{X}^k + r \nabla f(\mathbf{X}^k)]$$

r^k is the value of r maximizing $h(r)$. Since $h(r)$ is a single-variable function, the search method in Section 19.1.1 may be used to find the optimum provided that $h(r)$ is strictly unimodal.

The proposed procedure terminates when two successive trial points $\mathbf{X}^k$ and $\mathbf{X}^{k+1}$ are approximately equal. This is equivalent to having

$$r^k \nabla f(\mathbf{X}^k) \cong \mathbf{0}$$

Under the assumption that $r^k \neq 0$, which will always be true unless $\mathbf{X}_0$ happens to be the optimum of $f(\mathbf{X})$, this is equivalent to the necessary condition $\nabla f(\mathbf{X}^k) = \mathbf{0}$.

Example 19.1–2. Consider maximizing the function

$$f(x_1, x_2) = 4x_1 + 6x_2 - 2x_1^2 - 2x_1 x_2 - 2x_2^2$$

$f(x_1, x_2)$ is a quadratic function whose absolute optimum occurs at $(x_1^*, x_2^*) = (1/3, 4/3)$. It is shown how the problem is solved by the steepest ascent method. Figure 19–2 shows the successive points. The gradients at any two successive points are necessarily orthogonal (perpendicular).

Let the initial point be given by $\mathbf{X}^0 = (1, 1)$. Now

$$\nabla f(\mathbf{X}) = (4 - 4x_1 - 2x_2, 6 - 2x_1 - 4x_2)$$

First Iteration

$$\nabla f(\mathbf{X}^0) = (-2, 0)$$

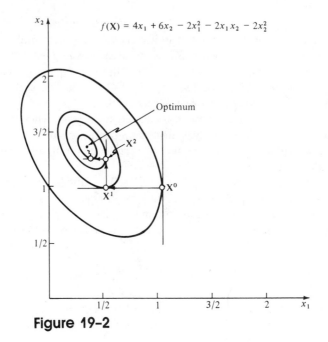

Figure 19-2

The next point X^1 is obtained by considering

$$X = (1, 1) + r(-2, 0) = (1 - 2r, 1)$$

Thus

$$h(r) = f(1 - 2r, 1) = -2(1 - 2r)^2 + 2(1 - 2r) + 4$$

The optimal step size yielding the maximum value of $h(r)$ is $r^1 = 1/4$. This gives $X^1 = (1/2, 1)$.

Second Iteration

$$\nabla f(X^1) = (0, 1)$$

Consider

$$X = (1/2, 1) + r(0, 1) = (1/2, 1 + r)$$

Thus

$$h(r) = -2(1 + r)^2 + 5(1 + r) + 3/2$$

This gives $r^2 = 1/4$ or $X^2 = (1/2, 5/4)$.

Third Iteration

$$\nabla f(X^2) = (-1/2, 0)$$

Consider

$$\mathbf{X} = \left(\frac{1}{2}, \frac{5}{4}\right) + r\left(-\frac{1}{2}, 0\right) = \left(\frac{1-r}{2}, \frac{5}{4}\right)$$

Thus

$$h(r) = -(1/2)(1-r)^2 + (3/4)(1-r) + 35/8$$

This gives $r^3 = 1/4$ or $\mathbf{X}^3 = (3/8, 5/4)$.

Fourth Iteration

$$\nabla f(\mathbf{X}^3) = (0, 1/4)$$

Consider

$$\mathbf{X} = \left(\frac{3}{8}, \frac{5}{4}\right) + r\left(0, \frac{1}{4}\right) = \left(\frac{3}{8}, \frac{5+r}{4}\right)$$

Thus

$$h(r) = -(1/8)(5+r)^2 + (21/16)(5+r) + 39/32$$

This gives $r^4 = 1/4$, or $\mathbf{X}^4 = (3/8, 21/16)$.

Fifth Iteration

$$\nabla f(\mathbf{X}^4) = (-1/8, 0)$$

Consider

$$\mathbf{X} = \left(\frac{3}{8}, \frac{21}{16}\right) + r\left(-\frac{1}{8}, 0\right) = \left(\frac{3-r}{8}, \frac{21}{16}\right)$$

Thus

$$h(r) = -\left(\frac{1}{32}\right)(3-r)^2 + \left(\frac{11}{64}\right)(3-r) + \frac{567}{128}$$

This gives $r^5 = 1/4$, or $\mathbf{X}^5 = (11/32, 21/16)$.

Sixth Iteration

$$\nabla f(\mathbf{X}^5) = (0, 1/16)$$

Since $\nabla f(\mathbf{X}^5) \cong \mathbf{0}$, the process can be terminated at this point. The *approximate* maximum point is given by $\mathbf{X}^5 = (.3437, 1.3125)$. Note that the exact optimum is $\mathbf{X}^* = (.3333, 1.3333)$.   ◀

19.2 CONSTRAINED NONLINEAR ALGORITHMS

The general constrained nonlinear programming problem may be defined as

$$\text{maximize (or minimize) } z = f(\mathbf{X})$$

subject to

$$\mathbf{g}(\mathbf{X}) \leq \mathbf{0}$$

The nonnegativity conditions $\mathbf{X} \geq \mathbf{0}$ are assumed to be part of the given constraints. Also, at least one of the functions $f(\mathbf{X})$ and $\mathbf{g}(\mathbf{X})$ is nonlinear. For the purpose of this presentation, these functions are continuously differentiable.

No general algorithm exists for handling nonlinear models, mainly because of the irregular behavior of the nonlinear functions. Perhaps the most general result applicable to the problem is the Kuhn–Tucker conditions. Section 18.2.2B shows that unless $f(\mathbf{X})$ and $\mathbf{g}(\mathbf{X})$ are well-behaved functions (convexity and concavity conditions), the Kuhn–Tucker theory yields only necessary conditions for optimum. This sets a big limitation on the application of the Kuhn–Tucker conditions to the general problem.

This section presents a number of algorithms, which may be classified generally as *indirect* and *direct* methods. Indirect methods basically solve the nonlinear problem by dealing with one or more *linear* problems that are extracted from the original program. Direct methods attack the nonlinear problem itself by determining successive search points. The idea is to convert constrained problems into unconstrained ones for which the gradient methods of Section 19.1.2 are applied with some modifications.

The indirect methods presented in this section include separable, quadratic, geometric, and stochastic programming. The direct methods include the method of linear combinations and a brief discussion of the sequential unconstrained maximization technique. Other important nonlinear techniques can be found in the Selected References at the end of the chapter.

19.2.1 SEPARABLE PROGRAMMING

A function $f(x_1, x_2, \ldots, x_n)$ is **separable** if it can be expressed as the sum of n single-variable functions $f_1(x_1), f_2(x_2), \ldots, f_n(x_n)$, that is,

$$f(x_1, x_2, \ldots, x_n) = f_1(x_1) + f_2(x_2) + \cdots + f_n(x_n)$$

For example, the linear function

$$h(x_1, x_2, \ldots, x_n) = a_1 x_1 + a_2 x_2 + \cdots + a_n x_n$$

(where the a's are constants) is separable. On the other hand, the function

$$h(x_1, x_2, x_3) = x_1^2 + x_1 \sin(x_2 + x_3) + x_2 e^{x_3}$$

is not separable.

Some nonlinear functions are not directly separable but can be made so by appropriate substitutions. Consider, for example, the case of maximizing $z = x_1 x_2$. Letting $y = x_1 x_2$, then $\ln y = \ln x_1 + \ln x_2$ and the problem becomes

$$\text{maximize } z = y$$

subject to

$$\ln y = \ln x_1 + \ln x_2$$

which is separable. The substitution assumes that x_1 and x_2 are *positive* variables; otherwise, the logarithmic function is undefined.

The case where x_1 and x_2 assume zero values (i.e., $x_1, x_2 \geq 0$) may be handled as follows. Let δ_1 and δ_2 be positive constants and define

$$w_1 = x_1 + \delta_1$$
$$w_2 = x_2 + \delta_2$$

This means that w_1 and w_2 are strictly positive. Now

$$x_1 x_2 = w_1 w_2 - \delta_2 w_1 - \delta_1 w_2 + \delta_1 \delta_2$$

Let $y = w_1 w_2$; then the problem is equivalent to

$$\text{maximize } z = y - \delta_2 w_1 - \delta_1 w_2 + \delta_1 \delta_2$$

subject to

$$\ln y = \ln w_1 + \ln w_2$$

which is separable.

Other functions that can be made readily separable (using substitution) are exemplified by $e^{x_1 + x_2}$ and $x_1^{x_2}$. A variant of the procedure just presented can be applied to such cases to effect separability.

Separable programming deals with nonlinear problems in which the objective function and the constraints are separable. This section shows how an approximate solution can be obtained for any separable problem by linear approximation and the simplex method of linear programming.

The single-variable function $f(x)$ can be approximated by a piecewise linear function using mixed integer programming (Chapter 8). Suppose that $f(x)$ is to be approximated over the interval $[a, b]$. Define a_k, $k = 1, 2, \ldots, K$, as the kth breaking point on the x-axis such that $a_1 < a_2 < \cdots < a_k$. The points a_1 and a_K coincide with the end points a and b of the interval under study. Thus $f(x)$ is approximated as follows:

$$f(x) \cong \sum_{k=1}^{K} f(a_k) t_k$$

$$x = \sum_{k=1}^{K} a_k t_k$$

where t_k is a nonnegative weight associated with the kth breaking point such that

$$\sum_{k=1}^{K} t_k = 1$$

Mixed integer programming ensures the validity of the approximation. Specifically, the approximation is valid if

1. At most two t_k are positive.
2. If t_{k^*} is a positive, then only an adjacent t_k (t_{k^*+1} or t_{k^*-1}) is allowed to be positive.

To show how these conditions are satisfied, consider the separable problem

$$\text{maximize (or minimize)} \ z = \sum_{i=1}^{n} f_i(x_i)$$

subject to

$$\sum_{i=1}^{n} g_i^j(x_i) \leq b_j, \qquad j = 1, 2, \ldots, m$$

This problem can be approximated as a mixed integer program as follows. Let the number of breaking points for the ith variable x_i be equal to K_i and let a_i^k be its kth breaking value. Let t_i^k be the weight associated with the kth breaking point of the ith variable. Then the equivalent mixed problem is

$$\text{maximize (or minimize)} \ z = \sum_{i=1}^{n} \sum_{k=1}^{K_i} f_i(a_i^k) t_i^k$$

subject to

$$\sum_{i=1}^{n} \sum_{k=1}^{K_i} g_i^j(a_i^k) t_i^k \leq b_j, \qquad j = 1, 2, \ldots, m$$

$$0 \leq t_i^1 \leq y_i^1$$

$$0 \leq t_i^k \leq y_i^{k-1} + y_i^k, \qquad k = 2, 3, \ldots, K_i - 1$$

$$0 \leq t_i^{K_i} \leq y_i^{K_i-1}$$

$$\sum_{k=1}^{K_i-1} y_i^k = 1$$

$$\sum_{k=1}^{K_i} t_i^k = 1$$

$$y_i^k = 0 \text{ or } 1, \qquad k = 1, 2, \ldots, K_i; \quad i = 1, 2, \ldots, n$$

The variables for the approximating problem are t_i^k and y_i^k.

This formulation shows how any separable problem can be solved, at least in principle, by mixed integer programming. The difficulty, however, is that the number of constraints increases rather rapidly with the number of breaking points. In particular, the computational feasibility of the procedure is highly questionable, since there are no efficient computer programs for handling large, mixed integer programming problems.

Another method for solving the approximate model is the regular simplex method (Chapter 3) under the condition of **restricted basis.** In this case the additional constraints involving y_i^k are disregarded. The restricted basis specifies that *no more* than two *positive* t_i^k can appear in the basis. Moreover, two t_i^k can be positive only if they are adjacent. Thus the strict optimality condition of the simplex method is used to select the entering variable t_i^k *only if* it satisfies the foregoing conditions. Otherwise, the variable t_i^k having the next best optimality indicator $(z_i^k - c_i^k)$ is considered for entering the solution. The process is repeated until the optimality condition is satisfied or until it is impossible to introduce new t_i^k without violating the restricted basis condition, whichever occurs first. At this point, the last tableau gives the approximate optimal solution to the problem.

Whereas the mixed integer programming method yields global optimum to the approximate problem, the restricted basis method can only guarantee a local optimum. Also, in the two methods, the approximate solution may not be feasible for the original problem. In fact, the approximate model may give rise to additional extreme points that do not exist in the original problem. This depends mainly on the degree of refinement of the linear approximation used. These inherent risks must be taken into consideration when using separable programming.

Example 19.2-1. Consider the problem

$$\text{maximize } z = x_1 + x_2^4$$

subject to

$$3x_1 + 2x_2^2 \leq 9$$
$$x_1, x_2 \geq 0$$

This example illustrates the application of the restricted basis method.

The exact optimum solution to this problem, obtained by inspection, is $x_1^* = 0$, $x_2^* = \sqrt{9/2} = 2.12$, and $z^* = 20.25$. To show how the approximating method is used, consider the separable functions

$$f_1(x_1) = x_1$$
$$f_2(x_2) = x_2^4$$
$$g_1^1(x_1) = 3x_1$$
$$g_1^2(x_2) = 2x_2^2$$

The functions $f_1(x_1)$ and $g_1^1(x_1)$ are left in their present form, since they are already linear. In this case, x_1 is treated as one of the variables. Considering $f_2(x_2)$ and $g_1^2(x_2)$, we assume that there are four breaking points ($K_2 = 4$). Since the value of x_2 cannot exceed 3, it follows that

k	a_2^k	$f_2(a_2^k)$	$g_1^2(a_2^k)$
1	0	0	0
2	1	1	2
3	2	16	8
4	3	81	18

This yields

$$f_2(x_2) \cong t_2^1 f_2(a_2^1) + t_2^2 f_2(a_2^2) + t_2^3 f_2(a_2^3) + t_2^4 f_2(a_2^4)$$
$$\cong 0(t_2^1) + 1(t_2^2) + 16(t_2^3) + 81(t_2^4) = t_2^2 + 16t_2^3 + 81t_2^4$$

Similarly,

$$g_1^2(x_2) \cong 2t_2^2 + 8t_2^3 + 18t_2^4$$

The approximation problem thus becomes

$$\text{maximize } z = x_1 + t_2^2 + 16t_2^3 + 81t_2^4$$

subject to

$$3x_1 + 2t_2^2 + 8t_2^3 + 18t_2^4 \le 9$$
$$t_2^1 + t_2^2 + t_2^3 + t_2^4 = 1$$
$$t_2^k \ge 0, \qquad k = 1, 2, 3, 4$$
$$x_1 \ge 0$$

together with the restricted basis condition.

The initial simplex tableau (with rearranged columns to give a starting solution) is given by

Basic	x_1	t_2^2	t_2^3	t_2^4	S_1	t_2^1	Solution
z	-1	-1	-16	-81	0	0	0
S_1	3	2	8	18	1	0	9
t_2^1	0	1	1	1	0	1	1

where S_1 (≥ 0) is a slack variable. (This problem happened to have an obvious starting solution. In general, one may have to use the artificial variables techniques, Section 3.2.3).

From the z-row coefficients, t_2^4 is the entering variable. Since t_2^1 is basic, it must be dropped first before t_2^4 can enter the solution (restricted basis condition). By the feasibility condition, S_1 must be the leaving variable. This means that t_2^4 cannot enter the solution. Next consider t_2^3 (next best entering variable). Again t_2^1 must be dropped first. From the feasibility condition, t_2^1 is the leaving variable as desired. The new tableau thus becomes

Basic	x_1	t_2^2	t_2^3	t_2^4	S_1	t_2^1	Solution
z	-1	15	0	-65	0	16	16
S_1	3	-6	0	10	1	-8	1
t_2^3	0	1	1	1	0	1	1

Clearly, t_2^4 is the entering variable. Since t_2^3 is in the basis, t_2^4 is an admissible entering variable. The simplex method shows that S_1 will be dropped. Thus

Basic	x_1	t_2^2	t_2^3	t_2^4	S_1	t_2^1	Solution
z	37/2	-24	0	0	13/2	-36	$22\frac{1}{2}$
t_2^4	3/10	$-6/10$	0	1	1/10	$-8/10$	1/10
t_2^3	$-3/10$	16/10	1	0	$-1/10$	18/10	9/10

The tableau shows that t_2^1 and t_2^2 are candidates for the entering variable. Since t_2^1 is not an adjacent point to the basic t_2^3 and t_2^4, it cannot be admitted. Also, t_2^2 cannot be admitted, since t_2^4 cannot be dropped. The process ends at this point and the solution given is the best feasible solution for the approximate problem.

To find the solution in terms of x_1 and x_2, we consider

$$t_2^3 = 9/10 \qquad \text{and} \qquad t_2^4 = 1/10$$

Thus

$$x_2 \cong 2t_2^3 + 3t_2^4 = 2(9/10) + 3(1/10) = 2.1$$

and $x_1 = 0$ and $z = 22.5$. The approximate optimum value of $x_2 (= 2.1)$ is very close to the true optimum value $(= 2.12)$. The value of z differs by about 10% error, however. The approximation may be improved in this case by using finer breaking points.   ◄

Separable Convex Programing

A special case of separable programming occurs when the functions $g_i^j(x_i)$ are convex so that the solution space of the problem is a convex set. In addition, the function $f_i(x_i)$ is convex in case of minimization and concave in case of maximization (see Table 18–2). Under such conditions, the following simplified approximation can be used.

Consider a minimization problem and let $f_i(x_i)$ be as shown in Figure 19–3. The k breaking point of the function $f_i(x_i)$ is determined by $x_i = a_{ki}$, $k = 0, 1, \ldots, K_i$. Let x_{ki} define the increment of the variable x_i in the range $(a_{k-1,i}, a_{ki})$, $k = 1, 2, \ldots, K_i$, and let ρ_{ki} be the corresponding slope of the line segment in the same range. Then

$$f_i(x_i) \cong \sum_{k=1}^{K_i} (\rho_{ki} x_{ki}) + f_i(a_{0i})$$

$$x_i = \sum_{k=1}^{K_i} x_{ki}$$

provided that

$$0 \le x_{ki} \le a_{ki} - a_{k-1,i} \qquad k = 1, 2, \ldots, K_i$$

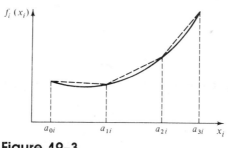

Figure 19–3

The fact that $f_i(x_i)$ is convex ensures that $\rho_{1i} < \rho_{2i} < \cdots < \rho_{K_i i}$. This means that in the minimization problem, for $p < q$, the variable x_{pi} is more attractive than x_{qi}. Consequently, x_{pi} will always enter the solution before x_{qi}. The only limitation here is that every x_{ki} must be restricted by the upper bound ($a_{ki} - a_{k-1,i}$).

The convex constraint functions $g_i^j(x_i)$ are approximated in essentially the same way. Let ρ_{ki}^j be the slope of the kth line segment corresponding to $g_i^j(x_i)$. It follows that the ith function is approximated as

$$g_i^j(x_i) \cong \sum_{k=1}^{K_i} \rho_{ki}^j x_{ki} + g_i^j(a_{0i})$$

The complete problem is thus given by

$$\text{minimize } z = \sum_{i=1}^{n} \left(\sum_{k=1}^{K_i} \rho_{ki} x_{ki} + f_i(a_{0i}) \right)$$

subject to

$$\sum_{i=1}^{n} \left(\sum_{k=1}^{K_i} \rho_{ki}^j x_{ki} + g_i^j(a_{0i}) \right) \leq b_j, \quad j = 1, 2, \ldots, m$$

$$0 \leq x_{ki} \leq a_{ki} - a_{k-1,i}, \quad k = 1, 2, \ldots, K_i, \quad i = 1, 2, \ldots, n$$

where

$$\rho_{ki} = \frac{f_i(a_{ki}) - f_i(a_{k-1,i})}{a_{ki} - a_{k-1,i}}$$

$$\rho_{ki}^j = \frac{g_i^j(a_{ki}) - g_i^j(a_{k-1,i})}{a_{ki} - a_{k-1,i}}$$

The maximization problem is treated essentially the same way. In this case $\rho_{1i} > \rho_{2i} > \cdots > \rho_{K_i i}$, which shows that, for $p < q$, the variable x_{pi} will always enter the solution before x_{qi} (see Problem 19–12 for the proof).

The new problem can be solved by the simplex method with upper bounded variables (Section 7.4). The restricted basis concept is not necessary here because the convexity (concavity) of the functions guarantees the proper selection of variables.

Example 19.2-2. Consider the problem

$$\text{minimize } z = x_1^2 + x_2^2 + 5$$

subject to

$$3x_1^4 + x_2 \leq 243$$
$$x_1 + 2x_2^2 \leq 32$$
$$x_1, x_2 \geq 0$$

The separable functions of this problem are

$$f_1(x_1) = x_1^2, \quad f_2(x_2) = x_2^2 + 5$$
$$g_1^1(x_1) = 3x_1^4, \quad g_2^1(x_2) = x_2$$
$$g_1^2(x_1) = x_1, \quad g_2^2(x_2) = 2x_2^2$$

These functions satisfy the convexity condition required for the minimization problems.

The range of the variables x_1 and x_2, calculated from the constraints, are given by $0 \le x_1 \le 3$ and $0 \le x_2 \le 4$. Thus x_1 and x_2 are partitioned in these ranges. Let $K_1 = 3$ and $K_2 = 4$ with $a_{01} = a_{02} = 0$. The slopes corresponding to the separable functions are as follows.

For $i = 1$,

k	a_{k1}	ρ_{k1}	ρ_{k1}^1	ρ_{k1}^2	x_{k1}
0	0	—	—	—	—
1	1	1	3	1	x_{11}
2	2	3	45	1	x_{21}
3	3	5	195	1	x_{31}

For $i = 2$,

k	a_{k2}	ρ_{k2}	ρ_{k2}^1	ρ_{k2}^2	x_{k2}
0	0	—	—	—	—
1	1	1	1	2	x_{12}
2	2	3	1	6	x_{22}
3	3	5	1	10	x_{32}
4	4	7	1	14	x_{42}

The complete problem then becomes

$$\text{minimize } z \cong x_{11} + 3x_{21} + 5x_{31} + x_{12} + 3x_{22} + 5x_{32} + 7x_{42} + 5$$

subject to

$$3x_{11} + 45x_{21} + 195x_{31} + x_{12} + x_{22} + x_{32} + x_{42} \le 243$$
$$x_{11} + x_{21} + x_{31} + 2x_{12} + 6x_{22} + 10x_{32} + 14x_{42} \le 32$$
$$0 \le x_{k1} \le 1, \qquad k = 1, 2, 3$$
$$0 \le x_{k2} \le 1, \qquad k = 1, 2, 3, 4$$

After solving this problem using upper bounding technique, let x_{k1}^* and x_{k2}^* be the corresponding optimal values. Optimal values of x_1 and x_2 are then given by

$$x_1^* = \sum_{k=1}^{3} x_{k1}^* \quad \text{and} \quad x_2^* = \sum_{k=1}^{4} x_{k2}^* \qquad \blacktriangleleft$$

19.2.2 QUADRATIC PROGRAMMING

A quadratic programming model is defined as follows:

$$\text{maximize (or minimize) } z = \mathbf{CX} + \mathbf{X}^T \mathbf{DX}$$

subject to

$$\mathbf{AX} \le \mathbf{P}_0, \qquad \mathbf{X} \ge \mathbf{0}$$

where

$$\mathbf{X} = (x_1, x_2, \ldots, x_n)^T$$
$$\mathbf{C} = (c_1, c_2, \ldots, c_n)$$
$$\mathbf{P}_0 = (b_1, b_2, \ldots, b_m)^T$$
$$\mathbf{A} = \begin{pmatrix} a_{11} & \cdots & a_{1n} \\ \vdots & & \vdots \\ a_{m1} & \cdots & a_{mn} \end{pmatrix}$$
$$\mathbf{D} = \begin{pmatrix} d_{11} & \cdots & d_{1n} \\ \vdots & & \vdots \\ d_{n1} & \cdots & d_{nn} \end{pmatrix}$$

The function $\mathbf{X}^T\mathbf{DX}$ defines a quadratic form (Section A.3) where $\mathbf{D}$ is symmetric. The matrix $\mathbf{D}$ is assumed negative definite if the problem is maximization, and positive definite if the problem is minimization. This means that z is strictly convex in $\mathbf{X}$ for minimization and strictly concave for maximization. The constraints are assumed linear in this case, which guarantees a convex solution space.

The solution to this problem is secured by direct application of the Kuhn–Tucker necessary conditions (Section 18.2.2B). Since z is strictly convex (or concave) and the solution space is a convex set, these conditions (as proved in Section 18.2.2B) are also sufficient for a global optimum.

The guadratic programming problem will be treated for the maximization case. It is trivial to change the formulation to minimization. The problem may be written as

$$\text{maximize } z = \mathbf{CX} + \mathbf{X}^T\mathbf{DX}$$

subject to

$$\mathbf{G(X)} = \begin{pmatrix} \mathbf{A} \\ -\mathbf{I} \end{pmatrix} \mathbf{X} - \begin{pmatrix} \mathbf{P}_0 \\ \mathbf{0} \end{pmatrix} \le \mathbf{0}$$

Let

$$\boldsymbol{\lambda} = (\lambda_1, \lambda_2, \ldots, \lambda_m)^T \qquad \text{and} \qquad \mathbf{U} = (\mu_1, \mu_2, \ldots, \mu_n)^T$$

be the Lagrange multipliers corresponding to the two sets of constraints $\mathbf{AX} - \mathbf{P}_0 \le \mathbf{0}$ and $-\mathbf{X} \le \mathbf{0}$, respectively. Application of the Kuhn–Tucker conditions immediately yields

$$\boldsymbol{\lambda} \ge \mathbf{0}, \qquad \mathbf{U} \ge \mathbf{0}$$
$$\nabla z - (\boldsymbol{\lambda}^T, \mathbf{U}^T)\nabla \mathbf{G(X)} = \mathbf{0}$$
$$\lambda_i \left(b_i - \sum_{j=1}^{n} a_{ij}x_j \right) = 0, \qquad i = 1, 2, \ldots, m$$
$$\mu_j x_j = 0, \qquad j = 1, 2, \ldots, n$$
$$\mathbf{AX} \le \mathbf{P}_0, \qquad -\mathbf{X} \le \mathbf{0}$$

Now

$$\nabla z = C + 2X^T D$$

$$\nabla G(X) = \begin{pmatrix} A \\ -I \end{pmatrix}$$

Let $S = P_0 - AX \geq 0$ be the slack variables of the constraints. The conditions reduce to

$$-2X^T D + \lambda^T A - U^T = C$$

$$AX + S = P_0$$

$$\mu_j x_j = 0 = \lambda_i S_i \qquad \text{for all } i \text{ and } j$$

$$\lambda, U, X, S \geq 0$$

Since $D^T = D$, the transpose of the first set of equations yields

$$-2DX + A^T\lambda - U = C^T$$

Hence the necessary conditions may be combined as

$$\left(\begin{array}{c|c|c|c} -2D & A^T & -I & 0 \\ \hline A & 0 & 0 & I \end{array}\right)\begin{pmatrix} X \\ \lambda \\ U \\ S \end{pmatrix} = \begin{pmatrix} C^T \\ P_0 \end{pmatrix}$$

$$\mu_j x_j = 0 = \lambda_i S_i \qquad \text{for all } i \text{ and } j$$

$$\lambda, U, X, S \geq 0$$

Except for the conditions $\mu_j x_j = 0 = \lambda_i S_i$, the remaining equations are linear functions in X, λ, U, and S. The problem is thus equivalent to solving a set of linear equations, while satisfying the additional conditions $\mu_j x_j = 0 = \lambda_i S_i$. Because z is strictly concave and the solution space is convex, the *feasible* solution satisfying all these conditions must give the optimum solution directly. From the conditions imposed on z and the solution space (i.e., z is strictly concave and the solution space is convex), the solution (when it exists) to the set of equations above must be unique.

The solution of the system is obtained by using phase I of the two-phase method (Section 3.2.3). The only restriction here is that the condition $\lambda_i S_i = 0 = \mu_j x_j$ should always be maintained. This means that if λ_i is in the basic solution at a *positive level*, S_i cannot become basic at positive level. Similarly, μ_j and x_j cannot be positive simultaneously. This is actually the same idea of the *restricted basis* used in Section 19.2.1. Phase I will end in the usual manner with the sum of the artificial variables equal to zero only if the problem has a feasible space. The feasibility of the solution space can be easily checked, however, by checking whether the system $AX \leq P_0$, $X \geq 0$ encloses a feasible space.

Example 19.2-3. Consider the problem

$$\text{maximize } z = 4x_1 + 6x_2 - 2x_1^2 - 2x_1x_2 - 2x_2^2$$

subject to

$$x_1 + 2x_2 \leq 2$$
$$x_1, x_2 \geq 0$$

This problem can be put in matrix form as follows:

$$\text{maximize } z = (4, 6) \begin{pmatrix} x_1 \\ x_2 \end{pmatrix} + (x_1, x_2) \begin{pmatrix} -2 & -1 \\ -1 & -2 \end{pmatrix} \begin{pmatrix} x_1 \\ x_2 \end{pmatrix}$$

subject to

$$(1, 2) \begin{pmatrix} x_1 \\ x_2 \end{pmatrix} \leq 2$$
$$x_1, x_2 \geq 0$$

This automatically defines all the information required to construct the following Kuhn–Tucker conditions.

$$\left(\begin{array}{cc|c|cc|c} 4 & 2 & 1 & -1 & 0 & 0 \\ 2 & 4 & 2 & 0 & -1 & 0 \\ \hline 1 & 2 & 0 & 0 & 0 & 1 \end{array} \right) \begin{pmatrix} x_1 \\ x_2 \\ \lambda_1 \\ \mu_1 \\ \mu_2 \\ S_1 \end{pmatrix} = \begin{pmatrix} 4 \\ 6 \\ 2 \end{pmatrix}$$

The initial tableau for phase 1 is obtained by introducing the artificial variables R_1 and R_2. Thus

Basic	x_1	x_2	λ_1	μ_1	μ_2	R_1	R_2	S_1	Solution
r_0	6	6	3	−1	−1	0	0	0	10
R_1	4	2	1	−1	0	1	0	0	4
R_2	2	4	2	0	−1	0	1	0	6
S_1	1	2	0	0	0	0	0	1	2

First Iteration

Since $\mu_1 = 0$, the most promising entering variable x_1 (minimization problem) can be made basic with R_1 as the leaving variable. This yields the following tableau:

Basic	x_1	x_2	λ_1	μ_1	μ_2	R_1	R_2	S_1	Solution
r_0	0	3	3/2	1/2	−1	−3/2	0	0	4
x_1	1	1/2	1/4	−1/4	0	1/4	0	0	1
R_2	0	3	3/2	1/2	−1	−1/2	1	0	4
S_1	0	3/2	−1/4	1/4	0	−1/4	0	1	1

Second Iteration

The most promising variable x_2 can be made basic since $\mu_2 = 0$. This gives

Basic	x_1	x_2	λ_1	μ_1	μ_2	R_1	R_2	S_1	Solution
r_0	0	0	2	0	−1	−1	0	−2	2
x_1	1	0	1/3	−1/3	0	1/3	0	−1/3	2/3
R_1	0	0	2	0	−1	0	1	−2	2
x_2	0	1	−1/6	1/6	0	−1/6	0	2/3	2/3

Third Iteration
Since $S_1 = 0$, λ_1 can be introduced into the solution. This yields

Basic	x_1	x_2	λ_1	μ_1	μ_2	R_1	R_2	S_1	Solution
r_0	0	0	0	0	0	−1	−1	0	0
x_1	1	0	0	−1/3	1/6	1/3	−1/6	0	1/3
λ_1	0	0	1	0	−1/2	0	1/2	−1	1
x_2	0	1	0	1/6	−1/12	−1/6	1/12	1/2	5/6

The last tableau gives the optimal solution for phase I. Since $r_0 = 0$, the solution given is feasible. Thus $x_1^* = 1/3$, $x_2^* = 5/6$. The optimal value of z can be computed from the original problem and is equal to 4.16. ◄

19.2.3 GEOMETRIC PROGRAMMING

A rather interesting technique for solving a special case of nonlinear problems is *geometric* programming. This technique, developed by R. Duffin and C. Zener in 1964, finds the solution by considering an associated dual problem (to be defined later). The advantage here is that it is usually much simpler computationally to work with the dual.

Geometric programming deals with problems in which the objective and the constraint functions are of the following type:

$$z = f(\mathbf{X}) = \sum_{j=1}^{N} U_j$$

where

$$U_j = c_j \prod_{i=1}^{n} x_i^{a_{ij}}, \quad j = 1, 2, \ldots, N$$

It is assumed that all $c_j > 0$, and that N is finite. The exponents a_{ij} are unrestricted in sign. The function $f(\mathbf{X})$ takes the form of a polynomial except that the exponents a_{ij} may be negative. For this reason, and because all $c_j > 0$, Duffin and Zener give $f(\mathbf{X})$ the name **posynomial.**

This section will present the unconstrained case of geometric programming. The objective is to familiarize the reader with this type of analysis. The treatment of the constrained problem is beyond the scope of this chapter. The interested reader may refer to the excellent presentation by Beightler and Wilde [1979, Chap. 6] for a more detailed treatment of the subject.

Consider the *minimization* of the function $f(\mathbf{X})$ as defined in the posynomial form given. This problem will be referred to as the *primal*. The variables x_i are assumed to be *strictly positive* so that the region $x_i \leq 0$ represents an infeasible space. It will be shown later that the requirement $x_i \neq 0$ plays an essential part in the derivation of the results.

The first partial derivative of z must vanish at a minimum point. Thus,

$$\frac{\partial z}{\partial x_k} = \sum_{j=1}^{N} \frac{\partial U_j}{\partial x_k} = \sum_{i=1}^{N} c_j a_{kj} (x_k)^{a_{kj}-1} \prod_{i \neq k} (x_i)^{a_{ij}} = 0, \qquad k = 1, 2, \ldots, n$$

Since each $x_k > 0$ by assumption,

$$\frac{\partial z}{\partial x_k} = 0 = \frac{1}{x_k} \sum_{j=1}^{N} a_{kj} U_j, \qquad k = 1, 2, \ldots, n$$

Let z^* be the minimum value of z. It follows that $z^* > 0$, since z is posynomial and each $x_k^* > 0$. Define

$$y_j = \frac{U_j^*}{z^*}$$

which shows that $y_j > 0$ and $\Sigma_{j=1}^{N} y_j = 1$. The value of y_j thus represents the relative contribution of the jth term U_j to the optimal value of the objective function z^*.

The *necessary* conditions can now be written as

$$\sum_{j=1}^{N} a_{kj} y_j = 0, \qquad k = 1, 2, \ldots, n$$

$$\sum_{j=1}^{N} y_j = 1, \qquad y_j > 0 \quad \text{for all } j$$

These are known as the **orthogonality** and **normality conditions** and will yield a unique solution for y_j if $(n + 1) = N$ and all the equations are independent. The problem becomes more complex when $N > (n + 1)$ because the values of y_j are no longer unique. It is shown later, however, that it is possible to determine y_j uniquely for the purpose of minimizing z.

Suppose now that y_j^* are the unique values determined from the equations above. These values are used to determine z^* and x_i^*, $i = 1, 2, \ldots, n$, as follows. Consider

$$z^* = (z^*)^{\sum_{j=1}^{N} y_j^*}$$

Since $z^* = U_j^*/y_j^*$, it follows that

$$z^* = \left(\frac{U_1^*}{y_1^*}\right)^{y_1^*} \left(\frac{U_2^*}{y_2^*}\right)^{y_2^*} \cdots \left(\frac{U_N^*}{y_N^*}\right)^{y_N^*} = \left\{ \prod_{j=1}^{N} \left(\frac{c_j}{y_j^*}\right)^{y_j^*} \right\} \left\{ \prod_{j=1}^{N} \left(\prod_{i=1}^{N} x_i^{*a_{ij}} \right)^{y_j^*} \right\}$$

$$= \left\{ \prod_{j=1}^{N} \left(\frac{c_j}{y_j^*}\right)^{y_j^*} \right\} \left\{ \prod_{i=1}^{N} (x_i^*)^{\sum_{j=1}^{N} a_{ij} y_j^*} \right\} = \prod_{j=1}^{N} \left(\frac{c_j}{y_j^*}\right)^{y_j^*}$$

This step is justified, since $\Sigma_{j=1}^N a_{ij} y_j = 0$. The value of z^* is thus determined as soon as all y_j^* are determined. Now, given y_j^* and z^*, $U_j^* = y_j^* z^*$ can be determined. Since

$$U_j^* = c_j \prod_{i=1}^n (x_i^*)^{a_{ij}}, \qquad j = 1, 2, \ldots, n$$

simultaneous solution of these equations should yield x_i^*.

The procedure described shows that the solution to the original posynomial z can be transformed into the solution of a set of linear equations in y_j. Observe that all y_j^* are determined from the necessary conditions for a minimum. It can be shown, however, that these conditions are also sufficient. The proof (under the given restrictions on z) is given in Beightler and Wilde [1979, p. 333].

The variables y_j actually define the dual variables associated with the z-primal problem. To see this relationship, consider the primal problem in the form

$$z = \sum_{j=1}^N y_j \left(\frac{U_j}{y_j} \right)$$

Now define the function

$$w = \prod_{j=1}^N \left(\frac{U_j}{y_j} \right)^{y_j}$$

Since $\Sigma_{j=1}^N y_j = 1$ and $y_j > 0$, by Cauchy's inequality,† we have

$$w \le z$$

The function w with its variables $y_1, y_2, \ldots, y_N$ defines the dual problem to the primal. Since w represents the lower bound on z and since z is associated with the minimization problem, it follows, by maximizing w, that

$$w^* = \max_{y_j} w = \min_{x_i} z = z^*$$

This means that the maximum value of w ($= w^*$) over the values of y_j is equal to the minimum value of z ($= z^*$) over the values of x_i.

Example 19.2–4. In this example a problem is considered in which $N = n + 1$ so that the solution to the orthogonality and normality conditions is unique. The next example will illustrate the other case, where $N > (n + 1)$. Consider the problem

$$\text{minimize } z = 7x_1 x_2^{-1} + 3x_2 x_3^{-2} + 5x_1^{-3} x_2 x_3 + x_1 x_2 x_3$$

---

† The Cauchy's inequality states that for $z_j > 0$,

$$\sum_{j=1}^N w_j z_j \ge \prod_{j=1}^N (z_j)^{w_j}, \qquad \text{where } w_j > 0 \quad \text{and} \quad \sum_{j=1}^N w_j = 1$$

This is also called the arithmetic–geometric mean inequality.

This function may be written as

$$z = 7x_1^1 x_2^{-1} x_3^0 + 3x_1^0 x_2^1 x_3^{-2} + 5x_1^{-3} x_2^1 x_3^1 + x_1^1 x_2^1 x_3^1$$

so that

$$(c_1, c_2, c_3, c_4) = (7, 3, 5, 1)$$

$$\begin{pmatrix} a_{11} & a_{12} & a_{13} & a_{14} \\ a_{21} & a_{22} & a_{23} & a_{24} \\ a_{31} & a_{32} & a_{33} & a_{34} \end{pmatrix} = \begin{pmatrix} 1 & 0 & -3 & 1 \\ -1 & 1 & 1 & 1 \\ 0 & -2 & 1 & 1 \end{pmatrix}$$

The orthogonality and normality conditions are thus given by

$$\begin{pmatrix} 1 & 0 & -3 & 1 \\ -1 & 1 & 1 & 1 \\ 0 & -2 & 1 & 1 \\ 1 & 1 & 1 & 1 \end{pmatrix} \begin{pmatrix} y_1 \\ y_2 \\ y_3 \\ y_4 \end{pmatrix} = \begin{pmatrix} 0 \\ 0 \\ 0 \\ 1 \end{pmatrix}$$

This yields the unique solution

$$y_1^* = 12/24, \quad y_2^* = 4/24, \quad y_3^* = 5/24, \quad y_4^* = 3/24$$

Thus

$$z^* = \left(\frac{7}{12/24}\right)^{12/24} \left(\frac{3}{4/24}\right)^{4/24} \left(\frac{5}{5/24}\right)^{5/24} \left(\frac{1}{3/24}\right)^{3/24} = 15.22$$

From the equation $U_j^* = y_j^* z^*$ it follows that

$$7x_1 x_2^{-1} = U_1 = (1/2)(15.22) = 7.61$$
$$3x_2 x_3^{-2} = U_2 = (1/6)(15.22) = 2.54$$
$$5x_1^{-3} x_2 x_3 = U_3 = (5/24)(15.22) = 3.17$$
$$x_1 x_2 x_3 = U_4 = (1/8)(15.22) = 1.90$$

The solution of these equations is given by

$$x_1^* = 1.315, \quad x_2^* = 1.21, \quad x_3^* = 1.2$$

which is the optimal solution to the primal. ◀

Example 19.2-5

Consider the problem

$$\text{minimize } z = 5x_1 x_2^{-1} + 2x_1^{-1} x_2 + 5x_1 + x_2^{-1}$$

The orthogonality and normality conditions are given by

$$\begin{pmatrix} 1 & -1 & 1 & 0 \\ -1 & 1 & 0 & -1 \\ 1 & 1 & 1 & 1 \end{pmatrix} \begin{pmatrix} y_1 \\ y_2 \\ y_3 \\ y_4 \end{pmatrix} = \begin{pmatrix} 0 \\ 0 \\ 1 \end{pmatrix}$$

Since $N > n + 1$, these equations do not yield the required y_j directly. Thus solving for y_1, y_2, and y_3 in terms of y_4 gives

$$\begin{pmatrix} 1 & -1 & 1 \\ -1 & 1 & 0 \\ 1 & 1 & 1 \end{pmatrix} \begin{pmatrix} y_1 \\ y_2 \\ y_3 \end{pmatrix} = \begin{pmatrix} 0 \\ y_4 \\ 1-y_4 \end{pmatrix}$$

or

$$y_1 = \frac{1 - 3y_4}{2}$$

$$y_2 = \frac{1 - y_4}{2}$$

$$y_3 = y_4$$

The dual problem may now be written as

$$\text{maximize } w = \left[\frac{5}{(1-3y_4)/2}\right]^{(1-3y_4)/2} \left[\frac{2}{(1-y_4)/2}\right]^{(1-y_4)/2} \left(\frac{5}{y_4}\right)^{y_4} \left(\frac{1}{y_4}\right)^{y_4}$$

Maximization of w is equivalent to maximization of $\ln w$. The latter is easier to manipulate, however. Thus,

$$\ln w = \frac{1-3y_4}{2}\{\ln 10 - \ln(1-3y_4)\} + \frac{1-y_4}{2}\{\ln 4 - \ln(1-y_4)\}$$
$$+ y_4\{\ln 5 - \ln y_4 + (\ln 1 - \ln y_4)\}$$

The value of y_4 maximizing $\ln w$ must be unique (since the primal problem has a unique minimum). Hence,

$$\frac{\partial \ln w}{\partial y_4} = \left(\frac{-3}{2}\right)\ln 10 - \left\{\left(\frac{-3}{2}\right) + \left(\frac{-3}{2}\right)\ln(1-3y_4)\right\}$$
$$+ \left(\frac{-1}{2}\right)\ln 4 - \left\{\left(\frac{-1}{2}\right) + \left(\frac{-1}{2}\right)\ln(1-y_4)\right\}$$
$$+ \ln 5 - \{1 + \ln y_4\} + \ln 1 - \{1 + \ln y_4\} = 0$$

This gives, after simplification,

$$-\ln\left(\frac{2 \times 10^{3/2}}{5}\right) + \ln\left[\frac{(1-3y_4)^{3/2}(1-y_4)^{1/2}}{y_4^2}\right] = 0$$

or

$$\frac{\sqrt{(1-3y_4)^3(1-y_4)}}{y_4^2} = 12.6$$

which yields $y_4^* \cong .16$. Hence $y_3^* = .16$, $y_2^* = .42$, and $y_1^* = .26$.
    The value of z^* is obtained from

$$z^* = w^* = \left(\frac{5}{.26}\right)^{.26}\left(\frac{2}{.42}\right)^{.42}\left(\frac{5}{.16}\right)^{.16}\left(\frac{1}{.16}\right)^{.16} \cong 9.661$$

Hence

$$U_3 = .16(9.661) = 1.546 = 5x_1$$
$$U_4 = .16(9.661) = 1.546 = x_2^{-1}$$

The solution here yields $x_1^* = .309$ and $x_2^* = .647$.   ◄

19.2.4   STOCHASTIC PROGRAMMING

Stochastic programming deals with situations where some or all parameters of the problem are described by random variables. Such cases seem typical of real-life problems, where it is difficult to determine the values of the parameters exactly. Chapter 4 shows that, in the case of linear programming, sensitivity analysis can be used to study the effect of changes in problem's parameters on optimal solution. This, however, represents only a partial answer to the problem, especially when the parameters are actually random variables. The objective of stochastic programming is to consider these random effects explicitly in the solution of the model.

The basic idea of all stochastic programming models is to convert the probabilistic nature of the problem into an equivalent deterministic situation. Several models have been developed to handle special cases of the general problem. In this section the idea of employing deterministic equivalence is illustrated with the introduction of the interesting technique of **chance-constrained programming.**

A chance-constrained model is defined generally as

$$\text{maximize } z = \sum_{j=1}^{n} c_j x_j$$

subject to

$$P\left\{\sum_{j=1}^{n} a_{ij} x_j \le b_i\right\} \ge 1 - \alpha_i, \qquad i = 1, 2, \ldots, m; \quad x_j \ge 0, \quad \text{for all } j$$

The name "chance-constrained" follows from each constraint

$$\sum_{j=1}^{n} a_{ij} x_j \le b_i$$

being realized with a minimum probability of $1 - \alpha_i$, $0 < \alpha_i < 1$.

In the general case it is assumed that c_j, a_{ij}, and b_i are all random variables. The fact that c_j is a random variable can always be treated by replacing it by its expected value. In what follows, three cases are considered. The first two correspond to the separate considerations of a_{ij} and b_i as random variables. The third case combines the random effects of a_{ij} and b_i. In all cases, it is assumed that the parameters are normally distributed with known means and variances.

Case 1: In this case, each a_{ij} is normally distributed with mean $E\{a_{ij}\}$ and variance var $\{a_{ij}\}$. Also, the covariance of a_{ij} and $a_{i'j'}$ is given by cov $\{a_{ij}, a_{i'j'}\}$.

Consider the ith constraint

$$P\left\{\sum_{j=1}^{n} a_{ij} x_j \le b_i\right\} \ge 1 - \alpha_i$$

and define

$$h_i = \sum_{j=1}^{n} a_{ij} x_j$$

Then h_i is normally distributed with

$$E\{h_i\} = \sum_{j=1}^{n} E\{a_{ij}\}x_j \qquad \text{and} \qquad \text{var}\{h_i\} = \mathbf{X}^T \mathbf{D}_i \mathbf{X}$$

where $\mathbf{X} = (x_1, \ldots, x_n)^T$

$$\mathbf{D}_i = i\text{th covariance matrix} = \begin{pmatrix} \text{var}\{a_{i1}\} & \cdots & \text{cov}\{a_{i1}, a_{in}\} \\ \vdots & & \vdots \\ \text{cov}\{a_{in}, a_{i1}\} & \cdots & \text{var}\{a_{in}\} \end{pmatrix}$$

Now

$$P\{h_i \le b_i\} = P\left\{ \frac{h_i - E\{h_i\}}{\sqrt{\text{var}\{h_i\}}} \le \frac{b_i - E\{h_i\}}{\sqrt{\text{var}\{h_i\}}} \right\} \ge 1 - \alpha_i$$

where $(h_i - E\{h_i\})/\sqrt{\text{var}\{h_i\}}$ is standard normal with mean zero and variance one. This means that

$$P\{h_i \le b_i\} = \Phi\left(\frac{b_i - E\{h_i\}}{\sqrt{\text{var}\{h_i\}}} \right)$$

where Φ represents the CDF of the standard normal distribution.

Let K_{α_i} be the standard normal value such that

$$\Phi(K_{\alpha_i}) = 1 - \alpha_i$$

Then the statement $P\{h_i \le b_i\} \ge 1 - \alpha_i$ is realized if and only if

$$\frac{b_i - E\{h_i\}}{\sqrt{\text{var}\{h_i\}}} \ge K_{\alpha_i}$$

This yields the following nonlinear constraint:

$$\sum_{j=1}^{n} E\{a_{ij}\}x_j + K_{\alpha_i}\sqrt{\mathbf{X}^T \mathbf{D}_i \mathbf{X}} \le b_i$$

which is equivalent to the original stochastic constraint.

For the special case where the normal distributions are independent,

$$\text{cov}\{a_{ij}, a_{ij}, a_{i'j'}\} = 0$$

and the last constraint reduces to

$$\sum_{j=1}^{n} E\{a_{ij}\}x_j + K_{\alpha_i}\sqrt{\sum_{j=1}^{n} \text{var}\{a_{ij}\}x_j^2} \le b_i$$

This constraint can now be put in the separable programming form (Section 19.2.1) by using the substitution

$$y_i = \sqrt{\sum_{j=1}^{n} \text{var}\{a_{ij}\}x_j^2}, \qquad \text{for all } i$$

Thus the original constraint is equivalent to

$$\sum_{j=1}^{n} E\{a_{ij}\}x_j + K_{\alpha_i}y_i \le b_i$$

and

$$\sum_{j=1}^{n} \text{var}\,\{a_{ij}\}x_j^2 - y_i^2 = 0$$

where $y_i \geq 0$.

Case 2: In this case only b_i is normal with mean $E\{b_i\}$ and variance var $\{b_i\}$. The analysis in this case is very similar to that of case 1. Consider the stochastic constraint

$$P\left\{ b_i \geq \sum_{j=1}^{n} a_{ij}x_j \right\} \geq \alpha_i$$

As in case 1,

$$P\left\{ \frac{b_i - E\{b_i\}}{\sqrt{\text{var}\,\{b_i\}}} \geq \frac{\sum_{j=1}^{n} a_{ij}x_j - E\{b_i\}}{\sqrt{\text{var}\,\{b_i\}}} \right\} \geq \alpha_i$$

This can hold only if

$$\frac{\sum_{j=1}^{n} a_{ij}x_j - E\{b_i\}}{\sqrt{\text{var}\,\{b_i\}}} \leq K_{\alpha_i}$$

Thus the stochastic constraint is equivalent to the deterministic linear constraint

$$\sum_{j=1}^{n} a_{ij}x_j \leq E\{b_i\} + K_{\alpha_i}\sqrt{\text{var}\,\{b_i\}}$$

Thus, in case 2, the chance-constrained model can be converted into an equivalent linear programming problem.

Example 19.2–6. Consider the chance-constrained problem

$$\text{maximize } z = 5x_1 + 6x_2 + 3x_3$$

subject to

$$P\{a_{11}x_1 + a_{12}x_2 + a_{13}x_3 \leq 8\} \geq .95$$
$$P\{5x_1 + x_2 + 6x_3 \leq b_2\} \geq .10$$

with all $x_j \geq 0$. Suppose that the a_{ij}'s are *independent* normally distributed random variables with the following means and variances:

$$E\{a_{11}\} = 1, \qquad E\{a_{12}\} = 3, \qquad E\{a_{13}\} = 9$$
$$\text{var}\,\{a_{11}\} = 25, \qquad \text{var}\,\{a_{12}\} = 16, \qquad \text{var}\,\{a_{13}\} = 4$$

The parameter b_2 is normally distributed with mean 7 and variance 9.

From standard normal tables,

$$K_{\alpha_1} = K_{.05} \cong 1.645, \qquad K_{\alpha_2} = K_{.10} \cong 1.285$$

For the first constraint, the equivalent deterministic constraint is given by

$$x_1 + 3x_2 + 9x_3 + 1.645\sqrt{25x_1^2 + 16x_2^2 + 4x_3^2} \le 8$$

and for the second constraint

$$5x_1 + x_2 + 6x_3 \le 7 + 1.285(3) = 10.855$$

If we let

$$y^2 = 25x_1^2 + 16x_2^2 + 4x_3^2$$

the complete problem then becomes

$$\text{maximize } z = 5x_1 + 6x_1 + 3x_3$$

subject to

$$x_1 + 3x_2 + 9x_3 + 1.645y \le 8$$
$$25x_1^2 + 16x_2^2 + 4x_3^2 - y^2 = 0$$
$$5x_1 + x_2 + 6x_3 \le 10.855$$
$$x_1, x_2, x_3, y \ge 0$$

which can be solved by separable programming.   ◄

Case 3: In this case all a_{ij} and b_i are normal random variables. Consider the constraint

$$\sum_{j=1}^{n} a_{ij} x_j \le b_i$$

This may be written

$$\sum_{j=1}^{n} a_{ij} x_j - b_i \le 0$$

Since all a_{ij} and b_i are normal, it follows from the theory of statistics that $\sum_{j=1}^{n} a_{ij} x_j - b_i$ is also normal. This shows that the chance constraint reduces in this case to the same situation given in case 1 and is treated in a similar manner.

19.2.5 LINEAR COMBINATIONS METHOD

This method deals with a constrained problem in which all constraints are linear. Specifically, the problem is given as

$$\text{maximize } z = f(\mathbf{X})$$

subject to

$$\mathbf{AX} \le \mathbf{b}, \qquad \mathbf{X} \ge \mathbf{0}$$

where $\mathbf{A}$ is a matrix and $\mathbf{b}$ is a vector.

The procedure is based on the general idea of the steepest ascent (gradient) method (Section 19.1.2). However, the direction specified by the gradient vector may not yield a feasible solution for the constrained problem. Also, the gradient vector will not necessarily be null at the optimum (constrained) point. The steepest ascent method must be modified to handle the constrained case.

Let $\mathbf{X}^k$ be the *feasible* trial point at the kth iteration. The objective function $f(\mathbf{X})$ can be expanded in the neighborhood of $\mathbf{X}^k$ using Taylor's series. This gives

$$f(\mathbf{X}) \cong f(\mathbf{X}^k) + \nabla f(\mathbf{X}^k)(\mathbf{X} - \mathbf{X}^k) = (f(\mathbf{X}^k) - \nabla f(\mathbf{X}^k)\mathbf{X}^k) + \nabla f(\mathbf{X}^k)\mathbf{X}$$

The procedure calls for determining a feasible point $\mathbf{X} = \mathbf{X}^*$ such that $f(\mathbf{X})$ is maximized subject to the (linear) constraints of the problem. Since $(f(\mathbf{X}^k) - \nabla f(\mathbf{X}^k)\mathbf{X}^k)$ is a constant, the problem of determining $\mathbf{X}^*$ becomes

$$\text{maximize } w_k(\mathbf{X}) = \nabla f(\mathbf{X}^k)\mathbf{X}$$

subject to

$$\mathbf{AX} \leq \mathbf{b}, \qquad \mathbf{X} \geq \mathbf{0}$$

This is a linear programming problem in $\mathbf{X}$ that can now be used to determine $\mathbf{X}^*$.

Since w_k is constructed from the gradient of $f(\mathbf{X})$ at $\mathbf{X}^k$, an improved solution point can be secured if and only if $w_k(\mathbf{X}^*) > w_k(\mathbf{X}^k)$. From Taylor's expansion, this does not guarantee that $f(\mathbf{X}^*) > f(\mathbf{X}^k)$ unless $\mathbf{X}^*$ is in the neighborhood of $\mathbf{X}^k$. However, given $w_k(\mathbf{X}^*) = w_k(\mathbf{X}^k)$, there must exist a point $\mathbf{X}^{k+1}$ on the line segment $(\mathbf{X}^k; \mathbf{X}^*)$ such that $f(\mathbf{X}^{k+1}) > f(\mathbf{X}^k)$. The objective is to determine $\mathbf{X}^{k+1}$. Define

$$\mathbf{X}^{k+1} = (1 - r)\mathbf{X}^k + r\mathbf{X}^* = \mathbf{X}^k + r(\mathbf{X}^* - \mathbf{X}^k), \qquad 0 < r \leq 1$$

This means that $\mathbf{X}^{k+1}$ is a **linear combination** of $\mathbf{X}^k$ and $\mathbf{X}^*$. Since $\mathbf{X}^k$ and $\mathbf{X}^*$ are two feasible points in a *convex* solution space, $\mathbf{X}^{k+1}$ is also feasible. By comparison with the steepest ascent method (Section 19.1.2), the parameter r may be regarded as a step size.

The point $\mathbf{X}^{k+1}$ is determined such that $f(\mathbf{X})$ is maximized. Since $\mathbf{X}^{k+1}$ is a function of r only, the determination of $\mathbf{X}^{k+1}$ is secured by maximizing

$$h(r) = f[\mathbf{X}^k + r(\mathbf{X}^* - \mathbf{X}^k)]$$

in terms of r.

The procedure just described is repeated until at the kth iteration the condition $w_k(\mathbf{X}^*) \leq w_k(\mathbf{X}^k)$ is satisfied. At this point, no further improvements are possible. The process is then terminated with $\mathbf{X}^k$ as the best solution point.

The linear programming problems generated at the successive iterations differ only in the coefficients of the objective function. The sensitivity analysis procedures presented in Section 4.5 thus may be used to carry out calculations efficiently.

Example 19.2-7. Consider the quadratic programming of Example 19.2–3. This is given by

$$\text{maximize } f(\mathbf{X}) = 4x_1 + 6x_2 - 2x_1^2 - 2x_1x_2 - 2x_2^2$$

subject to

$$x_1 + 2x_2 \leq 2$$

where x_1 and x_2 are nonnegative.

Let the initial trial point be $X^0 = (1/2, 1/2)$, which is feasible. Now

$$\nabla f(X) = (4 - 4x_1 - 2x_2, \ 6 - 2x_1 - 4x_2)$$

First Iteration

$$\nabla f(X^0) = (1, 3)$$

The associated linear program is to maximize $w_1 = x_1 + 3x_2$ subject to the same constraints as in the original problem. This gives the optimal solution $X^* = (0, 1)$. The values of w_1 at X^0 and X^* equal 2 and 3, respectively. Hence a new trial point must be determined. Thus

$$X^1 = (1/2, 1/2) + r[(0, 1) - (1/2, 1/2)] = \left(\frac{1-r}{2}, \frac{1+r}{2}\right)$$

Now, maximization of

$$h(r) = f\left(\frac{1-r}{2}, \frac{1+r}{2}\right)$$

yields $r^1 = 1$. Thus $X^1 = (0, 1)$ with $f(X^1) = 4$.

Second Iteration

$$\nabla f(X^1) = (2, 2)$$

The objective function of the new linear programming problem is $w_2 = 2x_1 + 2x_2$. The optimum solution to this problem yields $X^* = (2, 0)$. Since the values of w_2 at X^1 and X^* are 2 and 4 a new trial point must be determined. Thus

$$X^2 = (0, 1) + r[(2, 0) - (0, 1)] = (2r, 1 - r)$$

The maximization of

$$h(r) = f(2r, 1 - r)$$

yields $r^2 = 1/6$. Thus $X^2 = (1/3, 5/6)$ with $f(X)^2 \cong 4.16$.

Third Iteration

$$\nabla f(X^2) = (1, 2)$$

The corresponding objective linear function is $w_3 = x_1 + 2x_2$. The optimum solution of this problem yields the alternative solutions $X^* = (0, 1)$ and $X^* = (2, 0)$. The value of w_3 for both values of X^* equals its value at X^2. Consequently, no further improvements are possible. The *approximate* optimum solution is $X^2 = (1/3, 5/6)$ with $f(X^2) \cong 4.16$. This happens to be the exact optimum. ◀

19.2.6   SUMT ALGORITHM

In this section a more general gradient method is presented. It is assumed that the objective function $f(X)$ is concave and each constraint function $g_i(X)$ is convex. Moreover, the solution space must have an interior. This rules out both implicit and explicit use of *equality* constraints.

The SUMT (Sequential Unconstrained Maximization Technique) algorithm is based on transforming the constrained problem into an equivalent *unconstrained* problem. The procedure is more or less similar to the use of the Lagrange multipliers method. The transformed problem can then be solved using the steepest ascent method (Section 19.1.2).

To clarify the concept, consider the new function

$$p(X, t) = f(X) + t \left(\sum_{i=1}^{m} \frac{1}{g_i(X)} - \sum_{j=1}^{n} \frac{1}{x_j} \right)$$

where t is a nonnegative parameter. The second summation sign is based on the nonnegativity constraints, which must be put in the form $-x_j \leq 0$ to conform with the original constraints $g_i(X) \leq 0$. Since $g_i(X)$ is convex, $1/g_i(X)$ is concave. This means that $p(X, t)$ is concave in X. Consequently, $p(X, t)$ possesses a unique maximum. It is now shown that optimization of the original constrained problem is equivalent to optimization of $p(X, t)$.

The algorithm is initiated by arbitrarily selecting an initial *nonnegative* value for t. An initial point X^0 is selected as the first trial solution. This point must be an interior point; that is, it must not lie on the boundaries of the solution space. Given the value of t, the steepest ascent method is used to determine the corresponding optimal solution (maximum) of $p(X, t)$.

The new solution point will always be an interior point because if the solution point is close to the boundaries, at least one of the functions $1/g_i(X)$ or $(-1/x_i)$ will acquire a very large negative value. Since the objective is to maximize $p(X, t)$, such solution points are automatically excluded. The main result is that successive solution points will always be interior points. Consequently, the problem can always be treated as an unconstrained case.

Once the optimum solution corresponding to a given value of t is reached, a new value of t is generated and the optimization process (using the steepest ascent method) is repeated. Thus if t' is the current value of t, the next value, t'', must be selected such that $0 < t'' < t'$.

The SUMT procedure is terminated if, for two successive values of t, the corresponding *optimum* values of X obtained by maximizing $p(X, t)$ are approximately the same. At this point further trials will produce little improvement.

Actual implementation of SUMT involves more details than have been presented here. Specifically, the selection of an initial value of t is a very important factor which affects speed of convergence. Further, determination of an initial interior point may require special techniques. These details can be found in Fiacco and McCormick [1968].

19.3 SUMMARY

The solution methods of nonlinear programming can generally be classified as either *direct* or *indirect* procedures. Examples of direct methods are the gradient algorithms, wherein the maximum (minimum) of a problem is sought by following the fastest rate of increase (decrease) of the objective function at a point. In indirect methods, the original problem is first transformed into an auxiliary one from which the optimum is determined. Examples of these situations include quadratic programming, separable programming, and stochastic programming. We note that the auxiliary problems in these cases may yield an *exact* or an *approximate* solution of the original problem. For example, the use of the Kuhn–Tucker conditions with quadratic programming yields an exact solution, whereas separable programming yields an approximate solution only.

SELECTED REFERENCES

BAZARAA, M., and C. SHETTY, *Nonlinear Programming, Theory and Algorithms,* Wiley, New York, 1979.

BEIGHTLER, C., D. PHILLIPS, and D. WILDE, *Foundations of Optimization,* 2nd ed., Prentice-Hall, Englewood Cliffs, N.J., 1979.

FIACCO, A., and G. MCCORMICK, *Nonlinear Programming: Sequential Unconstrained Minimization Techniques,* Wiley, New York, 1968.

ZANGWILL, W., *Nonlinear Programming,* Prentice-Hall, Englewood Cliffs, N.J., 1969.

PROBLEMS

Section	Assigned Problems
19.1.1	19–1 to 19–3
19.1.2	19–4, 19–5
19.2.1	19–6 to 19–14
19.2.2	19–15, 19–16
19.2.3	19–17 to 19–20
19.2.4	19–21, 19–22
19.2.5	19–23

☐ **19–1** Solve Example 19.1–1 assuming that $\Delta = .01$. Compare the accuracy of the results with that in Table 19–1.

☐ **19–2** Find the maximum of each of the following functions by dichotomous search. Assume that $\Delta = .05$.

(a) $f(x) = 1/|(x-3)^3|$, $2 \le x \le 4$.

(b) $f(x) = x \cos x$, $0 \le x \le \pi$.

(c) $f(x) = x \sin \pi x$, $1.5 \le x \le 2.5$.

(d) $f(x) = -(x-3)^2$, $2 \le x \le 4$.

(e) $f(x) = \begin{cases} 4x, & 0 \le x \le 2. \\ 4-x, & 2 \le x \le 4. \end{cases}$

☐ **19–3** Develop an expression for determining the maximum number of iterations needed to terminate the dichotomous search method for a given value of Δ and an initial interval of uncertainty $I_0 = b - a$.

☐ **19–4** Show that, in general, the Newton–Raphson method (Section 18.1.2) when applied to a strictly concave quadratic function will converge in exactly one step. Apply the method to the maximization of

$$f(\mathbf{X}) = 4x_1 + 6x_2 - 2x_1^2 - 2x_1x_2 - 2x_2^2$$

☐ **19–5** Carry out at most five iterations for each of the following problems using the method of steepest descent (ascent). Assume that $\mathbf{X}^0 = \mathbf{0}$ in each case.

(a) $\min f(\mathbf{X}) = (x_2 - x_1^2)^2 + (1 - x_1)^2$

(b) $\max f(\mathbf{X}) = \mathbf{cX} + \mathbf{X}^T\mathbf{AX}$

where

$$\mathbf{c} = (1, 3, 5)$$

$$\mathbf{A} = \begin{pmatrix} -5 & -3 & -1/2 \\ -3 & -2 & 0 \\ -1/2 & 0 & -1/2 \end{pmatrix}$$

(c) $\max f(x) = 3 - x^2 - x^4$

(d) $\min f(\mathbf{X}) = x_1 - x_2 + x_1^2 - x_1x_2$

☐ **19–6** Formulate the following problem using the approximating mixed integer programming method.

$$\text{Maximize } z = e^{-x_1} + x_1 + (x_2 + 1)^2$$

subject to

$$x_1^2 + x_2 \le 3$$
$$x_1, x_2 \ge 0$$

☐ **19–7** Repeat Problem 19–6 using the restricted basis method. Then find the optimal solution.

☐ **19–8** Consider the problem

$$\text{maximize } z = x_1 x_2 x_3$$

subject to

$$x_1^2 + x_2 + x_3 \le 4$$
$$x_1, x_2, x_3 \ge 0$$

Approximate the problem as a linear programming model for use with the restricted basis method.

☐ **19–9** Show how the following problem can be made separable.

$$\text{Maximize } z = x_1 x_2 + x_3 + x_1 x_3$$

subject to

$$x_1 x_2 + x_2 + x_1 x_3 \le 10$$
$$x_1, x_2, x_3 \ge 0$$

☐ **19–10** Show how the following problem can be made separable.

$$\text{Minimize } z = e^{2x_1 + x_2^2} + (x_3 - 2)^2$$

subject to

$$x_1 + x_2 + x_3 \le 6$$
$$x_1, x_2, x_3 \ge 0$$

☐ **19–11** Show how the following problem can be made separable.

$$\text{Maximize } z = e^{x_1 x_2} + x_2^2 x_3 + x_4$$

subject to

$$x_1 + x_2 x_3 + x_3 \le 10$$
$$x_1, x_2, x_3 \le 0$$
$$x_4 \text{ unrestricted in sign}$$

☐ **19–12** Show that in separable convex programming (Section 19.2.1), it is never optimal to have $x_{ki} > 0$ when $x_{k-1,i}$ is not at its upper bound.

☐ **19–13** Solve as a separable convex programming problem.

$$\text{Minimize } z = x_1^4 + 2x_2 + x_3^2$$

subject to

$$x_1^2 + x_2 + x_3^2 \le 4$$
$$|x_1 + x_2| \le 0$$
$$x_1, x_3 \ge 0$$
$$x_2 \text{ unrestricted in sign}$$

☐ **19–14** Solve the following as a separate convex programming problem.

$$\text{Minimize } z = (x_1 - 2)^2 + 4(x_2 - 6)^2$$

subject to

$$6x_1 + 3(x_2 + 1)^2 \leq 12$$
$$x_1, x_2 \geq 0$$

☐ **19–15** Consider the problem

$$\text{maximize } z = 6x_1 + 3x_2 - 4x_1 x_2 - 2x_1^2 - 3x_2^2$$

subject to

$$x_1 + x_2 \leq 1$$
$$2x_1 + 3x_2 \leq 4$$
$$x_1, x_2 \leq 0$$

Show that z is strictly concave and then solve the problem using the quadratic programming algorithm.

☐ **19–16** Consider the problem:

$$\text{minimize } z = 2x_1^2 + 2x_2^2 + 3x_3^2 + 2x_1 x_2 + 2x_2 x_3 + x_1 - 3x_2 - 5x_3$$

subject to

$$x_1 + x_2 + x_3 \geq 1$$
$$3x_1 + 2x_2 + x_3 \leq 6$$
$$x_1, x_2, x_3 \geq 0$$

Show that z is strictly convex and then solve by the quadratic programming technique.

☐ **19–17** Solve the following problem by geometric programming.

$$\text{Minimize } z = 2x_1^{-1} + x_2^2 + x_1^4 x_2^{-2} + 4x_1^2$$
$$x_1, x_2 > 0$$

☐ **19–18** Solve the following problem by geometric programming.

$$\text{Minimize } z = 5x_1 x_2^{-1} x_3^2 + x_1^{-2} x_3^{-1} + 10x_2^3 + 2x_1^{-1} x_2 x_3^{-3}$$
$$x_1, x_2, x_3 > 0$$

☐ **19–19** Solve the following problem by geometric programming.

$$\text{Minimize } z = 2x_1^2 x_2^{-3} + 8x_1^{-3} x_2 + 3x_1 x_2$$

☐ **19–20** Solve the following problem by geometric programming.

$$\text{Minimize } z = 2x_1^3 x_2^{-3} + 4x_1^{-2} x_2 + x_1 x_2 + 8x_1 x_2^{-1}$$
$$x_1, x_2 > 0$$

☐ **19–21** Convert the following stochastic problem into an equivalent deterministic model.

$$\text{Maximize } z = x_1 + 2x_2 + 5x_3$$

subject to

$$P\{a_1x_1 + 3x_2 + a_3x_3 \leq 10\} \geq 0.9$$
$$P\{7x_1 + 5x_2 + x_3 \leq b_2\} \geq 0.1$$
$$x_1, x_2, x_3 \geq 0$$

Assume that a_1 and a_3 are independent and normally distributed random variables with means $E\{a_1\} = 2$ and $E\{a_3\} = 5$ and variances var $\{a_1\} = 9$ and var $\{a_3\} = 16$. Assume further that b_2 is normally distributed with mean 15 and variance 25.

☐ **19–22** Consider the following stochastic programming model:

$$\text{maximize } z = x_1 + x_2^2 + x_3$$

subject to

$$P\{x_1^2 + a_2x_2^3 + a_3\sqrt{x_3} \leq 10\} \geq 0.9$$
$$x_1, x_2, x_3 \geq 0$$

where a_2 and a_3 are independent and normally distributed random variables with means 5 and 2, and variance 16 and 25, respectively. Convert the problem into the (deterministic) separable programming form.

☐ **19–23** Solve the following problem by the linear combinations method.

$$\text{Minimize } f(\mathbf{X}) = x_1^3 + x_2^3 - 3x_1x_2$$

subject to

$$3x_1 + x_2 \leq 3$$
$$5x_1 - 3x_2 \leq 5$$
$$x_1, x_2 \geq 0$$

Appendix A

Review of Vectors and Matrices

A.1 VECTORS

A.1.1 DEFINITION OF A VECTOR

Let $p_1, p_2, \ldots, p_n$ be any n real numbers and $\mathbf{P}$ an ordered set of these real numbers, that is,

$$\mathbf{P} = (p_1, p_2, \ldots, p_n)$$

Then $\mathbf{P}$ is called an n-vector (or simply a vector). The ith component of $\mathbf{P}$ is given by p_i. For example, $\mathbf{P} = (1, 2)$ is a two-dimensional vector that joins the origin and the point $(1, 2)$.

A.1.2 ADDITION (SUBTRACTION) OF VECTORS

Let

$$\mathbf{P} = (p_1, p_2, \ldots, p_n) \quad \text{and} \quad \mathbf{Q} = (q_1, q_2, \ldots, q_n)$$

be two vectors in the n-dimensional space. Then the components of the vector $\mathbf{R} = (r_1, r_2, \ldots, r_n)$ such that $\mathbf{R} = \mathbf{P} \pm \mathbf{Q}$ are given by

$$r_i = p_i \pm q_i$$

In general, given the vectors $\mathbf{P}$, $\mathbf{Q}$, and $\mathbf{S}$,

$$\mathbf{P} \pm \mathbf{Q} = \mathbf{Q} \pm \mathbf{P} \quad \text{(commutative law)}$$
$$(\mathbf{P} + \mathbf{Q}) + \mathbf{S} = \mathbf{P} + (\mathbf{Q} + \mathbf{S}) \quad \text{(associative law)}$$
$$\mathbf{P} + (-\mathbf{P}) = \mathbf{0}, \quad \text{a zero (or null) vector}$$

A.1.3   MULTIPLICATION OF VECTORS BY SCALARS

Given a vector $\mathbf{P}$ and a scalar (constant) quantity θ, the new vector

$$\mathbf{Q} = \theta\mathbf{P} = (\theta p_1, \theta p_2, \ldots, \theta p_n)$$

is called the *scalar product* of $\mathbf{P}$ and θ.

In general, given the vectors $\mathbf{P}$ and $\mathbf{S}$ and the scalars θ and γ,

$$\theta(\mathbf{P} + \mathbf{S}) = \theta\mathbf{P} + \theta\mathbf{S} \qquad \text{(distributive law)}$$
$$\theta(\gamma\mathbf{P}) = (\theta\gamma)\mathbf{P} \qquad \text{(associative law)}$$

A.1.4   LINEARLY INDEPENDENT VECTORS

A set of vectors $\mathbf{P}_1, \mathbf{P}_2, \ldots, \mathbf{P}_n$ is said to be *linearly independent* if and only if, for all real θ_j,

$$\sum_{j=1}^{n} \theta_j \mathbf{P}_j = 0$$

implies that all $\theta_j = 0$, where the θ_j are scalar quantities. If

$$\sum_{j=1}^{n} \theta_j \mathbf{P}_j = 0$$

for some $\theta_j \neq 0$, the vectors are said to be *linearly dependent.* For example, the vectors

$$\mathbf{P}_1 = (1, 2,) \qquad \text{and} \qquad \mathbf{P}_2 = (2, 4)$$

are linearly dependent, since there exist $\theta_1 = 2$ and $\theta_2 = -1$ for which

$$\theta_1\mathbf{P}_1 + \theta_2\mathbf{P}_2 = 0$$

A.2   MATRICES

A.2.1   DEFINITION OF A MATRIX

A matrix is a rectangular array of elements. The (i, j)th element a_{ij} of the matrix $\mathbf{A}$ stands in the ith row and jth column of the array. The order (size) of a matrix is said to be $(m \times n)$ if the matrix includes m rows and n columns. For example,

$$\mathbf{A} = \begin{pmatrix} a_{11} & a_{12} & a_{13} \\ a_{21} & a_{22} & a_{23} \\ a_{31} & a_{32} & a_{33} \\ a_{41} & a_{42} & a_{43} \end{pmatrix} = \|a_{ij}\|_{4\times3}$$

is a (4×3)-matrix.

A.2.2   TYPES OF MATRICES

1. A *square* matrix is a matrix in which $m = n$.

2. An **identity** matrix is a square matrix in which all the diagonal elements are one and all the off-diagonal elements are zero; that is,

$$a_{ij} = 1, \quad \text{for } i = j$$
$$a_{ij} = 0, \quad \text{for } i \neq j$$

For example, a (3×3) identity matrix is given by

$$\mathbf{I}_3 = \begin{pmatrix} 1 & 0 & 0 \\ 0 & 1 & 0 \\ 0 & 0 & 1 \end{pmatrix}$$

3. A *row vector* is a matrix with one row and n columns.

4. A *column vector* is a matrix with m rows and one column.

5. The matrix $\mathbf{A}^T$ is called the **transpose** of $\mathbf{A}$ if the element a_{ij} in $\mathbf{A}$ is equal to element a_{ji} in $\mathbf{A}^T$ for all i and j. For example, if

$$\mathbf{A} = \begin{pmatrix} 1 & 4 \\ 2 & 5 \\ 3 & 6 \end{pmatrix}$$

then

$$\mathbf{A}^T = \begin{pmatrix} 1 & 2 & 3 \\ 4 & 5 & 6 \end{pmatrix}$$

In general, $\mathbf{A}^T$ is obtained by interchanging the rows and the columns of $\mathbf{A}$. Consequently, if $\mathbf{A}$ is of the order $(m \times n)$, $\mathbf{A}^T$ is of the order $(n \times m)$.

6. A matrix $\mathbf{B} = \mathbf{0}$ is called a **zero matrix** if every element of $\mathbf{B}$ is equal to zero.

7. Two matrices $\mathbf{A} = \|a_{ij}\|$ and $\mathbf{B} = \|b_{ij}\|$ are said to be *equal matrices* if and only if they have the same order and if each element a_{ij} is equal to the corresponding b_{ij} for all i and j.

A.2.3   MATRIX ARITHMETIC OPERATIONS

In matrices only addition (subtraction) and multiplication are defined. The division, although not defined, is replaced by the concept of inversion (see Section A.2.6).

Addition (Subtraction) of Matrices

Two matrices $\mathbf{A} = \|a_{ij}\|$ and $\mathbf{B} = \|b_{ij}\|$ can be added together if they are of the same order $(m \times n)$. The sum $\mathbf{D} = \mathbf{A} + \mathbf{B}$ is obtained by adding the corresponding elements. Thus,

$$\|d_{ij}\|_{m \times n} = \|a_{ij} + b_{ij}\|_{m \times n}$$

If one assumes that the matrices **A**, **B**, and **C** have the same order,

$$\mathbf{A} \pm \mathbf{B} = \mathbf{B} \pm \mathbf{A} \qquad \text{(commutative law)}$$
$$\mathbf{A} \pm (\mathbf{B} \pm \mathbf{C}) = (\mathbf{A} \pm \mathbf{B}) \pm \mathbf{C} \qquad \text{(associative law)}$$
$$(\mathbf{A} \pm \mathbf{B})^T = \mathbf{A}^T \pm \mathbf{B}^T$$

Product of Matrices

Two matrices $\mathbf{A} = \|a_{ij}\|$ and $\mathbf{B} = \|b_{ij}\|$ can be multiplied in the order $\mathbf{AB}$ if and only if the number of columns of **A** is equal to the number of rows of **B**. That is, if **A** is of the order $(m \times r)$, then **B** is of the order $(r \times n)$, where m and n are arbitrary sizes.

Let $\mathbf{D} = \mathbf{AB}$. Then **D** is of the order $(m \times n)$, and its elements d_{ij} are given by

$$d_{ij} = \sum_{k=1}^{r} A_{ik} b_{kj}, \qquad \text{for all } i \text{ and } j$$

For example, if

$$\mathbf{A} = \begin{pmatrix} 1 & 3 \\ 2 & 4 \end{pmatrix} \qquad \text{and} \qquad \mathbf{B} = \begin{pmatrix} 5 & 7 & 9 \\ 6 & 8 & 0 \end{pmatrix}$$

then

$$\mathbf{D} = \begin{pmatrix} 1 & 3 \\ 2 & 4 \end{pmatrix} \begin{pmatrix} 5 & 7 & 9 \\ 6 & 8 & 0 \end{pmatrix} = \begin{pmatrix} (1 \times 5 + 3 \times 6) & (1 \times 7 + 3 \times 8) & (1 \times 9 + 3 \times 0) \\ (2 \times 5 + 4 \times 6) & (2 \times 7 + 4 \times 8) & (2 \times 9 + 4 \times 0) \end{pmatrix}$$

$$= \begin{pmatrix} 23 & 31 & 9 \\ 34 & 46 & 18 \end{pmatrix}$$

Notice that, in general, $\mathbf{AB} \neq \mathbf{BA}$ even if **BA** is defined.

Matrix multiplication follows these general properties:

$$\mathbf{I}_m \mathbf{A} = \mathbf{A} \mathbf{I}_n = \mathbf{A}, \qquad \text{where } \mathbf{I} \text{ is an identity matrix}$$
$$(\mathbf{AB})\mathbf{C} = \mathbf{A}(\mathbf{BC})$$
$$\mathbf{C}(\mathbf{A} + \mathbf{B}) = \mathbf{CA} + \mathbf{CB}$$
$$(\mathbf{A} + \mathbf{B})\mathbf{C} = \mathbf{AC} + \mathbf{BC}$$
$$\alpha(\mathbf{AB}) = (\alpha\mathbf{A})\mathbf{B} = \mathbf{A}(\alpha\mathbf{B}), \qquad \alpha \text{ is a scalar}$$

Multiplication of Partitioned Matrices

Let **A** be an $(m \times r)$-matrix and **B** an $(r \times n)$-matrix. If **A** and **B** are partitioned into the following submatrices

$$\mathbf{A} = \begin{pmatrix} \mathbf{A}_{11} & \mathbf{A}_{12} & \mathbf{A}_{13} \\ \mathbf{A}_{21} & \mathbf{A}_{22} & \mathbf{A}_{23} \end{pmatrix} \qquad \text{and} \qquad \mathbf{B} = \begin{pmatrix} \mathbf{B}_{11} & \mathbf{B}_{12} \\ \mathbf{B}_{21} & \mathbf{B}_{22} \\ \mathbf{B}_{31} & \mathbf{B}_{32} \end{pmatrix}$$

such that the number of columns of $\mathbf{A}_{ij}$ is equal to the number of rows of $\mathbf{B}_{ji}$ and such that the number of columns of $\mathbf{A}_{ij}$ and $\mathbf{A}_{i+1,j}$ are equal, for all i and j, then

$$\mathbf{A} \times \mathbf{B} = \begin{pmatrix} \mathbf{A}_{11}\mathbf{B}_{11} + \mathbf{A}_{12}\mathbf{B}_{21} + \mathbf{A}_{13}\mathbf{B}_{31} & \mathbf{A}_{11}\mathbf{B}_{12} + \mathbf{A}_{12}\mathbf{B}_{22} + \mathbf{A}_{13}\mathbf{B}_{32} \\ \mathbf{A}_{21}\mathbf{B}_{11} + \mathbf{A}_{22}\mathbf{B}_{21} + \mathbf{A}_{23}\mathbf{B}_{31} & \mathbf{A}_{21}\mathbf{B}_{12} + \mathbf{A}_{22}\mathbf{B}_{22} + \mathbf{A}_{23}\mathbf{B}_{32} \end{pmatrix}$$

For example,

$$\begin{pmatrix} 1 & 2 & 3 \\ 1 & 0 & 5 \\ 2 & 5 & 6 \end{pmatrix} \begin{pmatrix} 4 \\ 1 \\ 8 \end{pmatrix} = \begin{pmatrix} (1)(4) + (2 \ \ 3)\begin{pmatrix} 1 \\ 8 \end{pmatrix} \\ \begin{pmatrix} 1 \\ 2 \end{pmatrix}(4) + \begin{pmatrix} 0 & 5 \\ 5 & 6 \end{pmatrix}\begin{pmatrix} 1 \\ 8 \end{pmatrix} \end{pmatrix} = \begin{pmatrix} 4 + 2 + 24 \\ \begin{pmatrix} 4 \\ 8 \end{pmatrix} + \begin{pmatrix} 40 \\ 53 \end{pmatrix} \end{pmatrix} = \begin{pmatrix} 30 \\ 44 \\ 61 \end{pmatrix}$$

The usefulness of partitioned matrices will come later in considering the inversion of matrices.

A.2.4 THE DETERMINANT OF A SQUARE MATRIX

Given the n-square matrix

$$\mathbf{A} = \begin{pmatrix} a_{11} & a_{12} & \cdots & a_{1n} \\ a_{21} & a_{22} & \cdots & a_{2n} \\ \vdots & \vdots & & \vdots \\ a_{n1} & a_{n2} & \cdots & a_{nn} \end{pmatrix}$$

consider the product

$$P_{j_1 j_2 \cdots j_n} = a_{1j_1} a_{2j_2} \cdots a_{nj_n}$$

the elements of which are selected such that each column and each row of $\mathbf{A}$ is represented exactly once among the subscripts of $P_{j_1 j_2 \cdots j_n}$. Next define $\epsilon_{j_1 j_2 \cdots j_n}$ equal to $+1$ if $j_1 j_2 \cdots j_n$ is an even permutation and -1 if $j_1 j_2 \cdots j_n$ is an odd permutation. Thus the scalar

$$\sum_{\rho} \epsilon_{j_1 j_2 \cdots j_n} P_{j_1 j_2 \cdots j_n}$$

is called the *determinant* of $\mathbf{A}$, where ρ represents the summation over all $n!$ permutations. The notation det $\mathbf{A}$ or $|\mathbf{A}|$ is usually used to represent the determinant of $\mathbf{A}$.

To illustrate, consider

$$\mathbf{A} = \begin{pmatrix} a_{11} & a_{12} & a_{13} \\ a_{21} & a_{22} & a_{23} \\ a_{31} & a_{32} & a_{33} \end{pmatrix}$$

Then

$$|\mathbf{A}| = a_{11}(a_{22}a_{33} - a_{23}a_{32}) - a_{12}(a_{21}a_{33} - a_{31}a_{23}) + a_{13}(a_{21}a_{32} - a_{22}a_{31})$$

The major properties of determinants can be summarized as follows:

1. If every element of a column or a row is zero, then the value of the determinant is zero.
2. The value of the determinant is not changed if its rows and columns are interchanged.
3. If $\mathbf{B}$ is obtained from $\mathbf{A}$ by interchanging any two of its rows (or columns), then $|\mathbf{B}| = -|\mathbf{A}|$.
4. If two rows (or columns) of $\mathbf{A}$ are identical, then $|\mathbf{A}| = 0$.
5. The value of $|\mathbf{A}|$ remains the same if a scalar α times a column (row) vector of $\mathbf{A}$ is added to another column (row) vector of $\mathbf{A}$.
6. If every element of a column (or a row) of a determinant is multiplied by a scalar α, the value of the determinant is multiplied by α.
7. If $\mathbf{A}$ and $\mathbf{B}$ are two n-square matrices, then

$$|\mathbf{AB}| = |\mathbf{A}|\,|\mathbf{B}|$$

Definition of the Minor of a Determinant. The minor M_{ij} of the element a_{ij} in the determinant $|\mathbf{A}|$ is obtained from the matrix $\mathbf{A}$ by striking out the ith row and jth column of $\mathbf{A}$. For example, for

$$\mathbf{A} = \begin{pmatrix} a_{11} & a_{12} & a_{13} \\ a_{21} & a_{22} & a_{23} \\ a_{31} & a_{32} & a_{33} \end{pmatrix}$$

$$M_{11} = \begin{vmatrix} a_{22} & a_{23} \\ a_{32} & a_{33} \end{vmatrix}, \quad M_{22} = \begin{vmatrix} a_{11} & a_{13} \\ a_{31} & a_{33} \end{vmatrix}, \ldots$$

Definition of the Adjoint Matrix. Let $A_{ij} = (-1)^{i+j}M_{ij}$ be defined as the **cofactor** of the element a_{ij} of the square matrix $\mathbf{A}$. Then, by definition, the adjoint matrix of $\mathbf{A}$ is given by

$$\text{adj } \mathbf{A} = \|A_{ij}\|^T = \begin{pmatrix} A_{11} & A_{21} & \cdots & A_{n1} \\ A_{12} & A_{22} & \cdots & A_{n2} \\ \vdots & \vdots & & \\ A_{1n} & A_{2n} & \cdots & A_{nn} \end{pmatrix}$$

For example, if

$$\mathbf{A} = \begin{pmatrix} 1 & 2 & 3 \\ 2 & 3 & 2 \\ 3 & 3 & 4 \end{pmatrix}$$

then, $A_{11} = (-1)^2(3 \times 4 - 2 \times 3) = 6$, $A_{12} = (-1)^3(2 \times 4 - 2 \times 3) = -2, \ldots$, or

$$\text{adj } \mathbf{A} = \begin{pmatrix} 6 & 1 & -5 \\ -2 & -5 & 4 \\ -3 & 3 & -1 \end{pmatrix}$$

A.2.5 NONSINGULAR MATRIX

A matrix is of a rank r if the largest *square* array in the matrix with nonvanishing determinant is of order r. A *square* matrix whose determinant does not vanish is called a **full-rank** or a **nonsingular** matrix. For example,

$$\mathbf{A} = \begin{pmatrix} 1 & 2 & 3 \\ 2 & 3 & 4 \\ 3 & 5 & 7 \end{pmatrix}$$

is a **singular** matrix, since

$$|\mathbf{A}| = 1(21 - 20) - 2(14 - 12) + 3(10 - 9) = 0$$

But $\mathbf{A}$ has a rank $r = 2$, since

$$\begin{pmatrix} 1 & 2 \\ 2 & 3 \end{pmatrix} = -1 \neq 0$$

A.2.6 THE INVERSE OF A MATRIX

If $\mathbf{B}$ and $\mathbf{C}$ are two n-square matrices such that $\mathbf{BC} = \mathbf{CB} = \mathbf{I}$, then $\mathbf{B}$ is called the inverse of $\mathbf{C}$ and $\mathbf{C}$ the inverse of $\mathbf{B}$. The common notation for the inverses is $\mathbf{B}^{-1}$ and $\mathbf{C}^{-1}$.

Theorem. *If* $\mathbf{BC} = \mathbf{I}$ *and* $\mathbf{B}$ *is* **nonsingular,** *then* $\mathbf{C} = \mathbf{B}^{-1}$, *which means that the inverse is unique.*

PROOF. By assumption,

$$\mathbf{BC} = \mathbf{I}$$

then

$$\mathbf{B}^{-1}\mathbf{BC} = \mathbf{B}^{-1}\mathbf{I}$$

or

$$\mathbf{IC} = \mathbf{B}^{-1}$$

or

$$\mathbf{C} = \mathbf{B}^{-1}$$

Two important results can be proved for nonsingular matrices:

1. If $\mathbf{A}$ and $\mathbf{B}$ are nonsingular n-square matrices, then $(\mathbf{AB})^{-1} = \mathbf{B}^{-1}\mathbf{A}^{-1}$.
2. If $\mathbf{A}$ is nonsingular, then $\mathbf{AB} = \mathbf{AC}$ implies that $\mathbf{B} = \mathbf{C}$.

The concept of matrix inversion is useful in solving n linearly independent equations. Consider

$$\begin{pmatrix} a_{11} & a_{12} & \cdots & a_{1n} \\ a_{21} & a_{22} & \cdots & a_{2n} \\ \vdots & \vdots & & \vdots \\ a_{n1} & a_{n2} & \cdots & a_{nn} \end{pmatrix} \begin{pmatrix} x_1 \\ x_2 \\ \vdots \\ x_n \end{pmatrix} = \begin{pmatrix} b_1 \\ b_2 \\ \vdots \\ b_n \end{pmatrix}$$

where x_i represent the unknowns and a_{ij} and b_i are constants. These n equations can be written in the form

$$\mathbf{AX} = \mathbf{b}$$

Since the equations are independent, it follows that $\mathbf{A}$ is nonsingular. Thus

$$\mathbf{A}^{-1}\mathbf{AX} = \mathbf{A}^{-1}\mathbf{b} \quad \text{or} \quad \mathbf{X} = \mathbf{A}^{-1}\mathbf{b}$$

gives the solution of the n unknowns.

A.2.7 METHODS OF COMPUTING THE INVERSE OF A MATRIX

Adjoint Matrix Method

Given $\mathbf{A}$ a nonsingular matrix of size n,

$$\mathbf{A}^{-1} = \frac{1}{|\mathbf{A}|} \operatorname{adj} \mathbf{A} = \frac{1}{|\mathbf{A}|} \begin{pmatrix} A_{11} & A_{21} & \cdots & A_{n1} \\ A_{12} & A_{22} & \cdots & A_{n2} \\ \vdots & \vdots & & \vdots \\ A_{1n} & A_{2n} & \cdots & A_{nn} \end{pmatrix}$$

For example, for

$$\mathbf{A} = \begin{pmatrix} 1 & 2 & 3 \\ 2 & 3 & 2 \\ 3 & 3 & 4 \end{pmatrix}$$

$$\operatorname{adj} \mathbf{A} = \begin{pmatrix} 6 & 1 & -5 \\ -2 & -5 & 4 \\ -3 & 3 & -1 \end{pmatrix} \quad \text{and} \quad |\mathbf{A}| = -7$$

Hence

$$\mathbf{A}^{-1} = \frac{1}{-7} \begin{pmatrix} 6 & 1 & -5 \\ -2 & -5 & 4 \\ -3 & 2 & -1 \end{pmatrix} = \begin{pmatrix} -6/7 & -1/7 & 5/7 \\ 2/7 & 5/7 & -4/7 \\ 3/7 & -3/7 & 1/7 \end{pmatrix}$$

Row Operations (Gauss–Jordan) Method

Consider the partitioned matrix $(\mathbf{A} \mid \mathbf{I})$, where $\mathbf{A}$ is nonsingular. By premultiplying this matrix by $\mathbf{A}^{-1}$, we obtain

$$(\mathbf{A}^{-1}\mathbf{A} \mid \mathbf{A}^{-1}\mathbf{I}) = (\mathbf{I} \mid \mathbf{A}^{-1})$$

Thus, by applying a sequence of row transformations only, the matrix A is changed to I and I is changed to A^{-1}.

For example, consider the system of equations of the form $AX = b$:

$$\begin{pmatrix} 1 & 2 & 3 \\ 2 & 3 & 2 \\ 3 & 3 & 4 \end{pmatrix} \begin{pmatrix} x_1 \\ x_2 \\ x_3 \end{pmatrix} = \begin{pmatrix} 3 \\ 4 \\ 5 \end{pmatrix}$$

The solution of X and the inverse of basis matrix can be obtained directly by considering

$$A^{-1}(A \mid I \mid b) = (I \mid A^{-1} \mid A^{-1}b)$$

Thus, by a row transformation operation, we get

$$\left(\begin{array}{ccc|ccc|c} 1 & 2 & 3 & 1 & 0 & 0 & 3 \\ 2 & 3 & 2 & 0 & 1 & 0 & 4 \\ 3 & 3 & 4 & 0 & 0 & 1 & 5 \end{array} \right)$$

Iteration 1:

$$\left(\begin{array}{ccc|ccc|c} 1 & 2 & 3 & 1 & 0 & 0 & 3 \\ 0 & -1 & -4 & -2 & 1 & 0 & -2 \\ 0 & -3 & -5 & -3 & 0 & 1 & -4 \end{array} \right)$$

Iteration 2:

$$\left(\begin{array}{ccc|ccc|c} 1 & 0 & -5 & -3 & 2 & 0 & -1 \\ 0 & 1 & 4 & 2 & -1 & 0 & 2 \\ 0 & 0 & 7 & 3 & -3 & 1 & 2 \end{array} \right)$$

Iteration 3:

$$\left(\begin{array}{ccc|ccc|c} 1 & 0 & 0 & -6/7 & -1/7 & 5/7 & 3/7 \\ 0 & 1 & 0 & 2/7 & 5/7 & -4/7 & 6/7 \\ 0 & 0 & 1 & 3/7 & -3/7 & 1/7 & 2/7 \end{array} \right)$$

This gives $x_1 = 3/7$, $x_2 = 6/7$, and $x_3 = 2/7$. The inverse of A is given by the right-hand-side matrix. This is the same as the inverse obtained by the method of adjoint matrix.

PARTITIONED MATRIX METHOD

Let the two nonsingular matrices A and B of size n be partitioned as shown here such that A_{11} is nonsingular.

$$A = \left(\begin{array}{c|c} \begin{matrix} A_{11} \\ (p \times p) \end{matrix} & \begin{matrix} A_{12} \\ (p \times q) \end{matrix} \\ \hline \begin{matrix} A_{21} \\ (q \times p) \end{matrix} & \begin{matrix} A_{22} \\ (q \times q) \end{matrix} \end{array} \right) \quad \text{and} \quad B = \left(\begin{array}{c|c} \begin{matrix} B_{11} \\ (p \times p) \end{matrix} & \begin{matrix} B_{12} \\ (p \times q) \end{matrix} \\ \hline \begin{matrix} B_{21} \\ (q \times p) \end{matrix} & \begin{matrix} B_{22} \\ (q \times q) \end{matrix} \end{array} \right)$$

If **B** is the inverse of **A**, from $\mathbf{AB} = \mathbf{I}_n$,

$$\mathbf{A}_{11}\mathbf{B}_{11} + \mathbf{A}_{12}\mathbf{B}_{21} = \mathbf{I}_p$$
$$\mathbf{A}_{11}\mathbf{B}_{12} + \mathbf{A}_{12}\mathbf{B}_{22} = 0$$

Also, from $\mathbf{BA} = \mathbf{I}_n$,

$$\mathbf{B}_{21}\mathbf{A}_{11} + \mathbf{B}_{22}\mathbf{A}_{21} = 0$$
$$\mathbf{B}_{21}\mathbf{A}_{12} + \mathbf{B}_{22}\mathbf{A}_{22} = \mathbf{I}_q$$

Since $\mathbf{A}_{11}$ is nonsingular, that is, $|\mathbf{A}_{11}| \neq 0$, solving for $\mathbf{B}_{11}$, $\mathbf{B}_{12}$, $\mathbf{B}_{21}$, and $\mathbf{B}_{22}$, we get

$$\mathbf{B}_{11} = \mathbf{A}_{11}^{-1} + (\mathbf{A}_{11}^{-1}\mathbf{A}_{12})\mathbf{D}^{-1}(\mathbf{A}_{21}\mathbf{A}_{11}^{-1})$$
$$\mathbf{B}_{12} = -(\mathbf{A}_{11}^{-1}\mathbf{A}_{12})\mathbf{D}^{-1}$$
$$\mathbf{B}_{21} = -\mathbf{D}^{-1}(\mathbf{A}_{21}\mathbf{A}_{11}^{-1})$$
$$\mathbf{B}_{22} = \mathbf{D}^{-1}$$

where

$$\mathbf{D} = \mathbf{A}_{22} - \mathbf{A}_{21}(\mathbf{A}_{11}^{-1}\mathbf{A}_{12})$$

To illustrate the use of these formulas, consider the example given previously,

$$\mathbf{A} = \begin{pmatrix} 1 & 2 & 3 \\ \hline 2 & 3 & 2 \\ 3 & 3 & 4 \end{pmatrix}$$

where

$$\mathbf{A}_{11} = (1), \quad \mathbf{A}_{12} = (2,\ 3), \quad \mathbf{A}_{21} = \begin{pmatrix} 2 \\ 3 \end{pmatrix}, \quad \text{and} \quad \mathbf{A}_{22} = \begin{pmatrix} 3 & 2 \\ 3 & 4 \end{pmatrix}$$

It is obvious that $\mathbf{A}_{11}^{-1} = 1$ and

$$\mathbf{D} = \begin{pmatrix} 3 & 2 \\ 3 & 4 \end{pmatrix} - \begin{pmatrix} 2 \\ 3 \end{pmatrix}(1)(2,\ 3) = \begin{pmatrix} -1 & -4 \\ -3 & -5 \end{pmatrix}$$

$$\mathbf{D}^{-1} = -1/7 \begin{pmatrix} -5 & 4 \\ 3 & -1 \end{pmatrix} = \begin{pmatrix} 5/7 & -4/7 \\ -3/7 & 1/7 \end{pmatrix}$$

Thus

$$\mathbf{B}_{11} = (-6/7) \quad \text{and} \quad \mathbf{B}_{12} = (-1/7 \quad 5/7)$$

$$\mathbf{B}_{21} = \begin{pmatrix} 2/7 \\ 3/7 \end{pmatrix} \quad \text{and} \quad \mathbf{B}_{22} = \begin{pmatrix} 5/7 & -4/7 \\ -3/7 & 1/7 \end{pmatrix}$$

which directly give $\mathbf{B} = \mathbf{A}^{-1}$.

A.3 QUADRATIC FORMS

Given

$$\mathbf{X} = (x_1, x_2, \ldots, x_n)^T$$

and

$$\mathbf{A} = \begin{pmatrix} a_{11} & a_{12} & \cdots & a_{1n} \\ a_{21} & a_{22} & \cdots & a_{2n} \\ \vdots & \vdots & & \vdots \\ a_{n1} & a_{n2} & \cdots & a_{nn} \end{pmatrix}$$

the function

$$Q(\mathbf{X}) = \mathbf{X}^T \mathbf{A} \mathbf{X} = \sum_{i=1}^{n} \sum_{j=1}^{n} a_{ij} x_i x_j$$

is called a *quadratic form*. The matrix $\mathbf{A}$ can always be assumed symmetric, since each element of every pair of coefficients a_{ij} and a_{ji} $(i \neq j)$ can be replaced by $(a_{ij} + a_{ji})/2$ without changing the value of $Q(\mathbf{X})$. This assumption has several advantages and hence is taken as a restriction.

To illustrate, the quadratic form

$$Q(\mathbf{X}) = (x_1, x_2, x_3) \begin{pmatrix} 1 & 0 & 1 \\ 2 & 7 & 6 \\ 3 & 0 & 2 \end{pmatrix} \begin{pmatrix} x_1 \\ x_2 \\ x_3 \end{pmatrix}$$

is the same as

$$Q(\mathbf{X}) = (x_1, x_2, x_3) \begin{pmatrix} 1 & 1 & 2 \\ 1 & 7 & 3 \\ 2 & 3 & 2 \end{pmatrix} \begin{pmatrix} x_1 \\ x_2 \\ x_3 \end{pmatrix}$$

Note that $\mathbf{A}$ is symmetric in the second case.

The quadratic form is said to be

1. *Positive-definite* if $Q(\mathbf{X}) > 0$ for every $\mathbf{X} \neq \mathbf{0}$.
2. *Positive-semidefinite* if $Q(\mathbf{X}) \geq \mathbf{0}$ for every $\mathbf{X}$ and there exist $\mathbf{X} \neq \mathbf{0}$ such that $Q(\mathbf{X}) = \mathbf{0}$.
3. *Negative-definite* if $-Q(\mathbf{X})$ is positive-definite.
4. *Negative-semidefinite* if $-Q(\mathbf{X})$ is positive-semidefinite.
5. *Indefinite* if it is none of these cases.

It can be proved that the necessary and sufficient conditions for the realization of the cases above are given by

1. $Q(\mathbf{X})$ is positive-definite (semidefinite) if the values of the principal minor determinants of $\mathbf{A}$ are positive (nonnegative).† In this case $\mathbf{A}$ is said to be positive-definite (semidefinite).

---

† The kth *principal minor* determinant of $\mathbf{A}_{n \times n}$ is defined by

$$\begin{vmatrix} a_{11} & a_{12} & \cdots & a_{1k} \\ a_{21} & a_{11} & \cdots & a_{2k} \\ \vdots & \vdots & & \vdots \\ a_{k1} & a_{k2} & & a_{kk} \end{vmatrix}, \qquad k = 1, 2, \ldots, n$$

+

2. $Q(X)$ is negative-definite if the value of kth principal minor determinant of **A** has the sign of $(-1)^k$, $k = 1, 2, \ldots, n$. In this case, **A** is called negative-definite.

3. $Q(X)$ is a negative-semidefinite if the kth principal minor determinant of **A** is either zero or has the sign of $(-1)^k$, $k = 1, 2, \ldots, n$.

SELECTED REFERENCES

HADLEY, G., *Matrix Algebra,* Addison-Wesley, Reading, Mass., 1961.
HOHN, F., *Elementary Matrix Algebra,* 2nd ed., Macmillan, New York, 1964.

PROBLEMS

☐ **A–1** Show that the following vectors are linearly dependent.

(a)
$$\begin{pmatrix} 1 \\ -2 \\ 3 \end{pmatrix} \quad \begin{pmatrix} -2 \\ 4 \\ -2 \end{pmatrix} \quad \begin{pmatrix} 1 \\ -2 \\ -1 \end{pmatrix}$$

(b)
$$\begin{pmatrix} 2 \\ -3 \\ 4 \\ 5 \end{pmatrix} \quad \begin{pmatrix} 4 \\ -6 \\ 8 \\ 10 \end{pmatrix}$$

☐ **A–2** Given

$$\mathbf{A} = \begin{pmatrix} 1 & 4 & 9 \\ 2 & 5 & -8 \\ 3 & 7 & 2 \end{pmatrix} \quad \text{and} \quad \mathbf{B} = \begin{pmatrix} 7 & -1 & 2 \\ 9 & 4 & 8 \\ 3 & 6 & 10 \end{pmatrix}$$

find
(a) $\mathbf{A} + 7\mathbf{B}$.
(b) $2\mathbf{A} - 3\mathbf{B}$.
(c) $(\mathbf{A} + 7\mathbf{B})^T$.

☐ **A–3** In Problem A–2, show that $\mathbf{AB} \neq \mathbf{BA}$.

☐ **A–4** Given the partitioned matrices

$$\mathbf{A} = \left(\begin{array}{c|cc} 1 & 5 & 7 \\ 2 & -6 & 9 \\ \hline 3 & 7 & 2 \\ 4 & 9 & 1 \end{array} \right) \quad \text{and} \quad \mathbf{B} = \left(\begin{array}{ccc|c} 2 & 3 & -4 & 5 \\ \hline 1 & 2 & 6 & 7 \\ 3 & 1 & 0 & 9 \end{array} \right)$$

find **AB** using partitioning.

☐ **A–5** In Problem A–2, find $\mathbf{A}^{-1}$ and $\mathbf{B}^{-1}$ using:
(a) The adjoint matrix method.
(b) The row operations method.
(c) The partitioned matrix method.

☐ **A–6** Verify the formulas given in the section "Partitioned Matrix Method" (pp. 805–806) for obtaining the inverse of a partitioned matrix.

☐ **A–7** Find the inverse of

$$A = \begin{pmatrix} 1 & G \\ H & B \end{pmatrix}$$

where **B** is a nonsingular matrix.

☐ **A–8** Show that the quadratic form

$$Q(x_1, x_2) = 6x_1 + 3x_2 - 4x_1 x_2 - 2x_1^2 - 3x_2^2 - 27/4$$

is negative-definite.

☐ **A–9** Show that the quadratic form

$$Q(x_1, x_2, x_3) = 2x_1^2 + 2x_2^2 + 3x_3^2 + 2x_1 x_2 + 2x_2 x_3$$

is positive-definite.

Appendix B

Review of Basic Theorems in Differential Calculus

B.1   DEFINITIONS

B.1.1   CONTINUOUS FUNCTION

A single-variable function $f(x)$ is said to be continuous at a point x_0 if for any $\epsilon > 0$, however small, there exists δ such that for $|h| < \delta$, $\delta > 0$,

$$|f(x_0 + h) - f(x_0)| \leq \epsilon$$

Similarly, an n-variable function $f(\mathbf{X})$, $\mathbf{X} = (x_1, x_2, \ldots, x_n)$, is continuous at a point $\mathbf{X}_0$ if for any $\epsilon > 0$, however small, there exists δ such that, for $|\mathbf{h}| < \delta$,

$$|f(\mathbf{X}_0 + \mathbf{h}) - f(\mathbf{X}_0)| \leq \epsilon$$

where

$$\mathbf{h} = (h_1, h_2, \ldots, h_n)$$
$$\delta = (\delta_1, \delta_2, \ldots, \delta_n) > 0$$

B.1.2   PARTIAL DERIVATIVE

For a single-variable function $f(x)$, the limit

$$\lim_{h \to 0} \frac{f(x_0 + h) - f(x_0)}{h}$$

at the point x_0, when it exists, defines the derivative of the function at x_0. This is usually written as $f'(x_0)$ or $df(x_0)/dx$.

For an n-variable function, an equivalent definition of the first *partial derivative* of the function with respect to any one of its n variables at a point $\mathbf{X}_0 = (x_1^0, x_2^0, \ldots, x_n^0)$ is given by

827

$$\frac{\partial f(\mathbf{X}_0)}{\partial x_i} = \lim_{h_i \to 0} \frac{f(x_1^0, \ldots, x_{i-1}^0, x_i^0 + h_i, x_{i+1}^0, \ldots, x_n^0) - f(\mathbf{X}_0)}{h_i}$$

provided that the limit exists.

The first partial derivatives of a function $f(\mathbf{X})$ at a certain point define the slopes of the tangent to the function with respect to the n coordinate axes. A useful notation for summarizing these first partial derivatives is to use the **gradient vector,** which is given by

$$\nabla f = \left(\frac{\partial f}{\partial x_1}, \ldots, \frac{\partial f}{\partial x_n} \right)$$

The second partial derivative is defined by taking the partial derivatives of the functions resulting from the first partial derivatives provided, of course, that they exist. This is written as

$$\frac{\partial^2 f}{\partial x_i \, \partial x_j}, \qquad \text{for all } i, j$$

A compact way for summarizing the second partial derivatives is to use the **Hessian** matrix. This is defined for $f(\mathbf{X})$ by

$$\mathbf{H} = \begin{pmatrix} \dfrac{\partial^2 f}{\partial x_1^2} & \dfrac{\partial^2 f}{\partial x_1 \, \partial x_2} & \cdots & \dfrac{\partial^2 f}{\partial x_1 \, \partial x_n} \\ \vdots & \vdots & & \vdots \\ \dfrac{\partial^2 f}{\partial x_n \, \partial x_1} & \dfrac{\partial^2 f}{\partial x_n \, \partial x_2} & \cdots & \dfrac{\partial^2 f}{\partial x_n^2} \end{pmatrix}$$

B.2 ROLLE'S THEOREM

Given the function $f(x)$ which is continuous in the closed interval $[a, b]$ with $f(a) = f(b)$, there exists at least one point ξ such that $a < \xi < b$ at which the first derivative of f vanishes, that is, $f'(\xi) = 0$.

The proof of this theorem can be found in Rudin [1964].

B.3 MEAN VALUE THEOREMS

B.3.1 FIRST MEAN VALUE THEOREM

Theorem. *Given a function $f(x)$ that is continuous in the closed interval $[a, b]$ and its first derivative exists at every interior point, there exists a point ξ in the open interval (a, b) that satisfies*

$$\frac{f(b)-f(a)}{b-a}=f'(\xi), \qquad a<\xi<b$$

Or, if b = a + h, then

$$\frac{f(a+h)-f(a)}{h}=f'(a+\theta h), \qquad 0<\theta<1$$

PROOF. Let $H(x) = f(x) - cx$, where c, is determined such that $H(a) = H(b)$. This means that

$$c=\frac{f(b)-f(a)}{b-a}$$

Since $H(x)$ satisfies Rolle's theorem (Section B.2), for some ξ in the open interval (a, b),

$$H'(\xi)=f'(\xi)-c=0$$

or

$$\frac{f(b)-f(a)}{b-a}=f'(\xi)$$

See Figure B–1.

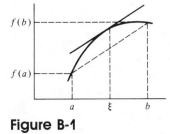

Figure B-1

B.3.2 CAUCHY'S SECOND MEAN VALUE THEOREM

Theorem. *Given the two functions $f(x)$ and $g(x)$ that are continuous at every point in the closed interval $[a, b]$ with $g(a) \neq g(b)$. Furthermore, their first derivatives exist at every interior point and do not vanish simultaneously at any interior point. There exists a point ξ, $a < \xi < b$, such that*

$$\frac{f(b)-f(a)}{g(b)-g(a)}=\frac{f'(\xi)}{g'(\xi)}, \qquad a<\xi<b$$

Or, if b = a + h, then

$$\frac{f(a+h)-f(a)}{g(a+h)-g(a)}=\frac{f'(a+\theta h)}{g'(a+\theta h)}, \qquad 0<\theta<1$$

PROOF. Let $H(x) = f(x) - cg(x)$ and determine c such that $H(a) = H(b)$. Thus,

$$c = \frac{f(b) - f(a)}{g(b) - g(a)}$$

By Rolle's theorem,

$$H'(\xi) = f'(\xi) - cg'(\xi) = 0$$

or

$$\frac{f(b) - f(a)}{g(b) - g(a)} = c = \frac{f'(\xi)}{g'(\xi)}, \qquad a < \xi < b$$

B.4 L'HÔPITAL'S RULE

Theorem. *Given two functions, $f(x)$ and $g(x)$, that both vanish $(= 0)$ at $x = a$. If $f(x)$ and $g(x)$ are continuous in the closed interval $[a, b]$ and are differentiable at every x, $a < x < b$, such that*

$$\lim_{x \to a} \frac{f'(x)}{g'(x)} = B$$

then

$$\lim_{x \to a} \frac{f(x)}{g(x)} = B$$

PROOF. Since $f(a) = g(a) = 0$, then for some x, $a < x \leq b$, by the second mean value theorem

$$\frac{f(x)}{g(x)} = \frac{f(x) - f(a)}{g(x) - g(a)} = \frac{f'(\xi)}{g'(\xi)}, \qquad a < \xi < x$$

Since $a < \xi < x$, then as $x \to a$, $\xi \to a$. Hence,

$$\lim_{x \to a} \frac{f(x)}{g(x)} = \lim_{\xi \to a} \frac{f'(\xi)}{g'(\xi)}$$

L'Hôpital's rule applies also to the undetermined forms of the type ∞/∞. The proof is not given here, however. Other forms, including $0 \cdot \infty$, 0^0, ∞^0, 1^∞, and $\infty - \infty$, can be reduced to the form ∞/∞ or $0/0$ using simple transformations. For example, if $f - g = \infty - \infty$, then

$$fg\left(\frac{1}{g} - \frac{1}{f}\right) \equiv \infty \cdot 0$$

or

$$\frac{1/g - 1/f}{1/fg} = \frac{0}{0}$$

for which the l'Hôpital rule is now applicable. The other undetermined forms, including 0^0, ∞^0, and 1^∞, are transformed to the forms indicated by taking their logarithms.

B.5 POLYNOMIAL APPROXIMATION

Given a function $f(x)$ which together with its first n derivatives are defined at the point $x = a$, then $f(x)$ can be approximated by the nth-order polynomial

$$p_n(x) = \sum_{i=0}^{n} c_i(x-a)^i$$

where c_i are constants to be determined as follows.

$$p_n(a) = f(a)$$

and

$$p_n^{(i)}(a) = f^{(i)}(a), \qquad i = 1, 2, \ldots, n$$

where the superscript (i) signifies the ith derivative. Thus

$$c_0 = p_n(a) = f(a)$$

and

$$c_i = \frac{f^{(i)}(a)}{i!}$$

This yields

$$p_n(x) = \sum_{i=0}^{n} \frac{f^{(i)}(a)}{i!}(x-a)^i$$

To illustrate, consider

$$f(x) = 5x^3 + 6x^2 - 3x + 5$$

For $a = 1$,

$$f(1) = 13, \quad f'(1) = 24, \quad f''(1) = 42, \quad f'''(1) = 30$$

and

$$f^{(i)}(1) = 0, \qquad i = 4, 5, \ldots$$

Thus

$$f(x) = p_3(x) = 13 + 24(x-1) + 21(x-1)^2 + 5(x-1)^3$$

Consider next $f(x) = \sin x$, for $a = 0$.

$$f(0) = 0$$

$$f^{(i)}(0) = \begin{cases} 1, & i = 1, 5, 9, \ldots \\ -1, & i = 3, 7, 11, \ldots \\ 0, & i = 2, 4, 6, \ldots \end{cases}$$

Thus

$$p_n(x) = x - \frac{x^3}{3!} + \frac{x^5}{5!} - \cdots + (-1)^{n+1} \frac{x^{2n-1}}{(2n-1)!}$$

It is noticed in the first example that $f(x) = p_3(x)$ because $f(x)$ itself is polynomial with a vanishing fourth derivative. In the second example $f^{(i)}(0)$ exists for odd $i \geq 1$, and hence $p_n(x)$ can only approximate $f(x)$.

The next step is to find an expression for $f(x) - p_n(x)$. This is called the **remainder** and may be written as

$$R_n(x) = f(x) - p_n(x)$$

By definition, since $f(a) = p_n(a)$ and $f^{(i)}(a) = p_n^{(i)}(a)$, for all i, then $R_n(a) = 0$ and $R_n^{(i)}(a) = 0$, for all i.

Assume that $f^{(n+1)}(x)$ exists at all points in the *closed interval* $[a, b]$. Thus $f^{(n)}(x)$ must exist also in the same interval and must be continuous. Consequently, $f^{(i)}(x)$, $i = 1, 2, \ldots, n - 1$, must also be continuous in $[a, b]$. Then it can be proved [Kaplan, (1952)] that

$$R_n(x) = \frac{f^{(n+1)}(\xi)}{(n+1)!} (x - a)^{n+1}, \qquad a < \xi < x$$

This result leads to Taylor's theorem, which is presented in the next section.

B.5.1 TAYLOR'S THEOREM

Given that the $(n + 1)$st derivative of $f(x)$ exists at every point of the closed interval $[a, b]$, for $0 < h \leq b - a$,

$$f(a + h) = f(a) + \sum_{i=1}^{n} \frac{f^{(i)}(a)}{i!} h^i + R_n(a + \theta h)$$

$$= f(a) + \sum_{i=1}^{n} \frac{f^{(i)}(a)}{i!} h^i + \frac{f^{(n+1)}(a + \theta h)}{(n+1)!} h^{n+1}, \qquad 0 < \theta < 1$$

This theorem follows directly from the development given previously.

If $R_n(a + \theta h) \to 0$ as $n \to \infty$, the theorem yields

$$f(a + h) = f(a) + \frac{f'(a)h}{1!} + \frac{f''(a)h^2}{2!} + \cdots$$

For example, for $f(x) = e^x, f^{(n)}(x) = e^x$. Let $a = 0, f^{(n)}(0) = 1$. Consequently,

$$f(0+h)=f(0)+\frac{f'(0)}{1!}h+\frac{f''(0)h^2}{2!}+\cdots+\frac{f^{(n+1)}(0+\theta h)}{(n+1)!}h^{n+1}$$

$$=1+h+\frac{h^2}{2!}+\cdots+\frac{h^n}{n!}+\frac{e^{\theta h}}{(n+1)!}h^{n+1}$$

It can be shown that

$$\lim_{n\to\infty}\frac{h^{n+1}}{(n+1)!}=0$$

Hence

$$e^x=1+x+\frac{x^2}{2!}+\cdots$$

Taylor's theorem may be extended to functions of n variables as follows. Assume that the second partial derivatives of $f(X)$ exist and are continuous. Let $h=(h_1, h_2, \ldots, h_n)^T$. Taylor's expansion around X_0 is given by

$$f(X_0+h)=f(X_0)+\nabla f(X_0)h+\tfrac{1}{2}h^T Hh\Big|_{X_0+\theta h}$$

where $\nabla f(X_0)$ is the gradient vector evaluated at X_0 and H is the Hessian matrix evaluated at $X_0+\theta h$ (see Section B.1.2). The third term is the expression representing the remainder.

In the case where the nth partial derivative of f exists and is continuous, Taylor's expansion can be generalized to $(n+1)$ terms. This generalization is not needed in this presentation and hence it is not included.

To illustrate the expansion of an n-variable function, consider

$$f(X)=f(x_1, x_2)=x_1^2+3x_1 e^{x_2}$$

It is required to expand $f(X)$ around $X_0=(1, 0)$. Thus

$$h=\binom{h_1}{h_2}=\binom{x_1}{x_2}-\binom{1}{0}=\binom{x_1-1}{x_2}$$

$$X_0+\theta h=\binom{1}{0}+\theta\binom{x_1-1}{x_2}=\binom{1-\theta+\theta x_1}{\theta x_2}$$

$$\nabla f(X_0)=\left(\frac{\partial f}{\partial x_1},\frac{\partial f}{\partial x_2}\right)=(2x_1+3e^{x_2}, 3x_1 e^{x_2})|_{X_0}=(5, 3)$$

$$H=\begin{pmatrix}\dfrac{\partial^2 f}{\partial x_1^2} & \dfrac{\partial^2 f}{\partial x_1\,\partial x_2}\\[2mm]\dfrac{\partial^2 f}{\partial x_2\,\partial x_1} & \dfrac{\partial^2 f}{\partial x_2^2}\end{pmatrix}=\begin{pmatrix}2 & 3e^{x_2}\\3e^{x_2} & 3x_1 e^{x_2}\end{pmatrix}$$

Hence

$$f(\mathbf{X}) = f(1, 0) + \nabla f(1, 0)\mathbf{h} + \tfrac{1}{2}\mathbf{h}^T \mathbf{H} \mathbf{h} \Big|_{\mathbf{X}_0 + \theta \mathbf{h}}$$

$$= 4 + (5, 3)\begin{pmatrix} x_1 - 1 \\ x_2 \end{pmatrix}$$

$$+ \tfrac{1}{2}(x_1 - 1, x_2)\begin{pmatrix} 2 & 3e^{\theta x_2} \\ 3e^{\theta x_2} & 3(1 - \theta + \theta x_1)e^{\theta x_2} \end{pmatrix}\begin{pmatrix} x_1 - 1 \\ x_2 \end{pmatrix}$$

Taylor's theorem will prove especially useful in developing the sufficiency conditions for identifying the maxima and minima of a differentiable function.

B.6 CONVEX AND CONCAVE FUNCTIONS

A function $f(\mathbf{X})$ is said to be strictly convex if, for any two other distinct points $\mathbf{X}_1$ and $\mathbf{X}_2$,

$$f(\lambda \mathbf{X}_1 + (1 - \lambda)\mathbf{X}_2) < \lambda f(\mathbf{X}_1) + (1 - \lambda)f(\mathbf{X}_2)$$

where $0 < \lambda < 1$. On the other hand, a function $f(\mathbf{X})$ is strictly concave if $-f(\mathbf{X})$ is strictly convex.

An important special case of the convex (concave) function is the quadratic form (see Section A.3)

$$f(\mathbf{X}) = \mathbf{C}\mathbf{X} + \mathbf{X}^T \mathbf{A} \mathbf{X}$$

where $\mathbf{C}$ is a constant vector and $\mathbf{A}$ is a symmetric matrix. It can be proved that $f(\mathbf{X})$ is strictly convex if $\mathbf{A}$ is positive-definite. Similarly, $f(\mathbf{X})$ is strictly concave if $\mathbf{A}$ is negative-definite.

SELECTED REFERENCES

BRAND, L., *Advanced Calculus*, Wiley, New York, 1955.

KAPLAN, W., *Advanced Calculus*, Addison-Wesley, Reading, Mass., 1952.

RUDIN, W., *Principles of Mathematical Analysis*, 2nd ed., McGraw-Hill, New York, 1964.

PROBLEMS

☐ **B–1** Given $f(x, y) = 0$, use Taylor's series to derive the expression for dy/dx in terms of the partial derivatives of $f(x, y)$.

☐ **B–2** Expand the following function around (1, 1, 0).

$$f(x, y, z) = 5x^2 \ln y + 3xye^z$$

☐ **B–3** Using Taylor's series, expand the function

$$f(x) = \cos x$$

around $x = 0$. Show that the remainder tends to zero as the number of terms n tends to infinity.

☐ **B–4** Show that the function $f(x) = e^x$ is strictly convex over all real values of x.

☐ **B–5** Show that the quadratic function

$$f(x_1, x_2, x_3) = 5x_1^2 + 5x_2^2 + 4x_3^2 + 4x_1x_2 + 2x_2x_3$$

is strictly convex.

☐ **B–6** In Problem B–5, show that $-f(x_1, x_2, x_3)$ is strictly concave.

Appendix C

General Program for LP Problems

The computer program in this appendix is designed to solve linear programs using the same format employed in the text. The program, which is totally interactive, is written in FORTRAN IV. It is self-documenting in the sense that it issues instructions for the preparation of the input data.

1. Program Limitations

The program handles any type of LP problem with any type of constraints, including unrestricted variables. The program will automatically add artificial variables where necessary. It will also convert the unrestricted variables into nonnegative variables. The augmentation of artificials and the conversion of the unrestricted variables is achieved in the exact manner presented in the text.

The program can handle up to 50 constraints and 100 variables. The limit on the variables is defined to include all the artificials as well as those resulting from the conversion of the unrestricted variables. These limits can be adjusted as desired by making a corresponding change in the DIMENSION statement of the program.

2. Output

The program will automatically print the initial and optimal tableaus (using the same format of the text) together with a summary of the optimal solution in terms of the original variables of the problem. At the request of the user, the program will also print all the intermediate tableaus. Special messages for the cases of unbounded and infeasible solutions are also printed out by the program.

3. Example of a Computer Session

In this section, we show how the problems listed are solved by the program in a typical interactive computer session.

Example 1

$$\text{Maximize } z = 2x_1 + x_2$$

subject to

$$x_1 - x_2 \le 10$$
$$2x_1 \quad\quad \le 40$$
$$x_1,\ x_2 \ge 0$$

Example 2

$$\text{Minimize } z = 10x_1 + 8x_2$$

Subject to

$$x_1 + 2x_2 \ge 5$$
$$2x_1 - x_2 \ge 12$$
$$x_1 + 3x_2 \ge 4$$
$$x_1 \ge 0, \quad\quad x_2 \text{ unrestricted}$$

```
DO YOU NEED INSTRUCTIONS (TYPE 1=YES OR 0=NO)
?
1
DATA MUST BE ENTERED AS FOLLOWS:
LINE 1: PROBLEM NAME, # CONSTRS, # VARS, # UNRESTRICTED VARS
        (IF NO UNRESTRICTED VARS, YOU MUST TYPE 0 "ZERO")
LINE 2: MAX OR MIN, OBJ COEFFS
FOLLOWING LINES: CONSTR TYPE(GE,LE OR EQ), CONSTR COEFFS, RHS
LAST LINE: INDICES OF UNRESTRICTED VARS. (IF NONE, DELETE LINE)
------------------------------------------------
EXAMPLE
    MAXIMIZE Z = 2X1      + 4X3 + 5X4
SUBJECT TO
               X1 + X2 - 3X3 - 2X4 <= 1
               5X1 + 7X2 + 2X3 - X4  = 8
               9X1 + X2        + 6X >= 9
               X1,X2 UNRESTR, X3 >=0
INPUT DATA ARE:
'EXAMPLE',3,4,2          <HIT RETURN>
'MAX',2,0,4,5            <HIT RETURN>
'LE',1,1,-3,-2,1         <HIT RETURN>
'EQ',5,7,2,-1,8          <HIT RETURN>
'GE',9,1,0,6,9           <HIT RETURN>
 1,2                     <HIT RETURN>
    ------------------------------------------

PLS ENTER DATA NOW
?
'EX1',2,2,0
?
'MAX',2,1
?
'LE',1,-1,10
?
'LE',2,0,40

 DATA SET IS NOW COMPLETE
```

```
DO YOU WANT TO PRINT ALL TABLEAUS (TYPE 1=YES OR 0=NO)
?
0
1                  PROBLEM-EX1 (MAX)

*** INITIAL TABLEAU ***
        OBJ COEFF     2.00      1.00      0.0       0.0
                        1         2         3         4
          X( 0)       0.0       0.0       0.0       0.0       0.0
          X( 3)       1.00     -1.00      1.00      0.0      10.00
          X( 4)       2.00      0.0       0.0       1.00      4.00

    UNBOUNDED SOLUTION  --  X( 2) CANNOT BE MADE BASIC

DO YOU WANT TO RUN A NEW PROBLEM (TYPE 1=YES OR 0=NO)
?
1

PLS ENTER DATA NOW
?
'EX2',3,2,1
?
'MIN',10,8
?
'GE',1,2,5
?
'GE',2,-1,12
?
'GE',1,3,4
?
2

  DATA SET IS NOW COMPLETE


DO YOU WANT TO PRINT ALL TABLEAUS (TYPE 1=YES OR 0=NO)
?
0
THE UNRESTR VARS RESULTED IN THE FOLLOWING SUBSTITUTION:
     ORIGINAL X( 1) = X( 1) =
     ORIGINAL X( 2) = X( 3) - X( 2) =

NOTE: VALUE OF "M" FOR ARTIFICIAL VARS IS "1E4".
                  PROBLEM-EX2 (MIN)

*** INITIAL TABLEAU ***
        OBJ COEFF    10.00      8.00     -8.00      0.0       0.0
                        1         2         3         4         5
          X( 0)       0.0       0.0       0.0       0.0       0.0       0.0
          X( 7)       1.00      2.00     -2.00     -1.00      0.0       5.00
          X( 8)       2.00     -1.00      1.00      0.0      -1.00     12.00
          X( 9)       1.00      3.00     -3.00      0.0       0.0       4.00
        OBJ COEFF     0.0    10000.00  10000.00  10000.00
                        6         7         8         9
          X( 0)       0.0       0.0       0.0       0.0       0.0
          X( 7)       0.0       1.00      0.0       0.0       5.00
          X( 8)       0.0       0.0       1.00      0.0      12.00
          X( 9)      -1.00      0.0       0.0       1.00      4.00

*** OPTIMUM TABLEAU (ITERATION # 4) ***
                        1         2         3         4         5
          X( 0)       0.0       0.0       0.0      -5.20     -2.40     54.80
          X( 6)       0.0       0.0       0.0      -1.40      0.20      0.60
          X( 3)       0.0      -1.00      1.00      0.40     -0.20      0.40
          X( 1)       1.00      0.0       0.0      -0.20     -0.40      5.80
```

```
                     6         7          8          9
         X( 0)      0.0    -9994.80   -9997.60  -10000.00    54.80
         X( 6)      1.00       1.40      -0.20      -1.00     0.60
         X( 3)      0.0       -0.40       0.20       0.0      0.40
         X( 1)      0.0        0.20       0.40       0.0      5.80
```

```
*** OPTIMAL SOLUTION ***
OBJECTIVE VALUE =     54.8000
     ORIGINAL X( 1) = X( 1) =       5.8000
     ORIGINAL X( 2) = X( 3) - X( 2) =     -0.4000


DO YOU WANT TO RUN A NEW PROBLEM (TYPE 1=YES OR 0=NO)
?
0
 STOP           0
```

4. Program Listing

```fortran
      DIMENSION A(50,100), B(50), CJ(100), NXI(50),
     *KODE(50),C(50,100),D(100),IN(50), IS(50,2)
      IIM=50
      IIN=100
C              SIMPLEX PROGRAM FOR LINEAR PROGRAMMING
      INTEGER GE,EQ
      DATA MIN/3HMIN/, MAX/3HMAX/,GE/2HGE/,LE/2HLE/, EQ/2HEQ/
   61 FORMAT (A4, 10I3)
   63 FORMAT (16F5.0)
   64 FORMAT (1H1, 20X, 'PROBLEM-',A4,1H(,A3,1H))
   65 FORMAT (/'  ITERATION NO.', I2)
   66 FORMAT (I24, 9I10)
   67 FORMAT (10X, 3H X(, I2, 1H), 10F10.2, 2X, F10.2)
   71 FORMAT(//' *** OPTIMUM TABLEAU (ITERATION #', I2,') ***')
   72 FORMAT(/' UNBOUNDED SOLUTION  --  X(',I2,') CANNOT BE MADE BASIC')
   73 FORMAT (10X, ' X( 0)', 10F10.2, 2X, F10.2)
   74 FORMAT (' NO FEASIBLE SOLUTION SINCE ARTIF. VAR. X(',I2,
     *  ') IS BASIC AND POSITIVE')
   75 FORMAT ('       OBJ COEFF',10F10.2, 2X, F10.2)
   80 FORMAT(//'DO YOU NEED INSTRUCTIONS (TYPE 1=YES OR 0=NO)')
   85 FORMAT('DATA MUST BE ENTERED AS FOLLOWS:'/
     *'LINE 1: PROBLEM NAME, # CONSTRS, # VARS, # UNRESTRICTED VARS'/
     *'         (IF NO UNRESTRICTED VARS, YOU MUST TYPE 0 "ZERO")'/
     *'LINE 2: MAX OR MIN, OBJ COEFFS'/
     *'FOLLOWING LINES: CONSTR TYPE(GE,LE OR EQ), CONSTR COEFFS, RHS'/
     *'LAST LINE: INDICES OF UNRESTRICTED VARS. (IF NONE, DELETE LINE)'/
     *  '----------------------------------------------'/
     * 'EXAMPLE'/'    MAXIMIZE Z = 2X1    + 4X3 + 5X4'/'SUBJECT TO '/
     * '              X1 + X2 - 3X3 - 2X4 <= 1'/
     * '              5X1 + 7X2 + 2X3 - X4  = 8'/
     * '              9X1 + X2       + 6X >= 9'/
     * '              X1,X2 UNRESTR, X3 >=0'/
     *  'INPUT DATA ARE:'/
     *  '''EXAMPLE'',3,4,2        <HIT RETURN>'/
     *  '''MAX'',2,0,4,5          <HIT RETURN>'/
     *  '''LE'',1,1,-3,-2,1       <HIT RETURN>'/
     *  '''EQ'',5,7,2,-1,8        <HIT RETURN>'/
     *  '''GE'',9,1,0,6,9         <HIT RETURN>'/
     * ' 1,2               <HIT RETURN>'/
     *  '----------------------------------------------')
   87 FORMAT(/'NOTE: VALUE OF "M" FOR ARTIFICIAL VARS IS "1E4".')
   90 FORMAT(/' DATA SET IS NOW COMPLETE'//)
   95 FORMAT('DO YOU WANT TO PRINT ALL TABLEAUS (TYPE 1=YES OR 0=NO)')
```

```
  500 FORMAT(5X,'ORIGINAL X(',I2,') = X(',I2,') - X(',I2,') = ',F10.4)
  510 FORMAT(5X,'ORIGINAL X(',I2,') = X(',I2,') = ', F10.4)
  520 FORMAT(/ '*** INITIAL TABLEAU ***')
  530 FORMAT('THE UNRESTR VARS RESULTED IN THE FOLLOWING SUBSTITUTION:')
  540 FORMAT(//'DO YOU WANT TO RUN A NEW PROBLEM (TYPE 1=YES OR 0=NO)')
  550 FORMAT(//'PLS ENTER DATA NOW')
  560 FORMAT(//'*** OPTIMAL SOLUTION ***'/'OBJECTIVE VALUE = ',F10.4)
  580 FORMAT('            X(',I2,') = ', F10.4)
      TOL=.00001
      WRITE(6,80)
      READ(9,*) INSTR
      IF(INSTR.EQ.0) GOTO 2323
      WRITE(6,85)
 2323 WRITE (6,550)
      CV=1E4
      READ (9,*) PROB, M, N, KUN
      READ(9,*) KODE(1),(D(J), J=1, N)
      N1=N
      NN=N
      NPR=0
      IF (KODE(1).EQ.MAX) GOTO 5
      KOD=1
      GOTO 6
    5 CV=-CV
      KOD=-1
    6 M1=M + 1
      DO 1000 I=2, M1
 1000 READ(9,*) KODE(I),(C(I,J), J=1, N), B(I)
      IF (KUN.EQ.0) GOTO 290
      READ(9,*)(IN(I), I=1, KUN)
  290 KNT=0
      WRITE (6,90)
      WRITE (6,95)
      READ(9,*) NPR
      DO 399 I=1, N
      IF (KUN.EQ.0) GOTO 305
      DO 300 J=1, KUN
      IF (IN(J).EQ.I) GOTO 320
  300 CONTINUE
  305 I1=I+KNT
      IS(I,1)=I1
      IS(I,2)=0
      CJ(I1)=D(I)
      DO 310 I2=2, M1
  310 A(I2,I1)=C(I2,I)
      GOTO 399
  320 KNT=KNT+1
      I1=KNT+I-1
      I2=I1+1
      IS(I,1)=I1
      IS(I,2)=I2
      CJ(I1)=D(I)
      CJ(I2)=-D(I)
      DO 330 L=2,M1
      A(L,I1)=C(L,I)
  330 A(L,I2)=-C(L,I)
  399 CONTINUE
      N=N+KUN
      DO 410 I=2,M1
      IF (KODE(I).NE.GE) GOTO 410
      N=N+1
      A(I,N)=-1
  410 CONTINUE
      AV=0
```

```
          DO 420 I=2,M1
          N=N+1
          A(I,N)=1
          NXI(I)=N
          IF (KODE(I).EQ.LE) GOTO 420
          AV=1
          CJ(N)=CV
   420 CONTINUE
  2222 ITER=0
          IF (KUN.EQ.0) GOTO 3333
          WRITE (6,530)
          DO 690 I =1, N1
          IF (IS(I,2).EQ.0) GOTO 600
          IF (IS(I,1).EQ.IS(I,2)) GOTO 600
          WRITE (6,500) I,IS(I,2),IS(I,1)
          GOTO 690
   600 WRITE(6,510)I, IS(I,1)
   690 CONTINUE
  3333 IF (AV.EQ.1) WRITE (6,87)
          WRITE(6,64) PROB, KODE(1)
          WRITE (6,520)
C        PRINT TABLEAU
          IF (ITER.EQ.0) GOTO 4444
          IF (NPR.EQ.0) GOTO 55
  1212 WRITE(6,65) ITER
  4444 N1=1
          N2=6
    43 IF (N2-N) 45,45,44
    44 N2=N
    45 IF (ITER.EQ.0) WRITE(6,75) (CJ(J), J=N1, N2)
          WRITE(6,66) (J, J=N1,N2)
          WRITE(6,73)(A(1,J), J=N1, N2), B(1)
          DO 48 I=2, M1
    48 WRITE(6,67) NXI(I), (A(I,J), J=N1, N2), B(I)
          IF (N2-N) 52,55,55
    52 N1=N1+6
          N2=N2+6
          GOTO 43
    55 CONTINUE
          IF (NPR.NE.2) GOTO 21
          WRITE (6,560) B(1)
          DO 800 J=1,NN
          D(J)=0
          DO 800 I=2, M1
          K=NXI(I)
          IF (K.EQ.IS(J,1)) D(J)=B(I)
   800 IF (K.EQ.IS(J,2)) D(J)=-B(I)
          DO 810 J=1, NN
          IF (KUN.EQ.0) WRITE (6,580) J, D(J)
          IF(KUN.GT.0.AND.IS(J,2).EQ.0) WRITE(6,510) J,IS(J,1), D(J)
          IF (KUN.GT.0.AND.IS(J,2).NE.0) WRITE(6,500) J,IS(J,2),IS(J,1),D(J)
   810 CONTINUE
          GOTO 1
C        COMPUTE Z AND ZC
    21 DO 25 J=1,N
          A(1,J)=0.
          DO 24 I=2, M1
          K=NXI(I)
    24 A(1,J)=A(1,J)+CJ(K)*A(I,J)
    25 A(1,J)=A(1,J)-CJ(J)
          B(1)=0.
          DO 28 I=2, M1
          K=NXI(I)
    28 B(1)=B(1)+CJ(K)*B(I)
```

```
C       DETERMINE PIVOT COLUMN
        ZCM=A(1,1)
        JM=1
        DO 109 J=2,N
        IF (KOD.EQ.1) GOTO 106
  105   IF (A(1,J)-ZCM) 107, 109, 109
  106   IF (A(1,J)-ZCM) 109, 109, 107
  107   ZCM=A(1,J)
        JM=J
  109   CONTINUE
C       CHECK FOR OPTIMAL
        CK=KOD*ZCM
        IF (CK.GT.TOL) GOTO 131
  123   DO 124 I=2, M1
        K=NXI(I)
        IF (CJ(K).NE.CV) GOTO 124
        IF (B(I).LE.TOL) GOTO 124
        WRITE(6,74) K
        GOTO 1
  124   CONTINUE
        WRITE(6,71) ITER
        NPR=2
        GOTO 4444
C       DETERMINE PIVOT ROW
  131   XM=1.0E38
        IM=0
        DO 139 I=2, M1
        IF (A(I,JM)) 139,139,135
  135   XX=B(I)/A(I,JM)
        IF (XX-XM) 137,139,139
  137   XM=XX
        IM=I
  139   CONTINUE
        IF (IM) 141,141,151
  141   WRITE(6,72) JM
        GOTO 1
C       PERFORM PIVOT OPERATION
  151   XX=A(IM,JM)
        B(IM)=B(IM)/XX
        ITER=ITER+1
        DO 154 J=1,N
  154   A(IM,J)=A(IM,J)/XX
        DO 161 I=1, M1
        IF (I-IM) 157,161,157
  157   XX=A(I,JM)
        B(I)=B(I)-XX*B(IM)
        DO 160 J=1,N
  160   A(I,J)=A(I,J)-XX*A(IM,J)
  161   CONTINUE
        NXI(IM)= JM
        IF (NPR.EQ.1)GOTO 1212
        GOTO 21

    1   WRITE (6,540)
        READ(9,*) INSTR
        IF(INSTR.EQ.0) STOP
        DO 700 I=1,IIM
        B(I)=0
        NXI(I)=0
        KODE(I)=0
        D(I)=0
        IN(I)=0
        IS(I,1)=0
        IS(I,2)=0
```

```
      DO 700 J=1,IIN
      A(I,J)=0
      CJ(J)=0
  700 C(I,J)=0
      ITER=0
      GOTO 2323
      END
```

Appendix D

General Program for Computing Poisson Queueing Formulas[†]

The computer program in this appendix is written in FORTRAN IV. It computes the basic steady-state results of any queueing model having the format $(M/M/c):(GD/N/K)$. The use of the program is now presented.

1. Input

Only five elements of information are needed for the input data of each model. These are

1. The arrival rate XLAM $(= \lambda)$.
2. The service rate XMU $(= \mu)$.
3. The number of parallel servers C $(= c)$.
4. The maximum number allowed in the system XN $(= N)$.
5. The maximum limit on the source XK $(= K)$.

These data are taken in the same order in which they appear in the Kendall notation, $(M/M/c):(GD/N/K)$, and are punched on one card according to the floating point format (5F10.0). If any of the elements c, N, and K is equal to ∞ (i.e., infinite number of servers, infinite system limit, or infinite source), this is entered in the input card as 9999.0. The program is coded such that any number of models can be computed in the same run.

The program may also be used to compute the pure birth and pure death Poisson probabilities for a given time interval t by using the following input data, again with the format (5F10.0).

---

† This program was written by Allen C. Schuermann, head of the Department of Industrial Engineering, Oklahoma State University.

Pure Birth
1. XLAM $= \lambda t.$
2. XMU $= 0.$
3. C $= 0.$
4. XN $= 9999.$
5. XK $= 9999.$

Pure Death
1. XLAM $= 0.$
2. XMU $= \mu t.$
3. C $= 1.$
4. XN $=$ initial number in system.
5. XK $=$ initial number in system.

Notice that the pure birth model can be used to generate any Poisson probabilities by replacing XLAM by the mean of the distribution.

2. Output

The output of the program summarizes the basic input data in addition to the values of λ_{eff} and ρ. The basic output information of the model includes W_s, W_q, L_s, and L_q. In addition, the values of p_n are computed for successive value of n until $p_n < 10^{-5}$ or until p_{1000} is computed, whichever occurs first. The limit $n = 1000$ is specified by the DIMENSION statement of the program. This can be increased as necessary if the computations terminate before $p_n < 10^{-5}$. Notice that the value of "M" as indicated by the second statement in the program must always be the same as the dimension of P and CP.

3. Error Message

The program checks automatically for invalid input data, which include (1) $\rho/c \geq 1$ for the $(M/M/c){:}(GD/\infty/\infty)$ and (2) $K > 1000$ in a *finite* source model. In both cases, the message INVALID DATA will be printed out. The first case leads to a situation where no steady-state results exist. The second case can be accounted for by increasing the dimension of P and CP to at least K.

Program Listing

```
      DIMENSION P(1000),CP(1000)
      M=1000
800 FORMAT(5F10.0)
900 FORMAT(//' *',21X,'(M/M/',I3,')-(GD/',I3,'/',I3,')')
901 FORMAT(/5X,'LAMBDA=',E12.5,5X,'LAMBDA EFF=',E12.5/
    *5X,'MU=',E12.5,9X,'RHO=',E12.5)
902 FORMAT(/5X,'WS=',E12.5,6X,'WQ=',E12.5/
    *5X,'LS=',E12.5,6X,'LQ=',E12.5)
903 FORMAT(/5X,'VALUES OF P(N) FOR N=0 TO',I4,
    *', OTHERWISE P(N) < 0.00001'/5X,'   P(0)=',F10.5/(5X,6F10.5))
```

```
    904 FORMAT(/5X,'INVALID DATA')
    905 FORMAT(//24X,'PURE BIRTH MODEL'//6X,'LAMBDA*T=',E12.5)
    906 FORMAT(//24X,'PURE DEATH MODEL'//6X,'MU*T=',E12.5,9X,'N=',I3)
    907 FORMAT (/5X,'CUMULATIVE VALUES OF P(N)'/8X,'P(0)=',F10.5/,(5X,
        *6F10.5))
    908 FORMAT (///' ENTER LAMBDA, MU,  # OF SERVERS, SYSTEM LIMIT, SOURCE
        * SIZE')
    909  FORMAT (6X,'E(N GIVEN T) =', E12.5)
        PTOL=1.E-5
      5 WRITE (6,908)
        READ(9,*,END=500)XLAM,XMU,REALNC,REALN,REALK
        NC=REALNC
        N=REALN
        K=REALK
        IT=1
        NN=MINO(M,N,K)
        DO 10 I=1,NN
     10 P(I)=0.
        IF(XLAM*XMU.EQ.0.) GO TO 210
        RHO=XLAM/XMU
        ELAM=XLAM
        WRITE(6,900) NC,N,K
        MBIG = MINO(K,N)
        IF (K.LT.9999)  IF (MBIG-M) 12,12,350
     12 RC=RHO/NC
        IF(NC.GT.1) GO TO 60
        IF(K.LT.9999) GO TO 30
        IF (RC.GE.1.)  IF (N-9999) 15,350,15
     15 CONTINUE
C       (M/M/1):(GD/*/*)
        IF(RHO.EQ.1.) GO TO 25
        PZ=1.-RHO
C       (M/M/1):(GD/N/*)
        IF(N.LT.9999) PZ=PZ/(1.-RHO**(N+1))
        P(1)=PZ*RHO
        DO 20 I=2,NN
        P(I)=P(I-1)*RHO
        IF(P(I).LT.PTOL) IF(IT) 20,150,20
        IT=0
     20 CONTINUE
        I=NN+1
        GO TO 150
     25 PZ=1./(N+1)
        DO 26 I=1,N
     26 P(I)=PZ
        I=N+1
        ELAM=XLAM*PZ*N
        QL=N/2.-ELAM/XMU
        GO TO 160
C       (M/M/1):(GD/K/K)
     30 P(1)=K*RHO
        PZ=1.+P(1)
        DO 40 I=2,NN
        P(I)=P(I-1)*RHO*(K-I+1)
     40 PZ=PZ+P(I)
     45 PZ=1./PZ
        DO 50 I=1,NN
        P(I)=PZ*P(I)
        IF(P(I).LT.PTOL) IF(IT) 50,51,50
        IT=0
     50 CONTINUE
        I=NN+1
     51 LL=I-1
        QL=0.
```

```
        DO 52 J=NC,LL
   52 QL=QL+P(J)*(J-NC)
        R=NC*PZ
        IF(I.GE.NC) GO TO 56
        DO 53 J=I,NC
   53 P(J)=0.
   56 DO 57 J=1,NC
   57 R=R+P(J)*(NC-J)
        ELAM=XMU*(NC-R)
        GO TO 160
   60 IF(NC.LT.9999) GO TO 80
C       (M/M/*):(GD/*/*)
        PZ=EXP(-RHO)
        P(1)=PZ*RHO
        DO 70 I=2,NN
        P(I)=P(I-1)*RHO/I
        IF(P(I).LT.PTOL) IF(IT) 70,75,70
        IT=0
   70 CONTINUE
        I=NN+1
   75 QL=0.
        GO TO 160
   80 PZ=1.
        IF(K.LT.9999) GO TO 120
        IF (RC.GE.1.)  IF (N-9999) 17,17,350
   17 CONTINUE
C       (M/M/C):(GD/*/*)
        P(1)=RHO
        DO 90 I=2,NC
        P(I)=P(I-1)*RHO/I
   90 PZ=PZ+P(I-1)
        IF(RC.EQ.1.) GO TO 115
        X=P(NC)/(1.-RC)
C       (M/M/C):(GD/N/*)
        IF(N.LT.9999) X=X*(1.-RC**(N-NC+1))
   95 PZ=PZ+X
        PZ=1./PZ
        DO 100 I=1,NC
        P(I)=PZ*P(I)
        IF(P(I).LT.PTOL) IF(IT) 100,145,100
        IT=0
  100 CONTINUE
        DO 110 I=NC,NN
        P(I)=P(I-1)*RC
        IF(P(I).LT.PTOL) IF(IT) 110,150,110
        IT=0
  110 CONTINUE
        I=NN+1
        IF(RC.EQ.1.) GO TO 117
        GO TO 150
  115 X=P(NC)*(N-NC+1)
        GO TO 95
  117 QL=P(NC)*(N-NC)*(N-NC+1)/2
        GO TO 155
C       (M/M/R):(GD/K/K)
  120 P(1)=K*RHO
        DO 130 I=2,NC
        P(I)=P(I-1)*RHO*(K-I+1)/I
  130 PZ=PZ+P(I-1)
        DO 140 I=NC,NN
        P(I)=P(I-1)*RC*(K-I+1)
  140 PZ=PZ+P(I)
        GO TO 45
  145 P(NC)=0.
```

```
    150 QL=RC*P(NC)/(1.-RC)**2
        IF(N.EQ.9999) GO TO 160
        QL=QL*(1.-RC**(N-NC)-(N-NC)*RC**(N-NC)*(1.-RC))
    155 ELAM=XLAM*(1.-P(NN))
    160 SL=QL+ELAM/XMU
        WS=SL/ELAM
        WQ=QL/ELAM
    200 WRITE(6,901) XLAM,ELAM,XMU,RHO
        WRITE(6,902) WS,WQ,SL,QL
    205 MAX=I-1
        CP(1)=PZ+P(1)
        VALM = P(1)
        DO 208 I=2,MAX
         VALM = VALM +I*P(I)
    208 CP(I)=CP(I-1)+P(I)
           IF (XLAM.EQ.0)   WRITE (6,909) VALM
        WRITE(6,903) MAX,PZ,(P(I),I=1,MAX)
        WRITE(6,907) PZ,(CP(I),I=1,MAX)
        GO TO 5
    210 IF(XMU.EQ.0.) GO TO 240
C       PURE DEATH MODEL
        WRITE(6,906) XMU,NN
        P(NN)=EXP(-XMU)
        PZ=P(NN)
        LL=NN-1
        DO 220 I=1,LL
        P(NN-I)=P(NN-I+1)*XMU/I
        PZ=PZ+P(NN-I)
        IF(P(NN-I).LT.PTOL) IF(IT) 220,230,220
        IT=0
    220 CONTINUE
    230 PZ=1.-PZ
        I=NN+1
        GO TO 205
C       PURE BIRTH MODEL
    240 WRITE(6,905) XLAM
        PZ=EXP(-XLAM)
        P(1)=PZ*XLAM
        DO 250 I=2,NN
        P(I)=P(I-1)*XLAM/I
        IF(P(I).LT.PTOL) IF(IT) 250,205,250
        IT=0
    250 CONTINUE
        I=NN+1
        GO TO 205
C       INVALID INPUT DATA
    350 WRITE(6,901) XLAM,ELAM,XMU,RHO
        WRITE(6,904)
        GO TO 5
    500 STOP
        END
/DATA
5.          0.          0.          9999.       9999.
0.          .5          1.          10.         10.
10.         5.          2.          10.         9999.
50.         12.         9999.       9999.       9999.
3.          4.          1.          5.          5.
20.         10.         1.          9999.       9999.
```

Examples of Program Output

PURE BIRTH MODEL

LAMBDA*T= 0.50000E+01

VALUES OF P(N) FOR N=0 TO 17, OTHERWISE P(N) < 0.00001
```
P(0)=    0.00674
0.03369    0.08422    0.14037    0.17547    0.17547    0.14622
0.10444    0.06528    0.03627    0.01813    0.00824    0.00343
0.00132    0.00047    0.00016    0.00005    0.00001
```

CUMULATIVE VALUES OF P(N)
```
P(0)=    0.00674
0.04043    0.12465    0.26503    0.44049    0.61596    0.76218
0.86663    0.93191    0.96817    0.98630    0.99455    0.99798
0.99930    0.99977    0.99993    0.99998    0.99999
```

PURE DEATH MODEL

MU*T= 0.50000E+01   N= 10
E(N GIVEN T) = 0.50222E+01

VALUES OF P(N) FOR N=0 TO 10, OTHERWISE P(N) < 0.00001
```
P(0)=    0.03183
0.03627    0.06528    0.10444    0.14622    0.17547    0.17547
0.14037    0.08422    0.03369    0.00674
```

CUMULATIVE VALUES OF P(N)
```
P(0)=    0.03183
0.06809    0.13337    0.23782    0.38404    0.55951    0.73497
0.87535    0.95957    0.99326    1.00000
```

*   (M/M/ 2)-(GD/ 10/***)

```
LAMBDA= 0.10000E+02     LAMBDA EFF= 0.90476E+01
MU= 0.50000E+01          RHO= 0.20000E+01

WS= 0.57895E+00     WQ= 0.37895E+00
LS= 0.52381E+01     LQ= 0.34286E+01
```

VALUES OF P(N) FOR N=0 TO 10, OTHERWISE P(N) < 0.00001
```
P(0)=    0.04762
0.09524    0.09524    0.09524    0.09524    0.09524    0.09524
0.09524    0.09524    0.09524    0.09524
```

CUMULATIVE VALUES OF P(N)
```
P(0)=    0.04762
0.14286    0.23810    0.33333    0.42857    0.52381    0.61905
0.71429    0.80952    0.90476    1.00000
```

*   (M/M/***)-(GD/***/***)

```
LAMBDA= 0.50000E+02     LAMBDA EFF= 0.50000E+02
MU= 0.12000E+02          RHO= 0.41667E+01

WS= 0.83333E-01     WQ= 0.0
LS= 0.41667E+01     LQ= 0.0
```

```
VALUES OF P(N) FOR N=0 TO  15, OTHERWISE P(N) < 0.00001
   P(0)=   0.01550
   0.06460   0.13458   0.18692   0.19471   0.16226   0.11268
   0.06707   0.03493   0.01617   0.00674   0.00255   0.00089
   0.00028   0.00008   0.00002

CUMULATIVE VALUES OF P(N)
   P(0)=   0.01550
   0.08010   0.21469   0.40160   0.59631   0.75857   0.87125
   0.93832   0.97325   0.98942   0.99616   0.99871   0.99960
   0.99988   0.99997   0.99999
```

```
*                        (M/M/  1)-(GD/  5/  5)

LAMBDA= 0.30000E+01    LAMBDA EFF= 0.39629E+01
MU= 0.40000E+01        RHO= 0.75000E+00

WS= 0.92838E+00    WQ= 0.67838E+00
LS= 0.36790E+01    LQ= 0.26883E+01

VALUES OF P(N) FOR N=0 TO   5, OTHERWISE P(N) < 0.00001
   P(0)=   0.00928
   0.03480   0.10440   0.23490   0.35235   0.26426

CUMULATIVE VALUES OF P(N)
   P(0)=   0.00928
   0.04408   0.14848   0.38338   0.73574   1.00000
```

```
*                        (M/M/  1)-(GD/***/***)

LAMBDA= 0.20000E+02    LAMBDA EFF= 0.20000E+02
MU= 0.10000E+02        RHO= 0.20000E+01

INVALID DATA
```

Answers to Selected Problems

Chapter 1

1–1 Use a semiautomatic machine for lot sizes between 0 and 120 units or between 200 and 240 units; otherwise, use an automatic machine. For lots greater than 400 units, no feasible solution exists.

1–2 Use a semiautomatic machine for lot sizes between 0 and 240; otherwise, use an automatic machine. For lots greater than 400 units, no feasible solution exists.

1–3 Use a semiautomatic machine for lots between 0 and 120 units. The old automatic machine is used for lot sizes between 120 and 150 units, and the new automatic machine is used for lots between 150 and 240 units. For lots greater than 240 units, no feasible solution exists.

1–4 Buy one ticket at A on Monday of the first week with return from B to A on Wednesday of the last week. All other tickets should be bought at B on Wednesday with return from A to B on Monday of the following week. In this manner each ticket spans a weekend.

1–5 Schools in communities 2 and 3 will minimize the total sum of student-miles.

Chapter 2

2–1 Produce 8 tables and 32 chairs.

2–2 Mix 52.94 lb of corn and 37.06 lb of soybean meal.

2–3 Car loans = $13,333 and personal loans = $6667.

2–4 Allocate 48,000 lb of tomato for juice and 12,000 lb for paste.

2–5 Produce approximately 36 units of HiFi-1 and 46 units of HiFi-2. (Exact values are 36.48 and 46.08.)

2–6 Produce 60 units of model 1 and 25 units of 2; $z = \$2300$.

2–7 Produce 54.55 units of product 1 and 10.9 units of 2; $z = \$141.8$.

2–8 Use 18.2 radio minutes and 9.1 TV minutes; $z = 245.7$.

2–9 Alternative optima at $(x_A, x_B) = (21.4, 14.3)$, or $(50, 0)$, or any point on the line segment joining the two points; $z = \$1000$.

2–10 Produce 150 hats of type 1 and 200 of type 2; $z = \$2200$.

2–11 System reduces to $4x_1 + 3x_2 \leq 12$, $-x_1 + x_2 \geq 1$, $x_1 \geq 0$.

2–12 $x_1 + x_2 \leq 5$, $x_1 - 2x_2 \leq 2$, $x_1 \geq 1$, $-x_1 + x_2 \leq 1$, $x_2 \geq 0$.

2–14 $2/7 \leq c_1/c_2 \leq 7/2$.

2–15 $(x_1, x_2) = (-10, -6)$; $z = -86$.

2–17 $(x_1, x_2) = (5, 5)$; $z = 35$.

2–18 $x_1 = 5$: $\{(5, 0), (5, 5)\}$; optimum at $(5, 5)$ with $z = 35$. $x_1 \leq 5$: $\{(0, 0), (5, 0),$ $(5, 5), (0, 10)\}$; optimum at $(5, 5)$ with $z = 35$. $x_1 \geq 5$: $\{(5, 0), (5, 5), (10, 0)\}$; optimum at $(10, 0)$ with $z = 50$.

2–19 (a) Minimum at $(1, 0)$; $z = 2$. (b) Maximum at $(2, 3)$; $z = 6$. (e) Minimum at $(1, 0)$ or $(1, 2)$; $z = 1$. (f) Maximum at $(4, 1)$; $z = 4$.

2–20 (a) Increase in resource $1 = 1$ unit and increase in $z = 5/3$. (b) Decrease in resource $2 = 2$ units and decrease in $z = 4/3$.

2-21 (a) 20 units. (b) 50 units. (c) 400 units. (d) Unit worths of line 1 and component resource are $5 and $2.5 per unit increase. Line 1 has higher priority.

2-22 (a) Unused capacity of machine 2 = .912 hour. (b) Unit worths of machines 1 and 3 are $.0547 and $.182 per additional minute. (c) Machine 3 has higher priority.

2-24 (a) $25 \leq c_1 < \infty$. (b) $0 \leq c_2 \leq 24$. (c) $c_1 = 25$. (d) $1.25 \leq c_1/c_2 < \infty$.

2-25 Ratio of sales due to radio to sales due to TV must be just above 1.25/25.

2-26 Maximize $z = 10x_2 - 5x_4$ subject to $2x_1 + 3x_2 + 4x_3 + 2x_4 \leq 500$, $3x_1 + 2x_2 + x_3 + 2x_4 \leq 380$, $x_1, x_2, x_3, x_4 \geq 0$.

2-27 Maximize $z = 30x_1 + 20x_2 + 50x_3$ subject to $2x_1 + 3x_2 + 5x_3 \leq 4000$, $4x_1 + 2x_2 + 7x_3 \leq 6000$, $x_1 + x_2/2 + x_3/3 \leq 1500$, $x_1/3 = x_2/2 = x_3/5$, $x_1 \geq 200$, $x_2 \geq 200$, $x_3 \geq 150$.

2-28 Let x_{ij} = amount invested in year i in plan j; $j = A, B$. Maximize $z = 3x_{2A} + 1.7x_{3A}$ subject to $x_{1A} + x_{1B} \leq 100,000$, $-1.7x_{1A} + x_{2A} + x_{2B} \leq 0$, $-3x_{1B} - 1.7x_{2A} + x_{3A} \leq 0$, all $x_{ij} \geq 0$.

2-33 Maximize $z = y$ subject to $-3x_1 + 4x_2 - 7x_3 + 15x_4 \geq y$, $5x_1 - 3x_2 + 9x_3 + 4x_4 \geq y$, $3x_1 - 9x_2 + 10x_3 - 8x_4 \geq y$, $x_1 + x_2 + x_3 + x_4 \leq 500$.

2-34 Maximize $z = 6x_1 + 4x_2 - 5(y_1'' + y_2'')$ subject to $x_1/5 + x_2/6 + y_1' - y_1'' = 8$, $x_1/4 + x_2/8 + y_2' - y_2'' = 8$, $y_1'' \leq 4$, $y_2'' \leq 4$, all variables ≥ 0.

Chapter 3

3-1 Maximize $z = 2x_1 + 3x_2 + 5x_3' - 5x_3''$ subject to $-x_1 - x_2 + x_3' - x_3'' + x_4 = 5$, $-6x_1 + 7x_2 - 9x_3' + 9x_3'' + x_5 = 4$, $x_1 + x_2 + 4x_3' - 4x_3'' = 10$, $x_1, x_2, x_3', x_3'' \geq 0$.

3-3 (a) (1) Yes. (2) No. (3) No. (4) No. (b) (1) Yes. (2) Yes. (3)No, C and I are not adjacent. (7) No, iterations cannot return to a previous extreme point.

3-4 A: Basic (s_1, s_2, s_3, s_4); nonbasic (x_1, x_2, x_3). E: Basic (x_1, x_2, s_3, s_4); nonbasic (s_1, s_2, x_3).

3-5 (a) $A \rightarrow B$: x_1 enters, s_2 leaves. (b) $E \rightarrow I$: x_3 enters, s_4 leaves.

3-6 (a) 15. (b) (8, 0, 3, 0, 0, 0); (2, 0, 0, 0, 0, 3); (0, 0, 1/3, 0, 8/3, 0); (0, 0, 0, 0, 2, 1); (0, 1/2, 0, 0, 0, 0); (0, 0, 0, 1/4, 0, 0). (c) (8, 0, 3, 0, 0, 0) $z = 31$.

3-7 (a) x_3 enters, improvement = 3. (b) x_1 enters, improvement = 5.

3-8 (a) $E = (5/2, 2)$, $z = 39/2$. (b) x_2 enters; $A \rightarrow G \rightarrow F \rightarrow E$. (c) Ratios = (2, 3, 5); x_1 enters at value 2. (d) Ratios = (1, 2, 4); x_2 enters at value 1. (e) Improvements are 8 when x_1 enters and 2 when x_2 enters.

3-9

Entering variable	x_1	x_2	x_3	x_4
Its value	3/2	1	0	0
Leaving variable	x_7	x_7	x_8	x_8

3-10

Entering variable	x_2	x_4	x_5	x_6	x_7
Its value	3	2	0	∞	2
Change in z	+15	-8	0	$+\infty$	0
Leaving variable	x_3	x_3	x_1	none	x_3

3-11 (a) (1.625, -1.114, 2.42). (b) ($-3/4$, 37/4, $-17/4$).

3-12 (a) Three iterations: (0, 1.71, 0, 4.86); $z = 26$. (b) Four iterations: (0, 2.2, 10.2, 0); $z = 43.8$. (c) Four iterations: (0, 2.2, 10.2, 0); $z = 28.4$. (d) Two iterations: (0, 3.33, 0, 0); $z = -13.33$. (e) Two iterations: (0, 0, 8, 0); $z = -16$.

3-13 $x_1 = 90$, all others = 0, $z = 450$.

3-14 (a) Four iterations: (3.12, 4.56, 1.1); $z = -12.76$. (b) Four iterations: (3.12, 4.56, 1.1); $z = -12.76$. (c) It is accidental that both methods required the same number of iterations. Computational experience shows that the criterion in part (a) is generally more efficient. (d) Number of iterations is the same. z-rows appear with opposite signs.

3–15 Three iterations: $(0, 0, 3/2, 0, 8, 0)$; $z = 3$.

3–16 (a) $z - (5 - 2M)x_1 - (6 + 3M)x_2 = -3M$. (b) $z - (2 + 6M)x_1 - (-7 + 16M)x_2 + Ms_2 + Ms_5 = -18M$. (c) $z - (3 - 4M)x_1 - (6 - 8M)x_2 - Ms_5 = 5M$.

3–17 (a) Three iterations: $(45/7, 4/7, 0)$; $z = 102/7$.

3–18 Two iterations: $(2, 0, 1)$; $z = 5$.

3–19 Three iterations: $(0, 2, 0, 2)$; $z = 16$.

3–20 Three iterations: $(0, 7/4, 0, 33/4)$; $z = 7/2$.

3–21 (a) Minimize R_1. (c) Minimize R_5.

3–23 $32/5 \le \max z \le 21$.

3–25 Four alternative basic optima are $(0, 0, 10/3, 0, 5, 1)$; $(0, 5, 0, 0, 0, 1)$; $(1, 4, 1/3, 0, 0, 0)$; $(1, 0, 3, 0, 4, 0)$.

3–28 Solution space is unbounded in the direction of x_2 and optimum z is unbounded also because of x_2.

3–30 (a) Resources 1 and 2 are scarce. (b) $y_1 = 5/8$, $y_2 = 1/8$, $y_3 = 0$. (d) $\Delta_1 \le 12$. (e) $\Delta_2 \le 2$. (f) For $\Delta_1 = 12$, z increases by $15/2$; and for $\Delta_2 = 2$, z increases by $1/4$. (g) $8/3 \le c_1 \le 8$. (h) $-1/4 \le c_2 \le 5/4$.

3–32 (a) (1) Infeasible. (2) $(0, 85, 230)$, $z = 1320$. (3) $(0, 102.5, 225)$, $z = 1330$. (b) (1) Solution does not change. (2) Solution changes, x_1 enters.

3–33 $(7/8, 7/2)$, $z = 77/8$.

3–34 $-5 \le \Delta_1 \le 3 + 2\Delta_2$.

3–37 $\sum_{j=1}^n a_{ij}x_j \le b_i$ and $\sum_{j=1}^n a_{ij}x_j \ge -b_i$.

3–38 Minimize $z = v$ subject to $\sum_{j=1}^n c_{ij}x_j \le v$, $\sum_{j=1}^n c_{ij}x_j \le -v$, for all i, $v \ge 0$.

Chapter 4

4–1 (a) Minimize $w = 3y_1 + 5y_2$ subject to $y_1 + 2y_2 \ge -5$, $-y_1 + 3y_2 \ge 2$, $-y_1 \ge 0$, $y_2 \ge 0$. (b) Maximize $w = 2y_1 + 5y_2$ subject to $6y_1 + 3y_2 \le 6$, $-3y_1 + 4y_2 \le 3$, $y_1 + y_2 \le 0$, $y_1 \ge 0$, $y_2 \ge 0$. (e) Minimize $w = 5y_1 + 6y_2$ subject to $2y_1 + 3y_2 = 1$, $y_1 - y_2 = 1$, y_1, y_2 unrestricted.

4–4 $y_1 = 5$, $y_2 = 0$, $w = 150$.

4–5 $y_1 = 3$, $y_2 = -1$, $w = 5$.

4–6 $y_1 = 4$, $y_2 = 0$, $w = 16$.

4–7 Optimal $z = 250/3$.

4–8 $x_1 = 36/13$, $x_2 = 28/13$.

4–9 Solve the dual in three iterations. The primal optimal solution is $x_1 = 0$, $x_2 = 20$, $x_3 = 0$; $z = 120$.

4–11 (a) Solutions feasible but not optimal. (b) Solutions infeasible. (c) Solutions feasible and optimal.

4–12 (a)

Basic	x_1	x_2	x_3	x_4	x_5	x_6	Solution
x_4	$-1/2$	2	0	1	$-1/2$	0	0
x_3	$3/2$	0	1	0	$1/2$	0	30
x_6	1	4	0	0	0	1	20

4–13 (a) $(x_2, x_4) = (3, 15)$; feasible. (d) $(x_1, x_4) = (21/2, -105/2)$; infeasible.

4–14 (a) $(y_1, y_2, y_3) = (0, 5/2, 0)$. (c) $(y_1, y_2, y_3) = (1, 2, 0)$.

4–15 (a)

Basic	x_1	x_2	x_3	x_4	x_5	Solution
z	0	0	$-2/5$	$-1/5$	0	$12/5$
x_1	1	0	$3/5$	$-1/5$	0	$3/5$
x_2	0	1	$-4/5$	$3/5$	0	$6/5$
x_5	0	0	1	-1	1	0

(b) Solution is optimal and feasible.

4–16 (a) Not optimal because the z-coefficients of x_1 and x_2 are $-7/3$ and $-40/3$.

4–17 Compute the objective value from the primal and dual. $z = w = 34$.

4–20 Dual infeasible and primal unbounded because primal has a feasible space.

4–22 (a) $(y_1, y_2, y_3) = (1, 2, 0)$. (b) Operation 2, then 1. (c) $\Delta_1 \leq 10$ minutes, $\Delta_2 \leq 400$ minutes. (d) Increase in z given that $\Delta_1 = 10$ is $10 and increase in z given that $\Delta_2 = 400$ is $800.

4–23 Either resource 1 or resource 2 can be considered for expansion. $\Delta_1 \leq 12$ and $\Delta_2 \leq 2$. New z for given Δ_1 is 8 and new z for given Δ_2 is $11/2$.

4–24 (a) $b_1 = 30$, $b_2 = 40$. (b) $y_1 = 5$, $y_2 = 0$. (c) $a = 23$, $b = 5$, $c = -10$. (d) $\Delta_1 \leq 10$.

4–25 (a) x_1 remains at zero level. (b) x_1 remains at zero level.

4–26 (a) Two iterations: $(x_1 = 0, x_2 = 5)$; $z = 15$. (b) Three iterations: $(x_1 = 2, x_2 = 0)$; $z = 10$.

4–27 Use artificial variables, dual simplex, and the solution obtained from the dual problem.

4–29 (a) $(x_1, x_2, x_3) = (0, 95, 230)$. (b) $(x_1, x_2, x_3) = (0, 150, 200)$. (c) Apply dual simplex $(x_1, x_2, x_3) = (0, 0, 300)$. (d) Apply dual simplex $(x_1, x_2, x_3) = (0, 75/2, 200)$.

4–30 $-30 \leq \theta \leq 5$.

4–31 Apply dual simplex $(x_1, x_2, x_3, x_4) = (75/2, 0, 5/4, 0)$.

4–32 (a) Constraint $4x_1 + x_2 + 2x_3 \leq 570$ is redundant. (b) Apply dual simplex $(x_1, x_2, x_3) = (0, 88, 230)$.

4–33 (a) Constraint redundant. (b) Apply dual simplex. No feasible solution. (c) Apply dual simplex. No feasible solution.

4–34 (a) Constraint is redundant. (b) Apply dual simplex. $(x_1, x_2, x_3) = (8/7, 12/7, 8/7)$; $z = 96/7$. (c) No feasible solution. (d) New solution is $(x_1, x_2, x_3) = (1, 7/4, 5/4)$; $z = 14$.

4–36 (a) x_1 enters the solution. New solution is $(x_1, x_2, x_3) = (460/3, 200/3, 0)$. (c) x_5 enters the solution. New solution is $(x_1, x_2, x_3) = (0, 105, 220)$. (d) Solution remains unchanged.

4–37 $\Delta c_1 > 4$.

4–38 x_2 enters the solution. New solution is $(x_1, x_2) = (11/5, 2/5)$.

4–40 Solution remains optimal for $\theta \leq 7/3$.

4–41 $4/3 \leq c_1/c_2 \leq 4$.

4–42 (a) x_7 will not improve solution. (b) x_7 improves solution. New solution is $(x_1, x_2, x_3) = (0, 0, 130)$.

4–43 (a) x_4 enters the solution. New solution is $(x_1, x_2, x_3, x_4) = (1, 0, 0, 1)$. (c) Solution remains unchanged and $x_4 = 0$.

4–45 x_2 remains nonbasic for $\theta \leq 23/5$.

4–46 Solution remains unchanged for $0 \leq \theta \leq \min (5, 23/6, 7/3) = 7/3$.

4–47 Solution remains unchanged for $0 \leq \theta \leq \min (5, 2.807, 7/3) = 7/3$.

4–48 (a) Minimize $w = 2y_1 + 6y_2$ subject to $y_1 + 2y_2 + y_3 \geq 0$, $y_2 - y_3 \geq 2$, $y_1 +$

$6y_2 + 3y_3 \geq -5$, $y_1 \leq 0$, $y_2 \geq 0$, y_3 unrestricted. (b) $y_1 = 0$, $y_2 = 2/3$, $y_3 = -4/3$. (c) $(x_1, x_2, x_3) = (5, 3, 0)$. (d) x_3 enters the solution. New solution is degenerate and remains the same $(x_1, x_2, x_3) = (2, 2, 0)$.

Chapter 5

5–1 Assume that supply and demand units are expressed in millions of gallons and that the unit transportation costs are given in thousand dollars per million gallons.

	Distribution Area			
	1	2	3	
1	12	18	M	6
Refinery 2	30	10	8	5
3	20	25	12	8
	4	8	7	

5–2 Use the same units as in Problem 5–1.

	Distribution Area			
	1	2	3	
1	12	18	M	6
Refinery 2	30	10	8	5
3	20	25	12	6
Dummy	M	50	50	2
	4	8	7	

5–3 Use the same units as in Problem 5–1.

	Distribution Area				
	1	2	3	Dummy	
1	12	18	M	15	6
Refinery 2	30	10	8	22	5
3	20	25	12	0	8
	4	8	4	3	

5–4 Supply and demand units are in truck loads and unit costs are in thousand dollars.

	1	2	3	4	5	
1	1	1.5	2	1.4	.35	23
2	.5	.7	.6	.65	.8	12
3	.4	.9	1	1.5	1.3	9
	6	12	9	9	8	

5–7 Let $a = \Sigma_{i=1}^{N} b_i$.

	Demand Day						
	1	2	3	$\cdots$	N	Dummy	
New Napkins	P_1	P_1	P_1	$\cdots$	P_1	0	a
Supply Day 1	M	P_2	P_3	$\cdots$	P_3	0	b_1
2	M	M	P_2	$\cdots$	P_3	0	b_2
3	M	M	M	$\cdots$	P_3	0	b_3
$\vdots$	$\vdots$	$\vdots$	$\vdots$		$\vdots$	$\vdots$	$\vdots$
N	M	M	M	$\cdots$	M	0	b_N
	b_1	b_2	b_3	$\cdots$	b_N	a	

5-10

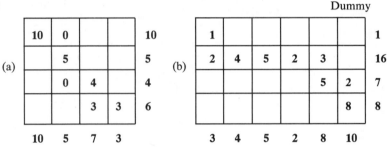

(a)

(b)

5-11 (a) x_{33} : $x_{33} \to x_{23} \to x_{22} \to x_{12} \to x_{11} \to x_{31} \to x_{33}$. (b) x_{33} enters at 5 and x_{31} leaves. (c) $\bar{c}_{33} = -24$ and change in $z = +120$.

5-12 (a) $x_{13} = x_{21} = x_{22} = x_{33} = 5$, $z = 35$.

5-13 (a) Starting solution $x_{11} = 4$, $x_{12} = 2$, $x_{22} = 5$, $x_{32} = 1$, and $x_{33} = 7$ is optimal. (d) Five iterations: optimal solution is $x_{11} = 6$, $x_{12} = 0$, $x_{14} = 9$, $x_{15} = 8$, $x_{22} = 3$, $x_{23} = 9$, and $x_{32} = 9$ with $z = 37$.

5-14 Five iterations: $x_{12} = x_{22} = x_{23} = 10$, $x_{21} = 60$, $x_{31} = 15$, and destination 3 will be 40 units short of its demand; $z = 595$.

5-16 Three iterations: $x_{13} = x_{34} = 20$, $x_{22} = x_{32} = 10$, $x_{21} = 30$; $z = 150$.

5-17 (a) $c_{11} = 0$, $c_{21} = 5$, $c_{22} = 8$, $c_{32} = 10$, $c_{33} = 15$; $z = 1475$. (b) $c_{12} \geq 3$, $c_{13} \geq 8$, $c_{23} \geq 13$, $c_{31} \geq 7$.

5-18 (a) $z = -100$. (b) $\theta = 1$ and x_{12} is the zero basic variable.

5-19 For each iteration, the z-equation coefficients for nonbasic x_{ij} will exactly equal the values of $u_i + v_j - c_{ij}$ in the corresponding transportation tableau. Also, the solutions for the corresponding iterations are exactly the same.

5-20 (a) $\Delta_{11} \leq 10$. (b) $\Delta_{21} \geq 0$. (c) $\Delta_{24} \leq 5$. (d) $\Delta_{34} = 0$.

5-21 (a) *Northwest corner:* $x_{11} = 7$, $x_{21} = 3$, $x_{22} = 9$, $x_{32} = 1$, $x_{33} = 10$, and $z = 94$. *Least cost:* $x_{13} = 7$, $x_{21} = 10$, $x_{23} = 2$, $x_{32} = 10$, $x_{33} = 1$, and $z = 40$, which is optimum. *VAM:* $x_{11} = 7$, $x_{21} = 2$, $x_{23} = 10$, $x_{31} = 1$, $x_{32} = 10$, and $z = 40$, which is optimum.

5-22 VAM has the best starting solution and yields the optimum in three iterations. $x_{13} = 10$, $x_{22} = 20$, $x_{31} = 30$, $x_{42} = 30$, $x_{44} = 10$, $x_{51} = 30$, $x_{52} = x_{53} = 10$, and $z = 820$.

5-23 The problem can be solved by one of three methods: (1) delete first column and reduce the supply at source 4 by 5 units; (2) assign $-M$ to c_{41}; or (3) assign $+M$ to c_{11}, c_{21}, and c_{31}. The first method is the simplest. Solution $x_{12} = 10$, $x_{12} = x_{13} = 5$, $x_{24} = 10$, $x_{34} = 15$, $x_{41} = 5$, $x_{43} = 10$, and $z = 55$.

5-25 (a) Four iterations. 1–5, 2–3, 3–2, 4–4, 5–1 with cost $= 21$. (b) Three iterations. 1–1, 2–2, 3–5, 4–4, 5–3 with cost $= 11$.

5–26 Five iterations. 1–4, 2–3, 3–2, 4–1 with cost = 14.

5–27 Machine 5 replaces machine 1.

5–29 $1 \rightarrow 3 \rightarrow 6 \rightarrow 7$ and total minimum distance = 8.

5–30 Ship 50 from factory (F)1 to store (S)1, 50 from F2 to S1, 200 from F2 to S2, and 50 from F2 to S3. Alternative solution: 50 from F1 to S1, 250 from F2 to S2, 50 from F2 to S3, and 50 from S2 to S1.

5–31 Assume a buffer $B = 110$.

	2	4	5	6	7	
1	20	M	M	M	3	**160**
3	M	30	M	M	9	**170**
5	M	2	M	4	10	**110**
6	8	M	4	M	M	**110**
7	40	M	M	M	M	**110**
	200	**130**	**110**	**110**	**110**	

Chapter 6

6–1 (a) Length = 14. Connect (1, 2), (2, 5), (5, 6), (4, 6), and either (1, 3) or (3, 4). (b) Length = 21. Connect (1, 2), (1, 3), (2, 4), (4, 5), and (4, 6). (e) Length = 13. Connect (1, 2), (2, 5), (5, 3), (2, 4), and (4, 6).

6–2 Length = 5080 miles connecting LA, SE, DE, DA, CH, NY, DC.

6–4 High pressure: connect (1, 2), (2, 3), (3, 4), and (4, 6); length = 33. Low pressure: connect (1, 5), (5, 7), (5, 9), and (9, 8); length = 20.

6–5 Replace at beginning of year 3 and abandon in year 5. Total cost = 12.5.

6–6 (a) Replace in year 2 and abandon in year 5. Total cost = 12.1. (b) Most reliable route is $1 \rightarrow 3 \rightarrow 5 \rightarrow 7$ with probability of not being caught = .0675.

6–10 (a)

	2	3	4	5	6	
1	4	6	7	∞	2	2
2	0	6	9	3	∞	1
3	4	0	2	9	5	1
4	5	6	0	11	6	1
6	∞	∞	3	3	0	1
	1	1	1	2	1	

6–12

Cut Set	Capacity
(1, 4), (2, 4), (2, 6), (2, 5), (3, 5)	115
(4, 7), (4, 6), (2, 6), (5, 6), (5, 8)	130
(4, 7), (6, 7), (6, 8), (5, 8)	110
(7, t), (8, t)	110

6–13 $S_1 \rightarrow D_1$: 20 units, $S_3 \rightarrow D_1$: 20, $S_2 \rightarrow D_4$: 20, $S_3 \rightarrow D_2$: 10, $S_3 \rightarrow D_3$: 10.

Chapter 7

7-1 Basis: B_1 and B_2, nonbasis: B_3 and B_4.

7-2 det $B_1 = -1$, det $B_2 = -3$, det $B_3 =$ det $B_4 = 0$.

7-3 (a) Unique solution with $x_1 > 0$ and $x_2 < 0$. (b) Unique solution with x_1 and $x_2 > 0$. (c) Unique solution with $x_1 < 0$ and $x_2 = 0$. (d) Infinity of solutions. (e) No solution. (f) No solution.

7-4 (a) det $= -4$, basis. (b) det $= -8$, basis. (c) det $= 0$, nonbasis.

7-5 Five basic feasible solution corresponding to the bases (P_1, P_2), (P_1, P_4), (P_2, P_3), (P_2, P_4), and (P_3, P_4), but a total of three feasible extreme points: (0, 2, 0, 0), (3, 0, 0, 1), and (0, 0, 6, 4).

7-6

Extreme Point	Basic Variables
(4, 0, 0, 0, 0)	(x_1, x_3), (x_1, x_4), (x_1, x_5)
(0, 2, 0, 0, 0)	(x_2, x_3), (x_2, x_4), (x_2, x_5)
(0, 0, 12/5, 0, 14/5)	(x_3, x_5)
(0, 0, 0, 12, 4)	(x_4, x_5)

7-11 (a) Number of extreme points $=$ number of basic solutions. (b) Number of extreme points $<$ number of basic solutions.

7-12 New $x_j = (1/\beta)$ old x_j.

7-13 New $x_j = (\gamma/\beta)$ old x_j.

7-15 New $(z_j - c_j) = (1/\beta)$ old $(z_j - c_j)$, which will not make x_j profitable. By making $z_j < c_j$, x_j becomes profitable.

7-22
$$A^{-1} = 1/26 \begin{pmatrix} -9 & -1 & 11 \\ 8 & -2 & -4 \\ 4 & 12 & -2 \end{pmatrix}$$

7-24 Three iterations: $(x_1, x_2, x_3) = (4, 6, 0)$ and $z = 12$.

7-25 Three iterations: $(x_1, x_2, x_3) = (3/2, 2, 0)$ and $z = 5$.

7-26 Three iterations: $(x_1, x_2) = (3/5, 6/5)$ and $z = 12/5$.

7-27 (a) Two iterations: $(x_1, x_2) = (0, 5)$ and $z = 15$. (b) Three iterations: $(x_1, x_2) = (2, 0)$ and $z = 10$.

7-28 (a) Phase I: three iterations with (x_1, x_2, x_4) as the basic feasible solution. Phase II: two iterations with the optimum $(x_1, x_2) = (2/5, 9/5)$ and $z = 17/5$.

7-29 Six iterations: $x_1 = x_5 = 0$, $x_3 = 3/4$, $x_2 = x_4 = x_6 = 1$, and $z = 20$.

7-30 Four iterations: $x_1 = 3$, $x_2 = 5$, $x_3 = 2$, and $z = 34$.

7-31 Four iterations: $x_1 = 2$, $x_2 = 8$, $x_3 = 0$, $x_4 = 12$, $x_5 = 28$, $x_6 = 0$, and $z = 156$.

7-32 Five iterations: $x_1 = 0$, $x_2 = 2$, $x_3 = 9$, $x_4 = 1$, and $z = 53$.

7-33 For each subproblem, select the extreme point associated with max $(z_j - c_j)$. Four iterations: $(x_1, x_2, x_3, x_4) = (5/3, 10/3, 0, 20)$ and $z = -245/3$.

7-34 Three iterations: $y_1 = 0$, $y_2 = 2$, $y_3 = 0$, $y_4 = 5$, $y_5 = 0$, and $z = 44$.

7-35 For $0 \leq t \leq 1/3$, $x_1 = 0$, $x_2 = 5$, $x_3 = 30$, and $z = 160 - 180t$. For $1/3 \leq t \leq 5/12$, $x_1 = 5$, $x_2 = 6.25$, $x_3 = 22.5$, and $z = 140 - 120t$. For $t \geq 5/12$, $x_1 = 20$, $x_2 = 2.5$, $x_3 = 0$, and $z = 65 + 60t$.

7-36 For $0 \leq t \leq 1/55$, $x_1 = 0$, $x_2 = 100 + 225t$, $x_3 = 230 + 50t$, and $z = 1350 + 700t$. For $1/55 \leq t \leq 2.1$, $x_1 = 0$, $x_2 = 105 - 50t$, $x_3 = 230 + 50t$, and $z = 1360 + 150t$. For $t > 2.1$, no feasible solution exists.

7-37 For $0 \leq t \leq 1/55$, $x_1 = 0$, $x_2 = 100 + 225t$, $x_3 = 230 + 50t$, and $z = 1350 - 680t - 300t^2$. For $1/55 \leq t \leq 5/12$, $x_1 = 0$, $x_2 = 105 - 50t$, $x_3 = 230 + 50t$, and $z = 1360 - 1230t - 300t^2$. For $5/12 \leq t \leq 8/7$, $x_1 = (460 + 100t)/3$,

$x_2 = (200 - 175t)/3$, $x_3 = 0$, and $z = (1780 + 1330t)/3 + 100t^2$. For $8/7 \le t \le 2.1$, $x_1 = 420 - 200t$, $x_2 = x_3 = 0$, and $z = 1260 + 660t - 600t^2$. For $t > 2.1$, no feasible solution exists.

7–38 For all $t \ge 0$, $x_1 = 2/5$, $x_2 = 9/5$, $x_5 = 0$, and $z = (17 - 29t)/5$.

7–39 For $0 \le t \le 3/2$, $x_1 = (2 + 2t)/5$, $x_2 = (9 - 6t)/5$, and $z = (17 - 43t + 2t^2)/5$. For $t > 3/2$, no feasible solution exists.

7–40 Same as Problem 7–39 except that $z = 18 - 7t - 9t^2$.

7–41 For $0 \le t \le 4/5$, $x_1 = 0$, $x_2 = 2 - t/4$, $x_3 = 2 - 3t/4$, $x_4 = 0$, and $z = 16 - 10t - 7t^2/4$. For $4/5 \le t \le 1$, $x_1 = 4 - 3t/2$, $x_2 = t/2$, $x_3 = x_4 = 0$, and $z = 8 + 3t - 2t^2$. For $1 \le t \le 4/3$, $x_1 = 4 - t$, $x_2 = x_3 = 0$, $x_4 = t$, and $z = 8 - t + 2t^2$. For $4/3 \le t \le 4$, $x_1 = x_2 = 0$, $x_3 = 4 - t$, $x_4 = 8 - t$, and $z = -8 + 15t - t^2$. For $t > 4$, no feasible solution exists.

7–42 Solution remains optimal and feasible for $0 \le t \le 1/2$.

7–43 $(6\beta - 12)/(\beta - 5) \le \alpha \le 6$.

7–44 For $0 \le t \le 1.3$, $x_1 = 0$, $x_2 = 9 - t^2$, and $z = 54 - 18t - 15t^2 + 2t^3 + t^4$. For $1.3 \le t \le 2.3$, $x_1 = 6 - 2t^2/3$, $x_2 = 0$, and $z = 18 + 6t - 8t^2 - 2(t^3 - t^4)/3$. For $2.3 \le t \le 3$, $x_1 = x_2 = 0$ and $z = 0$. For $t > 3$, no feasible solution exists.

7–45 For $-\infty < t \le -5$, $x_1 = 4$, $x_2 = 0$, and $z = 16 - 40t$. For $-5 \le t \le -1$, $x_1 = -(1 + t)$, $x_2 = 5 + t$, $z = 36 - 6t + 6t^2$. For $-1 \le t \le 2$, $x_1 = 0$, $x_2 = 3 - t$, and $z = 24 - 20t + 4t^2$. For $2 \le t \le 3$, $x_1 = x_2 = 0$ and $z = 0$. For $t > 3$, no feasible solution exists.

7–47 For $0 \le t \le 2/13$, $x_1 = 3/5$, $x_2 = 6/5$, and $z = (12 + 27t)/5$. For $t \ge 2/13$, $x_1 = 3/2$, $x_2 = 0$, and $z = (6 + 3t)/2$.

7–48 For $0 \le t \le 6/11$, $x_1 = (3 + 7t)/5$, $x_2 = (6 - 11t)/5$, and $z = (12 + 3t)/5$. For $t \ge 6/11$, $x_1 = (3 + 2t)/3$, $x_2 = 0$, and $z = (6 + 4t)/3$.

Chapter 8

8–1 The optimal LP solution is $x_1 = 10/3$, $x_2 = 0$, and $x_3 = 0$. The rounded solution $x_1 = 3$ and $x_2 = x_3 = 0$ satisfies the first constraint but never the second.

8–2 Minimize $\sum_{j=1}^{n} c_j x_j$, subject to $\sum_{j=1}^{n} a_{ij} x_j = 1$, $i = 1, 2, \ldots, m$, and $x_j = (0, 1)$, $j = 1, 2, \ldots, n$, where $a_{ij} = 1$ if the ith destination is reached on route j and zero otherwise.

8–5 (a) $(x_1 \le 1$ and $x_2 \le 2)$ or $(x_1 + x_2 \le 3$ and $x_1 \ge 2)$, which reduces to $x_1 - My \le 1$, $x_2 - My \le 2$, $x_1 + x_2 - M(1 - y) \le 3$, and $x_1 + M(1 - y) \ge 2$, where $y = (0, 1)$. The optimum solution is $(x_1, x_2) = (1, 2)$.

8–7 The optimal integer solution after multiplying the constraint by 4 is $x_1 = 0$, $x_2 = 6$, and the slack variable equals 1.

8–9 The rounded solution is $x_1 = 5$ and $x_2 = x_3 = 3$, which is infeasible. The optimal integer solution is $x_1 = 5$, $x_2 = 2$, and $x_3 = 2$.

8–10 Both cuts are the same in this case mainly because the problem has a small number of integer variables.

8–12 $x_1 = 5$, $x_2 = 2.75$, and $x_3 = 3$ with $z = 26.75$.

8–16 The alternative optima are (5, 0), (4, 1), and (3, 2) with $z = 5$.

8–18 $x_1 = 15$, $x_2 = 0$, $x_3 = 1$, $x_4 = 17$, and $x_5 = 0$ with $z = 151$. Alternative solution: $x_1 = 14$, $x_4 = 19$; all others $= 0$.

8–19 The starting solution is optimal and feasible, $x_1 = x_2 = x_3 = 1$.

8–21 $x_1 = x_2 = x_3 = x_4 = 1$ and $x_5 = 0$ with $z = 95$.

8–23 Treat the first constraint as an "either–or" constraint. The optimal solution is $x_1 = x_2 = 0$, $x_3 = 10$, and $z = 50$.

Chapter 9

9-2 (a) Optimum projects $(1, 3, 1)$ and optimum revenue $= 13$. (b) Optimum projects $(3, 2, 2, 1)$ and optimum revenue $= 14.1$.

9-3 Same as in Problem 9-2.

9-5 $(m_1, m_2, m_2, m_4) = (2, 3, 4, 1)$ with total points $= 250$.

9-6 (a) $(k_1, k_2, k_3) = (0, 0, 3), (0, 2, 2), (0, 4, 1),$ or $(0, 6, 0)$ with value $= 120$.
(b) $(k_1, k_2, k_3) = (0, 2, 0), (2, 1, 0),$ or $(4, 0, 0)$ with value $= 120$.

9-7 $x_4 = 4, 5, 6, 7; x_3 = 6, 7, 8, 9; x_2 = 9, 10, 11, 12; x_1 = 12, 13, 14, 15.$ $(m_1, m_2, m_3, m_4) = (1, 1, 2, 2)$ with $R = .432$.

9-8 $x_1 = 0, x_2 = 2, x_3 = 0,$ and $z = 6$.

9-10 (a) Successive decisions for weeks $1, 2, \ldots,$ and 5 are: hire one, fire one, fire two, hire three, and hire two. Total cost $= 24$.

9-11 Optimum route is $A \to 1 \to 3 \to 5 \to B$ with total distance $= 12$.

9-12 $x_1 = 5, x_2 = x_3 = x_4 = 0$ with $z = 74$.

9-13 $y_i = 8^{1/10}$, for all i.

9-14

Period	1	2	3	4	5
Type I investment	0	6,000	0	0	1,890
Type II investment	10,000	0	5,400	2,700	0

9-17 $(y_1, y_2, y_3) = (13/11, 7/11, 90/11)$.

9-18 Five alternative solutions: $(y_1, y_2, y_3) = (0, 0, 16), (1, 0, 12), (2, 0, 8), (3, 0, 4), (4, 0, 0)$ with total profit $= 64$.

9-19 $y_1 = y_2 = \cdots = y_{N-1} = 0, y_N = \alpha^N C$.

9-20 $y_i = \alpha^i C/(1 + \alpha + \cdots + \alpha^{N-1}), i = 1, 2, \ldots, N$.

9-21 Decisions: (keep, keep, replace) or (keep, replace, keep) with total return $= 13$.

9-25 $(x_1, x_2) = (0, 7)$ with $z = 49$.

9-27 $(x_1, x_2) = (9.6, .2)$ with $z = 707.72$.

Chapter 10

10-1 (a) $\{(H, H), (H, T), (T, H), (T, T)\}$. (b) $\{2, 3, 4, \ldots, 11, 12\}$. (c) $\{(H, 1), (H, 2), \ldots, (H, 6), (T, 1), (T, 2), \ldots, (T, 6)\}$. (d) $\{t | t \geq 0\}$, t is the life of the battery.

10-3 (a) $A\bar{B} = \{1, 2, 3, 4\}$. (b) $\bar{A} + B = \{5, 6, 7, 8, 9, 10\}$. (c) $AB + \bar{B}C = \{2, 4, 10\}$.
(d) $\bar{A}B + \bar{C} = \{1, 3, 5, 6, 7, 8, 9\}$. (e) $\overline{BCS} = \{1, 3\}$. (f) $A\bar{S} = \emptyset$.

10-7 (a) .25. (b) .15. (c) .6.

10-8 $k = 20$.

10-10 (a) $f(t) = 2e^{-2t}, t \geq 0$. (c) $P = 0$. (d) $P = 1 - e^{-10} + e^{-20}$.

10-13 (a) $f_x(x) = e^{-x}, x \geq 0$. (b) $F(x, y) = (1 - e^{-x})(1 - e^{-y})$.

10-15 (a) $k = 6$. (b) $g(x_1 | x_2) = 2(1 - x_1 - x_2)/(1 - x_2)^2$. Because $f_1(x_1) = 3(1 - x_1)^2 \neq g(x_1 | x_2), x_1$ and x_2 are dependent.

10-17 $E\{x_1 | x_2\} = (1 - x_2)/3$.

10-18 $E\{S^2\} = \sigma^2(n - 1)/n$.

10-21 $M_x(t) = 1/(1 - t)$, which is the mgf for $f(x) = e^{-x}, x > 0$.

10-24 $E\{x^n\} = 2^{-n/2}n!/(n/2)!$ if n is even and 0 otherwise.

10-27 $P = (.01, .04, .1, .2, .25, .24, .16)$.

10-28 (a) $p_1 = p_2 = p_3 = 1/3$. (b) $p_i = (p^{i-1}q)/(1 - p^5)$ for $i = 1, 2, \ldots, 5$.

10-31 $x_n = b(y_1 a^{n-1} + y_2 a^{n-2} + \cdots + y_n) + y_1(1 - b)a^{n-1}$.

Chapter 11

11-1 $T^* = 3$ with $EC(T^*) = 290$.

11-2 Expected costs for A, B, and C are \$190, 120, and 205. B has the highest priority.

11-3 Optimal stock $= 200$ loaves and the optimal expected profit is \$36.95.

11-4 $\alpha = 49$ pieces per day.

11-5 $d^* = \dfrac{1}{2}\left(t_L + t_u - \dfrac{2\sigma^2}{t_L - t_u}\ln\dfrac{c_2}{c_1}\right).$

11-9 $I \geq 2$, $2 \leq I \leq 4$.

11-10 $I \geq 4$.

11-11 Expected costs for A, B, and C are \$263.7, 93.3, and 168.2. B has the highest priority.

11-12 Posterior probabilities for A and B are .6097 and .3903.

11-13 (a) $E\{a_1\} = 3.2$, $E\{a_2\} = 9.8$, $E\{a_3\} = 10.6$. (b) $E\{a_1 \mid z_1\} = -3.7$, $E\{a_2 \mid z_1\} = 7.12$, $E\{a_3 \mid z_1\} = 13.17$.

11-15 Stock 130 loaves.

11-16 Order 130 loaves in day 1. In day 2, if demand on day 1 is 100 order 120. If it is 120, order 120. If it is 130, order 130.

11-17 Construct large plant.

11-18 Build a large plant.

11-19 (a) a_4. (b) a_2. (c) a_2. (d) a_4.

11-21 All criteria select machine 3.

11-22 (b) (1) $p \geq 5$ and $q \leq 5$. (2) $p \leq 7$ and $q \geq 7$.

11-23 (a) $2 < v < 4$. (b) $-1 < v < 0$.

11-25 The game has a saddle point at the strategy where both companies use TV, radio, and newspapers.

11-28 (a)$x_1 = x_2 = 1/2$, $y_1 = y_2 = 0$, $y_3 = 13/20$, $y_4 = 7/20$, and $v = 1/2$.

11-29 Blotto's optimal strategy is $(1/5, 3/5, 1/5)$. His enemy's strategy is $(1/3, 1/5, 0, 7/15)$.

Chapter 12

12-5 Critical path $(A, C, D, F, G, H, J, L, N, S, T)$ with duration $= 38$ days.

12-7 Four critical paths with duration $= 111$ days: (A, B, D, E, Q, R, S), (A, B, D, F, Q, R, S), (A, C, D, E, Q, R, S), (A, C, D, F, Q, R, S).

12-8 Critical paths: $(A, C, E, F, J, L, M, P, Q, S, T, U)$, $(A, C, E, F, J, L, N, P, Q, S, T, U)$. Duration $= 22.1$.

12-10 (a) Critical path $(1, 2, 3, 4, 6, 7)$. Duration $= 35$.

12-15 Minimum number of men is determined by the critical activities.

12-16 (a) Respective probabilities for events 2, 3, 4, 5, 6, and 7 are .5, .5, .5, 1., .5, and .5.

12-18 (a)

Duration	25	24	23	21	18	17	14
Cost	1150	1157	1170	1201	1276	1303	1403

Chapter 13

13-2 (a) $y = 346.6$, $t_0 = 11.55$, $\text{TCU}(y) = 17.3$.

13-3 (a) $y \cong 50$ units, $t_0 \cong 12$ days. (b) Excess annual cost $= \$540.20$.

13-4 Optimum number of orders per year $= 129$.

13-5 Let R = reorder point in number of units. (1-a) $t_0 = 11.55$, $R = 73.5$. (1-c) $t_0 = 22.36$, $R = 560$. (2-b) $t_0 = 8.16$, $R = 220.8$. (2-d) $t_0 = 15.8$, $R = 168$.

13–6 (1-a) $B \geq 8.74$. (1-c) $B \geq 23.12$. (2-a) $B \geq 12.36$. (2-d) $B \geq 17.48$.

13–8 (a) $y = 547.7$, $t_0 = 18.25$. (b) $y = 387.38$, $t_0 = 12.9$.

13–9 *Produce:* $y = 703.7$, total cost/day $= \$4.05$. *Buy:* $y = 326.87$, total cost/day $= \$6.54$.

13–11 $\dfrac{p}{p-1} \leq h \leq \dfrac{p^2}{100-p}$ and $p \geq 10$.

13–12 Let $p \rightarrow \infty$ and find the limit for y.

13–15 For $q = 300$, $y^* = 347$ and for $q = 500$, $y^* = 500$.

13–16 Order $q = 150$ units and take advantage of the discount.

13–17 There is no advantage in using the discount if it is $\leq 5.48\%$.

13–19 $y_i^* = \sqrt{(2K_i\beta_i - 2\lambda D_i)/h_i}$, $\lambda^* \cong -.103$.

13–21 $z_1 = 2$, $z_2 = 0$, and $z_3 = 3$ and total cost $= 65$.

13–23 $(z_1, z_2, z_3, z_4) = (5, 7, 14, 0)$ or $(6, 6, 14, 0)$.

13–25 $(z_1, z_2, \ldots, z_{10}) = (100, 120, 0, 200, 0, 0, 310, 0, 190, 0)$.

13–26 $z_1 = 50$, $z_2 = 260$, $z_3 = z_4 = z_5 = 0$.

13–27 $z_1 = 150$, $z_2 = 120$, $z_4 = 110$, $z_6 = 90$, $z_7 = 310$, $z_9 = 190$, all others $= 0$.

13–35 If $x = 2$, order .88; if $x = 5$, do not order.

13–37 If $x = 2$, order 6; if $x = 5$, order 3.

13–38 $19 \leq p \leq 35.7$.

13–39 If $x = 2$, order 21; if $x = 5$, order 18. Implied penalty is $p \leq 29$.

13–40 $P\{\xi \leq y^* - 1\} \leq \dfrac{r-c}{r-v} \leq P\{\xi \leq y^*\}$.

13–44 If $x < 3.78$, order $6.7 - x$; otherwise, do not order.

13–46 If $x < 1.25$, order $6.25 - x$; otherwise, do not order.

13–47 $P\{\xi \leq y^*\} = (r + p - c - h)/(r + p - v)$.

13–50 $y^* = 4.61$.

13–53 $y^* = E\{\xi\} - [(1 - \alpha)c/2p]$ when $h = p$.

Chapter 14

14–1 Years 1 and 2: advertise only if product is unsuccessful. Year 3: do not advertise.

14–2 Use radio advertisement if sales volume is poor; otherwise, use newspaper advertisement.

14–4 If beginning of month stock is zero, order two refrigerators; otherwise, order none.

14–5 Order 2 in state 0; otherwise, order none.

14–7 Advertise whenever in state 1.

Chapter 15

15–1

Customer	Server	Number of Servers	Queue Size	Source Size
Shopper	Checkout stand	>1	∞	∞
Car	Stoplight	1	∞	∞
Patient	Doctor	>1	Finite	∞
Plane	Runway	≥ 1	Finite	∞
Car	Toll gate	≥ 1	∞	∞
Program	Terminal	>1	∞	∞
Person	Teller	≥ 1	∞	∞

15–3 (a) .5 time unit. (b) $p_0(.5) = .3679$. (c) $P\{t \geq 3\} = .002479$. (d) $P\{t \leq 2\} = .9817$.

15–4 (a) 16 customers. (b) $P\{n \geq 1 \mid t = 1\} = .8647$.

15–5 (a) $P\{n = 2 \mid t = 5 \text{ minutes}\} = .2623$. (b) $P\{t \leq 2 \text{ minutes}\} = .4866$.

15–6 (a) $\lambda t = 7.5$. (b) $P\{n > 5000 \mid t = 30\} \cong 0$.

15–7 Jim's expected gain/8 hours = $1.27.

15–8 Jim's expected gain/8 hours = $1.96.

15–9 Input data: (a) 12.12, 0, 0, 9999., 9999. (b) 24.24, 0, 0, 9999., 9999.

15–10 (a) $q_{10}(2) = p_{70}(2) = .1251$. (b) $p_0(4) \cong .00001$. (c) $\Sigma_{n=0}^{80} n q_n(4)$.

15–11 Input data: (a) 0, 10, 1, 80, 80. (b) 0, 20, 1, 80, 80.

15–12 $p_0(5) = .00008$ and $p_0(2) \cong 0$.

15–13 (a) $\Sigma_{n=0}^{N} n q(6) \cong 17.89$. (b) $p_0(4) = .00069$. (c) $P\{n < 20 \mid t = 6\} = .99968$.

15–14 (a) $p_0 = .4$. (b) $L_q = .9$ customer. (c) $W_q = 2.25$ minutes. (d) $P\{n \geq 11\} = .00363$.

15–15 Input data: .4, .66667, 1., 9999., 9999.

15–16 New device is justified based on expected number of waiting customers in old system (= 19) but not on the basis of percent idle time in new system (= 25%).

15–17 (a) $p_0 + p_1 + p_2 \cong .42$. (b) .58. (c) $W_q = .417$ hour. (d) $n \geq 2$ spaces.

15–18 $P\{\tau > W_q\} = .549$.

15–19 Expected cost/day = $37.95.

15–20 (a) $p_0 = .40146$. (b) $L_q = .8615$ customer. (c) $W_q = 2.16$ minutes. (d) $p_{10} = .00243$.

15–21 Input data: .4, .66667, 1., 10., 9999.

15–22 (a) $p_{50} \cong .00002$. (b) $P\{n > 47\} = .00007$.

15–23 (a) $\lambda_{\text{eff}} = 19.98$. (b) $p_0 = .000762$. (c) $W_s = .652$ hour.

15–24 (a) $W_s = .618$ hour, $W_q = .451$ hour. (b) $W_s = .368$ hour, $W_q = .236$ hour.

15–25 $L_q = 7.04$ items.

15–26 $L_q = .333$ customer.

15–27

c	4	5	6	7
p_0	.0042	.01662	.02013	.0212
W_q	1.05	.081	.022	.0068

15–28 (a) .15. (b) .85. (c) .52.

15–29 (a) .70125. (b) $W_s = 1.202$ hours. (c) $L_q = 3.5$ programs. (d) .5. (e) $p_0 = .04494$. (f) 16.7%.

15–30 (a) Three counters. (b) Three counters. (c) At least five counters.

15–31 (a) Two tellers. (b) Two tellers.

15–32 (a) 8.33 vacant lots. (b) $p_{10} \cong 0$. (c) $\lambda_{\text{eff}} \cong 10$.

15–34 Input data: 20., 5.217, 4., 20., 9999.

15–36 (a) $L_s = 4.17$. (b) $p_0 = .0155$.

15–37 (a) .081%. (b) $L_q \cong 6$.

15–38 (a) $p_0 = .04305$, $p_1 = .16144$. (b) $L_q = .911$.

15–39 Pooling reduces waiting time.

15–40 Input data: (a) 2., 6., 1., 10., 10. (b) 3., 4., 2., 5., 5.

15–41 (a) $W_q^1 = 1.16$ hours, $W_q^2 = 7.27$ hours, $W_q^3 = 65.1$ hours. (b) $W_q = 17.4$ hours. (c) $L_q^1 = .194$ job, $L_q^2 = .909$ job, $L_q^3 = 5.42$ jobs. (d) $L_q = 6.5$ jobs.

15–49 (a) .833. (b) $L_s = 3.1$ customers. (c) $W_q = .454$ hour.

15–50 (a) $q_7 \cong 0$. (b) $p_0 = .42$. (c) $W_s = .234$ hour.

15–51 (a) $d_3 \cong .146$. (b) $d_1 + d_2 \cong .577$. (c) $L_s = .51$. (d) $W_s = .15$ hour. (e) $P\{n \geq 2\} = .114$.

15–55 $p_0'(t) = -\lambda p_0(t) + \mu p_1(t)$; $p_1'(t) = \lambda p_0(t) - \mu p_1(t)$; $\lambda < \mu$ is not essential for reaching steady state.

Chapter 16

16–2 $\bar{t} = 4.517$, χ^2-value $= 5.502 < \chi_5^2(.05) = 11.071$. Thus accept the hypothesis that interarrival time follows the distribution

$$f(t) = (1/4.517)e^{-t/4.517}, \qquad t > 0$$

16–3 $\bar{n} = 2.97$ customers/hour, χ^2-value $= 1.3494 < \chi_5^2(.05)$. Thus accept the hypothesis that the arrivals follow the distribution

$$p_n(t) = \frac{(2.97t)^n e^{-2.97t}}{n!}, \qquad n = 0, 1, 2, \ldots$$

16–4 $\mu = 15$ jobs/day.

16–6 (a) $C_1\rho/(1-\rho) + C_2(1-\rho)$. (b) $\rho = 1 - \sqrt{C_1/C_2}$.

16–7 Daily cost of model A $= \$554.67$ and of model B $= \$60.57$. Choose model B.

16–9 $c^* = 6$.

16–10 Hire the second repairman because total hourly cost of first repairman is $73 whereas that of the second is $69.

16–11 (a) $16.16. (b) $457.06. (c) $244.8.

16–12 (a) Monthly savings of one WATS line is $5920. (b) Additional monthly savings of two WATS lines $= \$2280$.

16–13 (b) $c = 5$. (c) $c = 4$.

Chapter 17

17–5 (a) 47%. (b) $\overline{W} = 1.2$ hour. (c) $\overline{Q} = 1.44$ customers.

17–6 Win if $0 \le R \le 1/3$ and lose if $1/3 < R \le 1$.

17–8

Day	1	2	3	4	5
Demand	0	2	1	3	2

17–9 $26.00.

17–10 Successive arrival times for customers 1, 2, and 3 are .283, .396, and .4501 time units.

17–11 $n = 5$ in first hour. Also $n = 5$ in the second hour.

17–13 $22.69.

17–14

			d				$p(L)$
		0	1	2	3	4	
L	1	1/23	7/23	2/23	0	0	10/23
	2	2/23	0	7/23	4/23	0	13/23
$p(d)$		3/23	7/23	9/23	4/23	0	

17–15 Estimated area $= 1.386$ and exact area $= 1.732$.

17–17 Estimated area $= 57.6$ and exact area $= 63$.

17–18 $(Q_1, Q_2, Q_3, Q_4, Q_5) = (.6, .6, 3.2, 2.8, 0)$, $\overline{Q} = 1.44$, and $S_Q^2 = 2.108$.

Chapter 18

18–1 (a) None. (b) Minimum at $x = 0$. (e) Inflection at $x = 0$, minimum at $x = .63$, and maximum at $x = -.63$.

18–2 (a) Minimum at $(1, 1)$.

18–3 Minimum occurs at $(1, 2, 0)$ only.

18–5 One root of $4x^4 - x^2 + 5 = 0$ occurs at $x \cong .353$ if starting point is $x = 1$.

18–8 $\partial f = 2\delta C^{(2-n)/n}$.

18–9 For $\partial g = -.01$, (a) $\partial f = -.0647$, (b) $\partial f = -.12$.

18–11 Minimum point, $(x_1, x_2, x_3, x_4) = (-5/74, -10/74, 155/74, 60/74)$. Sensitivity coefficients are $(-90/37, 85/37)$.

18–20 $\lambda_1 \le 0$, λ_2 unrestricted, $\lambda_3 \ge 0$. The necessary conditions are sufficient if f is concave, g_1 concave, g_2 linear, and g_3 convex.

Chapter 19

19–3 Maximum number of iterations $= 1.44 \ln \{(b - a)/\Delta - 1\}$.

19–7 $x_1 = 0$, $x_2 = 3$, $z = 17$.

19–11 Let $w_j = x_j + 1$, $j = 1, 2, 3$ and substitute for x_j in terms of w_j.

19–15 $x_1 = 1$, $x_2 = 0$, $z = 4$.

19–16 $x_1 = 0$, $x_2 = .4$, $x_3 = .7$.

19–17 Necessary conditions are not satisfied for $x_j > 0$. The problem has an infimum at $x_j = 0$; that is, $z \rightarrow 0$ as $x_j \rightarrow 0$ for all j.

19–18 $x_1 = 1.26$, $x_2 = .41$, $x_3 = .59$, $z = 10.28$.

19–20 $x_1 = 1.26$, $x_2 = 1.887$, $z = 13.07$.

Index

N

T